W9-CKH-035

IDEAL GAS

$h_2 - h_1 = C_p (T_2 - T_1)$

$u_2 - u_1 = c_v (T_2 - T_1)$

MIXTURES

AIR = DRY AIR + WATER VAPOR

specific humidity $= \omega = 0.6219 \dfrac{\phi}{P/P_g - \phi} =$

no. changes for different than air

ϕ = relative humidity $= \dfrac{\omega}{0.6219 + \omega} \dfrac{P}{P_g} = \omega \dfrac{R_v}{R_a} \dfrac{P_a}{P_g} = \omega (1.6079) \dfrac{P - P_v}{P_g}$

SECOND LAW

$\eta_{CAR} = 1 - \dfrac{T_L}{T_H} = \dfrac{\dot{W}_{OUT}}{\dot{Q}_{in}}$ $\quad \dfrac{Q_H}{Q_L} \overset{Rev}{=} \dfrac{T_H}{T_L}$

$S_2 - S_1 > 0$ IRREV

$S_2 - S_1 = 0$ REV

$COP_{carnot} = \dfrac{T_L}{T_H - T_L}$

* max. possible

if more impossible

if less irreversible

= reversible

$COP = \dfrac{Q_{OUT}}{W_{in}}$

CLOSED

$\Delta S_{UNIV} = \Delta S_{SYS} + \Delta S_{SURR}$ (crossed out, 0) Adiabatic

$= \Delta S_H + \Delta S_L$

$= mc \left[\ln \dfrac{T_2}{T_H} + \ln \dfrac{T_2}{T_L} \right]$

↑ FOR H ↑ FOR L

OPEN SYSTEM

$\Delta S_{UNIV} = \Delta S_{SYS} + \Delta S_{SURR}$

$\Delta S_{SURR} = \int \dfrac{\delta Q}{T} \Rightarrow \dfrac{1}{T_{SURR}} - Q_{in}$

carnot

$Q_{in} = T_H (S_2 - S_1)$

1st Law

$Q_{in} - P\Delta V = \Delta U$

$= m(h_2 - h_1)$

$S_2 - S_1 = m(s_2 - s_1)$

$S_2 - S_1 = C_v \ln \dfrac{T_2}{T_1} + R \ln \dfrac{v_2}{v_1}$ * use because $v_1 = v_1$ $\ln 1 = 0$

$S_2 - S_1 = C_p \ln \dfrac{T_2}{T_1} - R \ln \dfrac{P_2}{P_1}$

$$\eta = \frac{w_t' - w_p'}{h_3 - h_{2a}} \qquad \eta = \frac{w'_{isen}}{w'_{act.}}$$

BRAYTON CYCLE

$$\eta = \frac{\text{net } w'_{OUT}}{q_{in}}$$

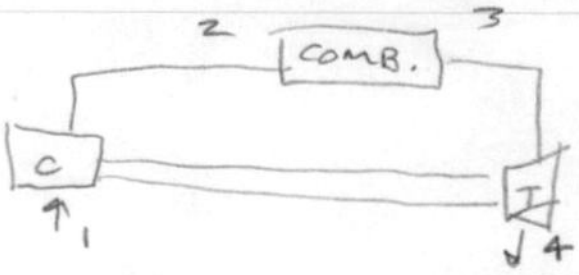

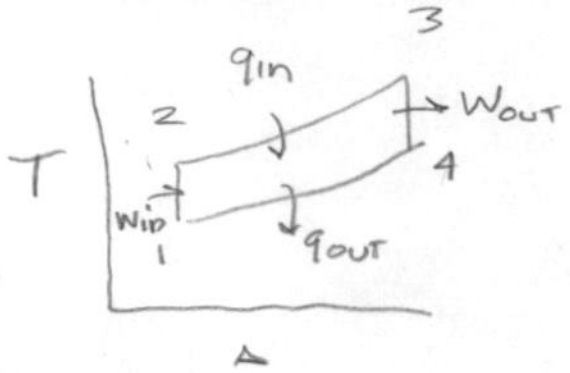

COMP: $\frac{T_2}{T_1} = \left(\frac{P_2}{P_1}\right)^{\frac{\gamma-1}{\gamma}}$ TURB.: $\frac{T_3}{T_4} = \left(\frac{P_3}{P_4}\right)^{\frac{\gamma-1}{\gamma}}$

$$\eta_{TURB} = \frac{w'_{OUT}\text{ (actual)}}{w'_{OUT}\text{ (isen.)}} = \frac{c_p(T_3 - T_4)}{c_p(T_3 - T_{4A})} \qquad \eta_c = \frac{c_p(T_{2A} - T_1)}{c_p(T_2 - T_1)}$$

$$\text{thrust} = \frac{\dot{m}}{g_c}(V_{exit} - V_{in})$$

Regenerated Brayton

$$e = \frac{T_3 - T_2}{T_5 - T_2} = \frac{q_{act}}{q_{max}} \qquad \eta = \frac{(T_4 - T_5) - (T_2 - T_1)}{T_4 - T_3}$$

for $e = 0$ $\quad \eta = 1 - \left(\frac{P_2}{P_1}\right)^{\frac{1-\gamma}{\gamma}}$ $\qquad$ for $e = 1$ $\quad \eta = 1 - \left(\frac{T_1}{T_4}\right)\left(\frac{P_2}{P_1}\right)^{\frac{\gamma-1}{\gamma}}$

Refrigeration Cycle

$$COP_{REF} = \frac{q_{in}}{\text{Net } w'_{in}}$$

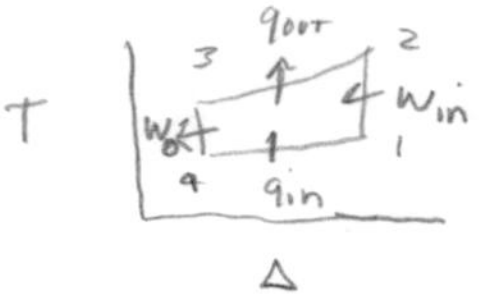

$$COP_{CAR} = \frac{T_L}{T_H - T_L}$$

OTTO CYCLE

$$\text{Power} = \text{net } w_{OUT}\left(N/2\right) \quad N = \text{rev/times}$$

net w_{out} = mep D ← Displacement/cylinder * #cyl

$$\text{Power} = \text{mep}\left(N/2\right)(D)$$

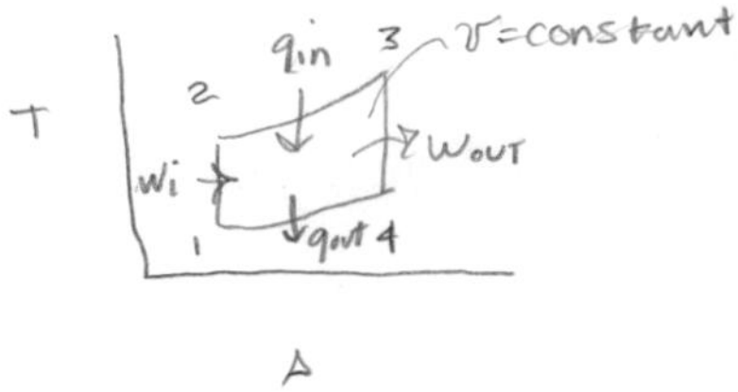

$$w_{in} = u_2 - u_1$$

$$q_{in} = u_3 - u_2$$

$$w_{OUT} = u_3 - u_4 \qquad \eta = 1 - \left(\frac{V_1}{V_2}\right)^{1-\gamma}$$

$$q_{OUT} = u_4 - u_1$$

$$Pv = RT$$

$$\left(\frac{V_1}{V_2}\right)^{\gamma-1} = \frac{T_2}{T_1}$$

ENGINEERING THERMODYNAMICS with Heat Transfer

Second Edition

WILLIAM L. HABERMAN
Engineering Consultant
Rockville, Maryland

JAMES E. A. JOHN
Dean, College of Engineering
University of Massachusetts, Amherst

ALLYN AND BACON

Boston London Sydney Toronto

This book is part of the

ALLYN AND BACON SERIES IN ENGINEERING

CONSULTING EDITOR: FRANK KREITH
UNIVERSITY OF COLORADO

A division of Simon & Schuster
160 Gould Street
Needham Heights, Massachusetts 02194

Library of Congress Cataloging-in-Publication Data

Haberman, William L., 1922–
Engineering thermodynamics with heat transfer / William L. Haberman, James E.A. John. — 2nd ed.
p. cm. — (Allyn and Bacon series in engineering)
Rev. ed. of: Engineering thermodynamics. c1980.
Includes bibliographies and index.
ISBN 0-205-11364-8 :
1. Thermodynamics. 2. Heat—Transmission. I. John, James E. A., 1933– . II. Haberman, William L., 1922– Engineering thermodynamics. III. Title. IV. Series.
TJ265.H24 1989
621.402′1—dc19 88-28297
CIP

Printed in the United States of America

10 9 8 7 6 5 4 3 2 1 93 92 91 90 89

TO THE MEMORY OF MY FATHER

William L. Haberman

TO MY WIFE

James E. A. John

Contents

chapter 4 Thermodynamic Properties of Substances 78

chapter 5 Gas and Gas-Vapor Mixtures 111

chapter 6 The Second Law of Thermodynamics 136

chapter 7 Second Law Analysis 177

Preface

This book is intended to provide the undergraduate engineering student with an understanding of the basic principles of thermodynamics and to introduce the student to the concepts of heat transfer. A course in thermodynamics is generally one of the first courses that an engineering student takes in his or her chosen field. This student lacks the sophistication of a senior-level student; yet too often the course in thermodynamics tends to be a series of very abstract concepts, with little application to the real world. For example, in many textbooks, the Rankine and Otto cycles are treated in a cursory manner, with no attempt to tie the cycles to real systems. Here, we have attempted, wherever possible, to use realistic values of pressures and temperatures; further, we have used the basic concepts of thermodynamics to explain why we are facing such problems as thermal pollution, air pollution, etc. In this way, the student should come to see thermodynamics as an exciting, relevant course to study.

In most available thermodynamics texts, the first several chapters and many pages of material are devoted to an almost endless series of definitions. Too often, the student becomes lost or loses motivation. From such an exposure, the student gets the impression that everything he or she is learning in thermodynamics is new; the series of definitions and words he or she is exposed to amplify this impression. However, the student has already been exposed to the first law and many thermodynamic principles in physics courses. We have tried to take full advantage of this exposure by first drawing on material already covered in physics courses dealing with the conservation of energy, then relating this to the thermodynamic system. Therefore, we have placed our discussion of the first law ahead of the material on thermodynamic properties in Chapters 4 and 5. This way, the student can see very clearly the need for tables of such properties and the necessity for learning how to use these tables.

To present the concepts of entropy and the second law, we utilize the classical approach, starting with the Kelvin–Planck and Clausius statements of the second law. Once the significance of the concept of reversibility has been established, we introduce reversible cycles, the Clausius inequality, and then entropy. Entropy is shown to represent a measure of irreversibility. In this way, the student is able to grasp the logic of introducing entropy as a thermodynamic property. This is reinforced with many examples of the application of entropy and the second law.

In this Second Edition, we have provided new material on second law analysis in Chapter 7. Here, we introduce the concepts of reversible work and availability and show how they can be used to provide an understanding of the potential utilization of energy for a system. Availability analysis is later applied to cycles in Chapter 10 and to combustion processes in Chapter 12.

The laws of thermodynamics and the basic thermodynamic properties are introduced in the first seven chapters. The next five chapters are devoted to engineering applications. Problems and examples have been carefully selected to reflect real engineering situations. Differences between theory and practice are discussed and difficulties involved in the possible achievement of theoretical efficiencies are analyzed. Chapter 9 is new in the Second Edition and covers the thermodynamics of fluid flow, including compressible fluid flow. The development of the equations of fluid flow, using the first and second laws, may be quite different from the way the same equations were developed in courses the student may have had in physics or engineering fluid mechanics. Chapter 10 presents a very thorough treatment of power and refrigeration cycles, including hardware, engineering, and environmental considerations, and the basic thermodynamics of refrigerators, heat pumps, stationary power plants, and automotive engines. Chapter 12 on future energy conversion systems should demonstrate to the student the continuing utility of thermodynamics in treating a wide variety of new problems and situations.

Chapters 13 and 14 present an introduction to heat transfer, with the fundamentals in Chapter 13 and applications in Chapter 14. These chapters are meant to broaden the coverage so as to make the text more suitable for a service course. For example, an electrical engineering curriculum may only provide room for one course in the thermal sciences, so it would be useful for these students to study heat transfer as well as basic thermodynamics. Our text fulfills this need and even offers a section on the cooling of electronic components.

Many worked out examples are provided throughout the text, and there are extensive problem sets at the end of each chapter. Since one of the goals of the text is to give the student an understanding of how to solve engineering problems, the examples and problems involve systems and devices that are familiar to the student and have application in the real world. In many cases, photographs and sketches of hardware are provided along with the discussion to bring the text more graphically to the student. It is recognized that many students have not previously had a great deal of exposure to engineering hardware; pains are taken to ensure that before a device is introduced in the text or in a problem, a description is given of the device and its function.

A dual system of units is used. It is felt that, although we are currently switching over to SI units (Systéme International d'Unités) in the United States, the student must still deal with English units. One has only to walk through a power plant to observe that virtually all pressures, temperatures, and dimensions are still measured and indicated in psi, °F, and feet; there are many other industries in which English units still predominate as well. However, most other countries in the world use SI units. Therefore, we feel that the engineering student should have a feel for both systems. With the use of dual units, the reader should be able to obtain a comparison of the two systems and an understanding of the magnitude of the units. Complete tables of thermodynamic properties for four substances, water, air, R-22, and carbon dioxide, are given in both SI units and English units in the Appendices.

We are pleased with the popularity that the First Edition of this text has enjoyed. We feel that the significant additions and changes we have made make the Second Edition even more suitable not only for use in a thermodynamics course for majors in mechanical engineering but also for use in a service course in thermal sciences.

We acknowledge the many useful suggestions and comments we have received from colleagues relative to this text. We also thank Oretta Taylor for her help in the preparation of portions of the manuscript.

The authors extend thanks to the following reviewers whose contributions have enriched the text:

Professor Theo G. Keithe, Jr.
University of Toledo

Dr. William S. Janna
University of New Orleans

Professor R. D. Slonneger
Morgantown, West Virginia

W. L. H.
J. E. A. J.

chapter

1

Introduction

1.1 WHAT IS THERMODYNAMICS?

Our modern technological society is based largely on the replacement of human and animal labor by inanimate, power-producing machinery. Examples of such machinery are steam power plants that generate electricity, locomotives that pull freight and passenger trains, and internal combustion engines that power automobiles. In each of these examples, working fluids such as steam and gases are generated by combustion of a fuel-air mixture and then are caused to act upon mechanical devices to produce power. Predictions of how much energy can be obtained from the working fluid and how well the extraction of energy from the working fluid can be accomplished are the province of an area of engineering called *thermodynamics*.

Thermodynamics is based on two experimentally observed laws. The first is the *law of conservation of energy*, familiar to the student from the study of classical mechanics. Whereas in mechanics only potential and kinetic energies are involved, in thermodynamics the law of conservation of energy is extended to include thermal and other forms of energy. When an energy transformation occurs, the same total energy must be present after the transformation as before; in other words, according to the first law, all the different types of energy must be accounted for and balanced out when a transformation occurs. For example, in an automobile engine, a specific quantity of thermal energy is released due to the combustion of gasoline in the engine cylinders. Some of this energy goes out the tailpipe as heated exhaust gases and is lost; some is converted to useful work in moving the car; and some is dissipated to the air via the cooling system. Whereas the distribution of these various types of energy is clearly of importance to the engineer, who wants to obtain as much useful work as possible from a given quantity of fuel, the first law merely states that energy can be neither created nor destroyed; it does not provide information as to the ultimate distribution of the energy in its various forms.

The second law provides further information about energy transformations. For example, it places a *limitation* on the amount of useful mechanical work that can be obtained from combustion of the fuel in an automobile engine. The first law states that energy must be conserved. Thus, according to the first law, all the thermal energy available from combustion of the fuel could be converted to useful mechanical work with no losses. Intuitively, however, we know that thermal and other losses are present in the engine. The second law provides a quantitative prediction of the extent of these losses.

An understanding of thermodynamics and the limitations it imposes on the conversion of energy from one to another is very relevant to what is going on in the world today. With limited supplies of conventional energy resources of oil and gas, and with increased demands for an improved standard of living and an accompanying increased demand for energy, it is important that we obtain the maximum utilization of our oil, gas, and coal reserves. Conversion of the chemical energy available in these fuels to usable form should be done as efficiently as possible. Further, we must examine the potential of new sources of energy, such as the sun and the oceans. Again, thermodynamics will be used to evaluate new energy sources and methods of converting the available energy to useful form.

1.2 THERMODYNAMIC SYSTEM

In treating problems in engineering mechanics, it is important first to define a free body, then analyze the forces acting on it, and finally apply the appropriate equations of motion. In thermodynamics, we are interested in the changes that take place either to a substance or within a device as the result of energy transfer or energy transformation. Analogous to the free body, we define a *thermodynamic system* to be the substance or volume in space that we wish to study. Everything outside the boundaries of the system is called the *surroundings*. In a thermodynamic analysis, we must carefully define the system boundaries and then study the changes that take place either to the working fluid within the system or to the working fluid as it crosses the system boundaries.

We can define three types of thermodynamic systems: closed, open, and isolated. In a *closed system*, no mass crosses the system boundaries. For example, in a certain situtiion, we might wish to define a closed system as the liquid water and water vapor contained within a rigid vessel (Figure 1.1). In this case, during a process in which thermal energy is added to the system, the volume of the system remains constant since the vessel is rigid. The addition of energy causes the relative amounts of liquid and vapor within the system to change.

In another case, we might define a closed system to be the gases contained within a piston-cylinder arrangement (Figure 1.2). As the piston is forced upward during a process, the system is reduced in volume; again, no mass crosses the system boundaries, so the system is still closed.

In an *open system*, mass does cross the system boundaries. Generally, we define an open system to be a fixed volume in space. For example, if we wish to analyze thermodynamically the operation of a water pump, we might use the system boundaries shown in Figure 1.3. In this case, the working fluid, water, flows into the system across the surface labeled *in*; then it flows out across surface *e* at higher pressure. Electrical power is supplied across the system boundary to drive the pump.

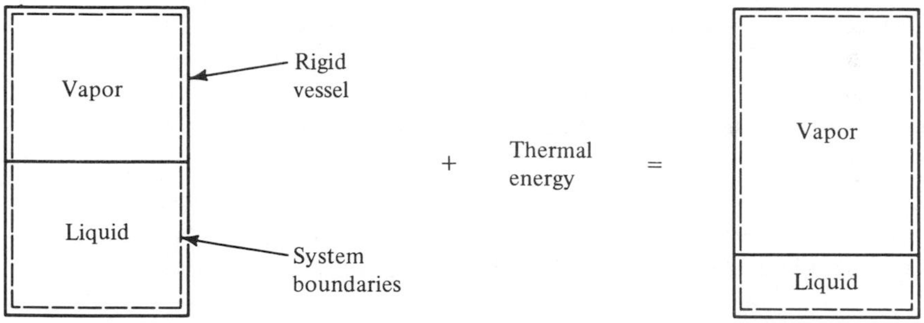

FIGURE 1.1 *Closed System with Constant Volume.*

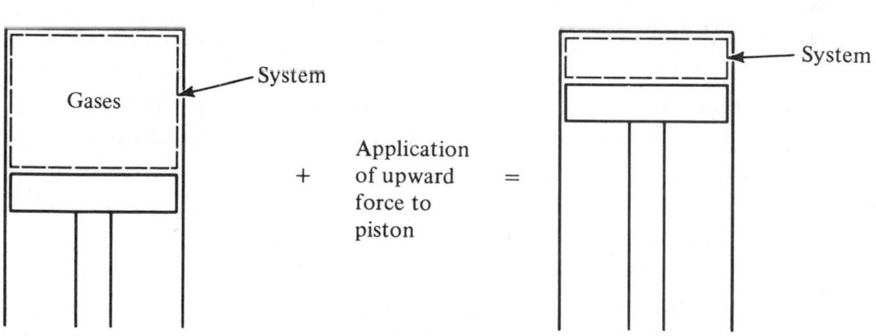

FIGURE 1.2 *Closed System with Nonconstant Volume.*

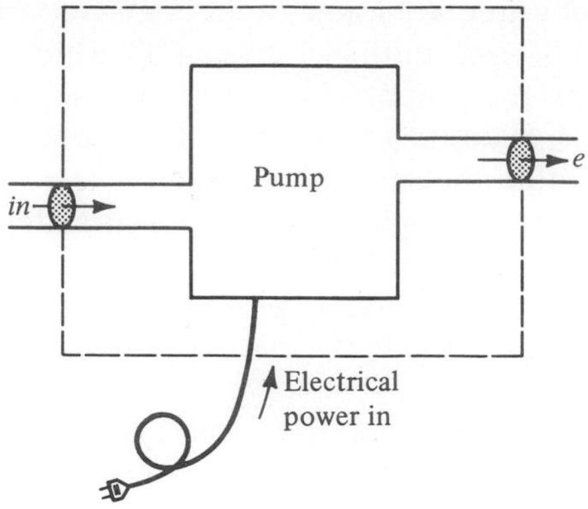

FIGURE 1.3 *Open System.*

An *isolated system* is one that is completely isolated from its surroundings. In other words, no energy crosses the boundaries of an isolated system.

In the next chapter, when we start working problems involving the first law of thermodynamics, we will show how to select the boundaries of a system for a given analytical situation.

1.3 THERMODYNAMIC PROPERTIES

The thermodynamic state of a substance is described by a specification of the thermodynamic properties of the substance at that state. For a given state of a substance, each of the thermodynamic properties has one and only one value. In thermodynamics we want to be able to evaluate quantitatively the change of state of a system as a result of energy transfer across system boundaries or energy transformation within a system. The state of a thermodynamic system is given by a specification of the properties of the substance comprising the system, such as temperature and pressure. Note that if we refer to the value of a property for a system, we imply that this property is the same at all points of the system. This situation, in which the value of a property has significance for an entire system, is called *thermodynamic equilibrium*. For example, if we say that the temperature of a system in thermodynamic equilibrium is 25°C, we mean that the temperature at each point throughout the entire system is 25°C. In this text we will deal with *equilibrium thermodynamics*, a study of the change of equilibrium state of a system due to energy transfer across system boundaries or energy transformation within the system.

There are two types of thermodynamic properties: extensive and intensive. *Extensive properties*, such as mass and total volume, depend on the total mass of the substance present. *Intensive properties* are definable at a point in a substance; if a substance is uniform and homogeneous, the value of the intensive property will be the same at each point in the substance. Specific properties are defined as properties per unit mass, and hence they are intensive. For example, we define *specific volume* as the volume of a substance per unit mass.

To illustrate further the difference between intensive and extensive properties, consider a system consisting of a mass of 10 pounds of a substance, with a system volume of 10 cubic feet. We will assume that the substance is homogeneous and is uniformly distributed over the system volume. This means that at each point in the system, the specific volume is 10/10 or 1 cubic foot per pound. Now divide the system into two equal parts, as shown in Figure 1.4. The mass of the left side is 1/2 the total mass, and the volume of the left side is 1/2 the total volume. However, the specific volume of the left side is still 5/5, or 1 cubic foot per pound. In other words, an extensive property of a system is equal to the sum of the values of the extensive properties of its parts. An intensive property of a system is equal to the value of the intensive property of each of its parts.

We will now discuss several important thermodynamic properties—pressure, density, and temperature—which may already be familiar to the student from introductory courses in physics. Later in the text, other thermodynamic properties, such as internal energy, enthalpy, and entropy, will be introduced.

The mean fluid *pressure*, p, over a plane area is defined to be the ratio of the normal component of force exerted by the fluid on that area to the area. The pressure at a point on the area is the limit that the mean pressure approaches as the area is reduced to a very small size around the point. Since pressure can be defined at a point on a surface, it is an intensive property.

Several methods are used to measure the pressure of a fluid. Usually a pressure gauge measures the difference between the local atmospheric

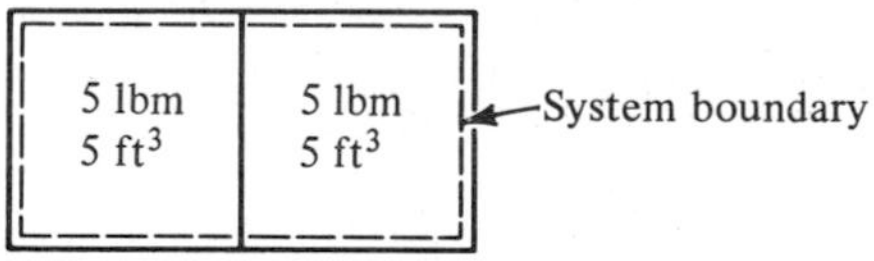

Total system mass = 10 lbm
Total system volume = 10 ft^3

FIGURE 1.4

pressure surrounding the gauge (ambient pressure) and the pressure in the region to which the gauge is attached (Figure 1.5). The gauge reading p_g is

$$p_g = p_{chamber} - p_{ambient}$$

The actual total pressure in the chamber is

$$p_{chamber} = p_g + p_{ambient}$$

The total pressure exerted is called *absolute pressure.*

Another method of measuring pressure is to use a U-tube manometer in which the pressure difference is used to support a liquid column. As shown in Figure 1.6(a), with the right leg open to atmospheric pressure, the magnitude of the distance d gives an indication of the pressure difference between the chamber and the atmosphere (see Example 1.2). Note that the manometer measures a gauge pressure, the difference between chamber pressure and local atmospheric pressure. The simple U-tube manometer shown in Figure 1.6(a) can be modified to measure absolute pressure. The right leg is sealed off and evacuated, so that the pressure at the top of the leg is zero as shown in Figure 1.6(b). Such a pressure-measuring device is called a *barometer*. The difference d in the

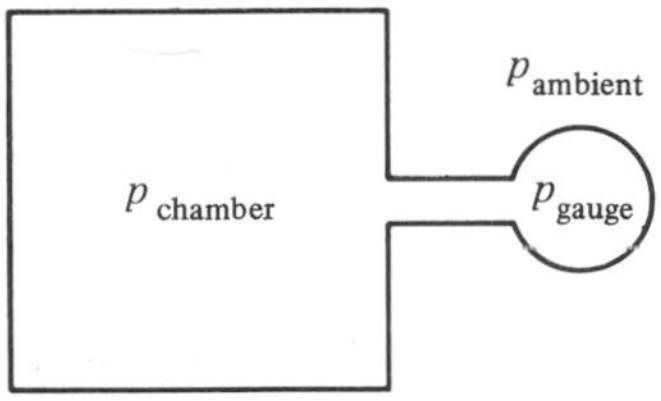

FIGURE 1.5 *Pressure Gauge.*

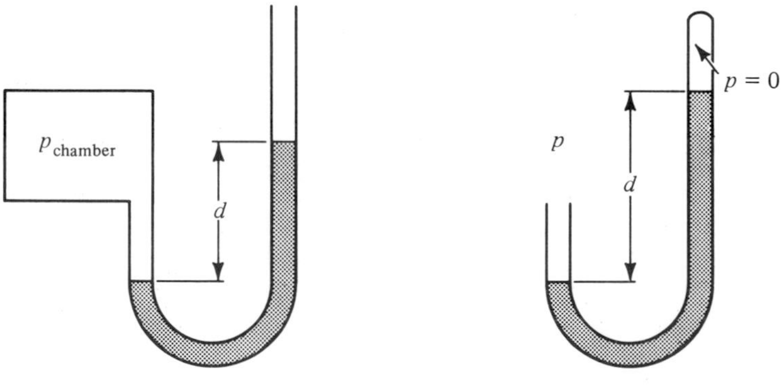

(a) U-tube manometer (b) Barometer

FIGURE 1.6 *Manometers.*

levels is proportional to the absolute pressure p. A barometer can thus be utilized to determine local atmospheric pressure.

Another thermodynamic property is *density*, ρ, which is defined as the mass of a substance per unit volume of that substance. The reciprocal of density has already been introduced and is called *specific volume*, v.

An important property is *temperature*, T. It is a manifestation of the average kinetic energy of the molecules of a substance and is a measure of the hotness or coldness of a substance. We know from experience that if two bodies at different temperature are brought into contact, energy (heat) will flow from the body of higher temperature (warmer body) to the body of lower temperature (cooler body). Further, if two bodies remain in contact over a sufficiently long period of time, they will reach the same temperature. For example, if we touch a container of hot water and then one of ice water, we say one feels hot, the other cold (relative to the temperature of our body). The reason is that the hot water releases energy to us while the ice water extracts energy from us. Note that the concept of temperature is based on the transfer of energy to or from the sensing body. The temperature of a body is a thermodynamic property and a measure of the ability of the body to transfer thermal energy (heat) to another body. When two bodies in contact have the same temperature, no energy transfer will take place.

A seemingly obvious conclusion that can be derived from the above is that when each of two bodies has the same temperature as a third body, the two bodies have equal temperatures. This statement, called the *zeroth law of thermodynamics*, is important in the measurement of temperature.

To detect changes in temperature of a substance, certain physical properties that change with temperature are used. Examples of such properties are the length of a rod, the volume of a liquid, the color of a glowing filament, and the electrical resistance of a wire. The most common temperature-sensing device is the liquid-in-glass thermometer. It consists of a glass tube attached to a glass bulb filled with a liquid (usually mercury or alcohol), as shown in Figure 1.7. The upper end of the tube is sealed off and evacuated. With increase of temperature, the volume of the liquid increases more rapidly than that of the glass container, and the liquid level rises. With decrease of temperature, the liquid level in the thermometer falls. The amount of expansion or contraction of the liquid is used as a measure of the temperature. To give the thermometer a scale for describing the position of the top of the liquid column, two reference points are selected and arbitrary values of temperature assigned to them. Usually, one reference point is chosen as the ice point (freezing point) of water, the other the steam point (boiling point) of water, with both taken at atmospheric pressure. Details of two commonly used temperature scales will be presented in Section 1.5.

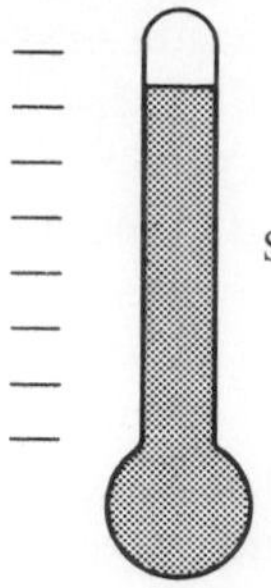

FIGURE 1.7 *Liquid-in-Glass Thermometer.*

The choice of values of temperature at the two reference points will determine the magnitude of each division of temperature and the zero value of temperature. If, for example, we choose water as the reference substance, we might assign a temperature of zero to the freezing point of water and some arbitrary value to the boiling point, with both at atmospheric pressure. The Celsius scale assigns 0° and 100° to the freezing and boiling points of water. The zero points for two different reference substances will not be the same, and hence, just as in the case of pressure, the values of temperature obtained are "gauge" temperature and will depend on the reference substance.

In utilizing temperature in thermodynamic relationships, we often require temperature scales having the same absolute reference point. Such an absolute reference point can be obtained by use of a constant-volume gas thermometer, for which the absolute reference point is independent of the gas used within the thermometer. In such a thermometer (Figure 1.8), gas is contained in a constant-volume vessel equipped with a gauge for measuring gas pressure. When the vessel is subjected to different external temperatures, the pressure of the gas within the vessel will vary, the pressure within the vessel becoming smaller as the temperature is reduced. After first reducing the vessel pressure to a very low value to eliminate the effects of intermolecular forces, we can define an absolute temperature scale such that

$$\frac{p_2}{p_1} = \frac{T_2}{T_1}$$

where p_2 and p_1 are the gas pressures attained when the vessel is exposed to temperatures T_2 and T_1, respectively. For example, if we expose a gas thermometer first to boiling water at atmospheric pressure and then to ice water at atmospheric pressure, the result will show that

$$\frac{T_{\text{boil}}}{T_{\text{ice}}} = 1.366$$

FIGURE 1.8 *Gas Thermometer.*

If we now divide the temperature range between boiling water and ice water into 100 increments, as with the Celsius scale, we obtain the absolute Celsius or Kelvin (K) scale with

$$T_{ice} = \frac{T_{boil}}{1.366}$$

$$= \frac{100 + T_{ice}}{1.366}$$

$$= \frac{100}{0.366} = 273 \text{ K}$$

This result is independent of the gas used in the vessel. In a later chapter, we will derive an absolute temperature scale from purely thermodynamic considerations.

Each of the properties considered—specific volume, pressure, density, and temperature—is a macroscopic property, representing the average effects of a large number of molecules. These properties are useful to the engineer since they can be easily measured, and hence changes in these properties can be readily determined. Throughout this text we shall utilize a macroscopic approach to thermodynamics, in which a thermodynamic state is described by macroscopic properties. This approach is in contrast to a microscopic approach, in which individual molecules and their motions are studied. Clearly, it would be impossible to measure or determine the trajectory of each molecule in a gas or other substance; for a microscopic approach, statistical mechanics would be required.

It is customary to define values of pressure, specific volume, temperature, and other macroscopic thermodynamic properties as existing at each point in a substance or system. This approach assumes that the substance is distributed continuously throughout space. Actually, the substance consists of a large number of molecules, with finite distances between molecules. Any dimensions considered in the macroscopic or continuum approach must greatly exceed intermolecular distances, so the smallest region of interest will include a large number of molecules. For example, suppose that we wish to define the specific volume v of a substance at a point in that substance. Let

$$v = \lim_{M \to M^*} \frac{\forall}{M}$$

where $\forall$ is the volume occupied by the mass M. Here, M^* represents the smallest mass surrounding the point consistent with the continuum approach, and therefore contains a large number of molecules. Note that in air at room temperature and pressure, a cube 0.01 millimeter on a side contains 3×10^{10} molecules. Thus, under these conditions, this dimension of 0.01 millimeter, smaller than that of most engineering measuring instruments or probes, is consistent with the continuum approach. As long as the density of the substance is not too low (in which case a physical dimension might be comparable to an intermolecular distance), the continuum approach can be used to provide an accurate representation of a substance and its properties.

1.4 EQUILIBRIUM

In order to assign one specific value of an intensive property to a thermodynamic system as a whole, the system must be in a state of thermodynamic equilibrium. In this section we will discuss mechanical and thermal equilibrium; in a later chapter we will be concerned with chemical equilibrium of thermodynamic systems. By *mechanical equilibrium*, we mean that there are no unbalanced forces present, either internal to the system or between the system and its surroundings. For example, consider the gas contained within a piston-cylinder arrangement as shown in Figure 1.9. Let the mass of the gas be the thermodynamic system. Take the external pressure applied on the gas to be 30 pounds per

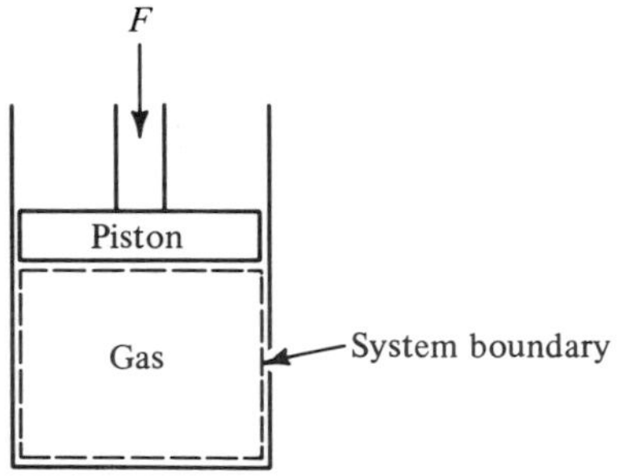

FIGURE 1.9

square inch absolute pressure (30 psia), caused by the action of atmospheric pressure (14.7 pounds per square inch) on the topside of the piston, the weight of the piston, and the force F applied to the piston rod. If the gas pressure is also 30 psia, there will exist a state of mechanical equilibrium, with no unbalanced forces. If now the external force F is removed, there will be an unbalanced force across the piston, and the piston will accelerate upward. During the period of acceleration, the gas pressure and density throughout the gas volume will not be uniform; for example, there will be a greater number of molecules near the power part of the cylinder than near the accelerating piston. It is hence impossible to specify a single pressure for the entire system. In fact, only when there is mechanical equilibrium can a single system pressure be specified.

Thermal equilibrium means that there is no temperature imbalance either within the system or between the system and its surroundings. For example, suppose that a hot flame is brought into contact with the gas in the cylinder, with the piston fixed, as shown in Figure 1.10. The flame temperature is much greater than the gas temperature. Heat therefore flows from the flame to the gas, bringing about an increase of gas temperature. The temperature of the gas will not be uniform, however, since the part of the gas nearest the flame, part A, is hotter than that near the piston, part B. Again, a single temperature cannot be specified for the system when the system is not in thermal equilibrium.

A thermodynamic state of a system is defined by the thermodynamic properties of the system at that state. Thus, a single thermodynamic state of a system can be defined only if that system is in thermodynamic equilibrium. It becomes evident that it is impossible to define the thermodynamic state of a substance at each point in a process if that process is brought about by unbalanced forces or unbalanced tempera-

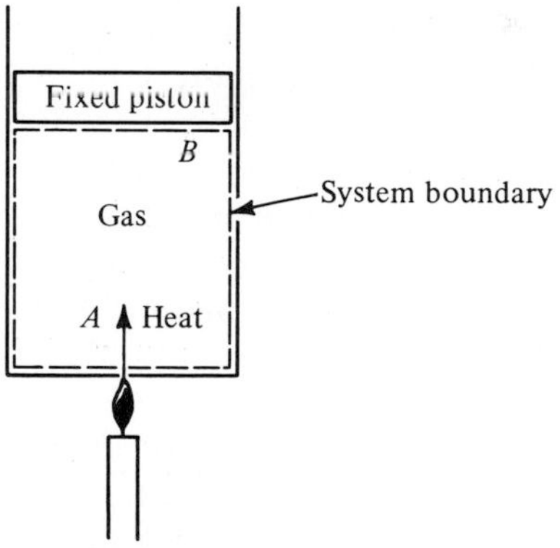

FIGURE 1.10

tures. Whereas this represents, in a sense, a limitation on the type of situation that we can deal with in equilibrium thermodynamics, there are many types of problems where we need be concerned only with the *end* states of a process, at which points equilibrium does exist. Further, in a desire to describe the states of a system during a process, we will use an ideal process, called a *quasi-static* or *quasi-equilibrium process*, in which changes are brought about by infinitesimal forces or infinitesimal temperature differences. In our example of the gas contained in the piston and cylinder of Figure 1.9, we showed that if the force F were completely removed suddenly, the piston would accelerate upward, and it would be impossible to specify a single thermodynamic state of the gas. Now, let us suppose that the force F is gradually released from the piston in small increments. We will perform the process very slowly, so that the gas has a chance to reach an equilibrium state after each increment is removed. We can now specify a thermodynamic state after each force increment has been released.

If these increments are made infinitesimally small, also implying that the piston is allowed to move upward infinitesimally slowly in the cylinder, then each state of the gas can be defined by thermodynamic properties. We will use this quasi-static process many times throughout the text, but remember that it is an idealization.

1.5 UNITS

In classical mechanics it is possible to describe all physical phenomena in terms of three fundamental physical quantities: mass, length, and time. From these quantities it is possible to define derived quantities, such as force, energy, momentum, acceleration, density, and pressure. In thermodynamics we need one additional quantity which cannot be derived from the three fundamental quantities used in mechanics. This additional quantity is *temperature*.

Two systems of units will be used throughout this text: the *English* system of units and the *International System* of Units (SI). In the English system, mass is expressed in pounds mass (lbm), distance in feet (ft), time in seconds (s), temperature in degrees Fahrenheit (°F), and force in pounds force (lbf). In the metric SI system, mass is expressed in kilograms (kg), distance in meters (m), time in seconds (s), temperature in degrees Celsius (°C) or degrees Kelvin (K), and force in newtons (N).

On the Celsius temperature scale (°C), the freezing point of water is 0°C and the boiling point of water is 100°C, with both at standard atmospheric pressure. On the Fahrenheit scale (°F), the freezing point of water is 32°F, while the boiling point is 212°F, again with both at standard atmospheric pressure. A comparison of the two scales is shown in Figure 1.11. Note that the magnitude of the degree Celsius is greater than the magnitude of the degree Fahrenheit by a factor of $(212-32)/100=1.8$. The conversion equations are

$$°F = 1.8°C + 32$$

$$\text{or} \quad °C = \frac{5}{9}(°F - 32) \tag{1.1}$$

For example, a temperature of $-40°F$ is equal to the Celsius temperature of $\frac{5}{9}(-40-32) = -40°C$. A temperature of 1000°C is equal to the Fahrenheit temperature of $1.8(1000)+32 = 1832°F$. The Celsius and Fahrenheit temperature scales have been defined in terms of the freezing and boiling points of water.

Associated with the Celsius scale is the *Kelvin* absolute temperature scale (K), with $K = °C + 273.15$ (usually rounded off to 273); note that the magnitude of the degree Kelvin is equal to the magnitude of the degree Celsius. Associated with the Fahrenheit scale is the *Rankine* absolute temperature scale (°R), with $°R = °F + 459.67$ (usually rounded off to 460); again the magnitude of the degree Rankine is equal to the magnitude of the degree Fahrenheit.

Absolute zero is at the same point; thus conversion is given by

$$K = \frac{5}{9}°R \quad \text{and} \quad °R + 1.8\,K$$

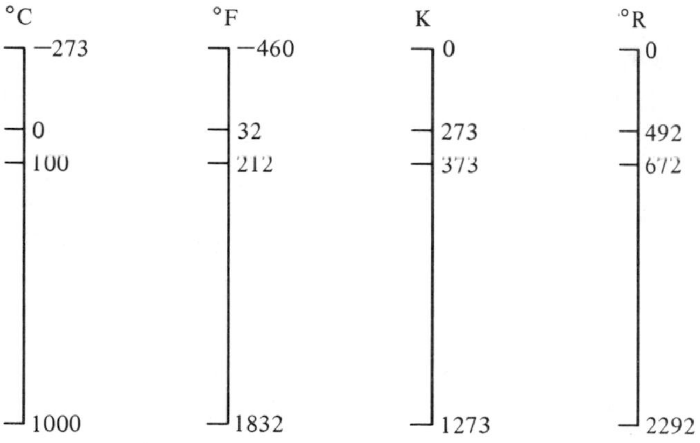

FIGURE 1.11 *Comparison of Temperature Scales.*

For example,

$$300\ \text{K} = 1.8(300) = 540°\text{R}$$

$$450°\text{R} = \frac{5}{9}(450) = 250\ \text{K}$$

The differentiation between units of mass and units of force is often confusing. Remember that force and mass are related by Newton's law:

$$F = kma$$

where k is a proportionality constant. In the English system, 1 pound mass weighs 1 pound on the surface of the earth, where the acceleration due to gravity is 32.2 ft/s^2. Since weight is a force, we obtain a value for k as follows:

$$1\ \text{lbf} = k(1\ \text{lbm} \times 32.2\ \text{ft/s}^2)$$

$$\text{or} \qquad k = \frac{1\ \text{lbf}}{32.2\ \text{lbm-ft/s}^2} = \frac{1}{g_c}$$

where g_c is the numerical value of the standard acceleration due to gravity, or

$$g_c = 32.2\ \frac{\text{lbm-ft/s}^2}{\text{lbf}}$$

According to the SI, the newton (N) is the force required to give a mass of 1 kilogram an acceleration of 1 m/s^2. In this case,

$$k = 1\ \frac{\text{newton}}{\text{kg} \cdot \text{m/s}^2} \qquad \text{or} \qquad g_c = 1\ \frac{\text{kg} \cdot \text{m/s}^2}{\text{N}}$$

Comparing units in the two systems, note that 1 pound mass is equal to 0.454 kilogram, 1 foot is equal to 0.305 meter, and 1 pound force is equal to 4.45 newtons. For example, a pressure of 1 standard atmosphere or 14.7 pounds per square inch is equal to

$$\left(14.7\ \frac{\text{lbf}}{\text{in}^2}\right)\left(144\ \frac{\text{in}^2}{\text{ft}^2}\right)\left(\frac{1}{0.305^2}\ \frac{\text{ft}^2}{\text{m}^2}\right) 4.45\ \frac{\text{N}}{\text{lbf}} = 1.013 \times 10^5\ \frac{\text{N}}{\text{m}^2}$$

It can be seen that N/m^2 is a relatively small unit of pressure; it is usually more appropriate to express pressure in terms of kN/m^2, where 1 kN/m^2 = 1000 N/m^2. Another unit of pressure used in the International System is the *pascal*, where 1 pascal (Pa) is equal to 1 N/m^2. Again, it is usually more convenient to express pressure in terms of kPa, where

1 kPa = 1000 Pa. Note that one standard atmosphere is equal to 101.3 kPa.

Finally, remember that energy has units of force times distance, that is, ft-lbf in the English system (1 Btu = 778 ft-lbf) and N · m in the SI, where 1 joule (J) = 1 N · m.

It is conventional in the SI to use the following prefixes:

k (kilo)	$10^3 \times$	(kPa, km)
M (mega)	$10^6 \times$	(MPa, MJ)
m (milli)	$10^{-3} \times$	(mm, mJ)
μ (micro)	$10^{-6} \times$	(μm)

A complete set of conversion factors is given in Appendix C.

The following examples are to provide familiarity with both the unit systems and the thermodynamic properties discussed in this chapter.

EXAMPLE 1.1

a. Find the weight of a body having a mass of 10 lbm, located at the surface of Mars, where the acceleration due to gravity is 39 percent of that on Earth.

b. Find the weight of a 1-kilogram body of mass on the surface of Mars.

Solution

a. At the surface of Mars, the acceleration due to gravity is $0.39 \times 32.2 = 12.56\ \text{ft/s}^2$. Hence,

$$F = \frac{ma}{g_c} = \frac{(10\ \text{lbm})(12.56\ \text{ft/s}^2)}{32.2\ \text{lbm-ft/lbf-s}^2} = 3.9\ \text{lbf}$$

The body has a weight of 3.9 lbf at the surface of Mars.

b. At the earth's surface, the standard acceleration due to gravity is $9.81\ \text{m/s}^2$. Therefore, on the surface of Mars, $g = 3.83\ \text{m/s}^2$. Again,

$$F = \frac{ma}{g_c} = \frac{(1.0\ \text{kg})(3.83\ \text{m/s}^2)}{1.0\ \text{kg}\cdot\text{m/N}\cdot\text{s}^2} = 3.83\ \text{N}$$

The body weighs 3.83 newtons on the surface of Mars.

EXAMPLE 1.2

The pressure p in an air tank is able to support a mercury column 500 millimeters high (Figure 1.12). Find the absolute tank pressure in pascals if the local atmospheric pressure is 95 kPa. Take the density of mercury to be $13.6 \times 10^3\ \text{kg/m}^3$, and assume that the pressure at point 1 is equal to tank pressure.

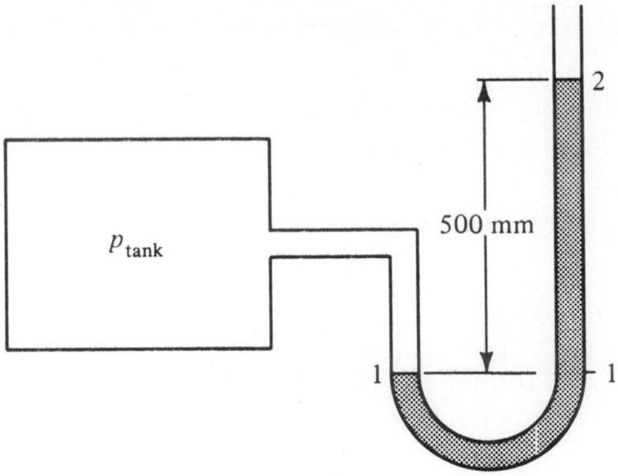

FIGURE 1.12 *Air Tank with Manometer.*

Solution

First write a force balance for the mercury. Note that the pressure at 1′ is equal to that at 1, since these points are at the same level in a static fluid (Figure 1.13).

$$\sum F_{vert} = 0$$

$$\text{or} \qquad p_2 A + W = p_1' A$$

where

$$W = \rho_{Hg} \frac{g}{g_c} AL \quad \text{and} \quad A = \text{cross-sectional area of tube}$$

Solving, we obtain

$$p_1' - p_2 = \rho_{Hg} \frac{g}{g_c} L$$

$$p_{tank} - p_{atm} = (13.6 \times 10^3) \frac{kg}{m^3} \frac{9.81 \text{ m/s}^2}{1 \text{ kg} \cdot \text{m/N} \cdot \text{s}^2} 0.50 \text{ m}$$

$$= (13.6 \times 10^3 \text{ kg/m}^3) 4.905 \text{ N} \cdot \text{m/kg}$$

$$= 6.67 \times 10^4 \text{ N/m}^2 = 66.7 \text{ kPa}$$

Thus

$$p_{tank} = 66.7 + 95 = 161.7 \text{ kPa}$$

We have shown that 500 millimeters of mercury measured on a U-tube manometer are equivalent to 66.7 kPa or 0.660 standard atmosphere. It follows that 1 standard atmosphere (101.3 kPa) is equivalent to 760 millimeters of mercury.

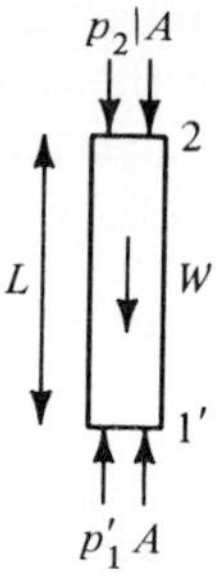

FIGURE 1.13

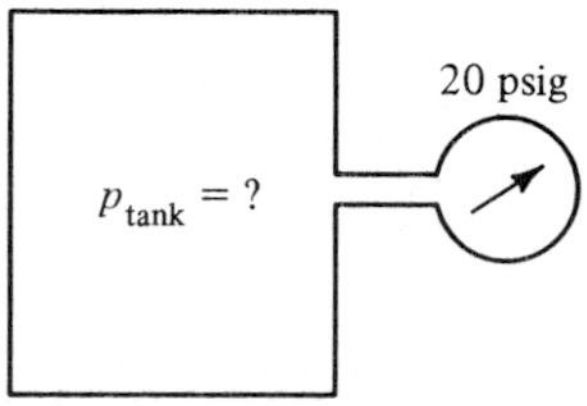

FIGURE 1.14 *Air Tank with Pressure Gauge.*

EXAMPLE 1.3

A pressure gauge connected to an air tank reads 20 pounds per square inch gauge pressure (20 psig). Find the absolute pressure (psia) of the air inside the tank (Figure 1.14). Assume an ambient pressure of 14.6 psi.

Solution

The pressure gauge reads the difference between the absolute pressure and the local ambient pressure. Thus

$$p_{tank} = 20 + 14.6 = 34.6 \text{ psia}$$

EXAMPLE 1.4

A 10-lbm mass of air is contained in a closed tank with rigid walls and an internal volume of 100 cubic feet. Heat is added to the air, raising its temperature from an initial value of 500°R to a final value of 1000°R. Determine the specific volume of the air at initial and final states.

Solution

Initially, the specific volume of the air is $100/10 = 10\ \text{ft}^3/\text{lbm}$. Since the internal volume of the tank does not change (rigid walls) and no air can enter or leave the tank (closed tank), the final specific volume must also be $10\ \text{ft}^3/\text{lbm}$.

The next two examples will introduce energy units, dimensions, and concepts that will be presented in the next chapter; they also serve as a review of some principles already learned in mechanics.

EXAMPLE 1.5

An automobile having a mass of 3200 lbm travels at a speed of 50 mph at sea level. What is the kinetic energy of the car? At what elevation above sea level would the automobile have a potential energy greater than that at sea level by the same amount?

Solution

$$50\text{ mph} = 50\ \frac{\text{miles}}{\text{h}}\ \frac{5280\text{ ft/mile}}{3600\text{ s/h}} = 73.3\text{ ft/s}$$

$$\text{Kinetic energy} = \frac{1}{2}\frac{m}{g_c}V^2 = \frac{1}{2}\ \frac{(3200\text{ lbm})}{(32.2\text{ lbm-ft/lbf-s}^2)}\ 73.3^2\ \frac{\text{ft}^2}{\text{s}^2}$$

$$= 267{,}000\text{ ft-lbf}$$

$$\text{Potential energy} = m\frac{g}{g_c}h = 267{,}000\text{ ft-lbf}$$

$$h = \frac{(267{,}000\text{ ft-lbf})}{(3200\text{ lbm})32.2\text{ ft/s}^2}\ 32.2\ \frac{\text{lbm-ft}}{\text{lbf-s}^2}$$

$$= 83.4\text{ ft above sea level}$$

EXAMPLE 1.6

A block having a mass of 50 kg starts from rest atop the inclined plane shown in Figure 1.15. Determine the speed at the bottom of the plane from energy considerations. Assume a coefficient of friction of 0.3 between block and plane.

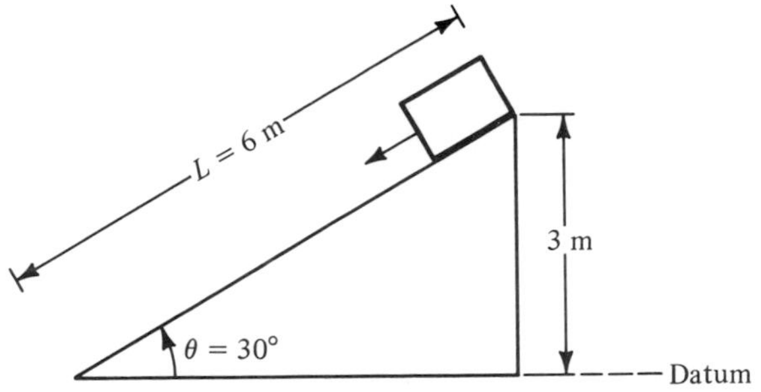

FIGURE 1.15

Solution

The initial potential energy of the body above the datum is

$$m\frac{g}{g_c}h = \frac{(50\,\text{kg})(9.81\,\text{m/s}^2)3\,\text{m}}{1\,\text{kg}\cdot\text{m/N}\cdot\text{s}^2} = 1471.5\,\text{N}\cdot\text{m} = 1471.5\,\text{J}$$

Work done against friction as the body moves down the plane is

$$\text{Friction force} \times L = \left[\mu\,\frac{mg}{g_c}\cos\theta\right]L$$

$$= \left[\frac{(0.3)(50\,\text{kg})(9.81\,\text{m/s}^2)(0.866)}{1.0\,\text{kg}\cdot\text{m/N}\cdot\text{s}^2}\right]6\,\text{m}$$

$$= 764.6\,\text{N}\cdot\text{m} = 764.6\,\text{J}$$

We can now equate the difference between the initial energy of the block and the work energy used up to overcome friction to the final kinetic energy of the block:

$$\frac{1}{2}\frac{m}{g_c}V^2 = \frac{mg}{g_c}h - (\text{friction force}) \times L$$

$$= 1471.5\,\text{J} - 764.6\,\text{J}$$

$$= 706.9\,\text{J}$$

$$V^2 = \frac{(706.9\,\text{N}\cdot\text{m})}{50\,\text{kg}}\left(1\,\frac{\text{kg}\cdot\text{m}}{\text{N}\cdot\text{s}^2}\right)2$$

$$= 28.28\,\frac{\text{m}^2}{\text{s}^2}$$

or $V = 5.32\,\text{m/s}$ at bottom of inclined plane

1.6 SUMMARY

In this first chapter, we have attempted to provide some idea of what thermodynamics is all about. Fundamentally, it is a study of the transformation of energy from one form to another, and the change in state of a system brought about by the transfer of energy in the form of heat and work. Since a single thermodynamic state of a system can be defined only if the system is in equilibrium, it follows that thermodynamics is a study of equilibrium states.

In the following chapters, we will study the two laws of thermodynamics and apply them to diverse physical situations. We will analyze power-producing systems, such as steam power plants and internal combustion engines, as well as power-absorbing devices, such as heat pumps and refrigerators.

PROBLEMS

1.1 The air temperature on a nice summer day is 80°F. What is this temperature in °C? What is the absolute air temperature in °R and in K?

1.2 The normal human body temperature is 98.6°F. What is it on the Celsius scale, on the Rankine scale, and on the Kelvin scale?

1.3 The reading of a thermometer using the Fahrenheit scale is the same as the reading on a Celsius thermometer. What is the temperature? This value is often used as a reference temperature for presenting thermodynamic properties of substances.

1.4 What will an astronaut's weight be on the moon if his weight on earth is 185 lbf? The acceleration due to the moon's gravity is 1/6 that of the earth's. What will his mass be on the moon?

1.5 The mass of the earth is approximately equal to 6×10^{24} kg. What is the average density of the earth, assuming the earth to be a sphere of radius 6.4×10^{6} m?

1.6 A cubical block of a material having a volume of 1 m^3 weighs 2000 newtons at the earth's surface. What is its density?

1.7 A rigid container having a volume of 3.5 ft^3 holds 12 pounds mass of a gaseous substance. Find the specific volume of the gas. If 6 pounds mass of the gas is discharged from the container, what will be the specific volume of the gas remaining in the container?

1.8 Gas at a pressure of 50 psia is contained in a cylinder. Calculate the force exerted on a 1-ft^2 plane surface located at the bottom of the cylinder.

1.9 One kilogram of a liquid having a density of 1000 kg/m^3 is mixed with 0.5 kilogram of a liquid having a density of 800 kg/m^3. Find the density of the mixture.

1.10 Two columns of liquid are connected to the same vacuum pump as shown in Figure 1.16. The column of water stands at 0.25 m and that of the unknown

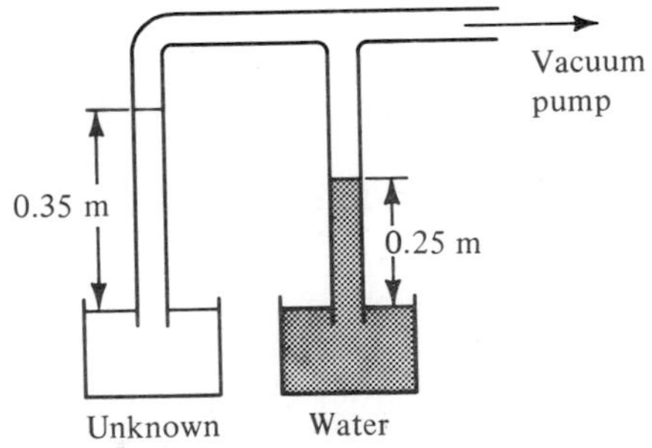

FIGURE 1.16

liquid at 0.35 m. Determine the density of the unknown liquid. The specific volume of the water is 0.001 m^3/kg.

1.11 A closed-tube manometer containing mercury is used to measure atmospheric pressure (Figure 1.17). What will the atmospheric pressure be if the height h of mercury column is 30 in.? The density of mercury is 810 lbm/ft^3.

1.12 A water manometer shown in Figure 1.18 is used to measure the low pressure in a natural gas main. The water stands 5 millimeters higher in the right-hand tube than in the left-hand tube. Determine the absolute pressure of the natural gas in Pa if a closed-tube barometer measuring local atmospheric pressure has a reading of 750 millimeters of mercury.

1.13 In Figure 1.19, the pressure within the air below the floating piston is 18 psig. The diameter of the piston is 4 in. What is the weight of the piston at equilibrium?

1.14 A 10-kg mass is accelerated at a constant rate from standstill along a frictionless horizontal plane. What is the work required to reach a velocity of 10 m/s? 20 m/s?

1.15 A wooden block having a mass of 20 kg is projected with a speed of 10 m/s up a plane inclined at 45 deg with the horizontal. For a coefficient of friction of 0.25 between block and plane, what height above initial position will the block reach?

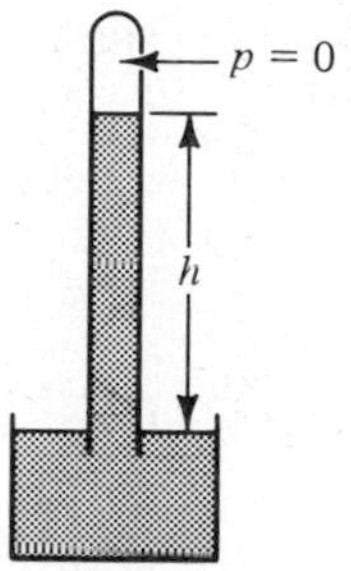

FIGURE 1.17

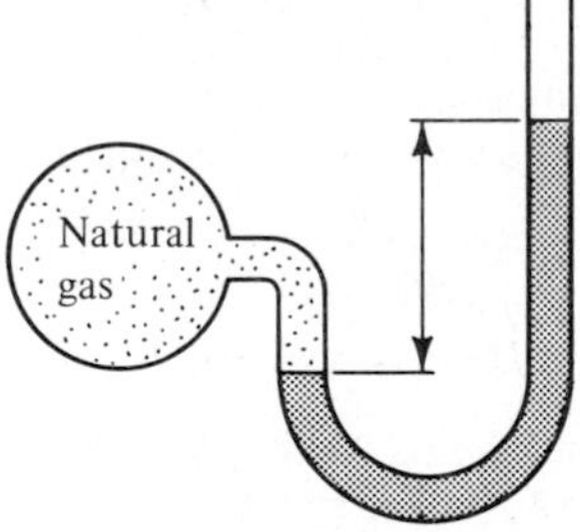

FIGURE 1.18

FIGURE 1.19

1.16 An elevator having a mass of 1000 kg is raised 100 m at a constant speed of 2 m/s by the action of a motor. Neglecting frictional effects, what is the power output of the motor? (Power is the rate of doing work.)

1.17 An automobile weighing 3000 lbf travels at a speed of 55 mph at the bottom of a hill. It is allowed to coast uphill to a standstill 70 ft above the bottom of the hill. Calculate the energy dissipated during the ascent.

1.18 The hammer of a pile driver is lifted 1.5 m in 5 seconds. What is the power of the engine of the pile driver for a hammer having a mass of 700 kg? (Power is the rate of doing work.)

1.19 If the engine of an automobile delivers 30 hp when the automobile is traveling along a level highway at a constant speed of 50 mph, what will be the resisting force? The automobile weighs 2500 pounds.

1.20 Determine the horsepower requirement of the automobile of Problem 1.19 if the automobile descends a grade of 5 percent; that is, the road descends vertically 1 ft in 20 ft horizontally.

1.21 Discuss the relation between the zeroth law and the measurement of temperature.

1.22 A force of 25 newtons accelerates a mass of 5 kg. Calculate the acceleration of the mass.

1.23 The cylinder of Figure 1.19 contains helium gas. The piston has a mass of 30 kg and a diameter of 200 mm. Determine the pressure within the cylinder for a gravitational acceleration of 9 m^2/s.

1.24 A tank, open to the atmosphere, is filled with a liquid to a depth of 30 ft. The density of the liquid is 50 lbm/ft^3. Calculate the pressure of the liquid at the tank bottom.

1.25 The tank of Problem 1.24 is 35 ft high. Determine the work required to raise all the liquid over the top rim of the tank.

1.26 Calculate the work done to accelerate a 10-kg body from a speed of 100 to 125 m/s in an interval of 20 s.

chapter

2

The First Law of Thermodynamics for Closed Systems

2.1 INTRODUCTION

In this chapter we will discuss the principle of conservation of energy as applied to thermodynamic systems. This is one of the most important principles of thermodynamics, and is called the *first law*.

In purely mechanical systems in which no frictional interactions occur (called *conservative* mechanical systems), two types of energy are distinguished: *kinetic energy* and *potential energy*. The total energy of a conservative mechanical system consists of the sum of the kinetic and potential energies of the particle or particles comprising the system; it remains constant if no work is done on the system by external forces. For example, if a ball is thrown vertically up into the air, the ball leaves the hand of the thrower with a certain velocity and hence a certain kinetic energy. When the ball reaches its maximum height, its velocity has vanished, and the ball possesses potential energy only. The initial kinetic energy that the ball had was transformed in the gravitational force field of the earth into gravitational potential energy.

In thermodynamics, the transformation of one energy form into another is of prime interest. Not only is the conversion of one mechanical energy form into another mechanical energy form of interest, but also important are both the conversion of chemical into mechanical energy, such as occurs in the gasoline engine, and the conversion of mechanical into electrical energy, such as occurs in an electric generator. In the latter, the kinetic energy of rotation of the drive shaft of the generator emerges as electric energy at the generator's output terminals, as shown in Figure 2.1(a). On the other hand, in an electric motor the reverse process occurs; that is, electric energy introduced to the motor at its terminals emerges as mechanical energy at the motor shaft, as Figure 2.1(b) demonstrates.

In addition to a transformation energy from one form to another within the boundaries of an energy device, there can occur a transfer of energy across the boundaries of a device. Such transfer can be of two forms: *performance of work* and *transfer of heat*. The throwing of a ball serves as an example of how energy is transferred as *work* from one system to another. In order to give the ball its initial velocity, the thrower applies a force to the ball and hence does work on the ball, increasing its kinetic energy from zero to its initial value as it leaves the thrower's hand. Another everyday example of the transfer of energy by the performance of work occurs in a kitchen blender, in which the blender motor does work on the liquid to be mixed through the blender shaft and blades (Figure 2.2).

The second form of energy transfer is *heat*. For example, when water is warmed on the stove, heat is transferred into the water in the kettle. For the transfer of heat to be effected, there must exist a difference in temperature between the gas flame (or heating coils) and the water. The temperature of the gas flame is several hundreds of degrees Fahrenheit, while that of the water is somewhat above room temperature.

These examples illustrate (1) the transformation of energy from one form to another within a system or device and (2) the transfer of energy across the boundaries of a system or device. The transfer of heat occurs solely because there exists a difference of temperature between two media and the performance of work occurred because a force was applied in the direction of motion. In either case, the energy of the system receiving the energy transfer is increased. If the system does work, or if heat is transferred from the system, the final energy level of the system will decrease.

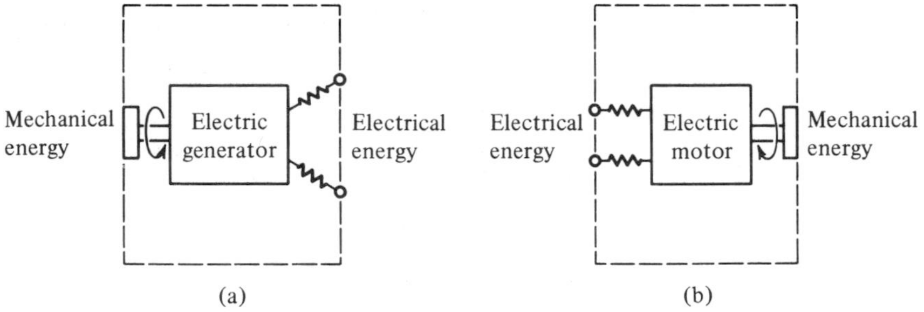

FIGURE 2.1 *Transformation of Energy.*

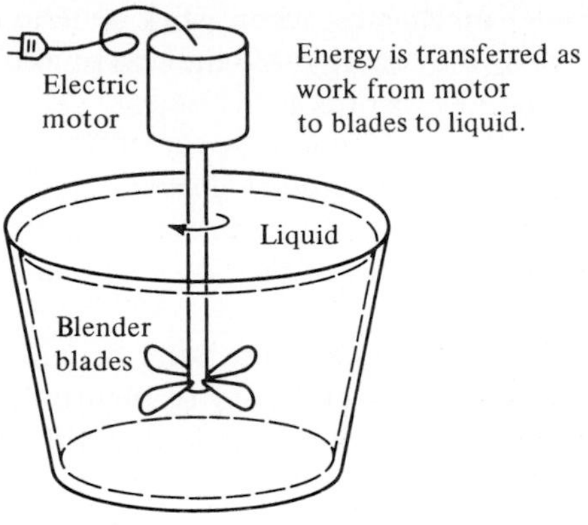

FIGURE 2.2 *Performance of Work.*

2.2 WORK

Now we will discuss in further detail the concept as encountered in mechanics. In mechanics *work* is defined as the product of the component of force along the line of motion of a body and the magnitude of the displacement of the body. For example, in Figure 2.3, force F is applied to the body, causing it to move a distance s, with the force applied at an angle θ to the body motion; thus the work (W) done on the body is

$$W = (F \cos \theta)s$$

where $F \cos \theta$ represents the component of force along the line of motion of the body. Thus work is done only when the force has a component along the line of motion of the body. If the force is at right angles to the displacement, the work is zero. In mechanics, one uses the convention that work is considered positive when the force component and the displacement are in the same direction, that is, when they have the same sign. For example, when a body is raised in an elevator, the work of the lifting force is positive, since force and displacement are in the same direction.

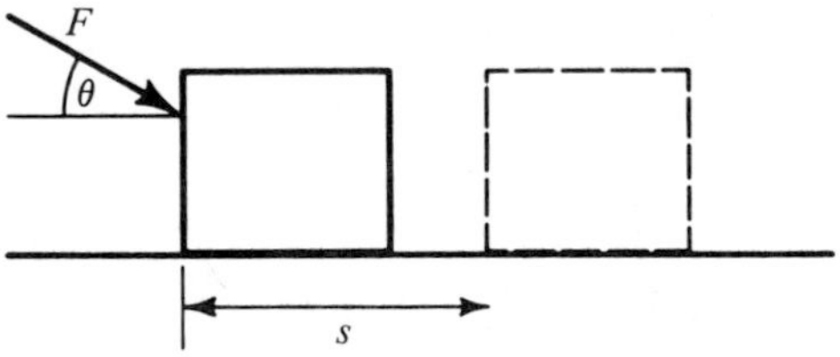

FIGURE 2.3 *Work Done on Body.*

In the convention of mechanics, when work is done on a body, it is considered positive, and the energy of the body is increased. The work-energy principle stated in words is

The work of all externally applied forces on a body, with the exception of the gravitational force, equals the change in total mechanical energy of the body (sum of the potential and kinetic energies).

As an example of this principle, let a body of mass m move along path s from point 1 to point 2, as indicated in Figure 2.4. Let F be the resultant of all external forces acting on the body, with the exception of gravity. The component of F along path s is

$$F_s = F \cos \theta$$

With the definition of work as the component of force along the line of motion times the magnitude of displacement, we obtain from conservation of energy

$$\int_1^2 F \cos \theta \, ds = m \frac{g}{g_c} (z_2 - z_1) + \frac{m}{2g_c} (V_2^2 - V_1^2) \tag{2.1}$$

Work due to forces other than gravity	Change of potential energy of body	Change of kinetic energy of body

When the only force acting on a body is the gravitational force (that is, when $F = 0$ and hence work $= 0$), the energy equation in this special case

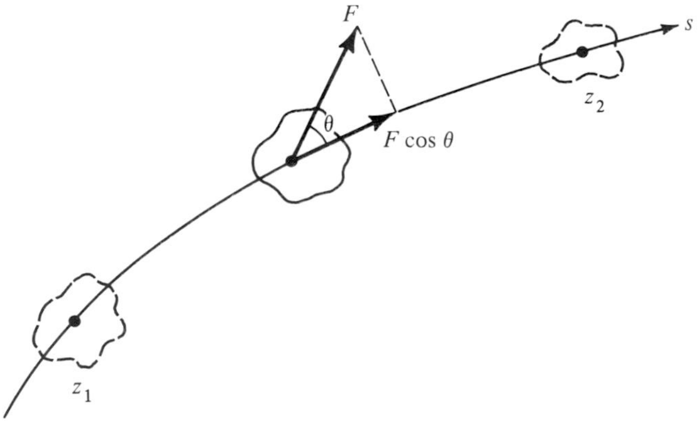

FIGURE 2.4

becomes

$$\frac{1}{2}mV_2^2 + mgz_2 = \frac{1}{2}mV_1^2 + mgz_1 \tag{2.2}$$

EXAMPLE 2.1

A ball is thrown vertically upward. During the throwing process with the ball in the thrower's hand, a ball having a mass of 1/3 lbm is moved upward from rest over a distance of 1.5 feet, leaving the thrower's hand with a velocity of 20 fps. Find the average force that the thrower exerts on the ball, the work done on the ball, and the height that the ball will reach.

Solution

To find the work W done on the ball, substitute into Equation (2.1) with $z_1 = 0$ (ground level) and $V_1 = 0$:

$$W_{\text{in}} = \frac{1}{3}\text{ lbm}\,\frac{32.2\text{ ft/s}^2}{32.2\text{ lbm-ft/lbf-s}^2}\,1.5\text{ ft} + \frac{1}{3}\text{ lbm}\,\frac{1}{2}\,\frac{20^2\text{ ft}^2/\text{s}^2}{32.2\text{ lbm-ft/lbf-s}^2}$$

$$= 0.50\text{ ft-lbf} + 2.07\text{ ft-lbf} = 2.57\text{ ft-lbf}$$

The average force that the thrower exerts is

$$F = \frac{W_{\text{in}}}{z_2 - z_1} = \frac{2.57}{1.5} = 1.713\text{ lbf}$$

Next consider the motion of the ball after it leaves the thrower's hand; neglect the resistance of the air during the motion. Since no external forces (other than gravity) are exerted, Equation (2.2) applies:

$$\frac{1}{2}\frac{m}{g_c}V_3^2 + m\frac{g}{g_c}z_3 = \frac{1}{2}\frac{m}{g_c}V_2^2 + m\frac{g}{g_c}z_2 = 2.57\text{ ft-lbf}$$

To find the maximum height (above the point at which the ball started in the thrower's hand) that the ball will reach, substitute $V_3 = 0$ into the above equation:

$$\frac{1}{3}\text{ lbm}\,\frac{(32.2\text{ ft/s}^2)z_3(\text{ft})}{32.2\text{ lbm-ft/lbf-s}^2} = 2.57\text{ ft-lbf} \quad \text{or} \quad z_3 = 7.71\text{ ft}$$

In thermodynamics, interest lies in devices that do work, and hence emphasis is on performance of work by a system. It thus has become customary in thermodynamics to refer to work done by a system as *positive*. In this convention, positive work causes the energy of a thermodynamic system to decrease. For clarity, we will use the symbol W_{out} to signify that work is done *by* the system, and W_{in} to signify that work has been done *on* the system.

There exists a number of ways in which work can be performed by or on a system. Of major interest in general thermodynamics is mechanical work, that is, work done by the movement of the boundary or portions of the boundary of the system. This movement may be normal to the boundary, for example, the movement of a piston in a cylinder (Figure 2.5); or it may be tangential to the boundary, such as rotating blades or paddles (Figure 2.6); or it may consist of the stretching of the boundary, such as the small extension of a wire under tension (Figure 2.7). The mechanical work involving large-scale motion of the boundary usually involves a shaft and is therefore also termed *mechanical shaft work*.

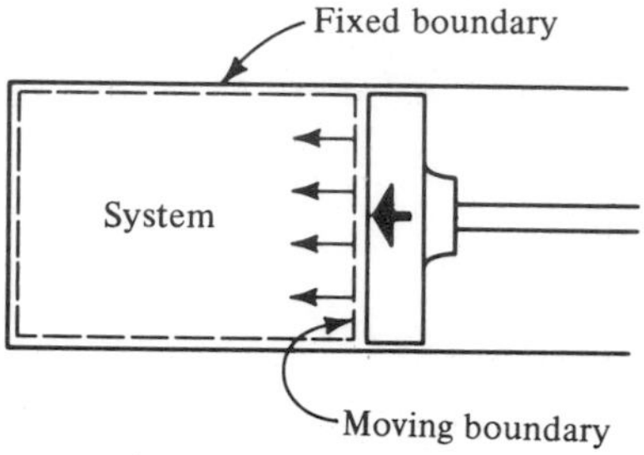

FIGURE 2.5 *Work Done by Piston.*

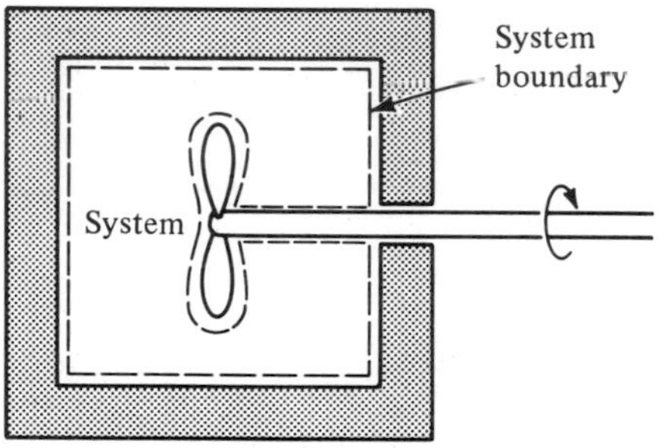

FIGURE 2.6 *Work Done by Rotor.*

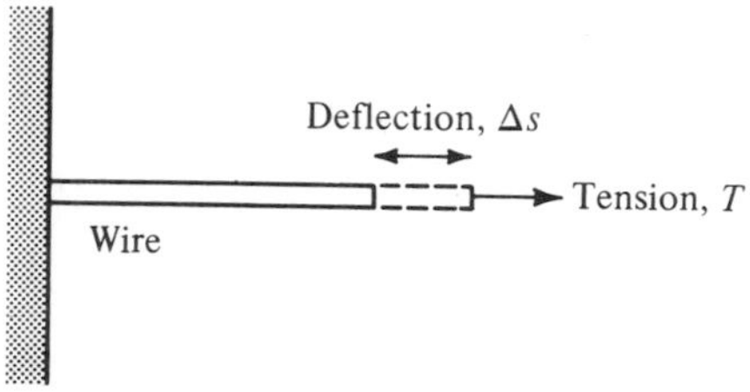

FIGURE 2.7 *Work Done in Extending Wire.*

EXAMPLE 2.2

A gas is contained in a cylinder and expands against a piston, as shown in Figure 2.8. Find an expression for the work done by the gas in terms of the thermodynamic properties of the gas. Assume that the expansion is a quasi-static process, so that the gas is in thermodynamic equilibrium at each point in the process.

Solution

Since the pressure p of the gas will decrease as the gas expands behind the piston, the work must be obtained by integration over the interval from position 1 to position 2. As the piston moves an incremental distance dL, the incremental work done by the gas, dW, is $pA\,dL$, where pA is the force exerted on the piston and dL is the displacement (Figure 2.9). If the gas is considered as the system, work is done by the gas; but $A\,dL = d\mathcal{V}$, the incremental change of volume of the gas. Therefore, $dW = p\,d\mathcal{V}$. The total work done by the gas is found by integration over the interval 1 to 2:

$$W_{\text{out}} = \int_{\mathcal{V}_1}^{\mathcal{V}_2} p\,d\mathcal{V} \tag{2.3}$$

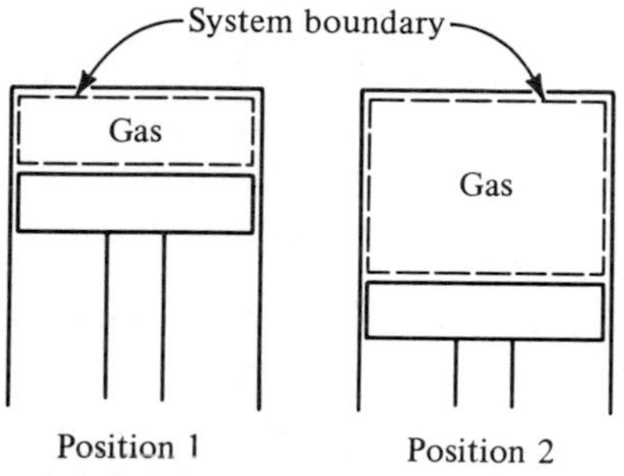

FIGURE 2.8 *Cylinder-Piston System.*

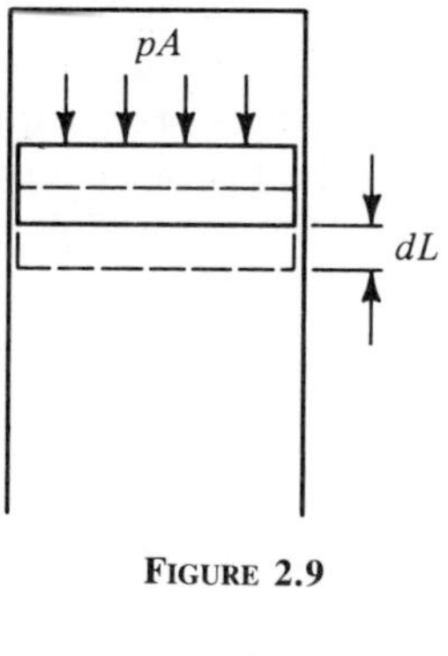

FIGURE 2.9

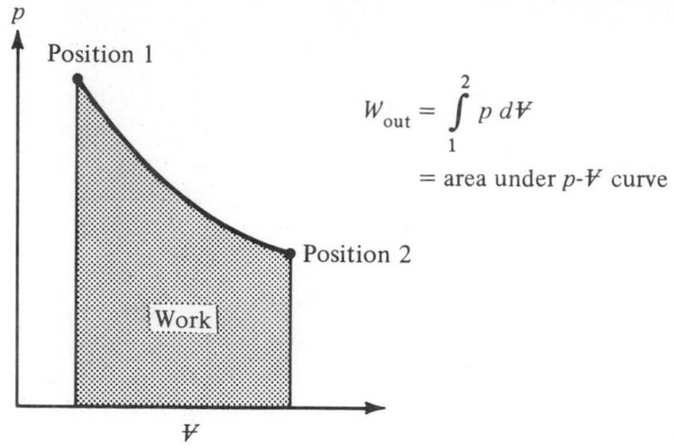

FIGURE 2.10 *p-$\mathcal{V}$ Integral.*

We therefore need to have a relationship between the pressure p and volume $\mathcal{V}$ during the expansion process in order to perform the actual integration (Figure 2.10). Note that for Equation (2.3) to apply, there must exist a state of thermodynamic equilibrium at each point in the expansion process, so that we can ascribe a single value of pressure to the gas in the cylinder at each point. For example, there can be no force imbalance either within the system or between the system and its surroundings. In other words, the external pressure due to externally applied forces, external pressure, and piston weight can be only infinitesimally less than the internal gas pressure at each point in the expansion.

The piston in an actual reciprocating internal combustion engine moves up and down very rapidly in the cylinder. It is clear that, in actuality, there are departures from thermodynamic equilibrium taking place within the cylinder. However, Equation (2.3) is used to obtain an approximation of the work done by the gases in the cylinder, with perhaps a correction factor applied if a more exact value is desired.

From the results of Example 2.2, it is clear that the amount of work performed by a system in going between two states depends on the path the system takes between the states. For example, consider the quasi-static expansion of gases behind a piston in a cylinder, with the gas going from pressure p_1, volume $\mathcal{V}_1$ to pressure p_2, volume $\mathcal{V}_2$. There are a multitude of paths available, some of which have been indicated on Figure 2.11. For each possible curve between state 1 and state 2,

$$W_{out} = \int_1^2 p\, d\mathcal{V}$$

will be different. In other words, we say that work is a path function.

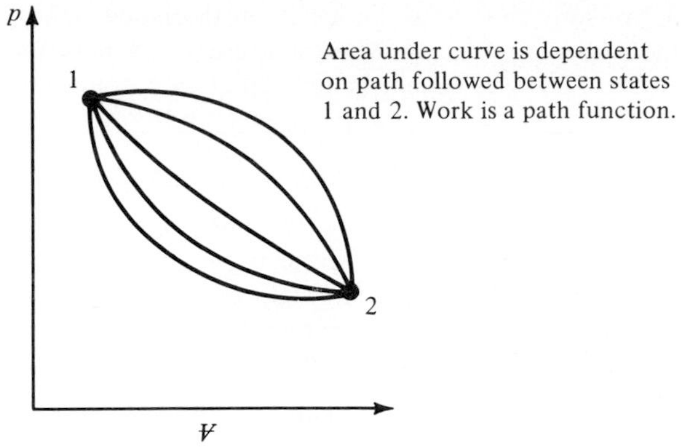

FIGURE 2.11 *p-V Integral Indicating Work Is a Path Function.*

Work is not a characteristic of a substance or a system of a given state, and therefore it is not a thermodynamic property.

EXAMPLE 2.3

A gas at pressure p_1 is contained behind a membrane in one half of a rigid vessel, as shown in Figure 2.12(a). The other half of the vessel is completely evacuated. The membrane is broken, allowing the gas to fill the entire volume, as shown in Figure 2.12(b). Calculate the work done by the gas.

Solution

Of importance here is the selection of system boundaries. Let the system consist of the entire contents of the rigid vessel. Since the boundaries of the system do not move, there is no work done by the system. However, the only substance present in the system is the gas; hence it follows that no work is done by the gas. Note that, in this case, the process consists of a sudden expansion of a gas, caused

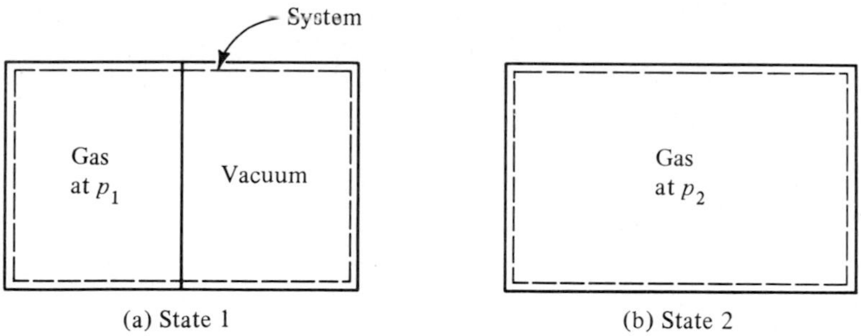

FIGURE 2.12

by imbalance of pressure, so there is not a state of thermodynamic equilibrium at each point in the expansion. Clearly, we cannot use $W = \int p\, d\mathcal{V}$ for this process, since a unique system pressure cannot be defined at each point in the expansion.

EXAMPLE 2.4

An elastic wire of length L is stretched by the application of a tensile force T to the end of the wire (Figure 2.13). The stretching is carried out quasi-statically. Determine the work done when the wire length is increased by length ΔL.

Solution

Take the wire as the thermodynamic system. The differential work done as the wire changes by length dL is $T\, dL$. If we assume that the wire is elastic, stress σ is proportional to strain ε, or

$$\sigma = \frac{T}{A} = E\varepsilon$$

where A is the cross-sectional area of the wire and E is Young's modulus. In this example,

$$d\varepsilon = \frac{dL}{L}$$

so that

$$dW_{\text{in}} = (AE\varepsilon)L\, d\varepsilon$$

Integrating, we obtain

$$W_{\text{in}} = \int_{\varepsilon=0}^{\varepsilon_{\text{final}}} AE\varepsilon L\, d\varepsilon$$

$$= AEL\, \frac{\varepsilon_{\text{final}}^2}{2}$$

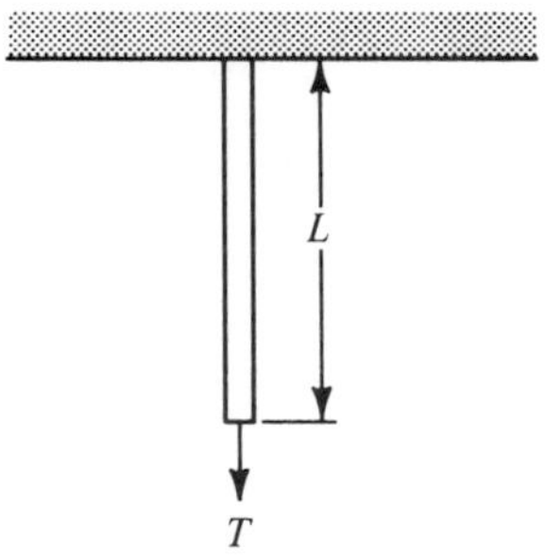

FIGURE 2.13

To determine $\varepsilon_{\text{final}}$, integrate the expression given above for $d\varepsilon$:

$$\varepsilon_{\text{final}} - 0 = \ln \frac{L_{\text{final}}}{L_{\text{initial}}}$$

For small ΔL, this result can be approximated by

$$\varepsilon_{\text{final}} = \frac{\Delta L}{L_{\text{initial}}}$$

Therefore,

$$W_{\text{in}} = AEL \frac{\varepsilon_{\text{final}}^2}{2}$$

$$= \frac{AE}{L} \frac{(\Delta L)^2}{2}$$

There are other forms of work that do not involve movement of system boundaries. An example is the work done as an electric current flows through a resistor. Let ΔV be the electrical potential developed across the resistance. Remember that electrical potential is the potential for doing work per unit charge, while current is the flow of charge per unit time. If the resistor is treated as the system, it follows that the work done on the system is

$$W_{\text{in}} = (\Delta V) I \, \Delta t$$

where Δt is the time during which the current flows and I is the current.

2.3 HEAT

Heat is defined as thermal energy transferred across the boundaries of a system solely because of a temperature difference between the system and its surroundings. Specifically, heat is transferred from a body at higher temperature to a body in contact at lower temperature.

By way of sign convention in thermodynamics, heat is taken to be positive if it is added to the system and is usually given the symbol Q. For clarity in the text, we will use the notation Q_{in} to signify heat transfer into the system, and Q_{out} to signify heat transfer from the system to the surroundings.

Heat is a form of energy, and therefore units for heat are energy units, or ft-lbf in the English system. Historically, heat has also been given in Btus (British thermal units), where 1 Btu is the amount of heat necessary to raise 1 pound mass of water from 59.5°F to 60.5°F. In the International System, joules are used. Historically, the calorie is also used in the metric system, where 1 calorie is the heat required to raise the temperature of 1 gram of water from 14.5°C to 15.5°C. Some of the conversion factors for the various units employed for heat are as follows:

$$1 \text{ Btu} = 778.17 \text{ ft-lbf (usually rounded off to 778 ft-lbf)}$$

$$1 \text{ calorie} = 4.187 \text{ J}$$

$$1 \text{ Btu} = 252 \text{ calories} = 1055 \text{ J} = 1.055 \text{ kJ}$$

It is extremely important to realize that just like work, heat is a path function, defined as it crosses the boundaries of a system. Heat is not a thermodynamic property; it is not a characteristic of a system in a given state.

The actual physical ways in which heat can be transferred from a system to its surroundings are threefold: conduction, convection, and radiation. In a *conduction* process, physical contact between media is required, as shown in Figure 2.14(a). Thermal energy is transferred by purely internal motion of the molecules of the media. No external motion of the media is involved in the transfer; neither the brick nor the cold table of Figure 2.14(a) needs to move for heat flow by conduction to take place between the brick and the table. In a *convection* process, heat is transferred by external movement of a liquid or a gas from one location to another, as Figure 2.14(b) demonstrates. In a *radiation* heat transfer process, thermal energy is transferred by means of electromagnetic waves, as Figure 2.14 shows; neither physical contact nor material media between the bodies is necessary. An example of radiative heat transfer is that between the sun and the earth, with space vacuum between the two bodies.

In a number of thermodynamic devices, the transfer of heat or work is accomplished while the substance undergoes a change of phase.

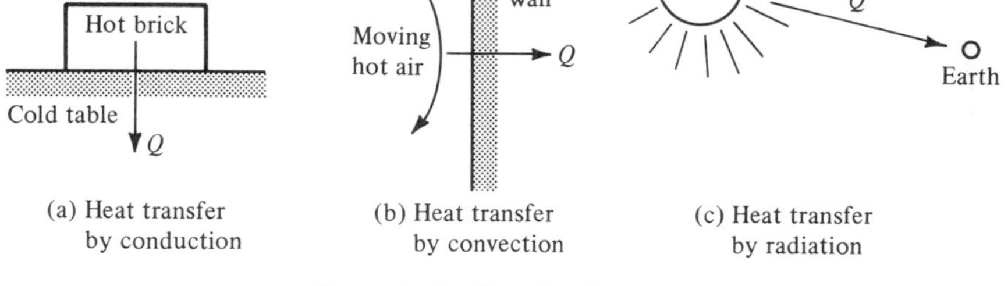

(a) Heat transfer by conduction

(b) Heat transfer by convection

(c) Heat transfer by radiation

FIGURE 2.14 *Transfer of Energy.*

Substances can exist in three phases: solid, liquid, and gas. For example, H_2O exists in the solid phase as ice, in the liquid phase as water, and in the gaseous phase as steam. The characteristics of the three phases are as follows: The *solid* phase retains its own shape and volume; the *liquid* phase retains its volume but takes the shape of its container; and the *gaseous* phase takes the volume and shape of its container. Changes from one phase to another are accompanied by the absorption or the release of heat.

As an illustration, consider a pot of water being heated on a gas burner. If the water is initially at room temperature, its temperature will increase steadily until it has reached 212°F (100°C). After this temperature has been reached and more heat has been added, some of the liquid water will vaporize; that is, it will turn into steam. The vaporizing process is one example of a change of phase, namely, from liquid phase into gaseous phase. Although heat continues to be added to the water, thermometers placed in the water will show no increase in temperature until after all the water is vaporized. When heat is removed from the steam vapor at atmospheric pressure and 212°F, the vapor returns to the liquid phase or condenses. In this process, it gives up the same amount of heat that was required to vaporize it.

When a substance changes from solid to liquid phase, the substance is said to *melt*; on the other hand, the change from liquid to solid phase is called *solidification*. Under special conditions of temperature and pressure, a substance can change directly from the solid to the gaseous phase. An example is dry ice (solid CO_2) at atmospheric pressure. This process is called *sublimation*.

The *latent heat of fusion* of a substance is the heat added per unit mass of a solid to change it to a liquid at the same pressure. The *latent heat of vaporization* of a substance is the heat added per unit mass of a liquid to change it to a vapor at the same pressure. The *latent heat of sublimation* of a substance is the heat added per unit mass of solid to change it to a vapor at the same pressure. No change in temperature takes place during any of these processes.

Since the latent heats of phase change are given for a unit mass, the heat released or absorbed by a mass m in the process of phase change is $Q = mLH$, where LH represents the appropriate latent heat.

A process that occurs without heat transfer to or from the system is called an *adiabatic process*. For example, a process taking place in a well-insulated container can be considered adiabatic. The combustion gases being accelerated in a rocket nozzle are at extremely high temperature. However, we can generally approximate the thermodynamic process that occurs in the nozzle as adiabatic since the gases are traveling at high velocity and there is little time for heat transfer from the gases.

2.4 INTERNAL ENERGY

We have discussed heat and work, types of energy that are defined as they cross the boundaries of a system. Heat and work are not the thermodynamic properties of a system. Clearly, a substance does not possess a given quantity of heat or work; rather, work can be done by a system or heat can be transferred to a system as the system undergoes a process between two states.

In addition to the potential and kinetic energies associated with the motion of an entire system or substance, a thermodynamic system in a given state also possesses internal energy due to the energies of the molecules that comprise the system. For example, the water in a stationary vessel or the gas in a cylinder consists of a large number of molecules moving constantly in a random fashion throughout the vessel or cylinder. Even though the vessel and cylinder boundaries are stationary, the molecules comprising the system possess both kinetic energy due to their translational, rotational, and vibratory motions and also potential energy due to intermolecular forces. The sum of these energies for all the molecules comprising the system is called the system *internal energy*. Internal energy is denoted by the symbol U, and internal energy per unit mass by u.

Internal energy is a thermodynamic property of a system in a given state; that is, associated with a substance at a given pressure, temperature, and specific volume will be one value of internal energy per unit mass. In other words, unlike heat and work, the value of internal energy of a substance at a given state is not a function of how the substance got to that state. The value of internal energy cannot be given in absolute terms; it is given as a difference in internal energy between the desired state and a reference state. This value is adequate for our purposes since we will be interested only in changes of internal energy of a system as it goes from one state to another.

The units of internal energy are the same as those of heat and work, although for historical reasons it is most often given in thermal units. In the English system, U is expressed in Btus, and u in Btu/lbm. In the International System, U is in joules, and u in J/kg.

2.5 THE FIRST LAW OF THERMODYNAMICS FOR CLOSED SYSTEMS

In Chapter 1, we discussed certain fundamental thermodynamic terms, including system, system boundaries, property, and state. In this chapter, we have introduced heat and work as types of energy that are transferred across system boundaries. We are also aware that a system can possess external kinetic and potential energies due to motion or position of the working fluid or system as a whole, as well as internal energy due to kinetic and potential energies of the molecules comprising the system. We are now in a position to apply the above concepts to the first law of thermodynamics for closed systems. For a closed system undergoing a thermodynamic process from state 1 to state 2,

Total energy in state 1 + heat in = work out + total energy in state 2

where

Total energy possessed by a system in a given state

= internal energy + potential energy + kinetic energy

We can therefore write

$$U_1 + KE_1 + PE_1 + Q_{in} = W_{out} + U_2 + KE_2 + PE_2 \qquad \textbf{(2.4)}$$

For a closed system, the external kinetic and potential energies are usually very small with respect to the changes of internal energy, and hence they can be neglected in the energy equation. For example, in the case of the movement of a piston in a cylinder, either the piston is at rest at the beginning and end of the process, or else the speed of the piston is small, as is the resulting motion of the gases in the cylinder at the beginning and end of the process. Neglecting potential and kinetic energy terms from Equation (2.4), we obtain for a closed system

$$U_1 + Q_{in} = W_{out} + U_2 \qquad \textbf{(2.5)}$$

Before proceeding to examples involving the first law for closed systems, we will review the concept of a thermodynamic system. As mentioned in Section 1.2, a thermodynamic system is a quantity of matter or a region of space that we wish to analyze thermodynamically. In working examples, one must first choose the boundaries of the system to be analyzed. Sometimes the choice will be quite obvious, but at other times it must be carefully chosen to provide the desired answer. The importance of the choice of system boundaries can be seen from the following examples.

Consider the case of two bricks, one hot and the other cold, that are placed so that two sides are in contact. Insulation material is wrapped around both bricks, as shown in Figure 2.15(a). First choose the system to include both bricks and insulation, as in Figure 2.15(b). Because of the insulation, no heat crosses the boundaries of the system. Next choose the system to include the hot brick and part of the insulation, as shown in Figure 2.15(c). In this case, heat is transferred across the system boun-

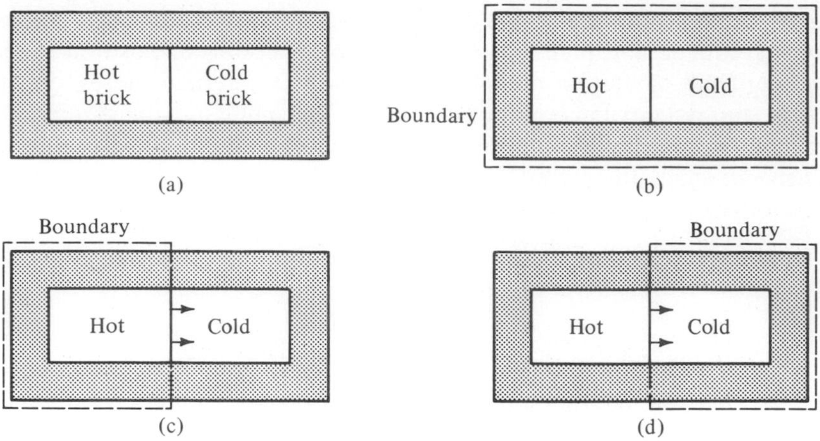

FIGURE 2.15 *Choice of System Boundary.*

dary out of the system. Finally, choose the system to include the cold brick and the insulation, as shown in Figure 2.15(d). In this case, heat crosses the system boundary and is transferred into the system.

Another example of how the selection of system boundaries may influence the analysis of a thermodynamic problem is shown in Figure 2.16. Part (a) depicts a physical system consisting of a small air tank which is heated by resistance coils with electric energy supplied from a storage battery. The equipment is enclosed in a heat-insulating container.

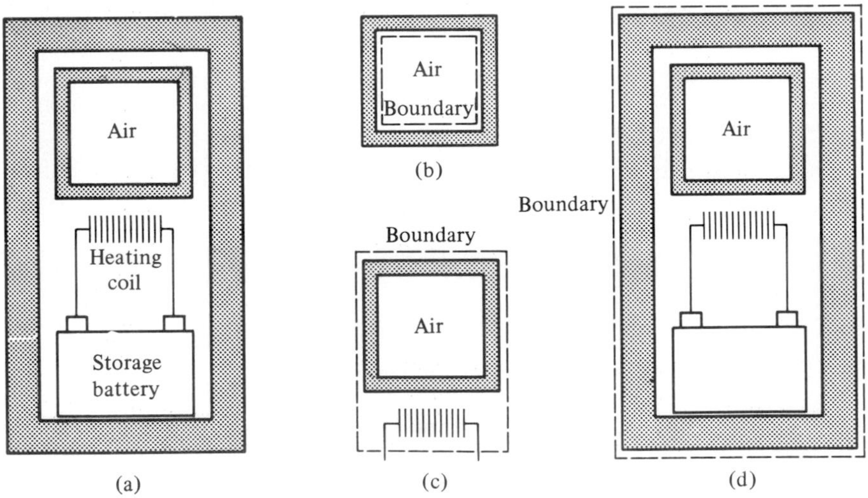

FIGURE 2.16 *Choice of System Boundary.*

If the system is chosen to include only the air within the tank, as in Figure 2.16(b), heat crosses the system boundary. If the system is chosen to consist of air, air tank, and heating coils, as in Figure 2.16(c), energy enters the system as electric work, with no heat crossing the system boundaries. Finally, if the system consists of air, air tank, heating coils, and storage battery, as in Figure 2.16(d), no energy crosses the boundary.

In order to work examples and problems involving the first law, one must know the values of thermodynamic properties of substances at given states. In this and the following chapter, such data will be provided in the specification of the problem or example. In actual practice, the engineer is not given values of thermodynamic properties at different states in a process. Rather, he must refer to tables of thermodynamic data that have been accumulated for substances of interest, or he must calculate properties from appropriate equations. The use of tables and equations to determine thermodynamic properties will be covered in detail in Chapter 4.

EXAMPLE 2.5

Two pounds of air inside a closed, rigid container are heated by the addition of 5 Btus of thermal energy. The internal energy of the air over the temperature range of this example can be written as

$$u - u_0 = 0.171(T - T_0)$$

where u is the internal energy per unit mass in Btu/lbm at temperature T (°F), and u_0 is the internal energy per unit mass at reference temperature T_0 (°F). Find the change in air temperature.

Solution

Select the system to be the air inside the tank as shown in Figure 2.17. The energy equation for the closed system is

$$Q_{in} - W_{out} + U_2 - U_1$$

Since the volume of the container does not change during the heat addition process, the work is zero, or $U_2 - U_1 = Q_{in} = 5$ Btu. Substituting the expression given for u, we obtain

$$\begin{aligned} Q_{in} &= m(u_2 - u_1) \\ &= m[u_0 + 0.171(T_2 - T_0)] - m[u_0 + 0.171(T_1 - T_0)] \\ &= m \times 0.171(T_2 - T_1) \end{aligned}$$

$$T_2 - T_1 = \frac{5\ \text{Btu}}{(2\ \text{lbm})(0.171\ \text{Btu/lbm-°F})} = 14.6\text{°F}$$

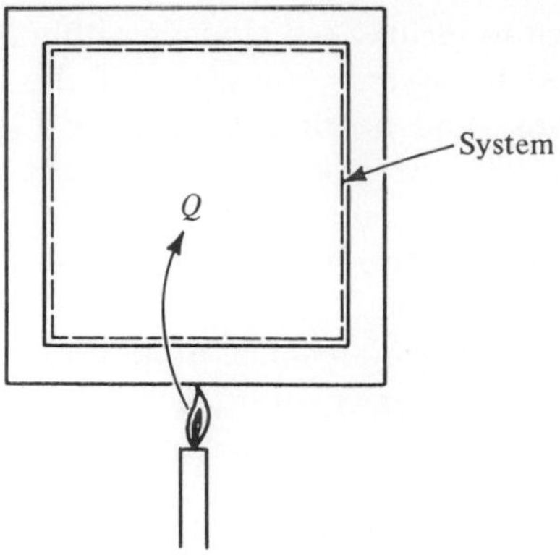

FIGURE 2.17

EXAMPLE 2.6

Air is contained within the cylinder and weightless floating piston shown in Figure 2.18. The air is heated by the addition of 120 joules of thermal energy, which causes the piston to rise in the cylinder. The initial volume of the air is 1000 cm^3, and the volume available to the air is doubled in the process. Determine the increase in internal energy of the air; assume a quasi-static process.

Solution

Select the air inside the piston-cylinder arrangement to be the system, since the air is the substance we wish to study (see Section 1.2). The energy equation for the closed system is given by Equation (2.5):

$$Q_{in} = W_{out} + U_2 - U_1$$

Since the volume of the container changes, the air does work. For a quasi-static process, the work is given by Equation (2.3):

$$W_{out} = \int_1^2 p\, d\forall = p_{atm}(\forall_2 - \forall_1)$$

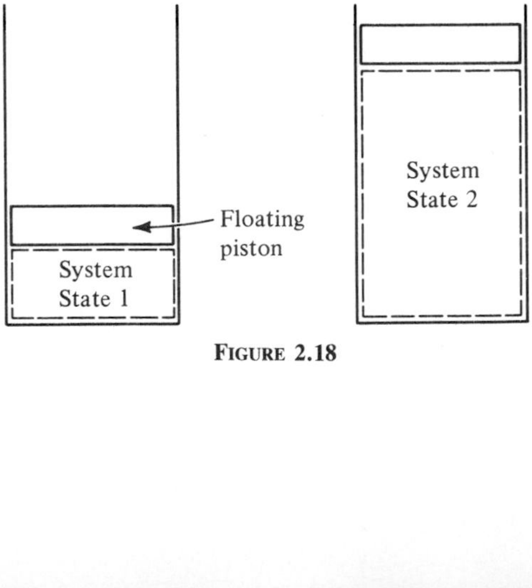

FIGURE 2.18

The pressure within the container must be atmospheric pressure, since the piston is weightless. Therefore,

$$W_{out} = (1.01 \times 10^5 \text{ N/m}^2)(1000 \text{ cm}^3)(10^{-6} \text{ m}^3/\text{cm}^3)$$
$$= 101 \text{ J}$$
$$U_2 - U_1 = Q_{in} - W_{out}$$
$$= 120 \text{ J} - 101 \text{ J}$$
$$= 19 \text{ J}$$

EXAMPLE 2.7

A gas is compressed by a piston in a cylinder from an initial pressure of 10 psia, initial volume at 40 ft^3, to a final pressure of 20 psia. During the compression process, the product $p\mathcal{V}$ remains a constant. Determine the work done on the gas for a quasi-static process. If there is no change of internal energy of the gas, determine the heat transfer to the gas.

Solution

Select the gas as the thermodynamic system to be analyzed. For a quasi-static process, $W_{out} = \int p\, d\mathcal{V}$, which is the area under the $p\mathcal{V}$ curve. For the situation under consideration, the $p\mathcal{V}$ curve appears in Figure 2.19. To calculate the area, integrate:

$$W_{out} = \int_1^2 p\, d\mathcal{V} \quad \text{where} \quad p\mathcal{V} = \text{constant} = C$$
$$= (10 \times 144 \text{ lbf/ft}^2)\, 40 \text{ ft}^3$$
$$= 57{,}600 \text{ ft-lbf}$$

$$W_{out} = \int_1^2 C \frac{d\mathcal{V}}{\mathcal{V}}$$
$$= C \ln \frac{\mathcal{V}_2}{\mathcal{V}_1}$$
$$= 57{,}600 \ln \frac{20}{40}$$
$$= -39{,}900 \text{ ft-lbf} \quad \text{or} \quad W_{in} = +39{,}900 \text{ ft-lbf}$$

From Equation (2.5),

$$Q_{in} = -39{,}900 \text{ ft-lbf} + U_2 - U_1$$
$$= (-39{,}900 \text{ ft-lbf}) \frac{\text{Btu}}{778 \text{ ft-lbf}} = -51.3 \text{ Btu}$$
$$\text{or} \qquad Q_{out} = +51.3 \text{ Btu}$$

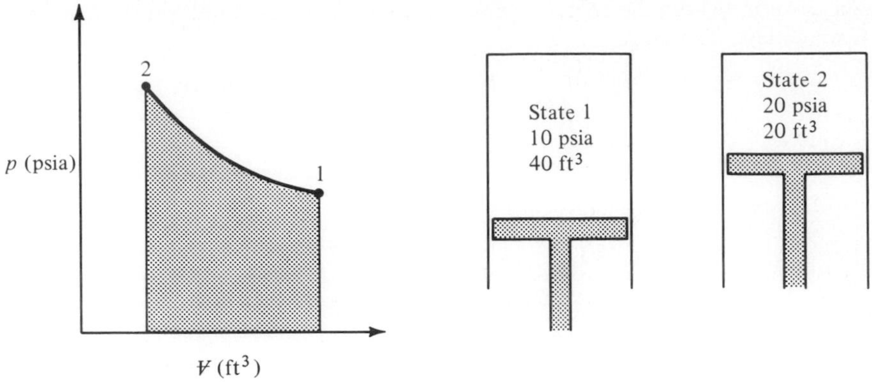

FIGURE 2.19 *p-V Curve.*

EXAMPLE 2.8

The voltage across a resistor is 115 volts, and the resistor draws a current of 20 amperes (Figure 2.20). How much heat must be carried away from the resistor in order that its internal energy (and hence temperature) remains constant?

Solution

Since we are concerned here with heat transfer from the resistor and the internal energy change of the resistor, let the resistor be the system. From Equation (2.5), $Q_{in} = W_{out} + U_2 - U_1$. From Section 2.2, we find the electric work:

$$W_{in} = (\text{voltage})(\text{current})\,\Delta t$$

Since $U_2 = U_1$ here, then

$$Q_{out} = (\text{voltage})(\text{current})\,\Delta t$$

or the time rate of heat removal $\dot{Q}_{out}$ is

$$\dot{Q}_{out} = (\text{voltage})(\text{current})$$

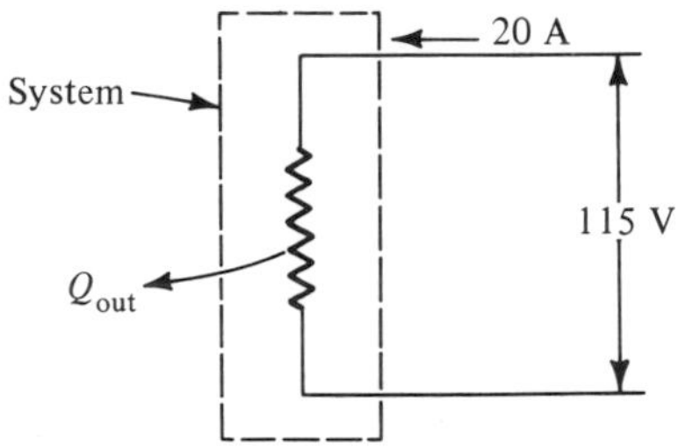

FIGURE 2.20

Hence,

$$\dot{Q}_{out} = (115 \text{ volts})\ 20 \text{ amperes}$$
$$= 2.3 \text{ kilowatts } (1 \text{ watt} = 1 \text{ volt} \cdot \text{ampere})$$

EXAMPLE 2.9

Steam (100 percent vapor) at atmospheric pressure and 212°F is contained inside a rigid vessel having a volume of 50 ft³. Heat is removed until the temperature drops to 180°F. Find the amount of heat removed and the masses of liquid and vapor at the final state. Take the properties of liquid water at 14.7 psia and 212°F to be as follows:

$$\text{Specific volume } (v_f) = 0.016719 \text{ ft}^3/\text{lbm}$$
$$\text{Specific internal energy } (u_f) = 180.12 \text{ Btu/lbm}$$

Take the properties of water vapor at 14.7 psia and 212°F to be as follows:

$$\text{Specific volume } (v_g) = 26.799 \text{ ft}^3/\text{lbm}$$
$$\text{Specific internal energy } (u_g) = 1077.6 \text{ Btu/lbm}$$

At 180°F,

$$v_f = 0.016510 \text{ ft}^3/\text{lbm} \qquad v_g = 50.22 \text{ ft}^3/\text{lbm}$$
$$u_f = 147.98 \text{ Btu/lbm} \qquad u_g = 1068.4 \text{ Btu/lbm}$$

Note: subscript f denotes liquid, subscript g denotes vapor.

Solution

Take the entire contents of the vessel as the system, since we are finding the heat removed from the entire vessel contents. This is also a convenient choice of system since, for the rigid vessel, the total volume of the system remains constant, so that $\mathcal{V}_1 = \mathcal{V}_2 = 50$ ft³. The mass inside the tank also remains the same, since no mass enters or leaves the tank. Therefore, the specific volumes are $v_1 = v_2 =$ 26.799 ft³/lbm. The total mass inside the tank is $m_{total} = 50/26.799 = 1.866$ lbm, which at state 1 is made up solely of vapor. Therefore, at state 2, $m_{liq\ 2} + m_{vap\ 2} =$ 1.866 lbm. Also, at state 2,

$$50 \text{ ft}^3 = v_{f2} m_{liq\ 2} + v_{g2} m_{vap\ 2}$$
$$= 0.01651 m_{liq\ 2} + (50.22)(1.866 - m_{liq\ 2})$$

Solving, we get

$$m_{liq\ 2} = 0.871 \text{ lbm}$$
$$m_{vap\ 2} = 1.866 - 0.871$$
$$= 0.995 \text{ lbm}$$

Since the system boundaries do not move, the work is zero, and from Equation (2.5), $Q_{in} = U_2 - U_1$. Now

$$U_2 = m_{liq\,2} u_{f2} + m_{vap\,2} u_{g2} = (0.871 \text{ lbm})\, 147.98 \text{ Btu/lbm} + (0.995 \text{ lbm})\, 1068.4 \text{ Btu/lbm}$$

$$= 128.89 \text{ Btu} + 1063.06 \text{ Btu} = 1191.95 \text{ Btu}$$

$$U_1 = m_1 u_1 = 1.866 \text{ lbm } (1077.6) \text{ Btu/lbm} = 2010.8 \text{ Btu}$$

Hence

$$Q_{in} = 1192.0 - 2010.8 = -818.8 \text{ Btu}$$

where the negative sign denotes heat removed from the system.

EXAMPLE 2.10

A cube of ice at 0°C is dropped into a glass of water at 20°C. The cube is initially 20 millimeters on a side, and the glass is 50 mm in diameter, filled to a depth of 120 mm. The latent heat of fusion of water is 333×10^3 J/kg; the specific volume of ice is 0.00109 m^3/kg; and the specific volume of water is 0.001 m^3/kg. The change of internal energy of water in the temperature range of interest is given by

$$u_2 - u_1 = 4.19 \times 10^3 (T_2 - T_1) \text{ J/kg with } T \text{ in °C}$$

Find the final equilibrium temperature of the water after the ice has melted. Neglect any transfer of heat or work across the system boundaries.

Solution

Consider the ice and water together as the system, since we are calculating the overall water mixture temperature. Since $Q = 0$ and $W = 0$, we get from Equation (2.5) that $U_2 - U_1 = 0$, or

$$(U_2 - U_1)_{ice} + (U_2 - U_1)_{water} = 0$$

The change in internal energy of ice and the water resulting from the melted ice is

$$(U_2 - U_1)_{ice} = m_{ice}[\text{LH} + 4.19 \times 10^3 (T_2 - T_1)_{ice}]$$

The change in internal energy of the water is

$$(U_2 - U_1)_{water} = m_{water} 4.19 \times 10^3 (T_2 - T_1)_{water}$$

The mass of the ice cube is

$$m_{ice} = \mathcal{V}/v = (0.02)^3 \text{ m}^3/(0.00109 \text{ m}^3/\text{kg})$$

$$= 0.00734 \text{ kg} = 7.34 \text{ g}$$

The initial mass of water in the glass is

$$m_{water} = \pi D^2 h/4 v_{water} = \pi (0.05^2 \text{ m}^2)(0.12 \text{ m})/4(0.001 \text{ m}^3/\text{kg})$$

$$= 0.236 \text{ kg} = 236 \text{ g}$$

Substituting numerical values in the above first law equation, we get

$$0.00734\ \text{kg}\ [333{,}000 + 4190(T_2 - 0)]\ \text{J/kg} + 0.236\ \text{kg}\ (4190)(T_2 - 20)\ \text{J/kg} = 0$$

$$T_2 = 17.0°\text{C}$$

PROBLEMS

2.1 A constant pressure of 50 psia acts on a floating piston having an area of 5 in^2, as shown in Figure 2.21. Calculate the work done by the gas during a piston displacement upwards of 3.0 in., neglecting the weight of the piston.

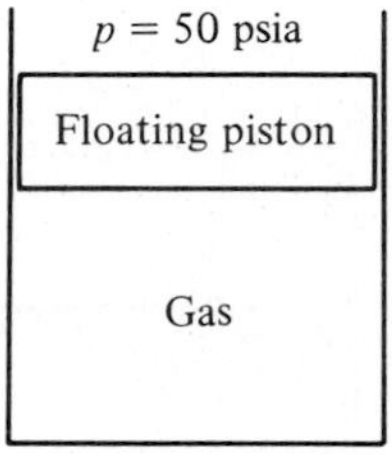

FIGURE 2.21

2.2 Repeat the Problem 2.1, assuming a piston weight of 1.5 pounds.

2.3 The pressure in the cylinder of Figure 2.22 provides a force sufficient to balance the weight of the piston. Atmospheric pressure of 101 kPa acts on top of the piston; the piston mass is 100 g, and piston diameter is 2.5 cm. Determine the gas pressure inside the cylinder. Determine also the work done by the gas during a vertical upward piston displacement of 2.0 cm.

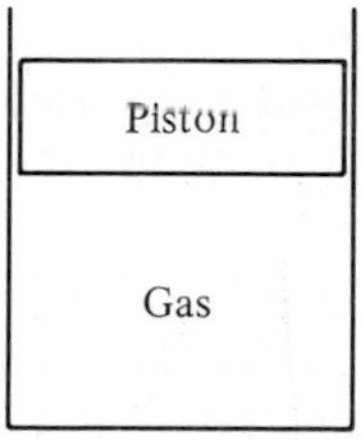

FIGURE 2.22

2.4 A gas is compressed by a piston from an initial pressure of 14.7 psia and an initial volume of 50 in^3 to a pressure of 100 psia and a volume of 5 in^3. The p-V relation is given by a straight line, as shown in Figure 2.23. Calculate the work required for the compression.

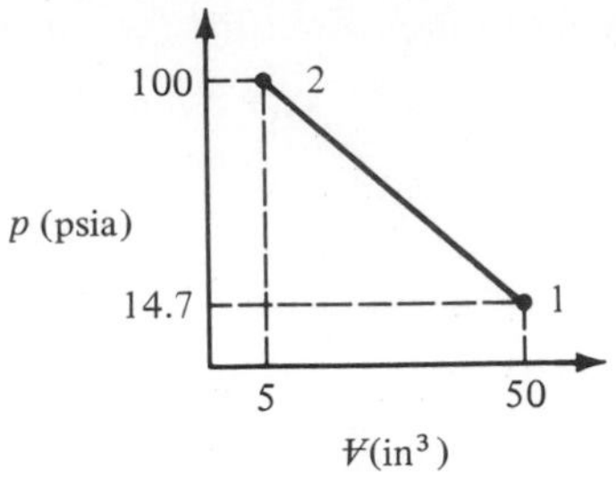

FIGURE 2.23

2.5 Calculate the work required for the compression of a gas in a piston-cylinder arrangement with the p-Ꝟ relation given in Figure 2.24.

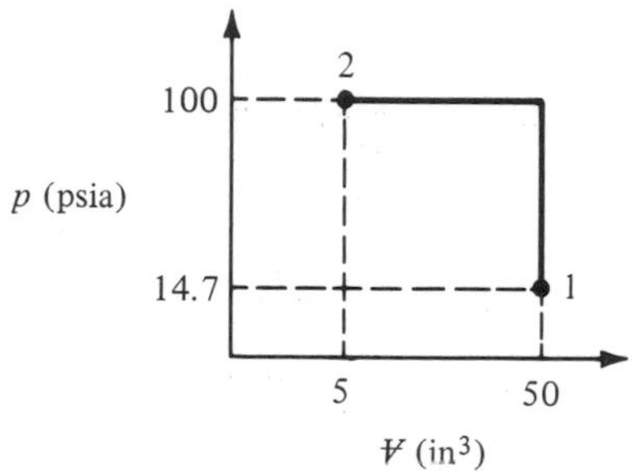

FIGURE 2.24

2.6 Calculate the work required to stretch a 6-in. length of steel wire by 0.1 in.; the wire diameter is 0.05 in. Take $E = 10^6$ psi.

2.7 Air expands against a piston of a hot air engine in such a way that the p-Ꝟ relation is described by a circular arc (one-quarter of a circle), as shown in Figure 2.25. Find the work done by the air and the change in internal energy; assume an adiabatic process.

2.8 A closed thermodynamic system undergoes a process change during which 2 kilojoules of thermal energy is supplied to the system, while 0.5 kilojoule of work is done on the system. Find the change in internal energy of the system.

2.9 A closed thermodynamic system is supplied with 100 Btu of heat during a quasi-static process in which the system expands against a constant external pressure of 50 psia. The internal energy at the beginning of the process is equal to that at the end of the process. Find the increase in volume of the system.

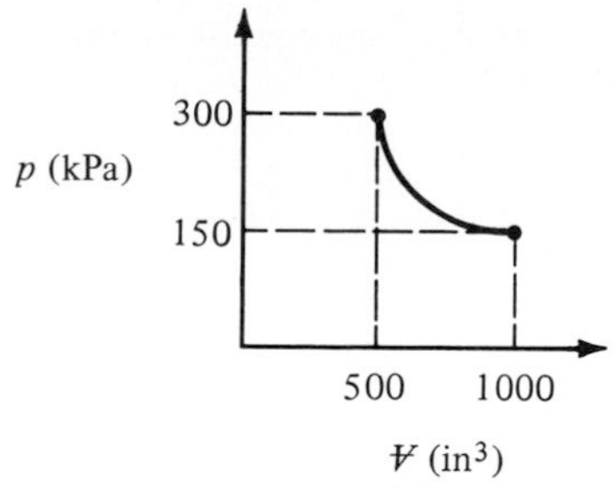

FIGURE 2.25

2.10 A gas is compressed slowly in a piston-cylinder system. The following volume-pressure relations are measured during the compression:

V (ft^3)	p (psia)
10.0	45.0
7.5	52.5
5.0	67.5
2.5	100.0

The change in internal energy is 4.5 Btu. Calculate the energies transferred as work and heat during the process.

2.11 An electrical current of 5 amperes flows through a 10-ohm resistor for a period of 10 seconds. Determine the change of internal energy of the resistor if it is well insulated (assume no heat transfer).

2.12 A container is divided into two compartments, one compartment containing a gas, the other evacuated. The partition between the two compartments is broken, allowing the gas to fill the entire container. Show that the internal energy of the gas does not change, assuming that the container is well insulated (see Figure 2.26).

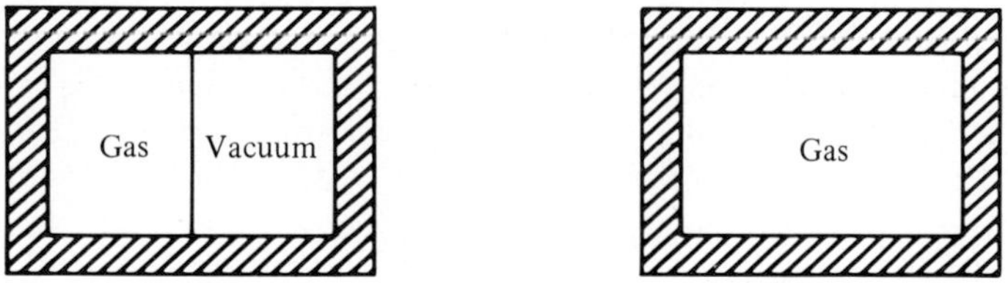

FIGURE 2.26

2.13 One gallon of water at 70°F is to be mixed with hot water at 140°F, to result in warm water at 120°F. Calculate the amount of hot water at 140°F to be added. The change in internal energy of water can be taken as $\Delta u = 1.0\,\Delta T$ Btu/lbm, with T in °F.

2.14 In an industrial process, 1000 kilograms of water must be heated from 25°C to 75°C. To accomplish this, steam at a temperature of 100°C will be mixed with water. Calculate the quantity of steam required. The latent heat of condensation of steam is 2256.9 kJ/kg, and the change in internal energy of water is given by $\Delta u = (4.19\,\Delta T)$ kJ/kg, with T in °C.

2.15 How long will it take a 1000-W electric immersion heat to bring 1/2 liter of water to a boil from a temperature of 20°C? The latent heat of evaporation is 2256.9 kJ/kg, and the change in internal energy of water is given by $\Delta u = (4.19\,\Delta T)$ kJ/kg, with T in °C. (Note: 1 liter = 10^{-3} m^3.)

2.16 An automobile having a mass of 1500 kg is traveling at a speed of 30 km/h. How many joules are developed in the brakes when the car is brought to rest? Assume that all of the energy is transferred as heat to the brakes.

2.17 A car weighing 2500 pounds is traveling at a speed of 50 mph. Calculate the heat that must be dissipated by the brakes as the car is brought to rest.

2.18 A 12-V battery is charged with a current of 10 amperes for a period of one hour. Calculate the change in internal energy of the battery if it loses 100 Btu during the charging process.

2.19 An outboard motor is operated in a drum containing 55 gallons of water. After the motor is run for a period of 20 minutes, it is found that the temperature of the water has risen by 10°F. Find the horsepower of the outboard motor. Assume that the change in internal energy of water is given by $\Delta u = 1.0\,\Delta T$ Btu/lbm, with T in °F.

2.20 An open diving bell, shown in Figure 2.27, is lowered to a depth of 30

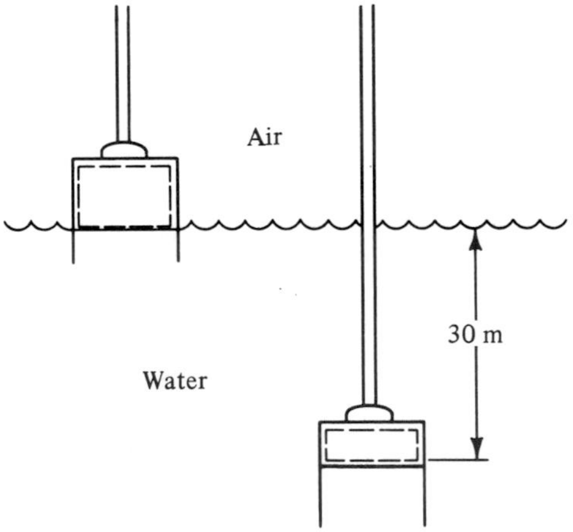

FIGURE 2.27 *Diving Bell.*

meters. How much work is done on the air in the diving bell as it is submerged to the 30-meter depth? The air has an initial volume of 3 m^3 at atmospheric pressure. The pressure-volume relation for the air is given by $p\mathcal{V}$ = constant. The density of water is 1000 kg/m^3.

2.21 A liquid is stirred with a paddle wheel driven by a 1/2-horsepower electric motor. It is assumed that 80 percent of the electric work of the motor goes into the liquid as mechanical work. Determine the rate of temperature rise of the liquid. Assume that the liquid container is well insulated, so that there is no heat transfer. Also, take the variation of internal energy with temperature of the liquid to be given by $du = 0.89\,dT$, with u in Btu/lbm and T in °F.

2.22 During an adiabatic thermodynamic process, air is expanded in a piston-cylinder arrangement from a volume of 20 cm^3 to a volume of 160 cm^3. During this process, the air obeys the relaton $p\mathcal{V}^{1.4}$ = constant. The initial pressure is 100 kPa. Determine the final air pressure, the work done by the air, and the change of internal energy of the air.

2.23 A gas is compressed by a piston in a cylinder from an initial pressure of 10 psia and an initial volume of 39.8 ft^3 to a final pressure and volume of 35 psia and 12.8 ft^3, respectively. The data available on the compression process are as follows:

p (psia)	$\mathcal{V}$ (ft^3)
10	39.8
15	27.5
20	21.4
25	17.3
30	14.7
35	12.8

Find the work that was required to compress the gas and the change of

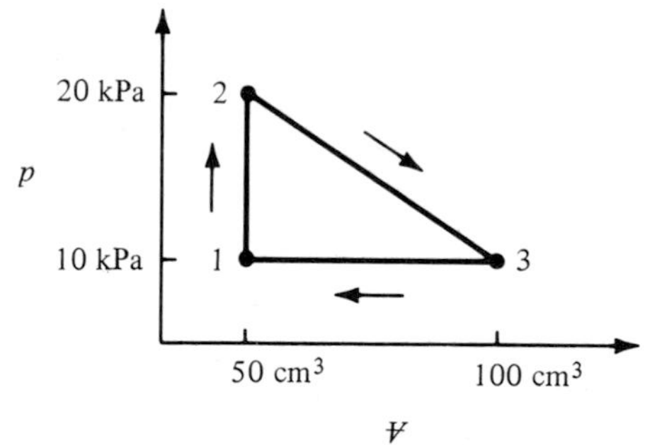

FIGURE 2.28

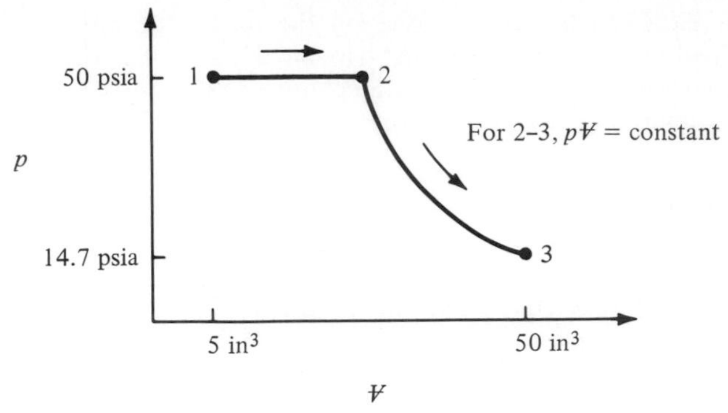

FIGURE 2.29

internal energy of the gas. Assume that the cylinder wall is insulated, so that there is no exchange of heat between the gas and its surroundings (adiabatic process), and that the process is quasi-static.

2.24 The gas in a piston-cylinder arrangement undergoes the three processes shown in Figure 2.28. Determine the work done by the gas in each process and in the cycle.

2.25 Determine the total work done by the gases in a piston-cylinder arrangement for the process 1–2 and 2–3 (see Figure 2.29).

chapter

3

The First Law of Thermodynamics for Open Systems

3.1 WHAT ARE OPEN SYSTEMS?

In the last chapter, we were concerned with closed thermodynamic systems, in which we took the system to consist of a fixed mass of a substance, with no mass crossing the system boundaries during the change of state under study. The volume of the closed system could increase or decrease (as in a piston-cylinder arrangement) or could remain constant (as in a closed, rigid container). In contrast, in an open system, mass flows do occur across the system boundaries. Thus we generally use a volume or region in space as our system and study the mass and energy flows crossing the system boundaries. In the open system, not only is energy added or removed during a process in the form of work or heat, but it is also carried into and out of the system by the mass flow entering and leaving the system. Remember that each particle of mass entering the system carries into the system its internal energy plus the kinetic energy associated with its velocity plus its gravitational potential energy.

In this chapter, we will discuss the conservation of energy and mass as applied to open thermodynamic systems.

3.2 THE CONSERVATION OF MASS FOR OPEN SYSTEMS

Until now we have utilized one of the basic principles of mechanics: the conservation of energy. This principle states that energy can be neither created nor destroyed, but merely transferred or transformed; that is, in any given situation, the energy is balanced. Another basic principle of physics concerns the conservation of mass. This principle states that mass can be neither created nor destroyed and is thus conserved.[1] In the case of the motion of a body as discussed in mechanics, this principle is inherently taken into account, and there is no need to discuss it specifically. Similarly, in a closed system, we are concerned with the energy balance of the system. Since the mass of the closed system remains constant during the process under consideration, no further specific statement has to be made.

On the other hand, in an open system we must account for the conservation of mass as well as the conservation of energy. Consider a mass balance for the open system shown in Figure 3.1. In the initial state, the system contains mass m_1 within its boundaries. During the process under study, mass m_{in} enters the system through an open portion of the boundary, with mass m_e exiting the system through another open portion of the boundary. In the final state, the mass inside the system is m_2. The conservation of mass merely states that the total mass must remain constant; that is, the initial mass m_1 inside the system, plus the mass m_{in} entering the system during a process, minus the mass m_e exiting from the system, is equal to the mass m_2 inside the system at the final state. Stated mathematically, this becomes for the system

$$m_1 + m_{\text{in}} - m_e = m_2 \tag{3.1}$$

---

1. Modern physics recognizes the convertibility of mass and energy. In problems considered in introductory thermodynamics, however, conversion of mass and energy is completely negligible. The relationship between the change in mass and the change in energy is given by

$$\Delta E = \frac{\Delta m}{g_c} c^2$$

where c is the speed of light (9.84×10^8 fps or 3×10^8 m/s). Since the speed of light is an invariant quantity, this relationship states that the mass of a system changes when its energy changes. From the above relation, we obtain that a change in energy of 1 Btu is equal to

$$\Delta m = \frac{(\Delta E) g_c}{c^2} = (778 \text{ ft-lbf/Btu}) \times \frac{32.2 \text{ lbm/lbf-ft/s}^2}{9.84^2 \times 10^{16} \text{ ft}^2/\text{s}^2} = 2.59 \times 10^{-14} \text{ lbm}$$

or a change of 1 joule is equivalent to 1.11×10^{-17} kg.

Next let us examine whether the magnitude of this change is significant for the problems in this text. For example, to change 1 pound of water at 212°F and atmospheric pressure from its liquid to vapor phase requires 930.3 Btu or an increase of 2.4×10^{-11} lbm in the mass of the water. (The boiling of 1 kilogram of water would result in an increase of 2.4×10^{-11} kg). Thus it can be concluded that for processes of interest here, the change in mass is certainly negligible. The two conservation laws can thus be considered as being completely independent.

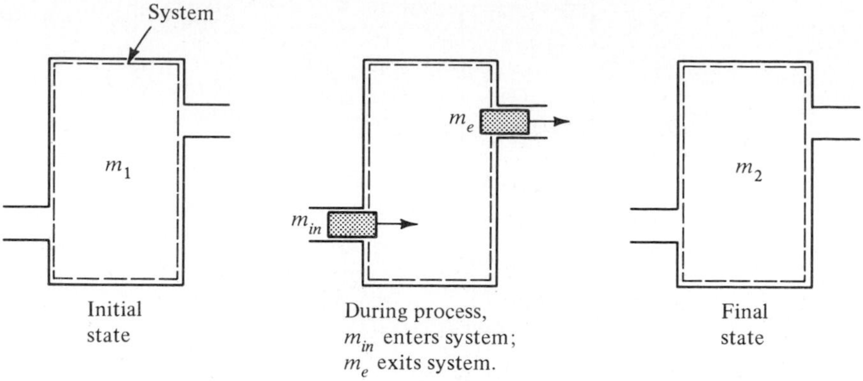

FIGURE 3.1 *Conservation of Mass for an Open System.*

In many applications, it is necessary to express Equation (3.1) on a rate basis, since devices such as water pumps, steam turbines, and air compressors operate on a continuous basis. In words, the equation for conservation of mass becomes

The rate at which mass flows into a system minus the rate at which mass flows out of a system yields the net rate of buildup of mass inside the system boundaries.

In other words,

$$\dot{m}_{\text{in}} - \dot{m}_e = \frac{dm_{\text{system}}}{dt} \qquad \textbf{(3.2)}$$

where

$\dot{m}_{\text{in}}$ = rate at which mass flows in across the system boundaries

$\dot{m}_e$ = rate at which mass flows out across the system boundaries

$\frac{dm_{\text{system}}}{dt}$ = rate of buildup of mass inside the system

As an example of Equation (3.2), if 5 pounds of water flows into a tank per second and 2 pounds of water flows out each second (Figure 3.2), it follows that the rate of increase of water inside the tank is 3 pounds per second. In order to keep the amount of water inside the tank at any time equal to a constant, the rate at which mass flows in must be balanced by the rate of outflow. In the above example, if $\dot{m}_{\text{in}} = \dot{m}_e = 5$ lbm/s, then $dm_{\text{system}}/dt = 0$, where the system is the contents of the tank.

Expressions for $\dot{m}_{\text{in}}$ and $\dot{m}_e$ can be written in terms of local fluid properties. Consider an element of fluid moving down a pipe and eventually crossing the boundaries of a system, as shown in Figure 3.3.

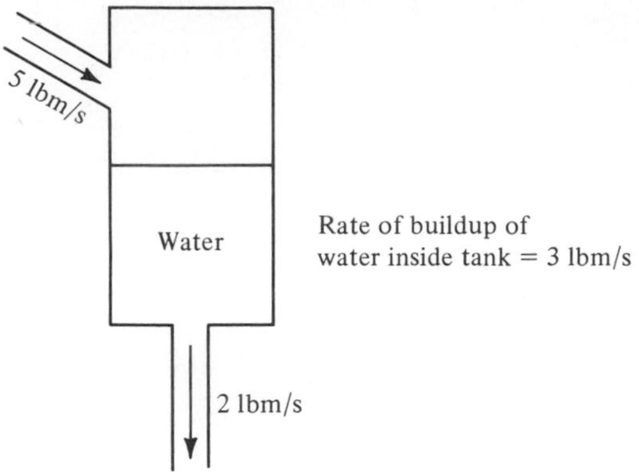

FIGURE 3.2

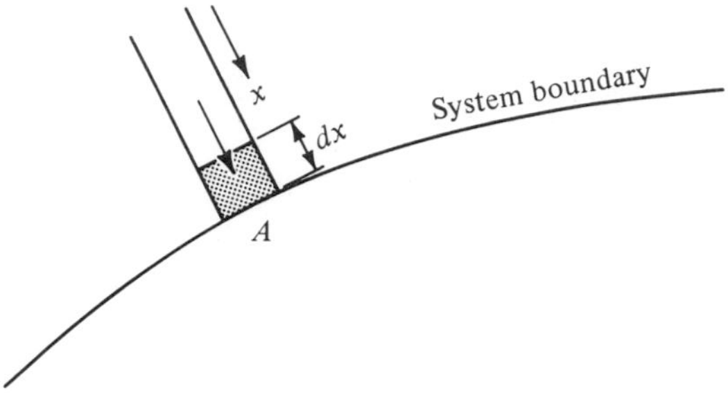

FIGURE 3.3

Let the element be moving at velocity V in the x-direction with the cross-sectional area A of the pipe normal to the direction of motion. The volume of the cross-hatched element is $A\,dx$, with its mass $A\,dx/v$. The rate dm/dt at which the mass of the element crosses the system boundaries and enters the system is $(A/v)(dx/dt)$, where dx/dt is simply the velocity of the element. In other words,

$$\dot{m} = \frac{AV}{v} \tag{3.3}$$

The conservation of mass equation, or continuity equation, can now be written in the following form for the system shown in Figure 3.4:

$$\frac{A_{\text{in}}V_{\text{in}}}{v_{\text{in}}} - \frac{A_e V_e}{v_e} = \frac{dm_{\text{system}}}{dt} \tag{3.4}$$

or, since $\rho = 1/v$,

$$\rho_{\text{in}}A_{\text{in}}V_{\text{in}} - \rho_e A_e V_e = \frac{dm_{\text{system}}}{dt}$$

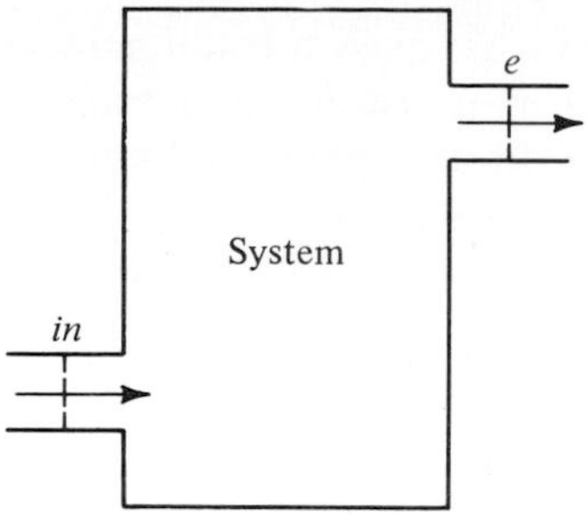

FIGURE 3.4

where ρ_{in}, A_{in}, v_{in}, V_{in}, for example, denote density, cross-sectional area, specific volume, and velocity. (Properties such as density, velocity, and specific volume have been assumed constant across each cross section.)

Since many devices that we are familiar with, such as pumps, turbines, heaters, and air conditioners, generally operate on a continuous basis for long periods of time, we are quite interested in the special case in which there is no buildup of mass inside the system, that is, when $dm_{system}/dt = 0$. This case, in which properties inside the system and at the system boundaries do not change with time, is called *steady flow*. It follows that, for steady flow,

$$\frac{A_{in}V_{in}}{v_{in}} = \frac{A_eV_e}{v_e} \tag{3.5}$$

or

$$\rho_{in}A_{in}V_{in} = \rho_eA_eV_e \tag{3.6}$$

3.3 THE ENERGY EQUATION FOR OPEN SYSTEMS

The energy equation for open systems can be obtained by appropriate modification of the energy equation for closed systems. The energy equation for a closed system was given by Equation (2.4):

$$Q_{in} - W_{out} = KE_2 - KE_1 + PE_2 - PE_1 + U_2 - U_1$$

This equation can be written in a simpler form as

$$Q_{in} - W_{out} = E_2 - E_1$$

where the total energy E of a system in a given state is the sum of its internal, kinetic, and potential energies:

$$E = U + KE + PE$$

In a closed system there is no mass transfer across system boundaries, but only energy flow. In an open system, mass transfer as well as energy flow occurs across openings in the physical boundaries. Thus in an open system we must account for energy entering, E_{in}, and energy leaving, E_e, the system through actual openings in the system boundary, as we did in the previous section for mass. The statement of energy conservation for open systems becomes

$$Q_{\text{in}} - W_{\text{out}} = E_e + E_2 - E_{\text{in}} - E_1 \tag{3.7}$$

This equation is shown pictorially in Figure 3.5. The work term W_{out} includes both the various forms of mechanical shaft work discussed in Section 2.2 for closed systems and the work done by or on the substance within the system at the physical openings. It is desirable to separate this latter form of work from the other forms of work included in the W_{out} term. Consider the work required to push the working fluid into or out of the system. As shown in Figure 3.6, a force is applied on the entering fluid element. This force, F_{in}, equals the product of pressure p_{in} and cross-sectional area A_{in}:

$$F_{\text{in}} = p_{\text{in}} A_{\text{in}}$$

During the push of the fluid element into the system, the element travels a distance l_{in}. The work done in pushing the fluid element across the system boundary is therefore

$$F_{\text{in}} l_{\text{in}} = p_{\text{in}} A_{\text{in}} l_{\text{in}} = p_{\text{in}} V_{\text{in}}$$

where V_{in} is the volume of the fluid element. This work is done on the system by the fluid outside the system; hence for the system it equals

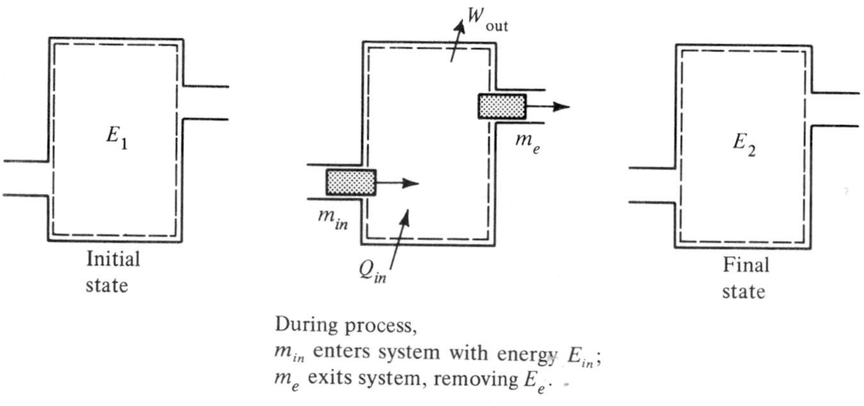

FIGURE 3.5 *Conservation of Energy in an Open System.*

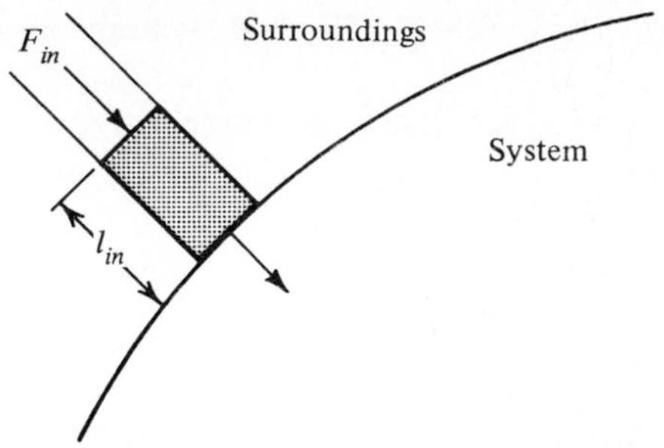

FIGURE 3.6 *Work to Push Substance into System.*

$-p_{\text{in}}V_{\text{in}}$, since work done on a system is negative. This work is done as a consequence of flow, so it is called *flow work*.

In a similar way, the system performs work in pushing out an element of fluid (Figure 3.7). This work amounts to p_eV_e. Separating flow work from W_{out}, and labeling the balance W'_{out}, we obtain

$$W_{\text{out}} = W'_{\text{out}} + p_eV_e - p_{\text{in}}V_{\text{in}}$$

Substituting back into Equation (3.7) yields

$$Q_{\text{in}} - (W'_{\text{out}} + p_eV_e - p_{\text{in}}V_{\text{in}}) = E_e - E_{\text{in}} + E_2 - E_1 \tag{3.8}$$

We obtain further modification by writing in terms of specific volume:

$$Q_{\text{in}} - W'_{\text{out}} = E_e + m_e p_e v_e - E_{\text{in}} - m_{\text{in}} p_{\text{in}} v_{\text{in}} + E_2 - E_1 \tag{3.9}$$

In Equation (3.8) expressions occur that involve the sum $E + pV$, where $E + pV = U + KE + PE + pV$. The internal energy U and the product pV are properties of a substance. Hence it is convenient to define a new property consisting of the sum of U and pV, namely, enthalpy, H:

$$H = U + pV \tag{3.10}$$

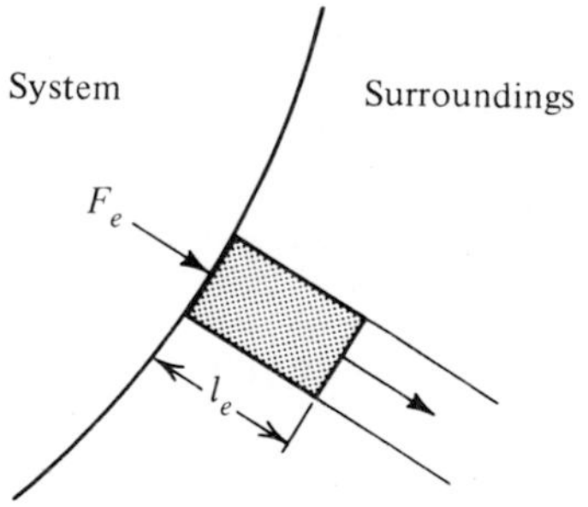

FIGURE 3.7 *Work to Push Substance out of System.*

The units of H are Btus in the English system and joules in the International System.

It is also appropriate to define an intensive property h equal to enthalpy per unit mass:

$$h = u + pv$$

Note that enthalpy, being a combination of thermodynamic properties, is itself a thermodynamic property of a substance in a given state. The units of h are Btu/lbm in English system and J/kg in the International System.

The energy equation for an open system can now be written as follows:

$$Q_{\text{in}} - W'_{\text{out}} = H_e - H_{\text{in}} + KE_e - KE_{\text{in}} + PE_e - PE_{\text{in}} + E_2 - E_1 \quad \textbf{(3.11)}$$

$$\text{or} \quad Q_{\text{in}} - W'_{\text{out}} = m_e h_e - m_{\text{in}} h_{\text{in}} + KE_e - KE_{\text{in}} + PE_e - PE_{\text{in}} + E_2 - E_1 \quad \textbf{(3.12)}$$

In Equations (3.11) and (3.12), W'_{out}, work done by the system, includes work done, for example, by a piston, an impeller, or a paddle wheel, and other mechanical shaft work; but it excludes the work required to push the fluid into or out of the system. Thus W'_{out} is consistent with the mechanical work, W_{out}, used in the closed system analysis of Chapter 2.

For steady flow, defined in the previous section as flow for which conditions within the system and at the system boundaries do not change with time, the energy within the system remains unchanged; that is, $E_2 = E_1$. From Equation (3.11), it follows that, for steady flow,

$$Q_{\text{in}} - W'_{\text{out}} = H_e - H_{\text{in}} + KE_e - KE_{\text{in}} + PE_e - PE_{\text{in}} \quad \textbf{(3.13)}$$

Using the expressions for KE $(=mV^2/2g_c)$ and for PE $(=mgz/g_c)$, together with the conservation of mass equation for steady flow ($m_{\text{in}} = m_e$), we obtain the following for $m_{\text{in}} = m_e = m$:

$$Q_{\text{in}} - W'_{\text{out}} = H_e - H_{\text{in}} + \frac{m}{2g_c}(V_e^2 - V_{\text{in}}^2) + m\frac{g}{g_c}(z_e - z_{\text{in}}) \quad \textbf{(3.14)}$$

The above equation represents the energy balance for a steady-flow open system. On a unit mass basis, the steady flow energy equation becomes

$$q_{\text{in}} - w'_{\text{out}} = h_e - h_{\text{in}} + \frac{1}{2g_c}(V_e^2 - V_{\text{in}}^2) + \frac{g}{g_c}(z_e - z_{\text{in}}) \quad \textbf{(3.15)}$$

To obtain an energy rate equation for steady flow, multiply Equation (3.15) by $\dot{m}$ (mass flow rate) to obtain

$$\dot{m}(q_{\text{in}} - w'_{\text{out}}) = \dot{m}(h_e - h_{\text{in}}) + \frac{\dot{m}}{2g_c}(V_e^2 - V_{\text{in}}^2) + \dot{m}\frac{g}{g_c}(z_e - z_{\text{in}}) \quad \textbf{(3.16)}$$

Letting $\dot{m}q_{\text{in}} = \dot{Q}_{\text{in}}$ and $\dot{m}w'_{\text{out}} = \dot{W}'_{\text{out}}$, we have

$$\dot{Q}_{\text{in}} - \dot{W}'_{\text{out}} = \dot{m}(h_e - h_{\text{in}}) + \frac{\dot{m}}{2g_c}(V_e^2 - V_{\text{in}}^2) + \dot{m}\frac{g}{g_c}(z_e - z_{\text{in}}) \tag{3.17}$$

The units for $\dot{Q}_{\text{in}}$ and $\dot{W}'_{\text{out}}$ are ft-lbf/s or Btu/s and J/s (watts). Note that in Equation (3.17), the term $\dot{W}'_{\text{out}}$ represents the rate of work done, or power.

3.4 APPLICATIONS OF STEADY-FLOW OPEN SYSTEMS

A large number of thermodynamic devices are continuous, steady-flow devices. Examples are steam and gas turbines, compressors and pumps, boilers, condensers, throttling devices, and nozzles. In this section, we present applications of the steady-flow energy equation to each of these examples.

1. Turbines

In a turbine, much of the energy content of the working medium passing through the turbine is converted into mechanical shaft work, as shown in Figure 3.8. Specifically, the working medium does work on the rotating member (rotor) of this machine, which carries a series of curved blades. The rotor is attached to a rotating shaft that transmits mechanical energy from the turbine. For example, in an electric power plant, steam at high pressure and temperature is used to turn a series of turbine wheels, each wheel consisting of curved blades mounted on a shaft. The

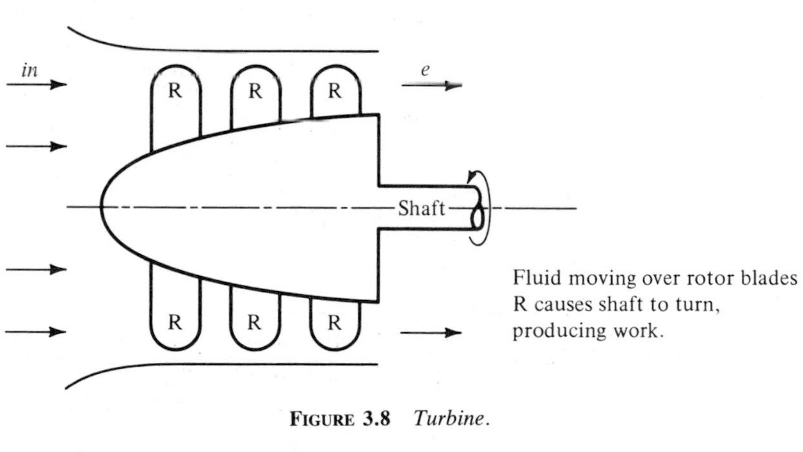

FIGURE 3.8 *Turbine.*

shaft power output is then utilized to drive an electric generator that provides electric power. Another example is the gas turbine on a jet engine, where combustion gases at high pressure and temperature are again used to turn a turbine wheel. Here the output turbine power is used to drive a compressor. Remember that in the turbine, power is delivered; there is a net positive work output.

The change between inlet and outlet potential energy in a turbine can usually be neglected as small in comparison to the enthalpy changes. Furthermore, in a well-insulated turbine, the heat losses through the turbine housing are also small. Under these assumptions, Equation (3.15) reduces to

$$w'_{\text{out}} = h_{\text{in}} - h_e + \frac{1}{2g_c}(V_{\text{in}}^2 - V_e^2) \tag{3.18}$$

The working fluid in a turbine can be either in a liquid phase (water turbine) or in a gas or vapor phase (gas or steam turbine).

2. *Compressors and Pumps*

Compressors and pumps are machines that are utilized to compress, or raise the pressure of, the fluid passing through them (Figure 3.9). The *compressor*, which uses a gas or vapor as its working fluid, can be a rotating or reciprocating type. The *rotating* type functions in a manner opposite to that of the turbine; the *reciprocating* type utilizes a piston-cylinder arrangement. In a household refrigerator, a compressor is used to raise the pressure of the refrigerant vapor; in a jet engine, a compressor is used to raise the pressure of the inlet air stream. A *pump* is used to raise the pressure of the liquid flowing through it. Note that work input is needed to operate a pump or compressor. For example, electric energy from a wall socket is used for a refrigerator compressor or water pump, and turbine work is used to drive the compressor in a jet engine.

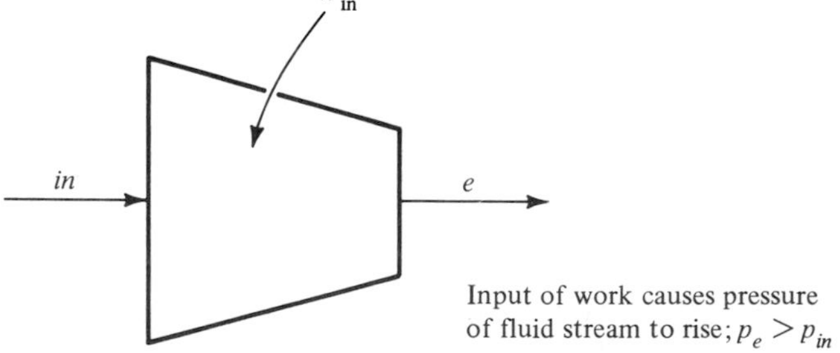

FIGURE 3.9 *Compressor or Pump.*

Again, the change between inlet and outlet potential energy is generally small and can be neglected. From Equation (3.15) we have

$$w'_{\text{in}} + q_{\text{in}} = h_e - h_{\text{in}} + \frac{V_e^2 - V_{\text{in}}^2}{2g_c} \tag{3.19}$$

For reciprocating compressors, cooling is usually provided. However, for rotating compressors, heat losses are usually small. Hence, for adiabatic flow,

$$w'_{\text{in}} = h_e - h_{\text{in}} + \frac{V_e^2 - V_{\text{in}}^2}{2g_c} \tag{3.20}$$

3. *Boilers*

A boiler is a vapor generator in which a liquid (for example, water) is converted into a vapor (for example, steam) by the addition of heat (Figure 3.10). Generally, changes between inlet and outlet kinetic and potential energies are very small in comparison with changes in enthalpy. Since no mechanical shaft work is done in a boiler, Equation (3.15) gives

$$q_{\text{in}} = h_e - h_{\text{in}} \tag{3.21}$$

4. *Condensers*

A condenser is an apparatus that condenses a substance from its vapor to its liquid phase by extracting heat from the substance (Figure 3.11). The coolant (substance that extracts heat) is usually water or air and generally is physically separated from the condensate (substance being condensed). In a steam power plant, steam from the turbine, flowing inside condenser tubes, is cooled by water flowing over the outer surfaces of the tubes. With the latent heat of the steam being transferred to the coolant water in the condenser, the inlet steam flow is converted to

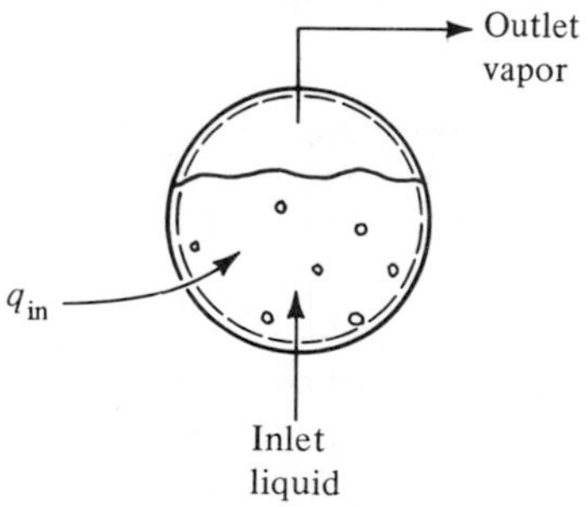

FIGURE 3.10 *Boiler.*

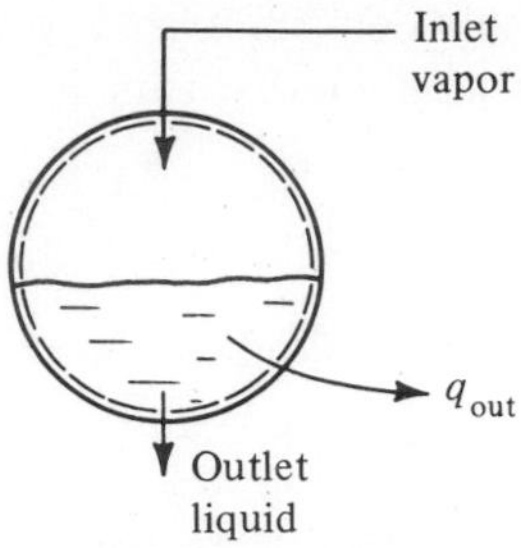

FIGURE 3.11 *Condenser.*

liquid water at the exit. In a household refrigerator, the condenser takes refrigerant vapor from the compressor and converts it into liquid. Here, the coolant medium is air. Again, changes in kinetic and potential energies of the cooling medium can be neglected. Since no mechanical shaft work is done in a condenser, Equation (3.15) gives

$$q_{out} = h_{in} - h_e \tag{3.22}$$

5. *Throttling Devices*

A throttling device is an apparatus which, by an obstruction in its throughflow (a valve or an orifice), reduces the pressure of the flow (Figure 3.12). It can be utilized to reduce the speed or power of an engine. In these devices, no mechanical shaft work is done, and heat transfer is negligible. Neglecting changes in potential and kinetic energies, we find that Equation (3.15) results in

$$h_{in} = h_e \tag{3.23}$$

6. *Nozzles*

A nozzle is a flow device, the purpose of which is to increase the flow speed of the substance passing through it (Figure 3.13). The

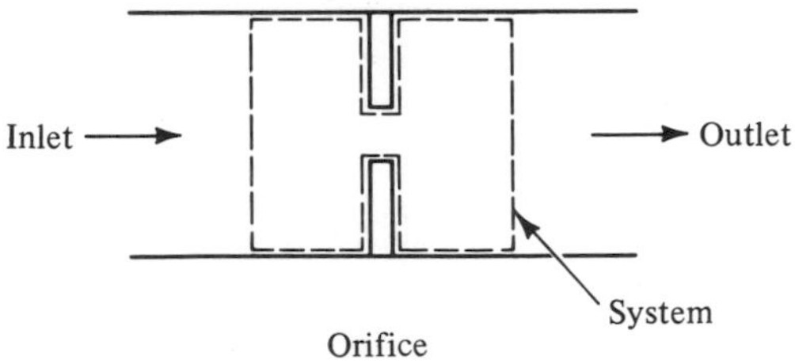

FIGURE 3.12 *Throttling Device.*

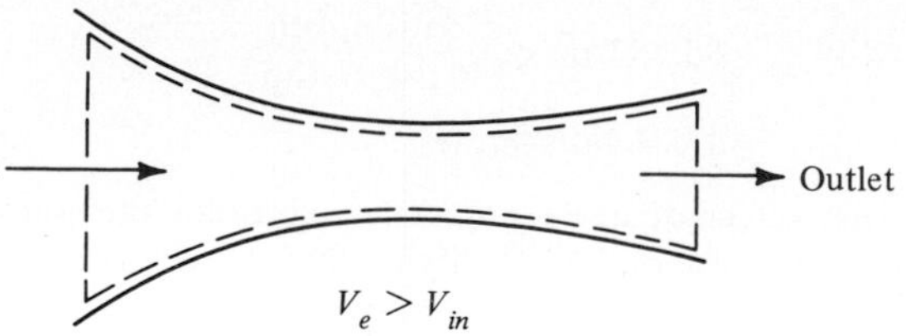

FIGURE 3.13 *Nozzle.*

substance may be a liquid, as with a garden hose nozzle, or a gas, as with an exit nozzle on a jet engine or rocket. In nozzles, there is no mechanical shaft work, and heat transfer is usually negligible. Therefore, for a horizontal nozzle, with $z_e = z_{in}$,

$$h_e - h_{in} = \frac{1}{2g_c}(V_{in}^2 - V_e^2)$$

$$\text{or} \qquad h_{in} + \frac{1}{2g_c}V_{in}^2 = h_e + \frac{1}{2g_c}V_e^2 \tag{3.24}$$

EXAMPLE 3.1

The flow rate of a steam turbine (Figure 3.14) is 0.5 lbm/s, with steam entering at a temperature of 600°F and a pressure of 200 psia. The steam leaves the turbine at a pressure of 5 psia and 162°F. What is the horsepower output of the turbine? Neglect changes in kinetic and potential energies and heat losses. The enthalpies of steam at these conditions can be found in tables of properties to be h_{in} = 1322.6 Btu/lbm and h_e = 1131.1 Btu/lbm.

Solution

In the absence of ΔKE and ΔPE, and when $\dot{Q}_{in} = 0$, the work rate output of the turbine is, from (3.18),

$$\dot{W}'_{out} = \dot{m}(h_{in} - h_e) = 0.5\ \text{lbm/s}\ (1322.6 - 1131.1)\ \text{Btu/lbm}$$
$$= 95.75\ \text{Btu/s}$$

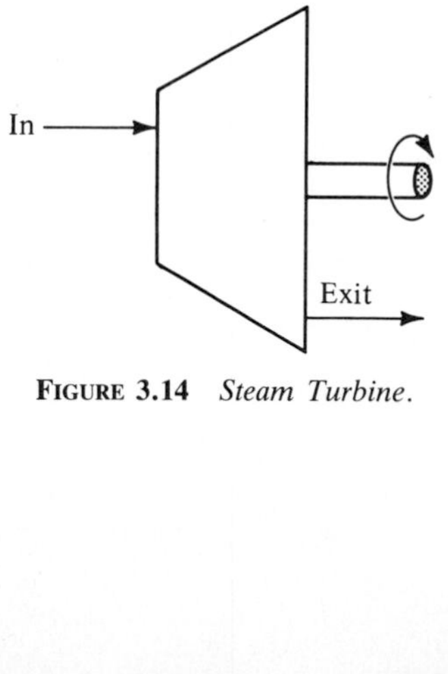

FIGURE 3.14 *Steam Turbine.*

$$\text{or} \qquad \dot{W}'_{\text{out}} = 95.75\ \text{Btu/s}\ \frac{778\ \text{ft-lbf/Btu}}{550\ \text{ft-lbf/s-hp}}$$

$$= 135.4\ \text{horsepower}$$

To examine the effect of neglecting ΔKE, consider the following: If the inlet velocity is 80 ft/s and the outlet velocity is 300 ft/s, what would the rate of change in KE be?

$$\Delta KE = \frac{\dot{m}}{2g_c}(V_{\text{in}}^2 - V_e^2) = \frac{0.5\ \text{lbm/s}}{64.4\ \text{lbm-ft/lbf-s}^2}(80^2 - 300^2)\ \text{ft}^2/\text{s}^2 = 649\ \text{ft-lbf/s}$$

$$= \frac{649}{550} = 1.18\ \text{hp}$$

To examine the effect of neglecting ΔPE, consider the following: If there is a 2-ft difference between the inlet and outlet locations, what would the change in PE be?

$$\Delta PE = \dot{m}\frac{g}{g_c}(z_e - z_{\text{in}}) = 0.5\ \text{lbm/s}\ \frac{(32.2\ \text{ft/s}^2)\text{lbf-s}^2}{32.2\ \text{lbm-ft}}\ 2\ \text{ft}$$

$$= 1\ \text{ft-lbf/s}$$

$$= \frac{1}{550} = 0.0018\ \text{hp}$$

EXAMPLE 3.2

A centrifugal air compressor has an air intake of 1.2 kg/min (Figure 3.15). The pressure and temperature conditions are inlet, 100 kPa, 0°C; and outlet, 200 kPa, 50°C. If the heat losses are negligibly small, what is the power input to the compressor? From tables of air properties, it is found that

$$u_{\text{in}} = 330.49\ \text{kJ/kg} \qquad v_{\text{in}} = 0.7841\ \text{m}^3/\text{kg}$$

$$u_e = 366.26\ \text{kJ/kg} \qquad v_e = 0.4640\ \text{m}^3/\text{kg}$$

Solution

The work input to the compressor is, from (3.20), if we neglect ΔKE,

$$w'_{\text{in}} = h_e - h_{\text{in}} = u_e - u_{\text{in}} + p_e v_e - p_{\text{in}} v_{\text{in}}$$

$$= 366.26\ \text{kJ/kg} - 330.49\ \text{kJ/kg} + (200\ \text{kN/m}^2)(0.4640\ \text{m}^3/\text{kg})$$

$$- (100\ \text{kN/m}^2)(0.7841\ \text{m}^3/\text{kg})$$

$$= 50.16\ \text{kJ/kg}$$

$$\text{Power input} = \dot{W}'_{\text{in}} = \dot{m}w'_{\text{in}}$$

$$= (50.16\ \text{kJ/kg})(1.2\ \text{kg/min})(\text{min}/60\ \text{s})$$

$$= 1.00\ \text{kJ/s}$$

$$= 1\ \text{kW}$$

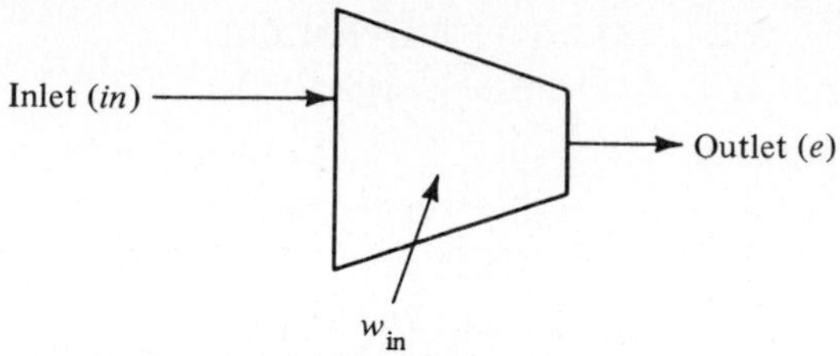

FIGURE 3.15 *Centrifugal Compressor.*

EXAMPLE 3.3

Air enters a nozzle at a pressure of 100 psia and a temperature of 200°F, and leaves at a pressure of 15 psia and a temperature of 420°R. The inlet velocity is 100 ft/s. Determine the exit velocity. For a flow of 1.1 lbm/s, find the inlet and exit areas required.

Assume that over the range of conditions in this example, the properties of air behave according to the expression $pv = RT$, with $R = 53.3$ ft-lbf/lbm-°R, and T in °R. Also assume that enthalpy change can be written as $\Delta h = 0.24\,\Delta T$, with h in Btu/lbm, and T in °R.

Solution

The applicable energy equation is (3.24),

$$h_{\text{in}} - h_e = \frac{1}{2g_c}(V_e^2 - V_{\text{in}}^2)$$

so that

$$V_e = \sqrt{2g_c(h_{\text{in}} - h_e) + V_{\text{in}}^2}$$

Now

$$h_{\text{in}} - h_e = 0.24[(200 + 460) - 420]$$
$$= 57.6 \text{ Btu/lbm}$$

so the exit velocity is

$$V_e = \sqrt{2\left(32.2\,\frac{\text{lbm-ft}}{\text{lbf-s}^2}\right)\left(57.6\,\frac{\text{Btu}}{\text{lbm}}\right)\left(778\,\frac{\text{ft-lbf}}{\text{Btu}}\right) + 10^4\,\frac{\text{ft}^2}{\text{s}^2}}$$
$$= 1702 \text{ ft/s}$$

From Equation (3.6),

$$\frac{A_{\text{in}}V_{\text{in}}}{v_{\text{in}}} = \dot{m} = \frac{A_eV_e}{v_e}$$

From the expression $pv = RT$,

$$v_{\text{in}} = \frac{RT_{\text{in}}}{p_{\text{in}}} = \frac{(53.3 \text{ ft-lbf/lbm-°R})660\text{°R}}{(100 \text{ lbf/in}^2)(144 \text{ in}^2/\text{ft}^2)} = 2.44 \text{ ft}^3/\text{lbm}$$

$$v_e = \frac{RT_e}{p_e} = \frac{(53.3\ \text{ft-lbf/lbm-°R})420\text{°R}}{(15\ \text{lbf/in}^2)(144\ \text{in}^2/\text{ft}^2)} = 10.36\ \text{ft}^3/\text{lbm}$$

The inlet and exit areas are

$$A_{\text{in}} = \frac{(1.1\ \text{lbm/s})2.44\ \text{ft}^2/\text{lbm}}{100\ \text{ft/s}} = 0.0268\ \text{ft}^2$$

$$A_e = \frac{(1.1\ \text{lbm/s})10.36\ \text{ft}^3/\text{lbm}}{1702\ \text{ft/s}} = 0.00670\ \text{ft}^2$$

EXAMPLE 3.4

A water pump is to be used in a water supply system as shown in Figure 3.16. The pressure at *in* is 70 kPa, and the pressure at *e* is 101 kPa. The velocity in the 60-mm pipe is 2 m/s. Find the pump power. Neglect friction, and assume no change of internal energy of the water and no heat transfer between *in* and *e*. Take the specific volume of the water to be constant at 0.001 m^3/kg.

Solution

Utilize a system bounded by surfaces at *in* and *e*, the pump casing, and the pipes. The work input to the fluid is, from Equation (3.15) with $q_{in} = 0$ and $h = u + pv$,

$$w'_{\text{in}} = u_e + p_e v_e - u_{\text{in}} - p_{\text{in}} v_{\text{in}} + (V_e^2 - V_{\text{in}}^2)/2g_c + (z_e - z_{\text{in}})g/g_c$$

The velocity at *e* is found from the continuity equation (3.5):

$$\frac{A_{\text{in}} V_{\text{in}}}{v_{\text{in}}} = \frac{A_e V_e}{v_e}$$

In this case, $v_{\text{in}} = v_e$, so that

$$V_e = \frac{A_{\text{in}} V_{\text{in}}}{A_e} = \frac{\frac{\pi}{4}(0.06)^2\ \text{m}^2\ (2\ \text{m/s})}{\frac{\pi}{4}(0.04)^2\ \text{m}^2}$$

$$= 4.5\ \text{m/s}$$

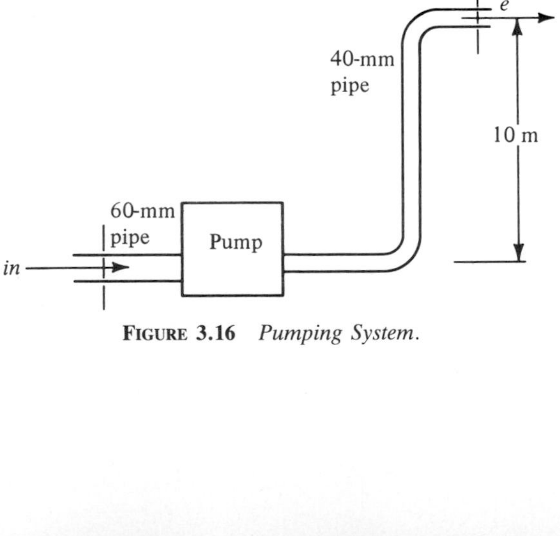

FIGURE 3.16 *Pumping System.*

$$w'_{in} = (1.01 \times 10^5\ \mathrm{N/m^2})(0.001\ \mathrm{m^3/kg}) - (0.7 \times 10^5\ \mathrm{N/m^2})(0.001\ \mathrm{m^3/kg})$$
$$+ \frac{4.5^2\ \mathrm{m^2/s^2} - 2^2\ \mathrm{m^2/s^2}}{2(1.0\ \mathrm{kg\cdot m/N\cdot s^2})} + \frac{(10\ \mathrm{m})(9.81\ \mathrm{m/s^2})}{1.0\ \mathrm{kg\cdot m/N\cdot s^2}}$$
$$= 31.0\ \mathrm{J/kg} + 8.13\ \mathrm{J/kg} + 98.1\ \mathrm{J/kg} = 137.2\ \mathrm{J/kg}$$

In order to find the power, multiply w'_{in} by mass flow $\dot{m}$, where

$$\dot{m} = \frac{A_{in}V_{in}}{v_{in}} = \frac{\frac{\pi}{4}(0.06^2\ \mathrm{m^2})(2\ \mathrm{m/s})}{0.001\ \mathrm{m^3/kg}} = 5.65\ \mathrm{kg/s}$$

$$\text{Power} = 137.2\ \mathrm{J/kg}\ (5.65\ \mathrm{kg/s}) = 776\ \mathrm{J/s} = 776\ \mathrm{W}$$

EXAMPLE 3.5

Steam is heated from a saturated vapor at 50 psia (281°F) to 300°F during a constant pressure process. How much heat must be added if the process takes place in

a. a container with a floating piston, and
b. a steady-flow process without external work?

From a thermodynamic properties table, it is found that at 50 psia and 281°F, $u = 1095.3$ Btu/lbm, $h = 1174.1$ Btu/lbm, and $v = 8.514\ \mathrm{ft^3/lbm}$, and at 50 psia and 300°F, $u = 1102.9$ Btu/lbm, $h = 1184.1$ Btu/lbm, and $v = 8.769\ \mathrm{ft^3/lbm}$.

Solution

a. From Equation (2.5), for a closed system,

$$Q_{in} - W_{out} = U_2 - U_1$$

or, on a unit mass basis,

$$q_{in} - w_{out} = u_2 - u_1$$

The work out is

$$w_{out} = p(v_2 - v_1) = \frac{(50 \times 144)\ \mathrm{lbf/ft^2}}{778\ \text{ft-lbf/Btu}}(8.769 - 8.514)\ \mathrm{ft^3/lbm}$$
$$= 2.36\ \mathrm{Btu/lbm}$$

Hence

$$q_{in} = 1102.9 - 1095.3 + 2.36 = 10\ \mathrm{Btu/lbm}$$

b. From Equation (3.15), for an open-flow system, if we neglect ΔKE and ΔPE,

$$q_{in} = h_e - h_{in} \qquad (\text{since } w'_{out} = 0)$$
$$= 1184.1\ \mathrm{Btu/lbm} - 1174.1\ \mathrm{Btu/lbm}$$
$$= 10\ \mathrm{Btu/lbm}$$

Note that q_{in} is identical in both cases.

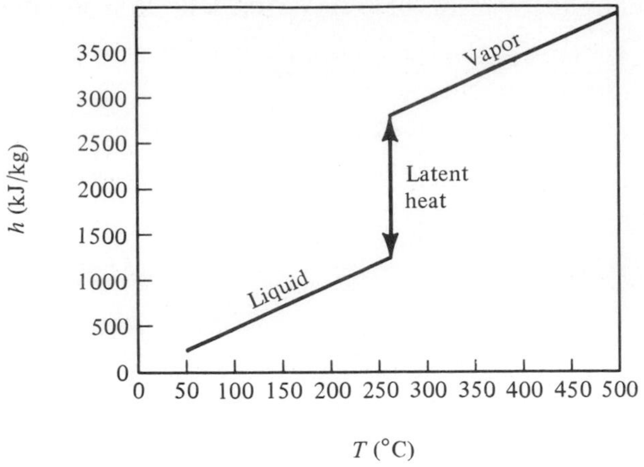

FIGURE 3.17 *Enthalpy-Temperature.*

EXAMPLE 3.6

Liquid water enters a boiler at a temperature of 50°C and a pressure of 5000 kPa, and leaves as steam at a temperature of 400°C and a pressure of 5000 kPa. A plot of h versus T for water at 5000 kPa is shown in Figure 3.17. For steady flow, what is the heat transferred to each kilogram of water in the boiler?

Solution

From the curve of Figure 3.17, at 50°C,

$$h_{\text{in}} \cong 215 \text{ kJ/kg}$$

At 400°C,

$$h_e \cong 3200 \text{ kJ/kg}$$

From Equation (3.21), for a boiler,

$$q_{\text{in}} = h_e - h_{\text{in}} = 2985 \text{ kJ/kg}$$

EXAMPLE 3.7

A stream of cold water is to be heated by a hot water flow in a steady-flow mixing process (Figure 3.18). In the process, 3.0 lbm/s of hot water at 180°F are to be mixed with 5.0 lbm/s of cold water at 40°F. What is the resultant temperature of the mixed stream? Assume that the enthalpy change of liquid water can be written as $\Delta h = 1.0\,\Delta T$, where Δh is in Btu/lbm and ΔT is in °F. Neglect changes of potential and kinetic energies of the water, and assume the mixer to be well insulated so that there is no external heat transfer.

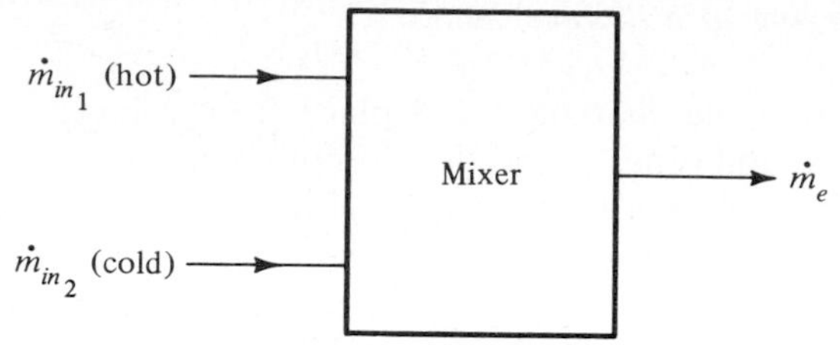

FIGURE 3.18 *Mixer.*

Solution

From the continuity equation, the total outflow from the mixer is $\dot{m}_e = 3\ \text{lbm/s} + 5\ \text{lbm/s} = 8\ \text{lbm/s}$. The energy equation becomes, with $Q_{\text{in}} = W'_{\text{out}} = 0$,

$$\dot{m}_{\text{in}_{\text{hot}}} h_{\text{in}_{\text{hot}}} + \dot{m}_{\text{in}_{\text{cold}}} h_{\text{in}_{\text{cold}}} = \dot{m}_e h_e$$

or

$$\dot{m}_{\text{in}_{\text{hot}}} (h_{\text{in}_{\text{hot}}} - h_e) = \dot{m}_{\text{in}_{\text{cold}}} (h_e - h_{\text{in}_{\text{cold}}})$$

$$(3.0\ \text{lbm/s})(1\ \text{Btu/lbm-°F})(180 - T_e)\text{°F} = (5\ \text{lbm/s})(1\ \text{Btu/lbm-°F})(T_e - 40)\text{°F}$$

$$T_e = 92.5\text{°F}$$

3.5 CHARGING AND DISCHARGING OF COMPRESSED GAS TANKS

Consider the flow that occurs in the charging and discharging of compressed gas tanks. During this process, the mass flow per unit time, and the pressure and the temperature of the gas within the storage tank, vary. The volume of the tank remains constant during the process, and there is no mechanical shaft work, $W'_{\text{out}} = 0$. Further, consider the case of insulated tanks or the case when the charging or discharging process is sufficiently fast so that only a negligible amount of heat is transferred across the walls of the storage tanks. From Equation (3.11), and neglecting changes in *KE* and *PE*, we have

$$H_e - H_{\text{in}} + U_2 - U_1 = 0$$

or

$$m_e h_e - m_{\text{in}} h_{\text{in}} + m_2 u_2 - m_1 u_1 = 0$$

The statement of the conservation of mass is from (3.1)

$$m_1 + m_{\text{in}} = m_2 + m_e$$

1. *Charging of a Storage Tank*

Assume that the charging takes place from a large supply tank so that its pressure and other properties do not change during the charging. However, within the tank being charged, the mass of compressed gas and other properties vary during the charging process. To simplify the analysis, we will assume uniformity of conditions at any given instant.

The storage tank is connected to the supply tank by a valve (Figure 3.19), which acts as a throttle and allows for the pressure difference between constant supply pressure and instantaneous pressure within the storage tank. Across the valve (throttling process), the enthalpy of the gas does not change. Thus the gas enters the tank with the supply gas enthalpy (h_{in}), but at a lower pressure. In the initial phases of charging, the mass flow per unit time is highest, decreasing in the latter phase as the pressure builds up within the tank. In this case, $m_e = 0$, $h_{in} =$ constant, and Equation (3.12) becomes $-m_{in}h_{in} + m_2u_2 - m_1u_1 = 0$, while Equation (3.1) becomes $m_1 + m_{in} = m_2$. Combining the two equations, we have

$$-(m_2 - m_1)h_{in} + m_2u_2 - m_1u_1 = 0$$
$$\text{or} \qquad m_1(u_1 - h_{in}) = m_2(u_2 - h_{in}) \tag{3.25}$$

In addition, the volume of the tank being charged remains constant. Thus

$$\forall = m_1v_1 = m_2v_2$$

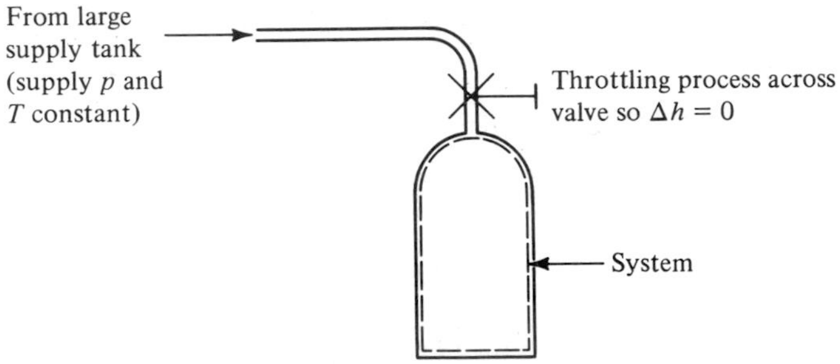

FIGURE 3.19 *Charging Process.*

EXAMPLE 3.8

A completely empty tank ($\forall = 0.5\ \text{ft}^3$) is charged with oxygen from a very large tank ($p_i = 100$ psia, $T_i = 80°\text{F}$) until the pressure in the tank reaches the supply pressure of 100 psia. Determine the temperature of the oxygen in the charged tank, which is well insulated.

From tables of thermodynamic properties of oxygen, it is found that $h_{in} = 117.1$ Btu/lbm. In addition, the following values of internal energy are found as a function of temperature:

T(°F)	u(Btu/lbm)
100	80.7
150	94.7
200	102.6
250	110.7
300	118.8

Solution

Since the supply tank is very large, we can assume that the conditions in it do not change during charging. The tank to be charged starts with $m_1 = 0$. From Equation (3.25), $u_2 = h_{in}$. Since the required value of internal energy lies between temperatures of 250°F and 300°F, interpolation is required. Using linear interpolation, we obtain

$$\frac{117.1 - 110.7}{118.8 - 110.7} = \frac{x - 250}{300 - 250}$$

$$0.790(50) = x - 250$$

$$x = T_2 = 289.5\text{°F}$$

EXAMPLE 3.9

Repeat Example 3.8 with $T_1 = 80$°F and residual mass $m_1 = 0.02$ lbm in the tank, which is charged until $m_2 = 0.16$ lbm. From tables of thermodynamic properties of oxygen, the internal energy u_1 is found to be 83.6 Btu/lbm.

Solution

In this case, Equation (3.25) applies:

$$m_1(u_1 - h_{in}) = m_2(u_2 - h_{in})$$

$$0.02\ \text{lbm}\ (83.6 - 117.1)\ \text{Btu/lbm} = 0.16\ \text{lbm}\ (u_2 - 117.1)\ \text{Btu/lbm}$$

$$u_2 = 112.9\ \text{Btu/lbm}$$

From the properties given in the statement of Example 3.8, we see that the value of internal energy lies between temperatures of 250°F and 300°F. Using linear interpolation, we get

$$\frac{112.9 - 110.7}{118.8 - 110.7} = \frac{x - 250}{300 - 250}$$

$$0.272(50) = x - 250$$

$$x = T_2 = 263.6\text{°F}$$

Note that in both examples the final temperature of oxygen in the tank is higher than that of the supply oxygen (which remains unchanged by the charging). This result is due to the flow work done by the supply oxygen on the oxygen in the tank.

---

2. *Discharging of a Storage Tank*

In this case we will assume that at any given instant, the conditions within the storage tank are uniform. The pressure difference between the inside of the tank and the environment is sustained by a throttling valve (Figure 3.20). Again, in the beginning of the discharge, the mass flow per unit time is greatest and decreases as the pressure within the tank drops.

Here, $m_{\text{in}} = 0$. However, h_e does not remain constant and decreases as the mass of gas within the tank decreases. Equation (3.12) becomes $m_e h_e + m_2 u_2 - m_1 u_1 = 0$, while Equation (3.1) becomes $m_1 = m_2 + m_e$. Combining the two equations, we have

$$(m_1 - m_2)h_e + m_2 u_2 - m_1 u_1 = 0 \quad \text{or} \quad m_1(u_1 - h_e) = m_2(u_2 - h_e) \tag{3.26}$$

The enthalpy h_e will vary from the initial value h_1 to the final value h_2. A good approximation in the solution of such problems can be obtained by assuming an average value of $h_e = (h_1 + h_2)/2$. Thus Equation (3.19)

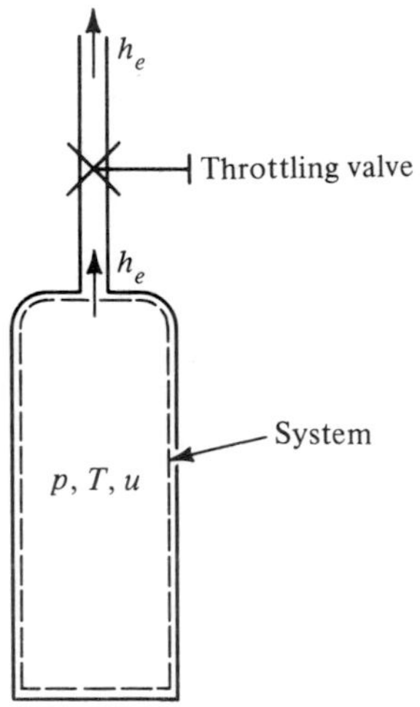

FIGURE 3.20 *Discharging Process.*

becomes

$$m_1\left(u_1 - \frac{h_1 + h_2}{2}\right) = m_2\left(u_2 - \frac{h_1 + h_2}{2}\right) \qquad \textbf{(3.27)}$$

In addition, the volume of the tank discharging remains constant:

$$V = m_1 v_1 = m_2 v_2$$

EXAMPLE 3.10

A tank is filled with 0.25 kilogram of air at 0°C and 300 kPa. It is discharged until 0.10 kilogram of air is left in the tank. If the tank is well insulated (no heat transfer), what will the final air temperature be in the tank?

From thermodynamic tables for air, it is found that at 0°C and 300 kPa, $u_1 = 330.0$ kJ/kg and $h_1 = 408.4$ kJ/kg. Further, it is to be noted in such tables that over the temperature and pressure ranges of interest in this problem, u and h are relatively insensitive to pressure. Values are found to be the following:

T(°C)	u(kJ/kg)	h(kJ/kg)
0	330.0	408.4
−50	294.6	358.6
−100	258.6	308.2

Solution

Using the internal volume of the tank as the thermodynamic system, apply Equation (3.27):

$$m_1\left(u_1 - \frac{h_1 + h_2}{2}\right) = m_2\left(u_2 - \frac{h_1 + h_2}{2}\right)$$

Substituting the values of thermodynamic properties, we obtain

$$0.25 \text{ kg } (330.0 - 204.2 - h_2/2) \text{ kJ/kg} = 0.10 \text{ kg } (-204.2 + u_2 - h_2/2) \text{ kJ/kg}$$

$$\text{or} \qquad 51.87 = 0.075h_2 + 0.1u_2$$

At $T = -50$°C, $0.075h_2 + 0.1u_2 = 56.36$ kJ/kg, while at $T = -100$°C, we get 48.98 kJ/kg. Interpolating linearly, we have

$$\frac{51.87 - 56.36}{48.98 - 56.36} = \frac{x - (-50)}{-100 - (-50)}$$

$$0.608(-50) = x + 50$$

$$x = T_2 = -80.4\text{°C}$$

It is evident that to solve the problems in this chapter, we needed to have information on the thermodynamic properties (u, h, v, etc.) of the various working substances undergoing changes. This information will be presented in detail in the next chapter.

PROBLEMS

3.1 The small, high-speed turbine in a dentist's drill is driven by compressed air and produces 50 W. The turbine inlet condition is 500 kPa, 25°C; the exit condition is 101.3 kPa, 0°C. The change in enthalpy of air is given by $\Delta h = 1.00\,\Delta T$ (kJ/kg) with T in °C. Calculate the mass flow rate of air through the turbine.

3.2 A small, liquid-water spray discharges steadily from a tank at 50 psia and 70°F through a nozzle into the atmosphere. Assuming that no heat is lost in the process, determine the nozzle exit velocity. The specific volume of water at 70°F is 0.01605 ft^3/lbm; neglect changes in the internal energy of the water.

3.3 Liquid water at 25°C and 101.3 kPa is compressed to 10,000 kPa in a steady-flow pump without heat losses. Find the work done in the compression process, neglecting changes in kinetic and potential energies between inlet and discharge. From tables of thermodynamic properties of water, we have $h_{\text{in}} = 104.8$ kJ/kg at 25°C and 101.3 kPa, and $h_e = 114.0$ kJ/kg at 10,000 kPa and 25°C.

3.4 Air at ambient atmospheric pressure and temperature enters the compressor of a jet power plant with an enthalpy of 450 kJ/kg; it discharges at a pressure of 400 kPa with an enthalpy of 600 kJ/kg. Calculate the power required to drive the compressor for a mass flow rate of 10 kg/s; assume adiabatic flow.

3.5 One cubic foot per second of water at 60°F passes through a nozzle. The inlet diameter of the nozzle is 4 in., and the outlet diameter is 1.8 in. What is the pressure drop between inlet and outlet sections? The specific volume of water at 60°F is 0.01603 ft^2/lbm; neglect changes of internal energy and potential energy.

3.6 The pressure of the working medium in a well-insulated steady-flow device changes from 75 to 25 psia, while the velocity changes from 100 to 1000 ft/s. The specific volume changes from 0.1 to 0.5 ft^3/lbm. The internal energy of the working medium decreases by 15 Btu/lbm. What is the work done by the working medium?

3.7 The enthalpy of high-pressure steam entering an insulated nozzle with a velocity of 50 m/s is 3000 kJ/kg. It leaves the nozzle with an enthalpy of 2200 kJ/kg. Calculate the nozzle exit velocity.

3.8 The pressure of air flowing through a reducing valve is changed from 250 to 75 psia. The initial temperature of the air is 100°F. The change in enthalpy of the air can be expressed by $\Delta h = 0.24\,\Delta T$ Btu/lbm with ΔT in °F. Find the temperature of the air after throttling.

3.9 The steam turbine of an electric power generating plant uses 5000 pounds of steam per hour. Steam enters the turbine at a pressure of 500 psia and a temperature of 900°F with an enthalpy of 1466.6 Btu/lbm. Steam leaves the turbine at a pressure of 5 psia with an enthalpy of 1031.0 Btu/lbm. Calculate the horsepower output of the turbine if (a) the turbine is completely insulated and (b) the heat loss from the turbine amounts to 10,000 Btu/hr.

3.10 Steam is discharged from the turbine of a steam plant at a pressure of 5 psia with an enthalpy of 1031.0 Btu/lbm. The turbine discharge enters a condenser and leaves the condenser as liquid water with an enthalpy of 130.2 Btu/lbm. Calculate the rate of energy removal in the condenser for a mass flow of 5000 pounds of steam per hour.

3.11 The boiler of a steam power plant receives 20,000 pounds of water per hour and discharges it as steam. The enthalpy of incoming water is 117.35 Btu/lbm, while that of the leaving steam is 1470.0 Btu/lbm. If the coal, which is burned in the boiler, releases 10,000 Btu/lbm of coal, find the mass of coal required per hour.

3.12 A steady-flow electric steam generator operating at atmospheric pressure uses 10 kg of water per minute at 20°C and transforms it into steam at 100°C. Calculate the electrical power requirement of the steam generator. The enthalpy of the water is 84 kJ/kg and that of the steam is 2676 kJ/kg.

3.13 A small container of carbon dioxide is at a temperature of 75°F, with a residual mass of 0.1 lbm. It is charged from a large storage tank (p_{in} = 150 psia, T_{in} = 75°F) until there is one pound of carbon dioxide in the container. From tables of thermodynamic properties of carbon dioxide, it is found that u_1 = 143.15 Btu/lbm and h_{in} = 165.62 Btu/lbm. Further, the following tabular values of internal energy are also found:

T(°F)	u(Btu/lbm)
100	147.7
200	165.3

Determine the temperature of carbon dioxide in the charged tank, which is well insulated.

3.14 The variation of enthalpy with temperature for steam at pressures of 1000 kPa and 100 kPa is shown in Figure 3.21. Steam is expanded in a nozzle from an entering pressure of 1000 kPa and an entering temperature of 700°C, to a final pressure and temperature of 100 kPa and 500°C. Find the exit velocity from the nozzle. Steam enters the nozzle at 20 m/s. Assume no heat transfer from the steam.

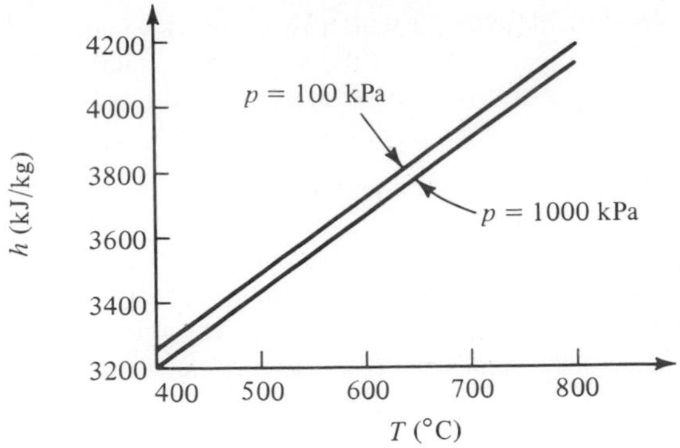

FIGURE 3.21

3.15 Find the pump horsepower required to pump 10 gallons of water per minute from the basement to the top floor of an 8-story building, 80 feet high. Assume the water temperature to be constant, with water pressure at the basement level of 20 psig and atmospheric pressure where the water flows out of the tap on the eighth floor. Take the density of water to be 62.4 lbm/ft^3. Neglect changes of kinetic energy of the water; assume adiabatic steady flow.

3.16 A steam turbine uses one kg of steam per second. Steam enters the turbine at 3500 kPa and 400°C with a velocity of 250 m/s; it leaves at a pressure of 50 kPa and a temperature of 95°C with a velocity of 30 m/s. Determine the power output of the turbine.

3.17 Water is compressed adiabatically at a steady flow rate of 2 lbm/s from a pressure of 50 psia and a temperature of 80°F to a pressure of 500 psia. What is the required horsepower? Neglect changes of kinematic and internal energies and assume v = constant.

3.18 Air enters an insulated nozzle at a pressure of 1000 kPa and a temperature of 150°C at negligible speed. Find the nozzle exit velocity. The nozzle exhausts to atmospheric pressure. Exit temperature is 0°C.

3.19 An insulated air turbine expands the air from 100 psia and 400°F to atmospheric pressure, 0°F. The turbine exit velocity is 300 ft/s. Determine the turbine work per lbm of air flow.

3.20 Water enters an insulated nozzle at a temperature of 50°C and a pressure of 1000 kPa and exits at atmospheric pressure. Find the nozzle exit velocity and flow rate if the exit area is 25 mm^2. Neglect changes of internal energy and specific volume

3.21 The following changes occur between the entrance and exit of an open system: pressure decreases from 250 to 30 psia; specific volume increases from 5 to 15 ft^3/lbm; internal energy decreases 50 Btu/lbm, and 25 Btu/lbm of heat is added to the fluid. No changes in entrance and exit velocities and in elevation occur. Find the work done by the fluid.

3.22 The inlet and exit conditions of an air blower are as follows:

	Inlet	Exit
Pressure, kPa	97	103
Specific volume, m^3/kg	0.890	0.850
Internal energy, kJ/kg	31.6	36.0
Velocity, m/s	3	15

Calculate the power consumption of the blower for a flow rate of 0.75 m^3/s.

3.23 Air flows through a circular tube, where it is heated by an electric coil from a temperature of 68°F to 300°F. The air enters the tube at a pressure of 100 psia and a velocity of 15 ft/s, and leaves the tube at a pressure of 93 psia. Determine the air velocity at tube exit.

chapter

4

Thermodynamic Properties of Substances

4.1 INTRODUCTION

In the examples and problems of the preceding chapters, it was necessary to provide information on the equilibrium thermodynamic properties of the working substance in order to obtain numerical results. It is clear that such information will not always be included in the definition of a problem, and thus the engineer must have means for determining such thermodynamic properties. The purposes of this chapter are to present the properties of and discuss behavior of four typical working substances.

4.2 THERMODYNAMIC PROPERTIES OF WATER (LIQUID AND VAPOR PHASES)

The thermodynamic properties of a substance that have been discussed in preceding chapters include pressure, temperature, specific volume (density), internal energy, and enthalpy. At this point, we would like to be able to determine relationships between these thermodynamic variables.

Let us start with an example familiar to all: the variation of specific volume of water with temperature at atmospheric pressure. Consider a sample of liquid water contained within a piston-cylinder arrangement, as shown in Figure 4.1, with a weightless piston such that the pressure on the water is maintained constant at one atmosphere. We will start with a unit mass of water at a temperature of 25°C (77°F) and slowly add heat to the water. Volume will be measured as a function of water temperature, with the water allowed to reach equilibrium prior to each measurement.

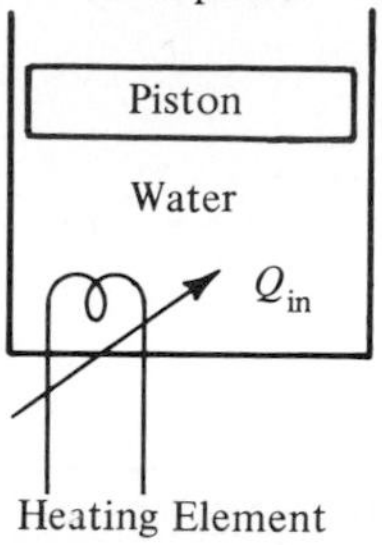

FIGURE 4.1

Remember that specific volume (v) is the ratio of volume and mass of water. The starting value for v is 0.001003 m^3/kg or 0.01606 ft^3/lbm at a room temperature of 25°C; these figures correspond to a density of 997.0 kg/m^3 or 62.27 lbm/ft^3. As the sample is heated at atmospheric pressure, the liquid water expands slightly, increasing its specific volume. For example, at 80°C (176°F), the specific volume is 0.00103 m^3/kg; at 93.3°C (200°F), the specific volume is 0.016637 ft^3/lbm. Eventually, as more heat is added, the water reaches 100°C (212°F), and with more heat input the water starts to boil. During the boiling process, two phases (liquid and vapor) coexist in equilibrium, with the temperature constant at 100°C (212°F).

The temperature at which vaporization (boiling) takes place at a given pressure is called the *saturation temperature*. In our example, the saturation temperature for water at one atmosphere is 100°C or 212°F. Similarly, the pressure at which vaporization takes place at a given temperature is called the *saturation pressure*; from our example, the saturation pressure at 100°C (212°F) is 101.3 kPa (kN/m^2) or 14.696 lbf/in^2.

With the addition of enough heat, our sample will pass from pure saturated liquid at 100°C to pure saturated vapor at 100°C; the specific volume of the sample will increase from 0.016719 ft^3/lbm to 26.799 ft^3/lbm, or from 0.0010437 m^3/kg to 1.673 m^3/kg, a factor of 1603! Further addition of heat to the pure saturated vapor at 100°C causes the temperature of the sample to increase above the saturation temperature; we call the vapor at a temperature above the saturation temperature a *superheated vapor*. Figure 4.2 shows a representation of the variation of v with T for water over the liquid-vapor region at atmospheric pressure.

Now let us examine the behavior of the sample at pressures other than atmospheric. We know that if the constant pressure on the sample is held below atmospheric pressure, the water will boil at a temperature below 100°C (212°F). If the pressure is maintained above atmospheric, as

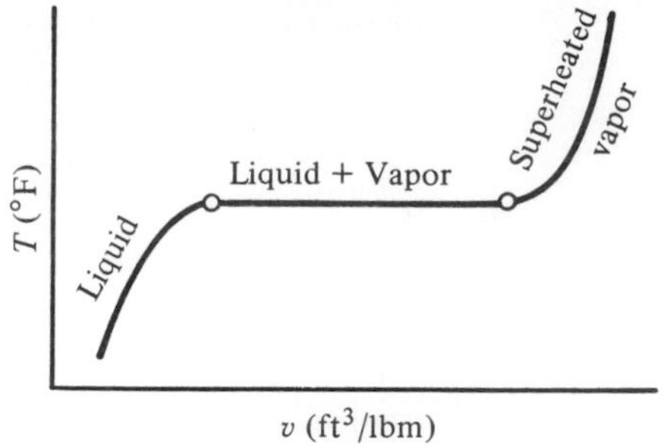

FIGURE 4.2

in a pressure cooker, the boiling temperature will be above atmospheric. Typical plots of v versus T for several pressures are given in Figure 4.3. Points A, B, and C correspond to the saturated liquid states; a, b, and c correspond to the saturated vapor states.

If we draw the locus of all saturation states, we have the saturation line shown in Figure 4.4. Again, to the right of the saturation line is the superheated vapor region, where the vapor temperature is greater than the saturation temperature at the specified pressure. To the left of the saturation line is the liquid region, called *compressed liquid region*, since the pressure on the liquid in this region is greater than the saturation pressure at a specified temperature. This region is also called the *sub-cooled liquid region*, since it corresponds to the case in which the temperature of the liquid is lower than the saturation temperature at a given pressure. From the shape of the saturation curve, it can be seen that the difference in specific volume between saturated liquid and vapor gets smaller as the pressure on the water is increased; at a high enough pressure (3208.2 psia, 22,120 kPa), liquid and vapor exist indistinguishably at 705.47°F (374.15°C). Above this pressure and temperature, there can be no liquid-vapor changes. The point labeled *C.P.* in Figure 4.4 is

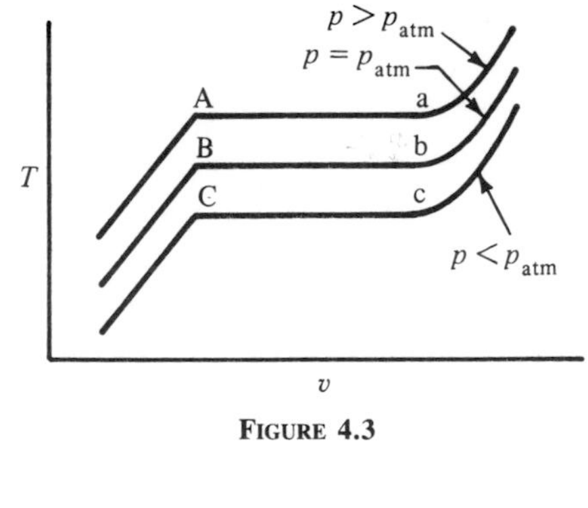

FIGURE 4.3

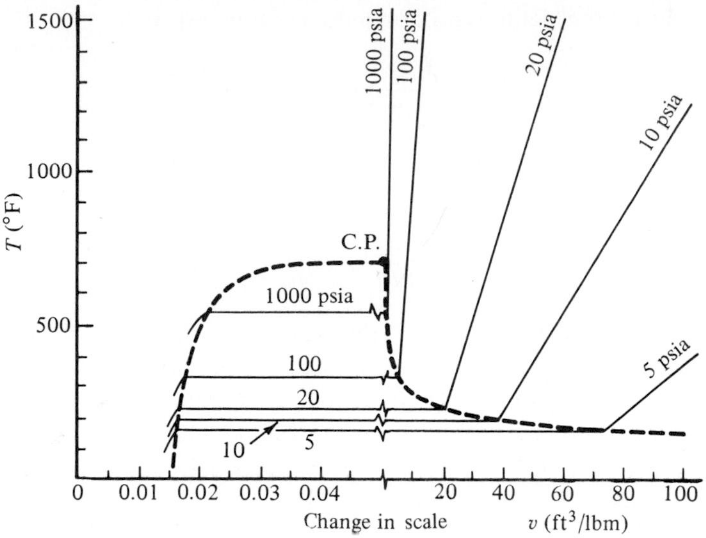

FIGURE 4.4 *Specific Volume-Temperature Diagram for Water.*

called the *critical point*, with pressure and temperature at this point called *critical pressure* and *critical temperature*.

Values of saturation pressure as a function of temperature are given in Table A.2 of Appendix A for water in English units and in Table B.2 of Appendix B in SI units. Also listed in the same tables are values of the specific volume of saturated liquid (v_f) at a given temperature, and specific volume of saturated vapor (v_g) at a given temperature. To provide for ease of calculation, the difference in specific volume, $v_g - v_f$ or v_{fg}, is also given.

Under the saturation curve of Figure 4.4, liquid and vapor coexist at equilibrium. The two phases of the mixture may exist in a container at separate locations (Figure 4.5) or as an intermingling of vapor and small liquid droplets, that is, as fog (Figure 4.6). To obtain a value of v in this region, we need to know how much vapor and liquid are present in the

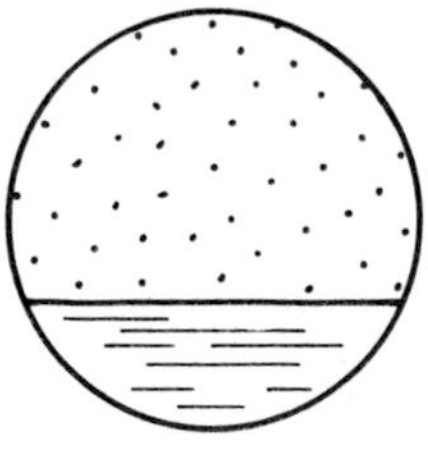

FIGURE 4.5

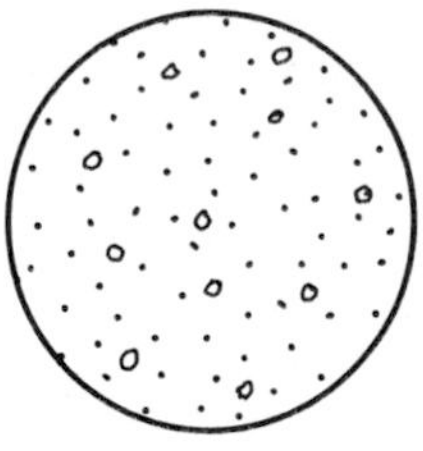

FIGURE 4.6

mixture. From the definition of specific volume, we have

$$v = \frac{V}{m} = \frac{V_f + V_g}{m_f + m_g} = \frac{m_f v_f + m_g v_g}{m_f + m_g}$$

where the subscript f refers to saturated liquid and g to saturated vapor. Simplifying the above, we obtain

$$v = \frac{m_f}{m_f + m_g} v_f + \frac{m_g}{m_f + m_g} v_g$$

The ratio of the mass of the vapor to the total mass of the liquid-vapor mixture, $m_g/(m_f + m_g)$, is called the *quality* of the mixture, denoted by the symbol x. The values of x can vary from zero (all saturated liquid; $m_g = 0$) to unity (all saturated vapor; $m_f = 0$). Recognizing that $m_f/(m_f + m_g) = 1 - x$, we have

$$v = (1 - x)v_f + xv_g$$

or

$$v = v_f + x(v_g - v_f)$$

$$v = v_f + xv_{fg} \tag{4.1}$$

In the liquid-vapor region, the value of v is completely specified by quality x and saturation pressure p, or by quality x and saturation temperature T.

EXAMPLE 4.1

A closed, rigid vessel contains 40 percent liquid water and 60 percent water vapor by volume in equilibrium at 300°F. Determine the pressure inside the vessel and the quality of the mixture.

Solution

The pressure in the vessel is equal to the saturation pressure at 300°F or 67.005 psia, from Table A.2, Appendix A. The volume of the vapor is given by $v_g m_g$; the volume of the liquid is $v_f m_f$. The ratio of liquid volume to vapor volume is $40/60 = v_f m_f / v_g m_g$. We find v_f and v_g in the saturated steam tables, Table A.2, Appendix A at 300°F, to be $v_f = 0.01745\ \text{ft}^3/\text{lbm}$ and $v_g = 6.4658\ \text{ft}^3/\text{lbm}$. Therefore

$$\frac{m_f}{m_g} = \frac{40}{60}\,\frac{6.4658}{0.01745} = 247 \qquad x = \frac{m_g}{m_g + m_f} = \frac{m_g}{m_g + 247m_g}$$

$$x = \frac{1}{248} = 0.00403$$

EXAMPLE 4.2

The mixture inside the tank of Example 4.1 is cooled to 200°F. Determine the percentage of liquid and the vapor by volume at the new state.

Solution

Since the vessel is closed, no mass enters or leaves the vessel. Since the vessel is rigid, the total volume of the contents is constant. Therefore, when the vessel is cooled, there is no change of specific volume. The specific volume at 200°F is the same as that at 300°F. At 300°F,

$$\begin{aligned} v &= v_f + xv_{fg} \\ &= 0.01745 + 0.00403(6.4483) \\ &= 0.01745 + 0.02599 \\ &= 0.04344 \text{ ft}^3/\text{lbm} \end{aligned}$$

At 200°F, v is also 0.04344 ft/lbm. At 200°F, $v_f = 0.016637 \text{ ft}^3/\text{lbm}$, $v_{fg} = 33.622 \text{ ft}^3/\text{lbm}$, and $v_g = 33.639 \text{ ft}^3/\text{lbm}$. The percent vapor by volume is $100v_g m_g/(vm)$, where $m_g/m = x$, and x is found from

$$0.04344 = 0.016637 + x33.622$$

$$x = \frac{0.02680}{33.622} = 0.000797$$

The percentage of vapor by volume is

$$100 \frac{33.639}{0.04344} 0.000797 = 61.7\%$$

EXAMPLE 4.3

Three kilograms of 30 percent quality steam at 12,056 kPa is contained behind a piston in a cylinder. The volume of the steam is doubled with no mass change, and the steam temperature is maintaned constant. Find the new steam pressure and quality.

Solution

At the initial condition, with a pressure of 12,056 kPa (kN/m^2), the saturation temperature is 325°C (see Table B.2, Appendix B). At this temperature, $v_f = 0.0015289 \text{ m}^3/\text{kg}$ and $v_{fg} = 0.01266 \text{ m}^3/\text{kg}$. Also, $v_1 = 0.00153 + x0.01266 = 0.00153 + 0.3(0.01266) = 0.00533 \text{ m}^3/\text{kg}$. Therefore, the total volume at state 1 is $0.00533(3) = 0.01599 \text{ m}^3$, and the volume at state 2 is 0.03198 m^3. Since no mass enters or leaves the system, $v_2 = (0.03198)/3 = 0.01066 \text{ m}^3/\text{kg}$. Again,

$$v_2 = v_f + x_2 v_{fg} = 0.0015289 + x_2 0.01266$$

$$x_2 = \frac{0.00913}{0.01266} = 0.721$$

$$p_2 = 12{,}056 \text{ kPa}$$

In the *superheated* region, specific volume at a given pressure is a function of temperature only. Values of specific volume for water vapor are given in Table A.3, Appendix A, and Table B.3, Appendix B. Likewise, in the *compressed liquid* region, specific volume at a given pressure is a function of temperature only. Values of specific volume of compressed liquid water are given in Tables A.4 and B.4. It can be seen that at a given temperature, the specific volume of compressed water has only a slight dependence on pressure. For example, at room temperature, the change in specific volume in going from one atmospheric pressure to 100 atmospheres is less than 1 percent. Generally, therefore, unless we are dealing with extremely high pressures, we say that water is incompressible, and we use the value of specific volume given for the saturated condition at the same temperature.

In handling problems involving the first law, one must determine the change of internal energy and enthalpy of substances undergoing thermodynamic processes. The internal energy of a substance is not directly measurable since it includes the effects of internal kinetic and potential energies of the atoms and molecules of a substance. However, as noted in Chapter 2, when energy is transferred into a closed system, there is an increase in internal energy of the system. Therefore, changes in internal energy are measurable, so values of internal energy can be given with respect to an arbitrarily assigned reference condition. In the thermodynamic tables for water, the reference condition for internal energy is taken as saturated liquid at 32°F ($u_f = 0$ at this point), (Tables A.2 and B.2.)

From the definition of enthalpy, $h = u + pv$, the value of enthalpy at the reference condition for internal energy is established. Values of changes in the enthalpy of water can be determined in the same setup used to determine the behavior of the specific volume of water. The piston-cylinder arrangement (Figure 4.1) is insulated, and heat is added by an electric coil; thus the energy addition is easily measured. Starting with liquid water at 0°C (32°F), we measure the heat added to the water to reach a liquid water temperature of 100°C (212°F). We find that the quantity of energy added to the water is 180.13 Btu/lbm (419.0 kJ/kg). To completely transform the saturated liquid water to saturated steam requires 970.3 Btu/lbm (2256.9 kJ/kg). To raise the temperature of the saturated steam from 212°F (100°C) to superheated steam at 1000°F (537.8°C) requires 384.0 Btu/lbm (893.2 kJ/kg). Since the water in the piston-cylinder arrangement constitutes a closed system, we have, from Equation (2.5),

$$q_{in} - w_{out} = u_2 - u_1$$

The only work done by the system is that done on the floating piston; that is, w_{out} equals $p(v_2 - v_1)$. Hence

$$q_{in} = u_2 + pv_2 - u_1 - pv_1 = h_2 - h_1$$

Thus, adding heat to the water at constant pressure results in an increase of enthalpy of the water. A representation of the change in enthalpy of water with temperature at atmospheric pressure is given in Figure 4.7.

Next, using the experimentally measured values of energy addition, we will determine values of enthalpy for water by using the reference condition for internal energy as saturated liquid at 32°F. At 32°F, the saturation pressure p_f is 0.08865 psia, and the specific volume v_f is 0.016022 ft^3/lbm. Therefore, with $h = u + pv$,

$$h_f = 0 + \frac{144 \text{ in}^2/\text{ft}^2}{778 \text{ ft-lbf/Btu}} (0.08865 \text{ lbf/in}^2)\, 0.016022 \text{ ft}^3/\text{lbm}$$
$$= 0.0003 \text{ Btu/lbm}$$

which can be taken as zero. This value of h_f is the enthalpy of liquid water at 32°F, 0.08865 psia. To evaluate the small change in enthalpy between saturation pressure and atmospheric pressure at a temperature of 32°F, we can assume that the change in specific volume of the liquid between the two pressures is negligibly small and that the change in internal energy is also very small, since any change in the internal energy of the liquid water is due largely to a change in temperature (which is zero here). But $\Delta h = \Delta u + p\,\Delta v + v\,\Delta p$, and $\Delta u = \Delta v = 0$. Thus, at 32°F,

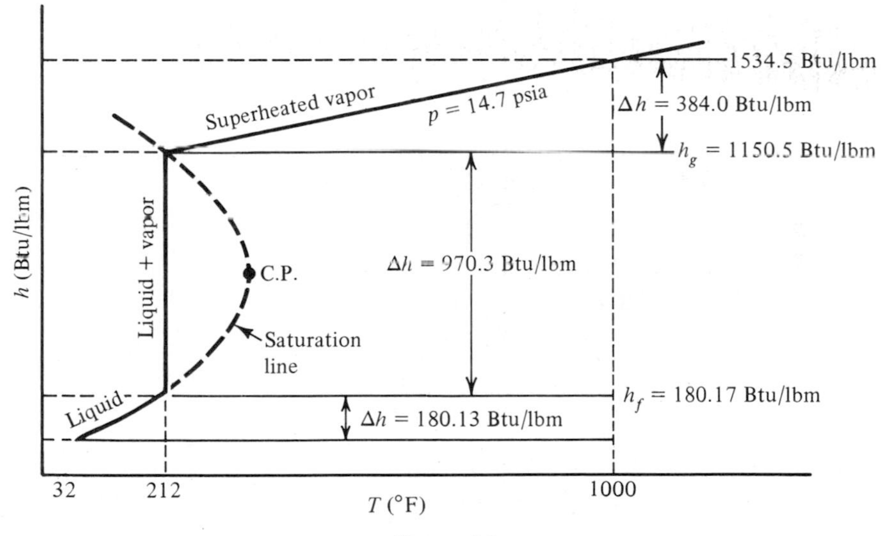

FIGURE 4.7

$$h_{1\ \text{atm}} - h_{0.08865\ \text{psia}} = v\,\Delta p$$

$$= (0.016022\ \text{ft}^3/\text{lbm})\ \frac{144\ \text{in}^2/\text{ft}^2}{778\ \text{ft-lbf/Btu}}$$

$$\times\ (14.696 - 0.08865)\ \frac{\text{lbf}}{\text{in}^2}$$

$$= 0.04\ \text{Btu/lbm}$$

Thus the enthalpy of saturated liquid water at 212°F becomes (with $u_f = 0$ at 32°F) 180.13 + 0.04 = 180.17 Btu/lbm. Similarly, the enthalpy of saturated steam at 212°F becomes 180.17 + 970.3 = 1150.5 Btu/lbm, while the value of enthalpy of superheated steam at 1000°F is 1150.5 + 384.0 = 1534.5 Btu/lbm. These values of enthalpy are at atmospheric pressure; they have also been indicated in Figure 4.7. The enthalpy-temperature behavior of water at pressures other than atmospheric is illustrated in Figure 4.8. From the relationship between enthalpy and internal energy, we can obtain the internal energy-temperature variation of water (Figure 4.9).

Values of internal energy and enthalpy of water (for the compressed liquid, saturated liquid, saturated vapor, and superheated vapor states)

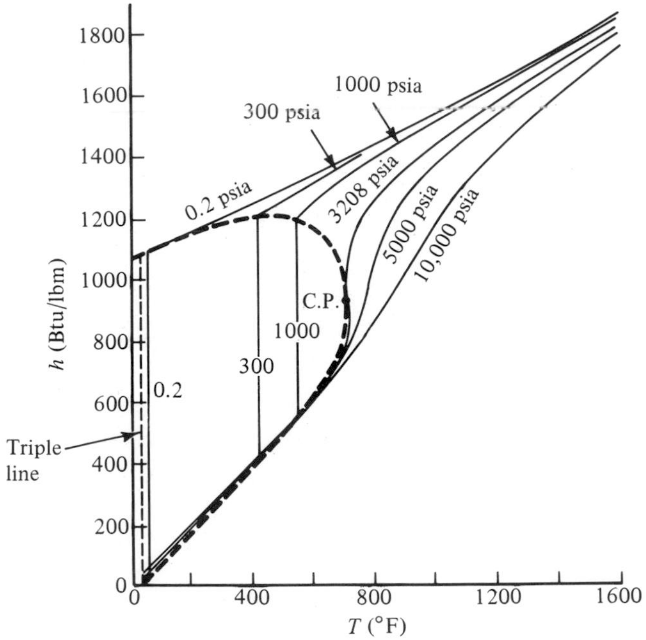

FIGURE 4.8 *Enthalpy-Temperature Diagram for Water.*

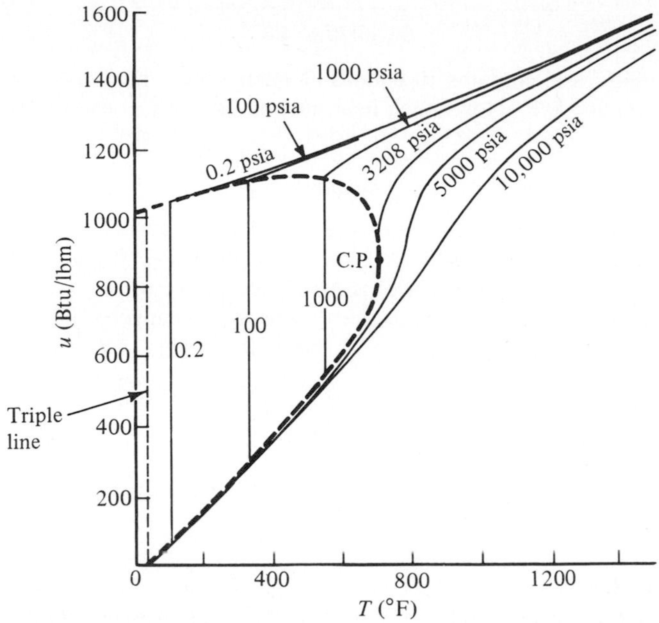

FIGURE 4.9 *Internal Energy-Temperature Diagram for Water.*

are given in Appendices A and B (Tables A.2, A.3, and A.4, or Tables B.2, B.3, and B.4) as a function of temperature and pressure. Appendix A also presents interpolation rules to obtain thermodynamic properties for values of temperature and pressure lying between values given in the tables.

In the saturation region, where liquid and vapor coexist at equilibrium, the value of h is obtained in a fashion similar to that given by Equation (4.1):

$$h = \frac{H}{m} = \frac{H_f + H_g}{m_f + m_g} = \frac{m_f h_f + m_g h_g}{m_f + m_g}$$

or $$h = h_f + x h_{fg} \tag{4.2}$$

where h_{fg} $(= h_g - h_f)$ is the amount of energy required to vaporize one unit mass of saturated liquid water to saturated steam at constant pressure. Similarly, for the internal energy u, we have

$$u = u_f + x u_{fg} \tag{4.3}$$

The following examples illustrate the use of thermodynamic tables for water.

EXAMPLE 4.4

A closed, rigid vessel contains 10 pounds of water vapor and 80 pounds of liquid in equilibrium at 300°F. How much heat must be added to raise the contents to 320°F?

Solution

Values of thermodynamic properties are found in Appendix A.2. At the initial state, $v_1 = v_f + x_1 v_{fg} = 0.01745\ \text{ft}^3/\text{lbm} + (10/90)(6.4483\ \text{ft}^3/\text{lbm}) = 0.7339\ \text{ft}^3/\text{lbm}$. Let the system be the vessel contents. Since the vessel is rigid, there is no change in volume of the system; further, there is no change of mass within the closed container since no mass enters or leaves the system. Therefore, $v_2 = v_1 = 0.7339\ \text{ft}^3/\text{lbm}$. Hence, $0.7339\ \text{ft}^3/\text{lbm} = v_f + x_2 v_{fg}$ at state 2, or $0.7339\ \text{ft}^3/\text{lbm} = 0.01766\ \text{ft}^3/\text{lbm} + x_2 4.8961\ \text{ft}^3/\text{lbm}$, and $x_2 = 0.146$.

With no work done on or by the system, $Q = U_2 - U_1$. At the initial state,

$$U_1 = m_f u_f + m_g u_g = m_1(1 - x_1)u_f + m_1 x_1 u_g = m_1[u_f + x_1(u_g - u_f)]$$

$$= 90\ \text{lbm}\left(269.5\ \text{Btu/lbm} + \frac{10}{90}\,830\ \text{Btu/lbm}\right) = 90\ \text{lbm}\ (361.7\ \text{Btu/lbm})$$

$$= 32{,}550\ \text{Btu}$$

$$U_2 = 90\ \text{lbm}\ (u_f + x_2 u_{fg}) = 90\ \text{lbm}\ [290.1\ \text{Btu/lbm} + 0.146\ \text{lbm}\ (813.6\ \text{Btu/lbm})]$$

$$= 90\ \text{lbm}\ (409\ \text{Btu/lbm}) = 36{,}800\ \text{Btu}$$

$$Q = 36{,}800 - 32{,}550 = 4250\ \text{Btu}$$

EXAMPLE 4.5

In a power plant, boiling water enters at 10,000 kPa and 50°C and is heated to 500°C in a steady-flow, constant-pressure process. Mass flow of steam is 10^6 kg/h. Determine the rate of heat input to the water, neglecting changes in kinetic and potential energies.

Solution

From Equation (3.21), $q_{\text{in}} = h_e - h_{\text{in}}$. From Table B.3, Appendix B, $h_e = 3374.6$ kJ/kg (superheated vapor), and from Table B.4, $h_{\text{in}} = 217.8$ kJ/kg (compressed liquid). Therefore,

$$q_{\text{in}} = 3374.6 - 217.8 = 3156.8\ \text{kJ/kg}$$

or

$$q_{\text{in}} = 3156\ \text{kJ/kg} \times 10^6\ \text{kg/h} = 3.16 \times 10^9\ \text{kJ/h}$$

4.3 INDEPENDENT PROPERTIES

Given certain thermodynamic properties, now we can determine other properties from tables and thereby solve thermodynamic problems. However, a fundamental question is, How many properties must be specified before the thermodynamic state of a pure substance can be defined?

We define a *pure substance* to be one that is both homogeneous and invariant in chemical composition. Thus, a mixture of liquid water and water vapor is still considered a pure substance, namely, all H_2O.

In the absence of magnetic, electric, surface, or gravity effects, one determines the state of a pure substance by specifying two independent, intensive thermodynamic properties. For example, as we have seen, the specification of p and T in the compressed liquid or superheated regions enables one to determine all other properties of the substance, such as v, u, and h. Thence, the specification of p and v, T and v, p and u, T and u, and so on, enables a determination of the state in these two regions. In the saturation region, however, care must be exercised, since p and T are no longer independent. For example, if the pressure of water is 14.7 psia and the temperature is 212°F, the water could exist as a saturated liquid, or a saturated vapor, or a mixture. Obviously, u and v or h could not be determined in this region from a specification of p and T, so p and T are not independent. However, we do have the additional property *quality* in the saturation region; x and T, x and p, v and p, and v and T are sets of two independent properties in the saturation region. If there is a chemical change or if magnetic or electric effects are present, additional properties would be required to completely specify the state. In this text, we will defer consideration of chemical change to Chapter 11; further, we will neglect magnetic, electric, and other such effects.

EXAMPLE 4.6

a. Superheated steam exists at 500 psia with a specific volume of 2.0 ft^3/lbm. Determine its temperature, internal energy, and enthalpy.

b. Superheated steam at a pressure of 500 psia has an internal energy of 1300 Btu/lbm. Determine its temperature, enthalpy, and specific volume.

c. The enthalpy of superheated steam at a temperature of 1000°F is 1495.0 Btu/lbm. Find the pressure and specific volume of the steam.

Solution

a. From the specific volume section of Table A.3 of Appendix A, we find that at a pressure of 500 psia, a specific volume of 2 ft^3/lbm lies between temperatures of 1200°F and 1300°F. Interpolating linearly between these temperatures,[1] we obtain

$$\frac{v - v_1}{v_2 - v_1} = \frac{T - T_1}{T_2 - T_1}$$

---

1. For a discussion on how to use the thermodynamic properties tables, see Appendix A.

Hence

$$\frac{2.0 - 1.9507}{2.0746 - 1.9507} = \frac{T - 1200}{1300 - 1200}$$

$$T = 1200 + 100\,\frac{0.0493}{0.01239}$$

$$= 1200 + 39.79 = 1239.8°\text{F}$$

From the internal energy section of Table A.3, we get

$$\frac{u - u_1}{u_2 - u_1} = \frac{T - T_1}{T_2 - T_1} = \frac{1239.8 - 1200}{1300 - 1200} = 0.3979$$

$$u = 1448.6 + 0.3979(1492.4 - 1448.6)$$

$$= 1448.6 + 17.4 = 1466.0 \text{ Btu/lbm}$$

From the enthalpy section of Table A.3, we get

$$\frac{h - h_1}{h_2 - h_1} = \frac{T - T_1}{T_2 - T_1} = 0.3979$$

$$h = 1629.1 + 0.3979(1684.4 - 1629.1)$$

$$= 1629.1 + 22.0 = 1651.1 \text{ Btu/lbm}$$

b. From Table A.3, we find that for $p = 500$ psia, $u = 1300$ Btu/lbm falls between $T_2 = 900°\text{F}$ and $T_1 = 800°\text{F}$. Thus

$$\frac{u - u_1}{u_2 - u_1} = \frac{T - T_1}{T_2 - T_1}$$

or

$$T = T_1 + \frac{u - u_1}{u_2 - u_1}(T_2 - T_1)$$

$$= 800 + \frac{1300 - 1279.5}{1321.3 - 1279.5}\,100 = 800 + 49.0$$

$$= 849.0°\text{F}$$

$$h = 1412.7 + 0.4904(1466.6 - 1412.7)$$

$$= 1412.7 + 26.43 = 1439.1 \text{ Btu/lbm}$$

$$v = 1.4397 + 0.4904(1.5708 - 1.4397)$$

$$= 1.4397 + 0.0643 = 1.5040 \text{ ft}^3/\text{lbm}$$

c. For enthalpy variation with pressure at a given temperature, use linear interpolation:

$$\frac{h - h_1}{h_2 - h_1} = \frac{p - p_1}{p_2 - p_1}$$

or

$$p = p_1 + \frac{h - h_1}{h_2 - h_1}(p_2 - p_1)$$

From the enthalpy tables, we see that $h = 1495.0$ Btu/lbm falls between

$p_1 = 1000$ psia and $p_2 = 1500$ psia. Hence

$$p = 1000 + \frac{1495.0 - 1505.4}{1490.1 - 1505.4}(1500 - 1000)$$
$$= 1000 + 0.6797(500)$$
$$= 1340 \text{ psia}$$

As discussed in Appendix A, use linear interpolation of the reciprocal of pressure for the determination of specific volume of superheated vapor. Thus

$$\frac{v - v_1}{v_2 - v_1} = \frac{\frac{1}{p} - \frac{1}{p_1}}{\frac{1}{p_2} - \frac{1}{p_1}}$$

or
$$v = v_1 + \frac{\frac{1}{p} - \frac{1}{p_1}}{\frac{1}{p_2} - \frac{1}{p_1}}(v_2 - v_1)$$
$$= 0.8295 + \frac{0.000746 - 0.001}{0.000667 - 0.001}(0.5394 - 0.8295)$$
$$= 0.8295 + 0.7628(-0.2901) = 0.6082 \text{ ft}^3/\text{lbm}$$

This value of v agrees exactly with the value obtainable from more extensive thermodynamic tables using smaller pressure increments than given in Appendix A. If we used linear interpolation of pressure to obtain v, we would get

$$v = 0.8295 + \frac{1340 - 1000}{1500 - 1000}(0.5394 - 0.8295) = 0.6322 \text{ ft}^3/\text{lbm}$$

which is a 4 percent difference in the value of v obtained.

EXAMPLE 4.7

a. Find the quality, internal energy, and enthalpy of saturated steam at a temperature of 120°C and specific volume of 0.7 m^3/kg.

b. The internal energy and enthalpy of saturated steam are $u = 1680$ kJ/kg and $h = 1780$ kJ/kg, respectively. Find the pressure, temperature, and quality of the steam.

Solution

a. From Table B.2, at $T = 120°C$ (saturation pressure = 198.54 kPa), $v_f = 0.0010607 \text{ m}^3/\text{kg}$ and $v_{fg} = 0.8904 \text{ m}^3/\text{kg}$. Hence from Equation (4.1), the quality x of the steam is

$$x = \frac{v - v_f}{v_{fg}} = \frac{0.7 - 0.0010607}{0.8904} = 0.785$$

From Equation (4.3),

$$u = u_f + xu_{fg} = 503.51 + 0.785(2025.5) = 2093.5 \text{ kJ/kg}$$

From Equation (4.2),

$$h = h_f + xh_{fg} = 503.72 + 0.785(2202.2) = 2232.4 \text{ kJ/kg}$$

b. From Equations (4.2) and (4.3), in the saturated region, $h = h_f + xh_{fg}$ and $u = u_f + xu_{fg}$; hence $x = (h - h_f)/h_{fg}$. Also $x = (u - u_f)/u_{fg}$. A trial-and-error solution is required here. Choosing different values of saturation temperature, we can find corresponding values of h_f, h_{fg}, u_f, and u_{fg}, and then calculate values of x. The value of temperature at which the two values of x obtained are identical is the required temperature. First select three temperatures and obtain values from Table B.2:

T(°C)	p(kPa)	h_f(kJ/kg)	h_{fg}(kJ/kg)	$x = \frac{h - h_f}{h_{fg}}$	u_f(kJ/kg)	u_{fg}(kJ/kg)	$x = \frac{u - u_f}{u_{fg}}$	Δx
150	476.0	632.15	2113.2	0.5432	631.63	1927.0	0.5440	−0.0008
200	1554.9	852.37	1938.6	0.4785	850.57	1742.5	0.4760	0.0025
250	3977.6	1085.8	1714.6	0.4049	1080.8	1520.5	0.3941	0.0108

Next try the following:

175	892.5	741.07	2030.7	0.5116	740.07	1838.5	0.5112	0.0003
170	792.02	719.12	2047.9	0.5180	718.24	1856.7	0.5180	0

Hence $T = 170$°C, $p = 792.02$ kPa, and $x = 0.518$.

4.4 THERMODYNAMIC PROPERTIES OF OTHER SUBSTANCES (LIQUID AND VAPOR PHASES)

The behavior of the thermodynamic properties of other pure substances is similar to that of water. For example, the shape of the saturation curve on a $T - v$ diagram for carbon dioxide and refrigerant R-22 (chlorodifluoromethane), shown in Figures 4.10 and 4.11, is similar to that for water. Naturally, the critical point is different for the different substances, as shown in Table 4.1. Also, the shape of the saturation curve on the h–T diagrams (shown in Figures 4.12 and 4.13), as well as the general behavior of the h–T curves, is similar to that of water.

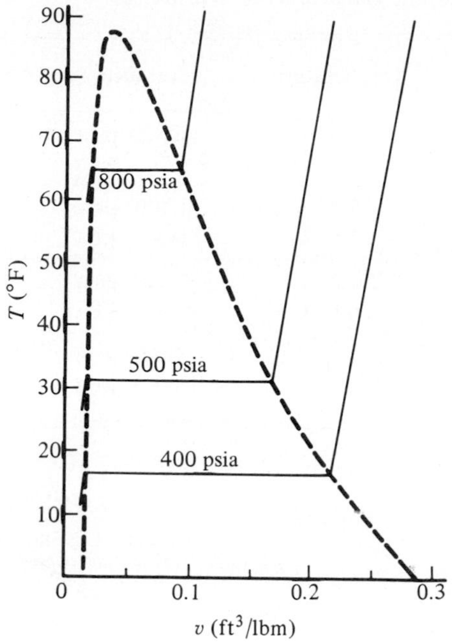

FIGURE 4.10 *Specific Volume-Temperature Diagram for Carbon Dioxide.*

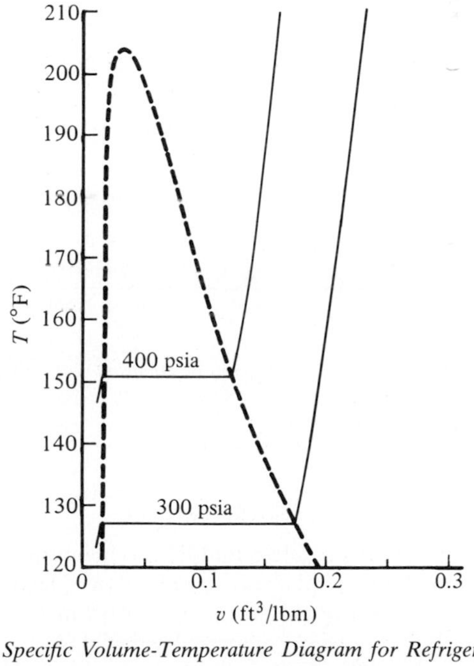

FIGURE 4.11 *Specific Volume-Temperature Diagram for Refrigerant-22.*

TABLE 4.1 *Critical Point Data for Four Substances*

	Temperature	Pressure	Specific Volume
Water	705.47°F 374.15°C	3208.2 psia 22,120 kPa	0.05078 ft^3/lbm 0.003170 m^3/kg
Carbon dioxide	87.87°F 31.04°C	1070.0 psia 7383 kPa	0.03423 ft^3/lbm 0.002137 m^3/kg
Refrigerant R-22	204.81°F 96.01°C	721.9 psia 4977 kPa	0.03053 ft^3/lbm 0.001906 m^3/kg
Dry air	−221.1°F −140.6°C	546.65 psia 3769.0 kPa	0.05006 ft^3/lbm 0.003125 m^3/kg

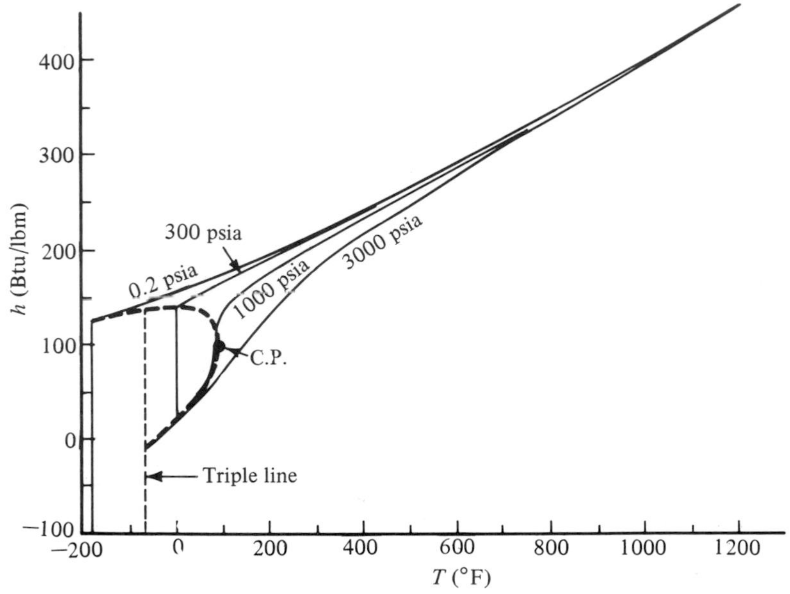

FIGURE 4.12 *Enthalpy-Temperature Diagram for Carbon Dioxide.*

EXAMPLE 4.8

In a refrigeration compressor, refrigerant R-22 is compressed from 15 psia, 0°F, to 100 psia, 200°F. Assuming adiabatic steady flow, determine the compressor horsepower required for a refrigerant flow of 350 lbm/h.

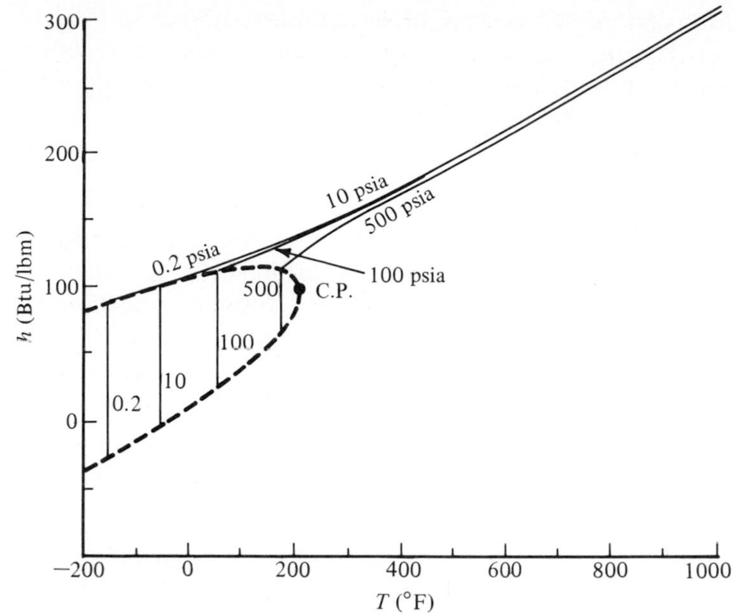

FIGURE 4.13 *Enthalpy-Temperature Diagram for Refrigerant-22.*

Solution

From Chapter 3, $q_{in} = w'_{out} + h_e - h_{in}$ for steady flow, where $q = 0$ for this adiabatic process. At state *in* (Figure 4.14), the refrigerant is superheated since the temperature is greater than saturation at 15 psia. From Table A.11, $h_{in} =$ 106.09 Btu/lbm. At state *e*, again the refrigerant is superheated; $h_e =$ 135.93 Btu/lbm.

Therefore, $w'_{out} = h_{in} - h_e = 29.84$ Btu/lbm, where the minus sign denotes work done on the fluid. The rate of work input, or power, is given by

$$W' = -29.84 \text{ Btu/lbm} \times 350 \text{ lbm/h}$$

$$= -10{,}440 \text{ Btu/h} = -4.1 \text{ horsepower}$$

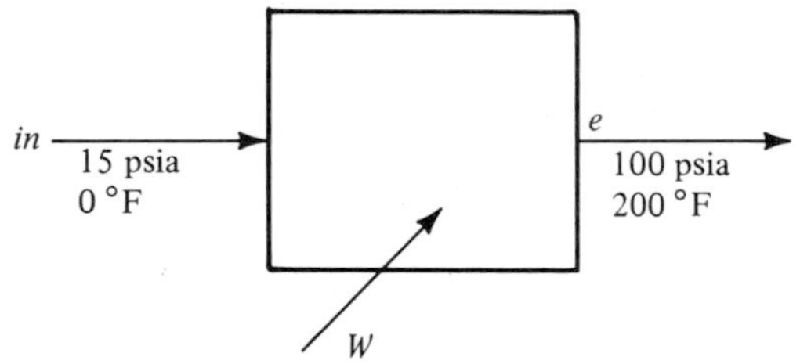

FIGURE 4.14

Although air is a mixture of several pure substances (mainly nitrogen and oxygen with some argon, carbon dioxide, and other trace components), it is itself not a pure substance. A *T-v* diagram for dry air is given in Figure 4.15. Note that in the saturation region, vaporization of air at constant pressure does not take place at constant temperature, but rather at a slight change in temperature. In a mixture of saturated liquid and vaporous air, the composition of the liquid air differs somewhat from that of the vapor phase. However, in the region where there is no change of phase (superheated air), the mixture exhibits the behavior of a pure substance. (This concept will be discussed further in Section 5.2.) An enthalpy-temperature diagram for dry air is shown in Figure 4.16. Values of thermodynamic properties for carbon dioxide, refrigerant R-22, and dry air are presented in the tables in Appendices A and B.

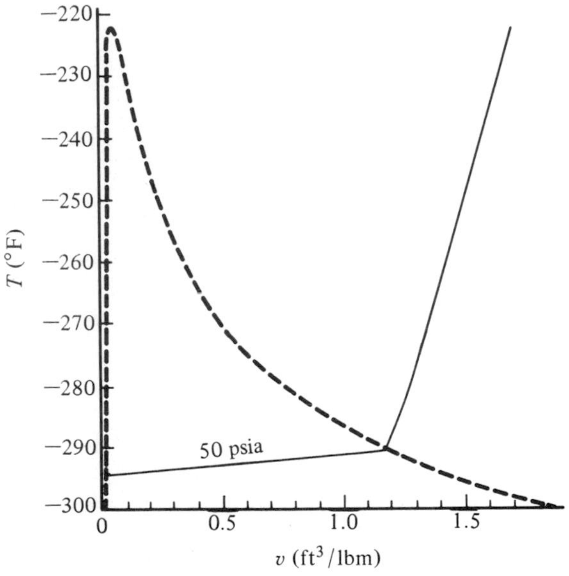

FIGURE 4.15 *T-v Diagram for Dry Air.*

EXAMPLE 4.9

In a hot air furnace, incoming air at 15°C and atmospheric pressure (100 kPa) is heated at essentially constant pressure to a temperature of 60°C. Determine the amount of heat that must be added to each kilogram of air passing through the furnace. Neglect changes of potential and kinetic energies.

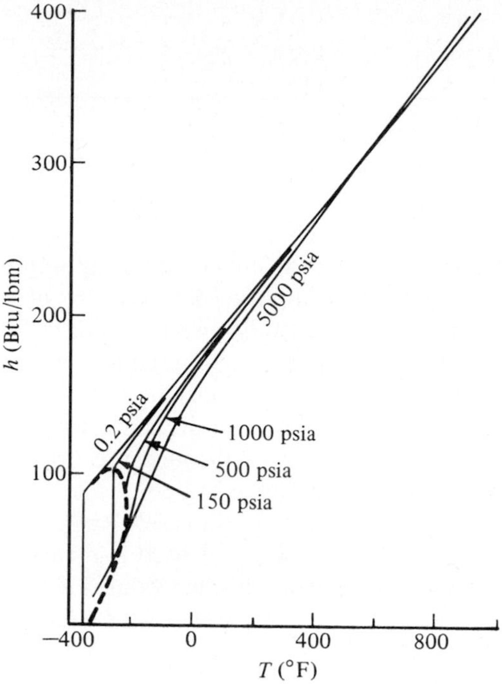

FIGURE 4.16 *h-T Diagram for Dry Air.*

Solution

For steady air flow through the furnace, we have, from Equation (3.15), $q_{in} - w'_{out} = h_e - h_{in}$, where $w'_{out} = 0$ for a furnace. The incoming air at 15°C and 100 kPa is superheated. The air leaving the furnace at 60°C and 100 kPa is also superheated. From Table B.13, we obtain, at a pressure of 100 kPa, enthalpy values of 414.48, 464.83, and 515.31 kJ/kg at temperatures of 0°, 50°, and 100°C, respectively. Using linear interpolation with temperature, we get at 15°C

$$\frac{h_{in} - 414.48}{464.83 - 414.48} = \frac{15 - 0}{50 - 0}$$

$$h_{in} = (15/50)(50.35) + 414.48 = 15.10 + 414.48$$

$$= 429.58 \text{ kJ/kg}$$

At 60°C,

$$\frac{h_e - 464.83}{515.31 - 464.83} = \frac{60 - 50}{100 - 50}$$

$$h_e = (10/50)(50.48) + 464.83 = 10.10 + 464.83$$

$$= 474.93 \text{ kJ/kg}$$

Therefore,

$$q_{in} = h_e - h_{in} = 474.93 - 429.58 = 45.35 \text{ kJ/kg}$$

4.5 IDEAL GASES

The variation of specific volume of steam with temperature was discussed in Section 4.2 and shown in Figure 4.4. Curves of v versus T for superheated water vapor are shown in further detail in Figure 4.17. Note that the curves become straight in the region somewhat beyond the saturation region. In this region, the specific volume v is thus linearly dependent on temperature T, at a given pressure. Further, if the straight lines are extended, they will all pass through the origin of the absolute temperature scale. Hence, in the linear region, $v = \text{constant} \times T$, where T is in absolute temperature units, °R (degrees Rankine) or K (Kelvin). Furthermore, if we multiply the slope of each v-T curve by the value of the pressure, we obtain an almost constant value for the product (Table 4.2). For pressures up to 50 psia, the value of the product changes only 1 percent of its lowest value. This product is given the symbol R and is called the *gas constant*.

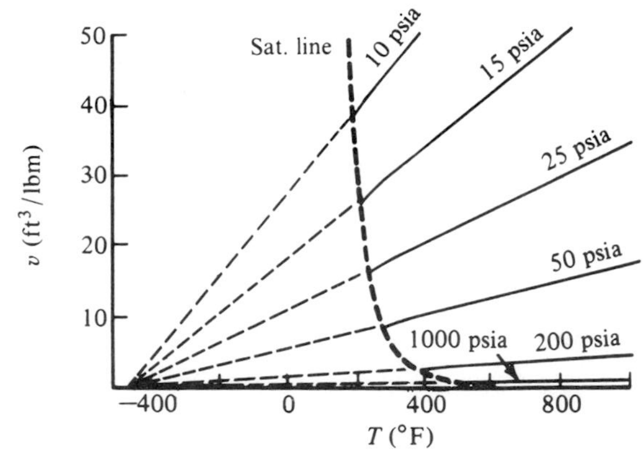

FIGURE 4.17 *v-T Diagram for Water Showing Linear Region.*

TABLE 4.2 *Slopes of the v-T Lines for Steam*

p (psia)	Slope ($ft^3/lbm°R$)	Slope × p (ft-lbf/lbm°R)
0.12	4.965	85.80
0.20	2.980	85.81
0.30	1.987	85.82
0.50	1.193	85.87
10	0.05968	85.94
15	3.982×10^{-2}	86.02
25	2.394×10^{-2}	86.18
50	1.203×10^{-2}	86.59
200	3.055×10^{-3}	87.97
1000	6.468×10^{-4}	93.14

As shown in Table 4.2, for low pressures, the value of the constant for water vapor in the linear region is independent of pressure and temperature. In this linear region, we therefore have the following relation between temperature, pressure, and specific volume (called an *equation of state*):

$$pv = RT \tag{4.4}$$

A substance exhibiting such a linear relation between p, v, and T is called an *ideal* or *perfect gas*.

The actual value of the gas constant R is obtained by extrapolation of the values of the slopes of the v-T curves times pressure to very low pressure, as illustrated in Figure 4.18. Values obtained in this fashion for the four substances are listed in Table 4.3. The units for the gas constant R are ft-lbf/lbm°R or J/kg K.

We can express the ideal gas law in terms of total volume as follows:

$$pV = mRT \tag{4.5}$$

TABLE 4.3 *Gas Constant R for Four Substances*

Substance	Gas Constant R (ft-lbf/lbm°R)	Gas Constant R (J/kg K)
Steam	85.78	461.5
Carbon dioxide	35.11	188.9
Refrigerant R-22	17.87	96.14
Dry air	53.35	287.0

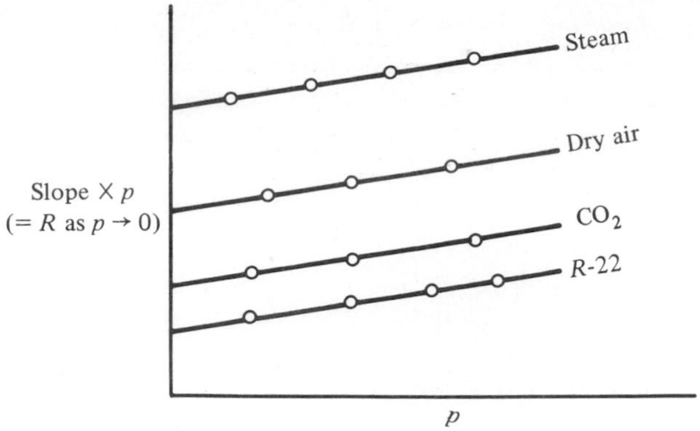

FIGURE 4.18 *Gas Constant R at Low Pressures.*

A mole of a substance is defined to be a mass equal to the molecular mass $\overline{M}$ of the substance. For example, the molecular mass of oxygen (O_2) is 32.00 (Appendix A.1); one kilogram-mole of oxygen has a mass of 32 kg. Likewise, one pound-mole of oxygen has a mass of 32 lbm. The mass m of a gas is thus equal to $N\overline{M}$, where N is the number of moles of the gas. The ideal gas law, (4.5), can now be written in terms of moles as:

$$p\forall = N\overline{M}RT$$

The product $R\overline{M}$ has the same value for all ideal gases and is called the *universal gas constant* R_u. In other words, for an ideal gas:

$$p\forall = NR_uT \tag{4.6}$$

From measurements of many gases at low pressure, it is found that R_u has the value:

$$R_u = 1545.33 \text{ ft-lbf/lbm-mole°R}$$
$$= 8314.3 \text{ J/kg} \cdot \text{mole K}$$

With a knowledge of R_u and the molecular mass of a gas, the gas constant R can be found from:

$$R = \frac{R_u}{\overline{M}} \tag{4.7}$$

For example, for molecular nitrogen with $\overline{M} = 28.014$,

$$R = \frac{1545.33 \text{ ft-lbf/lbm-mole°R}}{28.014 \text{ lbm/lbm-mole}} = 55.16 \text{ ft-lbf/lbm°R}$$

or

$$R = \frac{8314.3 \text{ J/kg} \cdot \text{mole K}}{28.014 \text{ kg/kg} \cdot \text{mole}} = 296.78 \text{ J/kg K}$$

Next, we will discuss the enthalpy and internal energy behavior of ideal gases. In general, the slopes of the h-T curves for a pure substance are a function of two independent thermodynamic variables (for example, p and T). For an ideal gas, the h-T curves at different low pressures coincide (see Figures 4.8, 4.9, 4.12, and 4.16), indicating h to be independent of pressure. Hence, for an ideal gas, enthalpy is a function only of temperature; that is, $h = h(T)$. In general, the slope of a constant pressure line on an h-T plot is called c_p, *specific heat at constant pressure*. For the ideal gas, we can therefore write $dh = c_p\,dT$, where c_p for an ideal gas is a function only of temperature.

A similar expression can be derived for the internal energy of an ideal gas. For an ideal gas, with $pv = RT$, it follows that $u = h - pv = h - RT$. Since h for an ideal gas is a function only of temperature, it follows that u is likewise a function only of temperature for an ideal gas. In general, for any substance, the *specific heat at constant volume* c_v is defined as the slope of a constant volume line on a plot of internal energy versus temperature. For an ideal gas, we can therefore write $du = c_v\,dT$, with c_v a function only of temperature. Note that for an ideal gas, with $du = dh - R\,dT$,

$$c_p - c_v = R$$

The change in enthalpy of an ideal gas in going from state 1 to state 2 by any arbitrary process is

$$h_2 - h_1 = \int_1^2 c_p(T)\,dT \tag{4.8}$$

If the temperature difference is not too large, c_p can be assumed constant, and

$$h_2 - h_1 = c_p(T_2 - T_1) \tag{4.9}$$

The change in internal energy of an ideal gas in going from state 1 to state 2 is

$$u_2 - u_1 = \int_1^2 c_v(T)\,dT \tag{4.10}$$

If the temperature difference is not too large, c_v can be assumed constant, and

$$u_2 - u_1 = c_v(T_2 - T_1) \tag{4.11}$$

It should be noted that for nonideal gases, c_p and c_v are the slopes of the enthalpy-temperature curves at constant pressure and the slopes of the internal energy–temperature curves at constant specific volume, respectively. However, for the special case of ideal gases, the changes in enthalpy, expressed by Equations (4.8) or (4.9), and the changes in internal energy, expressed by Equations (4.10) or (4.11), are valid for any process, including constant pressure and constant volume processes. Values of specific heat capacities for several ideal gases are tabulated as a function of temperature in Tables A.15 and B.15 of Appendices A and B.

EXAMPLE 4.10

Find the change in enthalpy and internal energy of superheated steam in going from a temperature of 1000°F to 1500°F at a pressure of 20 psia. Use the ideal gas approximation, and compare the results obtained from the steam tables. Also, find the specific volumes at the two temperatures.

Solution

From Appendix A, Table A.15 we find that at $T_1 = 1000°\text{F}$, $c_p = 0.515$ Btu/lbm°R and that at $T_2 = 1500°\text{F}$, $c_p = 0.559$ Btu/lbm°R. The average c_p over the temperature range is $(0.515 + 0.559)/2 = 0.537$ Btu/lbm°R. From Equation (4.9), we get

$$h_2 - h_1 = 0.537\ \text{Btu/lbm°R}\,(1500 - 1000)\text{°R} = 268.5\ \text{Btu/lbm}$$

From Table A.3, we get

$$h_2 - h_1 = 1803.3\ \text{Btu/lbm} - 1534.3\ \text{Btu/lbm} = 269.0\ \text{Btu/lbm}$$

Similarly, $c_{v1} = 0.400$ Btu/lbm°R and $c_{v2} = 0.454$ Btu/lbm°R; hence the average c_v is $(0.400 + 0.454)/2 = 0.427$ Btu/lbm°R. From Equation (4.11), we get

$$u_2 - u_1 = 0.427\ \text{Btu/lbm°R}\,(1500 - 1000)\text{°R} = 213.5\ \text{Btu/lbm}$$

From Table A.3, we obtain

$$u_2 - u_1 = 1587.3\ \text{Btu/lbm} - 1373.5\ \text{Btu/lbm} = 213.8\ \text{Btu/lbm}$$

From Equation (4.4), we get $v = RT/p$; hence

$$v_1 = 85.77\ \text{ft-lbf/lbm°R}\ \frac{(1000 + 460)\text{°R}}{20(144)\ \text{lbf/ft}^2} = 43.486\ \text{ft}^3\text{/lbm}$$

$$v_2 = 85.77\ \text{ft-lbf/lbm°R}\ \frac{1960\text{°R}}{20(144)\ \text{lbf/ft}^2} = 58.378\ \text{ft}^3\text{/lbm}$$

From Table A.3, $v_1 = 43.435\ \text{ft}^3\text{/lbm}$, $v_2 = 58.352\ \text{ft}^3\text{/lbm}$.

EXAMPLE 4.11

Dry air is at 80°F and atmospheric pressure, and c_p is 0.240 Btu/lbm°R. Find c_v. Using the ideal gas approximation, find the changes in enthalpy and internal energy if the air is heated to 200°F at atmospheric pressure.

Solution

$$c_v = c_p - R = 0.240\ \text{Btu/lbm°R} - \frac{53.35\ \text{ft-lbf/lbm°R}}{778\ \text{ft-lbf/Btu}} = 0.171\ \text{Btu/lbm°R}$$

From Table A.15, at 200°F, $c_p = 0.241$ and $c_v = 0.173$. Thus,

$$h_2 - h_1 = 0.2405\ \text{Btu/lbm°R}\ (200 - 80)\text{°F} = 28.86\ \text{Btu/lbm}$$

$$u_2 - u_1 = 0.172\ \text{Btu/lbm°R}\ (200 - 80)\text{°F} = 20.64\ \text{Btu/lbm}$$

The mathematical simplicity afforded by use of the ideal gas law can be seen from the examples shown. For air and many other gases, over the range of pressures and temperatures we commonly deal with, the assumption of ideal gas behavior yields a very excellent engineering approximation. However, as we get to higher and higher pressures, or as we get closer and closer to the saturation line, deviations from ideal gas behavior get larger and larger in magnitude. In these cases, use of the ideal gas law will depend on the degree of accuracy required for a particular problem.

4.6 THERMODYNAMIC PROPERTIES OF SUBSTANCES WITH A SOLID PHASE

In Section 4.2, we discussed regions in which pure substances exist in their liquid and vapor phases. Here we will discuss the regions in which one of the phases of a pure substance is solid. The behavior of the substance will be illustrated by *T-v* diagrams as shown in Figure 4.19 for water and in Figure 4.20 for carbon dioxide. The behavior of refrigerant R-22 is that of a pure substance; however, no data are available for this region, since the liquid and vapor phases of refrigerants are of primary interest in engineering applications. The phase diagram for air will be different since air behaves as a nonpure substance in this region, with the constituents nitrogen and oxygen freezing at different temperatures.

Water can exist in its solid phase (ice) only below temperatures of 32°F (0°C). The solid phase can exist by itself or in equilibrium with a liquid or vapor phase (or both), depending upon the pressure to which the water is subjected. To illustrate this concept, let us consider a sample of ice at atmospheric pressure and a temperature below 32°F (point 1 of Figure 4.19). Heat is slowly added at constant pressure. The temperature of the ice will rise with little accompanying change in specific volume until point 2 is reached (saturated ice). Further addition of heat results in the gradual melting of the ice at constant temperature until all of the ice turns

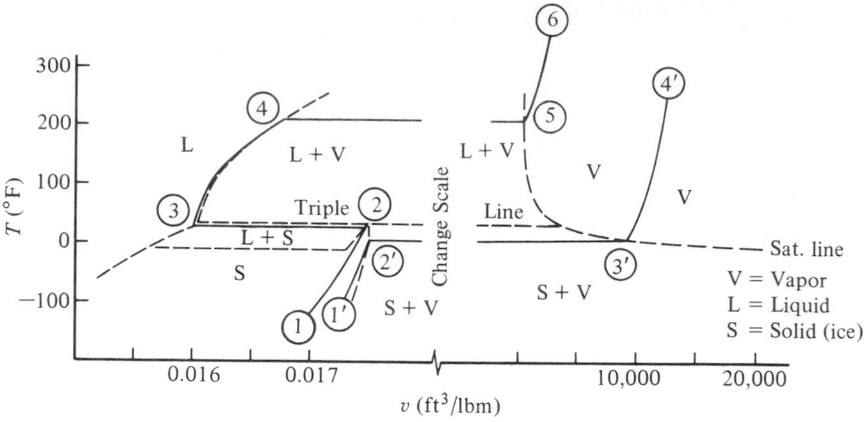

Figure 4.19 *T-v Diagram for Water.*

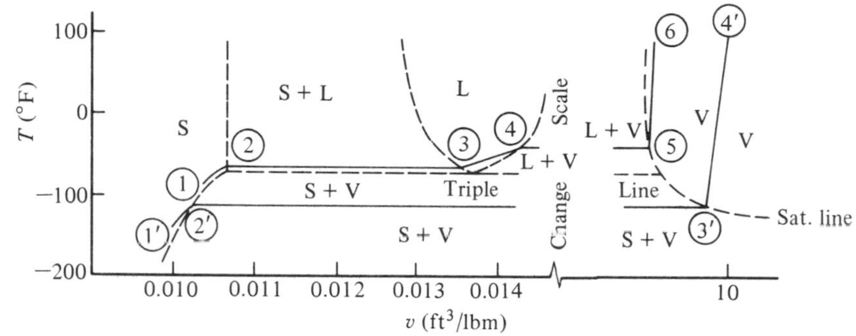

Figure 4.20 *T-v Diagram for Carbon Dioxide.*

into liquid water (point 3). During the melting process, there occurs an 8 percent decrease in the specific volume of the ice. In the melting region (2–3), ice and liquid water coexist in equilibrium. Additional heating will raise the temperature of the compressed liquid until the liquid saturation point 4 is reached at a temperature of 212°F (100°C). Further heat addition will result in a liquid-vapor mixture and finally in superheated steam, as previously described in Section 4.2 and illustrated in Figure 4.3. When the heating process of ice takes place at very low pressures (for example, 0.01 psia, 0.07 kPa), the *T-v* relationship will be represented by path 1′–2′–3′ in Figure 4.19. With the addition of heat, the sample of ice will change directly into water vapor. This process is called *sublimation*. Between points 2′ and 3′, ice and water vapor coexist in equilibrium.

Further addition of heat to the sample changes the saturated water vapor at point 3′ into superheated steam. If the pressure of the sample equals 0.08865 psia (0.6108 kPa), the solid, liquid, and vapor phases of water can coexist in equilibrium, as indicated in Figure 4.19 by the triple line.

Water is representative of pure substances for which the specific volume increases during freezing (that is, during change of phase from liquid to solid). On the other hand, carbon dioxide is representative of pure substances for which the specific volume decreases during freezing, as illustrated in the T-v diagram for carbon dioxide (Figure 4.20). Again, consider a sample of solid carbon dioxide (popularly called *dry ice*) at atmospheric pressure and a temperature of −120°F. When heat is added to the sample at constant pressure, the temperature of the dry ice will increase until a temperature of −109.4°F is reached and saturated ice (point 2′) is obtained. Upon further addition of heat, the dry ice will sublimate directly into carbon dioxide vapor. In the 2′–3′ region, dry ice and vapor coexist in equilibrium. The triple line of carbon dioxide occurs at a temperature of −69.9°F and a pressure of 75.146 psia. At a pressure above 75.146 psia, ice initially at point 1 will, when heated, become saturated ice at point 2, liquid and solid in equilibrium (2–3), compressed liquid (3–4), and saturated liquid at 4, saturated vapor at 5, and finally superheated carbon dioxide, 6.

Thermodynamic properties for saturated ice and saturated vapor in the solid-vapor region are given for water in Tables A.5 and B.5, and for carbon dioxide in Tables A.9 and B.9. The thermodynamic properties v, u, and h of a pure substance in this region are obtained in a fashion similar to that in the liquid-vapor region:

$$v = v_i + xv_{ig}$$

$$u = h_1 + xu_{ig}$$

$$h = h_i + xh_{ig}$$

where the subscript i denotes the saturated ice condition and the subscript ig denotes the difference between saturated ice (i) and saturated vapor (g) conditions, and $x = m_g/(m_i + m_g)$.

EXAMPLE 4.12

The specific volume of a saturated ice–water vapor mixture is 4000 ft^3/lbm. The quality of the mixture is 0.7. Determine the saturation pressure and temperature of the mixture.

Solution

From the expression $v = v_i + xv_{ig}$, we get 4000 ft^3/lbm = 0.0174 ft^3/lbm = $0.7v_{ig}$.

(Over the range of temperatures in Table A.5, v_i is relatively constant and, in any case, much less than v_{ig}.) Hence, $v_{ig} = 5714.3\ \text{ft}^3/\text{lbm}$. From Table A.5, we find that $v_{ig} = 5658\ \text{ft}^3/\text{lbm}$ at $T = 20°\text{F}$ and $v_{ig} = 9050\ \text{ft}^3/\text{lbm}$ at $T = 10°\text{F}$. Hence, using linear interpolation, we have

$$\frac{5714 - 5658}{9050 - 5648} = \frac{T - 20}{10 - 20}$$

and the saturation temperature is $T = 20 - 10(0.0165) = 19.83°\text{F}$. From Table A.5, we also find that the saturation pressure $p = 0.0505$ psia at $T = 20°\text{F}$ and $p = 0.0309$ psia at $T = 10°\text{F}$. Hence

$$\frac{p - 0.0505}{0.03039 - 0.0505} = 0.0165$$

and the saturation pressure is $p = 0.0505 - 0.0003 = 0.0502$ psia.

PROBLEMS

4.1 Find the specific volume of a mixture of liquid water and water vapor having a quality of 0.8 at 80°C.

4.2 A mixture of liquid water and water vapor is contained in a vessel of 20 ft^3 volume. If the pressure of the mixture is 9.34 psia, determine the mixture temperature, mass of liquid and mass of vapor, and specific volume. Mixture quality is 0.2.

4.3 A tank contains 30 kg of saturated CO_2 vapor at $-30°\text{C}$. What is the volume of the tank?

4.4 A storage vessel contains 40 pounds mass of saturated refrigerant R-22 at 158.33 psia. For a vessel volume of 1 ft^3, determine the percent mass and the percent volume of liquid and vapor in the vessel.

4.5 A spherical tank contains 25 pounds mass of saturated steam vapor at a temperature of 275°F. What is the diameter of the tank?

4.6 Superheated steam at 60 kPa, 200°C is contained in a spherical vessel of diameter 1.5 m. Determine the mass of steam in the vessel.

4.7 Calculate the internal energy and the specific volume of carbon dioxide at a pressure of 250 psia and a temperature of 340°F. Use thermodynamic properties given in Table A.7.

4.8 Using the thermodynamic properties given in Table B.11, calculate the enthalpy and the specific volume of refrigerant R-22 at a temperature of 150°C and a pressure of 375 kPa.

4.9 Find the internal energy and specific volume of superheated air at 100 kPa, 240°C.

4.10 The steam in a heating system of a tall skyscraper enters the pipe at the ground floor as saturated vapor at a pressure of 200 kPa. At the top floor, which is 200 m above the ground floor, the steam pressure in the pipe is 100 kPa. Assuming the steam pipe to be perfectly insulated, what will the quality of the steam be at the top floor?

4.11 A 1-kW electric heater is immersed in 2 kilograms of water at 15°C. Neglecting heat losses from the open container, calculate how long it will take to boil off 0.5 kilogram of water.

4.12 A 1-kW electric heater is immersed in a 1.0-ft^3 container of liquid water at 40°F until the water reaches 100°F. Determine the rate of rise of the water temperature and the time required to reach 100°F.

4.13 An automobile tire having a volume of 1.2 ft^3 is filled with air to a gauge pressure of 32 psi. What volume would the air in the tire occupy if it were released at atmospheric pressure of 14.7 psia with its temperature remaining constant? Assume air is an ideal gas.

4.14 Using the properties of dry air in Table A.13, determine the specific volume of air at 14.7 psia, 200°F. Compare with the value of specific volume found from the ideal gas equation of state (4.4). Use the value of R in Table 4.3. Repeat for air at 14.7 psia, 800°F.

4.15 Using the properties of superheated steam in Table B.3, determine the specific volume of steam at 100 kPa, 100°C. Compare with the value of specific volume found from the ideal gas equation (4.4), using the value of R in Table 4.3. Repeat for steam at 100 kPa, 500°C.

4.16 Water is supplied to a steam boiler at a rate of 5000 pounds per hour and a temperature of 65°F. It leaves as superheated steam at a pressure of 600 psia and a temperature of 800°F. Coal, which is used to fire the boiler, gives off 12,500 Btu/lbm during the combustion. Calculate the minimum rate at which coal must be fed to the boiler. Treat the process in the boiler as constant pressure.

4.17 Water at a temperature of 100°C is pumped at a pressure of 5000 kPa into a boiler. Determine the enthalpy of this compressed liquid. Compare with the enthalpy of saturated liquid water at 100°C.

4.18 Compressed liquid water fills a spherical vessel 3 ft in diameter. The water pressure and temperature are 3000 psia, 600°F. Determine the internal energy of the water in the vessel.

4.19 A Diesel engine begins its compression stroke with a volume of 100 cubic inches of air at atmospheric pressure and a temperature of 75°F. At the end of the compression stroke, the volume has been reduced to 5 in^3 with a pressure of 1000 psia. What is the temperature of the compressed air? Assume air is an ideal gas.

4.20 Air is compressed in a piston-cylinder arrangement from 100 kPa to 2000 kPa. If the temperature of the air prior to compression is 28°C, determine the temperature of the compressed air. Assume that the volume available to the air is reduced by a factor of 8 during the compression. Assume that the air behaves as an ideal gas.

4.21 To remove the superheat from steam which is at a pressure 400 psia and a temperature of 700°F, water having a temperature of 70°F will be added. Calculate the required amount of water per pound of steam.

4.22 The discharges from two boilers with identical mass flow rates are mixed. The discharge condition of the first boiler is 500 psia and 700°F, and that of the second boiler is 500 psia and 95 percent quality. Find the condition of the steam after mixing.

4.23 As a means of controlling the load of a steam turbine, the steam that leaves the boiler in a superheated condition at 300°C and 2000 kPa is throttled to a pressure of 1500 kPa. What will be the temperature and pressure of the steam after throttling?

4.24 A floating piston rests on one pound of 100 percent quality steam contained in a vertical cylinder such that the pressure of the steam is 140 psia with a volume of 3.2 ft^3. Energy is added slowly as heat to the steam until the piston reaches stops in the cylinder, with a resulting volume of steam of 5.3 ft^3. An additional 20 Btu are then added to the steam. Calculate the work done by and the total heat transferred to the steam.

4.25 Air at 26°C is flowing through a 2-cm-diameter pipe with an average velocity of 5.0 m/s. Find the mass flow of air for pressures of 100 kPa and 1000 kPa, using (a) the ideal gas approximation and (b) thermodynamic properties from tables.

4.26 Superheated steam at 100 psia, 400°F flows through a 3-in.-diameter pipe with a velocity of 2.6 ft/s. Determine the mass flow rate of steam.

4.27 The tube of a mercury barometer is 1 meter long and has a cross section of 100 mm^2. On a day when the column of mercury in the barometer is 770 mm and the room temperature is 25°C, a sufficient amount of argon gas is introduced into the evacuated space above the mercury to drop the height of the column to 650 mm. Calculate the mass of argon in the barometer (see Fig. 4.21).

4.28 Show that in a floating piston-cylinder arrangement, the temperature of an ideal gas within the cylinder is directly proportional to the distance between the bottom of the cylinder and the lower face of the piston.

4.29 An air bubble having a volume of 1000 mm^3 rises from the bottom of a mountain lake where the temperature and pressure are 10°C and 500 kPa, respectively. What will the volume of the bubble be when it reaches the surface of the lake, where the temperature is 20°C?

4.30 A well-insulated tank having a volume of 5 ft^3 contains 0.24 pound mass of steam at a pressure of 20 psia. Additional steam is supplied at a pressure of

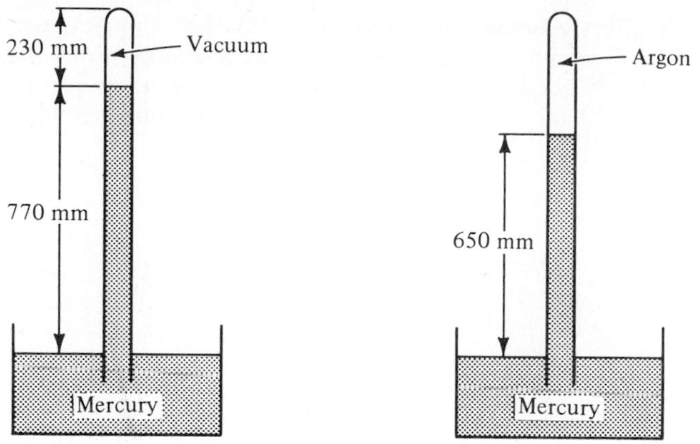

FIGURE 4.21

100 psia and a temperature of 500°F from a very large supply tank, until the pressure in the storage tank reaches 100 psia. Find the mass of steam added to the tank and the temperature of the steam in the tank.

4.31 A tank filled with 0.25 pound mass of carbon dioxide at 100°F, 10 psia is discharged until 0.15 pound mass of carbon dioxide is left in the tank. Determine the final temperature of the gas in the tank, which is well insulated.

4.32 The insulation of the tank of Problem 4.31 is removed and the tank is left standing until the carbon dioxide in the tank reaches the room temperature of 80°F. What will be the amount of heat transferred to the carbon dioxide?

4.33 A mixture of saturated liquid water and water vapor at 50°C is contained within a closed vessel having a total volume of 0.25 m^3. Initially, 80 percent of the total mass is in the vapor phase. Heat is added until all of the water in the container becomes saturated vapor. Calculate the heat input.

4.34 Ninety percent quality steam at 500°F is contained within a closed vessel of volume 10 ft^3. Heat is added until all the water is vapor. Calculate the heat input and the final pressure of the vapor.

4.35 Liquid water enters a steady flow boiler at 100°F. Steam is discharged from the boiler as saturated vapor at 500°F. For an output flow of 5000 pounds of steam per hour, determine the rate of heat input in the boiler in Btu/h.

4.36 Air enters a nozzle with a velocity of 25 m/s, $p = 500$ kPa, $T = 200$°C. The conditions at the nozzle discharge are $p = 100$ kPa, $T = 50$°C. Determine the nozzle exit velocity for adiabatic, steady flow. For your solution, first use the data given in Table B.13. Then assume the air to behave as an ideal gas with constant specific heat (use the average value of c_p over the temperature range of interest; see Table B.15).

4.37 Steam enters a turbine at a pressure of 1000 psia, temperature 600°F. The steam is discharged at 200°F with 95 percent quality. For a steady flow of 1000 pounds per hour, determine the power output of the turbine in kW. Assume adiabatic flow, and neglect charges of potential and kinetic energies.

chapter

5

Gas and Gas-Vapor Mixtures

5.1 INTRODUCTION

Since nonreacting gases can be mixed in any proportion, it becomes impractical to tabulate the thermodynamic properties of such mixtures. Therefore, we will develop a method for calculating the thermodynamic properties of a mixture from the properties of the component gases. We will apply the procedure to obtain the properties of gaseous mixtures as well as gas-vapor mixtures, such as moist atmospheric air.

5.2 THERMODYNAMIC PROPERTIES OF IDEAL GAS MIXTURES

In this section, we will show how to calculate the thermodynamic properties of mixtures of nonreacting ideal gases from a knowledge of the properties of the constituent gases. Each constituent gas is assumed to behave as an ideal gas; each will behave as if it resided in the container all by itself, occupying the entire volume of the container. This assumption is possible since, in an ideal gas, the distance between molecules is large (corresponding to low pressure), and intermolecular force small. For an ideal gas, the equation of state is given by Equation (4.4):

$$pv = RT$$

which, from (4.6), can be written as $p\forall = NR_uT$.

Each constituent gas (identified by subscripts A and B) in a mixture is at the same temperature, which is that of the mixture. Therefore,

$T = T_A = T_B$. Furthermore, each constituent gas occupies the entire volume $V\!\!\!/$, so that $V\!\!\!/ = V\!\!\!/_A = V\!\!\!/_B$.

Writing the ideal equation of state for each gas as well as that for the mixture, we have

$$p_A V\!\!\!/ = N_A R_u T \qquad p_B V\!\!\!/ = N_B R_u T \qquad p V\!\!\!/ = N R_u T$$

or

$$p_A/N_A = R_u T/V\!\!\!/ \qquad p_B/N_B = R_u T/V\!\!\!/ \qquad p/N = R_u T/V\!\!\!/$$

where p is the total pressure of the mixture and N is the number of moles of gas in the mixture. Hence $p_A/N_A = p_B/N_B = p/N$; therefore

$$p_A = \frac{N_A}{N} p \qquad \text{and} \qquad p_B = \frac{N_B}{N} p \tag{5.1}$$

The pressures p_A and p_B are termed the *partial pressure* of each constituent gas. The partial pressure is the pressure that each constituent would exert if it alone occupied the volume at the temperature of the mixture (Figure 5.1). The fractions N_A/N and N_B/N are called *mole fractions*. Due to conservation of mass, the sum of the moles of each constituent gas equals the total number of moles:

$$N = N_A + N_B$$

Therefore, from (5.1):

$$p_A + p_B = (N_A/N + N_B/N)p = p \tag{5.2}$$

This relation states that the sum of the partial pressures equals the total pressure of the mixture. The total pressure of the mixture is the pressure that a pressure-measuring device would sense.

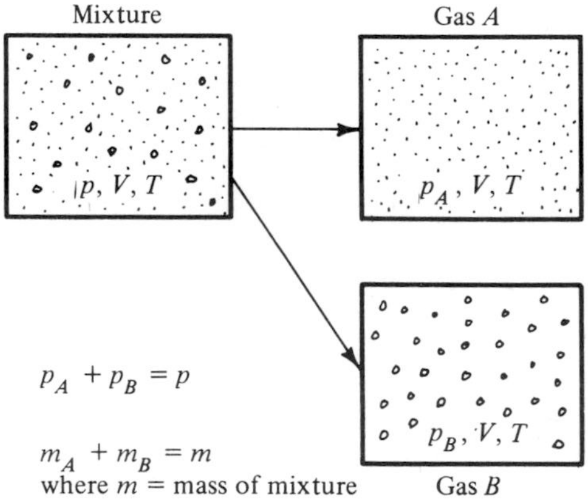

FIGURE 5.1

The molecular mass $\overline{M}$ of the mixture is obtained from the constituent molecular masses as follows:

$$m = N\overline{M}, \quad \text{with} \quad m = m_A + m_B, \quad \text{becomes} \quad m_A + m_B = N\overline{M}$$

$$\text{or} \qquad N_A\overline{M}_A + N_B\overline{M}_B = N\overline{M}$$

$$\text{and} \qquad \overline{M} = (N_A/N)\overline{M}_A + (N_B/N)\overline{M}_B \tag{5.3}$$

The molecular mass of the mixture can also be expressed in terms of the mass fractions m_A/m and m_B/m instead of mole fractions. Since $N/m = 1/\overline{M}$, we have $(N_A + N_B)/m = 1/\overline{M}$; or

$$\frac{1}{\overline{M}} = \frac{m_A}{m}\frac{1}{\overline{M}_A} + \frac{m_B}{m}\frac{1}{\overline{M}_B} \tag{5.4}$$

For ideal gases, the total internal energy U of a mixture is the sum of the internal energies of its constituents:

$$U = U_A + U_B$$

The specific internal energy u of the mixture is given by

$$u = \frac{U}{m} = \frac{m_A u_A + m_B u_B}{m}$$

Hence

$$u = (m_A/m)u_A + (m_B/m)u_B \tag{5.5}$$

In a similar fashion, we obtain the specific enthalpy h of the mixture:

$$h = (m_A/m)h_A + (m_B/m)h_B \tag{5.6}$$

For the specific heats of the mixture, we have

$$c_v = (m_A/m)c_{vA} + (m_B/m)c_{vB} \tag{5.7}$$

and

$$c_p = (m_A/m)c_{pA} + (m_B/m)c_{pB} \tag{5.8}$$

The relation between mole fraction and mass fraction is obtained from $N\overline{M} = m$. Thus, for example,

$$m_A/m = (N_A/N)(\overline{M}_A/\overline{M}) \tag{5.9}$$

From the definition of specific volume, we have

$$v = \forall/m = v_A m_A/m = v_B m_B/m \tag{5.10}$$

Since we are considering ideal gas mixtures only, we have, from the ideal equations of state (4.4) and (5.1),

$$v = \frac{R_A T}{p_A} \frac{m_A}{m} = \frac{R_A T}{pN_A/N} \frac{m_A}{m} = \frac{R_A T}{p} \frac{m_A}{N_A} \frac{N}{m} = \frac{R_A T}{p} \frac{\overline{M}_A}{\overline{M}}$$

Hence

$$v = R_A(\overline{M}_A/\overline{M})(T/p) = R_B(\overline{M}_B/\overline{M})(T/p) \tag{5.11}$$

Further, the gas constant of the mixture is

$$R = R_u/\overline{M} \tag{5.12}$$

EXAMPLE 5.1

Determine the gas constant of dry air by using information on properties of its constituents. In chemical formulas, the constituents of a mixture are usually given in moles. For dry air, assume the following composition:

$$1 \text{ mole of dry air} = 0.7809 \text{ mole } N_2 + 0.2095 \text{ mole } O_2 + 0.0093 \text{ mole A} + 0.0003 \text{ mole } CO_2$$

where N_2 is molecular nitrogen ($\overline{M} = 28.014$)
O_2 is molecular oxygen ($\overline{M} = 31.998$)
A is argon ($\overline{M} = 39.948$)
CO_2 is carbon dioxide ($\overline{M} = 44.009$)

Solution

The gas constant of the gas mixture is given by Equation (5.12). $R = R_u/\overline{M}$, with $\overline{M}$ given in Equations (5.3) as

$$\overline{M} = (N_A/N)\overline{M}_A + (N_B/N)\overline{M}_B + (N_C/N)\overline{M}_C + (N_D/N)\overline{M}_D$$

Thus the molecular mass of air becomes

$$\overline{M} = 0.7809 \times 28.014 + 0.2095 \times 31.998 + 0.0093 \times 39.948 + 0.0003 \times 44.009 = 28.964$$

Hence the gas constant R of air is

$$R = \frac{1545.3 \text{ ft-lbf/lbm-mole°R}}{28.964 \text{ lbm/lb-mole}} = 53.35 \text{ ft-lbf/lbm°R}$$

EXAMPLE 5.2

Determine the enthalpy of dry air at 44°F by using the enthalpy values of its constituent gases. Use ideal gas enthalpies as given in Table A.17 with a common reference temperature of 77°F. Argon has an enthalpy of −4.48 Btu/lbm at 44°F, using 77°F as reference.

Solution

From Equation (5.6), we have

$$h = (m_A/m)h_A + (m_B/m)h_B + (m_C/m)h_C + (m_D/m)h_D$$

where m_A/m, m_B/m, and so on, are the mass fractions of the constituents. The relation between mole fraction and mass fraction is $m_A/m = (N_A/N)(\overline{M}_A/\overline{M})$, and so on. Thus we have the following values:

Constituent	N_i/N	$\overline{M}_i/\overline{M}$	m_i/m	h (Btu/lbm)	$(m_i/m)h$
Nitrogen	0.7809	0.9672	0.7553	−8.19	−6.19
Oxygen	0.2095	1.1048	0.2315	−7.20	−1.67
Argon	0.0093	1.3792	0.0128	−4.48	−0.06
Carbon dioxide	0.0003	1.5194	0.0005	−6.30	−0.003
Air	1.000		1.0000		−7.92

$$h_{\text{air}} = -7.92 \text{ Btu/lbm}$$

To check, from Table A.17 we obtain the following for dry air at 44°F:

$$h = -18.5 + \frac{44-0}{77-0}[0-(-18.5)] = -7.93 \text{ Btu/lbm}$$

EXAMPLE 5.3

Find the specific volume of dry air at 62°F and 10.28 psia (0.7 atm) from property values of its constituent gases.

Solution

From Equation (5.11), the specific volume of the mixture is $v = R_A(\overline{M}_A/\overline{M})(T/p)$. First, use nitrogen with $R_A = 55.16$ ft-lbf/lbm°R, $\overline{M}_A/\overline{M} = 0.9672$ (from Example 5.2). Hence

$$v = (55.16 \text{ ft-lbf/lbm°R})(0.9672)\frac{522\text{°R}}{10.28 \text{ lbf/in}^2 \times 144 \text{ in}^2/\text{ft}^2} = 18.813 \text{ ft}^3/\text{lbm}$$

Second (to check), use carbon dioxide with $R_A = 35.11$ ft-lbf/lbm°R and $\overline{M}_1/\overline{M} = 1.5194$. Hence

$$v = (35.11 \text{ ft-lbf/lbm°R})(1.5194)\frac{522\text{°R}}{(10.28 \times 144) \text{ lbf/ft}^2} = 18.811 \text{ ft}^2/\text{lbm}$$

EXAMPLE 5.4

A gas mixture is composed of 40 pounds mass of carbon dioxide, 40 pounds mass of molecular nitrogen, and 20 pounds mass of carbon monoxide. Find the partial pressures of the constituents, given that the total pressure is 20 psia.

Solution

The partial pressure of each constituent is given in terms of the total pressure and the mole fractions of the constituents by Equation (5.1). The mole fractions are obtained from Equation (5.9), namely, $(m_i/m)(\overline{M}/\overline{M}_i) = N_i/N$, with the molecular mass given by Equation (5.4).

Constituent	Mass fraction (m_i/m)	$\overline{M}_i$	$(m_i/m)(1/\overline{M}_i)$	N_i/N	Partial pressure (psia)
CO_2	0.4	44.009	0.00909	0.2979	5.958
N_2	0.4	28.014	0.01428	0.4681	9.362
CO	0.2	28.010	0.00714	0.2340	4.680
			$1/\overline{M} = 0.030509$	1.0000	20.000

The molecular mass is $1/0.030509 = 32.78$. The partial pressure of each constituent is obtained from Equation (5.2); for example,

$$p_{CO_2} = 0.2979 \times 20 \text{ psia} = 5.958 \text{ psia}$$

EXAMPLE 5.5

Two kilograms of molecular nitrogen are stored at room temperature (20°C) in a tank at 100 atm (1.013×10^4 kPa). One kilogram of molecular oxygen is added to the tank, with the temperature of the mixture maintained at a constant temperature of 20°C. Find the total pressure of the gas mixture.

Solution

The mass of the nitrogen will not change as a result of the mixing process; the initial pressure of N_2 will be the partial pressure of nitrogen in the mixture. From Equation (5.1), we have for the nitrogen $p_{N_2} = (N_{N_2}/N)p$, and for the oxygen $p_{O_2} = (N_{O_2}/N)p$. Taking the ratio, we have

$$\frac{p_{N_2}}{p_{O_2}} = \frac{N_{N_2}}{N_{O_2}} = \frac{m_{N_2}}{\overline{M}_{N_2}} \frac{\overline{M}_{O_2}}{m_{O_2}} = \frac{2}{28.014} \frac{31.998}{1} = 2.2844$$

Thus, $p_{O_2} = 0.43775 p_{N_2} = 0.43775 \times 1.013 \times 10^4 \text{ kPa} = 0.4434 \times 10^4 \text{ kPa}$ (=43.78 atm). The total pressure is

$$p = p_{N_2} + p_{O_2} = 143.78 \text{ atm}$$

EXAMPLE 5.6

Find the specific heat capacity at constant pressure of the gas mixture of Example 5.5, and from it determine the change in enthalpy of the mixture due to a temperature rise of 20°C.

Solution

As indicated in Equation (5.8), the determination of the specific heat of the mixture requires a knowledge of the mass fractions and specific heats of the constituents (the c_p values are obtained from Table B.15):

Constituent	m_i/m	c_{pi} at 20°C (J/kg K)
N_2	0.667	1039
O_2	0.333	918

Thus

$$\begin{aligned} c_p &= \left(\frac{m_A}{m}\right)c_{pA} + \left(\frac{m_B}{m}\right)c_{pB} \\ &= 0.667 \times 1039 \text{ J/kg K} + 0.333 \times 918 \text{ J/kg K} \\ &= 998.7 \text{ J/kg K} \end{aligned}$$

For an ideal gas over a small temperature change,

$$\begin{aligned} h_2 - h_1 &= c_p(T_2 - T_1) = 998.7(20) \text{ J/kg K} \\ &= 19.97 \text{ kJ/kg} \end{aligned}$$

EXAMPLE 5.7

An insulated tank containing 3 cubic feet of molecular hydrogen at 50 psia and 60°F is connected to an insulated tank containing 2 cubic feet of molecular nitrogen at 25 psia and 90°F. A valve is opened in the line connecting the tanks, thereby allowing the gases to mix. Compute the resulting temperature and pressure of the gas mixture.

Solution

Since the tanks are insulated and the total volume of the tanks does not change, $Q_{in} = 0$ and $W_{out} = 0$, and the energy equation (2.5) for a closed system gives $U_2 - U_1 = 0$.

Let $U_1 = U_A + U_B$ be the total internal energy of the gases before mixing and $U_2 = U'_A + U'_B$ be the total internal energy of the mixture after mixing. Thus

$$U_A - U'_A + U_B - U'_B = 0$$

For an ideal gas, we have $\Delta U = mc_v \, \Delta T$. Thus, $m_A c_{vA}(T - T_A) + m_B c_{vB} (T - T_B) = 0$, where T is the temperature of the mixture. Solving for T, we obtain

$$T = \frac{m_A c_{vA} T_A + m_B c_{vB} T_B}{m_A c_{vA} + m_B c_{vB}}$$

Further, the volume and the mass of the mixture are the sums of volumes and masses, respectively, prior to mixing:

$$V = V_A + V_B$$

and $$m = m_A + m_B$$

From the ideal equation of state, we have

$$p = \frac{mRT}{V}$$

The mass of hydrogen in the tank prior to mixing is given by

$$m_{H_2} = \frac{p_{H_2}V_{H_2}}{R_{H_2}T_{H_2}} = \frac{(50 \times 144 \text{ lbf/ft}^2)(3 \text{ ft}^3)}{(766.5 \text{ ft-lbf/lbm°R})(520°\text{R})} = 0.0542 \text{ lbm}$$

while the mass of nitrogen is

$$m_{N_2} = \frac{p_{N_2}V_{N_2}}{R_{N_2}T_{N_2}} = \frac{(25 \times 144 \text{ lbf/ft}^2)(2 \text{ ft}^3)}{(55.16 \text{ ft-lbf/lbm°R})(550°\text{R})} = 0.2373 \text{ lbm}$$

$$V = \text{total mixture volume} = 3 + 2 = 5 \text{ ft}^3$$

$$m = \text{total mixture mass} = 0.0542 + 0.2373 = 0.2915 \text{ lbm}$$

For hydrogen, from Table A.15:

$$c_v = 2.43 \text{ Btu/lbm°R} \quad \text{and} \quad \overline{M} = 2.016$$

For nitrogen,

$$c_v = 0.177 \text{ Btu/lbm°R} \quad \text{and} \quad \overline{M} = 28.014$$

Hence,

$$T = \frac{0.0542 \text{ lbm} \times 2.43 \text{ Btu/lbm°R} \times 520°\text{R} + 0.2373 \text{ lbm} \times 0.177 \text{ Btu/lbm°R} \times 550°\text{R}}{0.0542 \text{ lbm} \times 2.43 \text{ Btu/lbm°R} + 0.2373 \text{ lbm} \times 0.177 \text{ Btu/lbm°R}}$$

$$= 527.3°\text{R} = 67.3°\text{F}$$

Now $R = R_u/\overline{M}$. From Equation (5.4),

$$\frac{1}{\overline{M}} = \left(\frac{m_A}{m}\right)\frac{1}{\overline{M}_A} + \left(\frac{m_B}{m}\right)\frac{1}{\overline{M}_B} = \frac{0.0542 \text{ lbm}}{0.2915 \text{ lbm}}\frac{1}{2.016} + \frac{0.2373 \text{ lbm}}{0.2915 \text{ lbm}}\frac{1}{28.014}$$

$$= 0.12129$$

$$R = \frac{R_u}{\overline{M}} = \frac{1545 \text{ ft-lbf/lb-mole°R}}{(1/0.12129) \text{ lbm/lb-mole}} = 187.4 \text{ ft-lbf/lbm°R}$$

Hence,

$$p = \frac{(0.2915 \text{ lbm})(187.4 \text{ ft-lbf/lbm°R})(527.3°\text{R})}{5 \text{ ft}^3} = 5761.0 \text{ lbf/ft}^2$$

$$= 40 \text{ psia}$$

EXAMPLE 5.8

If the two tanks of Example 5.7 are not insulated and the temperature of the mixture is measured to be 80°F, what amount of heat will be transferred?

Solution

From the energy equation for a closed system (2.5), we have

$$Q_{\text{in}} = U_2 - U_1 = (U'_A + U'_B) - (U_A + U_B)$$
$$= m_A c_{vA}(T - T_A) + m_B c_{vB}(T - T_B)$$

Thus

$$Q_{\text{in}} = (0.0542\,\text{lbm})2.43\,\text{Btu/lbm°R}(80 - 60)\text{°F}$$
$$+ (0.2373\,\text{lbm})0.177\,\text{Btu/lbm°R}(80 - 90)\text{°F}$$
$$= 2.634 - 0.420 = 2.214\,\text{Btu}$$

The system gains heat from the surroundings in the amount of 2.214 Btu.

5.3 THERMODYNAMIC PROPERTIES OF GAS-VAPOR MIXTURES

The analysis for obtaining properties of gas mixtures can easily be extended to mixtures of gases and condensable vapors when the partial pressure of the superheated vapor is sufficiently low so that it can be treated as an ideal gas. This analysis is demonstrated for water vapor in Figure 5.2, which shows that superheated water vapor at low pressures can be treated as an ideal gas up to saturation conditions. The additional phenomenon that occurs in the gas–superheated vapor mixture is that the saturation condition of the vapor can be reached. When the gas mixture becomes "saturated" with vapor, the addition of more vapor will result in the appearance of the condensable substance as liquid (fog).

The most important application for gas-vapor mixtures is the analysis of atmospheric air, which, in addition to gases, also contains water vapor. Before proceeding with a discussion of moisture in atmospheric air, we will illustrate the saturation of a gas by a condensable vapor in an example, using thermodynamic property tables for water.

EXAMPLE 5.9

A tank contains 1 pound mass of carbon dioxide and 1 pound mass of water vapor at 150°F. The pressure p in the tank is 2 psia.

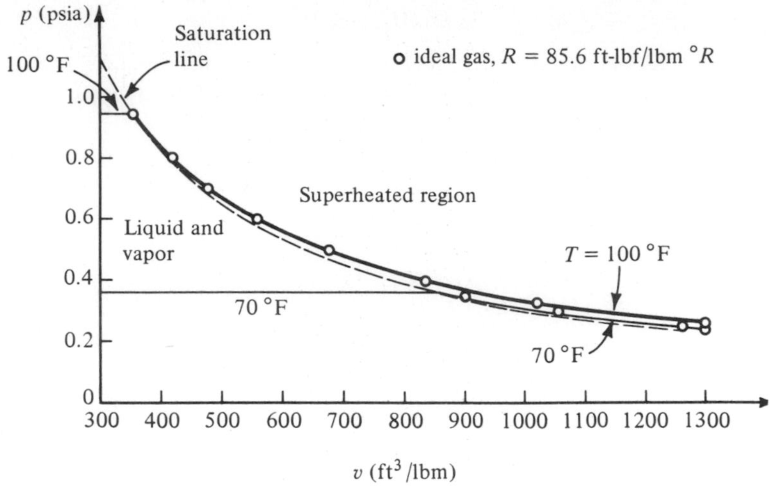

FIGURE 5.2 *p-v Relationship for Water Vapor.*

a. Find the partial pressures at 150°F.
b. Find the temperature and total pressure of the mixture at which the water vapor will be at the saturated condition.
c. Find the partial pressure of each constituent if the temperature is lowered to 50°F, 32°F, and 0°F.

Solution

a. For details, see Example 5.4.

Constituent	Mass fraction (m_i/m)	$(\bar{M}_i)$	$(m_i/m)(1/\bar{M}_i)$	N_i/N	Partial pressure (psia)
CO_2	0.5	44.009	0.01136	0.2905	0.581
H_2O	0.5	18.015	0.02775	0.7095	1.419

Since both CO_2 and H_2O occupy the entire volume of the tank, we can find the volume by using either CO_2 or H_2O. Thus

$$v_{CO_2} = \frac{R_{CO_2}T}{p_{CO_2}} = \frac{(35.11\ \text{ft-lbf/lbm°R})610°\text{R}}{(0.581\ \text{lbf/in}^2)144\ \text{in}^2/\text{ft}^2} = 256.0\ \text{ft}^3/\text{lbm}$$

or $\quad V = 256.0\ \text{ft}^3$ for CO_2 and for H_2O

From the steam tables, we find that at 1.419 psia and 150°F, the water vapor is in the superheated condition.

b. Since the volume and mass of water and CO_2 remain constant, the specific volume of each constituent also remains at 256.0 ft^3/lbm. From the steam tables (Table A.2), we find the following:

T_{sat}(°F)	p_{sat}(psia)	$v_{sat} = v_g$(ft^3/lbm)
110	1.2750	265.39
120	1.6927	203.26

By interpolation, with $v = 256.0$ ft^3/lbm, we find that

$$T_{sat} = 111.5°\text{F} \quad \text{and} \quad p_{H_2O} = 1.338 \text{ psia}$$

$$p_{CO_2} = \frac{(1 \text{ lbm})(35.11 \text{ ft-lbf/lbm°R})(571.5°\text{R})}{(256.0 \text{ ft}^3/\text{lbm})(144 \text{ in}^2/\text{ft}^2)} = 0.544 \text{ psia}$$

Total pressure $p = 1.338 + 0.544 = 1.882$ psia

c. At 50°F, the partial pressure of CO_2 is

$$p_{CO_2} = \frac{(1 \text{ lbm})(35.11 \text{ ft-lbf/lbm°R}) \times (510°\text{R})}{(256.0 \text{ ft}^3/\text{lbm})144 \text{ in}^2/\text{ft}^2} = 0.486 \text{ psia}$$

The specific volume of each constituent remains constant since the mass and volume remain constant. Hence we need to find p_{H_2O} as a function of v and T. A search of the steam tables locates steam at $T = 50°$F and $v = 256.0$ ft^3/lbm at a saturation pressure of 0.178 psia in the saturated vapor–liquid region. Hence by Equation (4.1),

$$v_{H_2O} = v_f + xv_{fg} = 256.0 = 0.016 + x1704.8$$

The assumption here is that the liquid-vapor behavior of the water is not affected by the presence of the carbon dioxide. The quality of water-vapor mixture becomes $x = 256.0/1704.8 = 0.15$.

The total pressure within the tank is 0.486 psia + 0.178 psia = 0.664 psia. Note that the pressure within the tank has dropped appreciably since most of the water has liquified.

Similarly, at $T = 32°$F, $p_{sat} = 0.08866$ psia, and $v_{fg} = 3302.4$ ft^3/lbm. Hence the quality is $x = 256.0/3302.4 = 0.0775$. The partial pressure of CO_2 is

$$p_{CO_2} = \frac{(1 \text{ lbm})(35.11 \text{ ft-lbf/lbm°R})(492°\text{R})}{(256.0 \text{ ft}^3/\text{lbm})(144 \text{ in}^2/\text{ft}^2)} = 0.469 \text{ psia}$$

The total pressure in the tank is

$$p = 0.08866 \text{ psia} + 0.0469 \text{ psia} = 0.558 \text{ psia}$$

At 0°F, water exists in its solid and vapor phase with $p_{sat} = 0.0185$ psia and $v_{ig} = 14{,}770$ ft^3/lbm. Thus, $x = 256.0/14{,}770 = 0.0173$. The partial pressure of CO_2 is

$$p_{CO_2} = \frac{(1 \text{ lbm})(35.11 \text{ ft-lbf/lbm°R})(460°\text{R})}{(256.0 \text{ ft}^3/\text{lbm})(144 \text{ in}^2/\text{ft}^2)} = 0.438 \text{ psia}$$

The total pressure in the tank is

$$p = 0.0185 \text{ psia} + 0.438 \text{ psia} = 0.457 \text{ psia}$$

Summarizing the partial and total pressure behavior with temperature, we have the following:

Temperature	150°F	111.5°F	50°F	32°F	0°F
p_{H_2O}/p	0.710	0.710	0.268	0.159	0.041
p_{CO_2}/p	0.290	0.290	0.732	0.841	0.959
p (psia)	2.000	1.882	0.664	0.558	0.457

5.4 THERMODYNAMIC PROPERTIES OF MOIST AIR (PSYCHROMETRICS)

The water vapor content of atmospheric air is usually given in somewhat different form than for other gas-vapor mixtures. Instead of using mass fractions (mass per total mass), the quantity *specific humidity* is used. *Specific humidity*, symbolized ω, is defined as the ratio of the mass of water vapor to the mass of air ($\omega = m_v/m_a$). The specific humidity is information required to properly design and size equipment used to condition (cool and dehumidify) indoor atmospheric air.

In addition to specific humidity, the quantity *relative humidity* is also used in air conditioning terminology. *Relative humidity*, symbolized ϕ, is defined as the ratio of the mass of water vapor to the mass of water vapor required to produce a saturated mixture at the same temperature ($\phi = m_v/m_{sat}$). The relative humidity together with temperature is a measure of human comfort (Figure 5.3). Next we will show there is a relationship between specific and relative humidity. To find this relation, proceed from the definition of relative humidity as follows:

$$\phi = \frac{m_v}{m_{sat}} = \omega \frac{m_a}{m_{sat}} = \omega \frac{\forall/v_a}{\forall/v_g} = \omega \frac{v_g}{v_a}$$

From the ideal equation of state, we have

$$v_a = R_a T/p_a \quad \text{and} \quad v_g = R_v T/p_g$$

Therefore

$$\phi = \omega \frac{R_v T}{p_g} \frac{p_a}{R_a T} = \frac{85.78}{53.35} \frac{p_a}{p_g} \omega = 1.6079 \frac{p_a}{p_g} \omega = \frac{1.6079(p - p_v)\omega}{p_g}$$

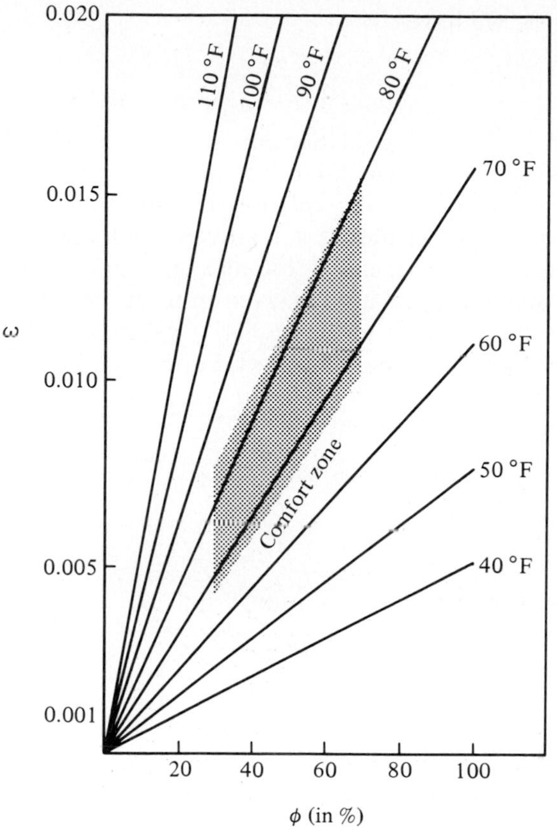

FIGURE 5.3 *Human Comfort Chart.*

Using $pV = mRT$, we have

$$\phi = \frac{p_v V / R_v T}{p_g V / R_v T} = \frac{p_v}{p_g}$$

Thus

$$\phi = 1.6079\left(\frac{p}{p_g} - \phi\right)\omega$$

The value of the saturation pressure p_g of water is obtainable as a function of the temperature from the steam tables. Solving the above equation for ω, we obtain

$$\omega = 0.6219 \frac{\phi}{p/p_g - \phi} \tag{5.13}$$

or solving for ϕ, we have

$$\phi = \frac{\omega}{(0.6219 + \omega)} \frac{p}{p_g} \tag{5.14}$$

Figure 5.3 shows the relationship between specific humidity and relative humidity at different air temperatures, at a constant pressure, say, 14.70 psia. Note that the relation is almost linear and that the amount of water vapor in the air is relatively small and rarely exceeds 3 percent of the mass of the air. In the air conditioning and refrigeration industry, the relationship is usually given in terms of ω versus T, with ϕ constant. This chart, shown in Figure 5.4, is called a *psychrometric chart* [from the Greek *psychro* (cold) + *meter*]. Further application of this chart will be discussed in Chapter 8 under air conditioning processes.

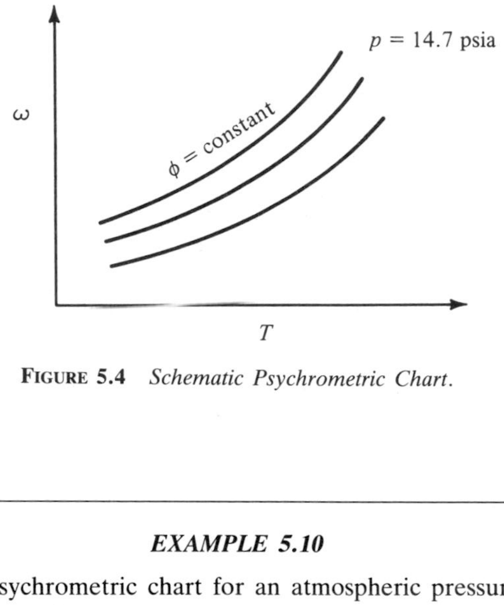

Figure 5.4 *Schematic Psychrometric Chart.*

EXAMPLE 5.10

Construct an air psychrometric chart for an atmospheric pressure of 14.70 psia.

Solution

From Equation (5.13), we have

$$\omega \text{ (lbm moisture/lbm dry air)} = 0.6219 \frac{\phi}{p/p_g - \phi}$$

Here, $p = 14.70$ psia. The values of p_g (the saturation pressure) at various temperatures are obtained from Table A.2. Hence, ω can be calculated as a function of temperature for different constant values of relative humidity ϕ as shown in the following table:

T(°F)	p_g(psia)	$\omega_{\phi=1}$	$\omega_{\phi=0.5}$	$\omega_{\phi=0.2}$
32.018	0.08865	0.0038	0.0019	0.0008
40	0.12163	0.0052	0.0026	0.0010
50	0.17796	0.0076	0.0038	0.0015
60	0.25611	0.0110	0.0055	0.0022
70	0.36292	0.0158	0.0078	0.0031
80	0.50683	0.0222	0.0109	0.0043
90	0.69813	0.0310	0.0151	0.0060
100	0.94924	0.0430	0.0208	0.0081

These results plus results of calculations at additional values of relative humidity are shown in Figure 5.5.

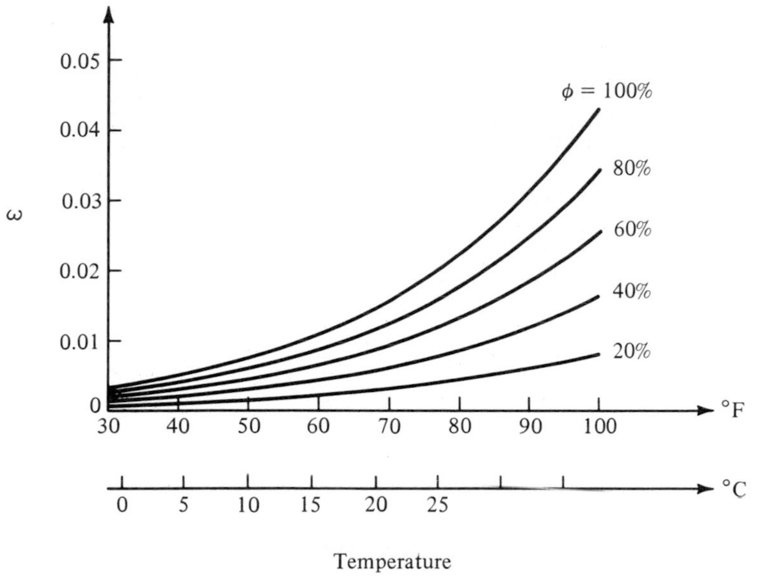

FIGURE 5.5 *Psychrometric Chart for Air at Standard Pressure of 14.7 psia (101.3 kPa).*

Summarizing the various conditions of water vapor in the atmospheric air, we have the following: For relative humidity *less than* 100 percent (usual condition), the water vapor is in its superheated condition. At 100 percent relative humidity, the vapor is in its saturated condition. When water vapor exists as a *wet mixture* (saturated vapor plus saturated liquid), a foggy condition in the atmospheric air will prevail.

EXAMPLE 5.11

When breathing of pure oxygen instead of atmospheric air becomes necessary, humidification of the oxygen prevents "oxygen burns" in the nostrils. Develop an oxygen psychrometric chart for a pressure of 14.70 psia.

Solution

The expression for the relative humidity ϕ uses the appropriate value of the gas constant for molecular oxygen:

$$\phi = \frac{R_v}{R_{O_2}} \frac{p_{O_2}}{p_g} \omega = \frac{85.78}{48.29} \frac{p_{O_2}}{p_g} \omega = 1.7764 \frac{p_{O_2}}{p_g} \omega$$

and $$\phi = 1.7764\left(\frac{p}{p_g} - \phi\right)\omega$$

or $$\omega = \frac{0.5629\phi}{p/p_g - \phi}$$

Hence, we can calculate the values shown in the following table:

T(°F)	p_g(psia)	$\omega_{\phi=1}$	$\omega_{\phi=0.5}$	$\omega_{\phi=0.2}$
32.018	0.08865	0.0034	0.0017	0.0007
40	0.12163	0.0047	0.0023	0.0009
50	0.17796	0.0069	0.0034	0.0014
60	0.25611	0.0100	0.0049	0.0020
70	0.36292	0.0143	0.0070	0.0028
80	0.50683	0.0201	0.0099	0.0039
90	0.69813	0.0281	0.0137	0.0054
100	0.94924	0.0389	0.0188	0.0074
110	1.27500	0.0535	0.0255	0.0099

EXAMPLE 5.12

Atmospheric air at a pressure of 14 psia and a temperature of 80°F has a relative humidity of 50 percent. Compute the specific humidity and compare it to the mass fraction of water vapor in the mixture.

Solution

From Equation (5.13), we have

$$\omega = 0.6219 \frac{\phi}{p/p_g - \phi} = 0.6219 \frac{0.5}{14/p_g - 0.5}$$

The saturation pressure p_g at 80°F is, from Table A.2, equal to 0.50683 psia. Thus

$$\omega = 0.6219 \frac{0.5}{14/0.50683 - 0.5} = 0.0115 \text{ lbm } H_2O/\text{lbm dry air}$$

The results of Example 5.10 indicate a specific humidity of 0.011 at an atmospheric pressure of 14.70 psia. The mass fraction of the water vapor is

$$\frac{m_{H_2O}}{m_{H_2O} + m_{air}} = \frac{m_{H_2O}/m_{air}}{1 + m_{H_2O}/m_{air}} = \frac{\omega}{1 + \omega}$$

$$= \frac{0.0115}{1.0115} = 0.0114 \text{ lbm } H_2O/\text{lbm mixture (moist air)}$$

In order to determine the change in enthalpy or change in internal energy of an ideal gas–water vapor mixture, we calculate the property change for each component separately. For example, if we are dealing with moist air, changes in enthalpy or internal energy of the air can be found either from ideal gas relations or from air tables. Changes in enthalpy or internal energy of the vapor can be found from steam tables.

EXAMPLE 5.13

Find the change in internal energy of air at atmospheric pressure as its temperature changes from 80°F to 70°F at a constant specific humidity ω of 0.0114.

Solution

The change in internal energy of the air alone is given by

$$u_{a2} - u_{a1} = (c_{va})(T_2 - T_1)$$

$$= (0.172 \text{ Btu/lbm°F})(-10\text{°F})$$

$$= -1.72 \text{ Btu/lbm dry air}$$

We can express the change of internal energy of the vapor as

$$u_{v2} - u_{v1} = c_{v_v}(T_2 - T_1)$$

$$= (0.335 \text{ Btu/lbm°F})(-10\text{°F}) = -3.35 \text{ Btu/lbm water vapor}$$

where we assume that the vapor also behaves as an ideal gas with $c_v = 0.335$ Btu/lbm°F.

Expressing the changes of internal energy of air and water vapor per pound of dry air, we have

$$u_2 - u_1 = -1.72 \text{ Btu/lbm dry air} + \omega(-3.35 \text{ Btu/lbm water vapor})$$

where

$$\omega \text{ (lbm vapor/lbm dry air)} = 0.0114$$

$$u_2 - u_1 = -1.72 \text{ Btu/lbm dry air} - 0.038 \text{ Btu/lbm dry air}$$

$$= -1.76 \text{ Btu/lbm dry air}$$

In actual practice, the relative humidity of an air–water vapor mixture is found from wet-bulb and dry-bulb temperatures. The *wet-bulb temperature* of a mixture is the temperature indicated by a thermometer with the bulb covered with a cotton wick saturated with water, as shown in Figure 5.6. *Dry-bulb temperature* is simply the temperature of the air. Although mass transfer and heat transfer processes are slightly different from those occurring in an adiabatic saturation apparatus, the wet-bulb temperature is very close to the adiabatic saturation temperature.

In an adiabatic saturation apparatus, an air–water vapor mixture is brought into contact with a reservoir of water, as shown in Figure 5.7. The mixture picks up moisture, leaving as saturated vapor at point 2, the process taking place at constant pressure. Enough makeup water is supplied to the reservoir to compensate exactly for that lost by evaporation to the mixture. The apparatus is well insulated, so that no heat is exchanged with the surroundings. By measuring T_1 and T_2, one can measure ω for the entering air–water vapor mixture. Let us analyze thermodynamically the apparatus of Figure 5.7, using the first law for steady flow. From Equation (3.17), neglecting changes of kinetic and potential energies, $\dot{m}(h_e - h_{\text{in}}) = 0$.

$$\dot{m}_{a1}h_{a1} + \dot{m}_{v1}h_{v1} + \dot{m}_{\text{liq.}}h_{f2} = \dot{m}_{a2}h_{a2} + \dot{m}_{v2}h_{v2}$$

For steady flow, applying conservation of mass for each constituent:

$$\dot{m}_{a1} = \dot{m}_{a2} \quad \text{and} \quad \dot{m}_{v1} + \dot{m}_{\text{liq.}} = \dot{m}_{v2}$$

so that

$$h_{a1} - h_{a2} = \omega_2 h_{v2} - \omega_1 h_{v1} - (\omega_2 - \omega_1)h_{f2}$$

However,

$$h_{v2} - h_{f2} = h_{fg2}$$

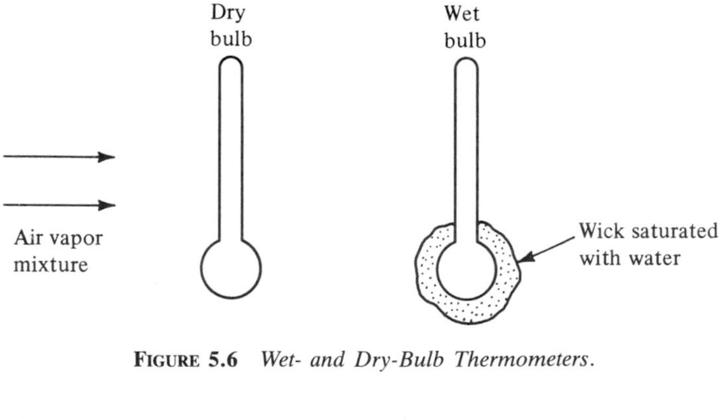

FIGURE 5.6 *Wet- and Dry-Bulb Thermometers.*

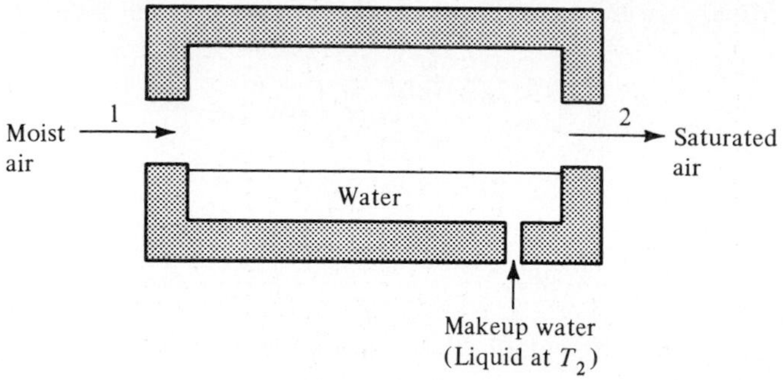

FIGURE 5.7 *Adiabatic Saturation Apparatus.*

Therefore

$$c_{pa}(T_1 - T_2) = \omega_2 h_{fg2} - \omega_1(h_{v1} - h_{f2})$$

and

$$\omega_1 = \frac{c_{pa}(T_2 - T_1) + \omega_2 h_{fg2}}{h_{v1} - h_{f2}} \tag{5.15}$$

EXAMPLE 5.14

A mixture of air and water vapor at atmospheric pressure enters an adiabatic saturation apparatus at 90°F and leaves at 70°F, the adiabatic saturation temperature. Calculate the relative humidity of the entering air.

Solution

Using Equation (5.15), with data from Tables A.2 and A.15, we have

$$\omega_1 = \frac{(0.24 \text{ Btu/lbm°F})(-20°\text{F}) + \omega_2(1054.0 \text{ Btu/lbm H}_2\text{O})}{1100.8 \text{ Btu/lbm H}_2\text{O} - 38.1 \text{ Btu/lbm H}_2\text{O}}$$

To solve for ω_2, use Equation (5.13) with $\phi = 1$ (saturated condition).

$$\omega_2 = 0.6219 \frac{1}{p_2/p_g - 1}$$

$$= 0.6219 \frac{1}{(14.696/0.36292) - 1}$$

$$= 0.01575 \text{ lbm vapor/lbm dry air}$$

Solving for ω_1, we obtain

$$\omega_1 = \frac{-4.8 \text{ Btu/lbm air} + 16.6 \text{ Btu/lbm air}}{1062.7 \text{ Btu/lbm H}_2\text{O}}$$

$$= 0.0111 \text{ lbm vapor/lbm dry air}$$

To solve for ϕ_1, use Equation (5.14) with $p_{g_1} = 0.6981$ psia at 90°F.

$$\phi_1 = \frac{\omega_1}{(0.6219 + \omega_1)} \frac{p}{p_g}$$

$$= \frac{0.0111}{(0.6219 + 0.0111)} \frac{14.696}{0.6981}$$

$$= 36.9\%$$

In calculating the relative humidity of moist air from the measurements of wet- and dry-bulb temperatures, the wet-bulb temperature is taken equal to adiabatic saturation temperature (T_2), with the dry-bulb temperature equal to the inlet temperature (T_1). The following example will illustrate the procedure.

EXAMPLE 5.15

Find the relative and specific humidities of moist air at an atmospheric pressure of 14.7 psia over a range of dry-bulb temperatures for a constant wet-bulb temperature of 70°F.

Solution

Using Equation (5.15) we have, with $T_1 = T_{DB}$ and $T_2 = T_{WB}$ and the values from Example 5.14,

$$\omega = \frac{(0.240 \text{ Btu/lbm°F})(70 - T_{DB}) + 16.60 \text{ Btu/lbm air}}{h_{v_1} \text{ Btu/lbm } H_2O - 38.05 \text{ Btu/lbm } H_2O}$$

where T_{DB} is the dry-bulb temperature and T_{WB} is the wet-bulb temperature.

The values of h_v at the various temperatures are obtained from Table A.2, with h_v taken as h_g.

By use of Equation (5.14), we obtain

$$\phi = \frac{\omega}{0.6219 + \omega} \frac{14.696}{p_g}$$

The values of ω and ϕ can be calculated as a function of dry-bulb temperature as shown in the following table:

T_{DB}(°F)	p_g(psia)	h_v(Btu/lbm H_2O)	ω	ϕ(%)
110	1.2750	1109.3	0.00653	12.0
100	0.94924	1105.1	0.00880	21.6
90	0.69813	1100.8	0.01109	36.9
80	0.50683	1096.4	0.01341	61.2
70	0.36292	1092.1	0.01574	100.0

The psychrometric chart in Figure 5.8 shows the calculated wet-bulb line for 70°F, as well as others which were calculated in a similar fashion.

Condensed moisture is called dew. Consequently, the *dew point* of an air–water vapor mixture is the temperature at which condensation occurs when the mixture is cooled at constant pressure. On a T-v diagram for water vapor, the dew point is shown as point D.P. on Figure 5.9. Measuring the temperature of the dew point is another method of determining the relative humidity of moist air. A brightly polished metal container is cooled and the temperature is observed at which the surface becomes clouded with condensed moisture. Suppose the dew point is measured in this way to be 15°C when the air temperature (dry-bulb) is 25°C. From Table B.2, the saturation pressure corresponding to 15°C is 1.7039 kPa. Thus, the water vapor in the air has a partial pressure of 1.7039 kPa, equal to the saturated vapor pressure at 15°C. The saturation pressure at 25°C is 3.1660 kPa. The relative humidity is therefore

$$\phi = \frac{p_v}{p_g} = \frac{1.7039}{3.1660} = 53.8\%$$

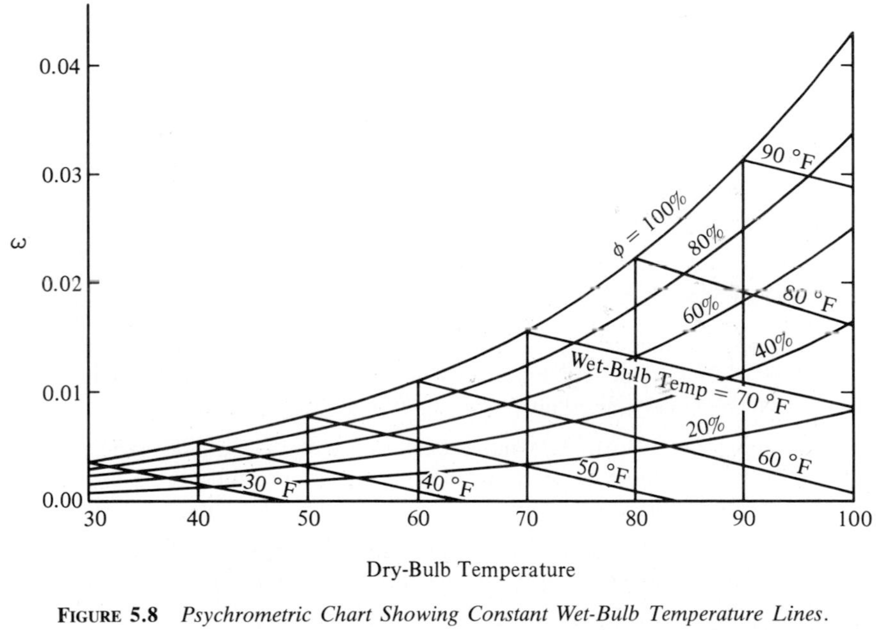

FIGURE 5.8 *Psychrometric Chart Showing Constant Wet-Bulb Temperature Lines.*

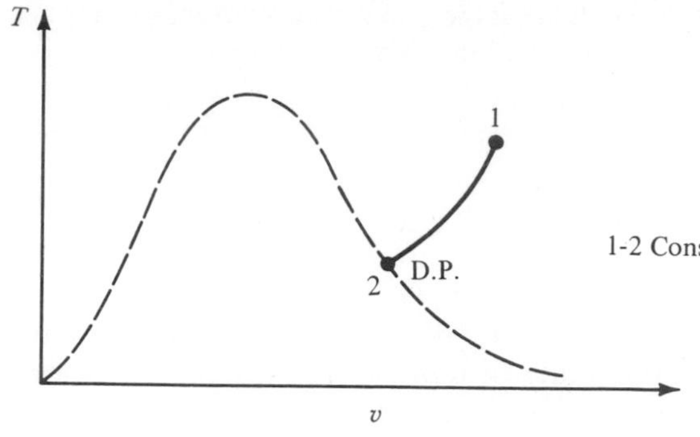

FIGURE 5.9 *Dew Point.*

PROBLEMS

5.1 An insulated container with 0.1 cubic meter of molecular oxygen at 25°C and 400 kPa is connected to one containing 0.2 cubic meter of molecular nitrogen at 200 kPa^2 and 25°C. Calculate the final pressure and temperature of the mixture.

5.2 An automobile exhaust analysis shows the following composition on a mole fraction basis:

N_2	0.808
CO_2	0.100
O_2	0.002
CO	0.065
H_2	0.025
	1.000

Calculate the gas constant of the mixture and the mass of 1 cubic meter of exhaust at a pressure of 101.3 kPa and a temperature of 25°C.

5.3 A humidifier is to increase the relative humidity of air from 40 to 50 percent when the air temperature is 70°F. The humidifier air flow is 1 cubic foot per second. Calculate the mass rate of water required in pounds per day.

5.4 An open container of water is placed in a closed room having a volume of 1500 ft^3 and an air temperature of 72°F. What will the specific humidity be after reaching equilibrium?

5.5 At a pressure of 101.3 kPa and a temperature of 28°C, atmospheric air has a relative humidity of 60 percent. Find the partial pressure of the dry air and the water vapor.

5.6 Atmospheric air at a pressure of 14.7 psia and a temperature of 60°F has a relative humidity of 70 percent. The atmospheric pressure drops to 14.5 psia while the temperature remains unchanged. Calculate the new relative humidity.

5.7 A mixture of water and air is contained in a 0.1-m^3 closed container at a temperature of 40°C, a pressure of 101.3 kPa, and 50 percent relative humidity. The tank is cooled. At what temperature will the water begin to condense, and what amount of heat will be removed?

5.8 A mixture of carbon dioxide and water vapor has a pressure of 14.7 psia and a temperature of 125°F. Find the partial pressure and the mole fraction of the two constituents, if the mass fraction of water vapor is 0.05.

5.9 Deverlop an air psychrometric chart for a pressure of 50 kPa.

5.10 An air–water vapor mixture at 100 kPa, 20°C, and 50 percent relative humidity is cooled at constant pressure. Find the temperature at which condensation will occur (dew point).

5.11 In a steady-flow, adiabatic process, 10 cubic feet per second of air at atmospheric pressure, 100°F, and 30 percent relative humidity are mixed with 5 cubic feet per second of air at atmospheric pressure, 60°F, and 80 percent relative humidity. Find the resultant temperature and relative humidity of the mixture.

5.12 Calculate the relative humidity of air with a dry-bulb temperature of 30°C and a wet-bulb temperature of 25°C.

5.13 A gas mixture has the following composition on a mass fraction basis:

N_2	0.40
CO	0.30
CO_2	0.15
CH_4	0.10
C_2H_6	0.05
	1.00

Find the composition of the gas mixture on a mole fraction basis.

5.14 In a mixture of nitric oxide (NO) and argon (A), at a temperature of 30°C and a pressure of 100 kPa, the partial pressure of argon is 30 percent of the total pressure. Calculate the mass fraction composition of the mixture.

5.15 The mole fraction composition of a furnace flue gas is

CO_2	0.10
O_2	0.06
H_2O	0.04
N_2	0.80
	1.00

Find the mass fraction composition and the gas constant of the flue gas.

5.16 The mass fraction composition of a gas mixture at 100 kPa is

CO_2	0.15
O_2	0.06
N_2	0.79
	1.00

Calculate the partial pressures and the gas constant of the mixture.

5.17 The products of combustion of coal mined in a certain region have the following mole fraction composition:

CO_2	0.100
O_2	0.070
N_2	0.830
	1.000

Find the gas constant, the molecular mass, and the specific heat at constant pressure of the gas mixture at a temperature of 100°F.

5.18 A tank has a volume of 50 ft^3 and contains 5 lbm of CO at a temperature of 120°F. Molecular oxygen at 120°F flows into the tank until a total pressure of 40 psia is reached. The final temperature of the mixture is also 120°F. Determine the mass fraction composition, the partial pressures of the constituent gases, and the specific heat at constant pressure of the mixture.

5.19 A mixture of one kg of O_2 and two kg of CO is contained in a tank at a pressure of 300 kPa and 20°C. Calculate the tank volume, the partial pressures of the constituent gases, and the volume each constituent gas alone would occupy at 300 kPa and 20°C.

5.20 An insulated container is divided into two sections by a partition. In one section, there is 0.1 cubic meter of molecular nitrogen at 750 kPa and 15°C and in the other there is one kg of molecular hydrogen at 350 kPa and 30°C. After removal of the partition, the two gases mix freely. Find the mass fraction composition, the gas constant, the molecular weight, and the total pressure and temperature of the mixture.

5.21 Show that, at a given temperature and barometric pressure, moist air is always slightly lighter than dry air.

5.22 Three pounds of carbon dioxide (CO_2) at a temperature of 120°F and a pressure of 25 psia are mixed in an adiabatic steady flow device with ten pounds of nitrogen (N_2) at 200°F and 20 psia. Determine the temperature of the mixture. Repeat if the same gases at the same conditions are mixed in a closed container.

5.23 The volumetric composition of a gas mixture is

$$N_2 = 60\%$$

$$O_2 = 15\%$$

$$CO_2 = 20\%$$

$$CO = 5\%$$

Calculate the mass fraction composition, the gas constant, the molecular mass, and the specific heat at constant pressure of the gas mixture.

5.24 Five kilograms of air at 150 kPa, 5°C, and 75 percent relative humidity are mixed with 10 kg of air at 150 kPa, 35°C, and 15 percent relative humidity in a steady flow adiabatic process. What will the mixture temperature be for a mixture pressure of 150 kPa?

5.25 Calculate the quantity of heat that must be added to air at 0°F and 95 percent relative humidity to result in a heated air temperature of 72°F. Also, determine the relative humidity of the heated air.

5.26 What is the relative humidity at which condensation will begin to form on a 10°C cold-water pipe in a 25°C warm room?

5.27 Develop an oxygen psychrometric chart (oxygen and water vapor mixture) for a pressure of 30 psia.

5.28 Determine the pressure and composition of a mixture of refrigerant R-22 and carbon dioxide such that both constituents will condense at a mixture temperature of −20°C.

5.29 If 5 percent of air (by mass) is trapped in a R-22 refrigeration system, what will be the R-22 saturation pressure at a temperature of 70°F?

chapter

6

The Second Law of Thermodynamics

6.1 INTRODUCTION

All thermodynamic processes obey the principle of conservation of energy (the first law of thermodynamics), which states that the total energy of any system and its surroundings is conserved. The second law of thermodynamics deals with the possible direction of a process, and it establishes restrictions with respect to certain processes. Let us briefly discuss the second law with two very simple examples familiar to all of us.

When a glass of ice water is placed in room air at normal temperatures, heat will flow from the air to the ice water, but not from the ice water to the surrounding air. When two pieces of sandpaper are rubbed together, the work used to overcome friction causes an increase in internal energy of the paper, and eventually heat flow to the surroundings. We know intuitively, however, that a process taking place in the reverse direction is impossible; that is, the thermal energy could not be used to generate an equivalent amount of work. Note that for both of the above examples, the process taking place in the reverse direction is allowable from conservation of energy principles. First law considerations do not prevent heat from flowing from ice to air, nor frictional heat from being converted entirely into work. Clearly, another fundamental law of thermodynamics is required to establish the possible direction of a process.

As stated above, the second law places such restrictions on certain processes. For example, engineers are interested in the maximum amount of work that can be obtained from the thermal energy provided by combustion of a fuel. They are also interested in the maximum amount of work that can be obtained from the combustion gases in an internal combustion engine as the piston moves between two positions in the cylinder, the maximum velocity that can be achieved by the expansion of high-pressure gas in a nozzle to ambient pressure, and the minimum amount of work necessary to maintain the cold space of a refrigerator. In the following sections we will show how the second law of thermodynamics helps to provide answers to the above and many other problems.

6.2 REVERSIBLE PROCESSES

The concept of a reversible process is fundamental to an understanding of the second law. A reversible process is one that can be reversed and leave no resultant change in either the system or the surroundings. We will show that the reversible process is an ideal process, with no natural reversible processes. It is an important and useful concept in that the engineer can calculate results and outputs that can be achieved during a reversible process and then strive to approach this upper limit as closely as possible in practice.

We will now present several cases of naturally occurring processes and examine them for reversibility. Consider first a process with *friction*. For example, a block is pushed across a horizontal surface as shown in Figure 6.1 with a force just adequate to overcome the frictional force between the block and the surface. If the block is considered as the system in this process, work is done on the system by the external force; also, as a result of friction between the block and the surface, the block heats up, increasing its internal energy. After the force is removed, heat will flow from the block to the surroundings, the block eventually reaching thermal equilibrium with its surroundings. At the conclusion of the entire process, the temperature of the block will be the same as that at the initial state, namely the temperature of the surroundings. As a result of the process, work has been done on the block and an equivalent amount of heat has flowed to the surroundings, with no internal energy change of the block. To completely reverse the process, heat would have

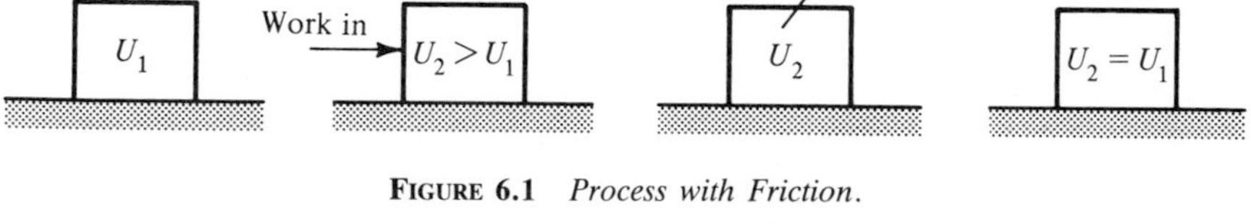

FIGURE 6.1 *Process with Friction.*

to flow into the system and an equivalent amount of work be done by the system. As we have observed, and might have guessed intuitively, such reversal is not possible.

As another example, consider heat transfer through a *finite temperature difference*. Suppose that two blocks, one at temperature T_1 and the other at a lower temperature T_2, are brought into direct contact (Figure 6.2). Heat will flow from the hotter block to the colder block. Now let us try to reverse the process, so that heat will have to flow from the colder block to the hotter block. As we know, this reversal is impossible without external machinery (as in a refrigerator) and work. But external work, from the surroundings, would leave a net change of the surroundings. Heat transfer through a finite temperature difference is thus an irreversible process.

As the temperature difference between the blocks gets less and less, this heat transfer process approaches a reversible one. If heat could be transferred between two bodies with only an infinitesimal temperature difference between them, the process would be reversible. In this situation, the bodies would be at thermal equilibrium with one another. Note that the transfer of a finite amount of heat through an infinitesimal temperature difference is an impractical situation, since infinite time or heat transfer area would be required. However, reversible heat transfer is an ideal situation that we may wish to approach to some degree in practice. It is important to note that a departure from equilibrium yields irreversibility.

A third type of process that we shall consider is the *nonequilibrium expansion of gases* behind a piston in a piston-cylinder arrangement, as shown in Figure 6.3. Gases initially at p_1 ($p_1 \gg p_0$) are allowed to push a weightless piston up to the stops at the top of the cylinder. This is an example of a nonequilibrium, non-quasi-static process (Section 1.3). For the process to be reversed, an external force must push down on the piston and compress the gases back to p_1. But more work must be done during the reverse process (in which the external force must push against

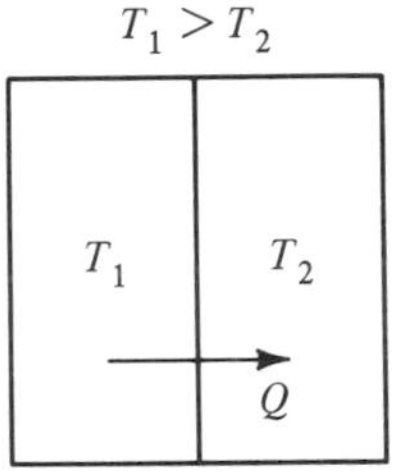

FIGURE 6.2 *Heat Transfer Through Finite Temperature Difference.*

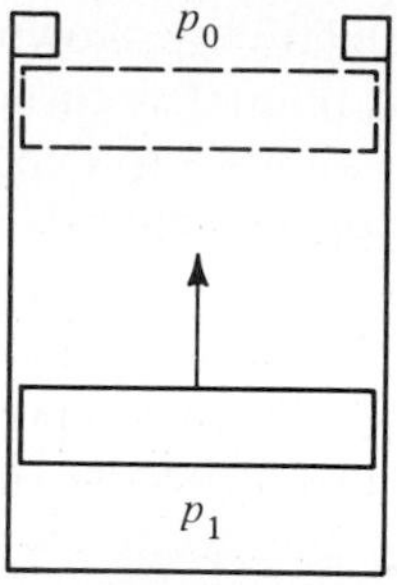

FIGURE 6.3 *Nonequilibrium Expansion.*

gas pressures that go up to p_1) than was done originally, with the cylinder gases pushing against a constant p_0. Since a net amount of work is required from the surroundings, a change is produced in the surroundings, rendering the process irreversible. Again, a departure from equilibrium yields irreversibility.

If the original expansion were quasi-static, the process would be reversible. For example, if weights were placed initially on top of the piston to maintain the pressure pushing down on the piston at p_1, and then if such weights were gradually removed, the difference between the external and internal pressures could be maintained, at least ideally, infinitesimally small. In this case, once the piston reached the top stops in the cylinder, the weights could be restored very gradually and the initial process reversed. Here, the work done by the force of the gases in the initial process would be the same as that done by the external forces in the reverse process, since the forces would be pushing against equivalent pressures.

The examples just given, depicting several processes of a specific nature, can now be generalized. All real processes are irreversible, since they all involve dissipative effects, such as the friction between the block and the surface of the first example, or departures from equilibrium, as with heat transfer through a finite temperature difference or with nonequilibrium expansion. The reversible process represents only a useful ideal; no such process can actually occur.

6.3 REVERSIBLE CYCLES

A *cycle* is a series of processes in which a system starts at a given thermodynamic state and is returned to that exact same state. The cycle is a very important concept for devices or machines that produce power or provide cooling for long periods of time. For example, in the steam power plant for producing electricity, shown schematically in Figure 6.4, high-pressure liquid water at point 1 is turned to hot vapor in the steam generator 1–2. The vapor is allowed to expand through the turbine 2–3, which is doing useful work driving the electric generator. The vapor is liquefied in the condenser 3–4, and the resultant liquid is returned by the pump to its initial state at 1. In this way, the same water flows around and around the cycle, continuing to produce power year after year.

Cycles that function to produce work output are called *power cycles*; cycles that function to pump heat from low- to high-temperature regions are called *refrigeration cycles*. A *heat engine* is a device that operates in a cycle in order to convert heat input into positive net work output. Examples of heat engines are the steam power plant and the internal combustion engine. A *heat pump* and a *refrigerator* are devices that operate in a cycle and transfer heat from a low-temperature region to a higher-temperature region with work input to the system.

There are two statements of the second law of thermodynamics—one that applies to heat engines and one that applies to heat pumps or refrigerators. Just as the first law is a naturally observed law and is not proven, so too these statements of the second law are not proven. Rather, their validity rests on the fact that there has never been an experiment with results that have contradicted them.

The following statement (called the *Kelvin-Planck statement*) applies to heat engines:

It is impossible to construct a cyclic engine that will take heat from a single reservoir and produce an equal amount of work.

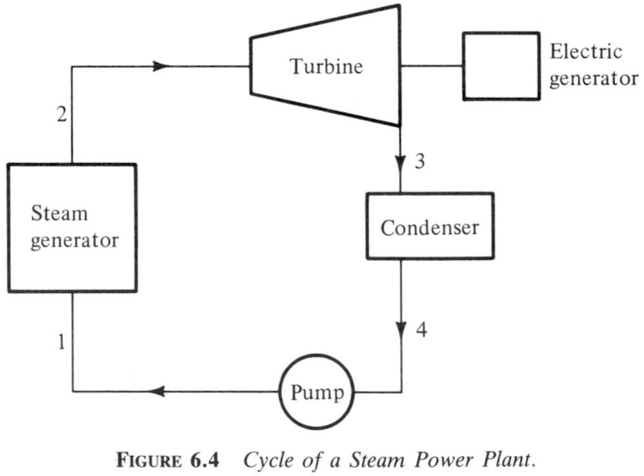

FIGURE 6.4 *Cycle of a Steam Power Plant.*

In other words, during a heat engine cycle, there must be heat rejection from the working fluid. The schematic drawing of a heat engine, shown in Figure 6.5, operating between two thermal reservoirs demonstrates this principle. A thermal reservoir is a body with large heat capacity such that when heat is removed or added to the reservoir, the temperature of the reservoir does not change. In actuality the high-temperature reservoir for a heat engine might be the hot gases generated by a combustion process; the low-temperature reservoir might be the atmosphere or a large body of water.

For a heat engine, we are interested in the amount of useful work that can be obtained for a given heat input. A measure of this conversion is called *cycle thermal efficiency*, η:

$$\eta_{\text{thermal}} = \frac{\text{net } W_{\text{out}}}{Q_{\text{in}}} \tag{6.1}$$

Applying the first law to the heat engine of Figure 6.5 yields, for each cycle,

$$Q_{\text{in}} = W_{\text{out}} + Q_{\text{out}}$$

since the energy E possessed by the system at beginning and end of the cycle must be the same. Therefore,

$$\eta_{\text{thermal}} = \frac{Q_{\text{in}} - Q_{\text{out}}}{Q_{\text{in}}}$$

$$= 1 - \frac{Q_{\text{out}}}{Q_{\text{in}}} \tag{6.2}$$

From the Kelvin-Planck statement of the second law, it follows that Q_{out} cannot be zero; the heat engine of Figure 6.6 is impossible. In other

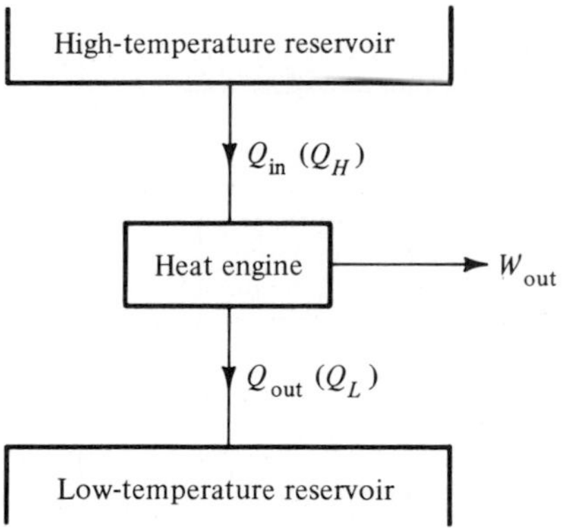

FIGURE 6.5 *Heat Engine.*

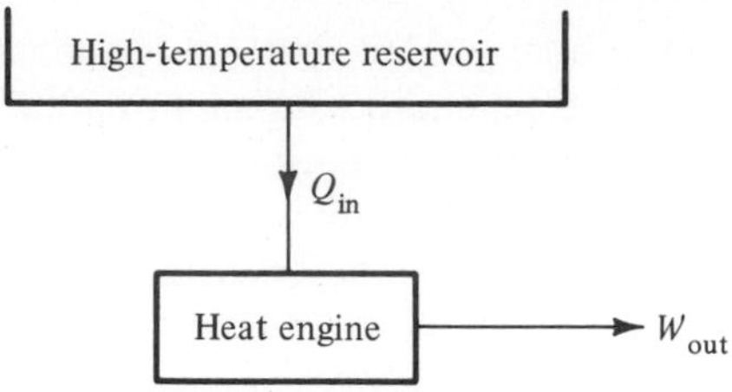

FIGURE 6.6 *Violation of Kelvin-Planck Statement of Second Law.*

words, the efficiency of a heat engine operating in a cycle cannot be 100 percent.

The second statement of the second law (called the *Clausius statement*) applies to heat pumps and refrigerators:

> *It is impossible to construct a cyclic device that will cause heat to be transferred from a low-temperature reservoir to a high-temperature reservoir without the input of work.*

The schematic drawing of a heat pump or refrigerator shown in Figure 6.7 demonstrates this principle. For example, in the household refrigerator, heat is removed from the low-temperature reservoir (inside the refrigerator) and transferred to the warm room air, with input of external electrical energy or work to drive the refrigerator compressor. The performance of a heat pump or refrigerator is measured by the *coefficient of performance* (COP):

$$\text{COP} = \frac{\text{useful output}}{\text{required work input}}$$

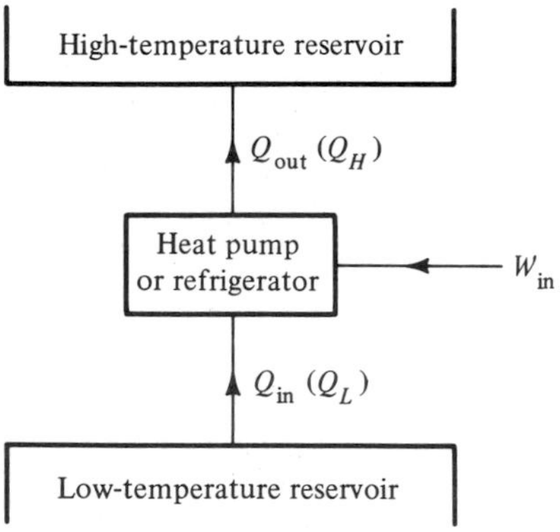

FIGURE 6.7 *Heat Pump or Refrigerator.*

For the refrigerator, useful output is Q_{in}, since it is desired to remove heat from the low-temperature region. Therefore,

$$\text{COP}_{ref} = \frac{Q_{in}}{W_{in}} \tag{6.3}$$

For the heat pump, useful output is Q_{out} since, for example, with a residential heat pump we desire to pump heat from the low-temperature outside air into the warm interior of the house. In this case,

$$\text{COP}_{heat\ pump} = \frac{Q_{out}}{W_{in}} \tag{6.4}$$

For both cases, the Clausius statement implies that the COP cannot be infinite (see Figure 6.8).

In light of the above discussion, we shall now consider two questions: What is the most efficient heat engine operating between two given reservoirs? What heat pump or refrigerator will provide the highest COP operating between two reservoirs?

For maximum efficiency of a heat engine cycle or maximum coefficient of performance of a heat pump or refrigeration cycle, all processes in the cycle must be reversible. From Section 6.2, this requirement implies frictionless motions, constant-temperature heat transfer only, and thermodynamic equilibrium of the system. Such a reversible cycle is the *Carnot cycle*, shown in Figure 6.9. In the *first* process of this cycle, the working fluid receives heat Q_H from the high-temperature reservoir at temperature T_H; during this process, the temperature of the working fluid must remain at T_H to allow for reversible heat transfer. In the *second* process, the temperature of working fluid is decreased to T_L; during this process, heat transfer is not allowed, since heat transfer through a finite

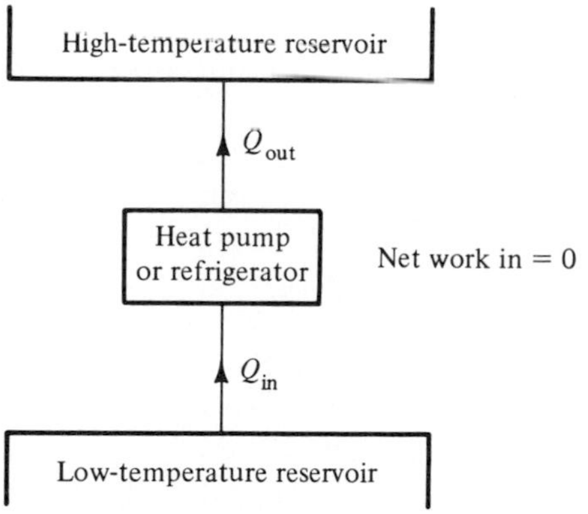

FIGURE 6.8 *Violation of Clausius's Statement of Second Law.*

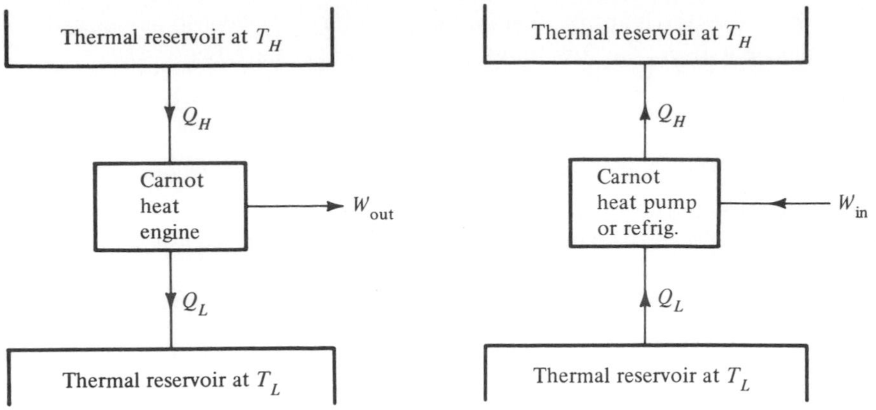

FIGURE 6.9 *Carnot Cycle.*

temperature difference is irreversible. Thus, the second process is adiabatic and reversible. For example, it could consist of expansion of the working fluid in an insulated piston-cylinder arrangement, with the temperature decreasing during this adiabatic process.

In the *third* process of the Carnot cycle, heat Q_L is rejected from the working fluid to the low-temperature reservoir. During this reversible process, the temperature of the working fluid is constant at T_L. Finally, to complete the cycle, the working fluid temperature is raised to T_H in an adiabatic reversible process. This process could consist of the compression of the working fluid in an insulated piston-cylinder arrangement.

A Carnot heat pump or refrigeration cycle operates in the opposite direction of the heat engine; heat is received from the low-temperature reservoir with the working fluid at T_L, and heat is rejected from the working fluid to the high-temperature reservoir at T_H.

Next we will show that all Carnot heat engines operating between the same thermal reservoirs have the same thermal efficiency. To prove this, we will first assume that one Carnot engine is more efficient than another and then show that this situation violates the second law. Consider two Carnot engines CE1 and CE2 operating between thermal reservoirs at T_H and T_L, as shown in Figure 6.10. We start by proposing that CE1 is more efficient than CE2. In other words, if Q_{H1} is set equal to Q_{H2}, it will follow that $W_1 > W_2$ and $Q_{L2} > Q_{L1}$. Since the Carnot cycle is a reversible cycle, we will now reverse CE2 and operate it as a refrigerator or heat pump, as shown in Figure 6.11. With Q_{H2} still equal to Q_{H1}, there is no net heat exchange with the high-temperature reservoir. The net result of the devices taken together is shown in Figure 6.12. But this is a violation of the Kelvin-Planck statement of the second law, since heat from the low-temperature reservoir is being converted entirely into

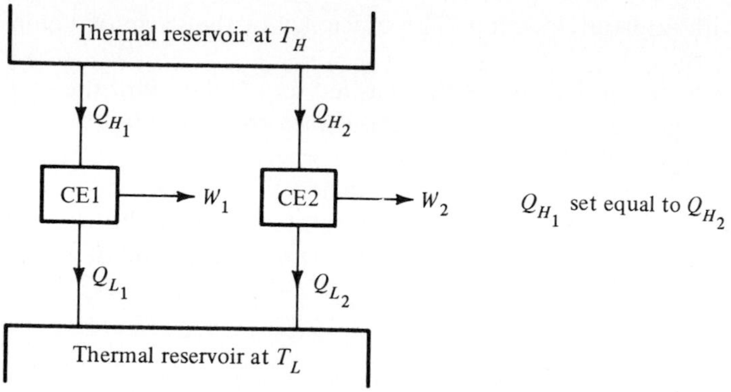

FIGURE 6.10 *Carnot Heat Engines Operating between Same Thermal Reservoirs.*

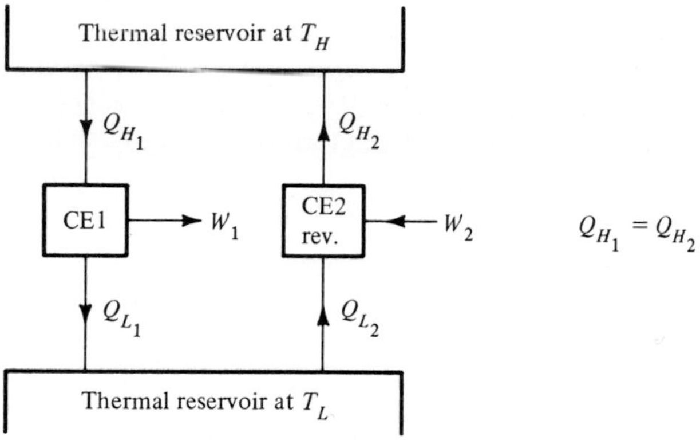

FIGURE 6.11 *Carnot Engine 2 Reversed.*

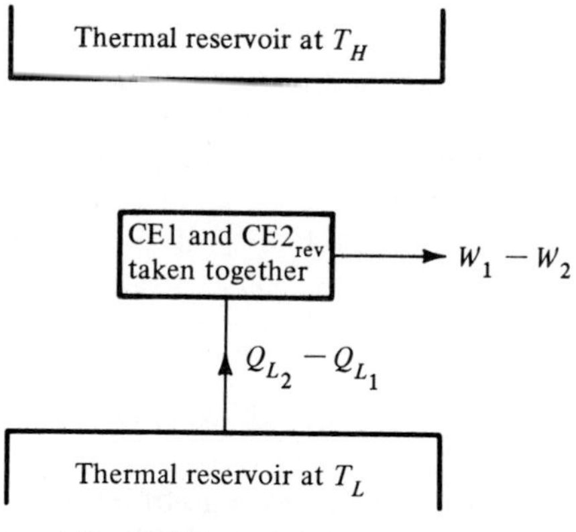

FIGURE 6.12 *Violation of Kelvin-Planck Statement.*

work with no heat rejection. The efficiency of the cycle of Figure 6.12 is 100 percent.

Since our initial proposition has led to a violation of the second law, we must conclude that the proposition was erroneous. In other words, we have the result that all Carnot engines operating between the same two thermal reservoirs have the same efficiency. Likewise, it can be shown that all Carnot heat pumps or refrigerators operating between the same two thermal reservoirs must have the same coefficient of performance.

6.4 THERMODYNAMIC TEMPERATURE SCALE

We have just shown that all Carnot heat engines operating between the same two thermal reservoirs have the same thermal efficiency. Thus, Carnot efficiency is independent of the working fluid used in the cycle. In fact, the efficiency of the Carnot engine can be dependent only on the two reservoirs themselves. But the only characteristic possessed by an infinite thermal reservoir is its temperature. It follows that

$$\eta_{\text{Carnot}} = f_1(T_H, T_L) \tag{6.5}$$

To this point in our study of thermodynamics, we have not defined a *thermodynamic temperature scale*. The zeroth law was useful for comparing temperatures of certain specified substances, and it could be used as a basis for temperature measurement. It is desirable, however, to have an absolute temperature scale independent of any substance. The Carnot engine gives us, at least theoretically, this benefit. For example, let us run a Carnot engine between the temperature of boiling water and the temperature of melting ice, both at atmospheric pressure. The efficiency of such an engine will always be 26.8 percent, independent of the working fluid used in the cycle. In other words,

$$0.268 = f_1(T_{\text{boiling water}}, T_{\text{ice}})$$

It is more convenient to define another function f such that

$$\frac{Q_L}{Q_H} = f\left(\frac{T_L}{T_H}\right)$$

By Equation (6.2), $\eta = 1 - (Q_L/Q_H) = 1 - f(T_L/T_H)$. In our example,

$$\frac{Q_L}{Q_H} = 1 - 0.268 = 0.732 = f\left(\frac{T_{\text{ice}}}{T_{\text{boil}}}\right)$$

The function f is used to define a temperature scale. The simplest case is the Kelvin scale, in which f is defined such that

$$\frac{Q_L}{Q_H} = \frac{T_L}{T_H} \qquad \textbf{(6.6)}$$

For the Kelvin scale, therefore,

$$\frac{T_{\text{ice}}}{T_{\text{boil}}} = 0.732$$

To complete our definition of temperature scale, we must assign a specific number of degrees to the temperature difference between boiling water and ice. For the Kelvin scale, this difference is set at 100°, or $T_{\text{boil}} - T_{\text{ice}} = 100°$. Solving, we have

$$(1 - 0.732)T_{\text{boil}} = 100$$

$$T_{\text{boil}} = 373\text{ K}$$

$$T_{\text{ice}} = 273\text{ K}$$

For the Rankine scale, we again set $Q_H/Q_L = T_H/T_L$, but here we assign a temperature difference of 180° between boiling water and ice. In this case,

$$T_{\text{boil}} = 672°\text{R}$$

$$T_{\text{ice}} = 492°\text{R}$$

Note that it would be possible to define other, possibly nonlinear, temperature scales using Carnot efficiency from Equation (6.5). However, the Kelvin and Rankine temperature scales are those in common usage. For these scales, we have

$$\eta_{\text{Carnot}} = 1 - \frac{T_L}{T_H} \qquad \textbf{(6.7)}$$

EXAMPLE 6.1

An engineer has invented an automobile engine that operates in a closed cycle, with the working fluid receiving heat from combustion gases at 2500°F and rejecting heat to ambient air. He claims that for a steady fuel flow of 10 pounds per hour, his engine can produce 60 horsepower. How do you evaluate this claim? Assume that for each pound of fuel burned, 18,000 Btu of heat are transferred into the working fluid.

Solution

The cycle efficiency claimed by the inventor is

$$\eta = \frac{W_{\text{out}}}{Q_{\text{in}}}$$

$$\text{or} \qquad \eta = \frac{\dot{W}_{\text{out}}}{\dot{Q}_{\text{in}}}$$

$$= \frac{60 \text{ horsepower} \times 2545 \text{ Btu/h-hp}}{10 \text{ lbm/h} \times 18{,}000 \text{ Btu/lbm}}$$

$$= 0.848 \quad \text{or} \quad 84.8\%$$

From the second law, the maximum efficiency for a cylic heat engine operating between two thermal reservoirs is Carnot efficiency, given by Equation (6.7):

$$\eta_{\text{Carnot}} = 1 - \frac{T_L}{T_H}$$

For this case, $\eta_{\text{Carnot}} = 1 - (T_{\text{ambient}}/2960°\text{R})$. For an ambient temperature of 40°F, $\eta_{\text{carnot}} = 1 - (500/2960) = 0.831$. Even for an ambient temperature of 0°F, $\eta_{\text{Carnot}} = 1 - (460/2960) = 0.845$. Since the inventor's claims exceed Carnot efficiency, and since Carnot efficiency itself can be only approached but not equaled in practice, the claims are faulty.

A Carnot heat engine operating in reverse is a *Carnot heat pump* or *Carnot refrigerator*. Again, the relationship $Q_H/Q_L = T_H/T_L$ applies for the Carnot heat pump or refrigerator, with T_H or T_L expressed in degrees Rankine or Kelvin. For any heat pump,

$$\text{COP} = \frac{Q_{\text{out}}}{W_{\text{in}}} = \frac{Q_{\text{out}}}{Q_{\text{out}} - Q_{\text{in}}} = \frac{Q_H/Q_L}{Q_H/Q_L - 1}$$

For a Carnot heat pump,

$$(\text{COP})_{\text{Carnot h.p.}} = \frac{T_H/T_L}{T_H/T_L - 1} = \frac{T_H}{T_H - T_L} \tag{6.8}$$

Likewise, for a refrigerator,

$$\text{COP} = \frac{Q_{\text{in}}}{W_{\text{in}}} = \frac{Q_{\text{in}}}{Q_{\text{out}} - Q_{\text{in}}} = \frac{1}{Q_H/Q_L - 1}$$

For a Carnot refrigerator,

$$(\text{COP})_{\text{Carnot refrig}} = \frac{1}{T_H/T_L - 1}$$

$$= \frac{T_L}{T_H - T_L} \tag{6.9}$$

EXAMPLE 6.2

Determine the maximum possible COP for a cyclic heat pump that is to pump heat from an ambient temperature of 0°C to the interior of a house at 20°C. Repeat for an ambient temperature of −20°C.

Solution

The maximum COP for a heat pump operating between two thermal reservoirs will be that of a Carnot heat pump. From Equation (6.8),

$$(\text{COP})_{\text{Carnot h.p.}} = \frac{T_H}{T_H - T_L}$$

For an ambient temperature of 0°C,

$$(\text{COP})_{\text{Carnot h.p.}} = \frac{293}{20-0} = 14.65$$

For an ambient temperature of −20°C,

$$(\text{COP})_{\text{Carnot h.p.}} = \frac{293}{20-(-20)} = 7.33$$

Once again, the COP of a Carnot heat pump can be only approached in practice, never reached. It is interesting to note that the rapid decrease of COP with ambient temperature for the Carnot heat pump is a characteristic or current commercially available heat pumps, which do not work on the Carnot cycle. Thus in cold climates, heat pumps for residential heating must be supplemented with heat from other sources, such as electrical heat.

6.5 CLAUSIUS' INEQUALITY

So far, our discussion of the second law has included reversible processes and reversible cycles. However, as engineers, we would like to have a quantitative measure of irreversibility in a process or cycle. In other words, we would like to establish a thermodynamic property from the second law, just as internal energy and enthalpy were obtained from first law considerations.

For a system undergoing a cycle, the value of each thermodynamic property at the endpoint must be the same as that at the initial point. Consider the Carnot heat engine cycle; characteristics of this cycle, as described in Sections 6.3 and 6.4, are the reservoir temperatures, T_H and T_L, and heat input and heat rejection, Q_H and Q_L. Now we will establish a thermodynamic property from these characteristics.

If we integrate heat transfer Q around the cycle, with Q_{in} positive, we get $\oint dQ > 0$, since $\oint dQ = Q_H - Q_L$ must be greater than zero if there is to be a positive work output.[1] Thus, as we are already aware, heat is *not* a thermodynamic property. However, for the Carnot cycle, if we consider the ratio of heat transfer to temperature and evaluate this

1. The symbol $\oint$ denotes an integral taken over all processes comprising the cycle.

ratio around the cycle, we obtain

$$\frac{Q_H}{T_H} - \frac{Q_L}{T_L} = 0$$

In other words, at least for the Carnot cycle, this ratio does have the property that its value is the the same as the initial points and the endpoints of the cycle. But what about other reversible cycles? In general, a reversible cycle may include heat input and heat rejection from and to many different thermal reservoirs. Consider the general reversible cycle shown in Figure 6.13 on *T-v* coordinates. Also shown is a Carnot cycle appearing on the same coordinates.

Now let us superimpose on the general reversible cycle a series of Carnot cycles, as shown in Figure 6.14. The adjacent adiabatic processes of the cycles coincide, so that the work of one cycle should, at least approximately, cancel the work of the adjacent cycle, since the adjacent processes occur in opposite directions. The greater the number of Carnot cycles used, the closer the result comes to the general reversible cycle. For each of the Carnot cycles, we have $Q_H/T_H - Q_L/T_L = 0$; or, following our sign convention of Q_{in} as positive heat transfer and Q_{out} as negative, we obtain

$$\sum \frac{Q}{T} = 0$$

where the summation is taken over all the Carnot cycles used to approximate the general reversible cycle.

If we were to use an infinite number of Carnot cycles, each involving infinitesimal heat transfer with thermal reservoirs, we could

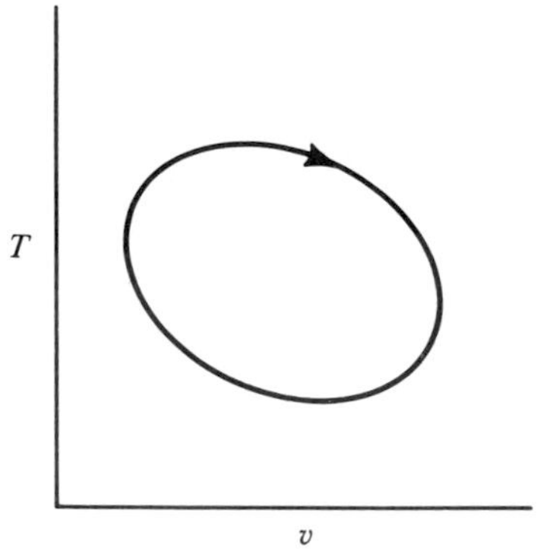

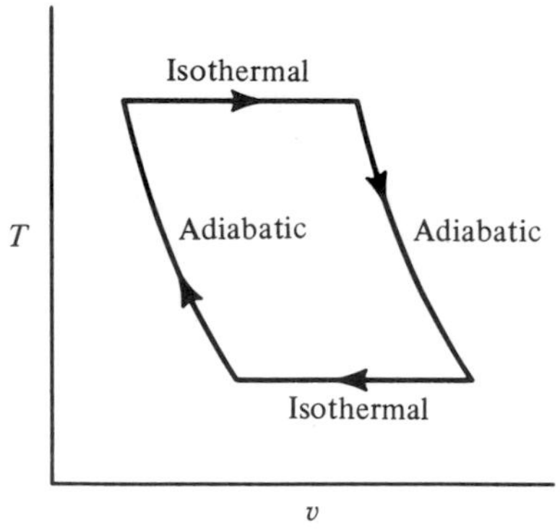

FIGURE 6.13 *T-v Diagram for Reversible Cycles.*

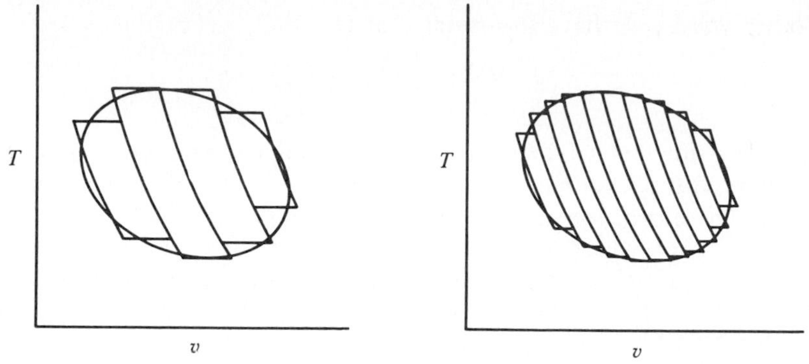

(a) Approximation of general reversible cycle with four Carnot cycles

(b) Closer approximation of general reversible cycle

FIGURE 6.14

duplicate the reversible cycle, and we would have, as a result,

$$\oint \left(\frac{dQ}{T} \right)_{\text{rev}} = 0 \tag{6.10}$$

where the subscript "rev" denotes "reversible cycle."

Now let us examine $\oint dQ/T$ for an irreversible cycle. Consider a reversible Carnot heat engine and an irreversible cyclic heat engine operating between the same two thermal reservoirs, one at T_H and the other at T_L, as shown in Figure 6.15. We will operate the two engines with the same energy input Q_H, so that $Q_{HC} = Q_{HI}$. The efficiency of the Carnot engine is greater than that of the irreversible engine, so that

$$W_C > W_I \quad \text{or} \quad Q_{HC} - Q_{LC} > Q_{HI} - Q_{LI}$$

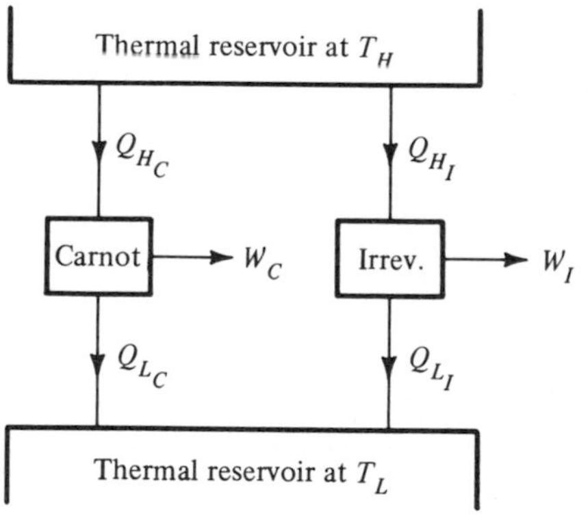

FIGURE 6.15

In other words, we have the result that $Q_{LI} > Q_{LC}$. For the cycle,

$$\oint \left(\frac{dQ}{T}\right)_I = \frac{Q_{HI}}{T_H} - \frac{Q_{LI}}{T_L}$$

since here the irreversible engine is exchanging heat with only two reservoirs. But $\oint (dQ/T)_C = Q_{HC}/T_H - Q_{LC}/T_L = 0$, as has been demonstrated. Therefore, for the irreversible regime,

$$\oint \left(\frac{dQ}{T}\right)_I < 0 \qquad \textbf{(6.11)}$$

since we have $Q_{HI}/T_H = Q_{HC}/T_H$.

Equation (6.11) can be taken as a general statement, true for any irreversible cycle operating between any number of thermal reservoirs. We have

$$\oint \frac{dQ}{T} \overset{\text{irrev}}{\underset{\text{rev}}{\leq}} 0 \qquad \textbf{(6.12)}$$

where the equality is true for a reversible cycle, and the inequality for an irreversible cycle. Equation (6.12) is called *Clausius' inequality*.

6.6 ENTROPY

We will now derive the thermodynamic property *entropy* from Clausius' inequality. Consider the reversible cycle shown on thermodynamic coordinates in Figure 6.16. The system proceeds from state 1 to state 2 along reversible path R_A, and it returns to its initial state along reversible path

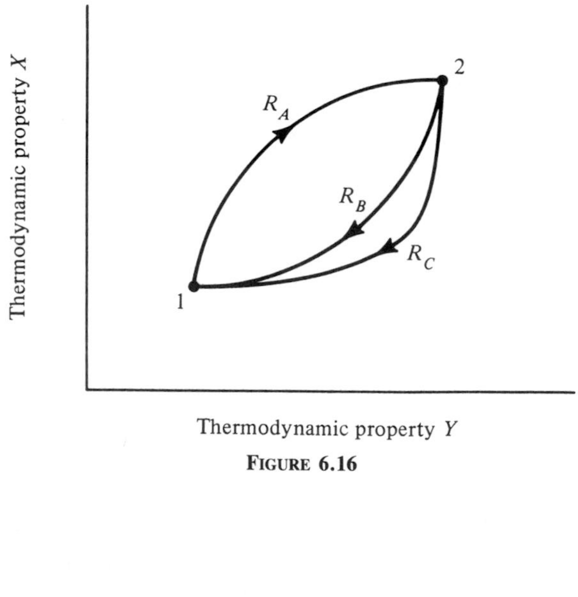

FIGURE 6.16

R_B. From Equation (6.12),

$$\int_1^2 \left(\frac{dQ}{T}\right)_{R_A} + \int_2^1 \left(\frac{dQ}{T}\right)_{R_B} = 0$$

Now consider another reversible cycle in which we again proceed from state 1 to state 2 along path R_A, but return from 2 to 1 along reversible path R_C. In this case,

$$\int_1^2 \left(\frac{dQ}{T}\right)_{R_A} + \int_2^1 \left(\frac{dQ}{T}\right)_{R_C} = 0$$

Subtracting the above results, we obtain

$$\int_2^1 \left(\frac{dQ}{T}\right)_{R_B} = \int_2^1 \left(\frac{dQ}{T}\right)_{R_C}$$

We could repeat the above exercise with reversible paths R_D, R_E, and so on, between 2 and 1, but we would always end up with the same result, namely, that

$$\int_2^1 dQ/T$$

is the same for all reversible paths between 2 and 1. Since the value of the integral is independent of the path taken between two state points, it follows that $\int (dQ/T)_{\text{rev}}$ is a thermodynamic property. We call this property *entropy* S:

$$S_2 - S_1 = \int_1^2 \left(\frac{dQ}{T}\right)_{\text{rev}} \tag{6.13}$$

If we follow an irreversible path between state 1 and state 2 and return from 2 to 1 along a reversible path (Figure 6.17), it follows that, from Clausius' inequality,

$$\int_1^2 \left(\frac{dQ}{T}\right)_{\text{irrev}} + \int_2^1 \left(\frac{dQ}{T}\right)_{\text{rev}} < 0$$

or, by the definition of entropy,

$$\int_1^2 \left(\frac{dQ}{T}\right)_{\text{irrev}} + S_1 - S_2 < 0$$

Rearranging, we obtain, for an irreversible process,

$$S_2 - S_1 > \int_1^2 \frac{dQ}{T} \tag{6.14}$$

We will now look at the meaning of *entropy* with respect to the reversible process and the Carnot cycle. Then, in the next section, we will examine entropy and irreversibility.

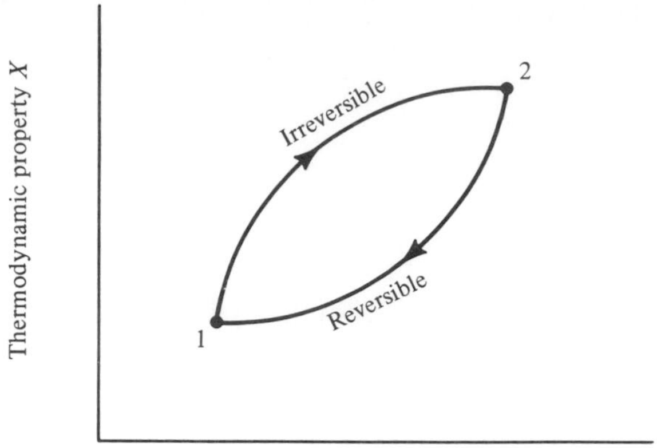

FIGURE 6.17

We can see from the definition of entropy that for a reversible process, $dQ = T\,dS$, or $Q = \int T\,dS$. Thus, for a reversible thermodynamic process, the area under a curve on a *T-S* diagram represents heat transfer. For process 1–2 on the *T-S* diagram of Figure 6.18, the shaded area represents heat transferred to the system.

Let us now return to the Carnot cycle and examine the significance of entropy with respect to this reversible cycle. In the first process of the cycle (Section 6.3), the working fluid receives heat Q_H reversibly from the high-temperature reservoir, with the working fluid remaining at T_H during the process (see 1–2 of Figure 6.19). From Equation (6.13), for a

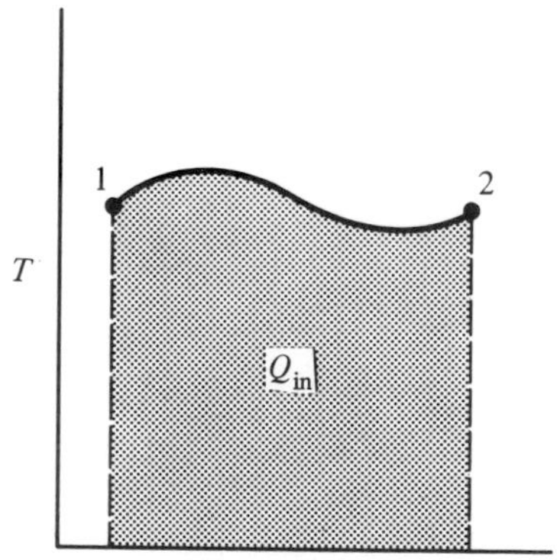

FIGURE 6.18 *T-S Diagram for Reversible Process.*

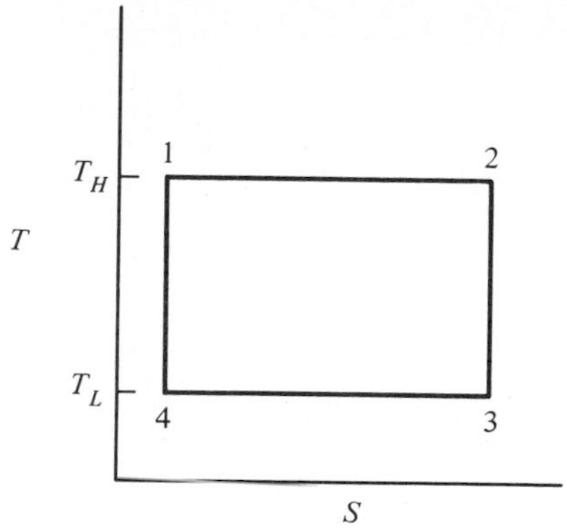

FIGURE 6.19 *T-S Diagram for Carnot Cycle.*

reversible process,

$$S_2 - S_1 = \int_1^2 \frac{dQ_{\text{in}}}{T} \quad \text{or} \quad S_2 - S_1 = \frac{Q_H}{T_H}$$

In the second process, the temperature of the working fluid is decreased to T_L, with the process occurring reversibly and adiabatically. For a reversible adiabatic process, from Equation (6.13), the entropy remains constant, or $S_2 = S_3$. The reversible adiabatic process is therefore called an *isentropic process*. In the third process, heat Q_L is rejected reversibly to the low-temperature reservoir, with the working fluid temperature constant at T_L, as shown by process 3–4. Finally, to complete the cycle, the working fluid is raised in temperature, reversibly and adiabatically, to T_H (process 4–1). The Carnot cycle appears as a rectangle on a *T-S* diagram, as Figure 6.19 shows.

In Chapter 10, when we discuss various types of power and refrigeration cycles, we will examine these cycles on temperature-entropy diagrams, thereby enabling a comparison with the ideal Carnot cycle.

6.7 ENTROPY AND IRREVERSIBILITY

In Section 6.1, we introduced the concepts of reversible and irreversible processes; in Section 6.6, we defined the thermodynamic property *entropy*. Now we will tie these concepts together and show how entropy can be used to indicate the degree of irreversibility in a thermodynamic process.

From Equations (6.13) and (6.14), we have

$$S_2 - S_1 \underset{\text{rev}}{\geq} \int_1^2 \frac{dQ_{\text{in}}}{T}$$

where the equality holds for a reversible process and the inequality for an irreversible process. This equation can be rewritten as

$$S_2 - S_1 = \left[\int_1^2 \frac{dQ_{\text{in}}}{T}\right] + [\text{positive term due to irreversibility}] \quad \textbf{(6.15)}$$

To illustrate the application of Equation (6.15), consider an example of an irreversible process. Oil in the insulated beaker shown in Figure 6.20 is being stirred by a paddle wheel. Let the oil be the thermodynamic system to be analyzed. Due to viscous, frictional forces between the paddle wheel and the oil, the temperature of the oil increases, so that temperature T_2 at the end of the process is greater than the initial temperature T_1. Since dissipative effects have caused the temperature increase, this process is irreversible.

We could take the same oil at temperature T_1 and have it attain the same final state with temperature T_2 as before, but by a different means. Place a large block with temperature $T_1 + \Delta T$ in contact with the oil at T_1, as shown in Figure 6.21. After the oil has been heated to $T_1 + \Delta T$, place another block at a slightly higher temperature in contact with the oil. Continue this procedure until the oil reaches temperature T_2. With no change of pressure, state 2 of the oil is exactly the same as state 2 of the oil in the paddle wheel process. We have now reached state 2 by a process that at least approaches a reversible process. For reversible heat

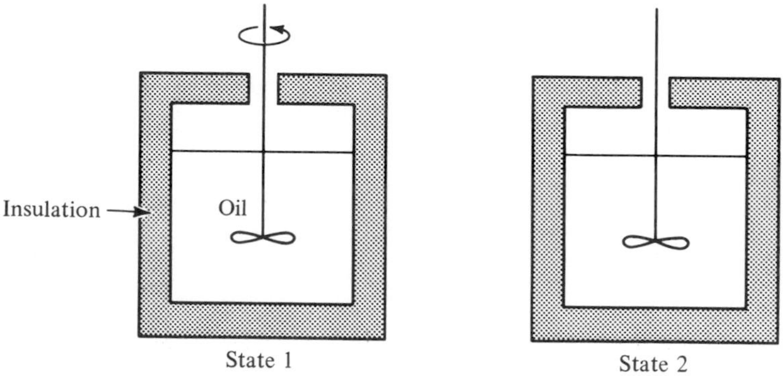

$T_2 > T_1$
$Q = 0$ (container insulated)

FIGURE 6.20

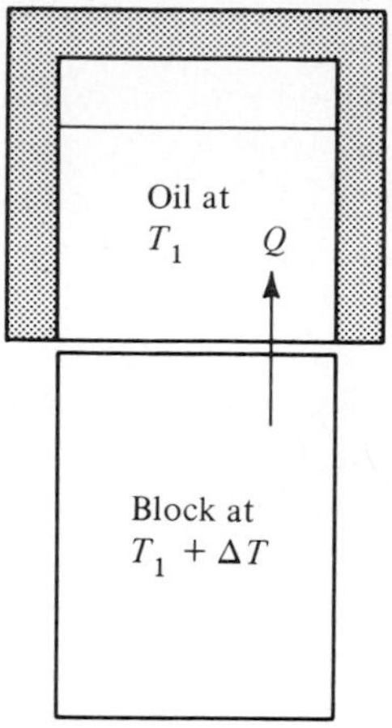

Heat transfer from block to oil

FIGURE 6.21

transfer, an infinite number of blocks would be necessary to allow for heat transfer only across an infinitesimally small temperature difference. Even though the two processes between states 1 and 2 are entirely different, the initial and final states are the same, so the change in any thermodynamic property evaluated between states 1 and 2 must be the same for the two processes.

To evaluate the entropy change of the oil between states 1 and 2, we would take the second process, which approaches a reversible process, determine the quantities of thermal energy added to the oil as a function of oil temperature, and evaluate the integral

$$\int_1^2 \frac{dQ_{\text{in}}}{T} \qquad \text{to obtain } S_2 - S_1$$

Since the processes have the same end states, the entropy change of the oil for the irreversible paddle wheel process must be the same as that of the reversible process. With no heat transfer for the paddle wheel process,

$$S_2 - S_1 > \int_1^2 \frac{dQ_{\text{in}}}{T}$$

since

$$\int_1^2 \frac{dQ_{\text{in}}}{T} = 0$$

for this process. In other words, the increase of entropy for the paddle wheel process is due to the fact that the process is irreversible.

From Equation (6.15), it can be seen that the entropy of a system can increase during a thermodynamic process due to heat transfer to the system or due to irreversibility. In other words, an increase of entropy of a system during a process does not necessarily indicate that the process is irreversible. However, by considering the entropy change of the system and its surroundings during a process, we can use entropy change as a measure of irreversibility. For the example just discussed, oil being stirred by a paddle wheel, with the oil increased in temperature due to viscous dissipation, there is no heat transfer to the surroundings since the beaker is insulated. Thus there is no change in either the temperature or any other thermodynamic property of the surroundings, so that $S_2 - S_1$ of the surroundings is zero. Adding the entropy change of the system to that of the surroundings, we have the result that the entropy change of the universe due to this irreversible process is positive, where *universe* is defined as the system and its surroundings taken together.

For the case of reversible heat transfer to the oil, with the oil heated by an infinite number of blocks, heat is transferred from the surroundings (blocks) to the system (oil). Assuming reversible heat transfer, the entropy of the system increases, but the entropy of the blocks decreases. During each incremental heat transfer, the block temperature is only infinitesimally greater than the oil temperature, so the entropy increase of the oil,

$$(S_2 - S_1)_{\text{oil}} = \int_1^2 \frac{dQ_{\text{to oil}}}{T_{\text{oil}}}$$

is balanced by the entropy decrease of the blocks,

$$\int_1^2 \frac{dQ_{\text{from blocks}}}{T_{\text{blocks}}}$$

Thus the total change of entropy of the universe (the system plus its surroundings) for the reversible process is zero. It appears that during a reversible process, the entropy of the universe remains constant. During an irreversible process, the entropy of the universe increases.

Let us now consider another example of the use of entropy to indicate irreversibility. Bring a hot block at temperature T_H in contact with a cold block of the same mass and specific heat at temperature T_L (Figure 6.22). Assume that the blocks are completely isolated from the surroundings, so that the only heat transfer is between the blocks. Left in contact, the blocks eventually reach a common final temperature T_2 halfway between T_L and T_H. The heat transfer process is irreversible since there is a finite temperature difference between the blocks during the process. To calculate the entropy change of the hot block, we must calculate the value of $\int dQ/T$ for a reversible heat transfer process between the initial and final states.

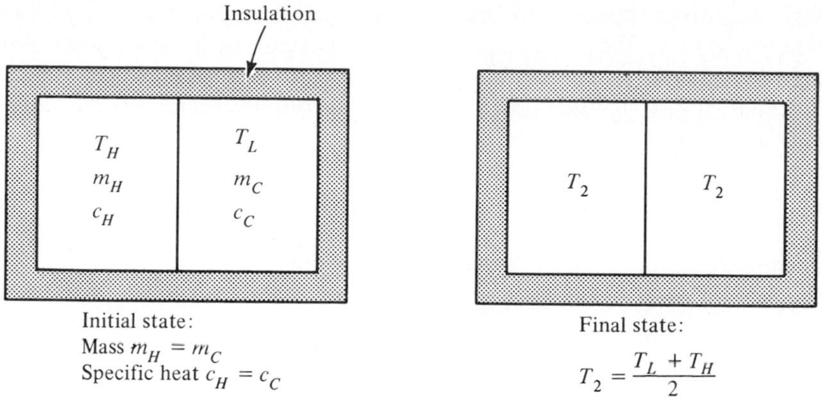

FIGURE 6.22

For the reversible process, bring a large body (a heat sink) at a temperature slightly less than T_H into contact with the hot block, and allow the heat sink to absorb an increment of heat from the block (Figure 6.23), lowering very slightly the block temperature. Now bring another heat sink into contact with the block, with this heat sink again at a slightly lower temperature than the block temperature. Repeat, gradually reducing the temperature of the heat sink and thus the temperature of the block. Eventually, heat sink and block temperatures are reduced to T_2. For an infinite number of heat sinks, each absorbing heat across an infinitesimally small temperature difference from the block, the process is reversible. The entropy change of the hot block during this reversible process is

$$(S_2 - S_1)_{\text{hot}} = \int_{T_H}^{T_2} \frac{dQ}{T} = m_H c_H \int_{T_H}^{T_2} \frac{dT}{T}$$

$$= m_H c_H \ln \frac{T_2}{T_H}$$

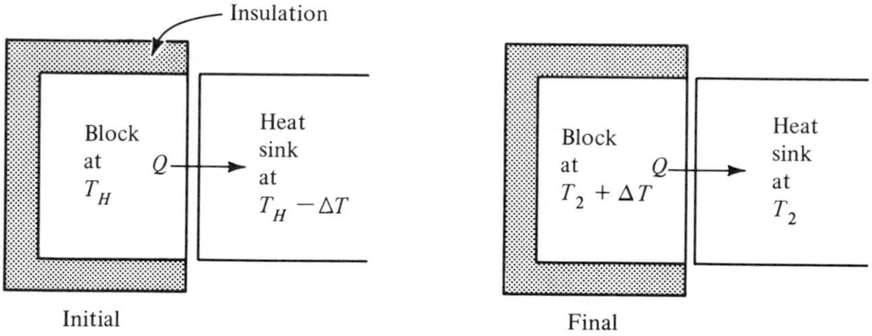

FIGURE 6.23

where m_H is the mass of the hot block, c_H is the specific heat of the block, and $m_H c_H\, dT$ is the infinitesimal heat carried away by each heat sink. Since this reversible process has been carried out between the same initial and final states as the actual process, it follows that, for the actual process,

$$(S_2 - S_1)_{\text{hot}} = m_H c_H \ln \frac{T_2}{T_H}$$

For the cold block, following similar reasoning,

$$(S_2 - S_1)_{\text{cold}} = m_C c_C \ln \frac{T_2}{T_L}$$

The total entropy change of the universe for this irreversible process is

$$(S_2 - S_1)_{\text{univ}} = m_C c_C \ln \frac{T_2}{T_L} + m_H c_H \ln \frac{T_2}{T_H} = mc \ln \frac{T_2^2}{T_L T_H}$$

where $m_c = m_H c_H = m_C c_C$ since the two blocks have the same mass and specific heat. Let $\Delta T = T_2 - T_L = T_H - T_2$ for this case in which $T_2 = (T_L + T_H)/2$. Substituting, we have

$$(S_2 - S_1)_{\text{univ}} = mc \ln \frac{T_2^2}{(T_2 - \Delta T)(T_2 + \Delta T)}$$

or $$(S_2 - S_1)_{\text{univ}} = mc \ln \frac{T_2^2}{T_2^2 - \Delta T^2}$$

It is clear that $(S_2 - S_1)_{\text{univ}}$ is positive. In other words, once again there is an increase in entropy of the universe due to this irreversible process of heat transfer across a finite temperature difference.

We have demonstrated another statement of the second law of thermodynamics:

> *During a thermodynamic process, the entropy of the universe can either increase or, in the limit of a reversible process, remain the same. The increase in entropy of the universe occurs due to irreversibility, brought about by factors such as dissipation and departure from thermodynamic equilibrium. It is impossible for any thermodynamic process to occur which involves a decrease in entropy of the universe.*

For a *closed* system, in which a mass undergoes a thermodynamic process that takes it from state 1 to state 2 (Figure 6.24), the change in entropy of the universe is simply

$$(S_2 - S_1)_{\text{univ}} = (S_2 - S_1)_{\text{system}} + (S_2 - S_1)_{\text{surround}} \tag{6.16}$$

Closed system at State 1 Closed system at State 2

FIGURE 6.24

For an *open* system, as shown in Figure 6.25, the effects of irreversibility and heat transfer across system boundaries during a process can contribute to an increase in entropy of the fluid passing through the system, as well as an increase in entropy of the mass within the system. In other words, for the open system, with s = entropy per unit mass,

$$(S_2 - S_1)_{\text{univ}} = m_e s_e - m_{\text{in}} s_{\text{in}} + (S_2 - S_1)_{\text{within system boundaries}} + (S_2 - S_1)_{\text{surround}} \tag{6.17}$$

For a *steady-flow* process, in which system properties are not a function of time, $(S_2 - S_1)_{\text{within system boundaries}} = 0$, and $m_e = m_{\text{in}} = m$, so that

$$(S_2 - S_1)_{\text{univ}} = m(s_e - s_{\text{in}}) + (S_2 - S_1)_{\text{surround}} \tag{6.18}$$

For a steady-flow, adiabatic reversible process, it follows that $s_e = s_{\text{in}}$.

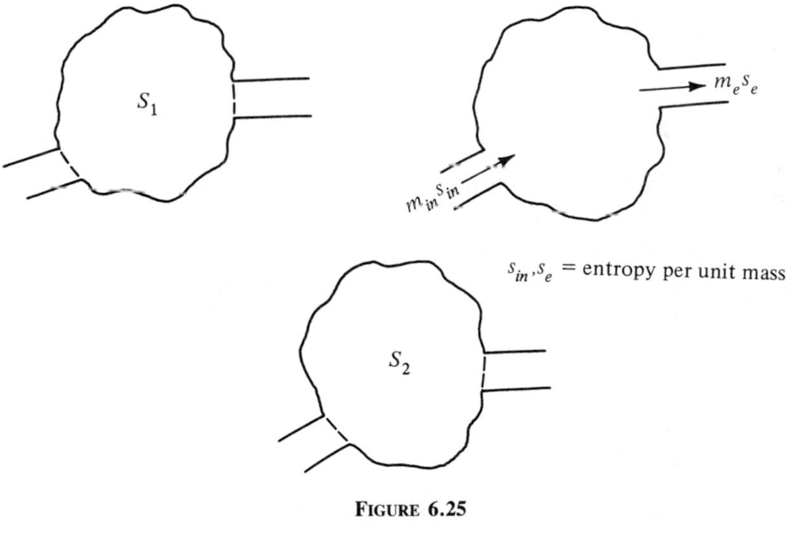

FIGURE 6.25

Since the increase in entropy of the universe measures irreversibility or losses, it follows that the ideal process is the reversible one. The maximum velocity that can be attained by the expansion of a flow from high pressure through an adiabatic nozzle will occur when the nozzle flow is reversible. The maximum work that can be obtained from the expansion of steam in an insulated turbine will occur during a reversible process. Thus, the second law tells us the best possible output from a thermodynamic process; we can hope to approach this ideal case, but we can never quite reach it since no actual process is completely reversible. The use of entropy is important to the engineer in helping to predict quantitatively the best possible output for a given system. A velocity through the nozzle greater than that of a reversible process would violate the second law and be impossible; a work output of the turbine greater than for a reversible turbine would similarly violate the second law. With good design, however, the engineer can produce a nozzle that will achieve over 95 percent of the velocity predictable from a second law analysis of a reversible process, and also a turbine that will yield over 90 percent of the work of a reversible turbine.

Thus, the second law places an upper limit on a thermodynamic process. In the following sections, we will show how entropy can be calculated and tabulated for substances, and then how entropy is used to determine this upper limit for a process.

The first and second laws of thermodynamics have now been stated. Note that we did not prove either of these laws. Rather, since no violation of either has been observed, we accept them as fundamental laws of nature.

In the next section, we will show how changes of entropy can be numerically evaluated for pure substances between given states. Then we will be able to tabulate entropy values and use entropy in the solution of engineering problems.

6.8 NUMERICAL EVALUATION OF ENTROPY

1. Entropy Change of a Pure Substance

Since entropy is a thermodynamic property of a substance, the entropy change occurring in a process is a function of the end states of the process. In other words, in evaluation of $S_2 - S_1$, how we get from state 1 to state 2 makes no difference. For a reversible process, the entropy change is given by Equation (6.13):

$$S_2 - S_1 = \int_1^2 \left(\frac{dQ_{\text{in}}}{T}\right)_{\text{rev}}$$

Note that in the evaluation of entropy change, only that part of the energy which is transferred in the form of heat during the reversible process enters into the calculation. Thus, if no heat is transferred in the reversible process from state 1 to state 2, the change of entropy will be zero, although there may be transfer of energy in the form of work.

For a reversible process taking place in a piston-cylinder arrangement, the work done going from state 1 to state 2 is, from Equation (2.3):

$$W_{\text{out}} = +\int_1^2 p\, d\forall$$

The energy equation (2.5) for a closed system becomes, in differential form,

$$dQ_{\text{in}} - dW_{\text{out}} = dU \quad \text{or} \quad dQ_{\text{in}} = dU + p\, d\forall \tag{6.19}$$

while the expression for the entropy change for a reversible process for a closed system is, in differential form,

$$dS = \frac{dQ_{\text{in}}}{T}$$

Substituting for dQ in Equation (6.19), we get

$$T\, dS = dU + p\, d\forall$$

or, on a per unit mass basis with $s = S/\text{mass}$,

$$T\, ds = du + p\, dv \tag{6.20}$$

where s is expressed in Btu/lbm°R or J/kg K.

Equation (6.20) has been derived for a specific type of process, namely, a reversible one taking place in a piston-cylinder arrangement. However, note that this equation involves only thermodynamic properties, which are independent of the path taken between thermodynamic states. It follows that Equation (6.20) must hold for any process, reversible or irreversible, in a closed or an open system. This equation is thus extremely important for evaluating entropy changes.

Since $h = u + pv$, it is sometimes convenient to rewrite Equation (6.20) as

$$T\, ds = dh - v\, dp \tag{6.21}$$

Again, Equation (6.21) involves only thermodynamic properties and is a general equation that can be used for evaluating entropy change. We will now work some examples that illustrate the evaluation of entropy change in several of the regions of thermodynamic behavior discussed in Chapter 4.

In the ideal gas region, we have the equation of state $pv = RT$, and the result that the internal energy and enthalpy are functions of temperature only. Simplifying Equation (6.20) with $du = c_v\, dt$, we obtain

$$ds = \frac{c_v\, dT}{T} + R\,\frac{dv}{v} \tag{6.22}$$

or $$s_2 - s_1 = \int_1^2 c_v(T)\,\frac{dT}{T} + R \ln \frac{v_2}{v_1}$$

For the constant c_v case (c_v independent of temperature), we can simplify Equation (6.22) by further integration:

$$s_2 - s_1 = c_v \ln \frac{T_2}{T_1} + R \ln \frac{v_2}{v_1} \tag{6.23}$$

Similarly, for an ideal gas, Equations (6.21) simplifies to

$$s_2 - s_1 = \int_1^2 c_p(T)\,\frac{dT}{T} - R \ln \frac{p_2}{p_1} \tag{6.24}$$

or, for constant c_p,

$$s_2 - s_1 = c_p \ln \frac{T_2}{T_1} - R \ln \frac{p_2}{p_1} \tag{6.25}$$

In a two-phase region, for example, the liquid-vapor region, we choose a reversible constant temperature and pressure process for evaluating entropy. Equation (6.20) can be integrated directly to obtain

$$s_2 - s_1 = (u_2 - u_1)/T + p(v_2 - v_1)/T = (h_2 - h_1)/T \tag{6.26}$$

EXAMPLE 6.3

Determine the entropy change when water is evaporated from a saturated liquid to a saturated vapor at a temperature of 430°F. The saturation pressure is 343.674 psia, and $h_{fg} = 796.0$ Btu/lbm.

Solution

In this case, state 1 is f and state 2 is g. Thus, from Equation (6.26)

$$s_2 - s_1 = s_g - s_f = s_{fg} = (h_g - h_f)/T_{\text{sat.}} = h_{fg}/T_{\text{sat.}}$$

$$= \frac{796.0 \text{ Btu/lbm}}{(430 + 459.69)\text{°R}}$$

$$= 0.8947 \text{ Btu/lbm°R}$$

In the liquid-vapor region, the enthalpy change of a mixture of liquid and vapor is given as $h - h_f = xh_{fg}$, where x is the quality of the mixture.

Thus the entropy change of a mixture is

$$s - s_f = (h - h_f)/T = xh_{fg}/T$$
$$\text{or} \qquad s - s_f = xs_{fg} \tag{6.27}$$

Similarly, in the solid-vapor region,

$$s - s_i = xs_{ig} \tag{6.28}$$

In the superheated vapor region, we can utilize a constant temperature or constant pressure process, since the other thermodynamic properties have been given as a function of temperature and pressure. For a constant temperature process, for example, we have, from Equation (6.20),

$$s_2 - s_1 = (u_2 - u_1)/T + \left(\int_1^2 p\,dv\right)\Big/T$$

or, since from integral calculus $p\,dv = d(pv) - v\,dp$,

$$s_2 - s_1 = (u_2 - u_1)/T + (p_2v_2 - p_1v_1)/T - \left(\int_1^2 v\,dp\right)\Big/T$$
$$s_2 - s_1 = (h_2 - h_1)/T - \left(\int_1^2 v\,dp\right)\Big/T \tag{6.29}$$

EXAMPLE 6.4

Water is heated from a saturated vapor at 430°F at constant temperature until a pressure of 240 psia is reached. Find the change in entropy. The saturation pressure is 343.647 psia, with $v_g = 1.3496\ \text{ft}^3/\text{lbm}$ and $h_g = 1203.9$ Btu/lbm. At 430°F and 240 psia, $h = 1222.3$ Btu/lbm.

Solution

Since this is a constant temperature process, Equation (6.29) is applicable. In order to evaluate the $\int v\,dp$ integral, we need to have the v-p relationship. From the Steam Tables, we find the specific volume of superheated steam at pressure increments of 10 psi at a temperature of 430°F. The values are given in the following table:

Pressure (psia)	Specific volume (ft^3/lbm)	Pressure (psia)	Specific volume (ft^3/lbm)
340	1.3668	280	1.7081
330	1.4153	270	1.7794
320	1.4667	260	1.8561
310	1.5214	250	1.9388
300	1.5796	240	2.0283
290	1.6417		

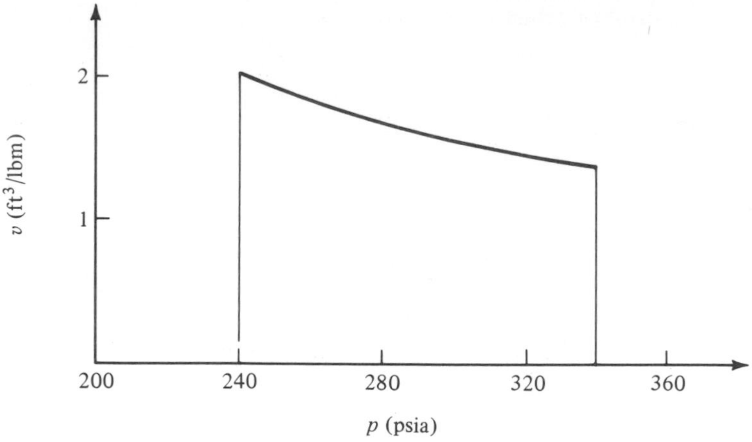

FIGURE 6.26

The v-p relation for steam at 430° is shown graphically in Figure 6.26. To evaluate the $\int v\,dp$ integral, we need to find the area under the curve.

By numerical integration, it is found that the area = −31.6501 Btu/lbm. Therefore, from (6.29),

$$s_2 - s_g = \frac{1222.3\ \text{Btu/lbm} - 1203.9\ \text{Btu/lbm}}{889.7°\text{R}} - \frac{(-31.6501)}{889.7°\text{R}}$$

$$= 0.02068\ \text{Btu/lbm°R} + 0.03557\ \text{Btu/lbm°R}$$

$$= 0.05625\ \text{Btu/lbm°R}$$

Using the procedures discussed above, one can calculate the entropy change of a pure substance as it passes from one state to another. Complete entropy tables for water, carbon dioxide, R-22, and dry air are presented in the appendices. In order to present such tables or values of entropy corresponding to a given state, it is necessary to select a datum and then express values of entropy relative to that datum. The values of entropy s tabulated in the appendices actually represent the difference between s at a given state and s at the datum. For water, the entropy datum used in the tables is $s = 0$ for saturated liquid water at 0°C or 32°F. For carbon dioxide and R-22, $s = 0$ for saturated liquid at −40°C or −40°F; for dry air, $s = 0$ for saturated liquid at −333.4°F or −203°C.

2. *Entropy Changes of Ideal Gas Mixtures*

In Section 5.2, a procedure was presented for the calculation of the thermodynamic properties of mixtures of nonreacting gases from a knowledge of the properties of the component gases. For example, the enthalpy of the mixture of two gases A and B was given by Equation (5.6) as

$$h = (m_A/m)h_A + (m_B/m)h_B$$

In a similar fashion, the entropy change of the mixture is

$$s_2 - s_1 = (m_A/m)(s_2 - s_1)_A + (m_B/m)(s_2 - s_1)_B \tag{6.30}$$

Since each component behaves as an ideal gas, we have, from Equation (6.22) for the entropy change of the gas mixture,

$$s_2 - s_1 = \int_1^2 [(m_A/m)c_{pA} + (m_B/m)c_{pB}]\, dT/T - (m_A/m)R_A \ln p_{A_2}/p_{A_1} - (m_B/m)R_B \ln p_{B_2}/p_{B_1}$$

For constant c_p values, the above equation simplifies to

$$s_2 - s_1 = [(m_A/m)c_{pA} + (m_B/m)c_{pB}] \ln T_2/T_1 - (m_A/m)R_A \ln p_{A_2}/p_{A_1} - (m_B/m)R_B \ln p_{B_2}/p_{B_1} \tag{6.31}$$

EXAMPLE 6.5

Find the change in entropy due to the mixing of the two gases (molecular hydrogen and nitrogen) in Example 5.7.

Solution

Before proceeding with the solution of the example, let us review some of the data and results obtained in Example 5.7:

State 1 (before mixing):

$$T_{H_2} = 520°R\,, \quad p_{H_2} = 50 \text{ psia}\,, \quad m_{H_2} = 0.0542 \text{ lbm}$$

$$R_{H_2} = 766.5 \text{ ft-lbf/lbm°R}\,, \quad c_{vH_2} = 2.44 \text{ Btu/lbm°R}$$

$$T_{N_2} = 550°R\,, \quad p_{N_2} = 25 \text{ psia}\,, \quad m_{N_2} = 0.2373 \text{ lbm}$$

$$R_{N_2} = 55.16 \text{ ft-lbf/lbm°R}\,, \quad c_{vN_2} = 0.177 \text{ Btu/lbm°R}$$

$$m = m_{H_2} + m_{N_2} = 0.2915 \text{ lbm}\,, \quad \frac{m_{H_2}}{m} = \frac{0.0542}{0.2915} = 0.1859$$

$$\frac{m_{N_2}}{m} = \frac{0.2373}{0.2915} = 0.8141$$

State 2 (after mixing):

$$T_2 = 527.3°R\,, \quad p = 40 \text{ psia}$$

Assuming constant c_p values, we can use Equation (6.31), for which we need c_{pH_2}

and c_{pN_2} and the partial pressures of the two constituents after mixing:

$$c_{pH_2} = 3.43 \text{ Btu/lbm°R} \qquad \text{(Appendix A.1)}$$

$$c_{pN_2} = 0.248 \text{ Btu/lbm°R} \qquad \text{(Appendix A.1)}$$

From the equation of state, we obtain the partial pressure after mixing:

$$p_{H_2} = \frac{m_{H_2} R_{H_2} T_2}{V} = \frac{(0.0542 \text{ lbm})(766.5 \text{ ft-lbf/lbm°R})(527.3\text{°R})}{(5 \text{ ft}^3)144 \text{ in}^2/\text{ft}^2} = 30.43 \text{ psia}$$

hence $p_{N_2} = p - p_{H_2} = 40 \text{ psia} - 30.43 \text{ psia} = 9.57 \text{ psia}$. Now from Equation (6.31),

$$\begin{aligned} s_2 - s_1 &= (m_{H_2}/m) c_{pH_2} \ln T_2/T_1 - (m_{H_2}/m) R_{H_2} \ln p_{H_22}/p_{H_21} \\ &\quad + (m_{N_2}/m) c_{pN_2} \ln T_2/T_1 - (m_{N_2}/m) R_{N_2} \ln p_{N_22}/p_{N_21} \\ &= (0.1859)(3.43 \text{ Btu/lbm°R}) \ln \frac{527.3}{520} \\ &\quad - (0.1859) \frac{766.5 \text{ ft-lbf/lbm°R}}{778 \text{ ft-lbf/Btu}} \ln \frac{30.43}{50} \\ &\quad + (0.8141)(0.248 \text{ Btu/lbm°R}) \ln \frac{527.3}{550} \\ &\quad - (0.8141) \frac{55.16 \text{ ft-lb/lbm°R}}{778 \text{ ft-lbf/Btu}} \ln \frac{9.57}{25} \\ &= 0.00889 + 0.09095 - 0.00851 + 0.05542 \\ &= 0.14675 \text{ Btu/lbm (mixture) °R} \end{aligned}$$

6.9 APPLICATIONS OF THE SECOND LAW

In the last section, we showed how to calculate the entropy change associated with a thermodynamic process. Further values of entropy associated with given thermodynamic states are available for four substances in the appendices. We are now in a position, therefore, to show how entropy and the second law can be used to calculate the best possible performance of a device.

EXAMPLE 6.6

Gaseous carbon dioxide is expanded through an insulated nozzle from 50 psia, 120°F, and zero velocity to 15 psia. What is the maximum exit velocity that can be achieved?

Solution

The best performance of the adiabatic nozzle will be obtained if the process is reversible. In this case for an adiabatic, reversible process, the change of entropy of the system (the gas within the nozzle) will be zero. Thus, $s_e = s_{in}$. Now from the thermodynamic tables given in Appendix A, $s_{in} = 0.458$ Btu/lbm°R, and hence $s_e = 0.458$ Btu/lbm°R. From the tables, we also obtain, for $s_e = 0.458$ Btu/lbm°R and $p = 15$ psia, $T_e = -18$°F.

From the energy equation for a nozzle given in Equation (3.24):

$$V_e^2 = 2g_c(h_{in} - h_e)$$

Again, from the thermodynamic tables, we have $h_{in} = 178.4$ Btu/lbm and $h_e = 151.3$ Btu/lbm. Thus

$$V_e^2 = 64.4 \frac{\text{lbm-ft/s}^2}{\text{lbf}} (178.4 - 151.3)\ \text{Btu/lbm}\ (778\ \text{ft-lbf/Btu})$$
$$= 1165\ \text{ft/s}$$

This velocity is the maximum that any insulated nozzle operating under the indicated conditions can achieve. In a real nozzle, there will be an increase of entropy; that is, for the insulated nozzle, $(s_e - s_{in})_{system} > 0$. For example, take the entropy increase to be 0.009 Btu/lbm. Thus, $s_e = 0.458 + 0.009 = 0.467$ Btu/lbm°R, $T_e = 0$°F, and $h_e = 154.8$ Btu/lbm. Therefore, $V_e = 1087.4$ ft/s, resulting in a velocity that is 93.3 percent of the ideal velocity.

EXAMPLE 6.7

A steam turbine operates between an inlet temperature of 400°C and pressure of 1.0 MPa and an exhaust pressure of 100 kPa. What is the maximum work output possible? Assume negligible heat losses from the turbine and negligible change of kinetic energy of the working fluid.

Solution

The maximum work output will be obtained for a reversible process. For a reversible, steady-flow, adiabatic process, $(s_e - s_{in})_{system} = 0$. From Table B.3 we obtain

$$h_{in} = 3059.8\ \text{kJ/kg} \qquad h_e = 2717.8\ \text{kJ/kg}$$
$$s_{in} = 7.4665\ \text{kJ/kg K} \qquad s_e = 7.4665\ \text{kJ/kg K}$$
$$T_e = 120.8°\text{C}$$

From Equation (3.18), we have

$$w'_{out} = h_{in} - h_e = 3059.8 - 2717.8$$
$$= 342.0\ \text{kJ/kg}$$

If the expansion process in the turbine were to take place irreversibly with an increase in entropy of 0.1 kJ/kg K, we would have $s_e = s_{in} + \Delta s = 7.4665 + 0.1 = 7.5665$ kJ/kg K. In this case, with $p_e = 100$ kPa and $s_e = 7.5665$ kJ/kg K, we find by interpolation from Table B.3 that $T_e = 140.6$°C and $h_e = 2757.5$ kJ/kg K. Therefore, $w'_{out} = 3059.8 - 2757.5 = 302.3$ kJ/kg. In this case, the actual (irreversible) work output is 88.4% of the ideal (reversible) work.

EXAMPLE 6.8

One pound of saturated R-22 vapor at 155°F is condensed at constant pressure to saturated liquid at 155°F by the transfer of heat to the surrounding air at a temperature of 95°F.

a. Determine the increase in entropy of the universe, that is, the system and its surroundings.

b. Find the highest temperature of the air permissible under the second law.

Solution

a. For the R-22, we have, from the property tables (A.10), $s_f = 0.11034$ Btu/lbm°R and $s_g = 0.19902$ Btu/lbm°R. Hence,

$$s_2 - s_1 = s_f - s_g = 0.11034 - 0.19902 = -0.08868 \text{ Btu/lbm°R}$$

For the surroundings, we have for the amount of heat transferred

$$(q_{\text{in}})_{\text{surround}} = -(h_f - h_g) = -(58.017 - 112.533) = 54.515 \text{ Btu/lbm(R-22)}$$

From the definition of entropy change, Equation (6.13), we have (since the air temperature remains constant)

$$(s_2 - s_1)_{\text{surround}} = q_{\text{in}}/T = 54.515/555 \text{ Btu/lbm°R} = 0.09822 \text{ Btu/lbm(R-22)°R}$$

Thus

$$(s_2 - s_1)_{\text{system}} + (s_2 - s_1)_{\text{surround}} = -0.08868 + 0.09822$$
$$= 0.00954 \text{ Btu/lbm(R-22)°R}$$

Since $(s_2 - s_1)_{\text{univ}} > 0$, this process is permissible by the second law.

b. The maximum temperature of air will occur when

$$(s_2 - s_1)_{\text{system}} + (s_2 - s_1)_{\text{surround}} = 0$$

This will occur when

$$(s_2 - s_1)_{\text{surround}} = 0.08868 \text{ Btu/lbm°R} = 54.515/T$$

$$\text{and} \quad T = 614.7\text{°R} \quad \text{or} \quad 154.7\text{°F}$$

The temperatures of the condensation process and the surrounding air are equal. This situation would require an infinite surface area to effect the transfer of heat reversibly.

EXAMPLE 6.9

Superheated steam in a closed can is heated from a temperature of 600°F and a pressure of 3 psia to a pressure of 5 psia. The heat is transferred from combustion gases which are at a constant temperature of 1500°F. How much heat is being transferred? Is the process irreversible?

Solution

This is a constant volume process. Take the steam as the system, the combustion gases as the surroundings. From Table A.3, we get $v_1 = 210.31$ ft^3/lbm, $u_1 =$

1219.25 Btu/lbm, and $s_1 = 2.1496$ Btu/lbm°R. By interpolation, we obtain the following for $p_2 = 5$ psia:

	T(°F)	v(ft^3/lbm)	u(Btu/lbm)	s(Btu/lbm°R)
	1300	209.62	1499.35	2.3509
State 2	1307	210.31	1502.56	2.3530
	1310	210.81	1503.75	2.3539

From the energy equation (2.5), we have, with $w_{out} = 0$ (constant volume process),

$$q_{in} = u_2 - u_1 = 1502.56 - 1219.25 = 283.31 \text{ Btu/lbm(steam)}$$

The change in entropy for the steam is $(s_2 - s_1)_{system} = 0.2034$ Btu/lbm°R. The entropy change of the surroundings is

$$(s_2 - s_1)_{surround} = q_{in}/T = -283.31/1960 = -0.1445 \text{ Btu/lbm(steam)°R}$$

The sum of the entropy changes is $0.2034 - 0.1445 = 0.0589$ Btu/lbm°R, which is greater than zero. Hence the process is irreversible.

PROBLEMS

6.1 Determine whether the following processes are reversible or irreversible, and explain why or why not:
a. A container of water is heated by an electrical immersion heater.
b. Water flows over a dam, turning a turbine to produce electrical energy.
c. A rubber ball is dropped onto a smooth pavement.
d. A wagon is caused to coast down a hill.
e. A hot metal sphere is quenched in a pail of water.

6.2 In a certain reversible thermodynamic process, the change of entropy of the system undergoing the process is measured to be 3.6 kJ/kg K. If the system were to undergo an irreversible process between the same end states, would the entropy change of the system be greater than, equal to, or less than 3.6 kJ/kg K? Explain.

6.3 Two pounds of water at 60°F are mixed adiabatically with 4.0 pounds of water at 120°F with all at 1 atmosphere pressure. Determine the final temperature of the mixture, the change of entropy of each batch of water, and the change of entropy of the universe due to this process. Take the specific heat of water to be 1.0 Btu/lbm/R°.

6.4 A 10-cm^3 block of copper at 200°C is brought into contact with a 20-cm^3 block of iron at 30°C. Determine the final equilibrium temperature of the two blocks. Assume no heat transfer with the surroundings. The specific heat of copper is 0.38 kJ/kg K; the specific heat of iron, 0.45 kJ/kg K; the density of copper, 8955 kg/m^3; and the density of iron, 7850 kg/m^3. Find the entropy change of the universe for this process.

6.5 A large heat reservoir at 130°F is brought into contact with 2.0 pounds of water at 60°F. Determine the change of entropy of the heat reservoir, the water, and the universe after the water has reached 130°F.

6.6 A large heat reservoir at 95°F is brought into contact with 2.0 pounds of water at 60°F until the water eventually reaches 95°F. Then a second reservoir at 130°F is brought into contact with the water until the water temperature reaches 130°F. Find the change of entropy of the water, the reservoirs, and the universe for the two processes.

6.7 A 50-ohm electrical resistor carries a current of 5 amperes. The resistor is maintained at a temperature of 30°C. Determine the rate at which heat must be transferred to the surroundings and the rate of entropy change of the resistor.

6.8 In a steady-flow mixing process, 10 lbm/s of hot water at 150°F and 10 lbm/s of cold water at 40°F enter a mixing chamber, as shown in Figure 6.27. Determine the rate of change of entropy for this process, assuming no heat transfer with the surroundings.

6.9 Three kilograms of liquid water are contained in a piston-cylinder arrangement at 20°C and 100 kPa. How much heat must be added to increase the temperature of the water to 500°C, while maintaining the pressure at 100 kPa? Calculate the change of entropy for the H_2O for this process.

6.10 The water of Problem 6.9 is heated from 20°C by bringing it into contact first with a heat reservoir at 100°C, and then with reservoirs at 200°C, 300°C, 400°C, and 500°C. Each time, the H_2O is allowed to reach equilibrium at the source temperature. The pressure is again maintained at 100 kPa. Find the total change of entropy of the H_2O and the change of entropy of each reservoir.

6.11 Air is contained in a rigid, closed vessel at 1 atmosphere and 60°F. Heat is added to the air until the air temperature reaches 200°F. The internal volume of the vessel is 3.5 ft^3. Find the change of entropy of the air, the heat added to the air, and the work done by the air. Treat the air as a perfect gas with constant specific heats (see Appendix A.1 for its properties).

6.12 Hydrogen is contained in a piston-cylinder arrangement, as shown in Figure 6.28. The hydrogen is allowed to expand, pushing the piston upward and

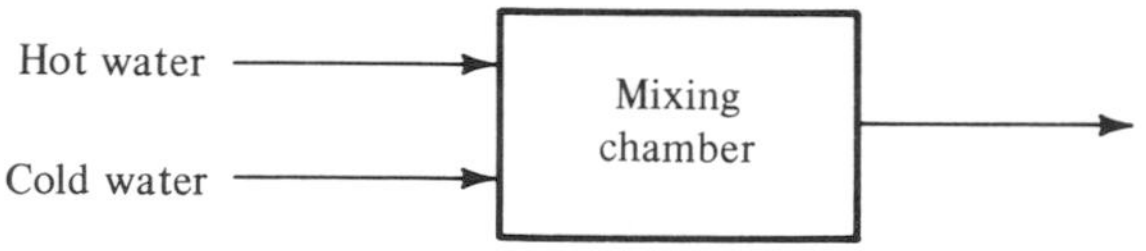

FIGURE 6.27

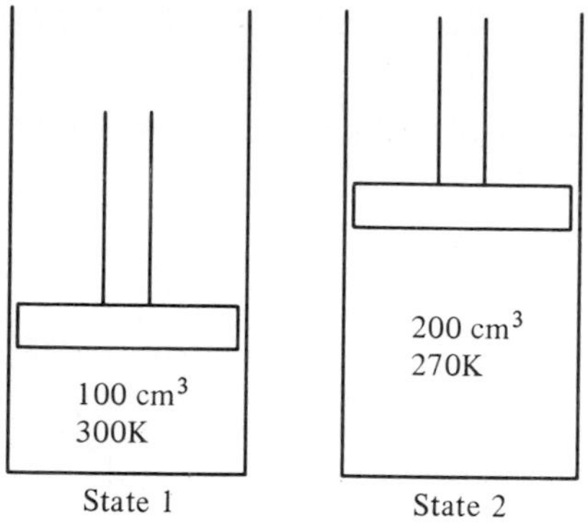

FIGURE 6.28

performing work. The process is carried out adiabatically. Find the change of entropy of the hydrogen and the work done by the hydrogen. Now suppose that the hydrogen is allowed to expand reversibly and adiabatically between the same volumes. Find the work done. Treat hydrogen as a perfect gas with constant specific heats; the properties are given in Table B.1.

6.13 A diaphragm divides a closed, well-insulated chamber into two equal volumes, each 1.0 liter. On one side of the diaphragm is nitrogen at 500 kPa and 300 K; on the other side is nitrogen at 1000 kPa and 400 K. The diaphragm is broken and the gases allowed to mix. Find the final temperature and pressure of the mixture. Calculate the total work done. Find the total entropy change of the universe. Treat nitrogen as a perfect gas with constant specific heats, given in Table B.1.

6.14 A block of dry ice (solid CO_2), at −109.4°F and having 1-ft^3 volume, is allowed to sublime to vapor at atmospheric pressure. The required heat transfer is from the surroundings at 100°F. Calculate the heat transfer to the CO_2 and the change of entropy of the universe when the CO_2 reaches 100°F.

6.15 A compressor is used to raise the pressure of an air stream at 14.7 psia, 100°F, to 100 psia. Assume a steady-flow process with negligible changes of potential or kinetic energies during the process. Determine the work required in Btu/lbm and the entropy change of the air stream for the following processes:

a. The process is reversible and adiabatic.

b. The work required is 10 percent more than that of (a) due to friction; the process is adiabatic.

c. The final temperature of the air is 100°F.

6.16 Steam is expanded in a well-insulated nozzle from 5000 kPa, 700°C, to 20 kPa. Determine the nozzle exit velocity for a reversible expansion in the nozzle. The actual nozzle exit velocity is measured to be 1200 m/s. Find the entropy change of the steam in kJ/kg K.

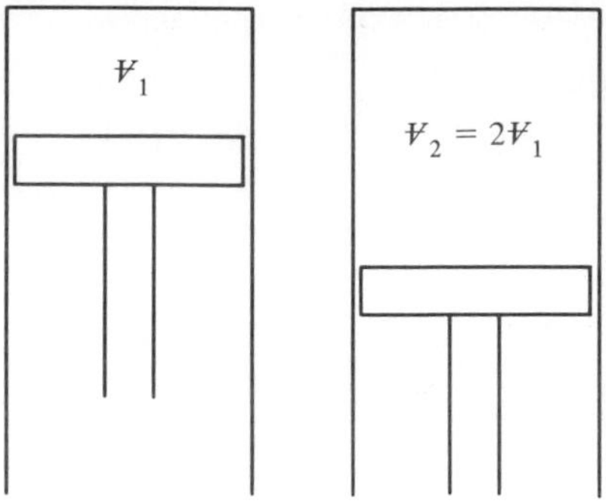

FIGURE 6.29

6.17 Air is contained in a piston-cylinder arrangement at 100 psia, 100°F. The air is allowed to push the piston outward in the cylinder, as shown in Figure 6.29. During this process, heat is received from the ambient air so as to maintain the compressed air in the cylinder at 100°F. The volume at state 2 is twice that at state 1. Find the entropy change of the air, assuming the process to be reversible. Repeat for an irreversible process between the same end states.

6.18 One mole of oxygen and one mole of nitrogen are contained in a rigid, well-insulated vessel; the gases are separated by a membrane, as shown in Figure 6.30. The membrane is broken, allowing the gases to mix. Find the final temperature and pressure in the vessel, and the change of entropy of the universe for this mixing process.

6.19 Repeat Problem 6.18 for the case shown in Figure 6.31.

6.20 Steam is expanded in a piston-cylinder arrangement from a volume of 10 in.^{3} to a volume of 20 in.^{3}. The expansion is reversible and isothermal, with

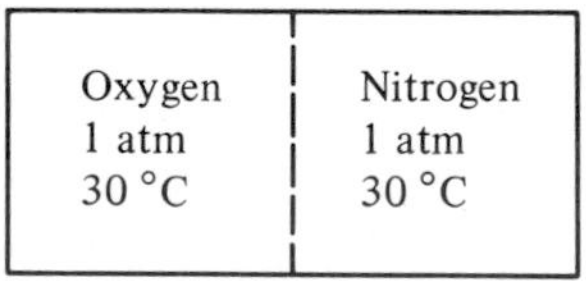

FIGURE 6.30

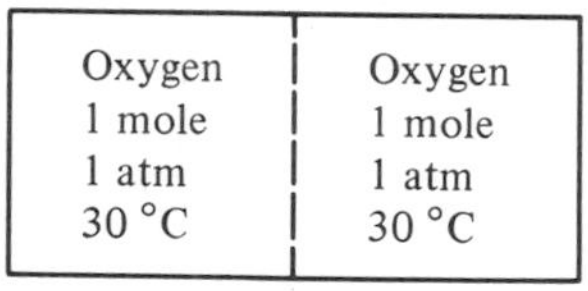

FIGURE 6.31

temperature maintained at 200°F and initial quality of 0.80. Determine the work done, the heat transferred, and the entropy change of the steam.

6.21 A container of 1.0 liter of hot water at 80°C, open to the atmosphere, is allowed to cool to an ambient air temperature of 20°C. Find the change of entropy of the universe for this process.

6.22 A container holds ice water at 32°F. There are 0.5 pint of liquid water and 3 ice cubes, each 0.75 inch on a side. The ice water is allowed to stand in ambient air at 80°F. Calculate the change of entropy of the universe when the water reaches 80°F.

6.23 A mixture of 0.1 liter of liquid R-22 and 10 liters of R-22 vapor at −40°C are contained in a closed rigid vessel. Heat is added until the temperature reaches 0°C. Calculate the change of entropy of the R-22 and the heat added.

6.24 Combustion gases are expanded in a steady-flow turbine from 200 psia, 500°F, to 20 psia. For a flow rate of 5 lbm/s, determine the maximum horsepower output of the turbine. Treat the gases as a perfect gas with $c_p = 0.20$ Btu/lbm°R and $R = 50$ ft-lbf/lbm°R.

6.25 Air is contained in a piston-cylinder arrangement at 500 kPa, 25°C, occupying a volume of 0.10 liter, as shown in Figure 6.32. The piston is restrained from moving by the upper stops. Suddenly, the upper stops are removed, causing the piston to jump to the lower position, restrained now by lower stops. The air is again allowed to reach an equilibrium temperature of 30°C, but at a volume of 0.20 liter. Determine the change of entropy of the air.

6.26 The piston of Problem 6.25 is pushed back upward until the air again occupies a volume of 0.10 liter. This process is carried out reversibly and isothermally. Calculate the work done, the heat transferred to the air, and the entropy change of the air for this process.

6.27 Scientists have proposed to use the natural thermal gradients existing in parts of the ocean to operate a heat engine. Maximum ocean temperature available near the surface is 30°C; minimum temperature near the ocean

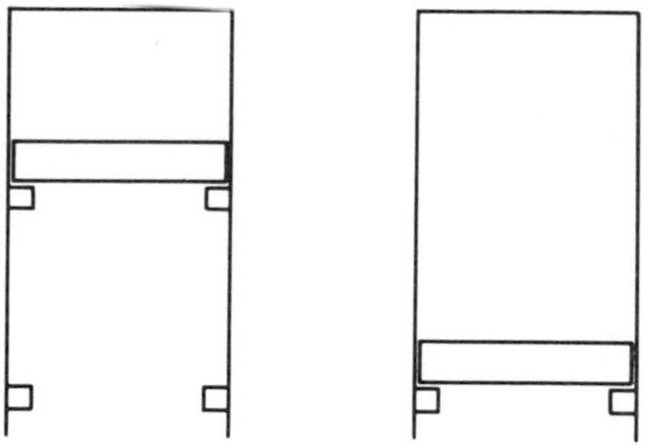

FIGURE 6.32

bottom is 5°C. What is the maximum possible thermal efficiency of such a heat engine?

6.28 Sketch a curve indicating the COP of a Carnot heat pump supplying heat to the interior of a house at 70°F over the ambient temperature range −20°F to 60°F.

6.29 In a Carnot heat engine, 3000 kilojoules of heat are added to the working fluid, with the working fluid temperature at 1000 K. Heat is rejected from the working fluid while the working fluid temperature is at 300 K. Determine the work output of the heat engine, the change of entropy of the working fluid during the heat addition process and during the heat rejection process, and the thermal efficiency of the engine. Sketch a *T-S* diagram for the cycle.

6.30 Determine the maximum coefficient of performance of a refrigerator designed to extract heat from a volume at 40°F and reject this heat to a volume of ambient temperature 80°F.

6.31 What is the maximum thermal efficiency of a cyclic heat engine operating between the temperature of combustion gases at 2000 K and ambient air at 25°C?

chapter

7

Second Law Analysis

7.1 INTRODUCTION

The second law of thermodynamics was introduced in Chapter 6. Whereas the first law, conservation of energy, can be written in terms of an equation expressing an energy balance such as (2.4) or (3.12), the second law places limits on the performance of a system. For example, the Carnot heat engine represents the most efficient cyclic heat engine operating between a high-temperature reservoir and a low-temperature reservoir. In Section 6.9 we used the second law concept of the reversible process to calculate the maximum exit velocity in a nozzle and the maximum work output of a steam turbine. Whereas the equation for the first law was expressed in terms of an equality, the second law is given by (6.13) and (6.14) in terms of an inequality–equality:

$$S_2 - S_1 \underset{\text{rev}}{\geq} \int_1^2 \frac{dQ_{\text{in}}}{T} \tag{7.1}$$

For an actual change of state of a system, it is important to be able to determine the maximum possible work that can be obtained. Second law analysis can provide us with this upper limit, called *reversible work*. Comparisons between the reversible work and actual work can then be used to determine irreversibilities and, hence, develop areas of energy waste. In current times, when we are anxious to use wisely our energy resources, concepts such as reversible work are useful tools to analyze how well we utilize our resources.

Availability is another property of importance in energy analysis; availability is the maximum work that can be done by a system in a given state. In this chapter, we will use the second law to devise expressions for reversible work and availability and then give examples of their calculations. We will also illustrate the use of second law analysis for several steady-flow devices and show how additional information on the performance of a device is obtainable from such analysis as compared to first law analysis. Second law analysis of various cycles will be discussed in Chapter 10, while Chapter 11 will provide second law analysis of combustion processes.

7.2 REVERSIBLE WORK

Reversible work is the maximum work done by a system in a given change of state. In Section 6.9, we considered some applications of the second law; in Example 6.6, we calculated the maximum exit velocity for a nozzle and in Example 6.7, we calculated the maximum work for a turbine. These maxima were ideal and based on the assumption of an isentropic process, one which is adiabatic and reversible. The final state in the actual process taking place in the nozzle or in the turbine would not be that of constant entropy, due to factors such as friction that take place in real processes. Let us assume, for the turbine, that the real process is adiabatic and irreversible. Schematically, we have the situation depicted in Figure 7.1. The actual final state $2a$ is not the same as the final state for an isentropic turbine, $2s$. With reversible work, we wish to calculate the maximum work between the actual states 1 and $2a$; isentropic work is not the same quantity, since $2s$ is a fictitious state based on an ideal, isentropic process. Let us now study how we can calculate reversible work. We will examine first an open, steady-flow system.

From Chapter 3, consider a generalized steady flow system, as shown in Figure 7.2. The first law is given by (3.15):

$$q_{\text{in}} - w'_{\text{out}} = h_e - h_{\text{in}} + \frac{1}{2g_c}(V_e^2 - V_{\text{in}}^2) + \frac{g}{g_c}(z_e - z_{\text{in}})$$

In order to calculate the maximum work between states in and e, assume that all processes are reversible and that any heat transfer from the

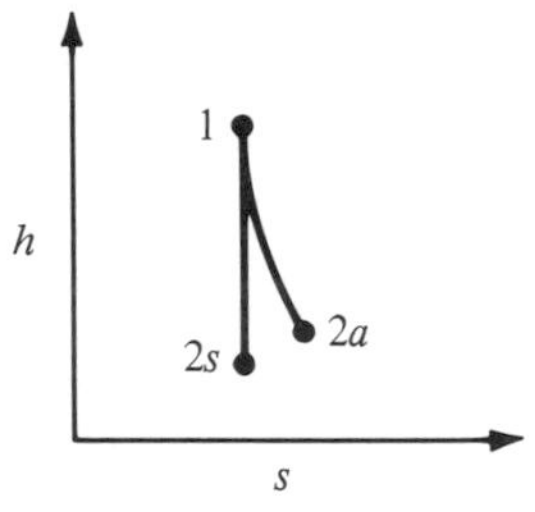

FIGURE 7.1 *Adiabatic and Isentropic Process.*

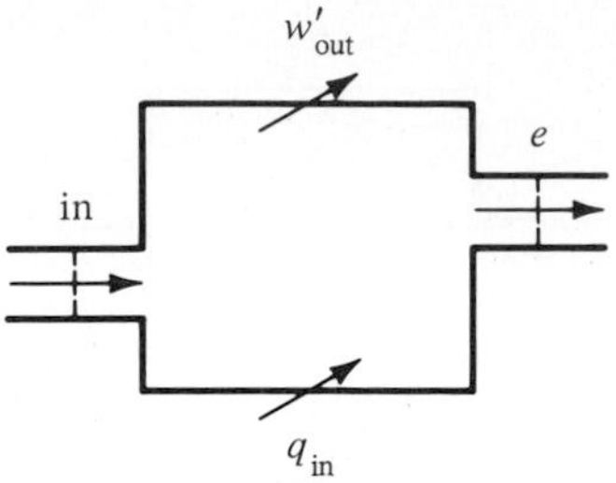

FIGURE 7.2 *Steady Flow System.*

surroundings is input to a Carnot engine, as shown in Figure 7.3. Note that, whereas states in and e are the same as those in Figure 7.2, the heat transfer and work output are different. T_o is the temperature of the surroundings; T is the system temperature at which heat q_c is received from the Carnot engine.

Since all processes depicted in Figure 7.3 are reversible, from (6.13),

$$q_c = T(s_e - s_{in})$$

Also, from (6.6),

$$\frac{q_0}{q_c} = \frac{T_o}{T}$$

From the first law, for the Carnot engine, $w'_c = q_0 - q_c$. Combining, we obtain

$$\begin{aligned} w'_c &= q_c \frac{T_o}{T} - q_c \\ &= T_o(s_e - s_{in}) - q_c \end{aligned}$$

The total work for the process of Figure 7.3 is the total of w' and w'_c, where, from (3.15),

$$q_c = w' + h_e - h_{in} + \frac{1}{2g_c}(V_e^2 - V_{in}^2) + \frac{g}{g_c}(z_e - z_{in})$$

$$w' + w'_c = q_c + h_{in} - h_e + \frac{1}{2g_c}(V_{in}^2 - V_e^2) + \frac{g}{g_c}(z_{in} - z_e) + T_o(s_e - s_{in}) - q_c$$

Therefore, the reversible work $w'_{rev} = w' + w'_c$

or $$w'_{rev} = h_{in} - h_e + \frac{1}{2g_c}(V_{in}^2 - V_e^2) + \frac{g}{g_c}(z_{in} - z_e) + T_o(s_e - s_{in}) \quad (7.2)$$

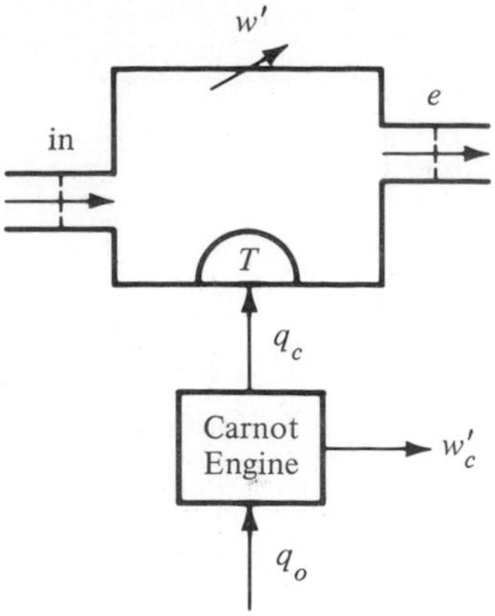

FIGURE 7.3 *Steady Flow System with Carnot Engine.*

For a closed system, as shown in Figure 7.4, we will assume no changes in potential or kinetic energy. Therefore, the first law becomes

$$q_{in} = w_{out} + u_2 - u_1$$

We will now make similar assumptions to that of the flow system, namely, that the heat transfer from the surroundings is allowed to be the heat input to a Carnot engine as in Figure 7.5, that all processes are reversible, and that the temperature of the surroundings is T_o.

Similar to the previous analysis,

$$w_c = q_0 - q_c$$

$$\frac{q_0}{q_c} = \frac{T_o}{T}$$

$$q_c = T(s_2 - s_1)$$

and

$$w_c = T_o(s_2 - s_1) - q_c$$

First law analysis yields

$$q_c = w + u_2 - u_1$$

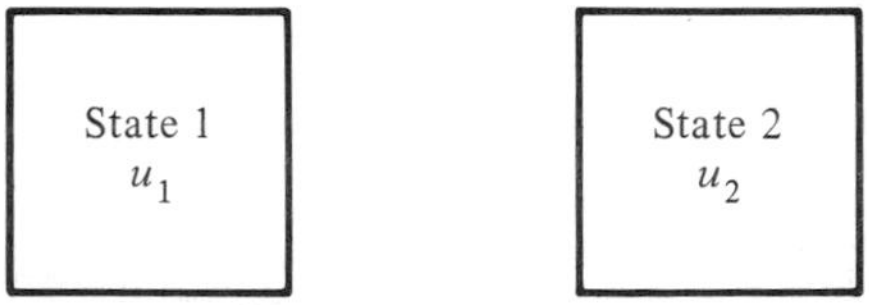

FIGURE 7.4 *Closed System.*

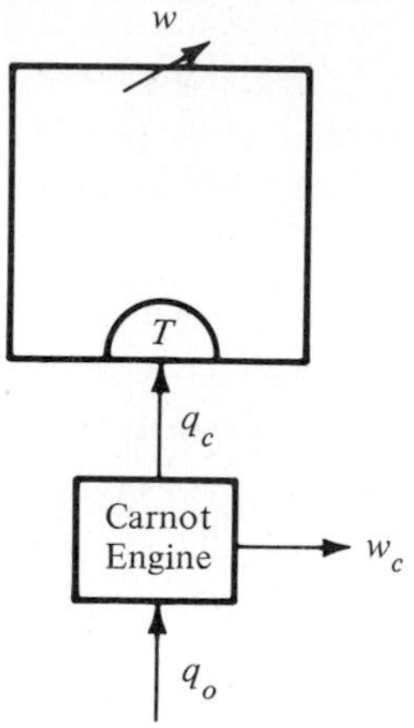

FIGURE 7.5 *Closed System with Carnot Engine.*

Now the reversible work

$$
\begin{aligned}
w_{\text{rev}} &= w + w_c \\
&= q_c + u_1 - u_2 + T_o(s_2 - s_1) - q_c \\
w_{\text{rev}} &= u_1 - u_2 + T_o(s_2 - s_1) \qquad \textbf{(7.3)}
\end{aligned}
$$

Another term of importance is irreversibility I (or i on a per unit mass basis) defined as the difference between reversible work and actual work for a given process. For the open, steady-flow system,

$$
\begin{aligned}
i &= w'_{\text{rev}} - w'_{\text{actual}} \\
&= h_{\text{in}} - h_e + \frac{1}{2g_c}(V_{\text{in}}^2 + V_e^2) + \frac{g}{g_c}(z_{\text{in}} - z_e) + T_o(s_e - s_{\text{in}}) \\
&\quad + h_e - h_{\text{in}} + \frac{1}{2g_c}(V_e^2 - V_{\text{in}}^2) + \frac{g}{g_c}(z_e - z_{\text{in}}) - q_{\text{in}} \qquad \textbf{(7.4a)} \\
i &= T_o(s_e - s_{\text{in}}) - q_{\text{in}} \qquad \textbf{(7.5a)}
\end{aligned}
$$

Similarly, for the closed system,

$$i = w_{\text{rev}} - w_{\text{actual}} \qquad \textbf{(7.4b)}$$

or

$$i = T_o(s_2 - s_1) - q_{\text{in}} \qquad \textbf{(7.5b)}$$

The following examples will illustrate the calculation of reversible work and irreversibility.

EXAMPLE 7.1

Steam enters a well-insulated turbine at 500 psia, 700°F and is discharged from the turbine at 14.7 psia. The work output of the turbine is measured to be 275 Btu/lbm. Determine the reversible work and irreversibility for this process. Neglect changes of potential and kinetic energies and assume that the temperature of the surroundings is 70°F.

Solution

First, we must find the actual final state. From the properties given,

$$h_{\text{in}} = 1357.7 \text{ Btu/lbm}$$

Using the first law, we have

$$w'_{\text{out}} = h_{\text{in}} - h_e$$

or

$$h_e = 1357.7 \text{ Btu/lbm} - 275 \text{ Btu/lbm} = 1082.7 \text{ Btu/lbm}$$

From Table A.2, at 14.7 psia, $h_f = 180.17$ Btu/lbm and $h_{fg} = 970.3$ Btu/lbm. Solving, we obtain

$$x_e = \frac{1082.7 - 180.17}{970.3} = 0.930$$

From (7.2),

$$w'_{\text{rev}} = h_{\text{in}} - h_e + T_o(s_e - s_{\text{in}})$$

For this case,

$$\begin{aligned}
s_e &= s_f + x_e s_{fg} \\
&= 0.3121 \text{ Btu/lbm°R} + 0.930(1.4447 \text{ Btu/lbm°R}) \\
&= 1.656 \text{ Btu/lbm°R} \\
s_{\text{in}} &= 1.6123 \text{ Btu/lbm°R} \\
w'_{\text{rev}} &= 275 \text{ Btu/lbm} + 530\text{°R}(1.656 - 1.612) \text{ Btu/lbm°R} \\
&= (275 + 23.3) \text{ Btu/lbm} \\
w'_{\text{rev}} &= 298.3 \text{ Btu/lbm}
\end{aligned}$$

$$i = w'_{\text{rev}} - w'_{\text{actual}} = 23.3 \text{ Btu/lbm}$$

EXAMPLE 7.2

Air is contained in a piston-cylinder arrangement at 20°C, 200 kPa. The temperature of the air is increased by heat exchange with the surroundings at 25°C. If the process is carried out at constant pressure until the air in the cylinder reaches 24°C, calculate the reversible work and irreversibility for the system consisting of the air in the cylinder. Treat the air as a perfect gas with constant specific heats.

Solution

For a closed system,

$$w_{\text{rev}} = u_1 - u_2 + T_o(s_2 - s_1)$$

For air, take

$$c_v = 0.715\ \text{kJ/kgK} \quad \text{and} \quad c_p = 1.004\ \text{kJ/kgK}$$

$$u_1 - u_2 = c_v(T_1 - T_2) = (0.715\ \text{kJ/kgK})(20 - 24)°\text{C} = -2.86\ \text{kJ/kg}$$

$$s_2 - s_1 = c_p \ln \frac{T_2}{T_1} - R \ln \frac{p_2}{p_1}$$

where

$$\ln \frac{p_2}{p_1} = 0$$

Therefore,

$$s_2 - s_1 = (1.004\ \text{kJ/kgK}) \ln \frac{24 + 273.15}{20 + 273.15}$$
$$= (1.004\ \text{kJ/kgK})(0.01355)$$
$$= 0.0136\ \text{kJ/kgK}$$

Now

$$w_{\text{rev}} = -2.86\ \text{kJ/kg} + 298.15\ \text{K}(0.0136\ \text{kJ/kgK})$$
$$= -2.8600\ \text{kJ/kg} + 4.0548\ \text{kJ/kg}$$
$$= +1.1948\ \text{kJ/kg}$$

The actual work for this process is

$$w = p(v_2 - v_1) = R(T_2 - T_1)$$

or

$$w_{\text{actual}} = (0.2870\ \text{kJ/kgK})4\ \text{K}$$
$$= 1.148\ \text{kJ/kg}$$
$$i = w_{\text{rev}} - w_{\text{actual}} = 0.0468\ \text{kJ/kg}$$

or, from (7.5),

$$i = T_o(s_2 - s_1) - q_{\text{in}}$$
$$q_{\text{in}} = w_{\text{actual}} + u_2 - u_1$$
$$= 1.148\ \text{kJ/kg} + 2.86\ \text{kJ/kg}$$
$$= 4.008\ \text{kJ/kg}$$

Therefore,

$$i = 4.0548\ \text{kJ/kg} - 4.008\ \text{kJ/kg}$$
$$= 0.0468\ \text{kJ/kg}$$

EXAMPLE 7.3

Ten lbm/s of hot water at 120°F are to be mixed with 5 lbm/s of cold water at 40°F. For a steady-flow process, determine the resultant temperature of the mixture and the rate of irreversibility for the mixing process. The temperature of the surroundings is 77°F. Take the specific heat of water as 1.0 Btu/lbm°F, and assume the mixing process to be adiabatic.

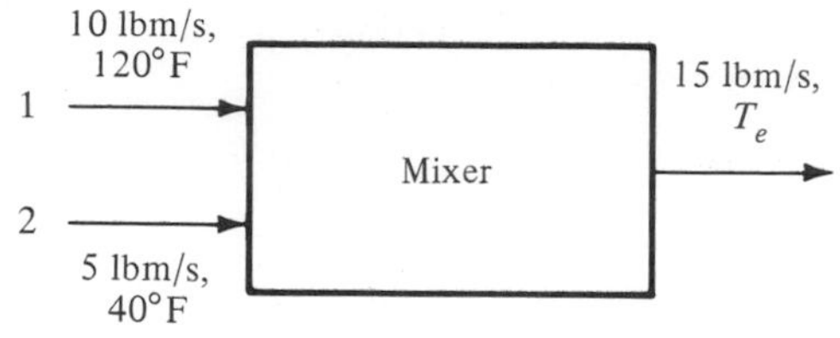

FIGURE 7.6 *Steady Flow Mixer.*

Solution

Writing an energy balance, we have

$$\dot{m}_1 h_1 + \dot{m}_2 h_2 = \dot{m}_e h_e$$

Taking the specific heat of water as constant over the temperature range of interest, we obtain

$$(5\ \text{lbm/s})(T_e - 40°\text{F}) = (10\ \text{lbm/s})(120°\text{F} - T_e)$$

or

$$T_e = 93.3°\text{F}$$

For the hotter water,

$$i = T_0(s_e - s_1) - q_{\text{in}}$$

$$= 537°\text{R}\left(c \ln \frac{T_e}{T_1}\right) - c(93.3°\text{F} - 120°\text{F})$$

$$= 537°\text{R}(1.0\ \text{Btu/lbm°F}) \ln \frac{553.3°\text{R}}{580°\text{R}} - (1.0\ \text{Btu/lbm°F})(-26.7°\text{F})$$

$$= -25.3\ \text{Btu/lbm} + 26.7\ \text{Btu/lbm}$$

$$= 1.4\ \text{Btu/lbm}$$

or $\dot{I} = (10\ \text{lbm/s})(1.4\ \text{Btu/lbm}) = 14\ \text{Btu/s}$

For the cooler water,

$$i = 537°\text{R}(1.0\ \text{Btu/lbm°F}) \ln \frac{553.3}{500} - (1.0\ \text{Btu/lbm°F})(53.3°\text{F})$$

$$= 54.4\ \text{Btu/lbm} - 53.3\ \text{Btu/lbm}$$

$$= 1.1\ \text{Btu/lbm}$$

or $\dot{I} = (5\ \text{lbm/s})(1.1\ \text{Btu/lbm}) = 5.5\ \text{Btu/s}$

Total rate of irreversibility = 19.5 Btu/s

7.3 AVAILABILITY

Availability is the maximum work that can be done by a given system in a given state. A system will continue to be able to do work until it reaches a state of equilibrium with its environment. Availability is thus the reversible work performed as the system goes from the given state to one of equilibrium with the surroundings. Return to (7.2), for a steady-flow system,

$$w'_{rev} = h_{in} - h_e + \frac{1}{2g_c}(V_{in}^2 - V_e^2) + \frac{g}{g_c}(z_{in} - z_e) + T_o(s_e - s_{in})$$

Now, to get an expression for availability, let state e be one of pressure and temperature equilibrium with the surroundings. In this case, at equilibrium, $V_e = 0$ and z_e will be a reference state z_0. Therefore, availability a is given by:

$$a = h - h_0 + \frac{V^2}{2g_c} + \frac{g}{g_c}(z - z_0) - T_0(s - s_0) \tag{7.6}$$

In the above, h_0 and s_0 correspond to enthalpy and entropy at surrounding temperature and pressure T_o and p_0.

Thus, for steady flow, the change in availability in going from state "in" to state "e" exactly equals the reversible work

$$a_{in} - a_e = w'_{rev} \tag{7.7}$$

For a closed system, we must subtract a term for the work done as the system volume changes to reach equilibrium with the surroundings. If the specific volume at a given state is v and v_0 is the specific volume at equilibrium, the system work against a surrounding pressure p_0 will be $p_0(v_0 - v)$. For a closed system, the availability is

$$a = u - u_0 + p_0(v - v_0) - T_o(s - s_0) \tag{7.8}$$

Thus, for a closed system,

$$a_1 - a_2 = w_{rev} + p_0(v_1 - v_2) \tag{7.9}$$

EXAMPLE 7.4

Combustion gases are expanded in a turbine from 1000 K, 500 kPa to 200 kPa. The process that takes place in the turbine is adiabatic, but the power output is only 95 percent of that of an isentropic turbine. Determine the availability of the gases at turbine inlet and outlet, the reversible work and the irreversibility for this

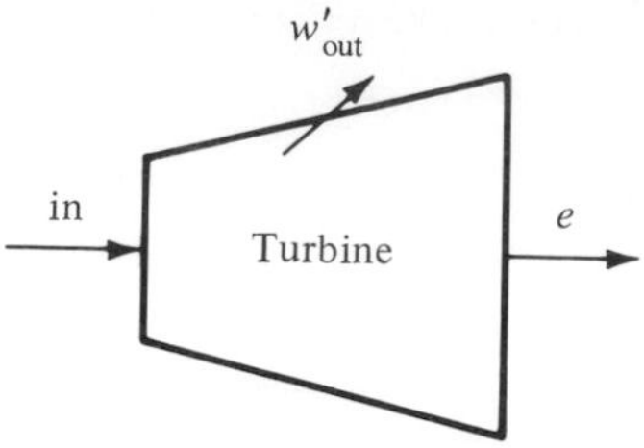

Figure 7.7 *Turbine.*

this process. Assume the combustion gases to behave as a perfect gas with constant specific heats, $c_p = 1.2\ \text{kJ/kgK}$, $\gamma = 1.35$. Surrounding temperature and pressure are 25°C, 101 kPa. Neglect changes in potential and kinetic energies.

Solution

First we must find the actual state at turbine outlet. For an isentropic turbine,

$$\frac{T_{e_s}}{T_{\text{in}}} = \left(\frac{p_e}{p_{\text{in}}}\right)^{\gamma-1/\gamma} = \left(\frac{200}{500}\right)^{0.35/1.35} = 0.7886$$

or
$$T_{e_s} = 788.6\ \text{K}$$

and
$$w'_s = h_{\text{in}} - h_{e_s} = c_p(T_{\text{in}} - T_{e_s})$$
$$= (1.2\ \text{kJ/kgK})(1000 - 788.6)\ \text{K}$$
$$- 253.7\ \text{kJ/kg}$$

$$w'_{\text{actual}} = 0.95(253.7\ \text{kJ/kg}) = 241.0\ \text{kJ/kg}$$

and
$$w'_{\text{actual}} = c_p(T_{\text{in}} - T_{e_a})$$
$$241.0\ \text{kJ/kg} = (1.2\ \text{kJ/kgK})(1000\ \text{K} - T_{e_a})$$

or
$$T_{e_a} = 799.2\ \text{K}$$

At inlet,

$$a_{\text{in}} = h_{\text{in}} - h_0 - T_o(s_{\text{in}} - s_0)$$
$$= c_p(T_{\text{in}} - T_o) - T_o(s_{\text{in}} - s_0)$$

$$s_{\text{in}} - s_0 = c_p \ln \frac{T_{\text{in}}}{T_o} - R \ln \frac{p_{\text{in}}}{p_0}$$

$$R = \frac{c_p(\gamma - 1)}{\gamma} = (1.2\ \text{kJ/kgK})\,\frac{0.35}{1.35} = 0.311\ \text{kJ/kgK}$$

$$s_{\text{in}} - s_0 = (1.2\ \text{kJ/kgK}) \ln \frac{1000}{298} - (0.311\ \text{kJ/kgK}) \ln \frac{500}{101}$$
$$= 1.453\ \text{kJ/kgK} - 0.497\ \text{kJ/kgK}$$
$$= 0.956\ \text{kJ/kgK}$$

$$a_{in} = (1.2 \text{ kJ/kgK})(1000 - 298) \text{ K} - 298 \text{ K}(0.956 \text{ kJ/kgK})$$
$$= 842.4 \text{ kJ/kg} - 284.9 \text{ kJ/kg}$$
$$= 557.5 \text{ kJ/kg}$$

At outlet,

$$a_e = c_p(T_e - T_o) - T_o(s_e - s_0)$$

$$s_e - s_0 = c_p \ln \frac{T_e}{T_o} - R \ln \frac{p_e}{p_0}$$
$$= (1.2 \text{ kJ/kgK}) \ln \frac{799.2}{298} - (0.311 \text{ kJ/kgK}) \ln \frac{200}{101}$$
$$= 1.184 \text{ kJ/kgK} - 0.212 \text{ kJ/kgK}$$
$$= 0.972 \text{ kJ/kgK}$$

$$a_e = (1.2 \text{ kJ/kgK})(799.2 - 298) \text{ K} - 298 \text{ K}(0.972 \text{ kJ/kgK})$$
$$= 601.4 \text{ kJ/kg} - 289.7 \text{ kJ/kg}$$
$$= 311.7 \text{ kJ/kg}$$

$$w'_{rev} = h_{in} - h_e + T_o(s_e - s_{in})$$
$$= c_p(T_{in} - T_e) + T_o(s_e - s_{in})$$

$$s_e - s_{in} = c_p \ln \frac{T_e}{T_{in}} - R \ln \frac{p_e}{p_{in}}$$
$$= (1.2 \text{ kJ/kgK}) \ln \frac{799.2}{1000} - (0.311 \text{ kJ/kgK}) \ln \frac{200}{500}$$
$$= -0.2690 \text{ kJ/kgK} + 0.2850 \text{ kJ/kgK} = 0.0160 \text{ kJ/kgK}$$

$$w'_{rev} = (1.2 \text{ kJ/kgK})(1000 - 799.2) \text{ K} + 298 \text{ K}(0.0160 \text{ kJ/kgK})$$
$$= 241.0 \text{ kJ/kg} + 4.8 \text{ kJ/kg}$$
$$= 245.8 \text{ kJ/kg}$$

Reversible work can also be found from

$$w'_{rev} = a_{in} - a_e = \underline{245.8 \text{ kJ/kg}}$$

$$i = w'_{rev} - w'_{actual}$$
$$= 245.8 \text{ kJ/kg} - 241.0 \text{ kJ/kg}$$
$$= 4.8 \text{ kJ/kg}$$

EXAMPLE 7.5

A 12-V auto battery is rated at 60 amp-hr. Determine the size of a compressed air tank that would have the same available energy as the battery. The air is stored at

300 psia, 77°F. Temperature and pressure of the surroundings are 77°F, 14.7 psia. Assume air to behave as a perfect gas with properties given in Table A.1.

Solution

$$a = u - u_0 + p_0(v - v_0) - T_o(s - s_0)$$

$$v = \frac{RT}{p} = \frac{(53.35 \text{ ft-lbf/lbm°R})(537\text{°R})}{(300 \times 144) \text{ lbf/ft}^2} = 0.663 \text{ ft}^3\text{/lbm}$$

$$v_0 = \frac{RT_o}{p_0} = \frac{(53.35 \text{ ft-lbf/lbm°R})(537\text{°R})}{(14.7 \times 144) \text{ lbf/ft}^2} = 13.534 \text{ ft}^3\text{/lbm}$$

$$s - s_0 = c_p \ln \frac{T}{T_o} - R \ln \frac{p}{p_0}$$

$$= 0 - (53.35 \text{ ft-lbf/lbm°R}) \ln \frac{300}{14.7} = -160.9 \frac{\text{ft-lbf}}{\text{lbm°R}}$$

$$u - u_0 = 0$$

so $$a = \left(14.7 \frac{\text{lbf}}{\text{in}^2}\right)\left(144 \frac{\text{in}^2}{\text{ft}^2}\right)\left(-12.871 \frac{\text{ft}^3}{\text{lbm}}\right) - 537\text{°R}\left(-160.9 \frac{\text{ft-lbf}}{\text{lbm°R}}\right)$$

$$a = -27{,}245 \text{ ft-lbf/lbm} + 86{,}403 \text{ ft-lbf/lbm}$$

$$= 59{,}158 \text{ ft-lbf/lbm}$$

$$\text{Available energy needed} = 12 \text{ V} \times 60 \text{ amperes} \times 3600 \text{ s}$$

$$= 2.592 \times 10^6 \text{ watt-s}$$

But 1 W = 0.7376 ft-lbf/s, so energy needed = 1.912×10^6 ft-lbf, or we need

$$\frac{1.912 \times 10^6 \text{ ft-lbf}}{59{,}158 \text{ ft-lbf/lbm}} = 32.32 \text{ lbm}$$

Since $p\forall = mRT$ for the air in the tank,

$$\forall = \frac{(32.32 \text{ lbm})(53.35 \text{ ft-lbf/lbm°R})537\text{°R}}{(300 \times 144) \text{ lbf/ft}^2}$$

$$= 21.43 \text{ ft}^3$$

7.4 FIRST AND SECOND LAW ANALYSIS APPLIED TO A TURBINE

In this section we will illustrate how additional information on the performance of a turbine is provided by a second law analysis. Such information is not obtainable from a first law analysis.

EXAMPLE 7.6

Two well-insulated turbines have the same work output (72 Btu/lbm). The high-pressure turbine operates between pressures of 600 psia and 300 psia, while the low-pressure turbine operates between pressures of 15 psia and 4.0 psia. Inlet temperature to the high-pressure turbine is 800°F, and the inlet quality to the low-pressure turbine is 0.945.

Calculate the changes in availability, irreversibilities, and efficiencies based on first and second law analysis and compare their performances. The second law efficiency is defined as the ratio of actual work and reversible work. Neglect changes in kinetic and potential energies between turbine inlet and outlet, and take the temperature of the surroundings to be 77°F (537°R).

Solution

From the thermodynamic properties tables of Appendix A, we obtain

Property	High-pressure Turbine	Low-pressure Turbine
p_{in}	600 psia	15 psia } $x = 0.945$
T_{in}	800°F	213°F } $x = 0.945$
h_{in}	1408.3 Btu/lbm	1097.6 Btu/lbm
s_{in}	1.6351 Btu/lbm°R	1.6759 Btu/lbm°R
p_e	300 psia	4.0 psia
s_{es}	1.6351 Btu/lbm°R	1.6759 Btu/lbm°R
h_{es}	1323.7 Btu/lbm	1012.9 Btu/lbm
h_e	1336.3 Btu/lbm	1025.6 Btu/lbm
s_e	1.6464 Btu/lbm°R	1.6968 Btu/lbm°R

Hence we obtain for the high-pressure turbine, $\Delta h_s = 1408.3$ Btu/lbm − 1323.7 Btu/lbm = 84.6 Btu/lbm and, similarly, for the low-pressure turbine, $\Delta h_s = 1097.6$ Btu/lbm − 1012.9 Btu/lbm = 84.7 Btu/lbm.

Neglecting changes in kinetic and potential energies between turbine inlet and exit, we obtain for the change in availability for the high-pressure turbine

$$\Delta a = a_{\text{in}} - a_e = h_{\text{in}} - h_e - T_0(s_{\text{in}} - s_e)$$

$$= 1408.3\ \text{Btu/lbm} - 1336.3\ \text{Btu/lbm}$$

$$- 537°\text{R}(1.6351\ \text{Btu/lbm°R} - 1.6464\ \text{Btu/lbm°R}) = 78.1\ \text{Btu/lbm}$$

Similarly, for the low-pressure turbine,

$$\Delta a = 1097.6\ \text{Btu/lbm} - 1025.6\ \text{Btu/lbm}$$

$$- 537°\text{R}(1.6759\ \text{Btu/lbm°R} - 1.6968\ \text{Btu/lbm°R}) = 83.2\ \text{Btu/lbm}$$

Hence, we can develop the following information for comparing the two turbines:

Quantity	High-pressure Turbine	Low-pressure Turbine
Isentropic work output $w'_s = \Delta h_s$	84.6 Btu/lbm	84.7 Btu/lbm
Actual work output $w'_{act} = \Delta h$	72.0 Btu/lbm	72.0 Btu/lbm
Change in availability $\Delta a = w'_{rev}$	78.1 Btu/lbm	83.2 Btu/lbm
Irreversibility $i = w'_{rev} - w'_{act}$	6.1 Btu/lbm	11.2 Btu/lbm
First law turbine efficiency $\eta_1 = \dfrac{w'_{act}}{\Delta h_s}$	85%	85%
Second law turbine efficiency $\eta_2 = \dfrac{w'_{act}}{w'_{rev}}$	92%	86.5%

The two turbines were chosen so that their actual work output was identical. It is seen that their first law efficiencies are identical, indicating no distinction in performance of the two turbines. On the other hand, second law analysis (availability balance) indicates a difference in their performance. The table indicates that the actual work output is less than the decrease in availability, since the change in availability equals the work output for a reversible change between the end states of the actual process. Further, it is seen that the decrease in availability is greater for the low-pressure turbine, i.e., there occurs a greater irreversibility in the low-pressure turbine. Hence, this turbine would be more sensitive to inefficiencies in design than the high-pressure turbine.

7.5 SECOND LAW ANALYSIS OF STEADY-FLOW DEVICES

In this section we will look at the changes in availability of other steady-flow devices, namely, the steam generator (boiler, superheater) and the condenser. In devices such as a turbine, a condenser, and in piping, the availability of the working medium *decreases* as it passes through the device. This decrease is associated with the production of work and friction in the turbine, and with friction in the piping. In a condenser, the decrease is due to the transfer of heat from the medium. The availability of the working medium *increases* as it passes through such devices as steam generators, reheaters, etc., due to the addition of heat to the working medium. Further, availability also increases in a pump or compressor due to the addition of work to the medium passing through these devices.

EXAMPLE 7.7

Compressed liquid water is supplied to a steam generator at a pressure of 500 psia and a temperature of 100°F. It leaves the steam generator at the same pressure and a temperature of 700°F. Calculate the change in availability of the water as it passes through the steam generator. Assume the temperature of the surroundings to be 77°F (537°R). Repeat the calculation for pressures of 250, 750, and 1000 psia, assuming the same heat input as for a pressure of 500 psia.

Solution

From Appendix A we obtain the enthalpy and entropy at the various conditions:

For $p = 500$ psia, $h_{in} = 69.3$ Btu/lbm, $s_{in} = 0.1292$ Btu/lbm°R and $h_e =$ 1357.7 Btu/lbm, $s_e = 1.6123$ Btu/lbm°R.

The change in availability becomes

$$\Delta a = a_e - a_{in} = h_e - h_{in} - T_o(s_e - s_{in}) = 1357.7\ \text{Btu/lbm} - 69.3\ \text{Btu/lbm}$$
$$- 537°\text{R}(1.6123\ \text{Btu/lbm°R} - 0.1292\ \text{Btu/lbm°R}$$
$$= 492.0\ \text{Btu/lbm}$$

an increase in availability.

From first law analysis (3.21), we get $q_{in} = h_e - h_{in} = 1357.7 - 69.3 =$ 1288.4 Btu/lbm.

For $p = 250$ psia, $h_{in} = 68.7$ Btu/lbm, $s_{in} = 0.1294$ Btu/lbm°R and $h_e = h_{in} +$ 1288.4 = 1357.1 Btu/lbm, $s_e = 1.6845$ Btu/lbm°R.

The increase in availability becomes

$$\Delta a = 1357.1\ \text{Btu/lbm} - 68.7\ \text{Btu/lbm}$$
$$- 537°\text{R}(1.6845\ \text{Btu/lbm°R} - 0.1294\ \text{Btu/lbm°R}) = 453.3\ \text{Btu/lbm}$$

For $p = 750$ psia, $h_{in} = 70.0$ Btu/lbm, $s_{in} = 0.1291$ Btu/lbm°R and $h_e =$ 1358.4 Btu/lbm, $s_e = 1.5713$ Btu/lbm°R.

The increase in availability becomes

$$\Delta a = 1358.4\ \text{Btu/lbm} - 70.0\ \text{Btu/lbm}$$
$$- 537°\text{R}(1.5713\ \text{Btu/lbm°R} - 0.1291\ \text{Btu/lbm°R}) = 513.9\ \text{Btu/lbm}$$

For $p = 1000$ psia, $h_{in} = 70.6$ Btu/lbm°R, $s_{in} = 0.1289$ Btu/lbm°R and $h_e =$ 1359.0 Btu/lbm, $s_e = 1.5424$ Btu/lbm°R.

The increase in availability becomes

$$\Delta a = 1359.0\ \text{Btu/lbm} - 70.6\ \text{Btu/lbm}$$
$$- 537°\text{R}(1.5424\ \text{Btu/lbm°R} - 0.1289\ \text{Btu/lbm°R}) = 529.4\ \text{Btu/lbm}$$

The following table summarizes the increase in availability obtained at the various pressure levels:

p (psia)	Δa (Btu/lbm)
250	453.3
500	492.2
750	513.9
1000	529.4

It is seen from the table that a higher "energy quality" steam is obtained at the higher pressure levels, although the same amount of energy (1288.4 Btu/lbm) was added in each case.

EXAMPLE 7.8

Saturated exhaust steam from a turbine enters a condenser at a temperature of 60°F and leaves as saturated liquid. Calculate the decrease in availability of the steam. Take the temperature of the surrounding to be 50°F (510°R).

Solution

From Appendix A we get the applicable properties:

$$h_g = 1087.7\text{ Btu/lbm}\,,\quad s_g = 2.0946\text{ Btu/lbm°R}\quad\text{and}\quad h_f = 28.1\text{ Btu/lbm}\,,$$
$$s_f = 0.0555\text{ Btu/lbm°R}$$

From first law analysis (3.22), we get $q_{\text{out}} = h_{\text{in}} - h_e = 1087.7\text{ Btu/lbm} - 28.1\text{ Btu/lbm} = 1059.6\text{ Btu/lbm°R}$.

The change in availability is

$$\begin{aligned}\Delta a = a_e - a_{\text{in}} &= 28.1\text{ Btu/lbm} - 1087.7\text{ Btu/lbm}\\ &\quad - 510\text{°R}(0.0555\text{ Btu/lbm°R} - 2.0946\text{ Btu/lbm°R})\\ &= 19.7\text{ Btu/lbm}\end{aligned}$$

a decrease in availability.

It is seen that the large amount of heat rejected in the condenser (1059.6 Btu/lbm) is associated with a small loss in availability (19.7 Btu/lbm) due to the relatively low temperature of the exhaust steam.

EXAMPLE 7.9

In a well-insulated boiler (heat exchanger), energy is transferred between hot products of combustion from a furnace and steam. The temperature of the products of combustion decreases by 500°C during its passage through the heat exchanger, while the pressure remains constant at 1 atm. Water enters the heat exchanger at 750 kPa, 150°C and exits at 750 kPa, 250°C.

Calculate the change in availability and irreversibility of the steam, of the products, and of the boiler, if the products enter the boiler at 1500°C, 1250°C, 1000°C, and 750°C, respectively. Take the specific heat of the products as 1.09 kJ/kgK.

Solution

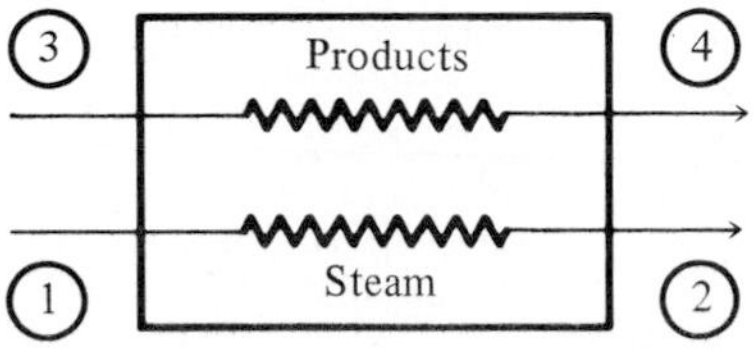

FIGURE 7.8 *Heat Exchanger (Boiler).*

From the first law, we get

$$\dot{m}_{\text{prod}}(h_4 - h_3) = \dot{m}_{\text{steam}}(h_2 - h_1)$$

From Tables B.3 and B.4, we obtain the enthalpy values of the steam. Hence,

$$\frac{\dot{m}_{\text{prod}}}{\dot{m}_{\text{steam}}} = \frac{h_2 - h_1}{h_4 - h_3} = \frac{2952.2 - 632.3}{1.09(500)} = 4.257$$

i.e., 4.257 kg of products of combustion pass through the boiler for every kilogram of steam. Neglecting changes in kinetic and potential energies, we get the change of availability from (7.6)

$$a_e - a_{\text{in}} = h_e - h_{\text{in}} - T_0(s_e - s_{\text{in}})$$

For the products of combustion, take $h_0 = 0$ and $s_0 = 0$ at a temperature of 25°C (298 K) and a pressure of 1 atm. For the steam, take, at 25°C, $h_0 = h_f =$ 104.77 kJ/kg and $s_0 = s_f = 0.3672$ kJ/kgK. All properties of the products of combustion listed in the table on page 194 are on a "per kilogram of steam" basis, e.g., at 750°C,

$$h = 4.257\ (1.09)\ (750 - 25) = 3364.1 \text{ kJ/kgK}$$

and from (6.25), at constant pressure,

$$s = c_p \ln \frac{T}{T_o} = 4.257\ (1.09) \ln \frac{750 + 273}{298} = 5.7231 \text{ kJ/kg(steam) K}$$

State	Medium	p (kPa)	T (°C)	h (kJ/kg)	s (kJ/kgK)
1	Steam	750	150	632.3	1.8413
2		750	250	2952.2	7.0732
3	Products	1 atm	750	3364.1	5.7231
4		1 atm	250	1044.0	2.6100
3	Products	1 atm	1000	4524.1	6.7376
4		1 atm	500	2204.1	4.4229
3	Products	1 atm	1250	5684.2	7.5696
4		1 atm	750	3364.1	5.7231
3	Products	1 atm	1500	6844.2	8.275
4		1 atm	1000	4524.1	6.7376

State	Medium	p (kPa)	T (°C)	q_{in} (kJ/kg)	Δa (kJ/kg)	i (kJ/kg)
1	Steam	750	150	2320	760.8	−760.8
2		750	250			
3	Products	1 atm	750	−2320	−1392.3	1392.3
4		1 atm	250			
			net	0	−631.5	631.5
3	Products	1 atm	1000	−2320	−1630.2	1630.2
4		1 atm	500			
			net	0	−869.4	869.4
3	Products	1 atm	1250	−2320	−1769.7	1769.7
4		1 atm	750			
			net	0	−1008.9	1009.1
3	Products	1 atm	1500	−2320	−1861.9	1861.9
4		1 atm	1000			
			net	0	−1101.1	1101.1

Note that the steam portion of the heat exchanger experiences a negative irreversibility (i), while the products portion experiences a positive irreversibility. Thus, the net change in irreversibility in the heat exchanger is 632 kJ/kg(steam). The irreversibility of an overall process is always greater than zero and becomes zero only when the process is reversible. The net change in irreversibility can also be calculated by considering the heat exchanger as a unit,

$$\text{net } i = \Sigma T_o(s_e - s_{\text{in}}) - \Sigma q_{\text{in}}$$

With $\Sigma q_{\text{in}} = 0$ (well-insulated), we get, for example, for the first case

$$298[(7.0732 - 1.8413) + (2.6100 - 5.7231)] = 631.4 \text{ kJ/kg}$$

It is possible to define a second law efficiency for the heat exchanger as the ratio of availability change of the steam (output) and of the combustion products (input)

$$\eta_2 = \frac{\Delta a_{\text{steam}}}{\Delta a_{\text{products}}}$$

For example, for the first case, we get

$$\eta_2 = \frac{760.8}{1392.3} = 54.6\%$$

Summarizing, we have

Inlet Temperature of Products (°C)	Net i (kJ/kg)	η_2
750	1392.3	54.6%
1000	1630.2	46.7%
1250	1769.7	43.0%
1500	1861.9	40.9%

We could also have defined a first law efficiency for the heat exchanger as the ratio of heat input and output

$$\eta_1 = \frac{q_{\text{out}}}{q_{\text{in}}} = \frac{2320}{2320} = 100\%$$

The value of 100 percent would be obtained for all four cases and hence shows that the first law indicates quantity of energy and not quality of energy as the second law does.

In this chapter we have discussed second law analysis of several devices separately. In power or refrigeration plants, several of these devices are combined to interact in a cycle. Second law analysis of a cycle will be discussed in Chapter 10, which deals with power and refrigeration cycles. Such second law analysis in conjunction with a first law analysis can indicate which devices in the cycle are most susceptible to improvement and hence deserve our effort to improve the overall performance of the cycle.

PROBLEMS

In all problems in this chapter, assume that the surroundings are at 25°C or 77°F and 101 kPa or 14.7 psia.

7.1 In an adiabatic compression, the pressure of a steady air flow is increased from 14.7 psia to 120 psia. The inlet air temperature is 40°F; outlet air temperature is 500°F. Determine the reversible work and irreversibility for this process per pound of air.

7.2 Superheated steam at 1500 psia, 600°F undergoes a throttling process to a pressure of 67 psia. Determine the availability change and irreversibility for this process.

7.3 In a refrigeration system, R-22 is throttled from saturated liquid at 50°C to a pressure of 500 kPa. Determine the availability change and irreversibility for this isenthalpic process (see Section 3.4).

7.4 Steam enters a well-insulated turbine at 650°C, 1000 kPa and is expanded to 1.0 kPa. If the work output of the turbine is 92 percent of that for an isentropic turbine, determine the actual work output per pound of steam, the reversible work for the actual change of state, and the irreversibility.

7.5 Repeat Problem 7.4 if the work output is 96 percent of that for an isentropic turbine. Repeat if the work output is equal to that of an isentropic turbine.

7.6 A glass of diameter 2.5 in. and height 5 in. is filled with water at 100°F. The water is allowed to cool to 77°F by means of heat transfer with the surroundings. Calculate the irreversibility for this process.

7.7 Air enters a nozzle with a velocity of 50 ft/s, pressure of 50 psia, and temperature of 200°F. The air flow is allowed to expand in the nozzle to a pressure of 20 psia, with the air temperature at the nozzle exit measured to be 80°F. There is a heat loss from the nozzle of 2.5 Btu per pound of air. Determine the actual velocity at nozzle exit, the reversible work, and the irreversibility for this nozzle process.

7.8 Combustion gases are contained in a rigid vessel at 1000 kPa, 1000°C. These gases are allowed to cool to 500°C by heat transfer to the surroundings. Determine the final pressure of the gases in the tank, the heat lost from the gases, the availability change, and the irreversibility for this process. Assume that the gases behave as a perfect gas with constant specific heats, $\gamma = 1.35$, $c_v = 1.6$ kJ/kgK.

7.9 A perfect gas is expanded in a piston-cylinder arrangement from 50 cm^3 to 500 cm^3. The process is reversible and adiabatic, with the initial gas pressure equal to 200 kPa. The gas obeys the relation $p\Psi^{\gamma}$ = constant, with $\gamma = 1.35$, $c_p = 1.2$ kJ/kgK. Determine the availability of the gas before and after the process and the actual work done by the gas.

7.10 Air at 100 psia, 100°F is contained in a chamber of volume 100 in^3, as shown. An adjacent chamber is also of volume 100 in^3, but it contains air at 100 psia, 20°F. The chambers are separated by a membrane, and the chambers are contained within a rigid, well-insulated vessel. The membrane is broken, allowing the two air masses to mix. Determine the total availability before mixing and after mixing and the irreversibility of this process.

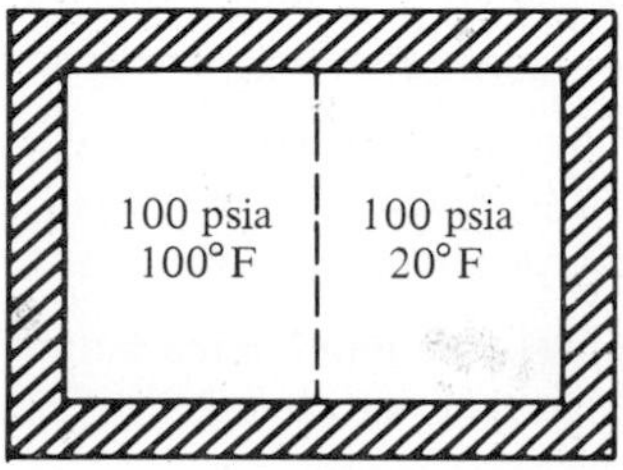

FIGURE 7.9

7.11 10,000 kg/h of steam with 10 percent quality are to be condensed to saturated liquid in a water-cooled condenser. Liquid water enters the condenser at 10°C and is to rise by no more than 8°C. The steam pressure is 7.5 kPa. For minimal cooling water flow, determine the irreversibility per unit time for this process and the total availability per unit time before and after the process. Assume that the condenser is well insulated so that there is no external heat loss.

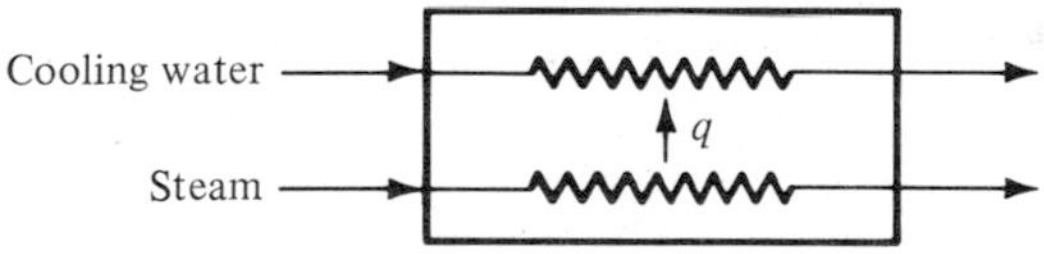

FIGURE 7.10

7.12 Determine the volume of a tank of superheated steam that would provide the same available energy as a 12-V car battery rated at 60 ampere-hours. Assume that the steam is at 100 psia, 600°F.

7.13 An ice cube one inch on a side is in a glass at a temperature of 0°C. The ice cube is allowed to melt and the resultant liquid water is warmed to 25°C by contact with the surroundings at 25°C. Determine the change of availability for the H_2O.

7.14 Air is heated in a heat exchanger from 0°F to 60°F by combustion gases that cool from 500°F to 350°F. The specific heat of the gases can be taken as 0.287 Btu/lbm°R. Determine the irreversibility for this process per pound of air, and the change of total availability per pound of air.

7.15 A 1.0-kg mass of metal at 500°C is quenched in a cylindrical tank of water 50 cm in diameter and 30 cm high, with the initial water temperature equal to 20°C. Determine the total change of availability and the irreversibility for this process. Take the specific heat of the metal as 0.5 kJ/kgK.

7.16 In a constant pressure, steady flow process, a stream of oxygen and a stream of nitrogen are combined to produce a mixture that contains 3.76 moles of nitrogen per mole of oxygen. Determine the reversible work per mole of mixture. Assume that the oxygen and nitrogen both enter the process at 1 atmosphere and 77°F.

7.17 Determine the minimum work in steady flow required to separate air into its components at 77°F, one atmospheric pressure. Assume that air is made up of 3.76 moles of nitrogen per mole of oxygen.

7.18 Steam at a pressure of 900 psia and a temperature of 1000°F is throttled to a pressure of 700 psia. Calculate the change in availability and the irreversibility for this throttling process.

7.19 Air enters a steady-flow compressor at a pressure of 14.7 psia and a temperature of 80°F and leaves it at a pressure of 75 psia. The compressor efficiency (ratio of work required for an isentropic compression process to that required for the actual compression process) is 85 percent. Find the required work input, the change in availability, and the irreversibility of the process. Consider air as an ideal gas with $c_p = 0.24$ Btu/lbm°R and $\gamma = 1.4$.

7.20 Flue gases are used to preheat the air supply required for combustion. The flue gases are cooled from 325°C to 200°C. The flue gas flow rate is 5 percent greater than the air supply rate. The air initially has a temperature of 30°C with $c_p = 1.01$ kJ/kgK. Calculate the availabilities of the flue gases and the air, and the irreversibilities of the heat exchange. Use $c_p = 1.10$ kJ/kgK for the flue gases.

chapter

8

Thermodynamic Processes

8.1 INTRODUCTION

In the preceding chapters, we studied the basic principles of equilibrium thermodynamics and expressed them in terms of the first and second laws of thermodynamics. In the remainder of this text, we will deal with applications of these principles. In this chapter, we will study several different types of thermodynamic processes and will indicate procedures and methods for solving engineering problems involving these processes.

In a thermodynamic process, the state of a substance comprising a system, or perhaps the state of a substance flowing across system boundaries, is changed due to the addition of heat or the performance of work. In many situations, there are constraints that cause one thermodynamic property to remain constant during a process. For example, if the process takes place in a closed, rigid vessel, such that no mass enters or leaves the vessel and the total volume of the vessel contents does not change, then the specific volume of the substance inside the vessel remains constant whatever process is taking place in the vessel. As another example, if a process occurs without heat transfer and is also reversible, then the entropy is constant. In a thermodynamics problem, we must be able to find the state of substance at the initial and final states of a process. To determine the state of a pure substance, in the absence of magnetic, surface, or gravity effects, one must specify two independent properties (see Section 4.3).

Thus it becomes very important in approaching a problem to be able to analyze the thermal and physical constraints so as to find which one, if any, of the thermodynamic properties of the substance comprising the system is maintained constant.

8.2 CONSTANT VOLUME PROCESS

A constant volume process will occur, for example, when a process takes place in a rigid container. If, in addition, the container is closed so that no mass can enter or leave the container, then the system comprising the contents of the container undergoes a process at constant specific volume.

EXAMPLE 8.1

Vapor of the refrigerant R-22 is sealed in a glass vial at 100°F. The vial is then cooled to a temperature at which tiny droplets of R-22 just begin to form on the inside walls of the vial. This temperature is measured to be 40°F. From this information, determine the pressure inside the vial at 100°F.

Solution

This process takes place at constant volume with the final state at saturation conditions. Since the mass of the system (R-22 inside the vial) remains constant, the specific volume of the system is constant during this process; that is, $v_2 = v_1$. From the thermodynamic tables for R-22 (Table A.10), we get, at 40°F, $v_2 = v_g = 0.65753\ \text{ft}^3/\text{lbm}$ and $p_2 = p_{\text{sat.}} = 83.206$ psia. At a temperature of 100°F, we have superheated R-22 with $v_1 = v_2 = 0.65753\ \text{ft}^3/\text{lbm}$. With two independent properties now known at state 1, namely, T and v, state 1 is determined and p_1 can be found from Table A.11. At 100°F and 80 psia, $v = 0.80477\ \text{ft}^3/\text{lbm}$; at 100°F and 100 psia, $v = 0.63003\ \text{ft}^3/\text{lbm}$. Interpolating, we obtain $p_1 = 96.85$ psia. Figure 8.1 illustrates the process on a p-T diagram.

EXAMPLE 8.2

A closed, insulated, rigid vessel having a volume of $10\ \text{m}^3$ contains saturated steam at 100°C and 50 percent quality. The mixture is stirred with a paddle wheel

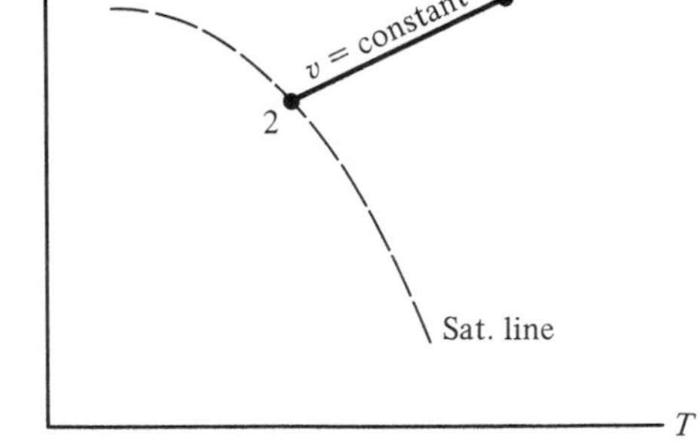

FIGURE 8.1 *Constant Specific Volume Process.*

until the temperature inside the vessel reaches 120°C, as shown in Figure 8.2(a). Find the work required.

Solution

At state 1, the temperature T and quality x are known. Since these two properties are independent, state 1 is determined. From Table B.2, we obtain

$$v_1 = v_f + 0.50v_{fg} = 0.0010437 + 0.50(1.672) = 0.8370 \text{ m}^3/\text{kg}$$

At state 2, $v_2 = v_1$ since the vessel is closed and rigid. Therefore, $v_2 = 0.8370 \text{ m}^3/\text{kg}$ and $T_2 = 120°\text{C}$. Again, with two independent properties, state 2 is determined. With $v_2 = 0.8370 \text{ m}^3/\text{kg}$,

$$x_2 = \frac{v_2 - v_f}{v_{fg}} = \frac{0.8370 - 0.00106}{0.8904} = 0.939$$

Figure 8.2(b) illustrates the process on a T-v diagram.

From the first law, $Q_{\text{in}} = W_{\text{out}} + U_2 - U_1$. For this case, with an insulated container, $Q_{\text{in}} = 0$. Therefore,

$$W_{\text{in}} = U_2 - U_1 = m(u_2 - u_1)$$

where

$$u_1 = u_f + 0.50u_{fg} = 418.95 + 0.50(2087.5) = 1462.7 \text{ kJ/kg}$$

and

$$m = \text{mass inside tank} = 10/0.8370 = 11.95 \text{ kg}$$

The internal energy at state 2 is

$$u_2 = u_f + x_2u_{fg} = 503.51 + 0.939(2025.5) = 2405.5 \text{ kJ/kg}$$

Therefore,

$$W_{\text{in}} = 11.95(2405.5 - 1462.7) = 11{,}270 \text{ kJ}$$

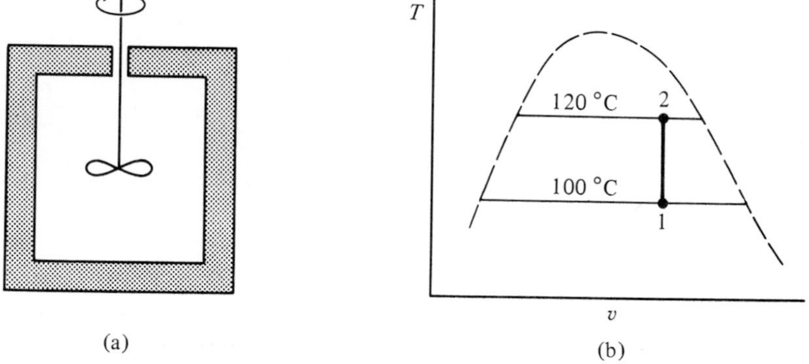

FIGURE 8.2 *Constant Specific Volume Stirring Process.*

8.3 CONSTANT PRESSURE (ISOBARIC) PROCESS

A constant pressure process might occur in a closed system when a portion of the boundary is moved (for example, by a piston or a diaphragm) sufficiently slowly so that the pressure within the system can equalize with that of the surroundings, which are at a constant value (for example, atmospheric pressure). A constant pressure process is illustrated in Figure 8.3 on *p-v* and *T-s* diagrams. Note that in the liquid-vapor regime, the constant pressure process is also a constant temperature process.

EXAMPLE 8.3

A tank having a volume of 10 ft^3 is charged with gaseous carbon dioxide at a temperature of 80°F and a pressure of 20 psia. Due to a leak in the tank's inlet valve, the tank continues to be charged. The pressure is kept constant at 20 psia by the extraction of heat. Determine the final state of the CO_2 in the tank. At the final state, three pounds of CO_2 are in the tank.

Solution

The mass of CO_2 within the tank is $m = \mathcal{V}/v = 10/v$. From Table A.7, at the initial state (80°F and 20 psia), specific volume v_1 is 6.545 ft^3/lbm, and hence $m_1 = 1.528$ lbm. Since the process proceeds at constant pressure, we can find, for different temperatures, the corresponding values of v_2 and m_2. Values are shown in the following table:

T(°F)	v(ft^3/lbm)	m(lbm)	
80	6.545	1.528	
60	6.297	1.588	
40	6.050	1.653	
20	5.802	1.724	
0	5.554	1.801	
−20	5.302	1.886	
−40	5.049	1.981	
−60	4.797	2.085	
−102.5	4.260	2.347	(Saturation)
−102.5	3.333	3.000	

When a temperature of −102.5°F is reached, the vapor-solid regime begins. Further addition of CO_2 does not change the temperature of the mixture, but rather only the quality. From Chapter 4, for the specific volume of a vapor-solid

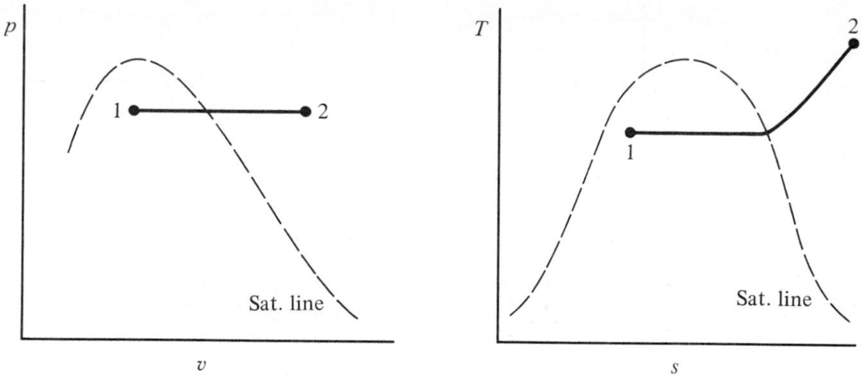

FIGURE 8.3 *Constant Pressure Process.*

mixture, $v = v_i + xv_{ig}$. Hence

$$x = \frac{v - v_i}{v_{ig}} = \frac{3.333 - 0.0103}{4.338} = 0.766 \text{ where } v \text{ at the final state is } 3.333 \text{ ft}^3/\text{lbm}.$$

The process is shown in Figure 8.4 on a p-T diagram.

EXAMPLE 8.4

Combustion gases in a piston-cylinder arrangement, as shown in Figure 8.5, occupy a volume of 100 cm^3 at a pressure of 1000 kPa and a temperature of 600 K. The gases are allowed to push the piston outward in the cylinder until the gases occupy a volume of 200 cm^3. During this process, heat is added to the gases so that the gas pressure remains constant. Assuming that the piston moves relatively slowly in the cylinder so that the gases remain very close to thermodynamic

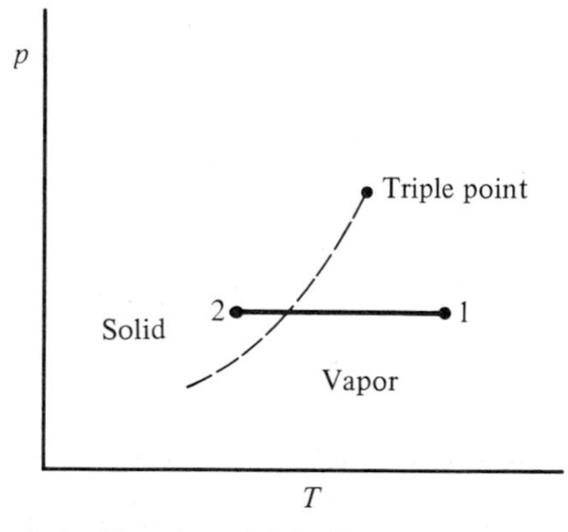

FIGURE 8.4 *Charging of CO_2 Tank at Constant Pressure.*

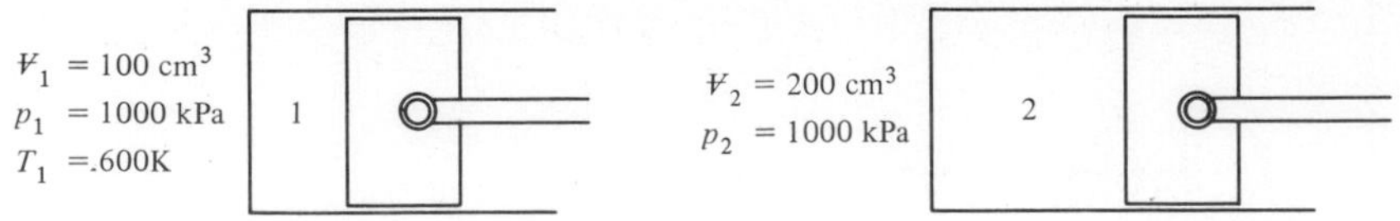

FIGURE 8.5 *Power Stroke of Piston-Cylinder.*

equilibrium during the entire (reversible) process, find the heat added to the gases and the work done. Assume that the combustion gases behave as an ideal gas with constant specific heat; take the gas constant R to be 300 J/kg K and the specific heat c_v to be 1.2 kJ/kg K.

Solution

From the first law, Equation (2.5), $Q_{in} = W_{out} + U_2 - U_1$. For this reversible constant pressure process,

$$W_{out} = \int_1^2 p\, dV = p(V_2 - V_1) = (1000 \times 10^3)\ \mathrm{N/m^2}(100 \times 10^{-6})\ \mathrm{m^3}$$
$$= 100\ \mathrm{J}$$

$$U_2 - U_1 = m(u_2 - u_1) = mc_v(T_2 - T_1)$$

The mass of the combustion gases is

$$m = \frac{pV}{RT} = \frac{(10^6\ \mathrm{N/m^2})(10^{-4}\ \mathrm{m^3})}{(300\ \mathrm{J/kg\ K})(600\ \mathrm{K})} = 0.0005556\ \mathrm{kg}$$

For this constant pressure process, $T_2/T_1 = V_2/V_1 = 2$; so $T_2 = 1200$ K. Substituting, we have

$$U_2 - U_1 = (0.0005556\ \mathrm{kg})(1200\ \mathrm{J/kg\ K})(1200\ \mathrm{K} - 600\ \mathrm{K})$$
$$= 400\ \mathrm{J}$$

Hence, $Q_{in} = W_{out} + U_2 - U_1 = 100 + 400 = 500$ J.

Alternatively, we can find Q_{in} for a reversible constant pressure process with an ideal gas with constant specific heat:

$$Q_{in} = W_{out} + U_2 - U_1 = p(V_2 - V_1) + U_2 - U_1$$

Since $H = U + pV$, the equation simplifies to

$$Q_{in} = H_2 - H_1$$
$$= mc_p(T_2 - T_1)$$

where $c_p = c_v + R = 1500$ J/kg K. Hence,

$$(0.0005556\ \mathrm{kg})(1500\ \mathrm{J/kg\ K})600\ \mathrm{K} = 500\ \mathrm{J} \quad \text{(as above)}$$

8.4 THROTTLING PROCESS (CONSTANT ENTHALPY)

A throttling process occurs when a fluid flows in a tube with a restriction (Figure 8.6) or through a valve. As discussed in Section 3.6, the heat transfer in this process is usually negligible and there is no mechanical shaft work. With negligible changes in potential and kinetic energies, the energy equation for steady flow indicates constant enthalpy. With a reduction in pressure due to frictional losses, the throttling process is a constant enthalpy irreversible process with no heat transfer; therefore the entropy of the fluid increases. The initial and final states for a pure substance undergoing a throttling process are shown on *p-h* and *h-s* diagrams in Figures 8.7 and 8.8.

EXAMPLE 8.5

Saturated liquid R-22 at a temperature of 30°C is throttled in a steady-flow process to a pressure of 354 kPa. Find the quality after throttling.

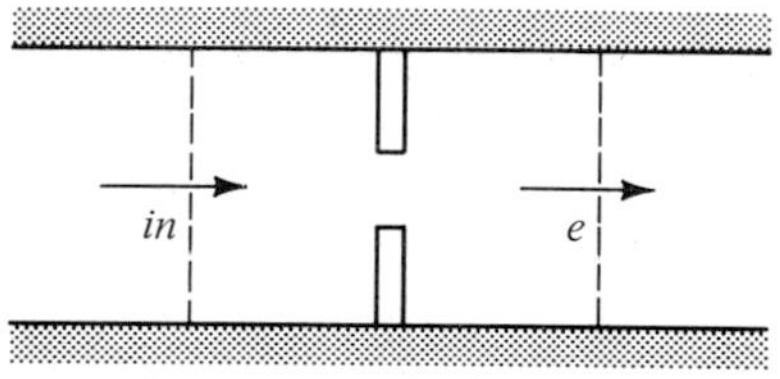

FIGURE 8.6 *Throttling Device.*

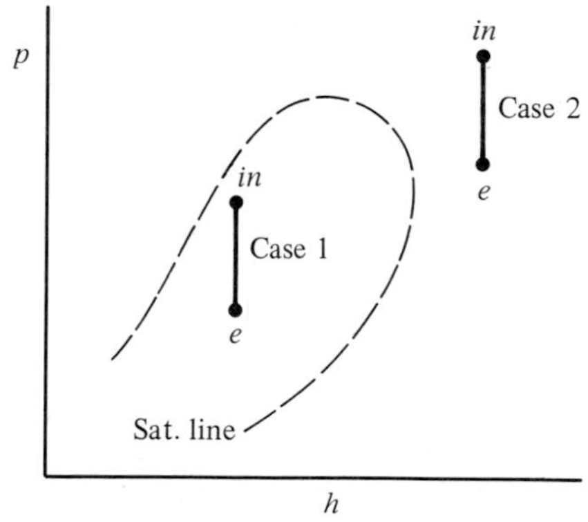

FIGURE 8.7 *p-h Diagram.*

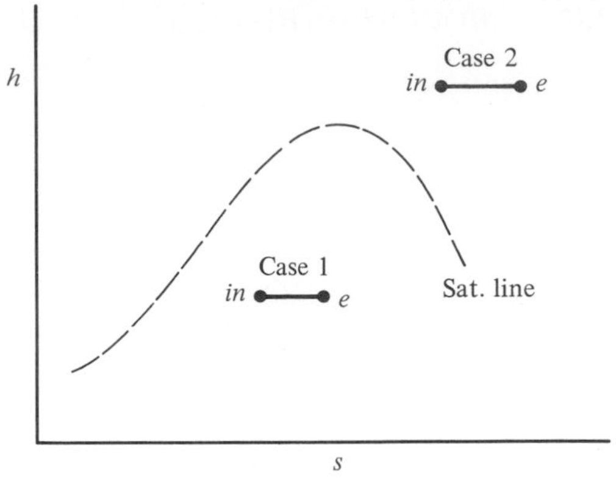

FIGURE 8.8 *h-s Diagram.*

Solution

For the throttling process, $h_e = h_{in}$. At *in*, the quality ($x = 0$) and temperature are known, so that $h_{in} = h_f$ at 30°C equals 81.25 kJ/kg (Table B.10). We know h_e and p_e at e, so state e is determined. At $p_e = 354$ kPa ($T_e = -10$°C), $h_f = 33.01$ kJ/kg and $h_{fg} = 213.13$ kJ/kg, so

$$x_e = \frac{h - h_f}{h_{fg}} = \frac{81.25 - 33.01}{213.13} = 0.226$$

EXAMPLE 8.6

Superheated steam at 100 psia and 600°F undergoes a throttling process to a pressure of 60 psia. Find the temperature after throttling and the entropy increase.

Solution

From Table A.3, we have, for the given inlet conditions, $h_{in} = 1329.6$ Btu/lbm and $s_{in} = 1.7586$ Btu/lbm°R. Since $h_e = h_{in}$ and the exit pressure is 60 psia, the state of the steam after throttling is determined. The values (from Table A.3) are given in the following table:

T(°F)	h(Btu/lbm)	s(Btu/lbm°R)	
500	1283.2	1.7681	
594.5	1329.6	1.8141	(By interpolation)
600	1332.3	1.8168	

Using these values, we obtain a temperature after throttling of 594.5°F. The change in entropy is

$$s_e - s_{in} = 1.8141 - 1.7586 = 0.0555 \text{ Btu/lbm°R}$$

EXAMPLE 8.7

In a throttling process of an ideal gas with constant specific heats, the pressure is reduced by 20 percent. Find the changes of temperature and entropy of the gas, assuming the gas constant R to be 287 J/kg K.

Solution

For an ideal gas, the change in enthalpy is given by Equation (4.9):

$$h_e - h_{in} = c_p(T_e - T_{in})$$

For this isenthalpic process, $h_e = h_{in}$, so that $T_e = T_{in}$; in other words, the temperature does not change.

From Equation (6.25), the entropy change for an ideal gas with constant specific heat is

$$s_e - s_{in} = c_p \ln T_e/T_{in} - R \ln p_e/p_{in}$$

or

$$s_e - s_{in} = -R \ln p_e/p_{in} \text{ (for } T_e = T_{in})$$

$$= -287 \ln 0.8 = 64.0 \text{ J/kg K}$$

8.5 ISOTHERMAL PROCESS

An isothermal process might occur, for example, during the compression of a substance in a piston-cylinder arrangement if sufficient heat is removed from the substance during the compression. In a steady-flow air compressor, if sufficient cooling is provided, the air temperature can be maintained constant. Further, if the changes of state occurring during a process are slow enough, the temperature of the substance comprising the system will have a chance to equalize with the constant temperature of the surroundings, as long as the substance is in thermal contact with the surroundings.

If a thermodynamic process is both *reversible* and *isothermal*, the expression for heat transfer from the second law, Equation (6.13), namely, $dQ_{in(rev)} = T\,dS$, becomes easy to integrate and results in

$$Q_{in} = T(S_2 - S_1) \quad \textbf{(8.1)}$$

If, in addition, the substance comprising the system is an ideal gas with constant specific heats, there is no change of either internal energy ($du = c_v\, dT$) or enthalpy ($dh = c_p\, dT$) for an isothermal process. For this case, the first law, in the absence of changes of kinetic and potential energies, becomes

$$Q_{in} = W_{out} = T(S_2 - S_1) \qquad \textbf{(8.2)}$$

From Equations (6.23) and (6.25), for an ideal gas with constant specific heats,

$$s_2 - s_1 = c_p \ln T_2/T_1 - R \ln p_2/p_1 = c_v \ln T_2/T_1 + R \ln v_2/v_1$$

If, an addition, the process is isothermal, then

$$s_2 - s_1 = -R \ln p_2/p_1 = R \ln v_2/v_1 \qquad \textbf{(8.3)}$$

Note that the above equation is consistent since $p_2/p_1 = v_1/v_2$ for an ideal gas ($pv = RT$) with T constant.

Combining Equations (8.2) and (8.3), we obtain, for an ideal gas undergoing a reversible, isothermal process

$$Q_{in} = W_{out} = -mRT \ln p_2/p_1 = mRT \ln v_2/v_1 \qquad \textbf{(8.4)}$$

As a check on the expression just derived, we know that for a piston-cylinder arrangement in which the process is reversible,

$$W_{out} = \int_1^2 p\, d\forall = m \int_1^2 p\, dv$$

For an ideal gas, $p = RT/v$, so

$$W_{out} = mR \int_1^2 T\, dv/v$$

For the isothermal process, we can simplify the above to $W_{out} = mRT \ln v_2/v_1$, consistent with Equation (8.4).

EXAMPLE 8.8

Steam is contained in a piston-cylinder arrangement at a temperature of 200°C, a pressure of 1000 kPa, and a total volume of 40 cm^3. The steam is allowed to expand slowly to a volume of 210 cm^3, pushing the piston outward and performing work. During the expansion, the steam is maintained at a constant temperature of 200°C. Assuming a reversible process, calculate the heat added to the steam and the work done by the steam.

Solution

At the initial state, the steam is superheated; with two independent properties, p and T, the initial state is determined. From Table B.3, $v_1 = 0.2059$ m^3/kg and

$u_1 = 2620.9$ kJ/kg. With no mass entering or leaving the cylinder, we can write

$$\frac{v_2}{v_1} = \frac{V_2/m_2}{V_1/m_1} = \frac{V_2}{V_1} = \frac{210}{40} = 5.25$$

Hence $v_2 = 1.081\ \text{m}^3/\text{kg}$. With T and v, two independent thermodynamic properties, the final state is determined. From Table B.3, $p_2 = 200$ kPa and $u_2 = 2654.4$ kJ/kg. From the first law for a closed system, Equation (2.5), we have

$$Q_{\text{in}} = W_{\text{out}} + U_2 - U_1$$

To find Q_{in}, we use Equation (8.1); $Q_{\text{in}} = T(S_2 - S_1) = mT(s_2 - s_1)$. For this example,

$$m = V/v = 40 \times 10^{-6}\ \text{m}^3/0.2059\ \text{m}^3/\text{kg} = 1.943 \times 10^{-4}\ \text{kg}$$

From Table B.3, $s_1 = 6.6922$ kJ/kg K and $s_2 = 7.5072$ kJ/kg K; also

$$Q_{\text{in}} = (1.943 \times 10^{-4}\ \text{kg})(473\ \text{K})(7.5072 - 6.6922)\ \text{kJ/kg K}$$
$$= 0.0749\ \text{kJ}$$

$$W_{\text{out}} = Q_{\text{in}} - m(u_2 - u_1) = 0.0749 + 1.943 \times 10^{-4}\ \text{kg}(2620.9 - 2654.4)\ \text{kJ/kg}$$
$$= 0.0684\ \text{kJ}$$

EXAMPLE 8.9

In a steady-flow air compressor (Figure 8.9), the air pressure is raised from 15 psia to 100 psia. Sufficient cooling is provided to maintain a constant air temperature of 100°F. For an intake air flow of 10.0 cfs, determine the horsepower required. Assume a reversible process, and neglect changes in potential and kinetic energies. Also assume that the air behaves as an ideal gas with constant specific heat, with $R = 53.3$ ft-lbf/lbm°R and $c_p = 0.24$ Btu/lbm°R.

Solution

For this case, the first law for steady flow, Equation (3.15), reduces to

$$q_{\text{in}} = w'_{\text{out}} + h_e - h_{\text{in}}$$

where, for an ideal gas with constant specific heat,

$$h_e - h_{\text{in}} = c_p(T_e - T_{\text{in}}) = 0$$

for this isothermal process. From the second law, we have using Equation (6.13):

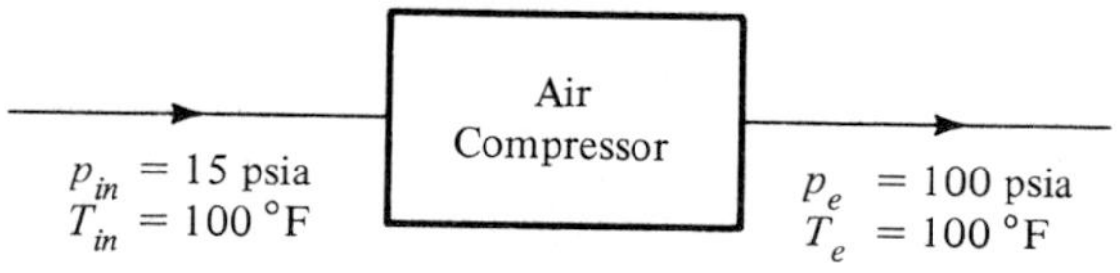

FIGURE 8.9 *Steady-flow Air Compressor.*

$q_{in} = T(s_e - s_{in})$, where from Equation (8.3):

$$s_e - s_{in} = -R \ln p_e/p_{in} = -53.3 \ln 6.67 = -101.1 \text{ ft-lbf/lbm°R}$$

Hence,

$$q_{in} = -(101.1 \text{ ft-lbf/lbm°R})(560°\text{R}) = -56{,}620 \text{ ft-lbf/lbm} = w'_{out}$$

$$\text{Power required} = (w'_{out}) \times \text{mass flow rate } (\dot{m})$$

where $\dot{m} = A_1 V_1 / v_1$. For this case,

$$v_1 = RT/p_1 = (53.3)560/15(144) = 13.82 \text{ ft}^2\text{/lbm}$$

$$\dot{m} = (10.0 \text{ ft}^3\text{/s})/13.82 \text{ ft}^3\text{/lbm} = 0.7236 \text{ lbm/s}$$

Therefore, the horsepower required is

$$\text{Required hp} = \frac{56{,}620 \text{ ft-lbf/lbm} \times 0.7236 \text{ lbm/s}}{550 \text{ ft-lbf/s hp}} = 74.5 \text{ hp}$$

8.6 ADIABATIC AND ISENTROPIC PROCESSES

A process during which no heat is transferred into or out of the system (that is, $Q = 0$) is called an *adiabatic process*. Such a process will occur, for example, in a well-insulated steam turbine in which high-pressure steam is expanded to do work on the turbine rotor without losing heat to the environment. The initial and final states of a pure substance undergoing an adiabatic process are shown on *T-s* diagrams in Figure 8.10.

EXAMPLE 8.10

A closed system of gaseous air is compressed adiabatically from a pressure of 15 psia and a temperature of 70°F to a pressure of 60 psia and a temperature of 350°F [see Figure 8.10(a)]. Using ideal gas approximations, compute the work input required and the entropy change of the process

---

Solution

The work input for this adiabatic process (no heat transfer) is, from Equation (2.5),

$$w_{in} = u_2 - u_1$$

which for an ideal gas with constant c_v becomes

$$w_{in} = c_v(T_2 - T_1)$$

From Table A.15, we obtain $c_v = 0.172$ Btu/lbm°R at 70°F and $c_v = 0.175$ Btu/lbm°R at 350°F. Using an average c_v of 0.174 Btu/lbm°R, we have

$$w_{in} = 0.174 \text{ Btu/lbm°R}(350°\text{F} - 70°\text{R}) = 48.72 \text{ Btu/lbm}$$

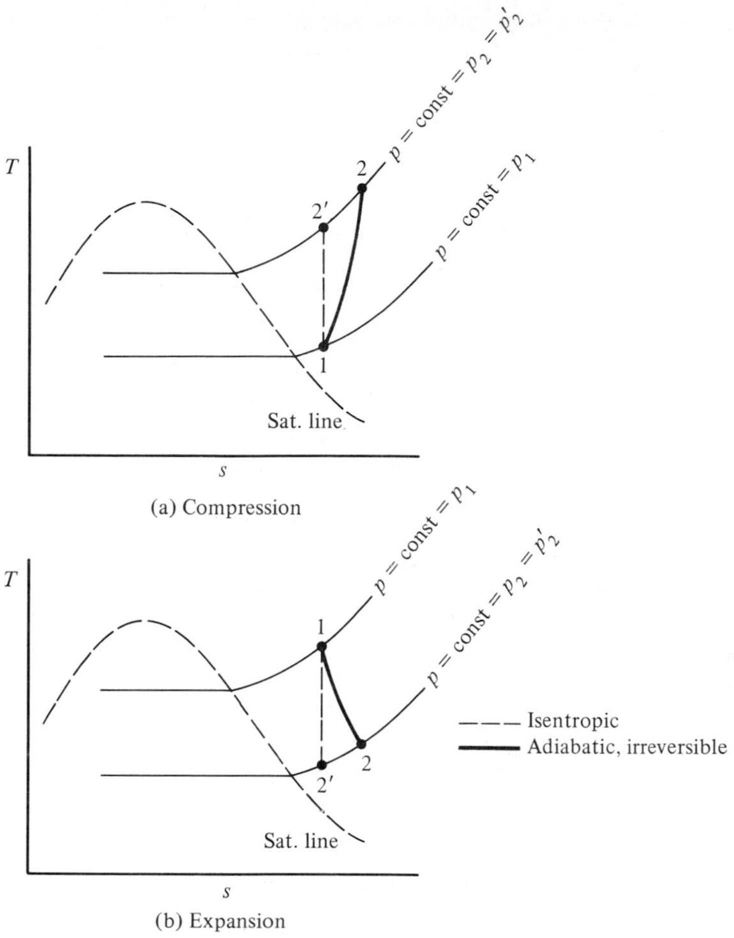

FIGURE 8.10 *Adiabatic and Isentropic Processes.*

The entropy change is obtained from Equation (6.25):

$$s_2 - s_1 = c_p \ln T_2/T_1 - R \ln p_2/p_1$$

$$= (0.242 \text{ Btu/lbm°R}) \ln (810/530) - (0.0686 \text{ Btu/lbm°R}) \ln (60/15)$$

where $c_p = 0.242$ Btu/lbm°R is the average value obtained from Table A.15, and $R = 53.35$ ft-lbf/lbm°R $= 0.0686$ Btu/lbm°R from Table A.1. Hence,

$$s_2 - s_1 = 0.1026 - 0.0951 = 0.0075 \text{ Btu/lbm°R}$$

Entropy was defined by Equation (6.15) as:

$$S_2 - S_1 = \int_1^2 \frac{dQ_{\text{in}}}{T} + (\text{irreversibility term})$$

Thus, if a process is both adiabatic and reversible, $S_2 - S_1 = 0$; in other words, entropy is constant during this process. Such a constant entropy process is called an *isentropic process*.[1] A comparison between the final states of isentropic and adiabatic irreversible processes with identical initial states is shown in Figure 8.10.

EXAMPLE 8.11

In a steady-flow refrigeration compressor, saturated R-22 vapor at −10°C is raised to 800 kPa. The inlet flow velocity is 3.0 m/s, and the outlet flow velocity 1.6 m/s. For a flow of 100 kg/h, find the required input power to the compressor. Assume that the compression process is isentropic.

Solution

The first law for this process is Equation (3.19),

$$q_{\text{in}} = w'_{\text{out}} + h_e - h_{\text{in}} + (V_e^2 - V_{\text{in}}^2)/2g_c = 0$$

since the process is isentropic (adiabatic and reversible).

The inlet state is determined from the temperature T and the quality ($x = 1.0$), so that $h_{\text{in}} = h_g$ at $-10°\text{C} = 246.14$ kJ/kg, and $s_{\text{in}} = 0.94224$ kJ/kg K. At the outlet, $p = 800$ kPa and $s_e = s_{\text{in}} = 0.94224$ kJ/kg K. Again, with two independent properties, state e is determined.

From Table B.11, at 800 kPa and 50°C, $s = 0.99155$ kJ/kg K. At the saturation line at 800 kPa, from Table B.11, $s_g = 0.90558$ kJ/kg K at $T_g = 15.47$°C. Since the actual s_e is 0.94224 kJ/kg K, it follows by interpolation that $T_e = 30.17$°C. Similarly, at 800 kPa,

$$h = 255.16 \text{ kJ/kg at } 15.47°\text{C} \quad \text{and} \quad h = 281.73 \text{ kJ/kg at } 50°\text{C}$$

Therefore, $h_e = 266.47$ kJ/kg. Substituting into the first law equation, we have

$$\begin{aligned} w'_{\text{out}} &= h_{\text{in}} - h_e + (V_{\text{in}}^2 - V_e^2)/2g_c \\ &= (246.14 - 266.47) \text{ kJ/kg} + \frac{(3^2 - 1.6^2)\text{ m}^2/\text{s}^2}{2(1)\text{ kg}\cdot\text{m/N}\cdot\text{s}^2} 10^{-3} \text{ kJ/N}\cdot\text{m} \\ &= -20.33 + 0.00322 = -20.33 \text{ kJ/kg} \end{aligned}$$

$$\begin{aligned} \text{Power in} &= 100 \text{ kg/h} \times \frac{1}{3600} \text{ h/s} \times 20.33 \text{ kJ/kg} \times 1 \text{ kW/kJ/s} \\ &= 0.565 \text{ kW} \end{aligned}$$

---

1. In principle, a constant entropy process could also be obtained in an irreversible process (increase in entropy) with sufficient cooling (decrease in entropy) to result in the same value of entropy at initial and final states of the process.

For an ideal gas undergoing an isentropic process, we can derive expressions for the variation of thermodynamic properties. From (6.20) and (6.21),

$$T\,ds = du + p\,dv = dh - v\,dp$$

For an isentropic process, $ds = 0$, so that

$$du + p\,dv = dh - v\,dp = 0$$

or, for an ideal gas,

$$c_v\,dT + p\,dv = c_p\,dT - v\,dp = 0$$

Substituting, we have:

$$dT = -\frac{p\,dv}{c_v}$$

or

$$v\,dp + \frac{c_p}{c_v}\,p\,dv = 0$$

For constant specific heats, integrate to obtain:

$$pv^{\gamma} = \text{constant, where } \gamma = \frac{c_p}{c_v}$$

This relation and its use will be discussed in more detail in Section 8.7.

As we discussed in Chapter 6, a reversible process, and hence an isentropic process, are idealizations. All actual processes involve dissipative effects and departures from equilibrium. However, the isentropic process is a useful concept in many cases, since it represents the best that can be achieved with a device. In an actual process, frictional losses and other irreversibilities result in an increase of entropy. To account for losses during the real process, the concept of efficiency of a device is introduced; that is, the efficiency of a device is a comparison between the actual, but adiabatic, performance of the device and its performance if it is operated isentropically between the same initial and final pressure conditions. For example, for a steam or gas turbine,

$$\eta_{\text{turbine}} = \frac{w'_{\text{out(actual)}}}{w'_{\text{out(isentropic)}}} \tag{8.5}$$

where the denominator is equal to the work performed by the turbine if it is operated isentropically between the same initial and final pressures of the actual turbine. Naturally, turbine efficiency is less than one; however, well-designed modern turbines have efficiencies as high as 90 percent. For a pump or compressor, work input to the device is required. Here the work input to an isentropic pump or compressor will be less than that of the actual device, operating between the same intake and exhaust pressures, due to the effects of friction and other irreversibilities. Therefore, defining an efficiency that will be less than 100 percent, we have

$$\eta_{\text{pump or compressor}} = \frac{w'_{\text{in(isentropic)}}}{w'_{\text{in(actual)}}} \tag{8.6}$$

where, again, $w'_{\text{in(isentropic)}}$ is the work input required for a device operating isentropically between the same pressures as the actual pump or compressor.

EXAMPLE 8.12

In an adiabatic steam turbine, steam is expanded from 1000 psia, 1000°F, to 1.0 psia (Figure 8.11). The efficiency of the turbine is 90 percent. Determine the flow required to produce an output of 100 megawatts. Also find the exhaust steam quality. Neglect changes of potential and kinetic energies as the steam flows through the turbine.

Solution

In this case, we will first calculate the isentropic work between the given pressures and then multiply by the efficiency to find the actual work. For inlet conditions of 1000 psia, 1000°F, from Table A.3:

$$s_{\text{in}} = 1.6530 \text{ Btu/lbm°R} \quad \text{and} \quad h_{\text{in}} = 1505.4 \text{ Btu/lbm}$$

For an isentropic turbine, $s_e = s_{\text{in}}$ (Figure 8.12). From Table A.2 at 1.0 psia

$$s_f = 0.1326 \text{ Btu/lbm°R}, \quad s_g = 1.9781 \text{ Btu/lbm°R}$$

$$h_f = 69.73 \text{ Btu/lbm}, \quad h_g = 1105.8 \text{ Btu/lbm}$$

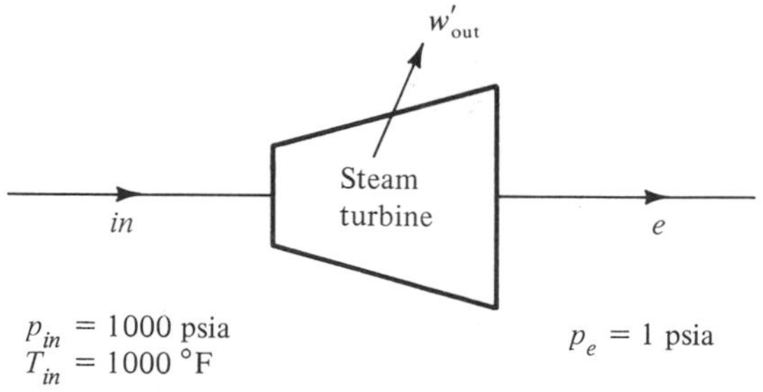

FIGURE 8.11

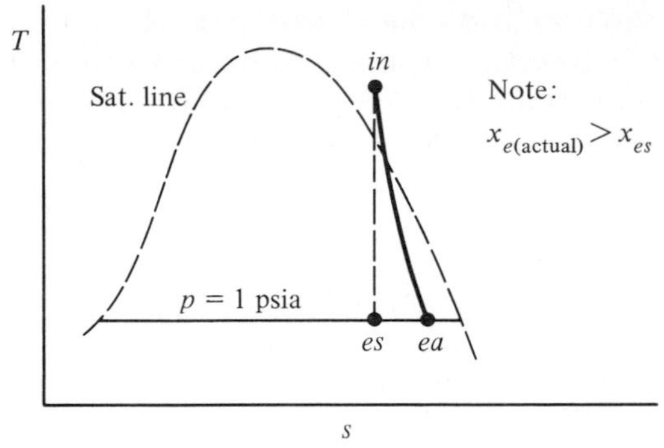

FIGURE 8.12

Solving for x_{es}, the exhaust quality for an isentropic turbine, we have

$$1.6530 = 0.1326 + x_{es}(1.9781 - 0.1326)$$

$$x_{es} = \frac{1.5204}{1.8455} = 0.824$$

Therefore,

$$h_{es} = 69.73 + 0.824(1105.8 - 69.73) = 69.73 + 853.72$$
$$= 923.5 \text{ Btu/lbm}$$

From Equation (3.18),

$$w'_{\text{out}(s)} = h_{\text{in}} - h_{es} = 1505.4 - 923.5 - 581.9 \text{ Btu/lbm}$$

From Equation (8.5),

$$w'_{\text{out(actual)}} = \eta w_{\text{out}(s)} = 0.9(581.9) = 523.7 \text{ Btu/lbm}$$

For an output of 100 megawatts (10^8 watts)

$$\dot{m} = \frac{(10^8 \text{ W})(0.9478 \times 10^{-3} \text{ Btu/s-W})}{523.7 \text{ Btu/lbm}}$$
$$= 181.0 \text{ lbm/s of steam}$$

We now determine $x_{e(\text{actual})}$.

For an adiabatic turbine,

$$h_{\text{in}} - h_{e(\text{actual})} = w'_{\text{out(actual)}}$$

Therefore,

$$h_{e(\text{actual})} = 1505.4 - 523.7 = 981.7 \text{ Btu/lbm}$$

$$x_{e(\text{actual})} = \frac{h_e - h_f}{h_g - h_f} = \frac{981.7 - 69.73}{1105.8 - 69.73} = 0.880$$

For a reversible process, the relationship for work for a steady-flow device can be reduced to a simplified, useful form.[2] Writing the first law for steady flow, Equation (3.15), in differential form, with no change of potential or kinetic energies, we have

$$dq_{\text{in}} = dw'_{\text{out}} + dh$$

From the second law, Equation (6.21), $T\,ds = dh - v\,dp$. Also, from (6.13), $dq_{\text{in}} = T\,ds$ for a reversible process. Therefore, substituting,

$$dw'_{\text{out}} = -v\,dp \qquad \textbf{(8.7)}$$

Repeating Equation (8.7) holds for a reversible steady-flow process in which changes of potential and kinetic energies are neglected. In order to integrate Equation (8.7), we must have a relationship between v and p. A simple case is that of an incompressible fluid, one for which v can be assumed constant. This assumption is a good one for liquids, except under extremely high pressures. For an incompressible fluid, Equation (8.7) reduces to

$$w'_{\text{out}} = -\int_{\text{in}}^{e} v\,dp = -v(p_e - p_{\text{in}}) \qquad \textbf{(8.8)}$$

Equation (8.8) holds for a reversible process; since an isentropic process is reversible (and adiabatic), (8.8) is valid for an isentropic process.

EXAMPLE 8.13

In a water pump, 100 gallons per minute are raised from atmospheric pressure, 60°F, to 100 psia; the pump efficiency is 80 percent. Determine the horsepower input to the pump.

Solution

Neglecting changes of potential and kinetic energies, we obtain, from Equations (8.6) and (8.8)

$$w'_{\text{out(actual)}} = \frac{w'_{\text{out}(s)}}{\eta} = -\frac{v(p_e - p_{\text{in}})}{\eta}$$

At 60°F, $v_f = 0.016033\ \text{ft}^3/\text{lbm}$ (the water is a slightly compressed liquid at this state), so that

$$w'_{\text{out}} = \frac{-(0.016033\ \text{ft}^3/\text{lbm})(100 - 14.7)\ \text{lbf}/\text{in}^2(144\ \text{in}^2/\text{ft}^2)}{0.80}$$

$$= -246\ \text{ft-lbf}/\text{lbm}$$

---

2. The corresponding expression for reversible work for a closed system was given in Equation (2.3).

(An alternate method of solution would have been to use $w'_{out(s)} = h_{in} - h_e$, with values of enthalpy determined from Table A.4 by interpolation. The latter solution is not as accurate as the first due to the complexities and approximations involved in interpolation.) Converting, we have 100 gpm = (100/7.48) cfm = 13.37 ft^3/min; therefore,

$$\dot{m} = \frac{13.37\ \text{ft}^3/\text{min}}{0.016033\ \text{ft}^3/\text{lbm}} = 833.8\ \text{lbm/min} = 13.9\ \text{lbm/s}$$

$$\begin{aligned}\text{Power in} &= \dot{m}w'_{\text{in(actual)}} = (13.9\ \text{lbm/s})(246\ \text{ft-lbf/lbm})\\ &= 3419\ \text{ft-lbf/s}\\ &= 6.22\ \text{hp}\end{aligned}$$

8.7 POLYTROPIC PROCESSES

A reversible process involving an ideal gas with heat added according to the relation

$$dq = c\,dT$$

is called a *polytropic* process (from the Greek meaning *multichange*). The variable c is the specific heat and thus is a function of the amount of energy transferred. We encountered and defined two special types of specific heat in Section 4.5. They were specific heats for a constant volume process and a constant pressure process. Here we place no such restriction on the process; rather, we require only that the amount of heat transferred be a function of c and the temperature difference.

For a reversible process, the second law has the form $dq_{in} = T\,ds$, where, from Equations (6.20) and (6.21),

$$T\,ds = du + p\,dv = dh - v\,dp$$

For an ideal gas, the internal energy and enthalpy are functions only of temperature; that is, $du = c_v\,dT$ and $dh = c_p\,dT$ for all processes, including nonconstant pressure, nonconstant volume processes. Combining the above, we obtain

$$\begin{aligned}dq_{in} &= c\,dT = c_v\,dT + p\,dv\\ &= c\,dT = c_p\,dT - v\,dp\end{aligned}$$

Dividing the two equations, we have

$$\frac{c - c_v}{c - c_p} = -\frac{p}{v}\frac{dv}{dp}$$

or

$$\frac{dp}{p} + \frac{c - c_p}{c - c_v}\frac{dv}{v} = 0 \qquad \textbf{(8.9)}$$

We have thus obtained a differential equation relating v and p. For brevity, let $n = (c - c_p)/(c - c_v)$. As can be proven by substitution, the solution of Equation (8.9) for constant specific heat is

$$pv^n = \text{constant} \tag{8.10}$$

or, for a polytropic process between states 1 and 2,

$$\frac{p_2}{p_1} = \left(\frac{v_1}{v_2}\right)^n \tag{8.11}$$

Equations (8.11) represent a relation between pressure and specific volume for a polytropic process. Using Equation (8.11) and the equation of state for an ideal gas, $pv = RT$, we can also get relationships between p and T or v and T:

$$Tv^{n-1} = \text{constant}$$

or

$$\frac{T_2}{T_1} = \left(\frac{v_1}{v_2}\right)^{n-1} \tag{8.12}$$

and

$$Tp^{n/(n-1)} = \text{constant}$$

or

$$\frac{T_2}{T_1} = \left(\frac{p_2}{p_1}\right)^{(n-1)/n} \tag{8.13}$$

Remember that Equations (8.11), (8.12), and (8.13) have all been derived for a reversible process, ideal gas with constant specific heats. Under these assumptions, the relationship between the three thermodynamic variables p, v, and T (equation of state) was reduced to a relationship between two variables (that is, p and v, p and T, or T and v). As long as the assumptions inherent in their derivation are valid, these equations provide very useful tools for the solution of thermodynamic problems, since they essentially take the place of property tables.

Several special polytropic processes occur, depending on the value of the specific heat c:

1. When no heat is transferred, $c = 0$. Hence, $n = c_p/c_v$ (call the ratio γ). As mentioned earlier, a process without heat transfer is called *adiabatic*. Hence this special process is the reversible adiabatic (*isentropic*) process.
2. When $c = c_p$, $n = 0$. Hence, p is constant and we have a *constant pressure* process.
3. When $c = c_v$, $n \to \infty$. Hence, v is constant and we have a *constant volume* process.
4. When $c \to \infty$, $n = 1$. Hence, from Equations (8.12), T is constant and we have a *constant temperature* process.

The variation of pressure with specific volume for polytropic processes having different values of the exponent n is given in Figure 8.13.

Next we will obtain expressions for the work output (w_{out}) of reversible polytropic processes. The work output for a closed system such as a piston-cylinder arrangement is, from Equation (2.3),

$$w_{out} = \int_1^2 p\,dv$$

From one of the expressions for dq_{in} obtained earlier,

$$(c - c_v)\,dT = p\,dv$$

or, for constant specific heats,

$$(c - c_v)(T_2 - T_1) = \int_1^2 p\,dv = w_{out}$$

with $n = (c - c_p)/(c - c_v)$ and $c_p - c_v = R$, we obtain

$$c - c_v = -\frac{R}{n-1}$$

Therefore, for a closed system,

$$w_{out} = -\frac{R}{n-1}(T_2 - T_1) \tag{8.14}$$

Using the equation of state of an ideal gas, $pv = RT$, we can give Equation (8.14) in an alternate form:

$$w_{out} = -\frac{1}{n-1}(p_2v_2 - p_1v_1) \tag{8.15}$$

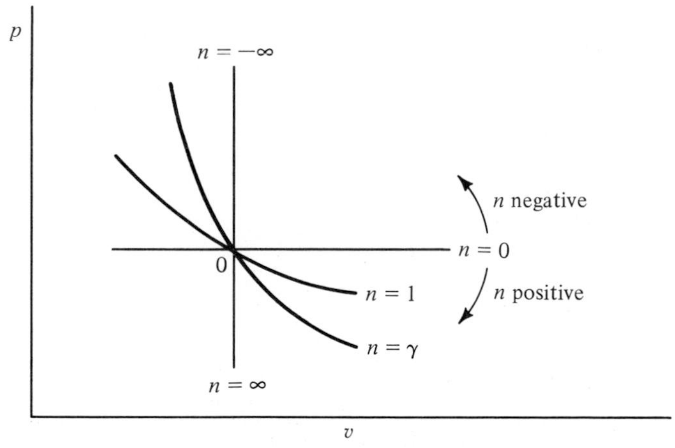

FIGURE 8.13 *Polytropic Processes.*

We could have obtained this result also by integrating the expression

$$\int_1^2 p\,dv$$

using the relation between p and v given in Equation (8.10).

For the special polytropic processes described above, we get the following for *closed systems*, from Equation (8.14) or (8.15):

For $n = 1$ (constant temperature), $w_{out} = 0/0$, which is indeterminate. However, from Equation (8.4), $w_{out} = RT \ln v_2/v_1$.
For $n \to \infty$ (constant volume), $w_{out} = 0$.
For $n = 0$ (constant pressure), $w_{out} = p(v_2 - v_1)$.
For $n = c_p/c_v = \gamma$ (constant entropy),

$$w_{out} = -\frac{R}{\gamma - 1}(T_2 - T_1)$$

$$= -\frac{1}{\gamma - 1}(p_2v_2 - p_1v_1)$$

The work output for a steady-flow reversible process with no changes of potential or kinetic energies is, from Equation (8.7),

$$w'_{out} = -\int_1^2 v\,dp$$

From one of the expressions for dq_{in} obtained earlier,

$$(c - c_p)\,dT = -v\,dp$$

or, for constant specific heats,

$$(c - c_p)(T_2 - T_1) = -\int_1^2 v\,dp = w'_{out}$$

For $n = (c - c_p)/(c - c_v)$ and $c_p - c_v = R$, we also obtain

$$c - c_p = -\frac{nR}{n - 1}$$

Therefore, for steady reversible flow,

$$w'_{out} = -\frac{nR}{n - 1}(T_2 - T_1) \qquad \textbf{(8.16)}$$

For this case of *reversible steady flow*, we obtain the following from Equation (8.16):

For $n = 1$ (constant temperature), $w'_{out} = 0/0$. However, from Equation (8.4), $w'_{out} = RT \ln v_2/v_1$.
For $n \to \infty$ (constant volume), $w'_{out} = -v(p_2 - p_1)$.
For $n = 0$ (constant pressure), $w'_{out} = 0$.

For $n = \gamma$ (constant entropy),

$$w'_{\text{out}} = -\frac{\gamma R}{\gamma - 1}(T_2 - T_1).$$

A comparison of the expressions for reversible work in a closed system and a steady-flow open system shows that for an isentropic process ($n = \gamma$), the work output is greater in steady flow by a factor of γ. For an isothermal process, the work outputs are identical.

EXAMPLE 8.14

In a rocket nozzle, gases are expanded adiabatically from a pressure of 4000 kPa and a temperature of 3000 K to a pressure of 50 kPa. The exhaust velocity is equal to 97 percent of that for an isentropic expansion. Determine the actual exhaust velocity. Assume the inlet velocity to the nozzle to be negligibly small, with the gases behaving as an ideal gas with constant specific heat, $R = 450$ J/kg K and $\gamma = 1.3$.

Solution

The first law for this process is, from Equation (3.24),

$$q_{\text{in}} = w'_{\text{out}} + h_e - h_{\text{in}} + (V_e^2 - V_{\text{in}}^2)/2g_c$$

where $q_{\text{in}} = 0$ for an adiabatic process, $V_{\text{in}} = 0$, and $w'_{\text{out}} = 0$. Hence

$$V_e^2/2g_c = h_{\text{in}} - h_e = c_p(T_{\text{in}} - T_e)$$

For an isentropic expansion, use Equation (8.13) to find T_{es}:

$$\frac{T_{es}}{T_{\text{in}}} = \left(\frac{p_e}{p_{\text{in}}}\right)^{(\gamma-1)/\gamma} = \left(\frac{50}{4000}\right)^{0.3/1.3} = 0.364$$

and $$T_{es} = 1092 \text{ K}$$

Now, since $\gamma = c_p/c_v$ and $R = c_p - c_v$,

$$c_p = \frac{R\gamma}{\gamma - 1} = 450(1.3)/0.3 = 1950 \text{ J/kg K}$$

Hence

$$V_{es} = \sqrt{1950 \frac{\text{N}\cdot\text{m}}{\text{kg}\cdot\text{K}}(3000 - 1092)\text{ K}(1 \text{ kg}\cdot\text{m/N}\cdot\text{s}^2)2}$$

$$= 2728 \text{ m/s}$$

and $$V_{e(\text{actual})} = 0.97(2728) = 2646 \text{ m/s}$$

EXAMPLE 8.15

In a water-cooled cylinder-piston device having a diameter of 10 cm, the following measurements of pressure vs. displacement were made:

Displacement (cm)	Pressure (kPa)
2	650
5	225
10	98
15	60

It is suggested that the process is polytropic. Determine the exponent (n) of the polytropic expansion, and the work done.

Solution

From Equation (8.10) and using the total volume $\forall$ in lieu of the specific volume v, we get

$$p\forall^n = \text{constant}$$

Here, $\forall = \frac{1}{4}\pi(10)^2 d$; therefore,

$$pd^n = \text{constant}/[\tfrac{1}{4}\pi(10)^2]^n \quad \text{and} \quad p = C \times d^{-n}$$

To determine the exponent n, plot the pressure p against the displacement d on log-log paper (Figure 8.14), and measure the actual slope in centimeters or inches. The slope is $-3.52/2.93 = -1.20$. Hence, $n = 1.2$ and $p\forall^{1.2} = \text{constant}$, and the process is polytropic.

By Equation (8.15),

$$\begin{aligned} W_{\text{out}} &= -\frac{1}{n-1}(p_2\forall_2 - p_1\forall_1) \\ &= -\frac{1}{1.2-1}\left[60\,\frac{\pi}{4}(0.1)^2 0.15 - 650\,\frac{\pi}{4}(0.1)^2 0.02\right] \\ &= 0.157\ \text{kJ} = 157\ \text{J} \end{aligned}$$

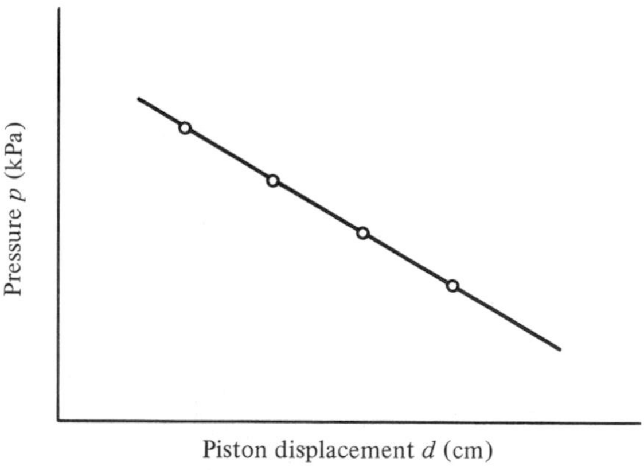

FIGURE 8.14 *Log-log Plot of Pressure vs. Piston Displacement.*

8.8 HUMIDITY PROCESSES

In Chapter 5 we discussed the behavior of gas-vapor mixtures and defined the concept of humidity in such mixtures. In particular, atmospheric air was shown to be a mixture of dry air and superheated water vapor. Here, we will examine processes that involve gas-vapor mixtures and that take place at constant humidity.

1. Constant Specific Humidity Process

When atmospheric air is heated (at constant pressure), the total amount of water vapor in the air remains constant while the temperature rises. The psychrometric chart (Figure 5.5) shows that the relative humidity of the air decreases, resulting in air that feels dry. Conversely, when atmospheric air is cooled at constant specific humidity, the relative humidity increases until a value of 100 percent is reached. Further cooling of the air is accompanied by a constant 100 percent relative humidity process.

2. Constant 100 Percent Relative Humidity Process

If atmospheric air containing water vapor at saturation pressure and temperature (100 percent R.H.) is cooled, a reduction in the temperature of the air results in the condensation of some of the water vapor since the air retains only a sufficient amount of water vapor to keep it at its 100 percent level. The resulting effect is a reduction in the specific humidity of the air.

EXAMPLE 8.16

Atmospheric air has a temperature of 10°C (50°F) and a relative humidity of 60 percent. It is heated at constant pressure to 20°C (68°F). Determine the resulting relative humidity.

Solution

Since the quantity of water vapor remains constant while the temperature of the mixture is raised, we can use Figure 5.3 or 5.5 to determine the resulting relative humidity. The two figures are reproduced in Figure 8.15. At a temperature of 10°C (50°F) and 60 percent relative humidity, we get a specific humidity of 0.0046 lbm(water vapor)/lbm(dry air). Proceeding to the left (increasing temperature) along the specific humidity line of 0.0046 to a temperature of 20°C (68°F), we obtain a relative humidity of 31 percent, a reduction of almost one half.

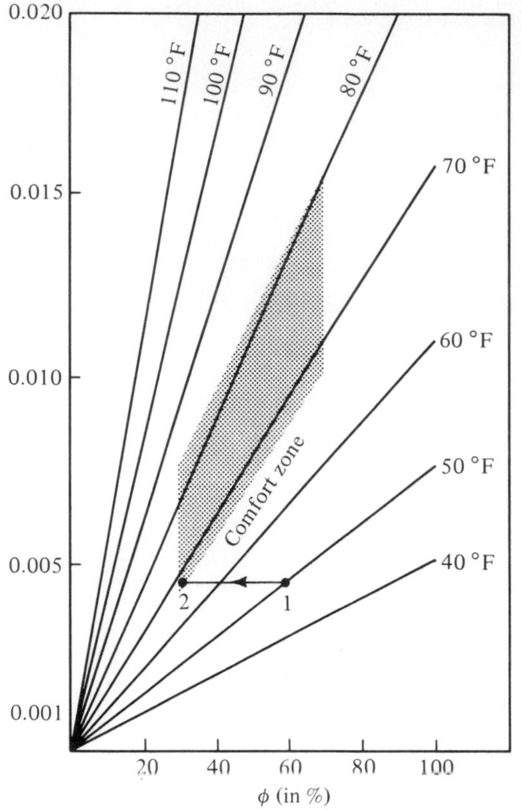

(a)

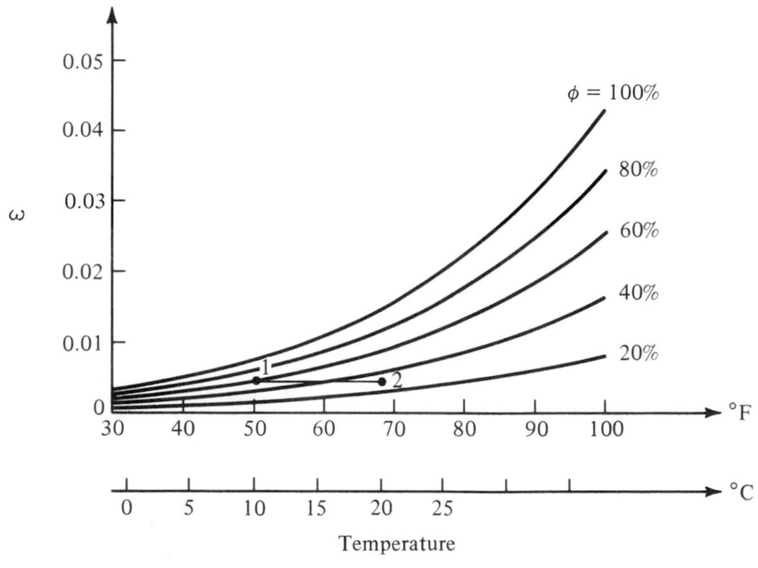

(b)

FIGURE 8.15 *Constant Specific Humidity Process.*

EXAMPLE 8.17

Atmospheric air at a temperature of 59°F (15°C) and a relative humidity of 100 percent is cooled to a temperature of 45°F (7.22°C). Find the amount of water condensed and the heat removed.

Solution

From Figure 8.16 we get $\omega_1 = 0.0106$ lbm(water vapor)/lbm(dry air), and $\omega_2 = 0.0063$ lbm(water vapor)/lbm(dry air). Hence the amount of water condensed in the process is $\omega_1 - \omega_2 = 0.0106 - 0.0063 = 0.0043$ lbm(water)/lbm(dry air). The amount of heat removed (if the air is in a closed room) is obtained from the energy equation (2.5):

$$q_{out} = (u_1 - u_2) \text{ Btu/lbm(dry air)}$$
$$= u_{a1} + \omega_1 u_{v1} - u_{a2} - \omega_2 u_{v2} - (\omega_1 - \omega_2) u_f$$

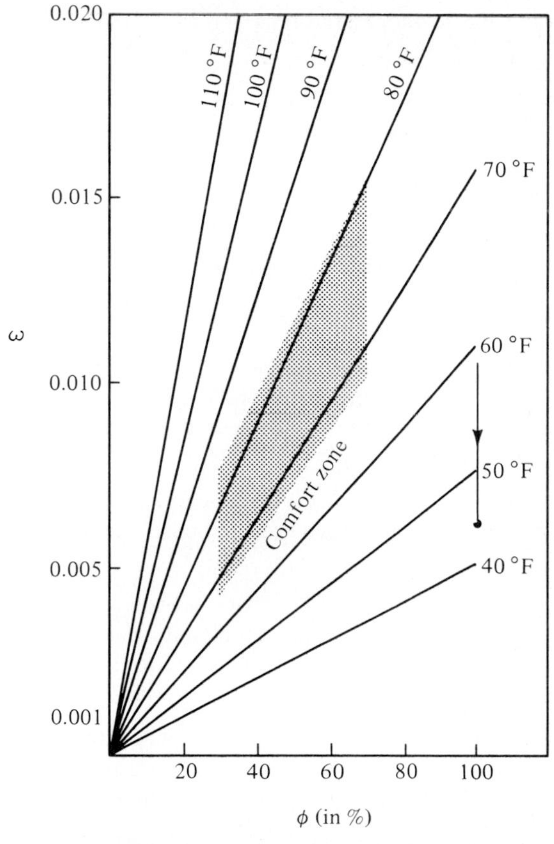

FIGURE 8.16 *Constant Relative Humidity Process.*

The last term accounts for the internal energy of the liquid condensate removed from the atmosphere. Since the temperature of the water condensate varies from T_1 to T_2, we will use an average temperature equal to $(T_1 + T_2)/2$ to find the value of u_f.

We evaluate the internal energy change of the dry air by using the ideal gas approximation, $c_v(T_2 - T_1)$. From Table A.2, we find by interpolation that u_f (at 52°F) = 20.06 Btu/lbm(water), u_g (at 45°F) = 1025.6 Btu/lbm and u_g (at 59°F) = 1030.2 Btu/lbm. Hence

$$\begin{aligned} q_{\text{out}} &= (0.171\ \text{Btu/lbm air°F})(59 - 45)\text{°F} \\ &\quad + (0.0106\ \text{lbm H}_2\text{O/lbm air})(1030.2\ \text{Btu/lbm H}_2\text{O}) \\ &\quad - (0.0063\ \text{lbm H}_2\text{O/lbm air})(1025.6\ \text{Btu/lbm H}_2\text{O}) \\ &\quad - (0.0043\ \text{lbm H}_2\text{O/lbm air})(20.06\ \text{Btu/lbm H}_2\text{O}) \\ &= 2.394 + 10.920 - 6.461 - 0.086 \\ &= 6.767\ \text{Btu/lbm(dry air)} \end{aligned}$$

Note that a major portion of heat removed is required to condense the vaporous water; a lesser portion, to reduce the temperature of the air.

PROBLEMS

8.1 Two kilograms of steam at 300 kPa and a quality of 70 percent are heated in a closed container until a pressure of 600 kPa is reached. Determine the final temperature of the steam and the heat input. Find the pressure at which saturated water vapor is obtained.

8.2 Five pounds of steam undergo a constant volume process from a temperature of 800°F and a pressure of 900 psia until a temperature of 600°F is reached. Determine the state of the steam, the heat transferred, and the change in internal energy.

8.3 Superheated steam at 100 kPa, 200°C is contained in a rigid vessel of volume 2.5 m^3. Determine the heat required to raise the steam temperature to 300°C.

8.4 Nitrogen is contained in a rigid vessel of volume 1.0 m^3; the nitrogen pressure is 200 kPa, and the nitrogen temperature is 100°C. The nitrogen is cooled until the nitrogen temperature is reduced to 0°C. Determine the nitrogen pressure after cooling and the heat withdrawn from the nitrogen.

8.5 A gas at constant atmospheric pressure is heated until it receives 20 Btu. What will the increase be in its internal energy if the volume of the gas increases from 1.0 ft^3 to 2.0 ft^3? The gas is contained in a piston-cylinder arrangement.

8.6 A mixture of carbon dioxide and argon is cooled at constant pressure from an initial temperature of 60°F and a pressure of 300 psia to a temperature of −50°F. The carbon dioxide begins to condense at −50°F. Find the composition of the mixture.

8.7 A small refrigerant-charging bottle contains R-22 in its vapor and liquid phases (Figure 8.17). Refrigerant is withdrawn slowly from the bottle so that the temperature of the refrigerant remains constant at the temperature of the surroundings or 70°F. What is the pressure within the bottle? Find the change in the height of the liquid during the discharge of 1 pound of R-22, and determine the entropy change of the system. Bottle diameter = 2.50 in.

8.8 Wet steam at a pressure of 1255 kPa and a quality of 95 percent is throttled to atmospheric pressure. Find the temperature after throttling and the entropy change of the steam.

8.9 Wet steam is throttled from a temperature of 320°F and a quality of 50 percent to a temperature of 200°F. Find the state of the steam after throttling, and determine the entropy change of the steam.

8.10 Two pounds of an ideal gas are contained within a piston-cylinder arrangement. The initial volume and pressure are 1 ft^3 and 75 psia, respectively. The final volume is 3 ft^3. Compute the work done by the gas for the following processes: (a) constant pressure, (b) constant temperature, (c) pressure proportional to the square root of volume, and (d) $pV^{1.3}$ = constant.

8.11 The temperature of an ideal gas is tripled in a constant volume process. The gas is then returned to its initial temperature during a constant pressure process. Find the expression for the work done by the gas, assuming the system is closed.

8.12 A closed system of vaporous refrigerant R-22 is compressed isentropically from a temperature of 40°F and a pressure of 60 psia to 250 psia. Find the work input and the change in internal energy.

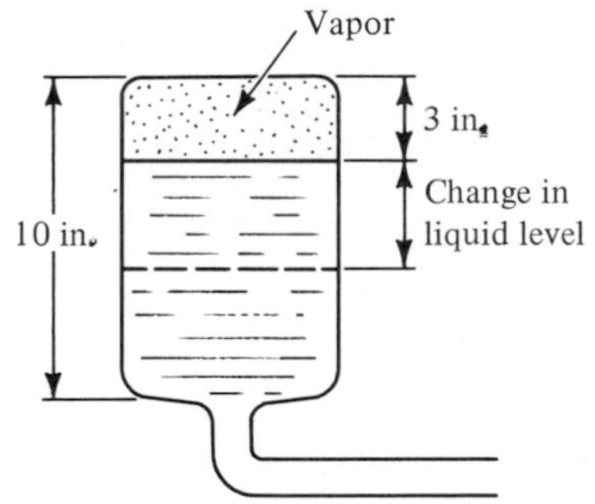

FIGURE 8.17

8.13 Steam expands isentropically in steady flow from a temperature of 200°C and a pressure of 1000 kPa to a pressure of 50 kPa. Find the work output if there is no change in potential and kinetic energies.

8.14 Steam at 500°C, 4000 kPa is expanded adiabatically to a pressure of 50 kPa and a temperature of 110°C; find the work output, assuming no change in potential and kinetic energies. What would the efficiency of the device be?

8.15 The pressure of carbon dioxide is increased threefold in a reciprocating compressor. The intake pressure is 15 psia with a temperature of 100°F. The compression is isentropic. Calculate the outlet temperature and the work required.

8.16 If the compression of Problem 8.15 is adiabatic with a compressor efficiency of 85 percent, what will be the outlet condition of the compressed carbon dioxide?

8.17 A polytropic process follows the pressure-volume relation $pv^{1.3}$ = constant. The initial pressure is 15 psia and the initial specific volume is 10 ft^3/lbm. Determine the specific volume when a pressure of 25 psia is reached. What will the temperature be at that pressure if the initial temperature is 40°F? Calculate the gas constant of the gas. Find the work done for closed and open systems.

8.18 Molecular nitrogen undergoes a polytropic process from a pressure of 20 psia and a volume of 10 ft^3, to a pressure of 100 psia and a volume of 2.9 ft^3. The system is closed, and the process is reversible. Determine the value of the exponent n, and calculate the work output. Also determine the heat transferred.

8.19 One cubic meter of dry air at 1000 kPa and 150°C is expanded in a steady-flow process according to the polytropic relation $pv^{1.35}$ = constant, until a pressure of 150 kPa is reached. Find the final specific volume and the temperature. Calculate the work output given that the process is reversible.

8.20 A cooling tower cools 1000 gallons of water per minute from a temperature of 120°F to 80°F. The air that is used to cool the water enters at a temperature of 96°F and a relative humidity of 60 percent and leaves as saturated air at a temperature of 102°F. Calculate the air flow rate and the water makeup rate.

8.21 Air having a relative humidity of 40 percent is expanded isentropically from a temperature of 35°C and a pressure of 500 kPa. What is the pressure in the nozzle at which condensation of water just begins?

8.22 Atmospheric air at 90°F and a relative humidity of 65 percent is heated to 110°F while the pressure remains constant. Find the new relative humidity and the total change in enthalpy of the air.

8.23 Atmospheric air at a temperature of 92°F, a pressure of 14.7 psia, and a relative humidity of 75 percent is cooled to 68°F. Find the relative humidity after cooling, and determine the change in enthalpy. After being cooled the air is reheated to 92°F. Find the new relative humidity.

8.24 A building is to be supplied with conditioned air at 20°C and a relative humidity of 50 percent. The outside atmospheric air is at 5°C and a relative humidity of 60 percent. Calculate the mass of water that must be added (or removed) per mass of outside air.

8.25 The composition of a mixture of gases on a mass fraction basis is N_2, 0.70; CO_2, 0.15; and H_2O, 0.15. At what temperature will the water vapor begin to condense if the total pressure of the mixture remains constant at 20 psia?

8.26 Develop a psychrometric chart for a mixture of nitrogen and water vapor at a total pressure of 35 kPa. Plot constant relative humidity lines of 100 percent, 60 percent, and 20 percent over a temperature range of 10°C to 40°C.

8.27 Due to leakage, air occasionally seeps into a refrigeration system. If 3 percent of air (on a mass basis) is found mixed with refrigerant R-22, what will the change in saturation pressure be for a temperature of 70°F?

8.28 A perfect gas with $\gamma = 1.3$ is expanded in a piston-cylinder arrangement from 50 cm^3 to 400 cm^3. Calculate the work done by the gas if the expansion is isentropic and the initial gas pressure is 200 kPa. Assume the gas has constant specific heats.

8.29 Nitrogen at 500 K is expanded isentropically from 300 kPa to 100 kPa in a nozzle. The velocity of the nitrogen at the entrance to the nozzle is 100 m/s. Find the velocity of the nitrogen at the exit from the nozzle, assuming that the nitrogen behaves as a perfect gas with constant specific heats (properties are given in Table B.1).

8.30 In a steady-flow compressor, 1.0 lbm/min of air is raised in pressure from 14.7 psia to 100 psia. Assuming that an isentropic compression process takes place on air entering at 50°F, determine the power required.

8.31 Helium is expanded in a nozzle from 200 psia, 500°F to a pressure of 10 psia. The velocity of the helium entering the nozzle is negligible; the exit velocity is 95 percent of that for isentropic flow. Determine the nozzle exhaust velocity and the temperature of the helium at nozzle exit. Also determine the increase in entropy of the helium. ($\gamma = 1.67$, $R = 386.0$ ft-lbf/lbm°R.)

8.32 Oxygen at 15 psia, 70°F is throttled to 10 psia in a steady-flow process. Determine the change of temperature of the oxygen as a result of the throttling process and the increase in entropy of the oxygen. Calculate the irreversibility for this process for an ambient temperature of 75°F.

chapter

9

The Thermodynamics of Fluid Flow

9.1 INTRODUCTION

Many of the thermodynamic cycles and processes we have discussed in previous chapters have involved the flow of a fluid, either liquid or gas. Thermodynamics enables us to develop a better understanding of these flow processes and the variation in property values that occurs as a result. In Section 3.4 we studied steady flow open systems using the first law of thermodynamics. We are now in a position to add second law analysis and thereby extend our study of flow processes.

9.2 STEADY, INCOMPRESSIBLE FLOW

The energy balance for a steady-flow open system on a unit mass basis is given by (3.15):

$$q_{in} - w'_{out} = h_e - h_{in} + \frac{1}{2g_c}(V_e^2 - V_{in}^2) + \frac{g}{g_c}(z_e - z_{in})$$

If the steady-flow process is reversible, from the second law,

$$\int_{in}^{e} T\,ds = q_{in}$$

Also, from (6.21), for any process,

$$T\,ds = dh - v\,dp$$

or

$$T\,ds = dh - \frac{dp}{\rho}$$

For a reversible process, therefore,

$$q_{in} = \int_{in}^{e} dh - \int_{in}^{e} \frac{dp}{\rho} = h_e - h_{in} - \int_{in}^{e} \frac{dp}{\rho}$$

Substituting back into (3.15), we obtain

$$-w'_{out} = \frac{1}{2g_c}(V_e^2 - V_{in}^2) + \frac{g}{g_c}(z_e - z_{in}) + \int_{in}^{e} \frac{dp}{\rho} \tag{9.1}$$

For flow of liquid water (except at extremely high pressures) it can be assumed that density ρ is constant; flow with constant density is called *incompressible flow*. For incompressible, reversible flow we now have:

$$-w'_{out} = \frac{1}{2g_c}(V_e^2 - V_{in}^2) + \frac{g}{g_c}(z_e - z_{in}) + \frac{p_e - p_{in}}{\rho} \tag{9.2}$$

If there is no shaft work ($w'_{out} = 0$), then we have:

$$\frac{V_{in}^2}{2g_c} + \frac{gz_{in}}{g_c} + \frac{p_{in}}{\rho} = \frac{V_e^2}{2g_c} + \frac{gz_e}{g_c} + \frac{p_e}{\rho} \tag{9.3}$$

or

$$\frac{V^2}{2} + gz + \frac{pg_c}{\rho} = \text{constant} \tag{9.4}$$

Equation (9.4) is called the Bernoulli equation and is valid for reversible, incompressible flow.

EXAMPLE 9.1

Water is to be discharged through a nozzle at the bottom of a large tank. The nozzle opening is circular, 5 cm in diameter; the nozzle discharges to atmospheric pressure. Determine the flow rate through the nozzle in m^3/s; assume steady reversible flow.

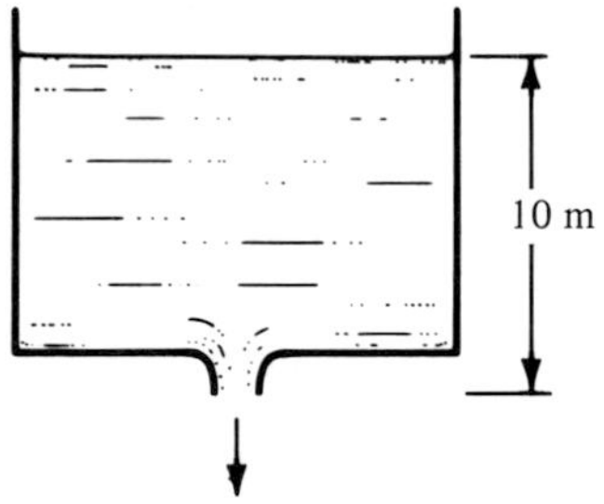

FIGURE 9.1

Solution

Select the system bounded by reservoir surface, tank, and nozzle exit plane. For a large tank, the velocity of the water level in the tank is negligible compared to the nozzle exit velocity. Applying (9.3), we obtain

$$\frac{V_{\text{in}}^2}{2g_c} + \frac{gz_{\text{in}}}{g_c} + \frac{p_{\text{in}}}{\rho} = \frac{V_e^2}{2g_c} + \frac{gz_e}{g_c} + \frac{p_e}{\rho}$$

Here,

$$V_{\text{in}} = 0$$

$$p_{\text{in}} = p_e = \text{atmospheric pressure}$$

and

$$z_{\text{in}} - z_e = 10\text{ m}$$

Therefore,

$$\frac{V_e^2}{2g_c} = \frac{g}{g_c}(z_{\text{in}} - z_e)$$

or

$$V_e = \sqrt{2g(z_{\text{in}} - z_e)}$$

$$= \sqrt{2(9.81\text{ m/s}^2)10\text{ m}}$$

$$= 14.0\text{ m/s}$$

and the volumetric flow Q out of the tank is

$$Q = A_e V_e$$

$$= \frac{\pi}{4}(0.05\text{ m})^2 14.0\text{ m/s}$$

$$= \underline{0.0275\text{ m}^3/\text{s}}$$

EXAMPLE 9.2

A Venturi meter is a converging-diverging section used to measure flow in a pipe. If the pressure difference between sections 1 and 2 in Figure 9.2 is measured to be 1.6 psi, determine the volumetric flow of water at 60°F through the horizontal pipe. The pipe diameter is 4 in., and the Venturi throat diameter is 3 in.; assume reversible flow.

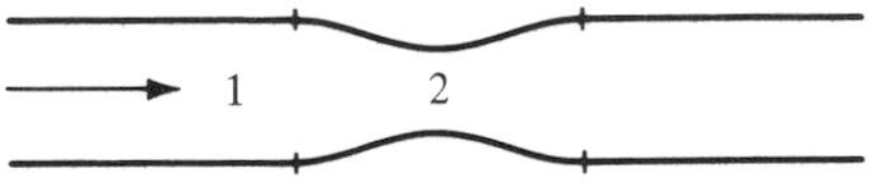

FIGURE 9.2 *Venturi.*

Solution

Applying Bernoulli's equation between sections 1 and 2, we obtain

$$\frac{p_1}{\rho} + \frac{V_1^2}{2g_c} = \frac{p_2}{\rho} + \frac{V_2^2}{2g_c}, \qquad z_1 = z_2 \text{ for a horizontal pipe}$$

From (3.6), the continuity equation for steady flow is

$$\rho_1 A_1 V_1 = \rho_2 A_2 V_2$$

where $\rho_1 = \rho_2$ for incompressible flow of water.

For our case,

$$V_2 = V_1 \frac{A_1}{A_2}$$

$$= V_1 \left(\frac{\frac{\pi}{4}\,16\text{ in}^2}{\frac{\pi}{4}\,9\text{ in}^2} \right) = 1.778V_1$$

and

$$\frac{V_2^2 - V_1^2}{2g_c} = \frac{p_1 - p_2}{\rho}$$

For this case,

$$p_1 - p_2 = 1.6\text{ lbf/in}^2$$

and

$$\rho = \frac{1}{v} = \frac{1}{0.01603\text{ ft}^3/\text{lbm}}$$

$$= 62.4\text{ lbm/ft}^3$$

so that

$$\frac{1.778^2 V_1^2 - V_1^2}{2(32.2\text{ lbm-ft/s}^2\text{lbf})} = \frac{(1.6\text{ lbf/in}^2)(144\text{ in}^2/\text{ft}^2)}{62.4\text{ lbm/ft}^3}$$

or

$$2.161V_1^2 = 237.8\text{ ft}^2/\text{s}^2$$

$$V_1 = 10.49\text{ ft/s}$$

and

$$Q = A_1 V_1$$

$$= \left(\frac{\pi}{4}\,\frac{16\text{ in}^2}{144\text{ in}^2/\text{ft}^2} \right) 10.49\text{ ft/s}$$

$$Q = 0.915\text{ ft}^3/\text{s}$$

EXAMPLE 9.3

It is desired to pump water at 60°F from a reservoir at a velocity of 25 ft/s through a 1.0-in.-diameter pipe. Find the pump horsepower required for frictionless, reversible flow. Assume that the exit flow is provided at atmospheric pressure.

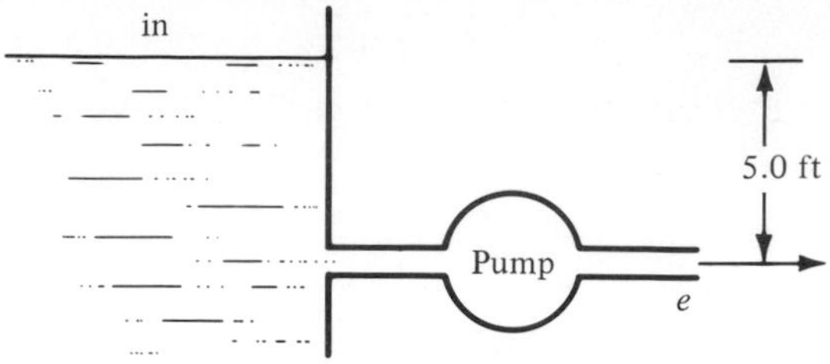

FIGURE 9.3

Solution

Locate sections in and e as shown, with $V_{in} = 0$ for the reservoir. From (9.2),

$$-w'_{out} = \frac{1}{2g_c}(V_e^2 - V_{in}^2) + \frac{g}{g_c}(z_e - z_{in}) + \frac{p_e - p_{in}}{\rho}$$

where $p_e = p_{in}$ = atmospheric pressure. Simplifying, we obtain

$$\begin{aligned}
-w'_{out} &= \frac{1}{2g_c}V_e^2 + \frac{g}{g_c}(z_e - z_{in}) \\
&= \frac{25^2\ \text{ft}^2/\text{s}^2}{2\left(32.2\ \dfrac{\text{lbm-ft}}{\text{lbf}\cdot\text{s}^2}\right)} + \frac{(32.2\ \text{ft/s}^2)(-5.0\ \text{ft})}{\left(32.2\ \dfrac{\text{lbm-ft}}{\text{lbf}\cdot\text{s}^2}\right)} \\
&= 9.705\ \frac{\text{ft-lbf}}{\text{lbm}} - 5.0\ \frac{\text{ft-lbf}}{\text{lbm}} \\
&= 4.705\ \frac{\text{ft-lbf}}{\text{lbm}}
\end{aligned}$$

Power $\mathscr{P}$ is a rate of doing work. In this case,

$$\mathscr{P} = -\left(4.705\ \frac{\text{ft-lbf}}{\text{lbm}}\right)\dot{m}$$

where

$$\begin{aligned}
\dot{m} &= \rho_e A_e V_e \\
&= (62.4\ \text{lbm/ft}^3)\left(\frac{\pi}{4}\frac{1^2}{12}\ \text{ft}^2\right)(25\ \text{ft/s}) \\
&= 8.508\ \text{lbm/s} \\
\mathscr{P} &= -\left(4.705\ \frac{\text{ft-lbf}}{\text{lbm}}\right)(8.508\ \text{lbm/s}) \\
&= -40.0\ \text{ft-lbf/s}
\end{aligned}$$

where

$$1.0 \text{ horsepower} = 550 \text{ ft-lbf/s}$$

so

$$\mathscr{P} = 0.0728 \text{ horsepower required}$$

9.3 COMPRESSIBLE FLOW

Section 9.2 has dealt with steady, incompressible flow in which density was assumed to be constant. In this and subsequent sections we will be studying compressible flow in which a significant variation in density occurs as a result of the flow. With a gas, for example, density variations can occur as a result of pressure changes. From the perfect gas law ($p = \rho RT$) we can see the dependence of density on pressure. It is true that, for certain low-velocity gas flows, pressure changes are small enough so that the assumption of incompressibility is a valid one. However, in the more general case, compressibility must be taken into account; the effect of compressibility may completely alter the characteristics of the flow. Some important applications of compressible flow are gas flows through jet engine and rocket nozzles, gas flows through jet engine diffusers, flow through turbine blade passages in gas and steam turbines, and high-speed flows over aircraft.

An important variable in compressible flow is the velocity of sound, the velocity at which an infinitesimal disturbance is propagated through a compressible medium. For example, a compressible fluid is contained in a cylinder to the right of a piston (Figure 9.4). The piston is suddenly given an infinitesimal velocity increment dV. At this instant, the fluid piles up next to the piston face and is slightly compressed; fluid further down the tube is at rest and does not immediately move with the piston. As time passes, the compressed fluid at the piston face compresses the layer of fluid next to it and so on; the disturbance caused by the motion of the piston is propagated through the fluid at finite velocity, the velocity of sound a.

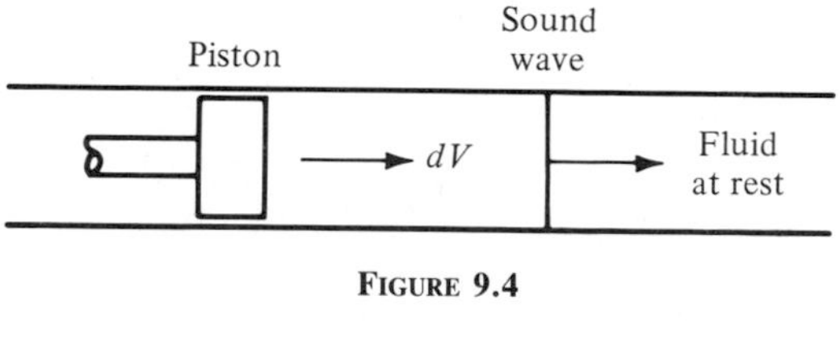

FIGURE 9.4

To analyze this flow, it must be realized that the fluid between piston face and sound wave in Figure 9.4 moves with the piston velocity dV. Let us now transform to a coordinate system fixed to the wave, as shown in Figure 9.5. Applying the continuity equation to a system consisting of the wave yields

$$\rho A_{\text{in}} a = (\rho + d\rho) A_e (a - dV) \quad \textbf{(9.5)}$$

or

$$\rho a = \rho a + a\, d\rho - \rho\, dV - d\rho\, dV$$

where $A_{\text{in}} = A_e$ for a very thin wave. Canceling terms and dropping those of second order, we have

$$a\, d\rho = \rho\, dV \quad \textbf{(9.6)}$$

Applying the energy equation, we obtain

$$h + \frac{a^2}{2g_c} = (h + dh) + \frac{(a - dV)^2}{2g_c}$$

Simplifying yields

$$dh = \frac{a\, dV}{g_c} \quad \textbf{(9.7)}$$

From Section 6.2, the sound wave process is reversible, since there are only infinitesimal departures from equilibrium. Further, the process is adiabatic; since there are no temperature gradients at the system boundaries, an infinitesimal temperature difference occurs inside the system. A process that if reversible and adiabatic is isentropic, $ds = 0$. But for any thermodynamic process,

$$T\, ds = dh - \frac{dp}{\rho}$$

For an isentropic process,

$$dh = \frac{dp}{\rho} \quad \textbf{(9.8)}$$

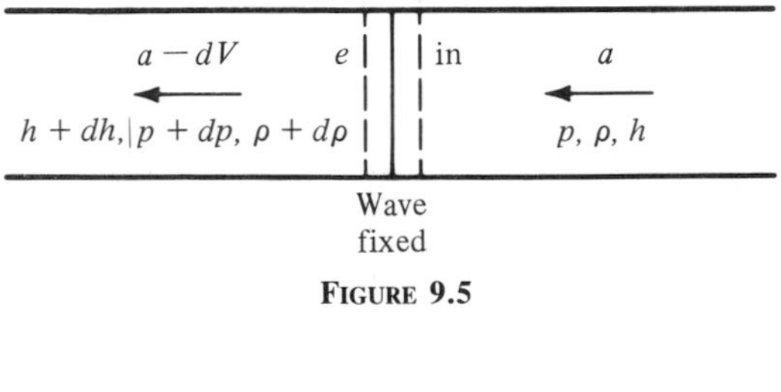

FIGURE 9.5

Substituting (9.8) into (9.7), we obtain

$$\frac{dp}{\rho} = \frac{a\,dV}{g_c}$$

Combining with (9.6), we have

$$a^2 = g_c \frac{dp}{d\rho}$$

or, more specifically,

$$a = \sqrt{g_c \left(\frac{\partial p}{\partial \rho}\right)_s} \tag{9.9}$$

For a perfect gas with constant specific heats undergoing an isentropic process, $p/\rho^{\gamma} = \text{constant}$ or

$$\left(\frac{\partial p}{\partial \rho}\right)_s = \frac{\gamma p}{\rho}$$

and

$$a = \sqrt{g_c \frac{\gamma p}{\rho}}$$

or, since $p = \rho RT$,

$$a = \sqrt{g_c\,\gamma RT} \tag{9.10}$$

EXAMPLE 9.4

Calculate the velocity of sound in air at 270 K.

Solution

From Table B.1, for air,

$$R = 0.2870\ \text{kJ/kg}\cdot\text{K}, \qquad \gamma = 1.40$$

Applying (9.10), we obtain

$$a = \sqrt{g_c \gamma RT}$$

$$= \sqrt{\left(1.0\,\frac{\text{kg}\cdot\text{m/s}^2}{\text{N}}\right)1.40(287.0\ \text{N}\cdot\text{m/kg}\cdot\text{K})300\ \text{K}}$$

$$a = 347.2\ \text{m/s}$$

The less compressible the medium, the higher the velocity of sound. For example, the velocity of sound in copper is more than 10 times that in air. If the fluid to the right of the piston in Figure 9.5 were truly incompressible, the entire fluid would move instantaneously to the right with incremental velocity dV as soon as the piston moves. The velocity of sound in an incompressible fluid is infinite.

An important dimensionless parameter for compressible flow is Mach number M, the ratio between flow velocity and the velocity of sound,

$$M = \frac{V}{a} \tag{9.11}$$

For $M < 1$, the flow is subsonic; for $M > 1$ we have supersonic flow. We have seen that an infinitesimal disturbance is propagated through a fluid at sonic velocity. If a plane is traveling subsonically through the atmosphere, the compression waves sent out into the air move ahead of the plane and signal the air ahead of it to adjust to the presence of the approaching object. For this reason, flow over a subsonic wing tends to be smooth, as shown in Figure 9.6. If a plane is traveling supersonically through the air, the plane is moving faster than the compression waves and therefore the flow ahead of the plane does not have a chance to adjust to the presence of the approaching object. In this case, adjustment of the flow occurs at the plane and is very sudden in nature; this sudden change in flow properties is called a *shock wave* (see Figure 9.7).

When an airplane is traveling supersonically near the ground, the sudden pressure change caused by the shock causes the occurrence of a sonic boom, which can be felt and heard by observers on the ground (see Figurc 9.8).

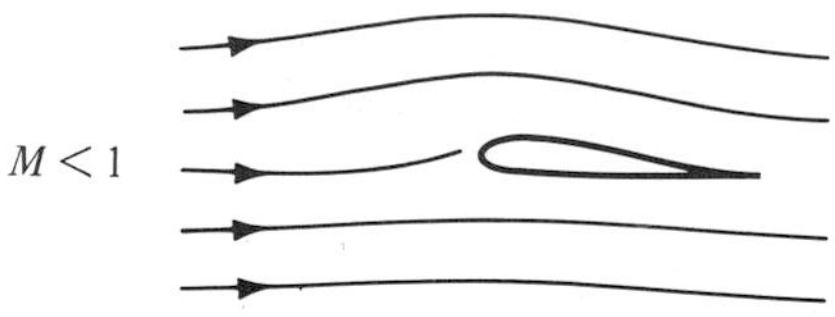

FIGURE 9.6 *Subsonic Flow.*

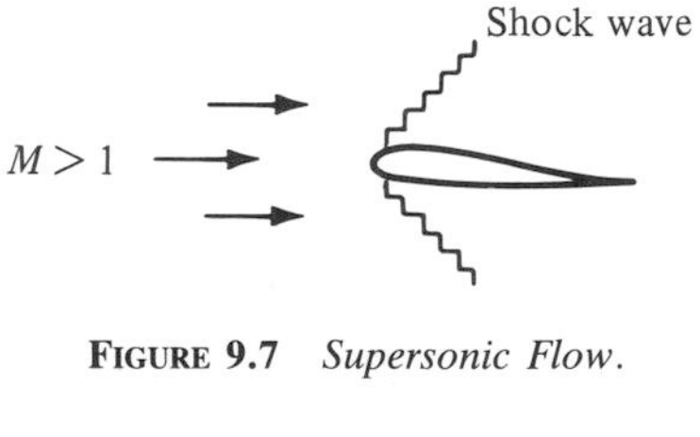

FIGURE 9.7 *Supersonic Flow.*

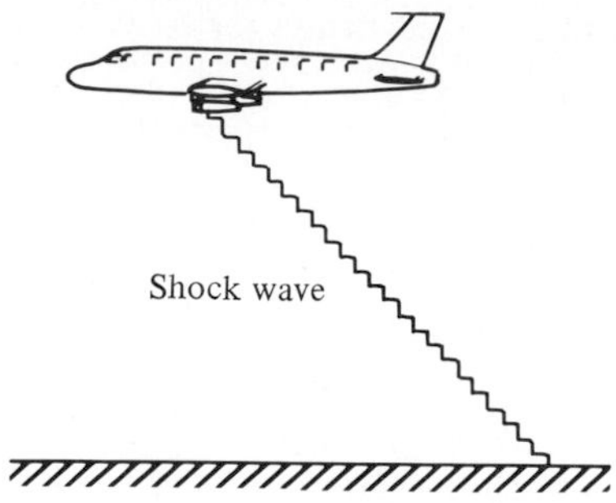

FIGURE 9.8 *Sonic Boom.*

9.4 ISENTROPIC, COMPRESSIBLE FLOW THROUGH A CHANNEL OF VARYING AREA

In this section, we will be primarily concerned with gas flow through variable area channels such as nozzles, diffusers, and turbine blade passages. We will assume the working fluid to be a perfect gas with constant specific heats, an assumption which is valid over a wide range of pressures and temperatures for air and other common gases. Further, we will assume that the flow is isentropic—in other words, adiabatic and reversible. Isentropic flow represents an ideal, frictionless flow; this analysis can be applied quite successfully and accurately to flows of hot, high-speed gases through rocket and turbine nozzles.

Consider steady flow through a varying area channel, as shown in Figure 9.9. We will define the boundaries of the open system as shown and write the continuity and energy equations as well as the second law for this system. We will take flows at inlet and outlet from the system to be one-dimensional, i.e., flow properties are uniform across each cross section of the flow. From Section 3.2, the continuity equation yields:

$$\rho AV = (\rho + d\rho)(A + dA)(V + dV)$$

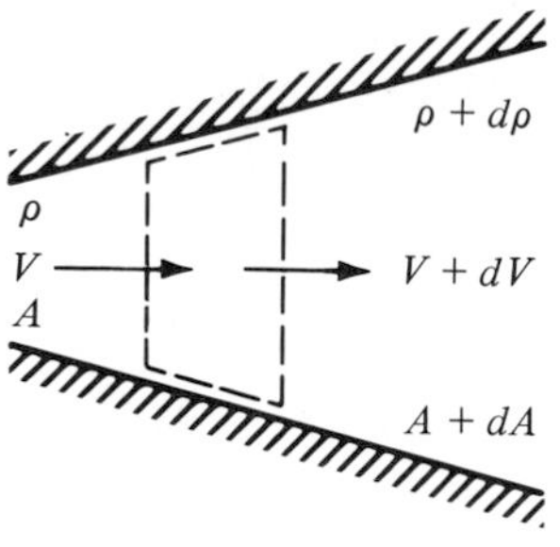

FIGURE 9.9

Combining and dropping second-order terms, we obtain

$$\frac{d\rho}{\rho} + \frac{dA}{A} + \frac{dV}{V} = 0 \tag{9.12}$$

The energy equation is given by (3.15), where we will neglect changes in gravitational potential energy for this gas flow, take $q_{\text{in}} = 0$ for isentropic flow, and also assume no shaft work, so $w'_{\text{out}} = 0$. There results

$$h_e + \frac{V_e^2}{2g_c} = h_{\text{in}} + \frac{V_{\text{in}}^2}{2g_c}$$

Applying this to the flow of Figure 9.6, we have

$$(h + dh) + \frac{(V + dV)^2}{2g_c} = h + \frac{V^2}{2g_c}$$

Simplifying and dropping second-order terms, we have

$$dh + \frac{V\,dV}{g_c} = 0 \tag{9.13}$$

Finally, from the second law,

$$T\,ds = dh - \frac{dp}{\rho}$$

For an isentropic process, $ds = 0$, so

$$dh = \frac{dp}{\rho} \tag{9.14}$$

Combining (9.13) and (9.14) yields

$$dp + \frac{\rho V\,dV}{g_c} = 0 \tag{9.15}$$

Finally, for an isentropic process, from (9.9)

$$a^2 = g_c \frac{dp}{d\rho} \tag{9.16}$$

Combining (9.12) and (9.15), we obtain

$$dp + \frac{\rho V^2}{g_c}\left(-\frac{d\rho}{\rho} - \frac{dA}{A}\right) = 0$$

We substitute for $d\rho$ from (9.16):

$$dp\left(1 - \frac{V^2}{a^2}\right) = \frac{\rho V^2}{g_c}\frac{dA}{A}$$

or, since $M = V/a$, we have

$$dp(1 - M^2) = \frac{\rho V^2}{g_c}\frac{dA}{A} \tag{9.17}$$

Equation (9.17) expresses the effect of Mach number on compressible flow through a varying area channel. For subsonic flow ($M < 1$), if area is increasing ($dA+$), then dp is positive and, from (9.15), dV is negative. Also, if area is decreasing in the flow direction, then dp is negative and dV is positive. This result is shown in Figure 9.10.

For supersonic flow, however, the reverse occurs. From (9.17), with $1 - M^2$ negative for supersonic flow, if area is increasing, the pressure will decrease and velocity will increase. If area decreases, the pressure will increase and velocity will decrease. This result is summarized in Figure 9.11.

It follows that a low-velocity flow cannot be accelerated to supersonic velocity in a converging nozzle; the highest velocity that can be obtained is sonic velocity. Acceleration to supersonic velocities can be achieved only in a converging-diverging nozzle, as shown in Figure 9.12.

In order to determine quantitatively the characteristics of a compressible flow through a variable area passage, it will prove useful to utilize the properties *stagnation temperature* and *stagnation pressure*. Stagnation or total temperature T_t is defined as the temperature attained by bringing a flow adiabatically to rest at a point in a steady-flow process. This differs with static temperature T, the temperature of a flowing stream measured by an observer moving at the stream velocity. Let us now derive an expression for T_t. For adiabatic steady flow with no shaft

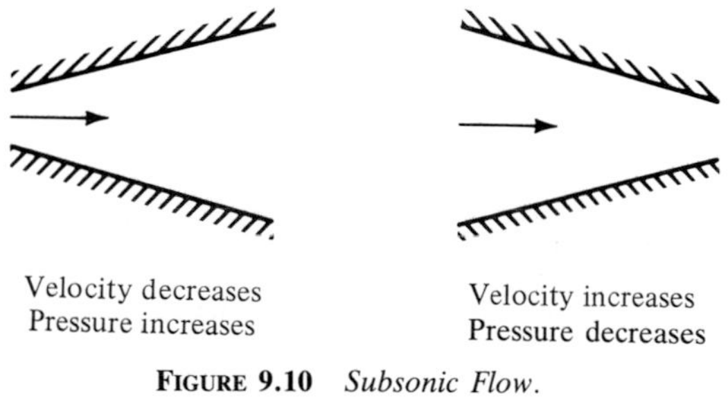

FIGURE 9.10 *Subsonic Flow.*

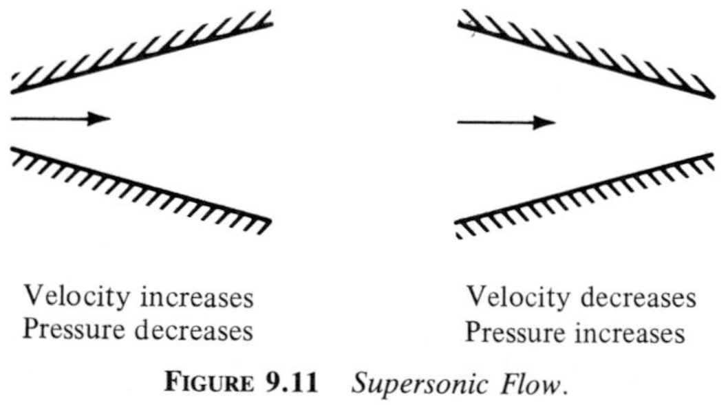

FIGURE 9.11 *Supersonic Flow.*

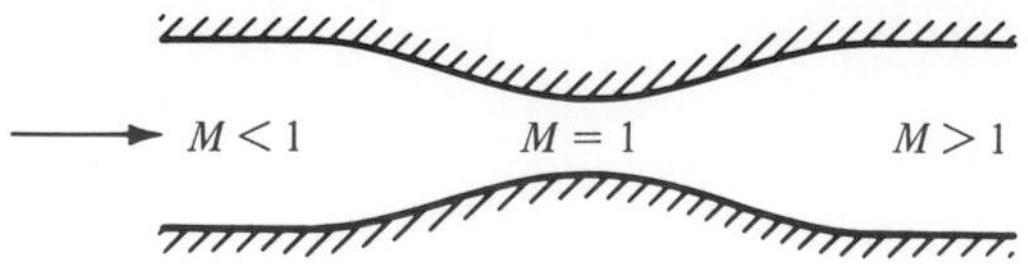

FIGURE 9.12 *Converging-Diverging Nozzle.*

work and negligible change in potential energy the energy equation (3.15) yields

$$h_e - h_{in} + \frac{1}{2g_c}(V_e^2 - V_{in}^2) = 0$$

If we let e refer to properties after the flow has been brought adiabatically to rest, we have

$$h_t - h = \frac{V^2}{2g_c} \tag{9.18}$$

where h_t is the total or stagnation enthalpy. Now we will assume the working fluid is a perfect gas with constant specific heats, so that $\Delta h = c_p \, \Delta T$ or, in our case,

$$c_p(T_t - T) = \frac{V^2}{2g_c}$$

or

$$T_t = T\left(1 + \frac{V^2}{2g_c c_p T}\right)$$

But

$$\gamma = \frac{c_p}{c_v} \quad \text{and} \quad c_p - c_v = R$$

so that

$$c_p = \frac{R\gamma}{\gamma - 1}$$

and

$$T_t = T\left(1 + \frac{V^2(\gamma - 1)}{2g_c \gamma RT}\right)$$

where

$$\gamma g_c RT = a^2$$

or

$$T_t = T\left(1 + \frac{\gamma - 1}{2} M^2\right) \tag{9.19}$$

For adiabatic steady flow, stagnation temperature is constant and therefore represents an important reference. Stagnation pressure in steady flow is defined as the pressure achieved by bringing a flow isentropically to rest. It was shown in Section 8.7 that, for a perfect gas with constant specific heats undergoing an isentropic process, $p/\rho^{\gamma} =$

constant, or

$$\frac{p_2}{p_1} = \left(\frac{T_2}{T_1}\right)^{\gamma/(\gamma-1)}$$

If we let state 2 be the stagnation state,

$$\frac{p_t}{p} = \left(\frac{T_t}{T}\right)^{\gamma/(\gamma-1)} \tag{9.20}$$

or

$$\frac{p_t}{p} = \left(1 + \frac{\gamma - 1}{2} M^2\right)^{\gamma/(\gamma-1)} \tag{9.21}$$

For isentropic flow, stagnation pressure is a constant and thus represents an important reference property. T/T_t and p/p_t have been plotted vs. M for $\gamma = 1.4$ in Appendix C. For example, suppose air is stored in a large reservoir and allowed to flow out through a converging nozzle (Figure 9.13). If the flow is adiabatic, then the stagnation temperature is constant and equal to 520°R at each point along the nozzle flow. If the flow is isentropic, then the stagnation pressure and stagnation temperature are both constant at each point along the nozzle flow. For example, for isentropic flow, if the Mach number at a point in the flow were 0.5, then the static temperature at that point would be

$$T = \frac{T_t}{1 + \dfrac{\gamma - 1}{2} M^2} = 495.2°\text{R}$$

or, from Appendix C, at $M = 0.5$,

$$\frac{T}{T_t} = 0.9524 \quad \text{and} \quad T = 495.2°\text{R}$$

At the same point, the static pressure would be

$$p = \frac{p_t}{\left(1 + \dfrac{\gamma - 1}{2} M^2\right)^{\gamma/(\gamma-1)}} = \frac{50 \text{ lbf/in}^2}{1.1862} = 42.15 \text{ lbf/in}^2$$

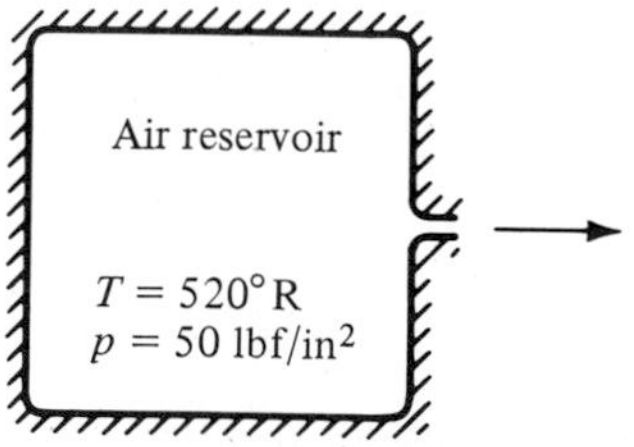

FIGURE 9.13

or, from Appendix C, at $M = 0.5$,

$$\frac{p}{p_t} = 0.8430 \quad \text{and} \quad p = 42.15 \text{ lbf/in}^2$$

For M^2 small, (9.21) can be expanded in a series to yield:

$$\frac{p_t}{p} = 1 + \frac{\gamma}{2} M^2 + \cdots$$

or, dropping higher-order terms, we obtain

$$p_t = p + \frac{\gamma p}{2} M^2$$

$$= p + \frac{\gamma p}{2} \frac{V^2}{\gamma g_c RT}$$

or

$$p_t = p + \frac{1}{2} \frac{\rho}{g_c} V^2$$

In other words, for M^2 very small, the expression for stagnation pressure is given by the incompressible Bernoulli equation, (9.4). Mach number is thus a measure of compressibility.

We can develop an equation for mass flow in terms of stagnation properties. For steady flow, the continuity equation yields

$$\rho AV = \text{constant} = \dot{m}$$

For a perfect gas, $\dot{m} = (p/RT)\, AM\sqrt{\gamma g_c RT}$. Substituting the expressions we have derived for stagnation properties, we obtain

$$\dot{m} = \frac{p_t}{\sqrt{RT_t}} AM\sqrt{\gamma g_c}\left(1 + \frac{\gamma - 1}{2} M^2\right)^{(1/2 - \gamma/\gamma - 1)}$$

But, for isentropic flow, p_t and T_t are constants. Using the area at which $M = 1$ as a reference area A^*, we obtain

$$\frac{A}{A^*} = \frac{1}{M} \frac{\left(1 + \frac{\gamma - 1}{2}\right)^{(1/2 - \gamma/\gamma - 1)}}{\left(1 + \frac{\gamma - 1}{2} M^2\right)^{(1/2 - \gamma/\gamma - 1)}} \qquad \textbf{(9.22)}$$

This equation is tabulated in Appendix C as A/A^* vs. M for $\gamma = 1.4$.

EXAMPLE 9.5

Air is expanded isentropically in a nozzle from $V_1 = 100$ m/s, $p_1 = 150$ kPa, $T_1 = 600$ K to $p_2 = 50$ kPa.

If the inlet area A_1 is 100 cm^2, determine A_2, the minimum nozzle area A_{min}, T_2, and V_2. Assume steady flow of a perfect gas with constant specific heats.

Solution

The Mach number

$$M_1 = \frac{V_1}{\sqrt{\gamma g_c R T_1}}$$

$$= \frac{100\ \text{m/s}}{\sqrt{1.4\left(1.0\ \dfrac{\text{kg}\cdot\text{m/s}^2}{\text{N}}\right)(287.0\ \text{J/kg}\cdot\text{K})(600\ \text{K})}}$$

$$= \frac{100\ \text{m/s}}{491\ \text{m/s}}$$

$$= 0.204$$

For this isentropic flow, p_t and T_t are constant. At $M_1 = 0.204$, from Appendix C, $p/p_t = 0.9714$. At state 2, with $p_{t2} = p_{t1}$,

$$\frac{p_2}{p_{t2}} = \frac{p_2}{p_1}\frac{p_1}{p_{t1}} = \frac{50}{150} \times 0.9714 = 0.3238$$

Therefore, $M_2 = 1.379$. We can now use A/A^* to find A_2. Remember that, since the flow goes from subsonic to supersonic, it must go through a minimum area at which $M = 1$ (Figure 9.14).

$$\frac{A_2}{A^*} = 1.103\,, \qquad \frac{A_1}{A^*} = 2.910$$

Therefore,

$$A_2 = \frac{1.103}{2.910}\ 100\ \text{cm}^2 = 37.9\ \text{cm}^2$$

$$A_{\text{min}} = A^* = \frac{100}{2.910} = 34.4\ \text{cm}^2$$

Since T_t is constant,

$$T_2 = T_1 \times \frac{T_{t1}}{T_1} \times \frac{T_2}{T_{t2}}$$

$$= 600 \times \frac{1}{0.9918} \times 0.7239$$

$$T_2 = 437.9\ \text{K}$$

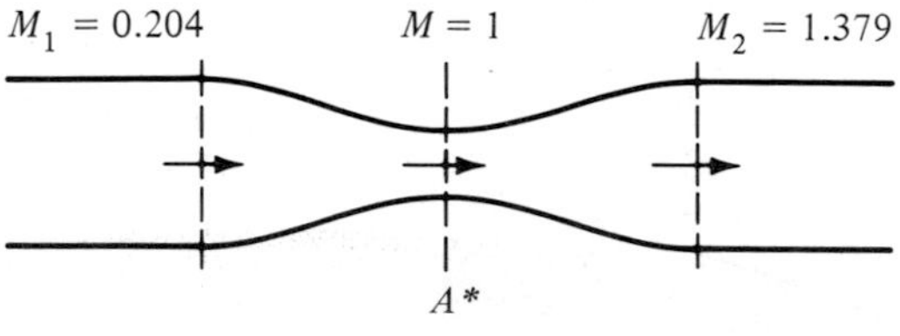

FIGURE 9.14

and $$V_2 = M_2\sqrt{\gamma g_c R T_2}$$

$$= 1.379\sqrt{1.4\left(1.0\ \frac{\text{kg}\cdot\text{m/s}^2}{\text{N}}\right)(287.0\ \text{J/kg}\cdot\text{K})(437.9\ \text{K})}$$

$$= 578.4\ \text{m/s}$$

Let us now consider compressible flow through a converging nozzle fed from a large reservoir. We will assume constant pressure and temperature in the reservoir with isentropic steady flow in the nozzle and we will examine characteristics of the flow as the back pressure is dropped below reservoir pressure. See Figure 9.15.

For back pressure equal to reservoir pressure, there will be no flow and hence no pressure change in the nozzle (curve ①). As the back pressure is lowered to values less than p_r, there will be flow through the nozzle (curves ② through ⑥). Since flow in the nozzle is subsonic, a decreasing area means increasing velocity and decreasing pressure (Figure 9.15). For a low enough back pressure, sonic velocity is attained at the nozzle exit (curve ④). This is the maximum velocity that can be achieved in the converging nozzle and thus corresponds to the maximum flow rate through the nozzle. Thus, with this limitation to flow, for all back pressure below that of ④, we say the nozzle is choked.

The phenomenon of choking goes along with the discussion at the end of Section 9.3. Changes in back pressure are felt in the reservoir by

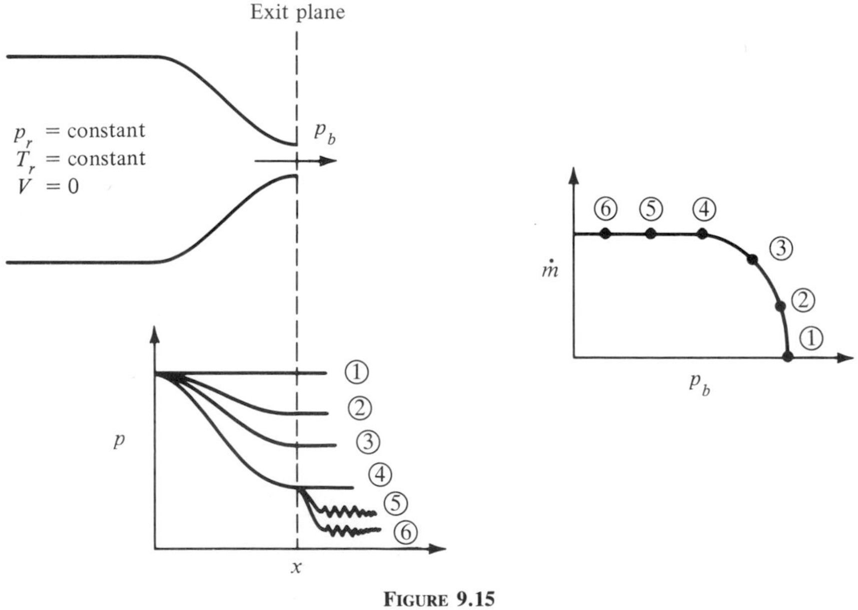

FIGURE 9.15

means of waves traveling at the velocity of sound with respect to the flow into which they are moving. If the flow itself is traveling at sonic velocity, the waves will not reach the reservoir; once sonic velocity has been reached, further changes in back pressure cannot be sensed by the reservoir. Therefore, the mass flow rates through the nozzle for ④, ⑤, and ⑥ are the same, as are the pressure variations in the nozzle. For ⑤ and ⑥, adjustment to the back pressure occurs via expansion waves occurring outside the nozzle. The maximum ratio of back pressure to reservoir pressure at which sonic velocity occurs is called the *critical pressure ratio*.

From (9.20),

$$\frac{p}{p_t} = \frac{1}{\left(1 + \frac{\gamma - 1}{2} M^2\right)^{\gamma/(\gamma-1)}}$$

For $M = 1$,

$$\left(\frac{p}{p_t}\right)_{\text{crit}} = \left(\frac{2}{\gamma + 1}\right)^{\gamma/\gamma - 1} \tag{9.23}$$

For example, for $\gamma = 1.4$, $(p/p_t)_{\text{crit}} = 0.5283$.

EXAMPLE 9.6

A large tank of nitrogen is exhausted through a converging nozzle as in Figure 9.15. If the tank pressure is 100 psia, tank temperature 600°R, determine the flow rate for back pressures of 80 psia, 50 psia, and 0 psia. Assume the tank pressure and temperature to be constant, and take the flow as steady, with a nozzle throat area of 2.0 in^2.

Solution

For a back pressure of 80 psia, the nozzle is not choked since $p_b/p_t > (p/p_t)_{\text{crit}}$. Therefore, flow at nozzle exit is subsonic with $M = 0.574$ at $p/p_t = 0.8$ (Appendix C); $T/T_t = 0.938$ so $T = 0.938(600) = 563$°R.

$$\begin{aligned}\dot{m} &= \rho A V \\ &= \frac{p}{RT} AM\sqrt{\gamma g_c RT}\end{aligned}$$

where $R = 55.16$ ft-lbf/lbm°R.

$$\dot{m} = \frac{(80\ \text{lbf/in}^2)(2.0\ \text{in}^2)0.574}{\left(55.16\ \frac{\text{ft-lbf}}{\text{lbm°R}}\right)(563°\text{R})} \times \sqrt{1.4\left(32.2\ \frac{\text{lbm-ft/s}^2}{\text{lbf}}\right)\left(55.16\ \frac{\text{ft-lbf}}{\text{lbm°R}}\right)563°\text{R}}$$

$$\dot{m} = 3.499\ \text{lbm/s} \qquad \text{for } p_b = 80\ \text{psia}$$

For $p_b = 50$ psia, the nozzle is choked ($M_{exit} = 1$) since $p_b/p_t < (p/p_t)_{crit}$. In this case, $M_{exit} = 1.0$, $p_{exit}/p_t = 0.5283$, and $T_{exit}/T_t = 0.8333$. Therefore, $p_{exit} =$ 52.83 psia, $T_{exit} = 500°R$.

$$\dot{m} = \frac{p}{RT} AM\sqrt{\gamma g_c RT}$$

$$= \frac{(52.83\text{ lbf/in}^2)(2.0\text{ in}^2)1.0}{(55.16\text{ ft-lbf/lbm°R})(500°\text{R})}$$

$$\times \sqrt{1.4\left(32.2\,\frac{\text{lbm-ft/s}^2}{\text{lbf}}\right)(55.16\text{ ft-lbf/lbm°R})(500°\text{R})}$$

$$= 4.272\text{ lbm/s} \qquad \text{for } p_b = 50\text{ psi}$$

For $p_b = 0$ psia, $\dot{m}$ will also be 4.272 lbm/s, since the nozzle is choked for this case also.

Now suppose the converging nozzle of Figure 9.15 is replaced by a converging-diverging nozzle. For constant reservoir conditions, we will examine flow characteristics as p_b is varied (Figure 9.16).

Clearly, for back pressure equal to reservoir pressure there is no flow and no pressure change ①. For a slight reduction in back pressure (curve ②), the pressure will decrease to the throat and then increase to the nozzle exit plane. For curve ②, all flow is subsonic. For a low enough

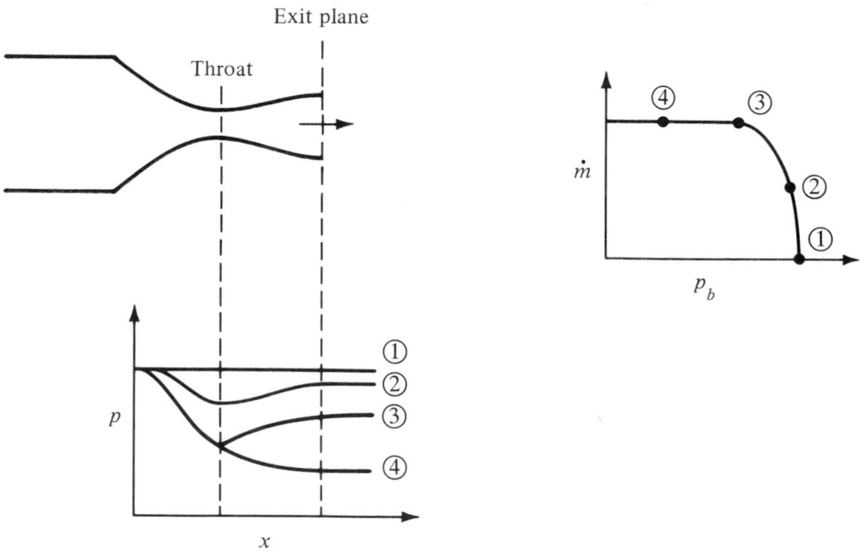

FIGURE 9.16

back pressure (curve ③), sonic flow occurs at the throat, with subsonic flow in the diverging section. Pressure again decreases to the throat, and then increases to the exit. Any further decrease in back pressure cannot be felt in the reservoir, so the nozzle is choked and $\dot{m}$ is therefore the same for all back pressures less than or equal to that of curve ③. For a low enough back pressure, we can have subsonic flow up to the throat followed by isentropic supersonic flow in the diverging section (curve ④). For expansion to back pressures between those of ③ and ④ and below ④, isentropic, one-dimensional solutions to the flow equations are not possible. We will refer later to these cases.

EXAMPLE 9.7

A rocket nozzle has a ratio of exit area to throat area of 3.37, and it is fed from a chamber in which conditions are maintained at 2500°F, 200 psia. Assuming the rocket exhaust gases behave as a perfect gas with $\gamma = 1.4$ and $R = 80$ ft-lbf/lbm°R, determine the exit plane velocity and the nozzle back pressure for supersonic, isentropic, perfectly expanded flow. (See Figure 9.17.)

Solution

For isentropic, supersonic flow at the exit plane, the Mach number at the nozzle throat = 1.0 and the throat area = A^*. For

$$\frac{A_e}{A^*} = 3.37, \quad \text{from Appendix C}, \quad M_e = 2.76$$

At $M_e = 2.76$,

$$\frac{p}{p_t} = 0.03917 \quad \text{and} \quad \frac{T}{T_t} = 0.3963$$

Therefore, $p_e = 7.834$ psia, which is equal to the back pressure for perfectly expanded flow.

$$T_e = 0.3963(2960°\text{R}) = 1173°\text{R}$$

and

$$\begin{aligned} V_e &= M_e\sqrt{\gamma g_c R T_e} \\ &= 2.76\sqrt{1.4\left(32.2\,\frac{\text{lbm-ft/s}^2}{\text{lbf}}\right)\left(80\,\frac{\text{ft-lbf}}{\text{lbm°R}}\right)(1173°\text{R})} \\ &= 5677 \text{ ft/s} \end{aligned}$$

FIGURE 9.17

EXAMPLE 9.8

The stagnation pressure of the air flow entering a converging-diverging nozzle is 200 kPa. If the area ratio of the nozzle is 2.0, determine the range of back pressures over which the nozzle will be choked.

Solution

Refer to Figure 9.16. The maximum back pressure for choked flow occurs under the conditions shown in Figure 9.18.

For $M = 1$ at throat, the throat area $= A^*$. The subsonic solution for $A/A^* = 2.0$ is $M_e = 0.306$ (Appendix C). At this value of M_e,

$$\frac{p_e}{p_t} = 0.937 \quad \text{or} \quad p_e = 187\ \text{kPa}$$

Therefore the nozzle is choked over the back pressure range 0 to 187 kPa.

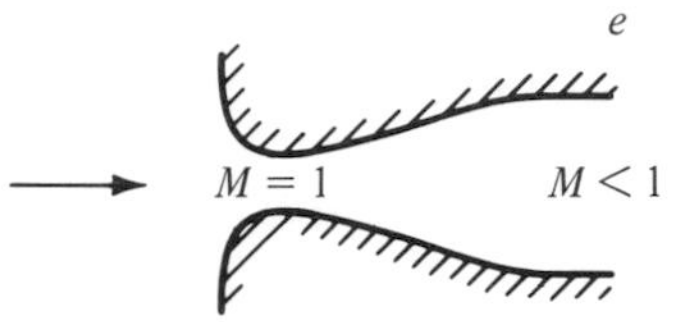

FIGURE 9.18

9.5 NORMAL SHOCK WAVES

Shock waves have been mentioned earlier in this chapter. The shock wave consists of a sudden change in flow properties in supersonic flow. We will now develop the appropriate equations for a normal shock wave, a wave normal to the flow direction. Consider a steady flow in which a normal shock occurs (see Figure 9.19). The shock is very thin, but there are finite changes in pressure and temperature across it. Select system

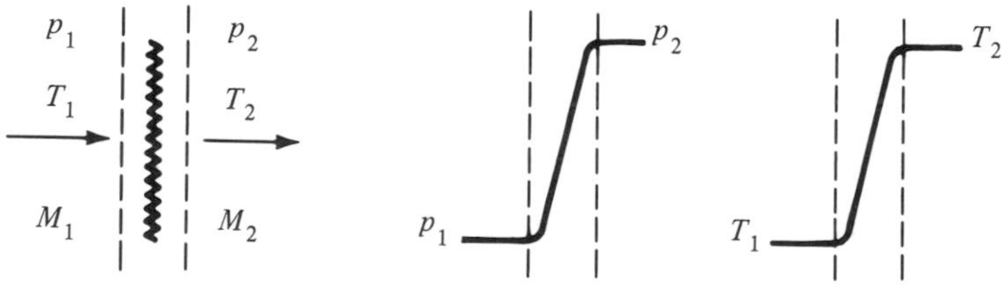

FIGURE 9.19

boundaries to include the wave, as shown, with temperature and pressure gradients internal to the wave. With no temperature gradient at system boundaries, we take the shock to be an adiabatic process. Further, since the shock represents a finite disturbance, it is irreversible.

The continuity equation for steady flow is $\rho_1 A_1 V_1 = \rho_2 A_2 V_2$ where, for a very thin shock, $A_1 = A_2$. Therefore,

$$\rho_1 V_1 = \rho_2 V_2 \tag{9.24}$$

or, for a perfect gas, with constant specific heats,

$$\frac{p_1 M_1}{\sqrt{T_1}} = \frac{p_2 M_2}{\sqrt{T_2}} \tag{9.25}$$

The energy equation for adiabatic flow of a perfect gas with constant specific heats is

$$T_{t_1} = T_{t_2}$$

or

$$T_1\left(1 + \frac{\gamma - 1}{2} M_1^2\right) = T_2\left(1 + \frac{\gamma - 1}{2} M_2^2\right) \tag{9.26}$$

Another equation from fluid flow is the momentum equation, which states that the sum of external forces in the x-direction acting on the fluid in a given volume is

$$\sum F_x = \frac{\dot{m}}{g_c} \Delta V_x$$

In our case, the only forces acting are pressure forces, so

$$p_1 A_1 - p_2 A_2 = \frac{\dot{m}}{g_c} (V_2 - V_1)$$

or, since $A_1 = A_2$,

$$p_1 - p_2 = \frac{\rho_2 V_2^2}{g_c} - \frac{\rho_1 V_1^2}{g_c} \tag{9.27}$$

Since $p = \rho RT$ and $a^2 = \gamma g_c RT$, we have

$$p_1(1 + \gamma M_1^2) = p_2(1 + \gamma M_2^2) \tag{9.28}$$

We now have three equations, (9.25), (9.26), and (9.28), in terms of three variables, M_2, p_2, and T_2, and we can solve for these variables in terms of M_1, p_1, and T_1. Solutions are provided in Appendix D for $\gamma = 1.4$ in terms of M_2 and for p_2/p_1 and T_2/T_1 in terms of M_1. For example, a sketch of M_2 vs. M_1 is given in Figure 9.20.

It can be seen that, for $M_1 > 1$, $M_2 < 1$ and vice versa. For $M_1 > 1$ it follows from (9.28) that $p_2/p_1 > 1$ and we have a compression shock. For $M_1 < 1$, $p_2/p_1 < 1$, and we have an expansion shock.

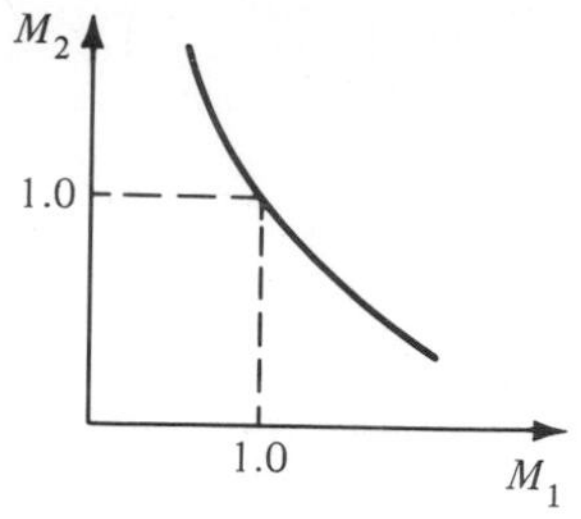

FIGURE 9.20

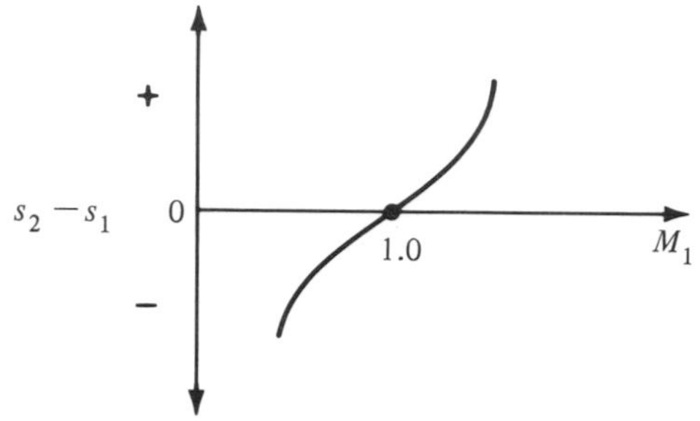

FIGURE 9.21

The change in entropy across a shock, $s_2 - s_1$, is given by

$$s_2 - s_1 = c_p \ln \frac{T_2}{T_1} - R \ln \frac{p_2}{p_1}$$

We can therefore solve for $s_2 - s_1$ in terms of M_1. The result is given in Figure 9.21.

For an adiabatic, irreversible process, the second law requires that $s_2 > s_1$, or $s_2 - s_1$ must be positive. The only solutions possible from Figure 9.21 are for $M_1 > 1$. In other words, an expansion shock is impossible; for a shock, we must have supersonic flow.

EXAMPLE 9.9

A stationary normal shock occurs in an air stream moving at Mach 3.0 (see Figure 9.22). The static pressure and temperature in the air stream ahead of the shock are 10 psia and 480°R. Determine the change in stagnation pressure and stagna-

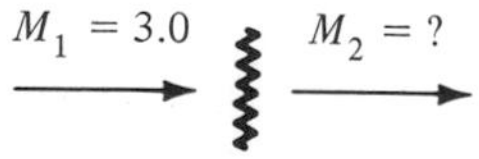

FIGURE 9.22

tion temperature across the shock and the increase in temperature and pressure of the air stream due to the shock.

Solution

From Appendix C.2, for $M_1 = 3.0$, $M_2 = 0.4752$, and $T_2/T_1 = 2.679$, $p_2/p_1 = 10.33$. Therefore,

$$T_2 = 1286°\text{R} \quad \text{and} \quad p_2 = 103.3\ \text{psia}$$

Then

$$T_2 - T_1 = 806°\text{R} \quad \text{and} \quad p_2 - p_1 = 93.3\ \text{psia}$$

Since the shock is an adiabatic process, $T_{t_2} = T_{t_1}$. At $M_1 = 3.0$,

$$\frac{p_1}{p_{t_1}} = 0.02722 \quad \text{and} \quad p_{t_1} = 367.4\ \text{psia}$$

$$\frac{p_2}{p_{t_2}} = 0.857 \quad \text{and} \quad p_{t_2} = \frac{103.3\ \text{psia}}{0.857} = 120.5\ \text{psia}$$

$$p_{t_2} - p_{t_1} = -246.9\ \text{psia}$$

EXAMPLE 9.10

A converging-diverging nozzle is fed from an air reservoir at 300 kPa. If the area ratio of the nozzle (exit area to throat area) is 2.5, determine the back pressure for a normal shock to appear at the nozzle exit. Assume isentropic flow up to the shock with $\gamma = 1.4$ (Figure 9.23).

Solution

Since the Mach number just before the shock must be supersonic, the Mach number at the nozzle throat = 1.0 and the throat area = A^* for the flow up to the shock. Thus, for $A_1/A^* = 2.5$, from Table C.1,

$$M_1 = 2.443 \quad \text{and} \quad \frac{p_1}{p_{t_1}} = 0.06396$$

$$p_1 = \frac{p_1}{p_{t_1}} \times p_1 = 0.06396 \times 300\ \text{kPa} = 19.2\ \text{kPa}$$

But from Table C.2, at $M_1 = 2.443$,

$$\frac{p_2}{p_1} = 6.796\,, \quad \text{so} \quad p_2 = p_b = 130.5\ \text{kPa}$$

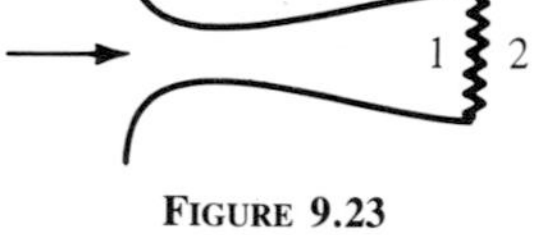

FIGURE 9.23

We can now fill in some of the gaps in our analysis of the flow through a converging-diverging nozzle (Figure 9.16). If the back pressure is reduced slightly below that of curve ③, a normal shock appears downstream of the nozzle throat ⓐ (see Figure 9.24).

For curve ⓐ, flow is sonic at the throat; supersonic flow occurs downstream of the throat with subsonic flow after the shock. Pressure decreases up to the shock, increases across the shock, and increases again for subsonic flow after the shock in an increasing area.

A lowering of back pressure below ⓐ moves the shock downstream until, at ⓑ, it appears at the nozzle exit. Pressure variations for ⓐ and ⓑ are shown in Figure 9.25.

For back pressures between those of curves ⓑ and ④, the adjustment to the back pressure occurs outside the nozzle in the form of oblique shock waves at the nozzle exit. Since, in this regime, the back pressure is greater than that of the exit plane, we say the flow is overexpanded.

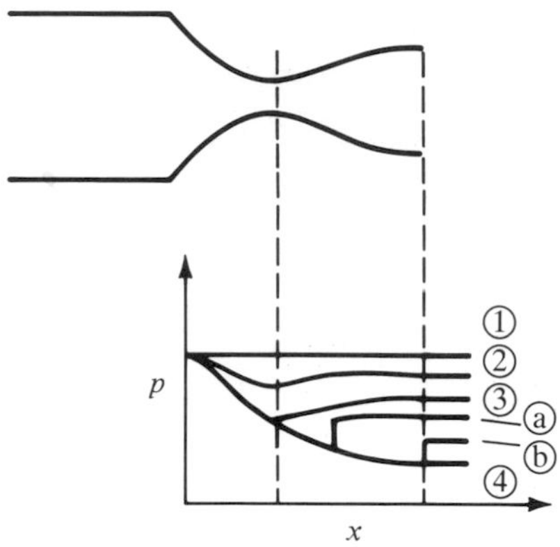

FIGURE 9.24

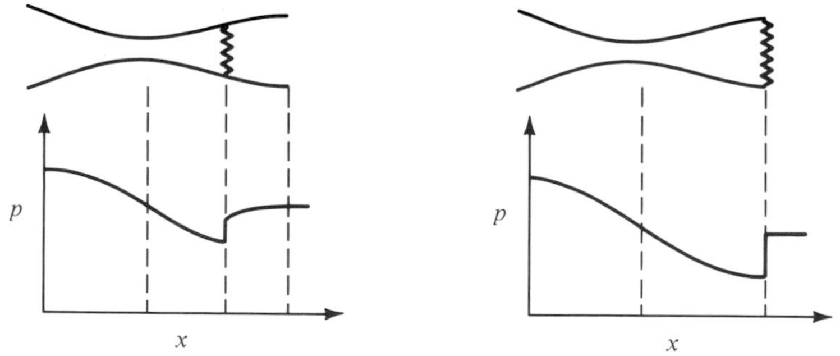

FIGURE 9.25

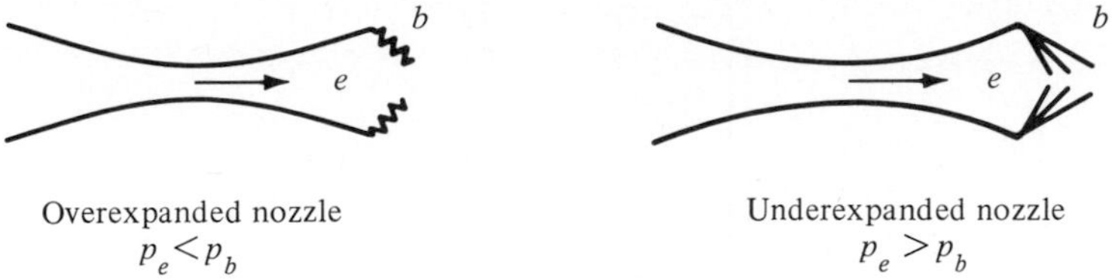

FIGURE 9.26

For back pressures below that of curve ④, adjustment to the back pressure occurs in the form of expansion waves at the nozzle exit. In this regime, the flow is underexpanded (see Figure 9.26).

For all back pressures less than that of ⓑ, the exit plane velocity and pressure are the same, equal to that for isentropic, supersonic flow.

EXAMPLE 9.11

A rocket nozzle has an area ratio of 4.0; the chamber pressure and temperature are 150 psia and 2000°R. Determine the rocket exit velocity in space for $\gamma = 1.4$ and $R = 100$ ft-lbf/lbm°R.

Solution

From Figure 9.26, with $p_b = 0$, it is clear that the nozzle is underexpanded. We thus have isentropic, supersonic flow up to the exit plane. For $A_e/A^* = 4.0$, $M_e = 2.94$ and $T_e/T_t = 0.3665$. Therefore,

$$T_e = 0.3665(2000°\text{R}) = 733°\text{R}$$

$$V_e = M_e a_e = 2.94\sqrt{\gamma g_c R T_e}$$

$$= 2.94\sqrt{1.4\left(32.2\ \frac{\text{lbm-ft/s}^2}{\text{lbf}}\right)\left(100\ \frac{\text{ft-lbf}}{\text{lbm°R}}\right)733°\text{R}}$$

$$V_e = 5344 \text{ ft/s}$$

PROBLEMS

9.1 A water tank is to be emptied through an aperture, as shown in Figure 9.27. When the height of the water level is 4 m above the aperture, determine the rate of flow of the water through the aperture in m³/s. Assume reversible flow, with an aperture diameter of 1.5 cm.

9.2 A 90-deg elbow is attached to the tank as shown in Figure 9.28. How high will the water jet reach?

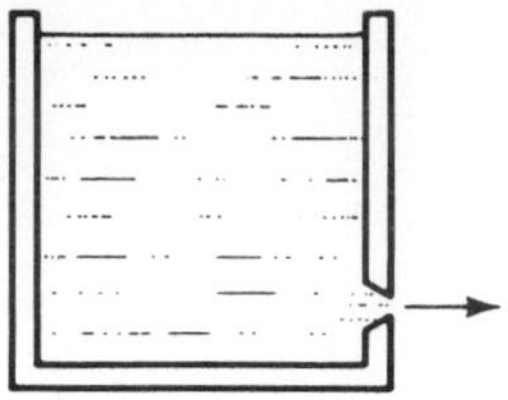

FIGURE 9.27

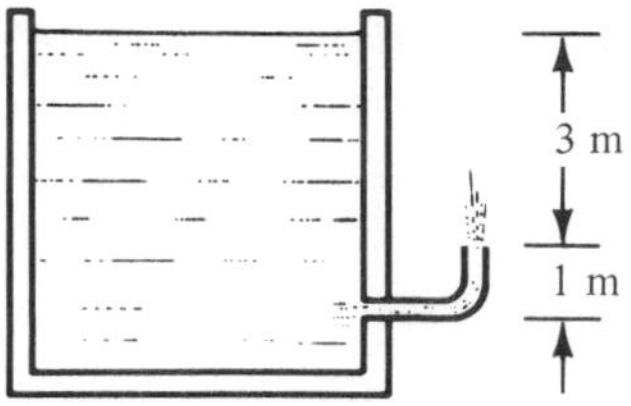

FIGURE 9.28

9.3 A pump is to be used to supply water from a reservoir to a location 100 feet above the reservoir (Figure 9.29). Water at 1.0 ft/s through a 1.0-in. diameter pipe is to be delivered at atmospheric pressure. Assuming reversible flow, determine the pump horsepower required.

9.4 Determine the pressure difference $p_1 - p_2$ for the reducing elbow shown. Assume reversible water flow, with the elbow in the horizontal plane, as shown in Figure 9.30. $V_1 = 0.8$ m/s.

9.5 An air nozzle exhausts to atmospheric pressure (see Figure 9.31). For $D_1 = 6$ in., $D_2 = 2$ in., and $V_1 = 10$ ft/s, determine the pressure p_1.

9.6 Determine the water flow out of the reservoir of Figure 9.32. Assume reversible flow, with a circular opening of diameter 2.2 cm.

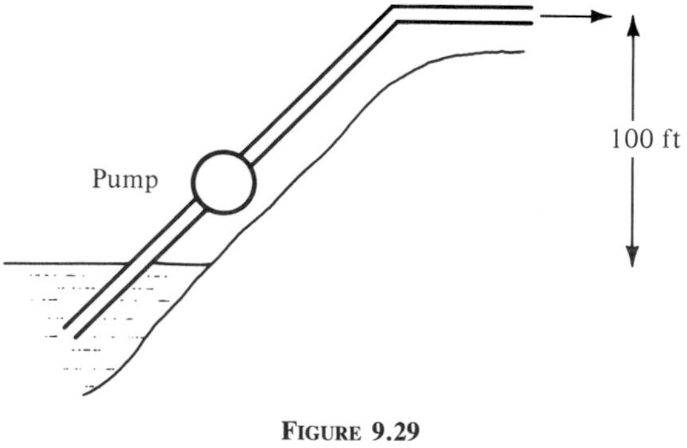

FIGURE 9.29

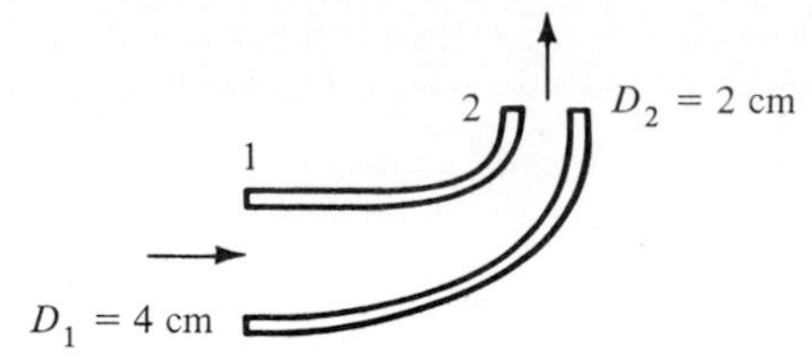

FIGURE 9.30

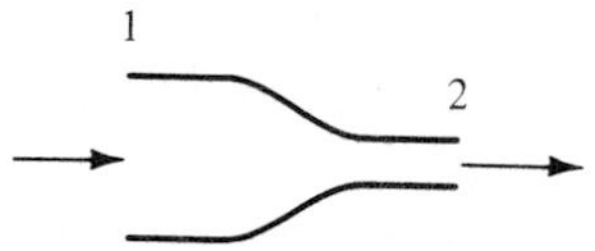

FIGURE 9.31

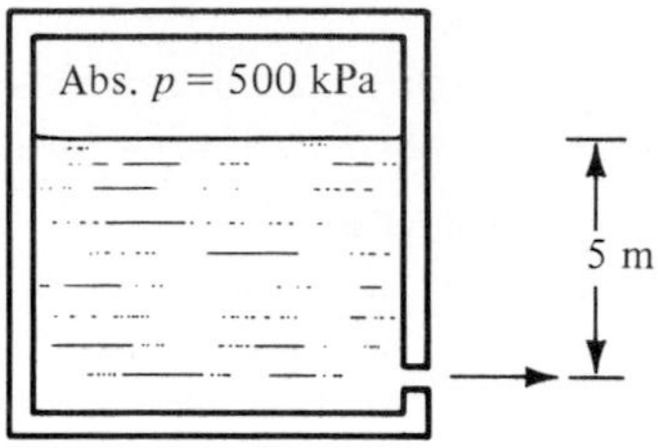

FIGURE 9.32

9.7 Determine the velocity of sound in hydrogen at atmospheric pressure and 60°F.

9.8 Determine the velocity of sound in helium at 2 atmospheres pressure and 50°C. (Molecular mass = 4.003 for helium.)

9.9 A jet plane is travelling at 500 mph at an altitude of 30,000 ft ($T = 411.7$°R). Determine the Mach number of the plane.

9.10 Air flows in a 3.0-cm-diameter duct with a Mach number of 0.6. The stagnation pressure of the flow is measured to be 200 kPa, the stagnation temperature is 400 K. Determine the air flow through the duct.

9.11 Air enters a diverging section of duct with a Mach number of 0.8, static pressure of 20 psia. Determine the Mach number and pressure at a cross-sectional area which is 20 percent larger than the inlet area. Assume isentropic flow.

9.12 A large reservoir tank contains air at 1.0 MPa, 30°C. A small leak develops in the tank, allowing air to escape to atmospheric pressure. Assume the leakage flow to be equivalent to isentropic flow through a converging nozzle of exit area 1 mm^2. Determine the rate of air flow out of the tank.

9.13 Plot mass flow rate vs. tank pressure for isentropic flow through the converging nozzle as shown in Figure 9.33. Converging nozzle exit diameter = 1.0 cm.

9.14 Hot gases are exhausted through a converging nozzle, with the back pressure on the nozzle maintained at 14.7 psia. The hot gases are supplied from a chamber at 500 psia and 2000°F. Determine the nozzle exhaust velocity for isentropic flow. Take $\gamma = 1.4$ and $R = 70$ ft-lbf/lbm°R. Repeat for a back pressure of 0 psia.

9.15 Air enters a converging-diverging nozzle with a velocity V_1 of 300 ft/s and temperature of 100°F (see Figure 9.34). Determine the velocities at sections 2 and 3, assuming isentropic flow.

9.16 A converging nozzle is fed from a large air tank at 1.0 MPa. Determine the range of back pressures over which the nozzle will be choked.

9.17 A converging-diverging nozzle is fed from a large air tank at 1.0 MPa. Determine the range of back pressures over which the nozzle will be choked. Take the nozzle to have a ratio of exit area to throat area of 2.3.

9.18 A converging-diverging rocket nozzle is supplied from a gas chamber at 1000 psia, 2500°F. Determine the rocket nozzle exhaust velocity in space. Assume the exhaust gases to behave as a perfect gas with constant $\gamma = 1.4$ and constant $R = 72$ ft-lbf/lbm°R. The nozzle area ratio is 4.0.

9.19 A stationary normal shock occurs in an air stream moving at Mach 2.0, static temperature 80°F. Determine the velocity of the air behind the shock.

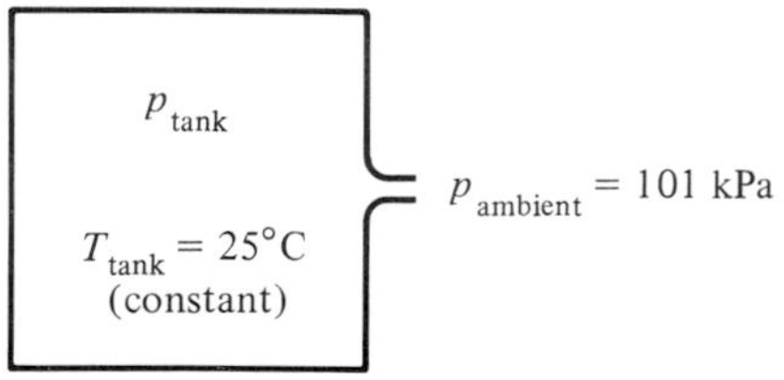

FIGURE 9.33

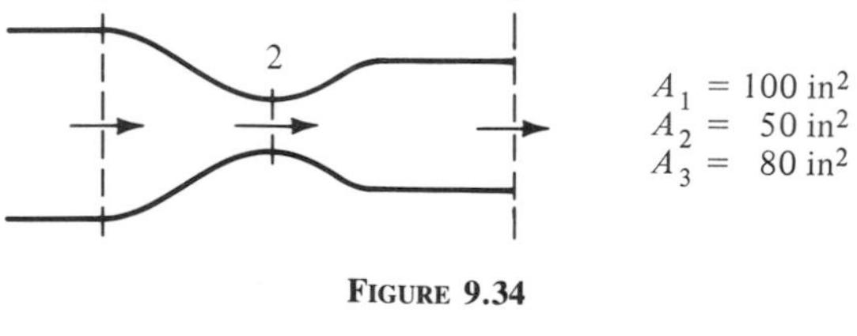

FIGURE 9.34

9.20 A normal shock moves into still air at 80°F. If the Mach number of the moving shock is 2.0, determine the velocity of the air behind the shock.

9.21 A converging-diverging nozzle has an area ratio of 2.5. The nozzle is supplied from a large tank containing air at 500 kPa, 30°C. Determine the range of back pressures over which a normal shock will appear in the nozzle. Over what range of back pressures will the nozzle be choked?

9.22 A converging-diverging rocket nozzle has an exit area of 320 cm^2, throat area of 100 cm^2. Determine the exit velocity if a normal shock appears in the diverging portion of the nozzle at an area of 200 cm^2. The nozzle is supplied from a chamber containing gases at 650 kPa, 1500 K. Find the exit plane pressure for this condition. Assume $\gamma = 1.4$, $R = 0.3$ kJ/kg K.

9.23 Determine the exhaust velocity for the rocket nozzle of the previous example if it is operating in space. Find the exit plane pressure.

9.24 For the conditions shown in Figure 9.35, determine the back pressure, the velocity at the throat, and V_1 and V_2.

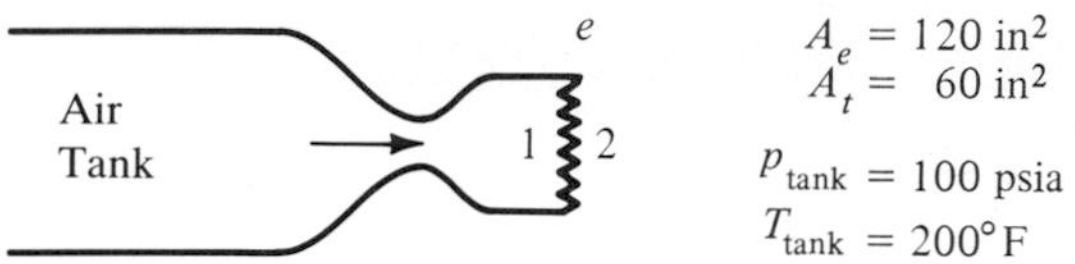

FIGURE 9.35

chapter

10

Power and Refrigeration Cycles

10.1 THERMODYNAMICS OF CYCLES

In this chapter, we will use the first and second laws of thermodynamics to analyze the performance of power and refrigeration cycles. As discussed in Chapter 6, a *cycle* is a series of processes in which a system starts at a given state and is returned to that exact same state.

A system undergoing cyclic changes can be either a closed system [Figure 10.1(a)] or an open system [Figure 10.1(b)], as defined in Chapters 2 and 3. Further, we can define open and closed loop cycles. In a *closed loop cycle* [Figure 10.1(c)], the same working fluid is utilized, circulating continuously within the system. An example of a closed loop cycle is an electric power generating station. Outputs as high as 1000 megawatts are obtained on a continual basis for many years. In such a stationary power generating plant, the working fluid (water) is recirculated. That is, the steam is cooled after it has expanded through the turbine to produce the required power; then it is reheated in the steam generator and allowed to pass again through the turbine. In this way, the same water flows around and around in the cycle.

In an *open loop cycle*, new working fluid is constantly furnished. An open loop cycle can be obtained when inlet and exit thermodynamic states are not equal and the working fluid is taken from a large reservoir and discharged into it [Figure 10.1(d)]. In this fashion, the inlet condition remains constant, while changes over part of the cycle take place outside the system. The mass contained within the system is very small compared to the mass contained within the large reservoir. Hence, the different thermodynamic state of this small mass has a negligibly small influence on

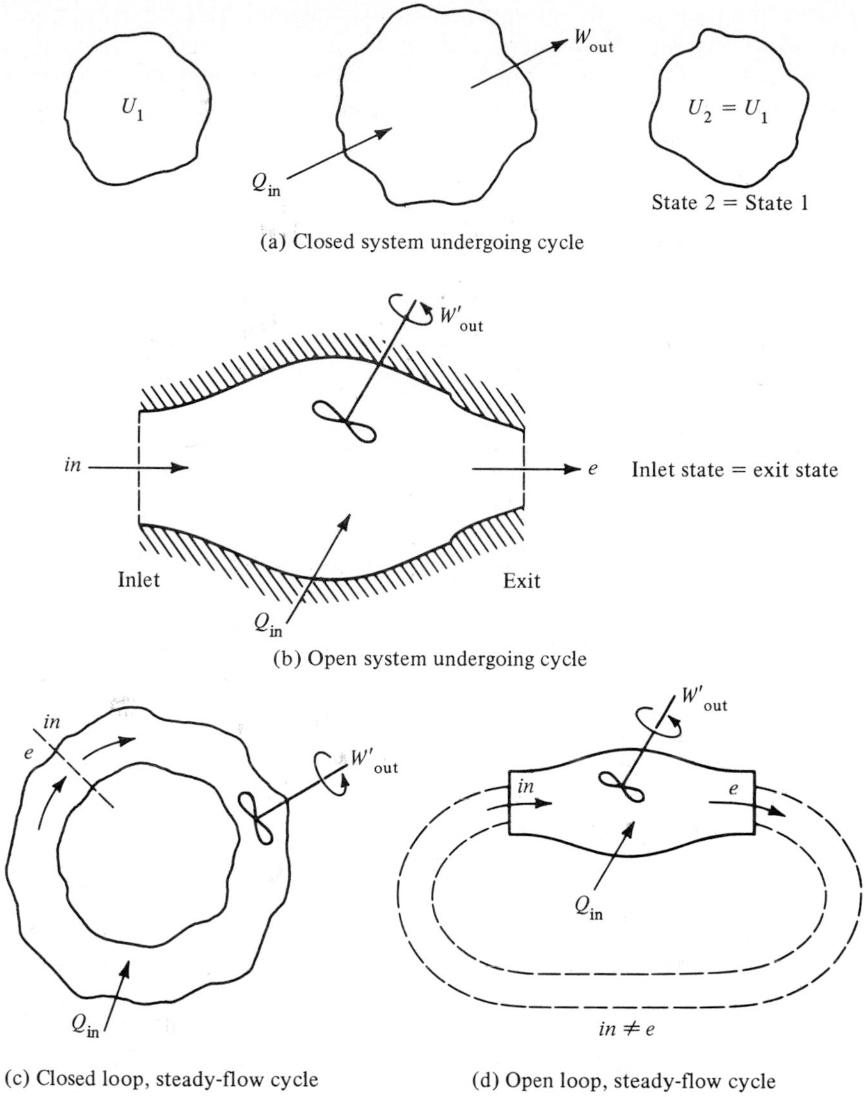

(a) Closed system undergoing cycle

(b) Open system undergoing cycle

(c) Closed loop, steady-flow cycle

(d) Open loop, steady-flow cycle

FIGURE 10.1

the thermodynamic state of the reservoir, when the exit mass mixes with that of the reservoir. An example of such a large reservoir is the atmosphere. An aircraft jet engine continually takes in fresh ambient air and discharges hot combustion gases to the atmosphere. Eventually, these gases cool down in the atmosphere to the same temperature as ambient temperature. This cycle can be treated as an open-loop cycle.

Also, in an automobile internal combustion engine, fresh air is continually taken in, and hot combustion gases are exhausted. Again, these combustion gases have negligible effect on the ambient temperature in the atmosphere, and this engine can be analyzed as an open loop cycle.

(*a*) *First Law Considerations*

If a system undergoing cyclic changes is a closed system, the first law becomes, with $U_2 - U_1 = 0$,

$$\text{net } Q_{\text{in}} = \text{net } W_{\text{out}} \tag{10.1}$$

This equation states that since the internal energy of the system remains unchanged as a result of the cyclic processes, an amount of energy equal to the net work output must have flowed into the system during the cycle as net heat input. Over part of the cycle, heat may be transferred into the system from the surroundings, while over another part of the cycle, heat may be transferred out of the system to the surroundings.

For a steady-flow system, the first law becomes, with $H_{\text{in}} = H_e$, $KE_{\text{in}} = KE_e$, and $PE_{\text{in}} = PE_e$:

$$\text{net } Q_{\text{in}} = \text{net } W'_{\text{out}} \tag{10.2}$$

(*b*) *Second Law Considerations*

The cycle must also satisfy the second law of thermodynamics. From Equation (6.16), for a closed system,

$$(S_2 - S_1)_{\text{univ}} = (S_2 - S_1)_{\text{system}} + (S_2 - S_1)_{\text{surround}} \geq 0$$

or, for an open steady-flow system, Equation (6.18),

$$(S_2 - S_1)_{\text{univ}} = m_e s_e - m_{\text{in}} s_{\text{in}} + (S_2 - S_1)_{\text{surround}} \geq 0$$

For a closed system undergoing a cycle, with 1 and 2 the same states.

$$(S_2 - S_1)_{\text{system}} = 0$$

For an open steady-flow system undergoing a cycle, as in Figure 10.1(c),

$$m_e s_e = m_{\text{in}} s_{\text{in}}$$

It follows that for a cycle in which irreversibilities occur,

$$(S_2 - S_1)_{\text{univ}} = (S_2 - S_1)_{\text{surround}} > 0$$

For a reversible cycle,

$$(S_2 - S_1)_{\text{univ}} = (S_2 - S_1)_{\text{surround}} = 0$$

Finally, the second law imposes a limitation on the performance of a cyclic device: The efficiency of a heat engine cannot be greater than that of a Carnot engine operating between the same maximum and minimum temperatures; the COP of a heat pump or refrigerator cannot be greater than that of a Carnot heat pump or refrigerator operating between the same maximum and minimum temperatures.

The concepts of availability and irreversibility were presented in Chapter 7. In this chapter, we will use availability equations in the second law analysis of cycles.

For an open, steady flow system,

$$a_{in} - a_e = w'_{rev} \tag{7.7}$$

Since, according to (7.4a), $i = w'_{rev} - w'_{out}$, we obtain

$$a_{in} - a_e = w'_{out} + i \tag{10.3}$$

Equation 10.3 states that changes in availability for a steady-flow system are equal to the sum of actual work output and irreversibility. The first law expresses the concept that all energy is conserved. Energy cannot be destroyed; only its form can change. Heat and work are forms of energy in transition across a system boundary. On the other hand, the second law expresses the concept that availability is not conserved, but decreases due to irreversibility. Thus, the first law, as expressed in Equation (3.15), gives the energy balance equation, while the second law, as expressed in Equation (10.3), presents the availability balance equation.

Applying Equation (10.3) to a steady-flow, open-loop system, we get for the change across the open-loop system

$$(a_{in} - a_e) = \text{net } w'_{out} + i \tag{10.3a}$$

This equation is also applicable to a system which consists of a closed-loop and an open-loop subsystem.

For a closed system, we have from (7.9):

$$a_1 - a_2 = w_{rev} + p_0(v_1 - v_2)$$

Since, from (7.4b), $i = w_{rev} - w_{out}$,

$$a_1 - a_2 = w_{out} + i - p_0(v_2 - v_1) \tag{10.4}$$

As mentioned earlier, a cycle consists of several processes in which a system starts at a given state and returns to that exact same state. Thus, for a steady-flow, closed-loop system undergoing cyclic changes, Equation (10.3a) becomes, with $a_{in} = a_e$,

$$0 = \text{net } w'_{out} + \Sigma i \tag{10.5a}$$

Similarly, for a closed system undergoing cyclic changes, we have from

Equation (10.4)

$$0 = \text{net } w_{\text{out}} + \Sigma i \tag{10.5b}$$

In this chapter, we will consider the thermodynamic cycles commonly used for power and refrigeration. We will study the systems used for the production of electrical power; the engines used to power automobiles, trucks, and planes; and the machinery used for home and industrial air conditioning and refrigeration. In each case, one of the main parameters of interest is output/input, that is, from first law considerations, thermal efficiency (η) for a heat engine and COP for a refrigerator and heat pump [see Equations (6.1), (6.2), and (6.3)]. Modifications to improve η or COP of the basic cycles will be presented.

We will also examine cycles and their components using second law availability considerations. From these considerations, we can define second law cycle efficiencies (output/input) as follows: for a *power* plant, where input is the availability input in the furnace-steam generator and output is the net work output of the cycle

$$\eta_2 = \text{net } w'_{\text{out}}/\Delta a_{\text{steam gen}} \tag{10.6}$$

for a *refrigeration* plant, where output is the availability change due to the cooling load and input is the net work input of the cycle

$$\eta_2 = \Delta a_{\text{C.L.}}/\text{net } w'_{\text{out}} \tag{10.7}$$

and for a *heat pump*, where output is the availability change due to the heating load and input is the net work input of the cycle

$$\eta_2 = \Delta a_{\text{H.L.}}/\text{net } w'_{\text{out}} \tag{10.8}$$

In each second law efficiency defined in this fashion, a maximum efficiency of unity is obtained for a reversible (Carnot) cycle. To obtain second law efficiencies of closed systems, replace w' by w.

The choice of the temperature of the surroundings T_0 may affect the value of second law efficiency obtained. In analyzing power cycles, the pressure of the surroundings p_0 is usually taken to be atmospheric (14.7 psia, 101.3 kPa). The temperature T_0 of the surroundings may vary depending on location and season. In plants using vapor power cycles, condenser cooling water is usually obtained from lakes or rivers, or from cooling towers. In such applications, the temperature of the cooling water is taken as reference temperature T_0 and may vary from 32°F to 90°F (0°C to 30°C) depending on geographical location and season. When the information is available, the actual value of temperature should be used as reference temperature; otherwise, use $T_0 = 77$°F (25°C), since this value is a reasonable average value. It also represents the usual reference temperature for thermochemical data (see Chapter 11).

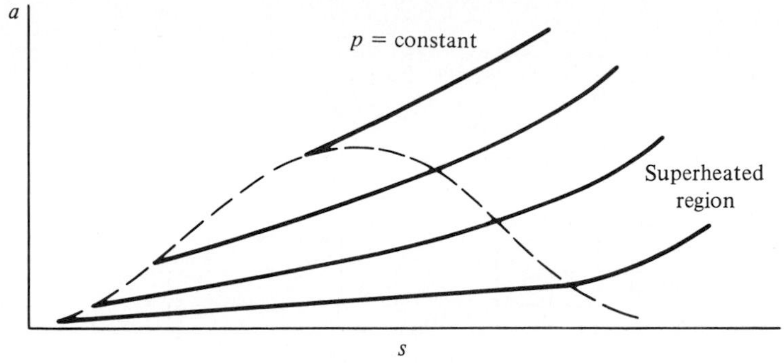

FIGURE 10.2 *a-s Diagram for Steam.*

In analyzing refrigeration cycles, the reference temperature T_0 should be chosen to be at least the lowest temperature in the cycle or the temperature of the cool room; otherwise, negative availability values and negative second law efficiency values will result.

Tables A.21 and A.22 and B.21 and B.22 present values of availability and entropy for water and refrigerant R-22, respectively. In calculating the values of availability, kinetic and potential energy contributions have been neglected, so that availability given by the expression $a = h - h_0 - T_0(s - s_0)$ becomes a state property. In this form, availability is also called *exergy*. For cases where the changes in kinetic and potential energies cannot be neglected, these changes are added to the changes in the values of availability given in the tables. For the tables, a reference temperature of 77°F (25°C) was chosen for water, while a reference temperature of 30°F (0°C) was selected for R-22.

Figure 10.2 shows an *a-s* diagram for steam. The shapes of the *a-s* curves are similar to those of the *h-s* curves. In the saturated region, the values of availability are obtained in a fashion analogous to (4.2), namely, $a = a_f + xa_{fg}$.

The following example will illustrate the use of first and second law analysis for a cycle and its components.

EXAMPLE 10.1

As an example of analyzing the performance of a cycle and its components, consider the reversible Carnot cycle shown in Figure 10.3 operating entirely in the saturated steam region with heat addition at a temperature (T_H) of 500°F and heat rejection at a temperature (T_L) of 60°F.

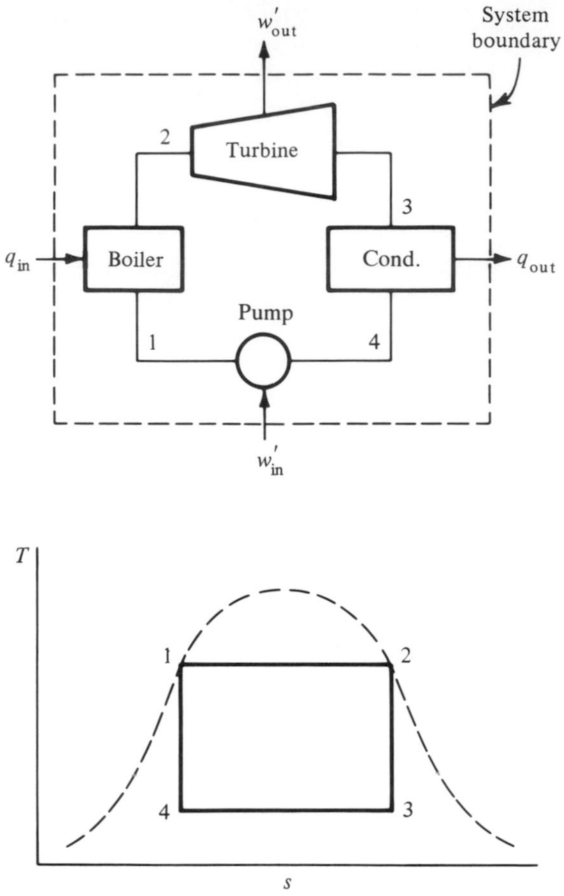

FIGURE 10.3 *Carnot Power Cycle Using Saturated Steam.*

Solution

The compression process (4-1) in the pump and the expansion process (2-3) in the turbine occur isentropically, i.e., without losses. In the heat addition process (1-2) in the boiler at constant temperature, an increase in entropy of the working fluid (steam) takes place, while in the heat rejection process (3-4) in the condenser at constant temperature, the entropy decreases.

Let us examine each component of the cycle and then determine the overall performance of the cycle using first law and second law analyses. Here, the compressor and turbine have an efficiency of 100 percent, i.e.,

$$\eta_{\text{turbine}} = \frac{w'_{\text{out}}}{w'_{\text{out isentropic}}} = 1 \quad \text{and} \quad \eta_{\text{pump}} = \frac{w'_{\text{in isentropic}}}{w'_{\text{in}}} = 1$$

From Table A.13, we obtain at $T = 500°F$, $h_1 = h_f = 487.9$ Btu/lbm and $s_1 = s_f = 0.6890$ Btu/lbm°R, $h_2 = h_g = 1202.2$ Btu/lbm, and $s_2 = s_g = 1.4333$ Btu/lbm°R. At $T = 60°F$, $h_f = 28.06$ Btu/lbm, $h_g = 1087.7$ Btu/lbm and $s_f = 0.0555$ Btu/lbm°R and $s_g = 2.0946$ Btu/lbm°R. Hence, the quality x of the steam at state 3 becomes with $s_2 = s_3$

$$1.4333 = 0.0555 + x_3 2.0391$$

$$x_3 = 0.6757$$

Thus, h_3 is 28.06 Btu/lbm + 0.6757(1059.7) Btu/lbm = 744.1 Btu/lbm. Similarly, we have at state 4 with $s_1 = s_4$

$$0.6890 \text{ Btu/lbm°R} = 0.0555 \text{ Btu/lbm°R} + x_4 2.0391 \text{ Btu/lbm°R}$$

$$x_4 = 0.3107$$

Thus, $h_4 = 28.06$ Btu/lbm + 0.3107(1059.7) Btu/lbm = 357.3 Btu/lbm.

We are now in a position to calculate the performance of each component. From Equation (3.21) we get for the boiler $q_{in} = h_2 - h_1 = 1202.2$ Btu/lbm 487.9 Btu/lbm = 714.3 Btu/lbm. From (3.18) we get for the turbine, neglecting changes in kinetic energy, $w'_{out} = h_2 - h_3 = 1202.2$ Btu/lbm − 744.1 Btu/lbm = 458.1 Btu/lbm. For the condenser, we get from Equation (3.22) $q_{out} = h_3 - h_4 =$ 744.1 Btu/lbm − 357.3 Btu/lbm = 386.8 Btu/lbm. Finally, from Equation (3.20) we get for the pump, neglecting changes in kinetic energy, $w'_{in} = h_1 - h_4 =$ 487.9 Btu/lbm − 357.3 Btu/lbm = 130.6 Btu/lbm.

The results are summarized in the following table:

State or component	T °F	h Btu/lbm	s Btu/lbm°R	q_{in} Btu/lbm	w'_{out} Btu/lbm
1	500	487.9	0.6890		
Boiler				714.3	
2	500	1202.2	1.4333		
Turbine					458.1
3	60	744.1	1.4333		
Condenser				−386.8	
4	60	357.3	0.6890		
Pump					−130.6
1	60	487.9	0.6890		
			Net	327.5 Btu/lbm	327.5 Btu/lbm

Since net $q_{in} = 327.5$ Btu/lbm equals net $w'_{out} = 327.5$ Btu/lbm, Equation (10.2) is satisfied.

The first-law efficiency of the cycle is from (6.1)

$$\eta = \frac{\text{net } w'_{out}}{\text{net } q_{in}} = \frac{327.5}{714.3} = 45.8\%$$

Since this is a Carnot cycle, we can also use Equation (6.7) to obtain the cycle efficiency, namely,

$$\eta = \frac{T_H - T_L}{T_H} = \frac{T_1 - T_3}{T_1} = \frac{500 - 60}{500 + 460} = 45.8\%$$

The first-law distribution of energy and work is:

Work output (turbine − pump)	327.5 Btu/lbm	45.8%
Energy rejection (condenser)	386.8 Btu/lbm	54.2%
Energy input (boiler)	714.3 Btu/lbm	100.0%

We see that a large quantity (54.2%) of the energy supplied to the cycle via the boiler is rejected from the cycle via the condenser. Next, we will analyze the cycle and its components using second-law analysis.

As pointed out in Chapter 7, the second law provides information on the quality of energy, whereas the first law accounts for the quantity of energy. For the purpose of second-law analysis, we need to obtain values of availability at the four states of the cycle. First, we select a temperature of 50°F and a pressure of 14.7 psia as our surroundings reference state 0. Condenser cooling water at a temperature of 50°F would typically be used in this case to condense the steam at a temperature of 60°F. For these conditions, we get for the steam at 50°F and 14.7 psia (a compressed liquid), $h_0 = 18.1$ Btu/lbm and $s_0 = 0.0361$ Btu/lbm°R. From (7.6) we get, neglecting kinetic and potential energies, an expression for availability, namely,

$$a = h - h_0 - T_0(s - s_0)$$

For example, for state 1 we get $a = 487.9$ Btu/lbm − 18.1 Btu/lbm − 510°R (0.6890 Btu/lbm°R − 0.0361 Btu/lbm°R) = 136.9 Btu/lbm. Values for availability for states 2, 3, and 4 were calculated in a similar fashion.

Next, we calculate the second law (availability) performance for each component. For the turbine, $a_{in} - a_e = a_2 - a_3 = 471.6$ Btu/lbm − 13.5 Btu/lbm = 458.1 Btu/lbm. From the first law calculation we obtained $w'_{out} =$ 458.1 Btu/lbm. For the reversible turbine used in this cycle, we have $i = 0$. Thus, the availability relation for steady flow (10.3), $a_2 - a_3 = w'_{out} + i$ is satisfied. In Example 7.9, we say that for a boiler (a heat exchanger), a "negative" irreversibility is obtained for the steam side of the boiler; however, the irreversibility in the combustion gas side is positive and sufficiently greater so that the combined irreversibility of the boiler becomes a positive quantity. Since we are not including the combustion side of the boiler in the system boundary of the cycle, we will assume perfect heat exchange so that, in magnitude, $\Delta a_{steam} = \Delta a_{combustion\ gases}$. Thus the steam side experiences an increase in availability, while the combustion side experiences an equal decrease in availability. In the condenser the unavailable part of the energy is rejected; hence the irreversibility is $i = -\Delta a$. The results of the calculations are presented in the following table:

State or component	a Btu/lbm	Δa Btu/lbm	i Btu/lbm
Steam (closed-loop cycle)			
1	136.9		
Boiler		334.7	−334.7
2	471.6		
Turbine		−458.1	0
3	13.5		
Condenser		−7.2	7.2
4	6.3		
Pump		130.6	0
1	136.9		
Furnace (combustion gases) (open-loop system)			
in			
		−334.7	334.7
e			
Combined net		−334.7	7.2

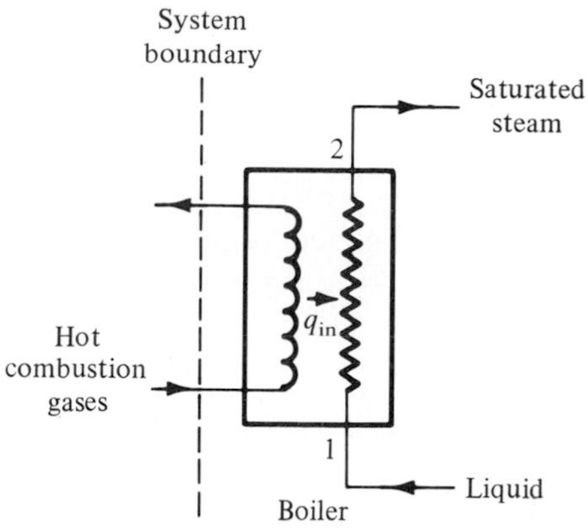

For a system that consists of a closed-loop cycle and an open-loop system, we get from Equation (10.3a)

$$-\Sigma(a_e - a_{in}) = \text{net } w'_{out} + \Sigma i$$

or

$$-(-334.7)\ \text{Btu/lbm} = 327.5\ \text{Btu/lbm} + 7.2\ \text{Btu/lbm}$$

The second-law efficiency of the cycle is from Equation (10.6)

$$\eta_2 = \frac{\text{net } w'_{out}}{\Delta a_{\text{steam gen}}} = \frac{327.5}{334.7} = 97.8\%$$

The second-law distribution of the pertinent quantities is:

	Btu/lbm		
Work output	327.5	97.8%	$(=\eta_2)$
Pump irreversibility	0	0%	
Turbine irreversibility	0	0%	
Condenser irreversibility	7.2	2.2%	
Total availability input	334.7	100.0%	

From the second-law analysis we see that the losses in the condenser are very small since the cooling process takes place at the low temperature of the cycle. In a subsequent example (Example 10.4), we will present a second law analysis of an actual power cycle which includes losses incurred in the combustion side of a boiler.

When considering a power system in which we assume that the availability input occurs by perfect heat exchange, we can, for brevity's sake, omit the irreversibility for the boiler (or equivalent heat exchanger) and deal with closed-loop cycles only.

We can classify cycles according to the nature of the working fluid. Thus, vapor cycles are those in which a phase change of the working fluid occurs, generally between liquid and vapor phases. Gas cycles involve gas or a mixture of gases as the working fluid, with no phase change.

10.2 VAPOR POWER CYCLES: THE RANKINE CYCLE

For maximum efficiency, it is logical to examine first the *Carnot* cycle. The Carnot cycle is shown in the *T-s* diagram in Figure 10.4 for a vapor cycle. Process 1–2 involves phase change from liquid to vapor (boiling); 3–4, phase change from vapor to liquid (condensation). It is relatively easy to design for both of these processes. However, process 4–1 involves an isentropic pressure increase, a pumping process in which a pump must handle a vapor-liquid mixture. Problems such as vapor lock have prevented the design of such a pump to handle a two-phase mixture. If, instead, the *T-s* diagram is as shown in Figure 10.5, with the pump now handling only liquid, excessively high pressures result at 1, many thousands of atmospheres. Further, if a phase change is to occur in process 1–2, the maximum temperature of the Carnot cycle is limited to a value below the critical temperature. Since $\eta_{\text{Carnot}} = 1 - (T_L/T_H)$, the cycle efficiency is severely limited. Assuming T_L is at an ambient tem-

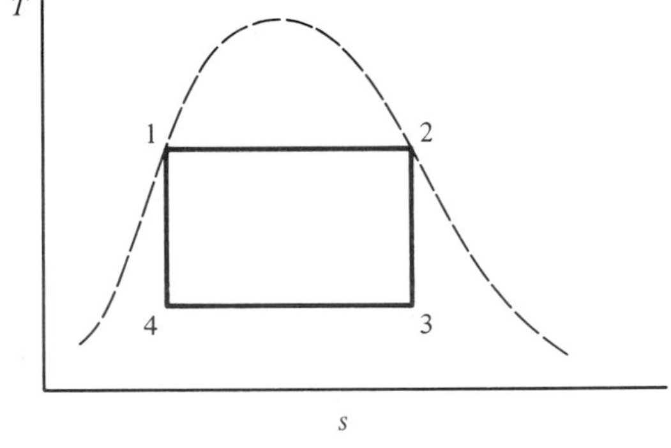

FIGURE 10.4 *Carnot Cycle.*

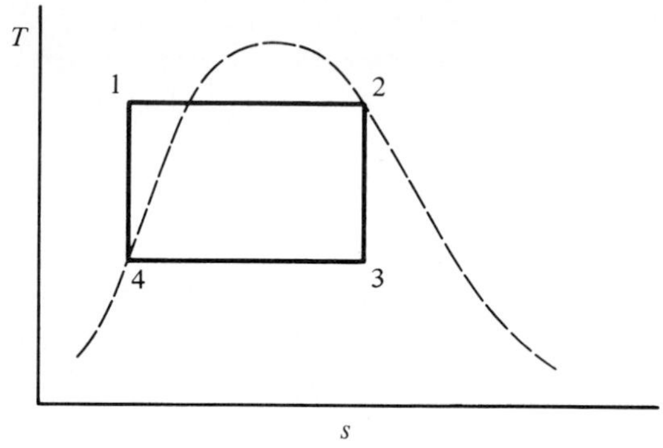

FIGURE 10.5

perature of 520°R, then at most we have

$$\eta = 1 - \frac{520}{705 + 460} = 0.554 \quad \text{or} \quad 55.4\%$$

If a Carnot cycle such as that shown in Figure 10.6 were tried, again excessive pressures would result, as would difficulty in transferring the necessary heat during the isothermal process 1–2 with varying pressure.

A practical version of the vapor power cycle is the *Rankine* cycle, currently used in a large majority of the power plants that generate our electrical power. The basic components consist of boiler, turbine, condenser, and pump, as shown in Figure 10.7(a). Water in the liquid and vapor phases is used as the working fluid. Energy input to the cycle is supplied by the combustion of a fossil fuel or by the heat from a nuclear reactor. In Figure 10.7(a), the hot products of combustion pass around the tubes of the boiler, raising the temperature of the water flowing

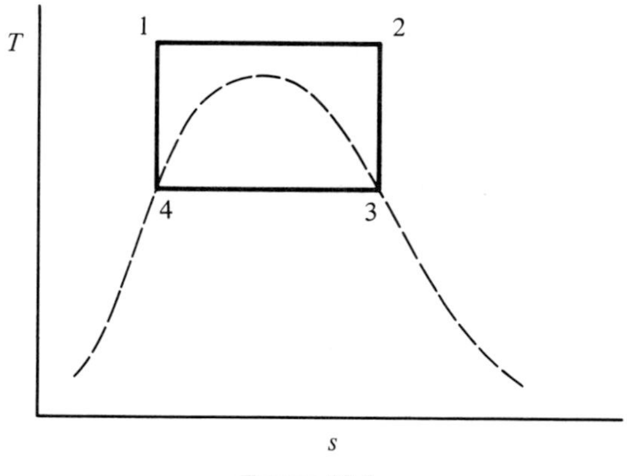

FIGURE 10.6

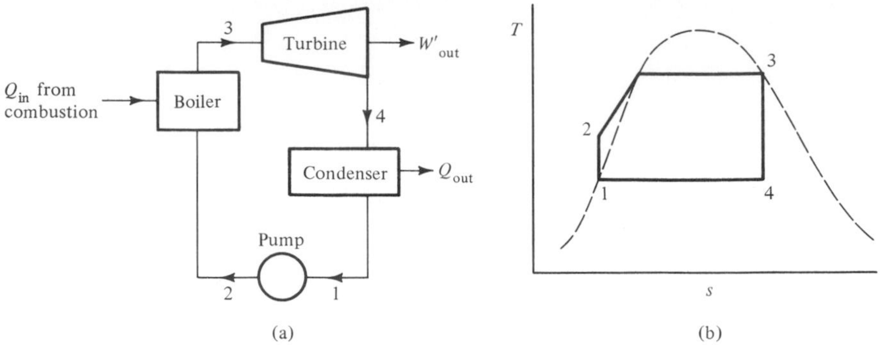

FIGURE 10.7 *Rankine Cycle.*

within the tubes to the boiling point. The water leaves the boiler as vapor and then enters the steam turbine, where it expands and does work on the rotating blades of the turbine. The turbine, in turn, is used to drive an electric generator. The vapor (or liquid–vapor mixture) from the turbine then passes across the outside of the condenser tubes. Next the pressure of the liquid formed in the condenser is raised back up to the boiler pressure by a pump, and the cycle is completed. The condenser tubes are kept cool by the passage of cooling water through the tubes.

The four processes involved in the basic Rankine cycle are shown in Figure 10.7(b). All processes in this cycle are assumed to be ideal and reversible, with an isentropic pump and turbine. All processes are steady flow, so that the boiler and condenser processes are taken as constant pressure processes. Note that the pump handles only liquid (process 1–2).

EXAMPLE 10.2

In a Rankine cycle, the maximum temperature in the boiler is 330°C, and the steam is condensed at 20°C. Find the cycle thermal efficiency, and compare it with the efficiency of a Carnot cycle operating between the same maximum and minimum temperatures. Assume that all processes are reversible, that the pump and turbine are isentropic, and that the substance leaving the boiler is saturated vapor (Figure 10.7).

Solution

At state 3, entering the turbine, the properties of saturated vapor, at 330°C are (from Table B.2)

$$s_3 = 5.4490 \text{ kJ/kg K} \quad \text{and} \quad h_3 = 2670.2 \text{ kJ/kg}$$

The turbine process is a steady-flow adiabatic process, so that

$$w'_{\text{out(turb)}} = h_3 - h_4$$

Next we find h_4. Since $s_4 = s_3 = 5.4490$ kJ/kg K and $T_4 = 20°C$, then

$$s_4 = s_f + x_4 s_{fg}$$

where, from Table B.2, $s_f = 0.2963$ kJ/kg K and $s_{fg} = 8.3721$ kJ/kg K. Therefore,

$$x_4 = \frac{5.4490 - 0.2963}{8.3721} = 0.615$$

and

$$\begin{aligned} h_4 &= h_f + x_4 h_{fg} \\ &= 83.861 + 0.615(2454.3) \\ &= 1593.3 \text{ kJ/kg} \end{aligned}$$

Hence,

$$w'_{\text{turb}} = 2670.2 - 1593.3 = 1076.9 \text{ kJ/kg}$$

At state 1, water leaves the condenser as saturated liquid at 20°C, so $h_1 = 83.861$ kJ/kg and $v_1 = 0.0010017\ \text{m}^3/\text{kg}$. To calculate pump work, use Equation (8.8): $w'_{\text{pump}} = -v(p_e - p_{\text{in}})$ with v assumed constant. In this case, $p_{\text{in}} =$ 2.3366 kPa, the saturation pressure at 20°C, and $p_e = 12\,863$ kPa, the boiler pressure or the saturation pressure at 330°C. Therefore,

$$\begin{aligned} w'_{\text{out(pump)}} &= -(0.001\ \text{m}^3/\text{kg})(12\,863 - 2)\ \text{kPa} \\ &= -12.86 \text{ kJ/kg} \end{aligned}$$

The first law for the steady-flow process in the pump yields, using (3.20),

$$w'_{\text{out}} + h_e - h_{\text{in}} = 0$$

or

$$h_e = h_2 = h_{\text{in}} - w'_{\text{out}} = 83.861 + 12.86 = 96.72 \text{ kJ/kg}$$

For the steady-flow process in the boiler, (3.21) gives

$$q_{\text{in}} = h_3 - h_2 = 2670.2 - 96.7 = 2573.5 \text{ kJ/kg}$$

For the condenser process, (3.22) yields

$$\begin{aligned} q_{\text{out}} &= h_4 - h_1 \\ &= 1593.3 - 83.861 \\ &= 1509.4 \text{ kJ/kg} \end{aligned}$$

Note that for the entire cycle,

$$\text{net } q_{\text{in}} = 2573.5 - 1509.4 = 1064.1 \text{ kJ/kg}$$

$$\text{net } w'_{\text{out}} = 1076.9 - 12.9 = 1064.0 \text{ kJ/kg}$$

As we showed in Section 10.1, for the cycle, $Q_{\text{in}} = W_{\text{out}}$; heat and work must balance.

The cycle efficiency is given by

$$\eta = \frac{\text{net } w'_{\text{out}}}{q_{\text{in}}} = \frac{1064.0}{2573.5}$$

$$= 41.3\%$$

For a Carnot cycle operating between the same maximum and minimum temperatures,

$$\eta_{\text{Carnot}} = 1 - \frac{T_L}{T_H} \quad \text{where } T_L \text{ and } T_H \text{ must be expressed as absolute temperatures}$$

$$= 1 - \frac{273 + 20}{273 + 330}$$

$$= 1 - 0.486$$

$$= 51.4\%$$

In order to increase the efficiency of the basic Rankine cycle, either we can increase the temperature and/or pressure at which heat is added to the working fluid, or we can decrease the temperature at which heat is rejected. To demonstrate, consider first the effect of reducing the condenser temperature. For this analysis, we will assume that all cycles are ideal, consisting of reversible processes. Starting with the base case 1–2–3–4 on a *T-s* diagram (Figure 10.8), we now lower the condenser temperature from $T_{\text{cond } A}$ to $T_{\text{cond } B}$, maintaining the same boiler pressure and temperature. The new cycle appears as 1′–2′–3–4′. For these ideal cycles, with all heat transfer reversible, $q = \int T\, ds$ or, for example, on a *T-s* diagram,

$$q_{\text{in(base cycle)}} = \text{area under 2–3}$$

$$q_{\text{out}} = \text{area under 4–1}$$

$$\text{net } w_{\text{out}} = q_{\text{in}} - q_{\text{out}} = \text{area 1–2–3–4}$$

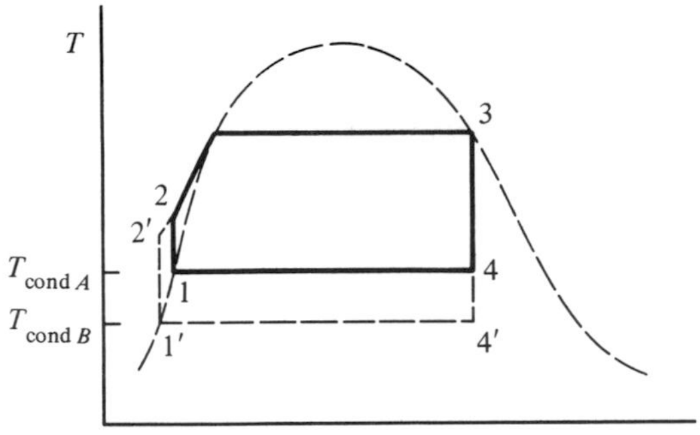

FIGURE 10.8 *Effect of Lowering Condenser Temperature.*

By lowering the condenser temperature, we increase net work by the area 1–2–2′–1′–4′–4–1; q_{in} is increased by the total area under 2′–2, down to the base line. Since these two areas are approximately equal, the net effect of a reduction of condenser temperature is an increase of cycle thermal efficiency.

Next, consider the effect of superheating the steam before it enters the turbine, as shown in Figure 10.9. At the same pressure, this superheating would increase the average temperature at which heat is added to the working fluid in the cycle. The boiler and superheater, taken together, will be called the *steam generator* (Figure 10.10). Note that q_{in} is increased by the area under 3–3′ on the *T-s* diagram, namely, 3–3′–0′–0–3. Net work is increased by the area 3–3′–4′–4–3. Since the ratio

$$\frac{\text{area } 3\text{–}3'\text{–}4'\text{–}4\text{–}3}{\text{area } 3\text{–}3'\text{–}0'\text{–}0\text{–}3}$$

is greater than the ratio of net work to heat input of the basic cycle 1–2–3–4, superheating the steam increases the cycle efficiency.

Finally, let us investigate the case in which we increase the steam generator pressure, holding the same maximum steam generator temperature. Such an example is shown in Figure 10.11. Again, take 1–2–3–4 as the base case and 1–2′–3′–4′ as the cycle with increased pressure, although it has the same maximum temperature as the base cycle. Now the net work has changed by an amount equal to the difference between area 2–2′–3′–x–2 and the area 3–4–4′–x–3. The heat input has decreased by an amount equal to the area under x–3 minus the area 2–2′–3′–x–2. By inspection, we find that the net work is approximately the same for both cases, but the heat input is less for the higher-pressure cycle.

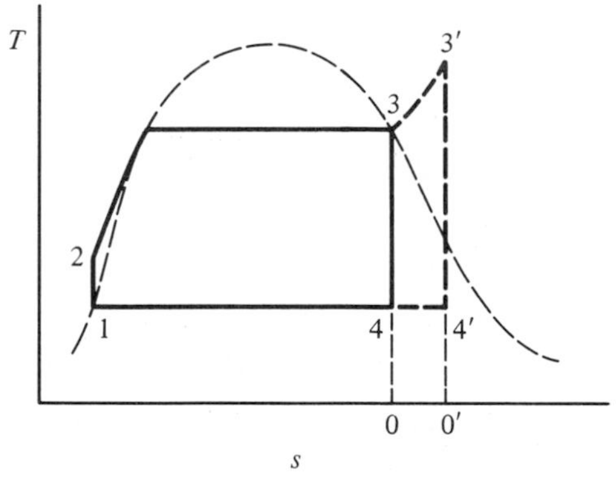

FIGURE 10.9 *Effect of Superheating Steam.*

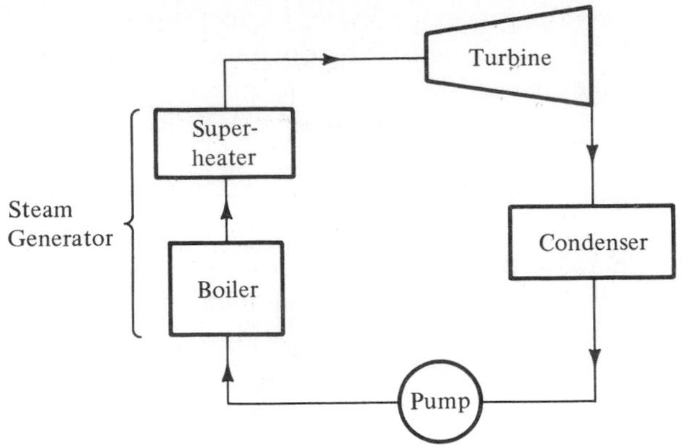

FIGURE 10.10 *Rankine Cycle with Superheat.*

Note that as with the Carnot cycle, in order to increase cycle efficiency we increase the temperature at which heat is added (in the steam generator) and decrease the temperature at which heat is rejected (in the condenser). In actual practice, the maximum pressure and temperature of the cycle are limited by metallurgical considerations—both the material of the boiler and superheater tubes and the material of the turbine blades. The condenser temperature is limited by the temperature of the heat sink available to absorb the heat rejection from the cycle, for example, a lake, river, or other large body of water. Typical maximum values are a steam generator pressure of 2000 psia and a steam generator temperature of 1000°F.

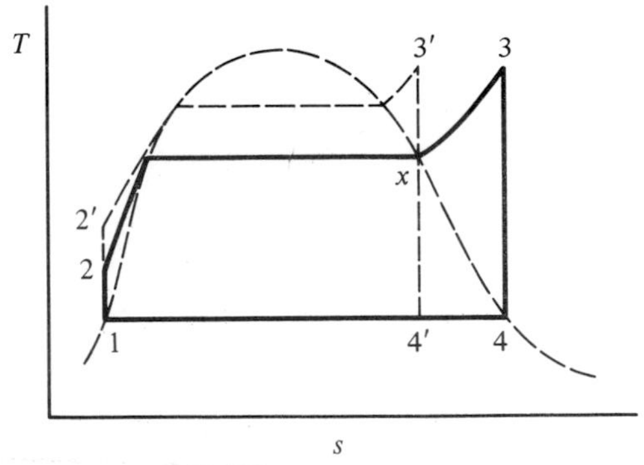

FIGURE 10.11 *Effect of Increased Boiler Pressure.*

EXAMPLE 10.3

In a 500-megawatt electric generating station, the steam temperature and pressure at the turbine inlet are 1000°F and 2000 psia (Figure 10.12). The temperature of the condensing steam in the condenser is maintained at 60°F. Assuming all processes to be ideal, determine the cycle thermal efficiency. The power plant is coal fired, with the heating value of the coal 12,000 Btus per pound of coal. Ninety percent of the heat generated by the combustion of the coal in the steam generator is available (the rest is lost up the stack). Determine the tons of coal required per hour to fuel the plant. The allowable temperature increase of the cooling water is 8°F. Determine the cooling water flow required.

Solution

At state 3, from the superheated steam tables (Table A.3),

$$h_3 = 1474.1 \text{ Btu/lbm}$$

$$s_3 = 1.5603 \text{ Btu/lbm°R}$$

For an isentropic turbine, $s_3 = s_4$. At a saturation temperature of 60°F, from Table A.2, $s_{4f} = 0.0555$ Btu/lbm°R and $s_{4fg} = 2.0391$ Btu/lbm°R.

Solving for quality at state 4, we have

$$s_4 = 1.5603 = 0.0555 + x_4 2.0391$$

$$x_4 = 0.738$$

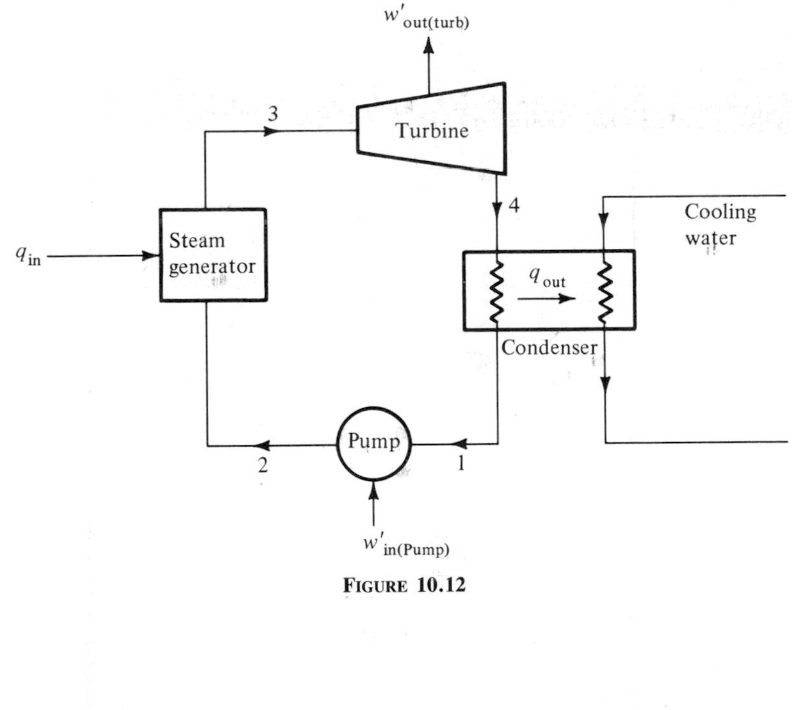

FIGURE 10.12

From Table A.2,

$$h_{4f} = 28.060 \text{ Btu/lbm} \qquad h_{4fg} = 1059.7 \text{ Btu/lbm}$$

Therefore,

$$\begin{aligned} h_4 &= 28.060 + 0.738(1059.7) \\ &= 810.12 \text{ Btu/lbm} \end{aligned}$$

Applying the first law for steady flow to the turbine yields

$$w'_{\text{turb}} = h_3 - h_4 = 664 \text{ Btu/lbm}$$

For the pump, Equation (8.8) gives

$$\begin{aligned} w'_{\text{out(pump)}} &= -v(p_e - p_{\text{in}}) \\ &= -(0.016033 \text{ ft}^3/\text{lbm})(2000 - 0.256) \text{ lbf/in}^2 \, \frac{144 \text{ in}^2/\text{ft}^2}{778 \text{ ft-lbf/Btu}} \\ &= -5.93 \text{ Btu/lbm} \end{aligned}$$

Applying the first law to the pump process, we have

$$\begin{aligned} h_2 &= h_1 + 5.93 \\ &= 28.060 + 5.93 \\ &= 34.0 \text{ Btu/lbm} \end{aligned}$$

For the steam generator, Equation (3.21) gives

$$\begin{aligned} q_{\text{in}} &= h_3 - h_2 \\ &= 1474.1 - 34.0 \\ &= 1440.1 \text{ Btu/lbm} \end{aligned}$$

For the condenser, Equation (3.22) gives

$$\begin{aligned} q_{\text{out}} &= h_4 - h_1 \\ &= 810.12 - 28.06 \\ &= 782.06 \text{ Btu/lbm} \end{aligned}$$

We can check the above calculations since, for the cycle, net q_{in} = net w'_{out}:

$$\text{net } q_{\text{in}} = 1440.1 - 782.1 = 658.0 \text{ Btu/lbm}$$

$$\text{net } w'_{\text{out}} = 664 - 5.9 = 658.1 \text{ Btu/lbm}$$

We now calculate cycle efficiency:

$$\begin{aligned} \eta &= \frac{w'_{\text{net}}}{q_{\text{in}}} \\ &= \frac{658.1 \text{ Btu/lbm}}{1440.1 \text{ Btu/lbm}} \\ &= 45.7\% \end{aligned}$$

For a power output of 500 megawatts,

$$(w'_{net})\dot{m} = (500)(3.412 \times 10^6 \text{ Btu/h-MW})$$

where $\dot{m}$ is the flow going around the cycle:

$$\dot{m} = \frac{1.7065 \times 10^9 \text{ Btu/h}}{658.1 \text{ Btu/lbm}}$$

$$= 2.59 \times 10^6 \text{ lbm/h}$$

Note the very large flow required to produce the desired power. Again, as long as there are no leaks, the same fluid will continue to flow around and around the cycle as long as the power plant is in operation.

The rate at which heat must be added in the steam generator is given by

$$\dot{q}_{in} = \dot{m}q_{in} = (2.59 \times 10^6)(1440.1) = 3.73 \times 10^9 \text{ Btu/h}$$

This heat input is supplied from combustion of coal. For each pound of coal burned, 0.90(12,000) or 10,800 Btus are available for heat input. Therefore,

$$\dot{m}_{coal} = \frac{3.73 \times 10^9 \text{ Btu/h}}{10.8 \times 10^3 \text{ Btu/lbm}} = 3.454 \times 10^5 \text{ lbm/h} = 173 \text{ tons/h}$$

The heat rejected from the power plant in the condenser goes into the cooling water. For the cooling water,

$$(\dot{m}_{cool.\ water})(c_p)\,\Delta T_{cool.\ water} = \dot{q}_{rej.} = \dot{m}q_{out(cycle)}$$

$$= (2.59 \times 10^6)782.1$$

$$= 2.026 \times 10^9 \text{ Btu/h}$$

For water at 60°F, $c_p = 1.0$ Btu/lbm°R. Therefore,

$$\dot{m}_{cool.\ water} = \frac{2.026 \times 10^9}{(1.0)8}$$

$$= 2.53 \times 10^8 \text{ lbm/h}$$

Thus one can see why a power plant is usually located near a large body of water.

Next, we will perform an availability (second law) analysis on the system of Example 10.3 and determine which system component has the best chance of performance improvement.

EXAMPLE 10.4

In the Rankine cycle power plant of Example 10.3, take the turbine and pump efficiencies as 85 percent and 90 percent, respectively. Analyze the performance of the cycle using (a) a first law analysis and (b) a second law analysis.

Solution

(*a*) *First law analysis*

In Example 10.3 we obtained the following thermodynamic properties of the steam at the various points in the cycle:

At point 3:

$T_3 = 1000°\text{F}$ $h_3 = 1474.1$ Btu/lbm

$p_3 = 2000$ psia $s_3 = 1.5603$ Btu/lbm°R

At point 4:

$T_4 = 60°\text{F}$ $h_{4s} = 810.1$ Btu/lbm

$s_{4s} = 1.5603$ Btu/lbm°R

$s_{4f} = 0.0555$ Btu/lbm°R

$s_{4g} = 2.0391$ Btu/lbm°R

$x_{4s} = 0.738$

In the turbine, $h_3 - h_{4s} = 1474.1$ Btu/lbm $- 810.1$ Btu/lbm $= 664.0$ Btu/lbm. Hence $w'_{\text{turb}} = \eta_t(664.0)$ Btu/lbm $= 0.85(664)$ Btu/lbm $= 564.4$ Btu/lbm $= h_3 - h_4$. Therefore, $h_4 = 1474.1$ Btu/lbm $- 564.4$ Btu/lbm $= 909.7$ Btu/lbm.

In the pump, $h_{2s} - h_1 = 5.9$ Btu/lbm, where $h_1 = 28.1$ Btu/lbm. From Table A.2, we have at 60°F, $h_f = 28.06$ Btu/lbm, $h_{fg} = 1059.7$ Btu/lbm, $h_g = 1087.7$ Btu/lbm, $s_f = 0.055$ Btu/lbm°R, $s_{fg} = 2.0391$ Btu/lbm°R, $s_g = 2.0946$ Btu/lbm°R; hencc $w'_{\text{pump}} - 5.9/0.90 = 6.6$ Btu/lbm $= h_2 - h_1$ and $h_2 = 34.7$ Btu/lbm.

In the steam generator, we have $q_{\text{in}} = h_3 - h_2 = 1474.1$ Btu/lbm $- 34.7$ Btu/lbm $= 1439.4$ Btu/lbm. Now $h_4 = 909.7$ Btu/lbm $= 28.06$ Btu/lbm $+ x_4(1059.7)$ Btu/lbm; hence $x_4 = 0.832$ and $s_4 = 0.0555$ Btu/lbm°R $+ 0.832(2.0391)$ Btu/lbm°R $= 1.7520$ Btu/lbm°R.

In the condenser, we have $q_{\text{out}} = h_4 - h_1 = 909.7$ Btu/lbm $- 28.1$ Btu/lbm $= 881.6$ Btu/lbm.

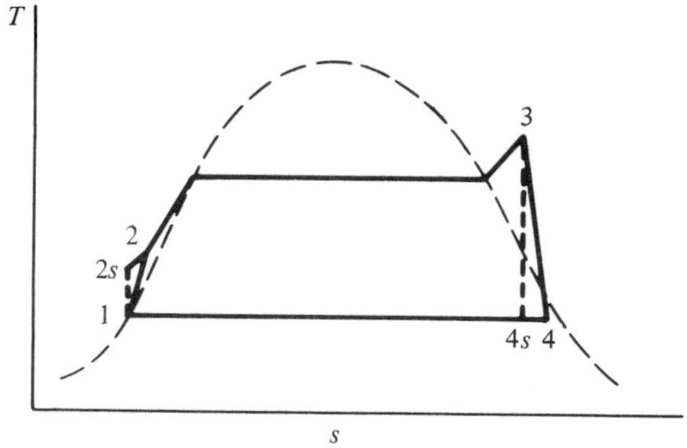

FIGURE 10.13 *T-s Diagram for Example 10.4.*

As check:

$$\text{net } q_{\text{in}} = 1439.4 \text{ Btu/lbm} - 881.6 \text{ Btu/lbm} = 557.9 \text{ Btu/lbm}$$

$$\text{net } w'_{\text{out}} = 564.4 \text{ Btu/lbm} - 6.6 \text{ Btu/lbm} = 557.8 \text{ Btu/lbm}$$

The cycle thermal efficiency is

$$\eta = \frac{\text{net } w'_{\text{out}}}{q_{\text{in}}} = \frac{564.4 - 6.6}{1439.4} = 38.8\%$$

as compared to a cycle efficiency of 45.7 percent (Example 10.3) for an isentropic turbine and pump.

The first law cnergy distribution of the cycle is:

Net work output	557.8 Btu/lbm	38.8%
Energy rejection	881.6 Btu/lbm	61.2%
Total energy input	1439.4 Btu/lbm	100.0%

Note that the calculations agree with the results of Equation 10.2 indicating that net $q_{\text{in}} = \text{net } w'_{\text{out}}$.

(b) Second-law analysis

First, let us exclude the furnace from the power plant system boundary and, second, let us include the furnace within the plant system boundary. If we exclude the furnace, the change in availability across the steam generator is considered as availability entering the system a_{in} (perfect heat exchanger).

For this power plant, using steam as the working fluid, let us choose the inlet temperature of the condenser cooling water as the temperature of the surroundings, i.e., take temperature $T_0 = 510°\text{R}$ (50°F) and p_0 as 14.7 psia; hence from Table A.4 we get for a compressed liquid $h_0 = 18.1$ Btu/lbm and $s_0 = 0.0361$ Btu/lbm°R.

State or component	p (psia)	T (°F)	h (Btu/lbm)	s (Btu/lbm°R)	a (Btu/lbm)
1	0.26	60	28.1	0.0555	0.2
Pump					
2	2000	61	34.7	0.0569	6.0
Steam generator					
3	2000	1000	1474.1	1.5603	678.7
Turbine					
4	0.26	60	909.7	1.7520	16.5
Condenser					
1	0.26	60	28.1	0.0555	0.2

State or component	q_{in} (Btu/lbm)	w'_{out} (Btu/lbm)	Δa (Btu/lbm)	i (Btu/lbm)
1				
Pump	0	−6.6	5.8	0.8
2				
Steam generator	1439.5	0	672.7	
3				
Turbine	0	564.4	−662.2	97.8
4				
Condenser	−881.6	0	−16.3	16.3
1				
Net	557.9	557.8	0	114.9

From (10.6) the second law efficiency is

$$\eta_2 = \frac{\text{net } w'_{out}}{\Delta a_{\text{ins.G.}}} = \frac{557.8}{672.7} = 82.9\%$$

(versus a first law efficiency of 45.7%).

The second law distribution of pertinent quantities in the cycle is:

Net work output	557.8 Btu/lbm	82.9% $(=\eta_2)$
Pump irreversibility	0.8 Btu/lbm	0.1%
Turbine irreversibility	97.8 Btu/lbm	14.6%
Condenser irreversibility	16.3 Btu/lbm	2.4%
Availability input	672.7 Btu/lbm	100.0%

It is noted that although a large quantity of heat is rejected in the condenser (q_{out} is 881.6 Btu/lbm and constitutes 61.2 percent of an energy input of 1439.5 Btu/lbm), the availability loss is relatively small. When the furnace is excluded from the analysis, the most likely improvement appears to be the turbine.

However, we also need to look at the furnace, since heat exchange between hot products of combustion and steam is a process which involves losses. When the furnace is included in our second-law analysis, a different picture of the loss distribution emerges. The availability of the fuel is taken to be the lower heating value of the coal (12,000 Btu/lbm coal). From Example 10.3, we have

$$\frac{\dot{m}_{steam}}{\dot{m}_{coal}} = \frac{0.9(12{,}000)}{h_3 - h_2} = \frac{0.9(12{,}000)}{1474.1 - 34.7} = 7.5 \text{ lbm steam/lbm coal}$$

Hence the availability of the fuel is

$$\frac{12{,}000 \text{ Btu/lbm coal}}{7.5 \text{ lbm steam/lbm coal}} = 1600 \text{ Btu/lbm steam}$$

The air supplied to the furnace for combustion is assumed to be at a temperature of 50°F.

State or component	p (psia)	T (°F)	h (Btu/lbm)	s (Btu/lbm°R)	a (Btu/lbm)
Steam (closed-loop cycle)					
1	0.26	60	28.1	0.0555	0.2
Pump					
2	2000	61	34.7	0.0569	6.0
Steam generator					
3	2000	1000	1474.1	1.5603	678.7
Turbine					
4	0.26	60	909.7	1.7520	16.5
Condenser					
1	0.26	60	28.1	0.0555	16.5
Furnace (combustion, heat exchange) (open-loop system)					
a					1600
Burner					
b (availability = 10% of 1600)					160

State or component		q_{in} (Btu/lbm)	w'_{out} (Btu/lbm)	Δa (Btu/lbm)	i (Btu/lbm)
1					
Pump		0	−6.6	5.8	0.8
2					
Steam generator		1439.5	0	672.7	−672.7
3					
Turbine		0	564.4	−662.2	97.8
4					
Condenser		−881.6	0	−16.3	16.3
1					
	Net	557.9	557.8	0	−557.8
Furnace (combustion, heat exchange) (open-loop system)					
a					
Burner			0	−1440	1440
b (availability = 10% of 1600)					
	Net			−1440	1440
Combined	net			−1440	882.2

In the table, all properties are given on a "per lbm of steam" basis. The net change in availability Δa in a closed-loop steam cycle is zero, since the working medium in a closed-loop system returns to its initial state ($a_{in} = a_e$). The cooled products of combustion (flue gases) are in an open-loop system and are continuously discharged to the environment; hence their irreversibility is 10 percent of 1600 Btu/lbm or 160 Btu/lbm. This is lost up the stack (Example 10.3).

The air supplied to the burner is at a temperature equal to T_0; hence its availability is zero.

The availability of the working medium in this closed cycle increases as it passes through the pump and steam generator, but decreases as it passes through the turbine and condenser. Since the working medium in a closed cycle returns to the same state at a given station in the cycle, the increase in availability of the working medim equals the decrease in availability, i.e., the net change in availability in the cycle is zero. In the turbine, the decrease in availability is accompanied by the production of work and the creation of irreversibility, i.e., the decrease in availability equals the sum of work output and irreversibility.

The second law efficiency of the cycle is from (10.6)

$$\eta_2 = \frac{\text{net } w'_{\text{out}}}{a_{\text{in}}} = \frac{557.8}{1600} = 34.9\%$$

The second law distribution in the cycle is:

Net work output	557.8 Btu/lbm	34.9%	$(=n_2)$
Pump irreversibility	0.8	0.1%	
Turbine irreversibility	97.8	6.1%	
Condenser irreversibility	16.3	1.0%	
Steam generator-burner irreversibility	767.3	48.0%	
Flue gas availability loss	160.0	10.0%	
Total availability input	1600 Btu/lbm	100.0%	

When the furnace is included in the analysis, the most likely improvement appears to be the steam generator-burner.

First-law analysis indicates that a large part of the initial energy in the coal is rejected to the condenser cooling water relative to other losses. Second-law analysis, on the other hand, also considers the quality of the energy and shows that the loss in the condenser is relatively small as compared to losses in the steam generator during combustion and heat transfer between the hot products of combustion and the steam. If the efficiency of the steam plant is to be improved, our effort should be directed to the high-temperature burner and steam generator area, where large losses in availability occur.

For this closed-loop cycle, the second-law efficiency of 34.9 percent is the same as that obtained from the first law, if the value energy input (q_{in}) in the thermal cycle efficiency is taken to be the heating value of the fuel (coal).

In order to prevent damage to the turbine blades, it is necessary to maintain the moisture content of the steam in the turbine below 10 percent. Hence, the steam quality at the turbine exit must be greater than 0.90. In the examples that we have worked, x_4 has been less than 0.90; further, from Figure 10.11, the increased thermal efficiency resulting from an increase in boiler pressure leads to a decrease in steam quality at the turbine exit. In order to obtain the advantages of high thermal efficiency with increased boiler pressure, with turbine steam quality maintained at acceptable levels, at least one stage of reheat is commonly used in a steam power plant. With reheat, the steam flow is extracted at an intermediate point in the turbine [4 of Figurc 10.14(a)] and then is returned to the steam generator, where it receives heat, increasing its temperature. The steam is then returned to the turbine, where it is further expanded to condenser pressure. From the T-s diagram of Figure 10.14(b), the effect of reheat is to raise the steam quality at the turbine exit from that of 4′ to that of 6. The following example will illustrate the advantage of reheat.

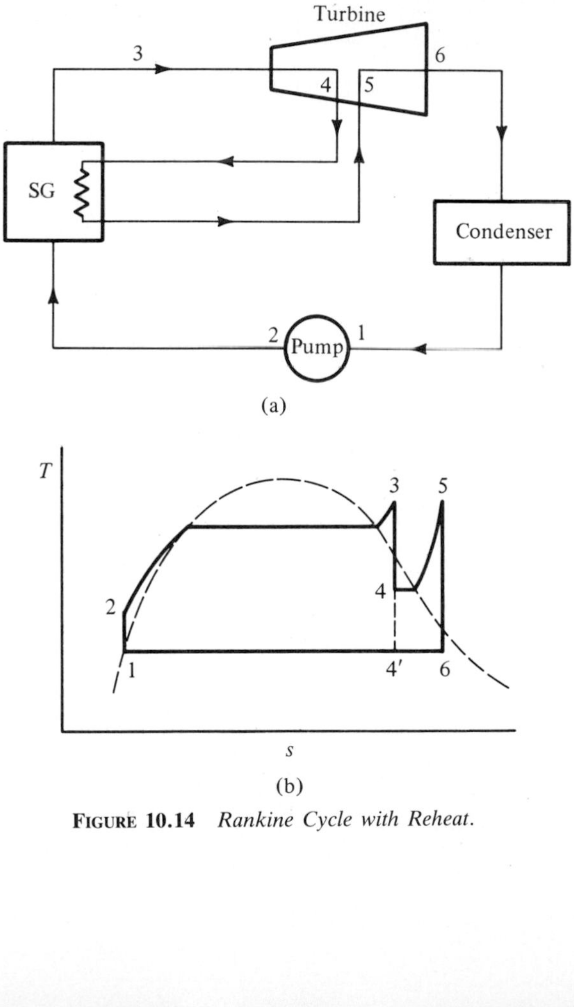

FIGURE 10.14 *Rankine Cycle with Reheat.*

EXAMPLE 10.5

In the power plant of Example 10.3, steam is extracted from the turbine at a pressure of 90 psia and is reheated in the steam generator to 1000°F (Figure 10.15). The steam is then returned to the turbine for expansion to the condenser temperature of 60°F. Assuming all processes to be ideal, calculate the quality of the steam at the turbine exit, and determine the cycle thermal efficiency.

Solution

From Example 10.3,

$$s_3 = 1.5603 \text{ Btu/lbm°R}$$

$$h_3 = 1474.1 \text{ Btu/lbm}$$

$$s_3 = s_4 = 1.5603 \text{ Btu/lbm°R}$$

$$p_4 = 90 \text{ psia}$$

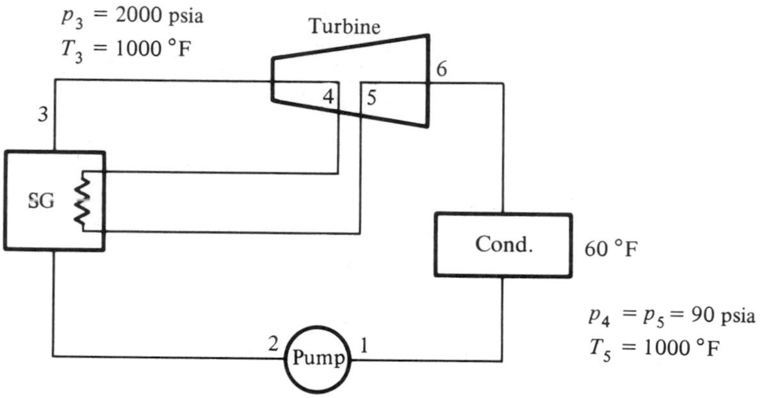

FIGURE 10.15

From Table A.2, at 90 psia,

$$s_f = 0.4640 \text{ Btu/lbm°R}$$

$$s_{fg} = 1.1477 \text{ Btu/lbm°R}$$

Therefore,

$$x_4 = \frac{1.5603 - 0.4640}{1.1477} = 0.955$$

$$h_4 = h_f + x_4 h_{fg}$$

$$= 290.4 + 0.955(894.8)$$

$$= 1144.9 \text{ Btu/lbm}$$

At state 5, $p_5 = 90$ psia, and $T_5 = 1000°F$. Interpolating from the superheat tables (Table A.3), we have

$$h_5 = 1532.3 \text{ Btu/lbm} \quad \text{and} \quad s_5 = 1.9330 \text{ Btu/lbm°R}$$

$$s_6 = s_5 = 1.9330 = s_{f6} + x_6 s_{fg}$$

$$= 0.0555 + x_6(2.0391)$$

Therefore, $x_6 = 0.921$:

$$h_6 = h_f + x_6 h_{fg} = 28.06 + (0.921)1059.7 = 976.0 \text{ Btu/lbm}$$

From Example 10.3, $w'_{\text{pump}} = 5.93$ Btu/lbm and $h_2 = 34.0$ Btu/lbm.

With reheat, heat is added to the working fluid at two locations in the cycle:

$$q_{\text{in}} = h_3 - h_2 + h_5 - h_4$$

$$= 1474.1 - 34.0 + 1532.3 - 1144.9$$

$$= 1827.5 \text{ Btu/lbm}$$

$$q_{\text{out}} = h_6 - h_1$$

$$= 976.0 - 28.06$$

$$= 947.9 \text{ Btu/lbm}$$

$$w'_{\text{net}} = h_3 - h_4 + h_5 - h_6 - w'_{\text{pump}}$$

$$= 1474.1 - 1144.9 + 1532.3 - 976.0 - 5.94$$

$$= 879.6 \text{ Btu/lbm}$$

Note that

$$\text{net } q_{\text{in}} = 1827.5 - 947.9 = 879.6 \text{ Btu/lbm} = \text{net } w'_{\text{out}}$$

The cycle thermal efficiency is

$$\eta = \frac{879.6}{1827.5} = 48.1\%$$

The effect of reheat has been to increase the quality of the steam leaving the turbine from 0.738 (see Example 10.3) to 0.921. There has also been a small change of cycle efficiency, from 45.7 percent to 48.1 percent.

One way to increase Rankine cycle efficiency is to increase the average temperature at which external heat is added to the working fluid. This temperature increase can be achieved with the use of feedwater heaters. In the feedwater heater, some hot steam extracted from an intermediate stage of the turbine is allowed to exchange heat with the cooler fluid before it enters the boiler. The effect of the heat exchange is shown in Figure 10.16. A fraction f of the total steam flow $\dot{m}$ is extracted at point 4 and is fed directly to the feedwater heater. The remainder of the steam flow $(1 - f)\dot{m}$ passes through the condenser and then is pumped to the feedwater heater. We do not want a liquid-vapor mixture at point 1, since the pump would not be able to handle it. Therefore, just enough steam is mixed with the liquid at 7 so that at point 1, the outlet of the feedwater heater, there will be saturated liquid.

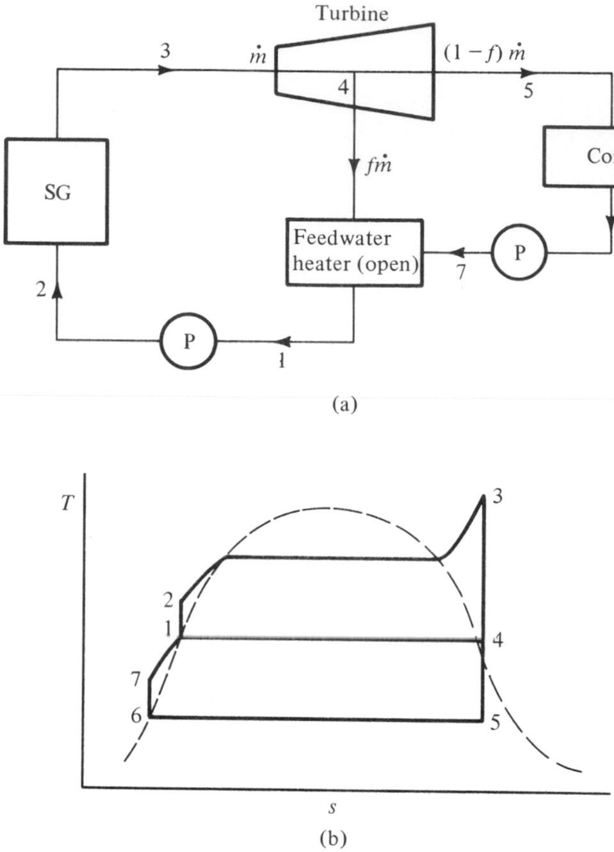

FIGURE 10.16 *Effect of Feedwater Heater.*

With a feedwater heater in the cycle, external heat is added during process 2–3. With no feedwater heater, the external heat addition process would occur over the wider temperature range T_7 to T_3. Thus, for the same maximum temperature, the average temperature at which external heat is added to the cycle with the feedwater heater is greater than that for the basic cycle without heater. It follows that the cycle of Figure 10.16 is able to provide a higher thermal efficiency than the basic cycle.

There are two types of feedwater heaters: open and closed. With the open heater, steam and liquid are intimately mixed together. With the closed heater, the fluids are kept separated, for example, the liquid inside a tube and the steam condensing on the outside of the tube, with heat transfer through the tube wall. For the type of open heater illustrated in Figure 10.16, mixing takes place at a constant pressure, with $p_4 = p_7 = p_1$. Therefore, an extra pump is required for each open feedwater heater.

For the closed heater, no extra pump is required. As shown in Figure 10.17, $p_2 = p_3$ = boiler pressure. Clearly, we would like to retain the hot condensate. In practice, either it is drained to a lower pressure feedwater heater (or to the condenser) through a trap or valve which allows only liquid to pass (Figure 10.18), or it is pumped to a higher-pressure region with a drip pump (Figure 10.19).

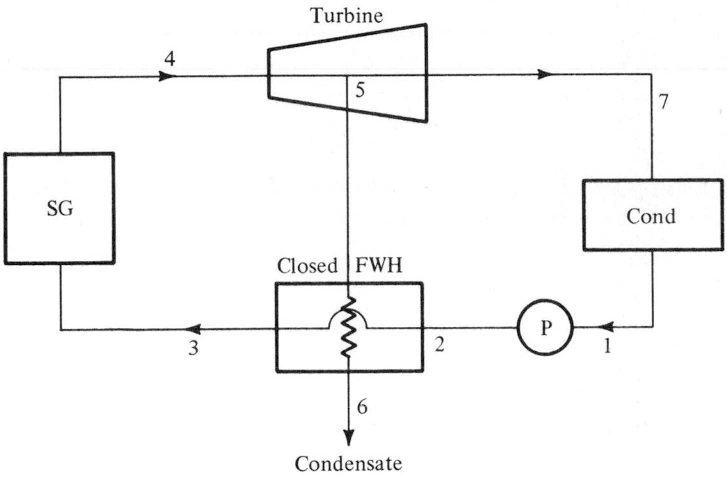

FIGURE 10.17 *Closed Feedwater Heater.*

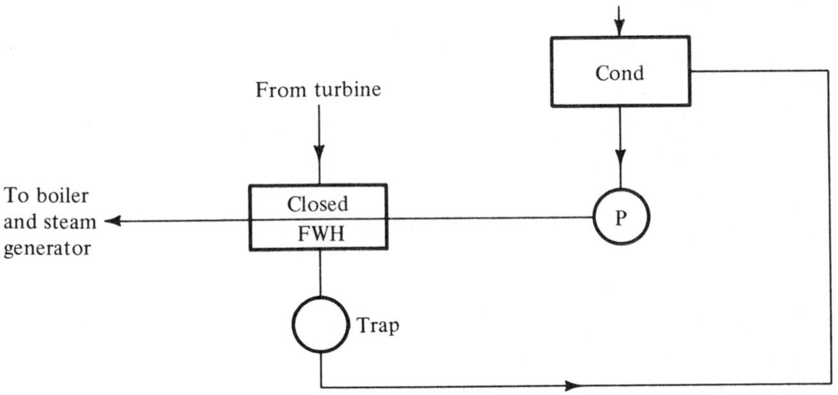

FIGURE 10.18 *Use of Trap.*

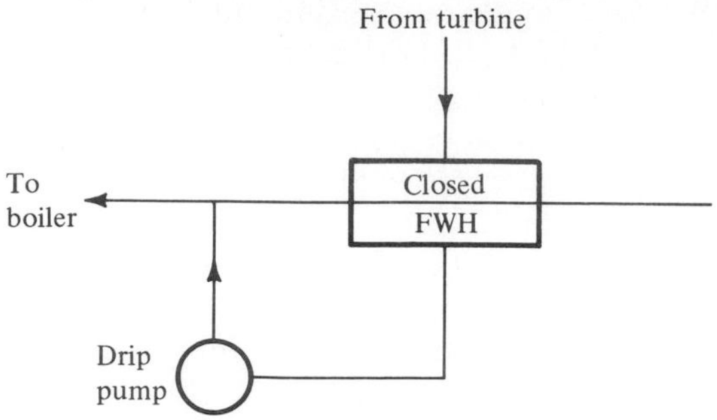

FIGURE 10.19 *Use of Drip Pump.*

In the following example, we will show how the addition of a feedwater heater can improve the efficiency of the Rankine cycle.

EXAMPLE 10.6

In the power plant of Example 10.3, a fraction of the steam flow is extracted from the turbine at 500 psia for the purpose of feedwater heating. Determine the cycle thermal efficiency with one open feedwater heater. Assume all processes to be ideal.

Solution

The cycle is indicated in Figure 10.20. From Example 10.3,

$$s_3 = 1.5603 \text{ Btu/lbm°R} \qquad h_3 = 1474.1 \text{ Btu/lbm}$$

$$h_5 = 28.06 \text{ Btu/lbm} \qquad h_4 = 810.1 \text{ Btu/lbm}$$

The expansion in the turbine is isentropic; therefore $s_7 = s_3$.

With $p_4 = 500$ psia, state 4 is determined. From the superheat tables, $h_4 = 1318.0$ Btu/lbm. The work input to each pump is equal to $-v\,\Delta p$. Thus,

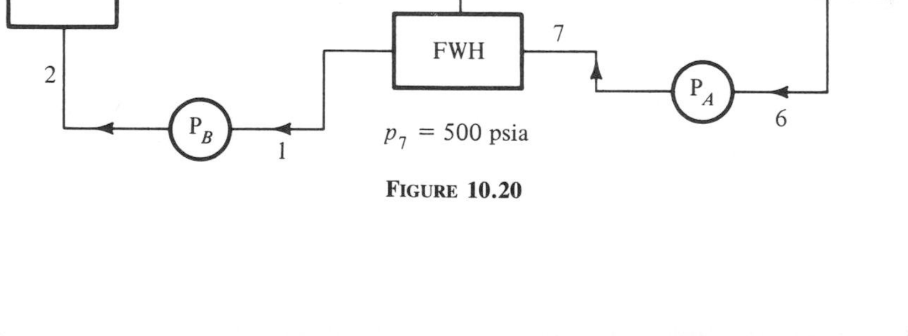

FIGURE 10.20

$$w'_{\text{pump }A} = \frac{(0.016033\ \text{ft}^3/\text{lbm})(500\ \text{lbf/in}^2 - 0.256\ \text{lbf/in}^2)(144\ \text{in}^2/\text{ft}^2)}{778\ \text{ft-lbf/Btu}}$$

$$= 1.48\ \text{Btu/lbm}$$

$$w'_{\text{pump }B} = \frac{(0.0197\ \text{ft}^3/\text{lbm})(2000\ \text{lbf/in}^2 - 500\ \text{lbf/in}^2)(144\ \text{in}^2/\text{ft}^2)}{778\ \text{ft-lbf/Btu}}$$

$$= 5.47\ \text{Btu/lbm}$$

Therefore,

$$h_7 = h_6 + w'_{\text{pump }A} = 28.06 + 1.48 = 29.54\ \text{Btu/lbm}$$

With saturated liquid at 1, $h_1 = h_f$ at 500 psi = 449.4 Btu/lbm. Therefore,

$$h_2 = 449.4 + 5.47 = 454.9\ \text{Btu/lbm}$$

In order to determine the fraction f of the total steam flow that is extracted at 4, we must run a mass balance and an energy balance on the feedwater heater. At steady state, the rate at which mass and energy enter the heater must be equal to the rate at which they leave. Therefore,

$$\dot{m}_4 + \dot{m}_7 = \dot{m}_1 \quad \text{and} \quad \dot{m}_4 h_4 + \dot{m}_7 h_7 = \dot{m}_1 h_1$$

But $\dot{m}_4 = f\dot{m}_1$ so that we have from the first equation (mass balance) $\dot{m}_7 = (1-f)\dot{m}_1$. Substituting into the second equation (energy balance), we have

$$1318.0f + 29.5(1-f) = 449.4$$

$$f = 0.326 \quad \text{and} \quad 1 - f = 0.674$$

$$w'_{\text{out(turb)}} = \dot{m}_1(h_3 - h_4) + \dot{m}_7(h_4 - h_5)$$

$$= \dot{m}_1(1474.1 - 1318.0) + 0.674\dot{m}_1(1318.0 - 810.1)$$

$$= 498.4\dot{m}_1$$

$$w'_{\text{in(pump)}} = 5.47\dot{m}_1 + (0.674)1.48\dot{m}_1$$

$$= 6.47\dot{m}_1$$

$$\dot{q}_{\text{in}} = \dot{m}_1(h_3 - h_2) = 1019.2\dot{m}_1$$

$$\eta = \frac{\text{net } \dot{w}_{\text{out}}}{\dot{q}_{\text{in}}} = \frac{498.4 - 6.47}{1019.2} = 48.3\%$$

Comparing Example 10.3 with Example 10.6, we find that the actual increase in efficiency is only 2.6 percent. However, power plants generally have several feedwater heaters, possibly as many as twelve. A typical arrangement is shown in Figure 10.21. As might be expected, the increase in cycle thermal efficiency per feedwater heater decreases as the number of feedwater heaters increases. Further, the addition of feedwater heaters adds to the initial cost of the plant and also involves frictional losses in

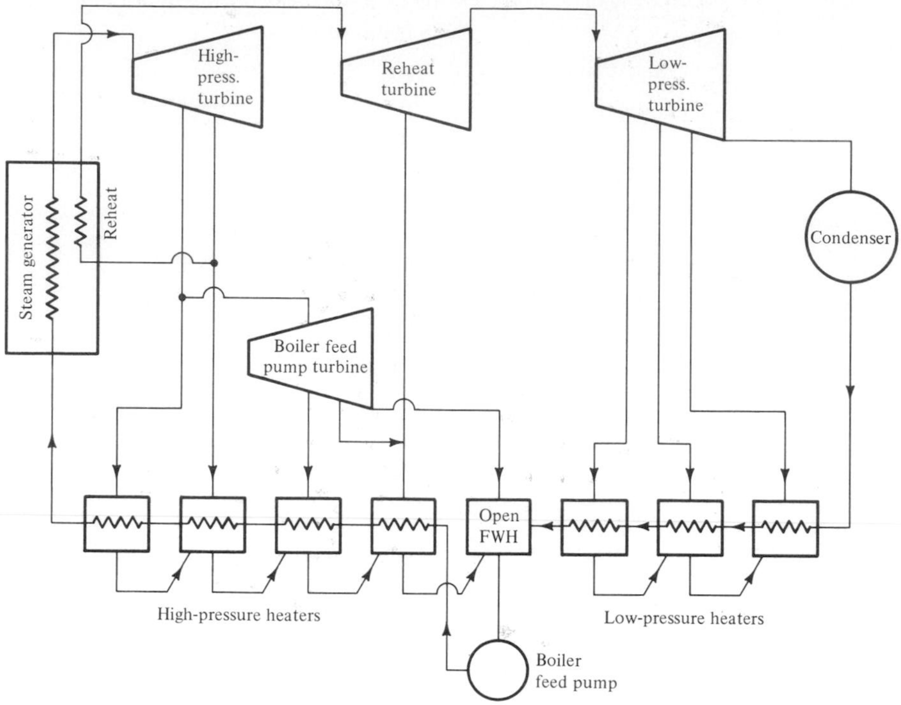

FIGURE 10.21 *Typical Power Plant Schematic: Eight-Stage Feedwater Heating.*

the piping and valving necessary for their utilization. Nevertheless, the use of several feedwater heaters is a common method of increasing plant efficiency.

Up to this point in our analysis of the Rankine cycle, we have mainly assumed all processes to be ideal. However, in an actual power plant, the turbine and pump are not isentropic. Further, there are pressure losses in the piping between the various components of the cycle, as well as heat losses. To account for irreversibilities in the turbine and pump, we use Equations (8.5) and (8.6) to determine turbine and pump efficiencies. To account for pressure and heat losses in the lines and valving, one must use methods from fluid mechanics which are outside the scope of this text. In the following example, therefore, we will assume values of pressure drop and determine how turbine and pump irreversibilities affect overall cycle performance.

EXAMPLE 10.7

A steam power plant (Figure 10.22) operates on a Rankine cycle with the following pressures and temperatures:

$p_1 = 1$ psia $\qquad T_{sat} = 101.74°F$ at 1 psia

$p_{1a} = 1$ psia $\qquad T_1 = 101.74°F = T_{1a}$

$p_{2'} = 850$ psia

$p_{2a} = 800$ psia $\qquad T_{2a} = 104°F$

$p_3 = 650$ psia $\qquad T_3 = 900°F$

$p_{3a} = 600$ psia $\qquad T_{3a} = 800°F$

$p_{4'} = p_1$

Turbine efficiency = 85%

Pump efficiency = 80%

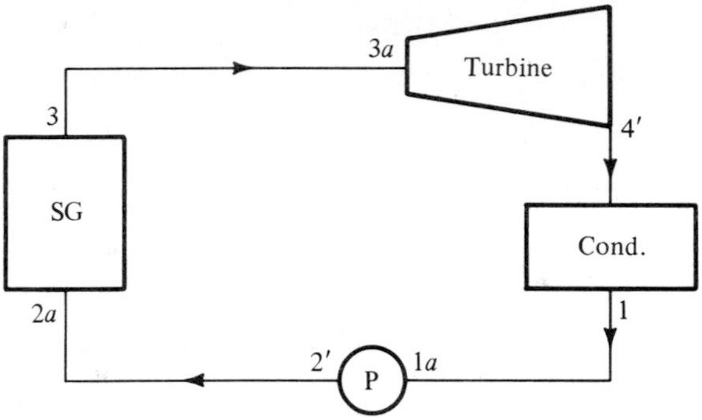

FIGURE 10.22

a. Calculate the cycle thermal efficiency and the net work output.
b. Compare your results with those for the ideal cycle shown in Figure 10.23, given the following values:

$p_1 = 1$ psia

$p_2 = 600$ psia

$p_3 = 600$ psia $\qquad T_3 = 900°F$

$p_4 = 1$ psia

Pump efficiency = turbine efficiency

= 100%

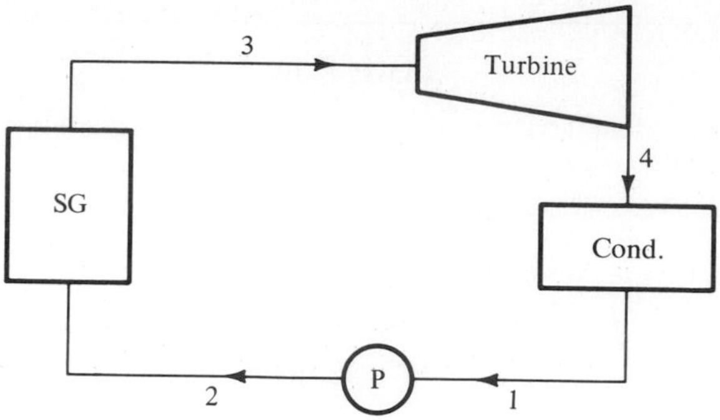

FIGURE 10.23

Solution

a. From the steam tables, Table A.2, $h_1 = h_f = 69.73$ Btu/lbm $= h_{1a}$.

$$w'_{\text{in(pump)}} = \frac{v_1(p_{2'} - p_{1a})}{\eta_{\text{pump}}} = \frac{0.01614\ \text{ft}^3/\text{lbm}}{0.80}(850-1)\ \text{lbf/in}^2\ \frac{144\ \text{in}^2/\text{ft}^2}{778\ \text{ft-lbf/Btu}}$$
$$= 3.17\ \text{Btu/lbm}$$

$$h_{2'} = h_{1a} + w'_{\text{in(pump)}} = 69.73 + 3.17 = 72.9\ \text{Btu/lbm}$$

From the compressed liquid tables, at $2a$,

$$h_{2a} = 74.1\ \text{Btu/lbm}$$

From the conditions given at 3,

$$h_3 = 1461.2\ \text{Btu/lbm}$$

The energy input to the boiler is

$$q_{\text{in}} = h_3 - h_{2a} = 1461.2 - 74.1 = 1387.1\ \text{Btu/lbm}$$

The work output of the turbine is

$$w'_{\text{out(turb)}} = h_{3a} - h_{4'} = 0.85(h_{3a} - h_{4's})$$

Thus,

$$h_{4's} = h_f + x_{4s}h_{fg} = h_1 + \frac{s_{3a} - s_1}{(s_{fg})_1}h_{fg} = 69.7 + \frac{1.6351 - 0.1326}{1.8455}1036.1$$
$$= 913.2\ \text{Btu/lbm}$$

$$h_{3a} = 1408.3\ \text{Btu/lbm}$$

Hence,

$$w'_{\text{out}} = (1408.3 - 913.2)0.85 = 420.8\ \text{Btu/lbm}$$
$$\text{net}\ w'_{\text{out}} = 420.8 - 3.2 = 417.6\ \text{Btu/lbm}$$
$$\eta_{\text{cycle}} = 417.6/1387.1 = 0.301$$

b. For the ideal cycle of Figure 10.23,

$$w'_{\text{pump}} = (0.01614\ \text{ft}^3/\text{lbm})(600 - 1)\ \text{lbf/in}^2\ \frac{14\ \text{in}^2/\text{ft}^2}{778\ \text{ft-lbf/Btu}} = 1.79\ \text{Btu/lbm}$$

$$h_2 = h_1 + w'_{\text{pump}} = 69.7 + 1.8 = 71.5\ \text{Btu/lbm}$$

$$\text{Energy input to boiler (steam generator)} = q_{\text{in}} = h_3 - h_2 = 1463 - 71.5 = 1391.5\ \text{Btu/lbm}$$

$$s_4 = s_3 = 1.6769\ \text{Btu/lbm°R}$$

$$= s_f + x_4 s_{fg} = 0.1326 + x_4(1.8455)$$

$$x_4 = 0.8368 \text{ and } h_4 = 69.73 + 0.8368(1036.1) = 936.7\ \text{Btu/lbm}$$

$$w'_{\text{out(turb)}} = h_3 - h_4 = 1463 - 936.7 = 526.3\ \text{Btu/lbm}$$

$$\text{net } w'_{\text{out}} = 526.3 - 1.8 = 524.5\ \text{Btu/lbm}$$

$$\eta_{\text{cycle}} = 524.5/1391.5 = 0.377$$

Several points should be stressed before we complete our study of the Rankine cycle power plant. First, the basic cycle with feedwater heaters and reheat is used in all large, central station power plants, using either a fossil fuel (such as natural gas, fuel oil, or coal) or a nuclear fuel. Unfortunately, due to the possibility of damage to the reactor, the peak cycle temperature of the nuclear power plant cannot be as high as that for a fossil fuel plant. Thus, whereas the fossil fuel plant may operate at temperatures of 1050°F, the maximum temperature of the working fluid (water) in a nuclear plant may be only 600°F. For this reason, the actual maximum thermal efficiency of fossil fuel plants today is about 40 percent, but the maximum thermal efficiency for a nuclear plant is only 33 percent.

With a maximum plant efficiency of 40 percent, 60 percent of the input energy to the boiler is rejected to the condenser cooling water. In other words, for each megawatt of electrical power produced, 1.5 megawatts are rejected, generally to a nearby natural body of water. This large rejection of heat into natural bodies of water has, in some cases, led to increases in natural water temperatures, which result in potential damage to aquatic life. This problem, called *thermal pollution*, has led certain states to impose regulations prohibiting significant changes in the temperature of natural bodies of water. These regulations have forced the power companies to resort to the use of cooling towers, in which the condenser cooling water is maintained in a closed loop and is cooled by evaporation to the surrounding air. Makeup water must still be provided to account for the loss due to evaporation. With the large cooling water flows required, the cooling tower must be a massive structure, hundreds

FIGURE 10.24 *Aerial Photo of Davis-Besse Nuclear Power Plant. (Courtesy of Toledo Edison Co.)*

of feet high. The cooling tower used in conjunction with the Davis-Besse nuclear power plant of Toledo Edison is 493 feet high with a base diameter of 415 feet (Figure 10.24).

The features of nuclear and fossil fuel power plants can be seen in the photos shown in Figures 10.24 through 10.28. Clearly visible in each nuclear power plant photo is the reactor containment vessel, generally composed of steel plate and concrete several feet thick. The reactors are designed to prevent radioactive fission products from being released to the environment, and to withstand disasters such as earthquakes, hurricanes, and tornadoes. Note that the fossil fuel plants have tall stacks to dissipate to the atmosphere the products of combustion of the fuel.

Figure 10.24 shows the Davis-Besse plant of Toledo Edison and Cleveland Electric Illuminating Companies. This nuclear plant is designed to generate 870 megawatts of electrical power. The containment vessel in the foreground, cylindrical in shape, contains the nuclear reactor and the steam generator. The containment vessel is dwarfed by the cooling tower in the background. Figures 10.25, 10.26, 10.27, and 10.28 depict other installations.

FIGURE 10.25 *Robinson Plant near Hartsville, South Carolina. This Plant, Which Has One Nuclear and One Coal-Fired Unit, Produces 903 Megawatts of Electrical Power. (Courtesy of Carolina Power and Light Co., Raleigh, North Carolina.)*

FIGURE 10.26 *Turkey Point Generating Plant. The Two Units at Left Are Fossil-Fuel Fired; the Two Units at Right, Nuclear Powered. Their Combined Capacity is 2.3 Million Kilowatts. (Photo Courtesy of Florida Power and Light Co.)*

FIGURE 10.27 *Karn Fossil Fuel Plant. (Courtesy of Consumers Power Co., Jackson, Michigan.)*

In Figure 10.28 it is of interest to follow the various fluids as they participate in the operation of the power plant. For the Rankine cycle, water is first pumped to the economizer where its temperature is raised via heat transfer from the hot gases leaving the furnace. This hot water is then passed through the boiler and superheater where its temperature is raised to that of the superheated vapor. The steam flows through the turbine to produce power and is discharged to the condenser where it is cooled to liquid phase, completing the cycle. Turbine power output is used to drive the electric generator.

Pulverized coal is fed into the furnace where it is combusted with outside air. The combustion gases are passed first to the economizer, transferring heat to the water of the main cycle, then to the fly ash collector, and finally up the chimney. Newer plants may require extensive, costly equipment, such as scrubbers, to remove sulfur oxide from exhaust gases. Finally, note the cooling water flow from the Saginaw River, which is used to carry away the heat from the condenser.

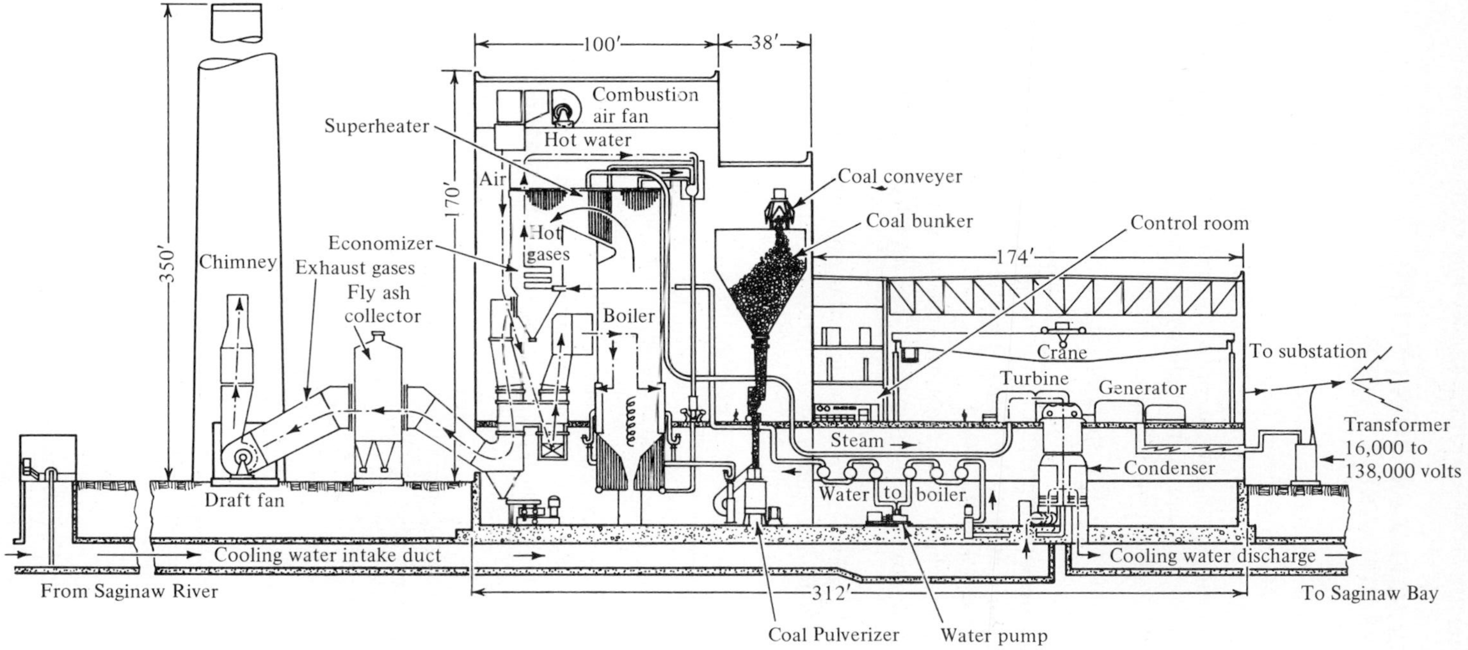

FIGURE 10.28 *Cross-Sectional View of Karn Coal-Fired Power Plant. (Courtesy of Consumers Power Co., Jackson, Michigan.)*

10.3 VAPOR REFRIGERATION CYCLES: THE REVERSED RANKINE CYCLE

In Chapter 6, we showed how a heat engine, operated in reverse, behaves as a heat pump or refrigerator. Likewise, by reversing the Rankine power cycle, we obtain a refrigerator or heat pump cycle.

A schematic arrangement of a refrigeration system utilizing the *reversed Rankine* (*vapor*) cycle is shown in Figure 10.29(a). The components of the system are as follows:

1. An *evaporator*, in which the working fluid is changed from wet vapor to dry (or superheated) vapor at constant pressure by the addition of heat from the low-temperature surroundings;
2. A *compressor* which, by work input, raises the temperature and pressure of the working fluid;
3. A *condenser* which, by rejecting heat to the high-temperature surroundings, changes the vapor to its liquid phase at constant pressure; and
4. An *expander engine* which reduces the pressure and temperature of the working fluid to those of its initial condition.

EXAMPLE 10.8

A reversed Rankine refrigeration cycle operates between the conditions indicated on Figure 10.29(b). States 1 and 3 are saturated conditions with $T_1 = 35°C$ and $T_3 = 150°C$. Assume that the expansion and compression processes are isentropic, with no pressure or temperature changes in the lines between the various components. Find the cycle coefficient of performance as a refrigerator, and determine the cooling capacity (q_{in}) of the cycle if the working fluid is water.

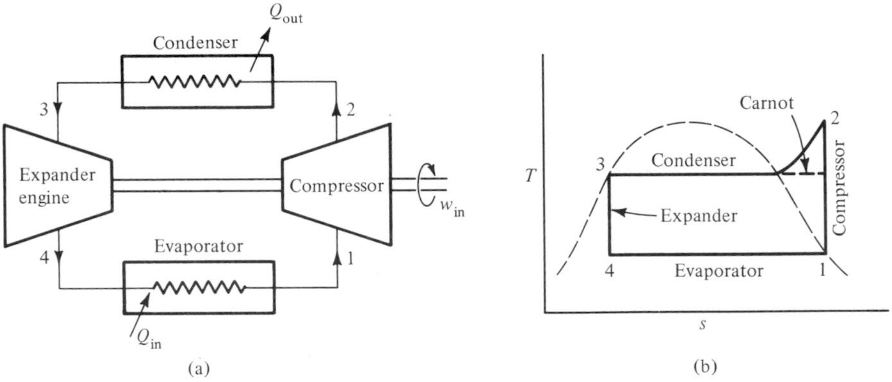

FIGURE 10.29 *Reversed Rankine Cycle.*

Solution

From the saturated steam tables B.2, at a temperature of 35°C,

$$p_1 = 5.6216\,\text{kPa} \qquad h_f = 146.6\,\text{kJ/kg} \qquad h_{fg} = 2418.8\,\text{kJ/kg}$$

$$h_g = 2565.3\,\text{kJ/kg} \qquad s_f = 0.5049\,\text{kJ/kg K}$$

$$s_{fg} = 7.8494\,\text{kJ/kg K} \qquad s_g = 8.3543\,\text{kJ/kg K}$$

Therefore, $h_1 = 2565.3\,\text{kJ/kg}$ and $s_1 = 8.3543\,\text{kJ/kg K}$. Further,

$$p_2 = p_3 = \text{saturation pressure at } 150°\text{C} = 476.0\,\text{kPa}$$

Also, $s_2 = s_1 = 8.3543\,\text{kJ/kg K}$. From the superheat tables B.3, by interpolation,

$$T_2 = 591.0°\text{C} \quad \text{and} \quad h_2 = 3682.0\,\text{kJ/kg}$$

From Table B.2,

$$h_3 = h_f \text{ at } 150°\text{C} = 632.15\,\text{kJ/kg}$$

$$s_3 = s_f = 1.8416\,\text{kJ/kg K}$$

Further, $s_4 = s_3 = 1.8416$, where state 4 has a quality x_4. Thus,

$$1.8416 = s_f + x_4 s_{fg} = 0.5049 + x_4(7.8494)$$

so that $x_4 = 0.170$.

Now $h_4 = h_f + x_4 h_{fg} = 146.6 + 0.170(2418.8) = 557.8\,\text{kJ/kg}$. Thus,

$$w'_{\text{out(comp)}} = h_2 - h_1 = 3682.0 - 2565.3 = 1116.7\,\text{kJ/kg}$$

$$w'_{\text{out(exp)}} = h_3 - h_4 = 632.15 - 557.8 = 74.35\,\text{kJ/kg}$$

$$q_{\text{in(evap)}} = h_1 - h_4 = 2565.3 - 557.8 = 2007.5\,\text{kJ/kg}$$

$$= \text{cooling capacity}$$

$$q_{\text{out(cond)}} = h_2 - h_3 = 3682.0 - 632.15 = 3049.8\,\text{kJ/kg}$$

Note that for a cycle, net $q_{\text{out}} = 3049.8 - 2007.5 = 1042.3\,\text{kJ/kg}$, net $w'_{\text{in}} = 1116.7 - 74.4 = 1042.3\,\text{kJ/kg}$. From Chapter 6, we get

$$\text{COP} = \frac{q_{\text{in}}}{\text{net } w_{\text{in}}} = \frac{2007.5}{1042.3} = 1.926$$

For a Carnot refrigeration cycle operating between temperature bounds $T_H = T_3 = 150°\text{C}$ and $T_L = T_4 = 35°\text{C}$, the coefficient of performance is, from Equation (6.9),

$$\text{COP}_{\text{Carnot refrig.}} = \frac{T_L}{T_H - T_L} = \frac{35 + 273}{150 - 35} = 2.678$$

Figure 10.30 shows the coefficient of performance of a basic reversed Rankine cycle, using water as the working fluid. The COP, calculated by the procedures of Example 10.8, is shown over a range of

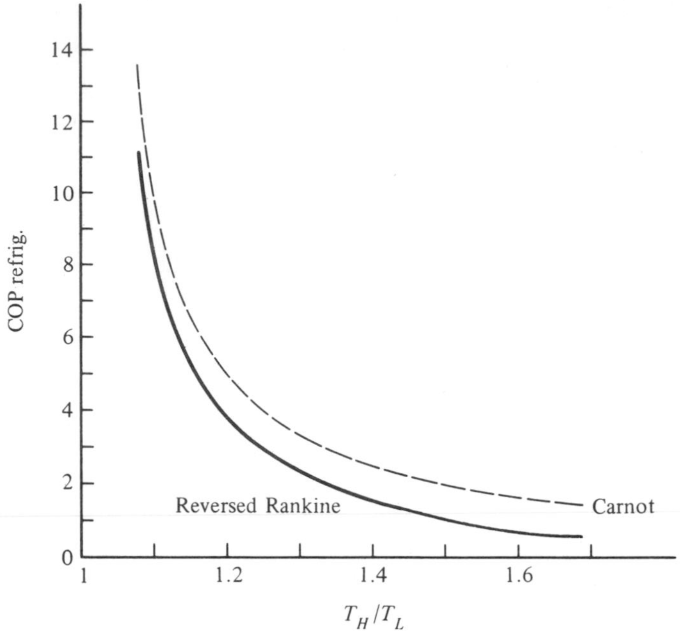

FIGURE 10.30 *Comparison of COPs of Reversed Rankine and Carnot Cycles. The Temperature of the Evaporator Water Is 40°F.*

temperature ratios for an evaporator temperature of 40°F. The COPs of the corresponding Carnot refrigeration cycles are also included in the figure. For example, for a condenser temperature of 115°F (a typical value in air conditioning application),

$$\frac{T_H}{T_L} = \frac{115 + 460}{40 + 460} = 1.15$$

The COP of the reversed Rankine cycle using water as working fluid is 5.4; the COP of the corresponding Carnot cycle is 6.67.

From Example 10.8, we can see that the expander engine work output is only a small part (less than 7 percent) of the compressor work required in a vapor refrigeration cycle. In most applications, therefore, the engine is replaced with a simpler, less expensive device: a *throttle*. A throttle can take the form of either a valve, which can also be used to control the flow, as in an automobile air conditioner, or a capillary tube, as in a household refrigerator. In either case, the throttling process, as described in Section 8.4, is isenthalpic and irreversible, thus involving an increase of entropy of the working fluid. Further, the compressor operates only in the superheat region, to avoid the deleterious effects of liquid-vapor mixture.

A schematic of the system components used in the great majority of automobile and home air conditioning systems, as well as household refrigerators and heat pumps, is shown in Figure 10.31. Corresponding T-s and p-h diagrams are shown in Figure 10.32, where the compressor process has been assumed isentropic.

For steady flow with negligible changes in kinetic and potential energies, from the energy equation (3.8), we obtain the following for the four processes of a vapor refrigeration cycle with a throttle:

(1–2) Isentropic compression process:

$$q_{in} = 0 \qquad s_2 - s_1 = 0$$

$$w'_{in} = h_2 - h_1$$

(2–3) Constant pressure, condensing (energy rejection) process:

$$w'_{out} = 0$$

$$q_{out} = h_2 - h_3$$

(3–4) Throttling process:

$$q_{in} = 0 \qquad s_4 > s_3$$

$$h_3 = h_4$$

(4–1) Constant pressure, evaporation (energy input) process:

$$w'_{out} = 0$$

$$q_{in} = h_1 - h_4$$

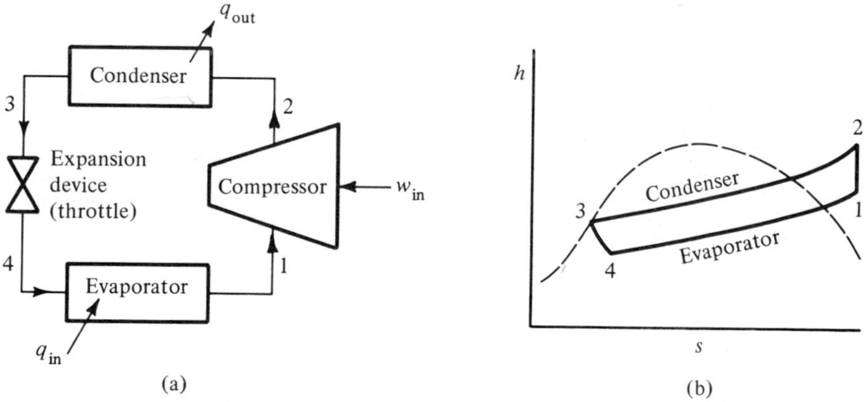

FIGURE 10.31 *Air Conditioning and Refrigerator Systems.*

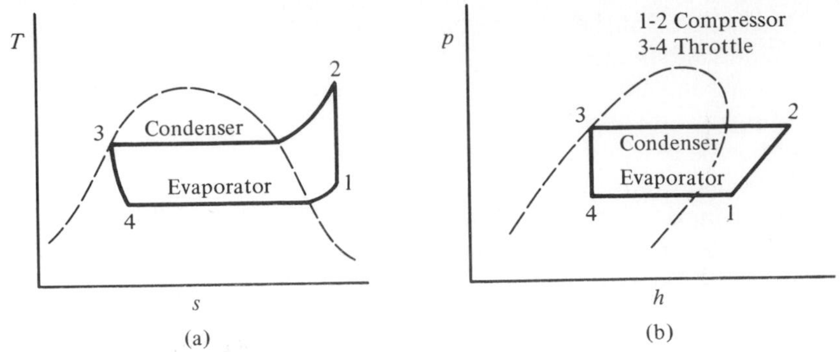

Figure 10.32 *T-s and p-h Diagrams for Vapor Refrigerator Cycle.*

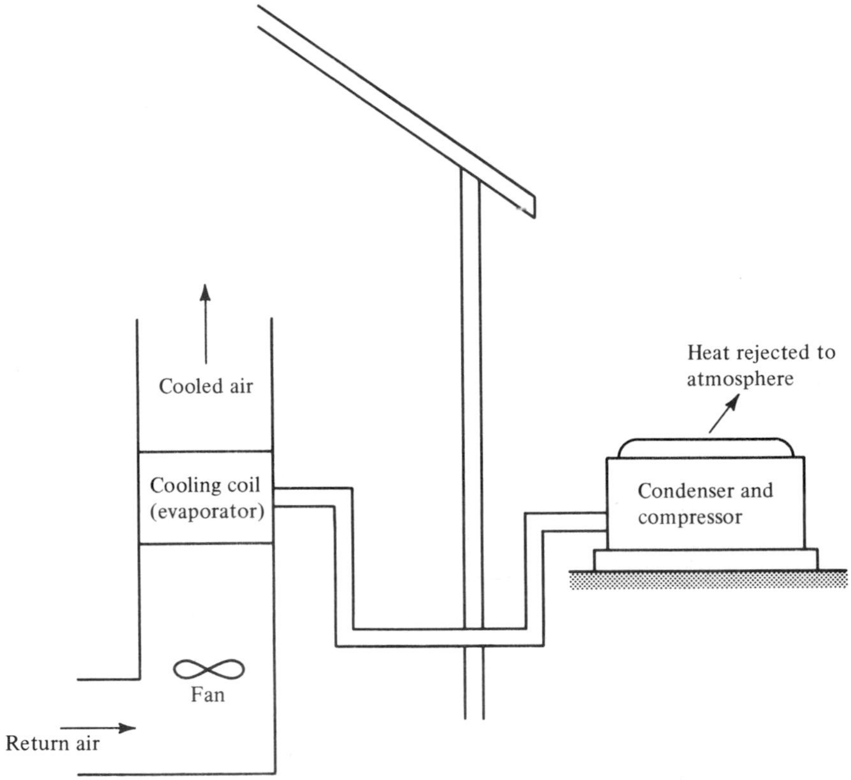

Figure 10.33 *Home Air Conditioning Installation.*

The net work input to the cycle is

$$\text{net } w'_{\text{in}} = h_2 - h_1$$

From Chapter 6, the coefficient of performance for a refrigeration cycle is

$$\text{COP}_{\text{refrig.}} = \frac{q_{\text{in}}}{\text{net } w'_{\text{in}}} = \frac{h_1 - h_4}{h_2 - h_1} \tag{10.9}$$

If water were used as the working fluid in a refrigeration cycle, the pressure inside the evaporator would have to be maintained at 0.1 psia (0.7 kPa) for heat to be withdrawn from a cold space temperature of 35°F (1.67°C). Clearly, it is not easy to hold such a vacuum. Further, with water, no temperature below 32°F (0°C) could be maintained, yet we require such low temperatures for many refrigeration applications. For a working fluid in refrigeration cycles, therefore, we require substances with appreciable saturation pressure at low temperatures.

In the early days of refrigeration, substances such as ammonia, sulfur dioxide, and carbon dioxide were utilized. In more recent years, however, special working fluids (called *refrigerants*) have been developed for refrigeration cycles. They have low toxicity and possess thermodynamic properties that make them suitable for use in vapor refrigeration cycles. One of these refrigerants is R-22, the thermodynamic properties of which are given in Tables A.10, A.11, B.10, and B.11. R-22 is chlorodifluoromethane, which is marketed as Freon (Du Pont), Genetron 22 (Allied Chemical), and Ucon 22 (Union Carbide). With R-22, for

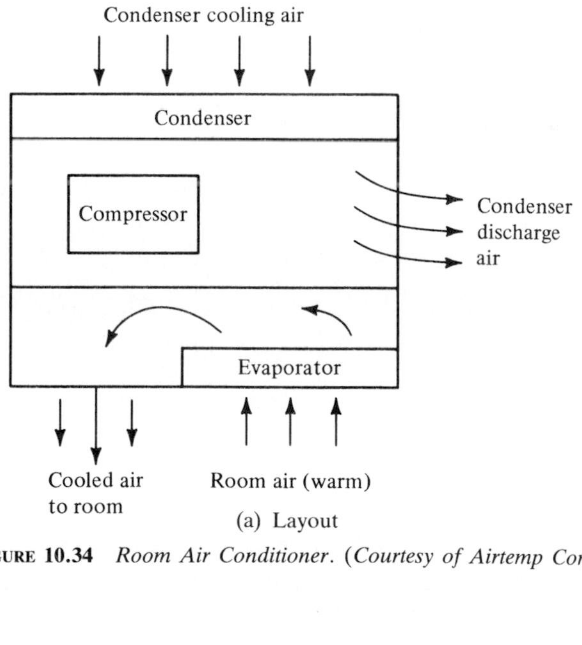

(a) Layout

FIGURE 10.34 *Room Air Conditioner. (Courtesy of Airtemp Corp.)*

(b) Window unit

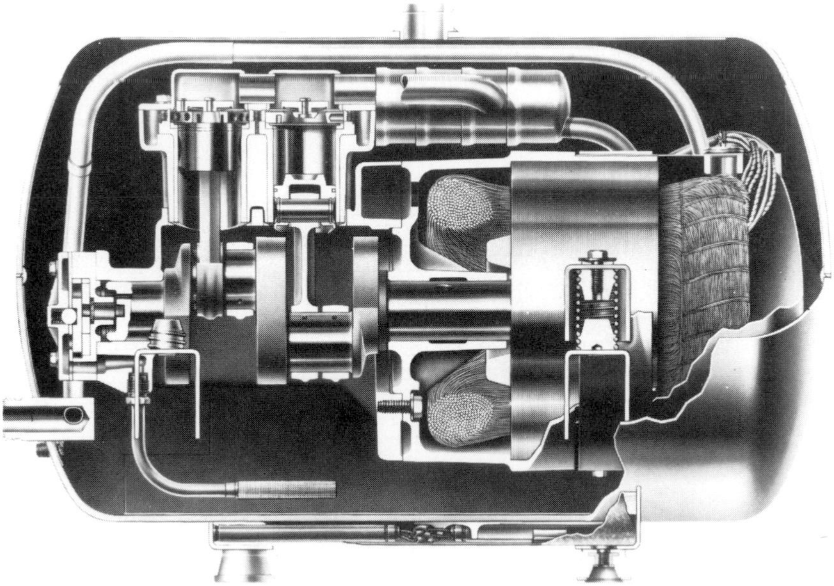

(c) Compressor

Figure 10.34 (*continued*)

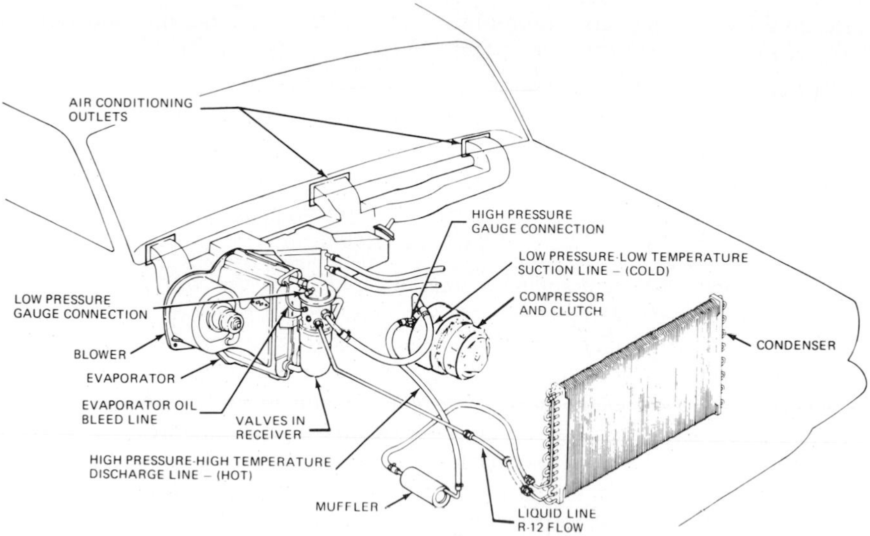

FIGURE 10.35 *Automobile Air Conditioning System.* (*Courtesy of Harrison Radiator Division, General Motors Corporation, Lockport, New York.*)

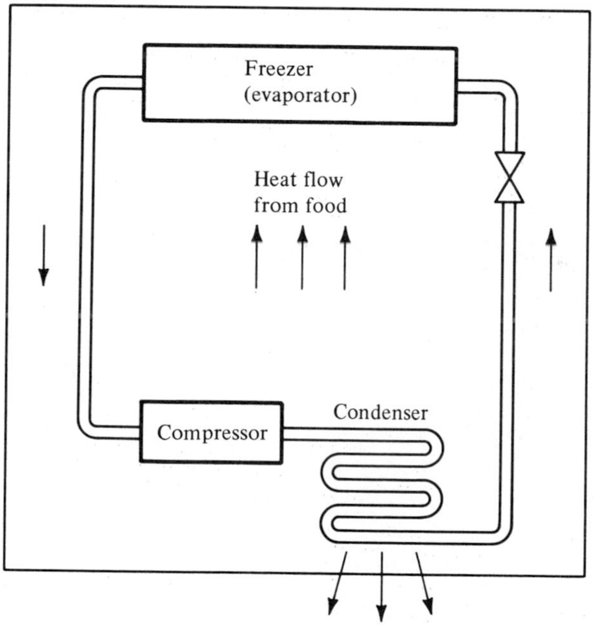

FIGURE 10.36 *Location of Components in Household Refrigerator.*

example, an evaporator temperature of −20°C can be reached with a pressure of 244.814 kPa, a pressure that should be easy to maintain in a refrigeration system.

Typical photographs and layouts of home and automobile air conditioning systems and household refrigerators are shown in Figures 10.33 through 10.36.

EXAMPLE 10.9

An automobile air conditioning system uses R-22 as the working fluid. The evaporator pressure is 80 psia, and the condenser outlet pressure is 250 psia. The temperature of the R-22 leaving the evaporator is 5° above saturation. Assume an isentropic compressor with no changes of pressure or temperature in the lines connecting the components of the cycle. The maximum heat load is 22,000 Btu/h; that is, 22,000 Btu/h must be removed to lower the air temperature of the interior of the car to a comfortable level. Determine the compressor horsepower required and the coefficient of performance of the cycle. Then repeat the problem for a compressor efficiency of 85 percent.

Solution

As shown on the *T-s* diagram of a vapor refrigeration cycle in Figure 10.31, $h_3 = h_f$ at 250 psia. By interpolation from Table A.10, we find that the corresponding saturation temperature is 112.76°F and h_f is 43.32 Btu/lbm. For the throttling process 3–4, $h_4 = h_3 = 43.32$ Btu/lbm. From Table A.11, the saturation temperature corresponding to 80 psia is 37.76°F. Therefore, $T_1 = 37.76 + 5 = 42.76$°F. State 1 is in the superheat region, with $p_1 = 80$ psia and $T_1 = 42.76$°F. From Table A.11, $h_1 = 108.80$ Btu/lbm and $s_2 = 0.22190$ Btu/lbm°R. For the compressor process, $s_2 = s_1 = 0.22190$ Btu/lbm°R. At $p_2 = 250$ psia and $s_2 = 0.22190$ Btu/lbm°R, $T_2 = 153.9$°F and $h_2 = 121.49$ Btu/lbm. For the cycle,

$$q_{\text{in}} = h_1 - h_4 = 108.80 - 43.32 = 65.48 \text{ Btu/lbm}$$

$$q_{\text{out}} = h_2 - h_3 = 121.49 - 43.32 = 78.17 \text{ Btu/lbm}$$

$$w'_{\text{in}} = h_2 - h_1 = 121.49 - 108.80 = 12.69 \text{ Btu/lbm}$$

Note that net $q_{\text{in}} = 65.48 - 78.17 = -12.69 =$ net w'_{out}. From Equation (6.3),

$$\text{COP} = \frac{q_{\text{in}}}{\text{net } w_{\text{in}}} = \frac{65.48}{12.69} = 5.16$$

The corresponding Carnot cycle COP is, from Equation (6.9),

$$\text{COP}_{\text{Carnot}} = \frac{T_L}{T_H - T_L} = \frac{42.76 + 459.67}{112.76 - 42.76} = 7.18$$

For a heat load ($\dot{q}_{\text{in}}$) of 22,000 Btu/h,

$$\dot{w}_{\text{in}} = \frac{22{,}000 \text{ Btu/h}}{\text{COP}} = 4264 \text{ Btu/h} = \frac{4264 \text{ Btu/h}}{2544.4 \text{ Btu/h/horsepower}}$$

$$= 1.676 \text{ compressor horsepower}$$

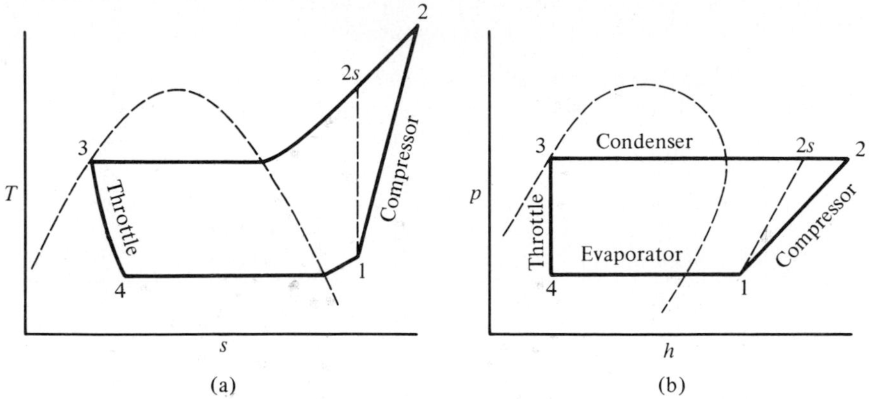

FIGURE 10.37 *T-s and p-h Diagrams of Vapor Compression Cycle with Nonisentropic Compression.*

The refrigerant flow rate is $\dot{q}_{in} = \dot{m}q_{in}$, or

$$\dot{m} = \frac{\dot{q}_{in}}{q_{in}} = \frac{22{,}000\ \text{Btu/h}}{3600\ \text{h/s}(65.48\ \text{Btu/lbm})} = 0.0933\ \text{lbm/s}$$

For nonisentropic compression, the cycle is illustrated in Figure 10.37. For a compressor efficiency of 0.85,

$$w'_{in(actual)} = \frac{(h_2 - h_1)_s}{0.85} = \frac{12.69}{0.85} = 14.93\ \text{Btu/lbm}$$

Therefore,

$$h_{2a} = 108.80 + 14.93 = 123.73\ \text{Btu/lbm}$$

For the cycle,

$$q_{in} = 65.48\ \text{Btu/lbm}$$

$$q_{out} = 123.73 - 43.32 = 80.41\ \text{Btu/lbm}$$

$$w'_{in} = 123.73 - 108.80 = 14.93\ \text{Btu/lbm}$$

$$\text{COP} = \frac{65.48}{14.93} = 4.38$$

$$\dot{w}_{in} = \frac{22{,}000}{\text{COP}} = 5016\ \text{Btu/h} = 1.972\ \text{hp}$$

or directly,

$$\dot{w}_{in(a)} = \dot{w}_{in}/0.85 = 1.676/0.85 = 1.972\ \text{hp}$$

EXAMPLE 10.10

Perform a first- and second-law analysis on the automobile air conditioning system of Example 10.9 if the compressor operates at an efficiency of 85 percent.

Solution

Working medium: Refrigerant R-22
Reference conditions: $T_0 = 490°R$ (30°F), $p_0 = 14.7$ psia, $h_0 = 110.59$ Btu/lbm, $s_0 = 0.26242$ Btu/lbm°R

State or component	p (psia)	T (°F)	h (Btu/lbm)	s (Btu/lbm°R)	a (Btu/lbm)
1	80	42.76	108.80	0.22190	18.07
Compressor					
2	250	164.18	123.73	0.22552	31.22
Condenser					
3	250	112.76	43.32	0.08643	18.97
Throttle					
4	80	37.76	43.32	0.09037	17.04
Evaporator (cooling load)					
1	80	42.76	108.80	0.22190	18.07

State or component	q_{in} (Btu/lbm)	w'_{out} (Btu/lbm)	Δa (Btu/lbm)	i (Btu/lbm)
1				
Compressor	0	14.93	13.16	1.77
2				
Condenser	−80.41	0	−12.26	12.26
3				
Throttle	0	0	−1.93	1.93
4				
Evaporator (cooling load)	65.48	0	1.03	
1				
Net	−14.93	−14.93	0	15.96

Values of T, h, and s at states 1, 3, and 4 are the same as given in Example 10.9. Compressor work $w'_{in} = 12.96$ Btu/lbm/0.85 = 14.93 Btu/lbm; hence $h_2 = 108.80$ Btu/lbm + 14.93 Btu/lbm = 123.73 Btu/lbm. Heat rejection in the condenser is $q_{out} = h_2 - h_3 = 123.73$ Btu/lbm − 43.32 Btu/lbm = 80.41 Btu/lbm. For this cycle, we choose $T_0 = 490°R$ (30°F) and $p_0 = 14.7$ psia, so that $h_0 = 110.59$ Btu/lbm and $s_0 = 0.26242$ Btu/lbm°R.

As explained in Example 10.1, we omit the value of irreversibility for the evaporator (availability input), since we are assuming perfect heat transfer in the evaporator.

The first law distribution of energy in the cycle is

Work input	14.93 Btu/lbm	(1.0)
Energy input	65.48 Btu/lbm	(5.386 = COP)
Energy rejection	80.41 Btu/lbm	(5.386)

The second law efficiency of this refrigeration cycle is from Equation (10.7)

$$\eta_2 = \frac{\Delta a_{\text{C.L.}}}{\text{net } w'_{\text{in}}} = \frac{1.03 \text{ Btu/lbm}}{14.93 \text{ Btu/lbm}} = 0.069$$

The second-law distribution in the cycle is:

Work input	14.93 Btu/lbm	(1.00)
Availability input (evaporator)	1.03 Btu/lbm	($0.069 = \eta_2$)
Irreversibilities:		
Compressor	1.77 Btu/lbm	
Condenser	12.26 Btu/lbm	
Throttle	1.93 Btu/lbm	
Irreversibilities	15.96 Btu/lbm	(1.069)

Note that the condenser irreversibility is a significant part of the availability losses, since the condensing process occurs at the high temperatures of the cycle. The condenser is therefore a likely candidate for improvement.

The cooling capacity of a refrigeration or air conditioning system is generally specified in terms of Btu/h or kJ/h. The cooling capacity of such systems is also expressed in terms of tons of refrigeration, where one ton is equal to 12,000 Btu/h. One ton is the amount of heat that must be removed to freeze one ton of water at 32°F and one atmosphere in one day. For a latent heat of water of 144 Btu/lbm, one ton of refrigeration becomes

$$\frac{(144 \text{ Btu/lbm})(2000 \text{ lbm})}{24 \text{ h}} = 12{,}000 \text{ Btu/h}$$

In Example 10.9, the maximum cooling capacity of the air conditioning unit was 22,000 Btu/h or 22,000/12,000 = 1.83 tons of refrigeration. With an isentropic compressor, there are 1.83/1.676 or 1.09 tons of refrigeration available per compressor horsepower. For an 85 percent efficient compressor, there is 1.83/1.972 or 0.93 ton of refrigeration available per compressor horsepower.

A refrigerant commonly used in smaller refrigerating machines is R-12 (dichlorodifluoromethane). It has a lower latent heat of evaporation than does R-22; the resulting larger refrigerant flow rate permits easier control and operation of the refrigeration system. Application of R-12 is illustrated in the next example.

EXAMPLE 10.11

A typical household refrigerator uses refrigerant R-12 as the working fluid. The refrigeration saturation temperature in the evaporator is 5°F; the condenser saturation temperature is 104.35°F. The refrigerant leaves the evaporator at 15°F above the saturation temperature (15° superheat). A capillary tube is used to throttle.

For a compressor efficiency of 75 percent, find the compressor horsepower required, the cooling capacity, and the coefficient of performance of the unit. The mass flow of the refrigerant is 0.01 lbm/s.

The thermodynamic properties of R-12 required for the calculations are given in the following table:

T (°F)	p (psia)	h_f (Btu/lbm)	h_g (Btu/lbm)	h (Btu/lbm)	s_f (Btu/lbm°R)	s_g (Btu/lbm°R)	s (Btu/lbm°R)
5	26.48	9.60	77.81		0.0216	0.1684	
20	26.48			79.99			0.1730
104.35	140.0	32.10	87.39		0.0649	0.1630	
136	140.0			93.20			0.1730

Solution

The thermodynamic representation of this refrigeration cycle is given in Figure 10.37. The temperature of refrigerant entering the compressor is $T_1 = 5° + 15° = 20°F$, with $p_1 = p_4$ (saturation pressure at 5°F) = 26.48 psia; hence, $h_1 =$ 79.99 Btu/lbm. The saturation pressure of R-12 corresponding to a temperature of 104.35°F is 140.0 psia; hence $h_3 = h_f = 32.10$ Btu/lbm. For an isentropic compressor, $s_2 = s_1 = 0.1730$ Btu/lbm°R; hence, at $p_2 = 140$ psia, $h_{2s} = 93.20$ Btu/lbm ($T_{2s} = 136°F$). Thus,

$$h_2 - h_1 = \frac{h_{2s} - h_1}{0.75} = \frac{93.20 - 79.99}{0.75}$$
$$= \frac{13.21}{0.75} = 17.62 \text{ Btu/lbm}$$

The compressor power required is

$$\dot{m}(h_2 - h_1) = (0.01 \text{ lbm/s})(17.62 \text{ Btu/lbm}) = 0.176 \text{ Btu/s}$$
$$= 0.176(1.415 \text{ hp/Btu/s}) = 0.25 \text{ hp}$$

In the capillary tube, $h_3 = h_4 = 32.10$ Btu/lbm. The cooling capacity $\dot{q}_{in}$ of the unit is

$$\dot{m}(h_1 - h_4) = (0.01 \text{ lbm/s})(79.99 \text{ Btu/lbm} - 32.10 \text{ Btu/lbm})$$
$$= 0.479 \text{ Btu/s} = 1725 \text{ Btu/h}$$
$$= \frac{1725}{12{,}000} = 0.143 \text{ ton (refrigeration)}$$

The coefficient of performance becomes, from Equation (10.9),

$$\mathrm{COP} = \frac{q_{\text{in}}}{\text{net } w'_{\text{in}}} = \frac{h_1 - h_4}{h_2 - h_1} = \frac{47.89}{17.62} = 2.72$$

In a household refrigerator, the compressor is usually driven by an electric motor. For an electric motor efficiency of 80 percent, the motor power requirement would become (0.25/0.80) (0.7457 kW/hp) = 0.233 kW.

The reversed Rankine cycle is also used as a basic heat pump cycle with the system components shown in Figure 10.31. Most commercially available heat pumps can be used for cooling as well as heating. In summer, the heat pump functions as a conventional cooling system. The indoor coil (acting as an evaporator) extracts heat from the indoor air, transferring it via the outdoor coil (acting as a condenser) to the outside air. In winter, the heat pump functions as a heating unit after a reversing valve changes the direction of the flow of the refrigerant in the coils of the two heat exchangers (evaporator and condenser). The outside coil (now acting as an evaporator) extracts heat from the outside air and transfers it through the indoor coil (now acting as a condenser) to the indoor air. The heating capacity of the system depends on the temperature of the outside air. If the building to be heated has a heat loss greater than the amount of heat that a pump can supply, supplementary heat is supplied by electric heating coils. Since electric energy is usually more expensive than energy obtained from natural gas or oil, heat pumps are usually utilized only in areas where the winter temperature remains moderate (50°F and above).

A schematic diagram of the cooling and heating operations of a heat pump is shown in Figure 10.38. Note that the flow reversing valve changes the flow direction in the two heat exchangers only, not in the compressor. Capillary tubes or expansion valves are used to control the refrigerant flow.

EXAMPLE 10.12

When a heat pump is operating in the cooling mode, the pressure in the evaporator is 100 psia and the condenser pressure is 300 psia. The refrigerant (R-22) leaves the evaporator at 65°F. The cooling capacity of the unit is 44,000 Btu/h.

When the pump is operating in the heating mode, the pressure in the evaporator is 60 psia, and in the condenser 200 psia. The heating capacity of the heat pump is then 49,000 Btu/h. The refrigerant leaves the evaporator at a temperature of 40°F.

Determine the compressor power requirement, the coefficient of performance, and the mass flow of refrigerant in

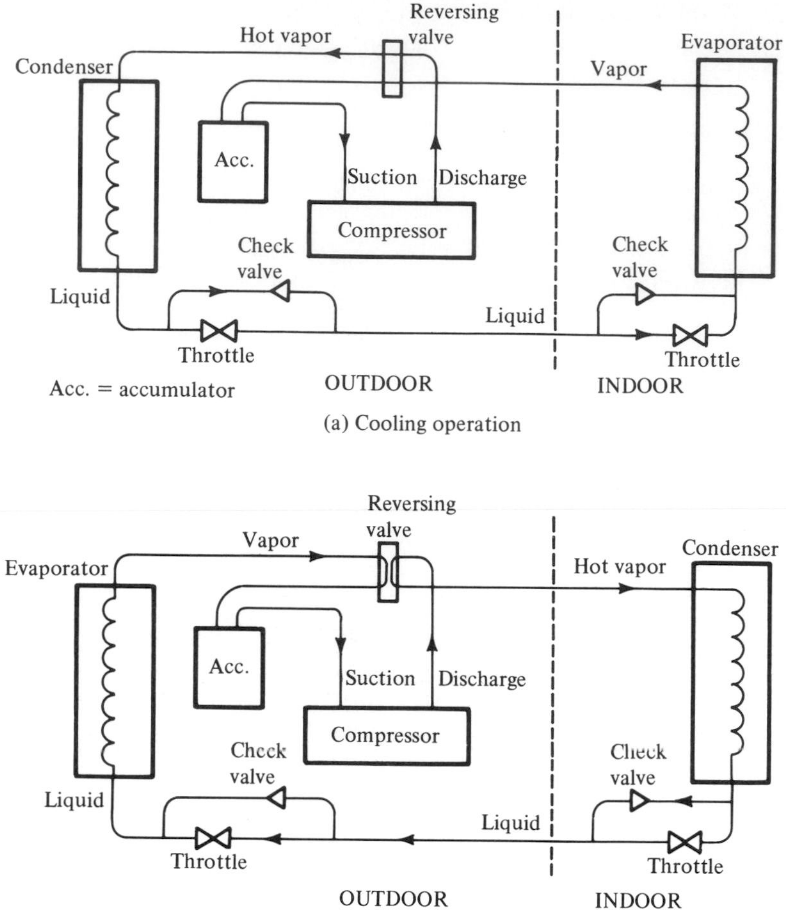

(b) Heating operation

FIGURE 10.38 *Schematic of Heat-Pump Operation.*

a. the cooling mode, and
b. the heating mode.

Assume a compressor efficiency of 70 percent.

A cooling capacity of 44,000 Btu/h will maintain a given building at an indoor temperature of 75°F when the outside temperature is 95°F (that is, the heat gain of the building from the outdoors is 44,000 Btu/h). The heat gain or loss to the building can be taken as proportional to the temperature difference between indoors and outdoors; 49,000 Btu/h are supplied by the heat pump in winter, when the outside temperature is 45°F. What will be the additional electric heater requirements to maintain the building at an indoor temperature of 75°F?

Solution

a. *Cooling mode*:

Referring to Figure 10.37, we find that $T_1 = 65°F$. Hence, by interpolation from Table A.11, at 100 psia,

$$h_1 = 111.58 \text{ Btu/lbm} \quad \text{and} \quad s_1 = 0.22271 \text{ Btu/lbm°R}$$

For an isentropic compressor, $s_{2s} = s_1 = 0.22271$ Btu/lbm°R. At $p_2 = 300$ psia, $T_{2s} = 174.93°F$ and $h_{2s} = 123.98$ Btu/lbm. Hence,

$$w'_{in} = (h_{2s} - h_1)/0.70 = (123.98 - 111.58)/0.70 = 17.72 \text{ Btu/lbm}$$

Thus, $h_2 = h_1 + 17.72 = 129.30$ Btu/lbm. From Table A.10, by interpolation, $h_3 = h_f$ at 300 psia = 48.04 Btu/lbm with $T_3 = 127°F$. With $h_3 = h_4$, h_4 is 48.04 Btu/lbm. The cooling capacity of the unit is

$$\dot{q}_{in} = \dot{m}(h_1 - h_4)$$

or

$$\dot{m} = \frac{44{,}000 \text{ Btu/h}}{(3600 \text{ s/h})(111.58 - 48.04) \text{ Btu/lbm}} = 0.192 \text{ lbm/s}$$

The compressor power requirement becomes

$$\dot{w}'_{in} = \dot{m}(h_2 - h_1) = (0.192 \text{ lbm/s})(17.72 \text{ Btu/lbm}) = 3.40 \text{ Btu/s}$$

$$= (3.40 \text{ Btu/s})(1.415 \text{ hp-s/Btu}) = 4.81 \text{ hp} = 3.59 \text{ kW}$$

The coefficient of performance as a refrigeration cycle is, from Equation (6.3),

$$\text{COP}_{\text{refrig}} = \frac{q_{in}}{\text{net } w'_{in}} = \frac{h_1 - h_4}{h_2 - h_1} = \frac{63.54}{17.72} = 3.59$$

b. *Heating mode*:

The cycle is represented again by Figure 10.37 with $T_1 = 40°F$. From Table A.11, at 60 psia,

$$h_1 = 109.58 \text{ Btu/lbm} \quad \text{and} \quad s_1 = 0.22950 \text{ Btu/lbm°R}$$

For an isentropic compressor, $s_2 = s_1 = 0.22950$ Btu/lbm°R. At p_2 of 200 psia, we obtain $h_{2s} = 123.13$ Btu/lbm. Hence,

$$w'_{in} = (123.13 - 109.58)/0.70 = 19.36 \text{ Btu/lbm}$$

Thus, $h_2 = h_1 + 19.36 = 128.94$ Btu/lbm. From Table A.10, $h_3 = h_f$ at 200 psia = 38.09 Btu/lbm with $T_3 = 96.27°F$. The heating capacity of the unit is

$$\dot{q}_{out} = \dot{m}(h_2 - h_3)$$

or

$$\dot{m} = \frac{49{,}000 \text{ Btu/h}}{(128.94 \text{ Btu/lbm} - 38.09 \text{ Btu/lbm})3600 \text{ s/h}} = 0.150 \text{ lbm/s}$$

The compressor power requirement is

$$\dot{w}'_{in} = \dot{m}(h_2 - h_3) = (0.150 \text{ lbm/s})(19.36 \text{ Btu/lbm}) = 2.90 \text{ Btu/s} = 4.10 \text{ hp}$$

$$= 3.06 \text{ kW}$$

The coefficient of performance as a heat pump is, from Equation (6.4),

$$\text{COP}_{\text{heat pump}} = \frac{q_{\text{out}}}{\text{net } w'_{\text{in}}} = \frac{h_2 - h_3}{h_2 - h_1} = \frac{90.85}{19.36} = 4.69$$

or

$$\text{COP}_{\text{heat pump}} = \frac{q_{\text{in}}}{\text{net } w'_{\text{in}}} + 1 = \frac{h_1 - h_4}{h_2 - h_1} + 1 = \frac{109.58 - 38.09}{19.36} + 1$$
$$= 3.69 + 1 = 4.69$$

The heat loss from the building is

$$\dot{q} = \frac{44{,}000 \text{ Btu/h}}{(95 - 75)°\text{F}} (75 - 45)°\text{F} = 66{,}000 \text{ Btu/h}$$

Hence the electric heater must supply

$$66{,}000 - 49{,}000 = 17{,}000 \text{ Btu/h} = 4.98 \text{ kW}$$

Thus, to maintain the inside of the building at a temperature of 75°F, when the outside temperature is 45°F, there must be 3 kW for the compressor and 5 kW for the supplementary electric heater. No supplementary heating would be required if the outside temperature were 52.7°F, $\dot{q} = 44{,}000(95 - 75)/(75 - 52.7) = 49{,}000$ Btu/h. Therefore, as stated previously, since electric energy is usually more expensive than energy obtained from natural gas or oil heat, the heat pump will operate most economically when confined to geographical regions where the outdoor temperature does not often drop below 50°F.

10.4 GAS POWER AND REFRIGERATION CYCLES: THE BRAYTON CYCLE

There are many cycles of interest in which the working fluid is always in the gaseous phase. An example is the *Brayton* cycle (also called the *Joule* cycle), used in such applications as stationary power plants, jet engines, and special air conditioning systems. Like the ideal Rankine cycle, (1–2–3–4) of Figure 10.39, the ideal Brayton cycle consists of two isentropic processes and two constant pressure processes, as shown by 1′–2′–3′–4′ of Figure 10.39.

1. Brayton Power Cycles

In our study of maximum efficiency of a gas power cycle, it is logical to look first, as we did for the vapor power cycle, at a Carnot power cycle operating between given maximum and minimum temperatures. As shown in Figure 10.40, to reach point $2a$ of the Carnot cycle would require going to much higher pressures than in the Brayton cycle.

Further, to reach point $4a$ of the Carnot cycle would involve large specific volumes and hence large specific volume changes during a cycle. Such volumes and volume changes would give low work output per specific volume, coupled with high maximum pressures. High pressures result in heavy, thick-walled machinery and piping, certainly unsuitable for light-weight application in aircraft. Thus, in the Brayton gas power cycle, an isentropic compression (process 1–2 of Figure 10.40) is used which does not reach as high a pressure as in the Carnot cycle (process 1–$2a$). Similarly, the isentropic expansion (power) process does not reach as low a pressure as in the Carnot cycle, since such low pressure would result in

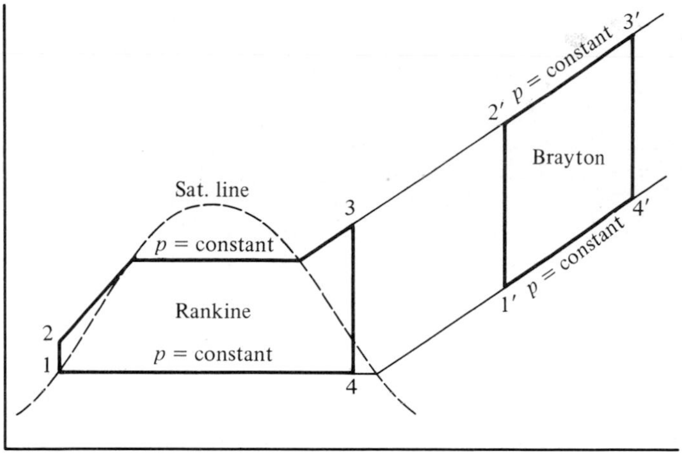

FIGURE 10.39 *T-s Diagram of Ideal Rankine and Brayton Cycles.*

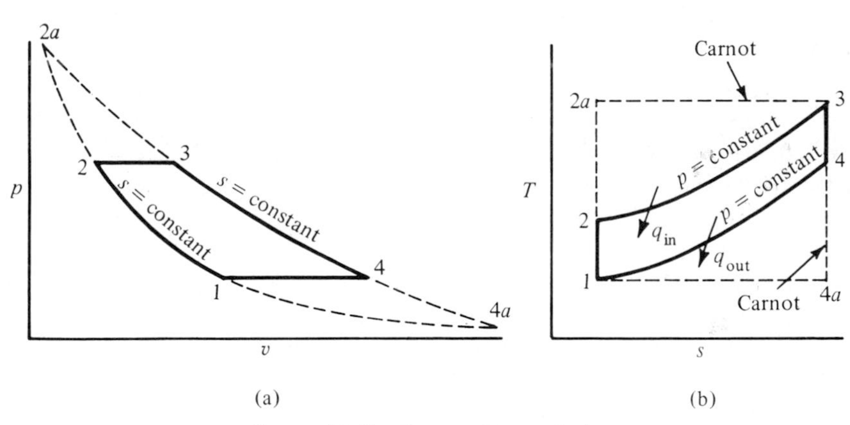

FIGURE 10.40 *Brayton Power Cycle.*

high specific volume at point 4*a* and would require large engine sizes. Hence, the heat rejection process in a Brayton cycle (process 4–1) is at constant pressure rather than at constant temperature. By comparison, in the Rankine cycle, the heat rejection process is at both constant temperature and constant pressure, since it occurs in the vapor-liquid region.

A schematic arrangement of a typical Brayton power generating system is shown in Figure 10.41. The components of the system are as follows:

1. a *compressor*, in which the inlet air flow is raised in pressure;
2. a *combustion chamber*, in which fuel is added to the air flow and the resultant fuel-air mixture is burned, increasing cycle temperature; and

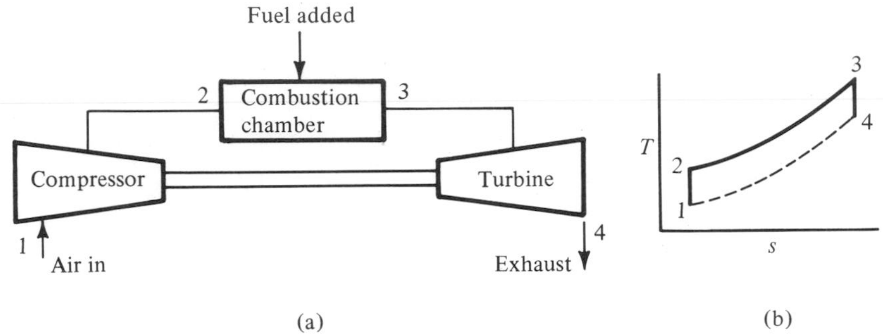

FIGURE 10.41 *Open-Loop Brayton Cycle.*

3. a *turbine*, in which the high-temperature, high-pressure gases from the combusion chamber are allowed to expand, producing work. Some of this work is used to drive the compressor, and the remainder is useful work output. Flow from the turbine is then allowed to discharge to the atmosphere.

This is an open loop cycle, in which fresh air from the atmosphere is continually drawn into the compressor, with the hot products of combustion at the exhaust discharged from the system. In a closed loop cycle, the heat exchanger (a condenser in a Rankine power cycle) removes heat from the working fluid, restoring the temperature of the gas exhausting from the turbine to the inlet temperature T_1 at the compressor. In an open loop cycle, this function of the heat exchanger is performed by the atmosphere (see Section 10.1).

The thermal efficiency of the Brayton cycle is

$$\eta = \frac{\text{net } w'_{\text{out}}}{q_{\text{in}}} = \frac{q_{\text{in}} - q_{\text{out}}}{q_{\text{in}}}$$

$$\eta = 1 - \frac{h_4 - h_1}{h_3 - h_2}$$

The net work output of the cycle is

$$\text{net } w'_{\text{out}} = h_3 - h_4 - (h_2 - h_1)$$

The energy input to the cycle, which takes place in the combustion chamber, is

$$q_{\text{in}} = h_3 - h_2$$

To simplify the analysis of the ideal Brayton cycle, we will assume the following:

1. The working fluid is air alone. That is, the products of combustion, obtained when air is burned with a small amount of fuel in the combustion chamber, retain the properties of air.
2. There is a fixed flow of air, with the small amount of fuel flow neglected in comparison to the air flow.
3. Air can be treated as an ideal gas with constant specific heats.

Such a cycle is called an *air-standard* cycle.

For the isentropic compression in the compressor, we obtain, from Equation (8.13),

$$\frac{T_2}{T_1} = \left(\frac{p_2}{p_1}\right)^{(\gamma-1)/\gamma}$$

For the isentropic expansion in the turbine, we have

$$\frac{T_3}{T_4} = \left(\frac{p_3}{p_4}\right)^{(\gamma-1)/\gamma}$$

Since $p_4 = p_1$ and $p_3 = p_2$, it follows that $(T_2/T_1) = (T_3/T_4)$ or $T_4 = T_3(T_1/T_2)$. Therefore, the thermal cycle efficiency for the air-standard Brayton cycle is

$$\eta = 1 - \frac{h_4 - h_1}{h_3 - h_2} = 1 - \frac{c_p(T_4 - T_1)}{c_p(T_3 - T_2)} = 1 - \frac{T_1(T_3/T_2 - 1)}{T_2(T_3/T_2 - 1)}$$

$$\eta = 1 - \frac{T_1}{T_2}$$

or $$\eta = 1 - \left(\frac{p_2}{p_1}\right)^{(1-\gamma)/\gamma} \qquad \textbf{(10.10)}$$

In other words, the thermal efficiency of the ideal Brayton cycle is dependent only on the compressor pressure ratio. In Figure 10.42, Brayton cycle efficiency has been plotted as a function of the compressor pressure ratio for $\gamma = 1.4$ and $\gamma = 1.35$.

From the results presented in Figure 10.42, it would seem advantageous to operate at as high a compressor pressure ratio as possible. However, the temperature of the cycle is limited; that is, due to metallurgical considerations, maximum permissible turbine blade temperatures at the present time must be limited to between 1800°F and 2000°F. Figure 10.43 depicts this limitation on a *T-s* diagram. For this ideal cycle,

$$q_{in} = \text{area under } 2\text{–}3$$

$$q_{out} = \text{area under } 4\text{–}1$$

$$w'_{out} = \text{area } 1\text{–}2\text{–}3\text{–}4$$

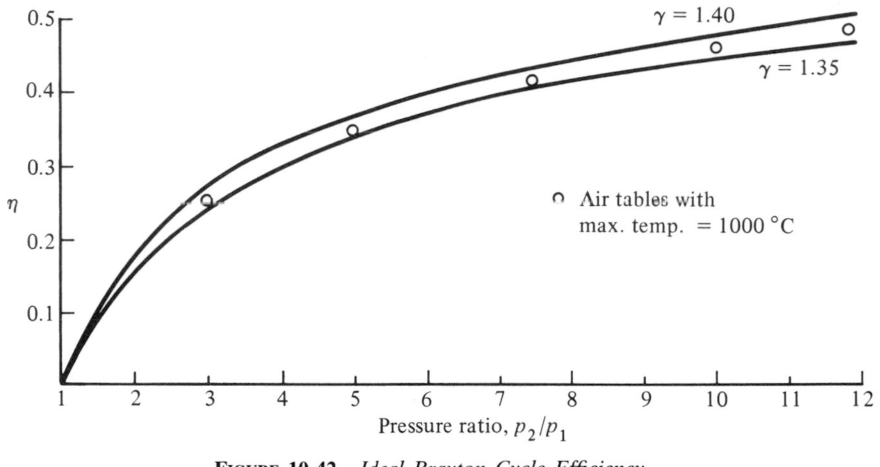

FIGURE 10.42 *Ideal Brayton Cycle Efficiency.*

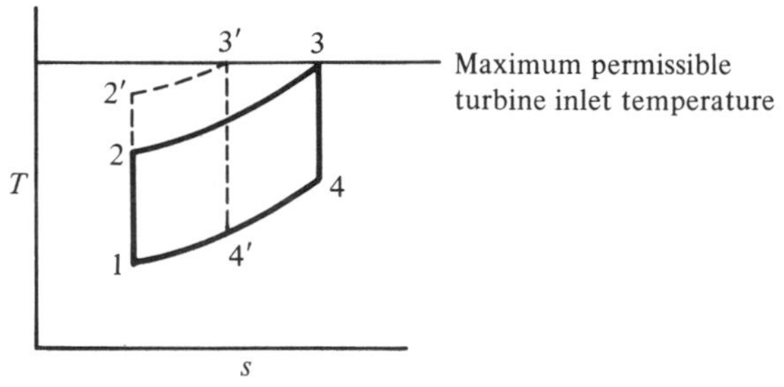

FIGURE 10.43 *T-s Diagram of Brayton Cycle with Temperature Limitation.*

For an increase of compressor pressure ratio from p_2/p_1 to $p_{2'}/p_1$, the temperature $T_{2'}$ will be greater than T_2. Hence although the efficiency of 1–2′–3′–4′ is greater than that of 1–2–3–4, less heat can be added to the working fluid before the maximum allowable turbine inlet temperature is reached. In turn, less work output results for the cycle, since the area 1–2′–3′–4′ is less than 1–2–3–4. In an actual engine, the work output for a given size is an important consideration. Further, some of the work output must be used to overcome friction. The frictional work required for 1–2′–3′–4′ would be a greater fraction of the total work output than 1–2–3–4. Thus, compressor pressure ratios for gas turbine power plants are in the range from 4 : 1 to 8 : 1.

EXAMPLE 10.13

A gas turbine power plant is to produce 10 megawatts. Air enters the compressor at one atmosphere and 20°C. The compressor pressure ratio is 5, and the maximum cycle temperature (turbine inlet temperature) is 1000°C. The turbine outlet pressure is one atmosphere.

Find the total turbine power output, the power required to drive the compressor, the cycle thermal efficiency, and the fuel flow and fuel-air ratio required for a fuel heating value of 50 000 kJ per kilogram of fuel.

Assume an ideal cycle, with isentropic compressor and turbine, and no pressure drop in the combustion chamber. Carry out the calculations

a. under the assumption that the working fluid can be taken as air behaving as an ideal gas with constant specific heats, $c_p = 1.01$ kJ/kg K and $\gamma = 1.4$; and

b. by use of the air tables given in Appendix B.

Solution

a. *For constant specific heats*

For the conditions given, $p_2/p_1 = 5$. Since the compressor process is isentropic, Equation (8.13) gives

$$T_2/T_1 = (5)^{(\gamma-1)/\gamma} = 1.584 \quad \text{and} \quad T_2 = 1.584(293) = 464.3 \text{ K}$$

The turbine exhausts to atmospheric pressure; hence, $p_4 = 1$ atmosphere. With no pressure drop in the combustion chamber, $p_3 = p_2$. For an isentropic turbine,

$$\frac{T_4}{T_3} = \left(\frac{p_4}{p_3}\right)^{(\gamma-1)/\gamma} = \left(\frac{p_1}{p_2}\right)^{(\gamma-1)/\gamma} = 1/1.584$$

or

$$T_4 = \frac{1273}{1.584} = 803.8 \text{ K}$$

For the cycle,

$$w'_{\text{out(turb)}} = c_p(T_3 - T_4) = (1.01 \text{ kJ/kg K})(1273 \text{ K} - 803.8 \text{ K})$$
$$= 473.9 \text{ kJ/kg}$$

$$w'_{\text{in(comp)}} = c_p(T_2 - T_1) = 1.01\ \text{kJ/kg K}(464.3\ \text{K} - 293\ \text{K})$$
$$= 172.9\ \text{kJ/kg}$$
$$\text{net } w'_{\text{out}} = 473.9 - 172.9 = 301.0\ \text{kJ/kg}$$

For an output of 10 megawatts,

$$\dot{m}_{\text{air}} = \frac{10 \times 10^6\ \text{J/s}}{3.01 \times 10^5\ \text{J/kg}} = 33.2\ \text{kg/s}$$

$$\text{Total turbine power out} = (33.2\ \text{kg/s})(473.9\ \text{kJ/kg}) = 15.73\ \text{MW}$$

$$\text{Power to drive compressor} = (33.2\ \text{kg/s})(172.9\ \text{kJ/kg}) = 5.74\ \text{MW}$$

From Equation (10.9), $\eta = 1 - 5^{(1-\gamma)\gamma} = 0.369$. The rate of energy input in the combustion chamber is

$$\dot{q}_{\text{in}} = \dot{m}c_p(T_3 - T_2) = (33.2\ \text{kg/s})(1.01\ \text{kJ/kg K})(1273\ \text{K} - 464.3\ \text{K}) = 27.1\ \text{MW}$$

$$\dot{m}_{\text{fuel}} = \frac{27\,100\ \text{kJ/s}}{50\,000\ \text{kJ/kg}} = 0.542\ \text{kg/s}$$

Hence

$$\frac{\dot{m}_{\text{fuel}}}{\dot{m}_{\text{air}}} = \frac{0.542}{33.2} = 0.0163$$

The thermal efficiency of the equivalent Carnot cycle is, with $T_H = T_3 = 1000°\text{C}$ and $T_L = T_1 = 20°\text{C}$,

$$\eta_{\text{Carnot}} = 1 - \frac{T_L}{T_H} = 1 - \frac{20 + 273}{1000 + 273} = 0.770\ (\text{vs. } 0.369)$$

b. *Using air tables*

We obtain the following from Table B.13, at a pressure of one atmosphere:

	0°C	50°C
h	414.48	464.83 kJ/kg
s	3.9752	4.1445 kJ/kg K

Hence, using linear interpolation, we obtain

$$s_1 = 3.9752 + \frac{20}{50}(4.1445 - 3.9752) = 4.0429\ \text{kJ/kg K}$$

$$h_1 = 414.48 + \frac{20}{50}(464.83 - 414.48) = 434.62\ \text{kJ/kg}$$

We get the following from Table B.13, at a pressure of 500 kPa:

	150°C	200°C
h	565.66	616.85 kJ/kg
s	3.9539	4.0682 kJ/kg K

For the isentropic compression process 1–2 we get, with $s_2 = s_1$,

$$T_2 = 150 + \frac{4.0429 - 3.9539}{4.0682 - 3.9539} 50 = 188.93°\text{C (vs. } 191.1°\text{C)}$$

$$h_2 = 565.66 + 0.7787(616.85 - 565.66) = 605.52 \text{ kJ/kg}$$

For the isentropic expansion process 3–4, we have $s_3 = s_4$. From Table B.13, we obtain, at a pressure of 500 kPa and a temperature of 1000°C,

$$h_3 = 1506.4 \text{ kJ/kg} \quad \text{and} \quad s_3 = 5.1556 \text{ kJ/kg K}$$

We get the following from Table B.13, at a pressure of 100 kPa:

	550°C	600°C
h	989.41	1044.91 kJ/kg
s	5.1183	5.1838 kJ/kg K

Hence

$$T_4 = 550 + \frac{5.1556 - 5.1183}{5.1838 - 5.1183} 50 = 578°\text{C (vs. } 530.8°\text{C)}$$

$$h_4 = 989.41 + 0.5695(1044.91 - 98.41) = 1021.02 \text{ kJ/kg}$$

Thus, the turbine work output is

$$w'_{\text{out(turb)}} = h_3 = h_4 = 1506.4 - 1021.0 = 485.4 \text{ kJ/kg}$$

The compressor work input is

$$w'_{\text{in(comp)}} = h_2 - h_1 = 605.52 - 434.62 = 170.90 \text{ kJ/kg}$$

Thus,

$$\text{net } w'_{\text{out}} = 485.4 - 170.9 = 314.5 \text{ kJ/kg}$$

From Equation (6.1), the thermal efficiency is

$$\eta = \frac{h_3 - h_4 - (h_2 - h_1)}{h_3 - h_2} = \frac{314.5}{1506.4 - 605.5} = 0.349 \text{ (vs. } 0.369)$$

Thus, if actual thermodynamic properties of air are used, a slightly lower value of efficiency is obtained. In addition, the changes of temperature across the compressor and turbine are smaller. However, these changes are relatively small and hence generally do not justify the additional calculation work required.

Using the procedure outlined in Example 10.13, part (b), one can calculate the thermal cycle efficiency of an ideal Brayton cycle over a range of pressure ratios; these results (for a turbine inlet temperature of 1000°C) have been included in Figure 10.42. The efficiencies calculated in this manner are slightly lower than those based on the air-standard cycle, since, as shown in Table B.15, γ decreases with increasing temperature.

The actual Brayton cycle involves deviations from the ideal cycle just discussed. There are, for example, pressure drops in the ducting connecting the components, as well as a pressure drop in the combustion chamber. The most significant deviation from ideal behavior, however, occurs in the compressor and turbine due to flow irreversibilities. The processes taking place in the actual turbine and compressor can be made adiabatic, with sufficient insulation to restrict heat losses. However, frictional and flow losses in both devices render the processes irreversible and, hence, nonisentropic. Just as with the Rankine cycle, we must associate component efficiencies with each of these devices. Ideal and actual processes that occur in the compressor and turbine are shown on an *h-s* diagram in Figure 10.44. Typical present-day turbine efficiencies range from 85 percent to 90 percent; typical compressor efficiencies range from 80 percent to 85 percent.

EXAMPLE 10.14

Calculate the thermal efficiency of the Brayton power plant of Example 10.13 with a turbine efficiency of 90 percent and a compressor efficiency of 80 percent. Assume the working fluid to be an ideal gas, as in Example 10.13(a).

Solution

From Equation (8.5), the turbine efficiency is

$$\eta_{turb} = \frac{w'_{out(actual)}}{w'_{out(isentropic)}}$$

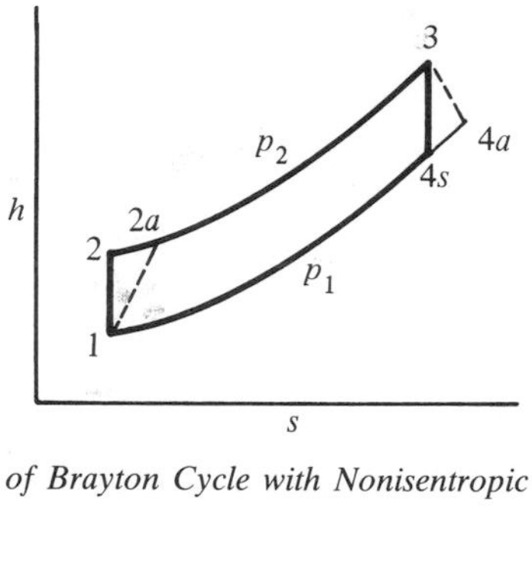

FIGURE 10.44 *h-s Diagram of Brayton Cycle with Nonisentropic Compressor and Turbine.*

while from Equation (8.6), the compressor efficiency is

$$\eta_{comp} = \frac{w'_{in(isentropic)}}{w'_{in(actual)}}$$

For the compressor, with p_2/p_1 still 5,

$$w'_{in(actual)} = \frac{172.9}{0.80} = 216.1 \text{ kJ/kg} = c_p(T_{2a} - T_1)$$

Therefore,

$$T_{2a} - T_1 = \frac{216.1}{1.01} = 214.0 \text{ K} \quad \text{and} \quad T_{2a} = 293 + 214 = 507 \text{ K}$$

For the turbine, p_2/p_1 is still 5; hence,

$$w'_{out(actual)} = 0.90(473.9) = 426.5 \text{ kJ/kg} = c_p(T_3 - T_{4a})$$

Therefore,

$$T_3 - T_{4a} = 426.5/1.01 = 422.3 \text{ K} \quad \text{and} \quad T_{4a} = 1273 - 422.3 = 850.7 \text{ K}$$

The energy input to the cycle is

$$q_{in} = c_p(T_3 - T_{2a}) = 1.01 \text{ kJ/kg K}(1273 \text{ K} - 507 \text{ K}) = 773.7 \text{ kJ/kg}$$

The cycle thermal efficiency is

$$\eta_{cycle} = \frac{\text{net } W'_{out}}{q_{in}} = \frac{426.5 - 216.1}{773.7} = 0.272$$

compared to 0.369 for $\eta_{turb.} = 100\%$ and $\eta_{comp.} = 100\%$.

The change in entropy due to the irreversibilities in the compressor is, from Equation (6.25),

$$s_{2a} - s_1 = c_p \ln \frac{T_{2a}}{T_1} - R \ln \frac{p_2}{p_1} = (1.01 \text{ kJ/kg K}) \ln \frac{507}{293} - (0.2870 \text{ kJ/kg K}) \ln 5$$
$$= 0.0919 \text{ kJ/kg K}$$

The entropy increase in the turbine is

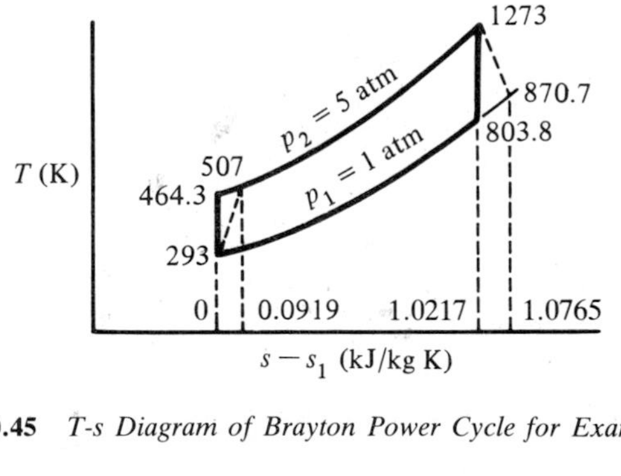

FIGURE 10.45 *T-s Diagram of Brayton Power Cycle for Example 10.14.*

$$s_{4a} - s_3 = cp \ln \frac{T_{4a}}{T_3} - R \ln \frac{p_4}{p_3} = (1.01 \text{ kJ/kg K}) \ln \frac{850.7}{1273} - (0.2870 \text{ kJ/kg K}) \ln \frac{1}{5}$$
$$= 0.0548 \text{ kJ/kg K}$$

The change in entropy during heat addition is

$$s_3 - s_{2a} = c_p \ln \frac{T_3}{T_{2a}} = 1.01 \text{ kJ/kg K} \ln \frac{1273}{507} = 0.9298 \text{ kJ/kg K}$$

Figure 10.45 illustrates the results of Example 10.14 on a *T-s* diagram.

EXAMPLE 10.15

Perform a second-law analysis on the Brayton power plant of Example 10.14.

Solution

Working medium: air
Reference conditions: $T_0 = 293$ K (20°C), $p_0 = 1$ atm (101.3 kPa), $h_0 = 0$ Btu/lbm, $s_0 = 0$ Btu/lbm°R

State or component	p (atm)	T (K)	h (kJ/kg)	s (kJ/kg K)	a (kJ/kg)
1	1	293	0	0	0
Compressor					
2a	5	507	216.1	0.0919	189.2
Energy input					
3	5	1273	989.8	1.0217	690.4
Turbine					
4a	1	851	563.3	1.0765	247.9
Energy rejection					
1	1	293	0	0	0

State or component		q_{in} (kJ/kg)	w'_{out} (kJ/kg)	Δa (kJ/kg)	i (kJ/kg)
1					
Compressor		0	−216.1	189.2	26.9
2a					
Energy input		773.7	0	501.2	
3					
Turbine		0	426.5	−442.5	16.0
4a					
Energy rejection		−563.3	0	−247.9	247.9
1					
	Net	210.4	210.4	0	290.8

The values of T were taken directly from Example 10.14. Values of enthalpy were calculated from $h = c_p(T - T_0) = 1.01\ \text{kJ/kg K}(T - 293)$. Values of entropy were calculated using (6.25) with $c_p = 1.01\ \text{kJ/kg K}$, $R = 0.287\ \text{kJ/kg K}$, and $s_0 = 0\ \text{kJ/kg K}$ at 20°C. For example, for state 2a, $s = (1.01\ \text{kJ/kg K}) \ln (507/293) - 0.287\ \text{kJ/kg K} \ln 5 = 0.0919\ \text{kJ/kg K}$. From (7.6) we obtain the availability, e.g., for state 2a, neglecting kinetic and potential energies, $a = h - h_0 - T_0(s - s_0) = 216.1\ \text{kJ/kg} - 293\ \text{K}(0.0919)\ \text{kJ/kg K} = 189.2\ \text{kJ/kg}$.

As with previous second law analysis examples, we will assume perfect heat transfer in the availability input component (combustion chamber) and omit the value of irreversibility in this component. In a more refined analysis, the irreversibility of this heat exchange must be included (see, e.g., Example 10.4).

From Equation (10.6) we get the second-law efficiency of the Brayton power cycle:

$$\eta_2 = \frac{\text{net } w'_{\text{out}}}{a_{\text{in}}} = \frac{210.4\ \text{kJ/kg}}{501.2\ \text{kJ/kg}} = 42.0\%$$

The second-law distribution of pertinent quantities of the cycle is:

Net work output	210.4 kJ/kg	42.0%	$(=\eta_2)$
Compressor irreversibility	26.9 kJ/kg	5.4%	
Turbine irreversibility	16.0 kJ/kg	3.2%	
Energy rejection irreversibility	247.9 kJ/kg	49.4%	
Availability input	501.3 kJ/kg	100.0%	

It is seen that a likely candidate for improvement is the energy rejection heat exchange. To recoup some of the losses in this heat exchanger, a regenerator is used to preheat the air entering the combustion chamber. The use of regenerators in Brayton cycles will be discussed later in Section 10.4.

In an open-loop, steady-flow cycle, the working fluid which was ingested at state 1 would be discharged at state 4. The availability of 247.9 kJ/kg would be lost directly to the environment instead of the energy rejection heat exchanger.

Figure 10.46 gives values of cycle efficiency for several different combinations of compressor and turbine efficiencies and a compressor pressure ratio of 5.

From the above example and discussion, one can see that Brayton cycle performance and efficiency are very must dependent on compressor and turbine efficiencies. In fact, the utilization of Brayton cycle engines to any great extent did not occur until the 1940s and 1950s with the jet engine for aircraft, even though the cycle had been proposed and some

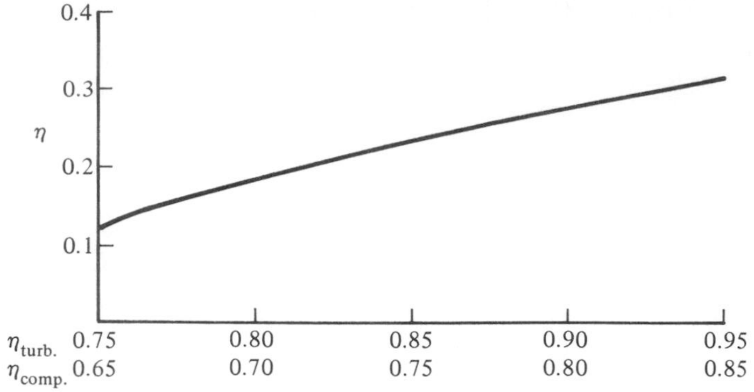

FIGURE 10.46 *Effect of Component Efficiencies on Brayton Cycle Performance (Pressure Ratio = 5).*

preliminary versions tried in the 1800s. The reasons for the long delay were the difficulties in finding materials for the turbine blades that would withstand the high temperatures and the problems in developing compressors and turbines with adequate efficiencies.

Let us now compare the Rankine and Brayton cycles as power sources. The working fluid on the Brayton cycle is a gas; the Rankine cycle involves a phase change. In the Rankine cycle, the pump handles pure liquid, so that the pump work is usually only a small fraction of the turbine work output. However, the compressor of the Brayton cycle handles a gas; the compressor work is an appreciable fraction of the turbine work output. This factor tends to give the Rankine cycle a higher thermal efficiency. However, the Brayton cycle has several practical advantages for certain applications. For electric power plants, it is desirable in peak loading situations to have a source of power that can be put on line immediately. Likewise, if there were an emergency plant shutdown of a conventional Rankine power plant, it would be desirable to have an auxiliary source of power to restart the plant. With a Rankine cycle, once the plant has been shut down, a long period of time is required to reheat the water in the boiler and get the plant running again. With the Brayton cycle, almost immediate startup is available. Another benefit of the Brayton power plant for stationary applications is that no cooling water is required. For this reason, Brayton power plants are seeing increased application for stationary power plants, especially to ease peaking electrical loads as might occur in midsummer with widespread use of air conditioning.

Another advantage of the Brayton cycle is that it has a better power-to-weight ratio than does the Rankine. Internal pressures in the Rankine cycle are high, over 1000 psia. Such pressure requires heavy,

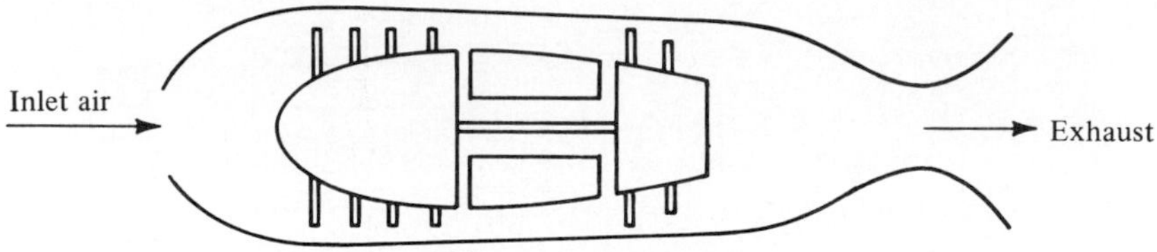

FIGURE 10.47 *Schematic of Jet Engine.*

thick-walled piping. The Brayton cycle involves no such high pressures. Thus, the Brayton or gas turbine cycle is widely used as the jet engine power plant for aircraft.

Again, the Rankine cycle currently has advantages over the Brayton cycle in basic cycle thermal efficiency, and it is therefore still the cycle used for the great majority of stationary electric power plants.

The turbojet engine, operating on the Brayton cycle, is used as the source of propulsive power for aircraft. With this engine, a forward thrust is provided by the acceleration of the air stream passing through the engine. As shown in Figure 10.47, ambient air is taken in through the engine inlet or diffuser. In the usual case, the engine or plane is flying into still air. In this analysis, we will take all velocities relative to the engine so that the inlet velocity will be equivalent to the forward velocity of the engine or plane. In the diffuser, the flow is slowed down, with a resultant increase in pressure and temperature of the air. The pressure and temperature are increased still further in the compressor. In this application, unlike the stationary gas turbine power plant, all the turbine work output is used to drive the compressor. After leaving the com-

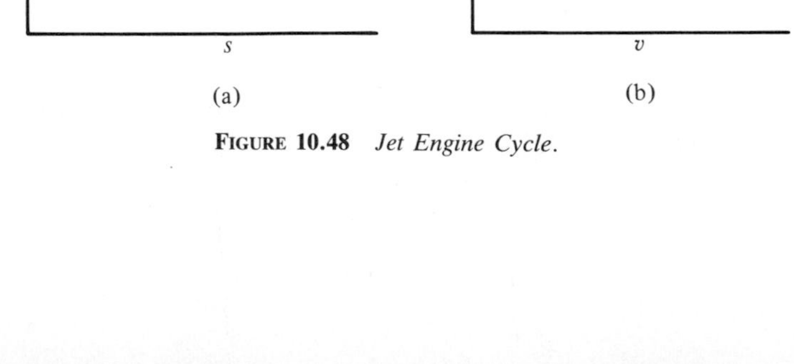

FIGURE 10.48 *Jet Engine Cycle.*

(a)

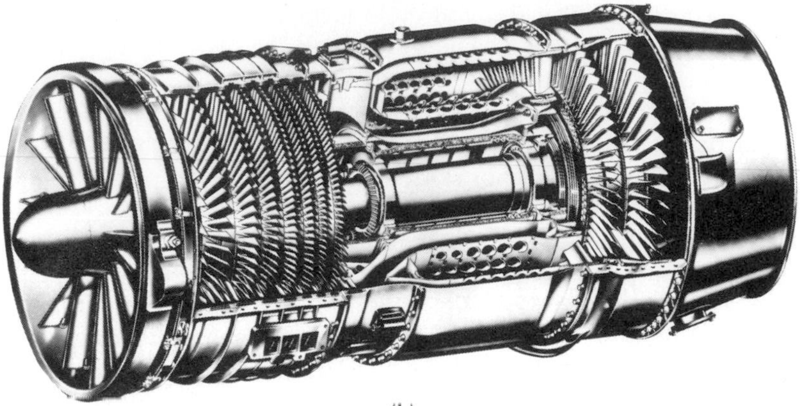

(b)

(c)

FIGURE 10.49 *Turbojet Engines. (a) Cutaway View of JT3C-7 Turbojet Engine. (Courtesy of Pratt & Whitney Aircraft Group, East Hartford, Conn.) (b) Cutaway View of CJ-610 Turbojet Engine. (Courtesy of General Electric Co.) (c) Components of CJ-610 Turbojet Engine. (Courtesy of General Electric Co.).*

pressor, the high-pressure air flow enters the combustion chamber, where fuel is added and the resultant mixture burned. The hot products of combustion are expanded through the turbine, doing work. Finally the combustion gases are expanded further through a nozzle to high velocity. For the ideal case in which the nozzle expands the gases perfectly to local ambient pressure, so that there are no unbalanced pressure forces acting on the flow, the forward thrust due to acceleration of the flow from inlet to exhaust is given by

$$\text{Thrust} = \frac{\dot{m}}{g_c}(V_{\text{exhaust}} - V_{\text{inlet}}) \qquad \textbf{(10.11)}$$

The ideal jet engine cycle, with isentropic compressor and turbine and no pressure drops in the combustion chamber or between components, is shown on a *T-s* diagram in Figure 10.48(a). A cutaway view of an actual turbojet engine is shown in Figure 10.49.

EXAMPLE 10.16

A turbojet engine flies at a steady forward speed of 150 m/s at an altitude where the atmospheric pressure is 96 kPa and the atmospheric temperature is 10°C. The compressor pressure ratio is 7:1, and the turbine inlet temperature is 800°C. Assume reversible processes within the turbojet engine and a constant specific heat c_p of 1.00 kJ/kg K. Neglect the effect of fuel flow rate. Find the pressures and temperatures at various locations in the jet engine, and determine the engine exhaust velocity and the thrust per unit mass flow.

Solution

From the energy equation (3.18), between stations 0 and 1 (see Figure 10.47), we get with $q_{\text{in}} = 0$ and no shaft work ($w'_{\text{out}} = 0$),

$$\frac{1}{2g_c}(V_1^2 - V_0^2) = h_0 - h_1$$

or, for constant specific heat c_p,

$$\frac{1}{2g_c}(V_1^2 - V_0^2) = c_p(T_0 - T_1)$$

Since V_1 is very small in comparison to V_0, we have

$$T_1 = T_0 + \frac{V_0^2}{2g_c c_p} = 10°\text{C} + \frac{150^2\ \text{m}^2/\text{s}^2}{2(1\ \text{kg}\cdot\text{m/N}\cdot\text{s}^2)(1000\ \text{J/kg K})}$$

$$= 10°\text{C} + 11.25°\text{C} = 21.25°\text{C}$$

For adiabatic flow in the inlet, which acts to slow down the incoming air flow, we have from Equation (8.13),

$$\frac{p_1}{p_0} = \left(\frac{T_1}{T_0}\right)^{\gamma/(\gamma-1)} = \left(\frac{21.25 + 273.15}{10 + 273.15}\right)^{3.50} = 1.146$$

or

$$p_1 = (96 \text{ kN/m}^2)(1.146) = 110.03 \text{ kPa}$$

The pressure ratio in the compressor is 7, so that $p_2 = 7p_1 = 7(110.03) = 770.2$ kPa. For the isentropic compression process 1–2, we have

$$\frac{T_2}{T_1} = \left(\frac{p_2}{p_1}\right)^{(\gamma-1)/\gamma} = 7^{0.4/1.4} = 7^{0.286} = 1.744$$

so that $T_2 = 1.744(294.40) = 513.33$ K $= 240.18$°C. The energy equation in the combustion chamber is

$$\begin{aligned} q_{\text{in}} &= h_3 - h_2 = c_p(T_3 - T_2) \\ &= (1.0 \text{ kJ/kg K})(800°\text{C} - 240.18°\text{C}) = 559.82 \text{ kJ/kg} \end{aligned}$$

The work output of the turbine is $w'_{\text{out}} = h_3 - h_4$, while the work input to the compressor (which in a jet engine equals the work output of the turbine) is $w'_{\text{in}} = h_2 - h_1$. Hence

$$c_p(T_2 - T_1) = c_p(T_3 - T_4)$$

so that $T_4 = T_3 - T_2 + T_1 = 800 - 240.18 + 21.25 = 581.07$°C.

For the expansion process of the turbine 3–4, we get

$$\frac{p_4}{p_3} = \left(\frac{T_4}{T_3}\right)^{\gamma/(\gamma-1)} = \left(\frac{581.07 + 273.15}{800 + 273.15}\right)^{3.50} = 0.450$$

so that with $p_3 = p_2$,

$$p_4 = 0.450(770.2) = 346.6 \text{ kPa}$$

For the expansion process in the exhaust nozzle 4–5, we get, using Equation (8.13),

$$\frac{T_5}{T_4} = \left(\frac{p_5}{p_4}\right)^{(\gamma-1)/\gamma} = \left(\frac{96}{346.6}\right)^{0.286} = 0.693$$

so that $T_5 = 0.693(581.07 + 273.15) = 591.95$ K $= 318.8$°C. The energy equation for the nozzle is, with V_4 very small relative to V_5,

$$\frac{V_5^2}{2g_c} = c_p(T_4 - T_5) = (1.0 \text{ kJ/kg K})(581.07°\text{C} - 218.80°\text{C}) = 262.27 \text{ kJ/kg}$$

Solving for V_5, we have

$$V_5 = \sqrt{2(1 \text{ kg} \cdot \text{m/N} \cdot \text{s}^2)1000 \text{ J/kJ}(262.27 \text{ kJ/kg})} = 724.3 \text{ m/s}$$

Note that the exhaust velocity V_5 is in excess of the inlet velocity of 150 m/s. The energy added to the cycle during process 2–3 is used to increase the kinetic energy of the flow through the jet engine.

From Equation (10.11), we obtain the thrust per unit mass flow:

$$\frac{\text{Thrust}}{\dot{m}} = \frac{V_{\text{exhaust}} - V_{\text{inlet}}}{g_c} = \frac{(724.3 - 150) \text{ m/s}}{1 \text{ kg} \cdot \text{m/N} \cdot \text{s}^2} = 574.3 \text{ N/kg/s}$$

To this point, with both the stationary power plant and the jet engine applications of the Brayton cycle, we have considered only cycles involving the basic components: compressor, combustion chamber, and turbine. As with the Rankine power cycle, however, it is possible to modify the Brayton cycle to increase its thermal efficiency.

From Example 10.13, one can see that the temperature of the gases at the turbine exit is 530.7°C (803.8 K). These hot gases are discharged from the engine and are therefore not utilized in the cycle and represent losses. It is desirable to try to recover some of this energy. Such heat recovery can be accomplished with the use of a heat exchanger, or *regenerator*, as shown in Figure 10.50. In the heat exchanger, hot gases from the turbine exhaust are utilized to heat the cooler air inflow before it enters the combustion chamber. Thus, less heat needs to be supplied in the combustion chamber to reach a certain maximum turbine inlet temperature. The regenerator can thus be used very effectively to increase cycle efficiency for stationary power plants and for automotive applications. However, the additional ducting and space and weight requirements prevent its use for the aircraft jet engine application. The improvement in cycle efficiency due to the use of the regenerator clearly depends on how much heat can be transferred between the two streams. The effectiveness e of a heat exchanger or regenerator is given by the ratio between the actual heat transfer and the maximum possible heat transfer. From Figure 10.50, the maximum possible value of T_3 is equal to T_5. Therefore,

$$e = \frac{q_{\text{actual}}}{q_{\text{max.}}} = \frac{c'_p(T_3 - T_2)}{c''_p(T_5 - T_2)}$$

where c'_p is the mean specific heat between T_2 and T_3, and c''_p is the mean specific heat between T_2 and T_5. For constant and equal specific heats,

$$e = \frac{T_3 - T_2}{T_5 - T_2} \tag{10.12}$$

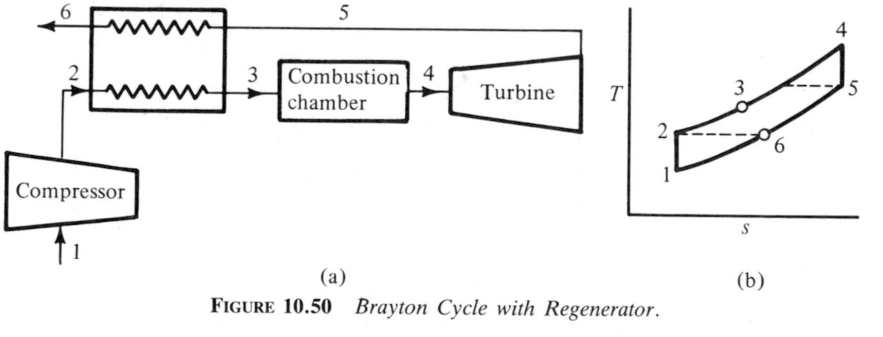

FIGURE 10.50 *Brayton Cycle with Regenerator.*

EXAMPLE 10.17

Perform a first- and second-law analysis on the Brayton gas power cycle of Example 10.14 (Figure 10.44) if a reheat regenerator having an effectiveness of 0.8 is added to the system (see Figure 10.50).

Solution

From Equation (10.12), for constant specific heats, regenerator effectiveness $e = (T_3 - T_2)/(T_5 - T_2) = 0.8$. From an energy balance, again with constant specific heats, $T_3 - T_2 = T_5 - T_6$ or $T_3 = T_2 + T_5 - T_6$.

From Example 10.14, $T_5 = 850.7$ K and $T_2 = 507$ K.

Solving, $T_6 = 576$ K and $T_3 = 782$ K.

Values of the properties of the working medium were calculated as described in Example 10.15, with the reference conditions again chosen as $T_0 =$ 293 K (20°C) and $p_0 = 1$ atm.

Working medium: air
Reference conditions: $T_0 = 293$ K (20°C), $p_0 = 1$ atm (101 kPa), $h_0 = 0$ kJ/kg, $s_0 = 0$ kJ/kg K

State or component	p (atm)	T (K)	h (kJ/kg)	s (kJ/kg K)	a (kJ/kg)
1	1	293	0	0	0
Compressor					
2	5	507	216.1	0.0919	189.2
Regenerator					
3	5	782	493.8	0.5295	338.7
Combustor					
4	5	1273	989.8	1.0217	690.4
Turbine					
5	1	851	563.3	1.0765	247.9
Regenerator					
6	1	576	285.6	0.6822	85.7
Energy rejection					
1	1	293	0	0	0

State or component		q_{in} (kJ/kg)	w'_{out} (kJ/kg)	Δa (kJ/kg)	i (kJ/kg)
1					
Compressor		0	−216.1	189.2	26.9
2					
Regenerator		277.7	0	149.5	−149.5
3					
Combustor		496.0		351.7	
4					
Turbine		0	426.5	−442.6	16.1
5					
Regenerator		−277.7	0	−162.2	162.2
6					
Energy rejection		−285.6	0	−85.7	85.7
1					
	Net	210.4	210.4	0	141.4

From Equation (6.1) we get the thermal cycle efficiency

$$\eta = \frac{\text{net } w'_{\text{out}}}{q_{\text{in}}} = \frac{210.4}{496.0} = 42.4\%$$

The first law distribution of pertinent quantities in the cycle is:

Work output	210.4 kJ/kg	42.4%	$(=\eta)$
Energy rejection	285.6	57.6%	
Energy input	496.0 kJ/kg	100.0%	

From Equation (10.6) we get the second-law efficiency of the cycle

$$\eta_2 = \frac{\text{net } w'_{\text{out}}}{a_{\text{in}}} = \frac{210.4}{351.7} = 59.7\%$$

The second-law distribution of the pertinent quantities in the cycle is:

Net work output	210.4 kJ/kg	59.8%	$(=\eta_2)$
Compressor irreversibility	26.9 kJ/kg	7.6%	
Turbine irreversibility	16.1 kJ/kg	4.6%	
Regenerator irreversibility (162.2 – 149.5)	12.7 kJ/kg	3.6%	
Energy rejection irreversibility	85.7 kJ/kg	24.4%	
Availability input	351.8 kJ/kg	100.0%	

From a first-law analysis we find that the thermal cycle efficiency increased from 27.2 to 42.4 percent due to the utilization of a regenerator in the Brayton cycle. Similarly, the second-law analysis indicates an increase in efficiency from 42.0 to 59.8 percent for the regenerator chosen. The second-law analysis also indicates the possibility of further increase in system performance by use of a more effective regenerator.

We will now calculate the thermal efficiency of the Brayton cycle with a regenerator. We will assume that the compressor and turbine are isentropic, that there is no pressure drop in the combustion chamber, and that the working fluid is air which behaves as an ideal gas with constant specific heats. Referring to the sketch and *T-s* diagram of Figure 10.50, we have, for the energy input to the cycle,

$$q_{\text{in}} = h_4 - h_3 = c_p(T_4 - T_3)$$

and, for work output of the turbine,

$$w'_{\text{out(turb)}} = h_4 - h_5 = c_p(T_4 - T_5)$$

The work input to the compressor is

$$w'_{\text{in(comp)}} = h_2 - h_1 = c_p(T_2 - T_1)$$

Hence, the cycle efficiency becomes

$$\eta = \frac{\text{net } w'_{\text{out}}}{q_{\text{in}}} = \frac{T_4 - T_5 - (T_2 - T_1)}{T_4 - T_3}$$

From Equation (10.12), we have $T_3 = e(T_5 - T_2) + T_2$, so that

$$\eta = \frac{T_4 - T_5 - (T_2 - T_1)}{T_4 - eT_5 + (e-1)T_2} = \frac{1 - T_5/T_4 - (T_2/T_1)(T_1/T_4) + T_1/T_4}{1 - eT_5/T_4 + (e-1)(T_2/T_1)(T_1/T_4)}$$

With $T_2/T_1 = (p_2/p_1)^{(\gamma-1)/\gamma}$ and $T_5/T_4 = (p_2/p_1)^{(1-\gamma)/\gamma}$, we obtain, after simplification, the efficiency of a Brayton power cycle with regeneration:

$$\eta = \frac{[(p_2/p_1)^{(1-\gamma)/\gamma} - T_1/T_4][(p_2/p_1)^{(\gamma-1)/\gamma} - 1]}{1 - e(p_2/p_1)^{(1-\gamma)/\gamma} + (e-1)(T_1/T_4)(p_2/p_1)^{(\gamma-1)/\gamma}} \qquad \textbf{(10.13)}$$

For $e = 0$ (no regenerator), Equation (10.13) reduces to Equation (10.10), namely, $\eta = 1 - (p_2/p_1)^{(1-\gamma)/\gamma}$, while for $e = 1$ (ideal, 100 percent effective regenerator), Equation (10.13) simplifies to $\eta = 1 - (T_1/T_4)(p_2/p_1)^{(\gamma-1)/\gamma}$.

From Equation (10.13), the effect of incorporating regenerators with various effectiveness values ($e = 1$, 0.8, 0.6 and 0) into the Brayton cycle was calculated and is shown in Figure 10.51 for a maximum cycle temperature ratio $T_4/T_1 = 4$ and $\gamma = 1.4$. Note that the use of a regenerator provides a significant improvement in Brayton cycle efficiency, especially at low compressor pressure ratios. This is particularly important for automobile applications. Whereas a stationary power plant may be able to operate near an optimum performance point on a continual basis, the automobile must accelerate, decelerate, cruise, and idle, and it operates at or near an optimum performance or efficiency point only a small portion of the time. Thus the use of a regenerator greatly benefits the overall fuel economy of a gas turbine-powered car. A cutaway view of an automotive gas turbine with a ceramic regenerator is shown in Figure 10.52(a): external views are shown in Figure 10.52(b) and 10.52(c).

2. *Brayton Refrigeration Cycles*

Like the Rankine cycle, the Brayton cycle can be operated "in reverse" to produce cooling. Although less efficient than cooling units based on the reversed Rankine cycle, the reversed Brayton cycle offers the advantage of lower weight of the unit. Hence, this cycle is of interest in applications where the weight of the cooling unit must be kept to a minimum, such as aircraft cabin cooling.

A schematic arrangement of a typical refrigeration system utilizing the *reversed Brayton (gas)* cycle is shown in Figure 10.53. The components of the system are as follows:

1. a *compressor* which raises the pressure of the working fluid from its lowest to its highest value;
2. an *energy rejection heat exchanger*, in which the high temperature of the working fluid is lowered;
3. an *expander engine*, in which the pressure and temperature of the working fluid are reduced (the work output of the engine can be used to supply part of the work input required for the compressor); and
4. an *energy input heat exchanger* which raises the temperature of the working fluid at constant pressure.

The four processes involved in an ideal reversed Brayton cycle are illustrated in Figure 10.54. For these reversible processes, for steady flow

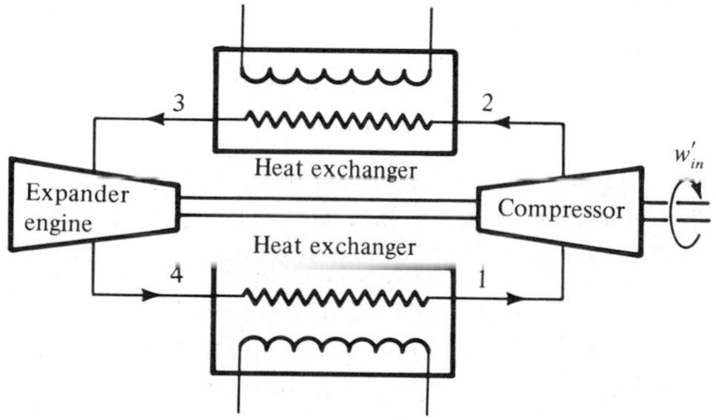

FIGURE 10.53 *Reversed Brayton Cycle Components.*

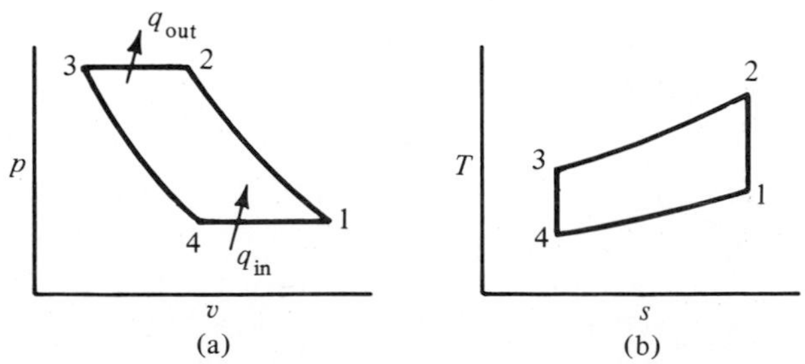

FIGURE 10.54 *Ideal Reversed Brayton Cycle.*

in a closed loop cycle, we have the following:

(1–2) Isentropic compression process:

$$q_{in} = 0 \qquad s_2 = s_1$$
$$w'_{in} = h_2 - h_1$$

(2–3) Constant pressure process with energy rejection:

$$w'_{out} = 0$$
$$q_{out} = h_2 - h_3$$

(3–4) Isentropic expansion process:

$$q_{in} = 0 \qquad s_3 = s_4$$
$$w'_{out} = h_3 - h_4$$

(4–1) Constant pressure process with energy input:

$$w'_{out} = 0$$
$$q_{in} = h_1 - h_4$$

The net work input to the cycle is

$$\text{net } w'_{in} = w'_{in(1-2)} - w'_{out} = h_2 - h_1 - (h_3 - h_4)$$

The coefficient of performance as a refrigerator is, from Equation (6.3),

$$\text{COP}_{\text{refrig}} = \frac{q_{in(4-1)}}{\text{net } w'_{in}} = \frac{h_1 - h_4}{h_2 - h_1 - (h_3 - h_4)} \qquad \textbf{(10.14)}$$

EXAMPLE 10.18

A closed loop refrigeration unit uses nitrogen as the working fluid and operates on a reversed Brayton cycle (Figure 10.53). The inlet conditions to the compressor are $p_1 = 200$ kPa and $T_1 = 30°\text{C}$. There is a fivefold increase in pressure across the compressor, and the inlet temperature to the heat exchanger T_4 is 10°C. Determine the coefficient of performance and the cycle work input for ideal performance of the compressor and expander engine. Assume that $c_p = 1.04$ kJ/kg K and $\gamma = 1.40$.

Solution

In the absence of pressure drops in the heat exchangers, we have $p_2 = p_3$ and $p_4 = p_1$. For the isentropic compression process 1–2, we get

$$T_2 = T_1(p_2/p_1)^{(\gamma-1)/\gamma} = (30 + 273.15)(5)^{0.286} = 303.15(1.584)$$
$$= 480.13 \text{ K} = 206.98°\text{C}$$

The work input to the compressor is

$$w'_{in} = h_2 - h_1 = c_p(T_2 - T_1) = (1.04 \text{ kJ/kg K})(206.98°\text{C} - 30°\text{C})$$
$$= 184.06 \text{ kJ/kg}$$

The isentropic expansion in the expander engine gives

$$T_3 = T_4(p_3/p_4)^{(\gamma-1)/\gamma} = (10 + 273.15) \text{ K}(5)^{0.286} = 283.15(1.584)$$
$$= 448.5 \text{ K} = 175.3°\text{C}$$

The work output of the engine is

$$w'_{out} = h_3 - h_4 = c_p(T_3 - T_4) = (1.04 \text{ kJ/kg K})(175.3°\text{C} - 10°\text{C})$$
$$= 171.91 \text{ kJ/kg}$$

The energy input to the cycle is

$$q_{in} = h_1 - h_4 = c_p(T_1 - T_4) = (1.04 \text{ kJ/kg K})(30°\text{C} - 10°\text{C})$$
$$= 20.80 \text{ kJ/kg}$$

The net work input to the cycle is

$$\text{net } w'_{in} = 184.06 - 171.91 = 12.15 \text{ kJ/kg}$$

The coefficient of performance as a refrigerator is, from Equation (10.14),

$$\text{COP}_{refrig} = \frac{q_{in}}{\text{net } w'_{in}} = \frac{h_1 - h_4}{h_2 - h_1 - (h_3 - h_4)} = \frac{20.80}{12.15} = 1.713$$

The ratio of energy rejection and energy input is

$$\frac{q_{out}}{q_{in}} = \frac{h_2 - h_3}{h_1 - h_4} = \frac{c_p(T_2 - T_3)}{c_p(T_1 - T_4)} = \frac{1.04(206.98 - 175.3)}{20.80}$$
$$= \frac{32.94}{20.80} = 1.584$$

The COP of the corresponding Carnot cycle is, with $T_H = T_3 = 175.3°\text{C}$ and $T_L = T_1 = 30°\text{C}$,

$$\text{COP}_{\text{Carnot refrig}} = \frac{T_L}{T_H - T_L} = \frac{30 + 273}{175.3 - 30} = 2.085 \text{ (vs. 1.713)}$$

EXAMPLE 10.19

Perform a second-law analysis on the reversed Brayton cycle described in Example 10.18, again assuming ideal performance of compressor and turbine.

Solution

Working medium: nitrogen
Reference conditions: $T_0 = 273$ K (0°C), $p_0 = 200$ kPa, $h_0 = 0$ kJ/kg,
$s_0 = 0$ kJ/kg K

State or component	p (kPa)	T (°C)	h (kJ/kg)	s (kJ/kg K)	a (kJ/kg)
1	200	30	31.2	0.1084	1.6
Compressor					
2	1000	207	215.3	0.1084	185.7
Energy rejection					
3	1000	175	182.3	0.0380	171.9
Expander					
4	200	10	10.4	0.0376	0.1
Energy input (cooling load)					
1	200	30	31.2	0.1084	1.6

State or component		q_{in} (kJ/kg)	w'_{out} (kJ/kg)	Δa (kJ/kg)	i (kJ/kg)
1					
Compressor		0	−184.1	184.1	0
2					
Energy rejection		−33.0	0	−13.8	13.8
3					
Expander		0	171.9	−171.8	0
4					
Energy input (cooling load)		20.8	0	1.5	
1					
	Net	−12.2	−12.2	0	13.8

Values of T, w'_{out}, and q_{in} were taken directly from Example 10.18. The values of enthalpy were calculated from $c_p(T - T_0) = 1.04$ kJ/kg K$(T - T_0)$. The values of entropy were calculated using Equation (6.25) with $R = 0.2971$ kJ/kg K, namely, $s - s_0 = c_p \ln(T/T_0) - R \ln(p/p_0)$.

Work input	12.2 kJ/kg	1.00	
Energy input	20.8 kJ/kg	1.71	(= COP)
Energy rejection	33.0 kJ/kg	2.71	

From Equation (10.7) we get the second-law efficiency of the cycle

$$\eta_2 = \frac{a_{in}}{\text{net } w'_{out}} = \frac{1.5}{12.2} = 0.123$$

The second-law distribution of pertinent quantities of the cycle is:

Work input		12.2 kJ/kg	(1.00)
Availability input		1.5 kJ/kg	($0.0123 = \eta_2$)
Irreversibilities:			
Compressor	0 kJ/kg		
Energy rejection	13.8 kJ/kg		
Expander	0 kJ/kg		
	13.8 kJ/kg	13.7 kJ/kg	(1.123)

From the second-law analysis, it is seen that the most likely candidate for improvement is the energy rejection process.

EXAMPLE 10.20

An air conditioning unit used in aircraft operates on a reversed Brayton cycle using air as working fluid. Since air is the refrigerant, it can be blown directly into the aircraft cabin without passing through a heat exchanger; that is, it operates as an open loop cycle [Figure 10.55(a)]. The following data are given:

$$p_1 = 14.7 \text{ psia} \qquad T_1 = 85°\text{F} \qquad p_2 = 100 \text{ psia}$$

$$T_4 = 50°\text{F} \qquad p_4 = 14.7 \text{ psia}$$

Assuming ideal performance, calculate the cycle work input and coefficient of performance,

a. assuming that $c_p = 0.24$ Btu/lbm°R and $\gamma = 1.40$; and
b. using thermodynamic properties from the tables for air in Appendix A.

Solution

a. Assuming constant specific heats:

The *T-s* diagram of this open loop is shown in Figure 10.55(b). In an ideal heat exchanger, there is no pressure drop; hence, $p_2 = p_3$. For the isentropic compression 1–2, we have

$$T_2 = T_1\left(\frac{100}{14.7}\right)^{0.286} = (85 + 460)°\text{R}(1.730) = 942.55°\text{R} = 482.55°\text{F}$$

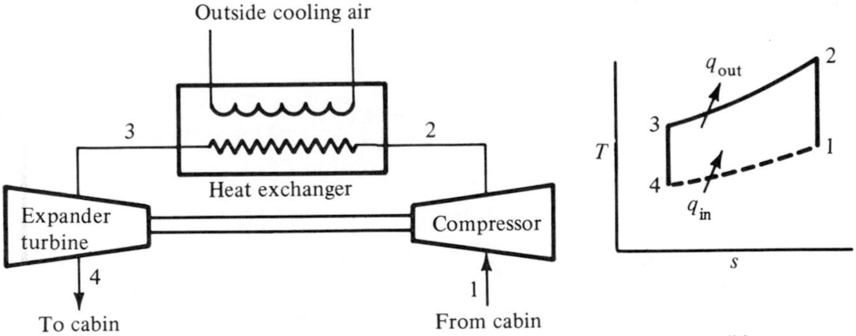

FIGURE 10.55 *Open-Loop Reversed Brayton Cycle.*

The work input to the compressor is

$$w'_{in} = h_2 - h_1 = c_p(T_2 - T_1) = (0.24 \text{ Btu/lbm°R})(482.55°\text{F} - 85°\text{F})$$
$$= 95.4 \text{ Btu/lbm}$$

For the isentropic expansion process 3–4, we get

$$T_3 = T_4\left(\frac{100}{14.7}\right)^{0.286} = (50 + 460)°\text{R}(1.730) = 882.0°\text{R} = 422.0°\text{F}$$

The work output of the engine is

$$w'_{out} = h_3 - h_4 = c_p(T_3 - T_4) = (0.24 \text{ Btu/lbm°R})(422.0°\text{F} - 50°\text{F})$$
$$= 89.3 \text{ Btu/lbm}$$

The energy input to the cycle is

$$q_{in} = h_1 - h_4 = c_p(T_1 - T_4) = (0.24 \text{ Btu/lbm°R})(85°\text{F} - 50°\text{F})$$
$$= 8.4 \text{ Btu/lbm}$$

The net work input to the cycle becomes

$$\text{net } w'_{in} = 95.4 - 89.3 = 6.1 \text{ Btu/lbm}$$

Note that in contrast to the vapor compression cycle, in this cycle, the expander engine work output is almost equal to compressor work input; hence expander engine work output is utilized in a Brayton refrigeration cycle. The coefficient of performance of the cycle is

$$\text{COP}_{\text{refrig}} = \frac{q_{in}}{\text{net } w'_{in}} = \frac{8.4}{6.1} = 1.38$$

The energy rejection of the cycle is

$$q_{out} = h_2 - h_3 = c_p(T_2 - T_3) = (0.24 \text{ Btu/lbm°R})(482.55°\text{F} - 422.0°\text{F})$$
$$= 14.53 \text{ Btu/lbm}$$

Hence

$$\frac{q_{out}}{q_{in}} = \frac{14.53}{8.4} = 1.73$$

The COP of an equivalent Carnot cycle would be, with $T_H = T_3 = 422°\text{F}$ and $T_L = T_1 = 85°\text{F}$,

$$\text{COP}_{\text{Carnot}} = \frac{T_L}{T_H - T_L} = \frac{85 + 460}{422 - 85} = 1.62 \text{ (vs. 1.38)}$$

b. Using air tables:
We obtain the following from Table A.13, at a pressure of 14.7 psia:

	0°F	100°F	
s	0.9357	0.9831	Btu/lbm°R
h	170.03	194.06	Btu/lbm

Hence, by linear interpolation, we have

$$s_1 = 0.9357 + 0.85(0.9831 - 0.9357) = 0.9759 \text{ Btu/lbm°R}$$

$$h_1 = 170.03 + 0.85(194.06 - 170.03) = 190.46 \text{ Btu/lbm}$$

We get the following from Table A.13, at a pressure of 100 psia:

	400°F	500°F
s	0.9552	0.9823 Btu/lbm°R
h	266.59	291.28 Btu/lbm

In the isentropic compression process 1–2, we have $s_2 = s_1$. Hence

$$T_2 = 400 + \frac{0.9759 - 0.9552}{0.9823 - 0.9552} 100 = 476.38\text{°F}$$

$$h_2 = 266.59 + 0.7638(291.28 - 266.59) = 285.45 \text{ Btu/lbm}$$

The work input to the compressor is

$$w'_{\text{in}} = h_2 - h_1 = 285.45 - 190.46 = 94.99 \text{ Btu/lbm}$$

Now $T_4 = 50$°F, and hence

$$h_4 = 170.03 + 0.5(194.06 - 170.03) = 182.05 \text{ Btu/lbm}$$

$$s_4 = 0.9357 + 0.5(0.9831 - 0.9357) = 0.9594 \text{ Btu/lbm°R}$$

In the isentropic expansion process 3–4, we have $s_4 = s_3$. Hence

$$T_3 = 400 + \frac{0.9594 - 0.9552}{0.9823 - 0.9552} 100 = 415.50\text{°F}$$

$$h_3 = 266.59 + 0.155(291.28 - 266.59) = 270.42 \text{ Btu/lbm}$$

The work output of the engine is

$$w'_{\text{out}} = h_3 - h_4 = 270.42 - 182.05 = 88.37 \text{ Btu/lbm}$$

The energy input to the cycle is

$$q_{\text{in}} = h_1 - h_4 = 190.46 - 182.05 = 8.41 \text{ Btu/lbm}$$

The coefficient of performance of the cycle as a refrigerator is

$$\text{COP}_{\text{refrig}} = \frac{8.41}{94.99 - 88.37} = \frac{8.41}{6.62} = 1.27$$

The ratio of energy input and rejection is

$$\frac{q_{\text{out}}}{q_{\text{in}}} = \frac{h_2 - h_3}{h_1 - h_4} = \frac{285.45 - 270.42}{8.41} = \frac{15.03}{8.41} = 1.79$$

Note the slightly lower value of COP obtained when values of thermodynamic properties from the air tables are used instead of constant specific heats.

Figure 10.56 shows a typical lightweight air conditioning unit for business aircraft and helicopter cabin cooling, based on the reversed Brayton cycle. As indicated in this schematic diagram, the refrigerant (air) is compressed in the compressor section of the main aircraft engine; that is, instead of a separate compressor being used, hot, pressurized bleed air is extracted from the aircraft engine. Heat rejection from the refrigerant takes place in the heat exchanger, which is cooled by flow of outside ambient air. The flow is induced by a fan driven by the expander turbine. The exhaust from the expander turbine provides cold air for distribution to the cabin. The weight of the entire air conditioning unit, which has a cooling capacity of one ton of refrigeration, is only 27 pounds.

Other lightweight air conditioning units have been built for applications where the engine bleed capability is limited. The schematic of such a system is illustrated in Figure 10.53(a), with the compressor and turbine on the same shaft. The cabin air is recirculated with fresh makeup air supplied by small bleed-off from the main engine.

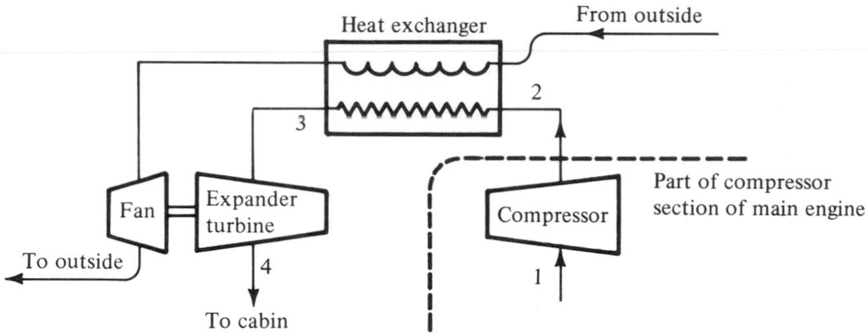

FIGURE 10.56 *Lightweight Aircraft and Helicopter Air Conditioning Unit.*

It is seen from the three preceding examples that gas compression refrigeration (reversed Brayton) cycles achieve lower coefficients of performance than comparable vapor compression refrigeration (reversed Rankine) cycles. However, the light weight and compactness of reversed Brayton cycle equipment offset its inherently low efficiency. Hence, such equipment is utilized in applications such as aircraft or mobile, temporary installations. Recent development of lightweight vapor refrigeration equipment using high-speed compressors, however, has decreased the use of air compression refrigeration equipment in aircraft systems.

10.5 POWER CYCLES: THE OTTO CYCLE

The engine used to power virtually all the automobiles in the world today is the reciprocating, spark-ignited, internal combustion engine. In this type of engine, combustion takes place inside a piston-cylinder arrangement, with the products of combustion used to drive the piston and produce power. The engine has a high power output per unit weight, is durable, and gives good performance over a wide range of operating conditions with a reasonably high thermal efficiency. It is therefore ideal for transportation applications.

The spark-ignited internal combustion engine operates according to the Otto cycle. On this cycle, a fuel-air mixture, prepared, for example, in a carburetor, is aspirated into the cylinder during the downstroke of the piston (Figure 10.57). During this process, with the intake valve open, the pressure in the cylinder is constant, with the cylinder volume available to the mixture increased (process 0–1 in Figure 10.58). At the bottom of the piston travel, called the *bottom dead center position* (BDC), the intake valve is closed. The piston is then pushed upward in the cylinder by the crankshaft (process 1–2), with the fuel-air mixture compressed, until the piston reaches the top of its travel, called the *top dead center position* (TDC). A spark is fired with the piston at TDC, causing ignition and combustion of the fuel-air mixture (process 2–3). Ideally, the combustion of the mixture and the resultant heat release occur instantaneously, raising the temperature and pressure of the combustion gases to T_3 and p_3. The gases at high pressure and temperature then push the piston down during the power stroke 3–4, performing work on the crankshaft. When the piston reaches BDC, the exhaust valve is opened, reducing cylinder pressure to the initial pressure, with a corresponding decrease of temperature (process 4–1). Finally, to complete the cycle, with the exhaust valve open, the piston is pushed upward (process 1–0), clearing the cylinder of the combustion gases. Note that for each cycle, there are four strokes of the piston; as described, this cycle is commonly referred to as the *four-stroke cycle*.

In order to analyze this open loop cycle thermodynamically, we will approximate it by a closed system cycle, in which the actual combustion process is replaced by an equivalent heat input process, and the heat rejection process is assumed to take place at constant specific volume [Figure 10.59(a)]. Note that in the cycle depicted in Figure 10.58, the heat rejection process, although at constant volume, proceeds with diminishing mass of the combustion gases in the cylinder. In the approximate closed system cycle, the system mass, consisting solely of air, remains constant during all processes of the cycle.

For comparison, the equivalent Carnot power cycle has also been

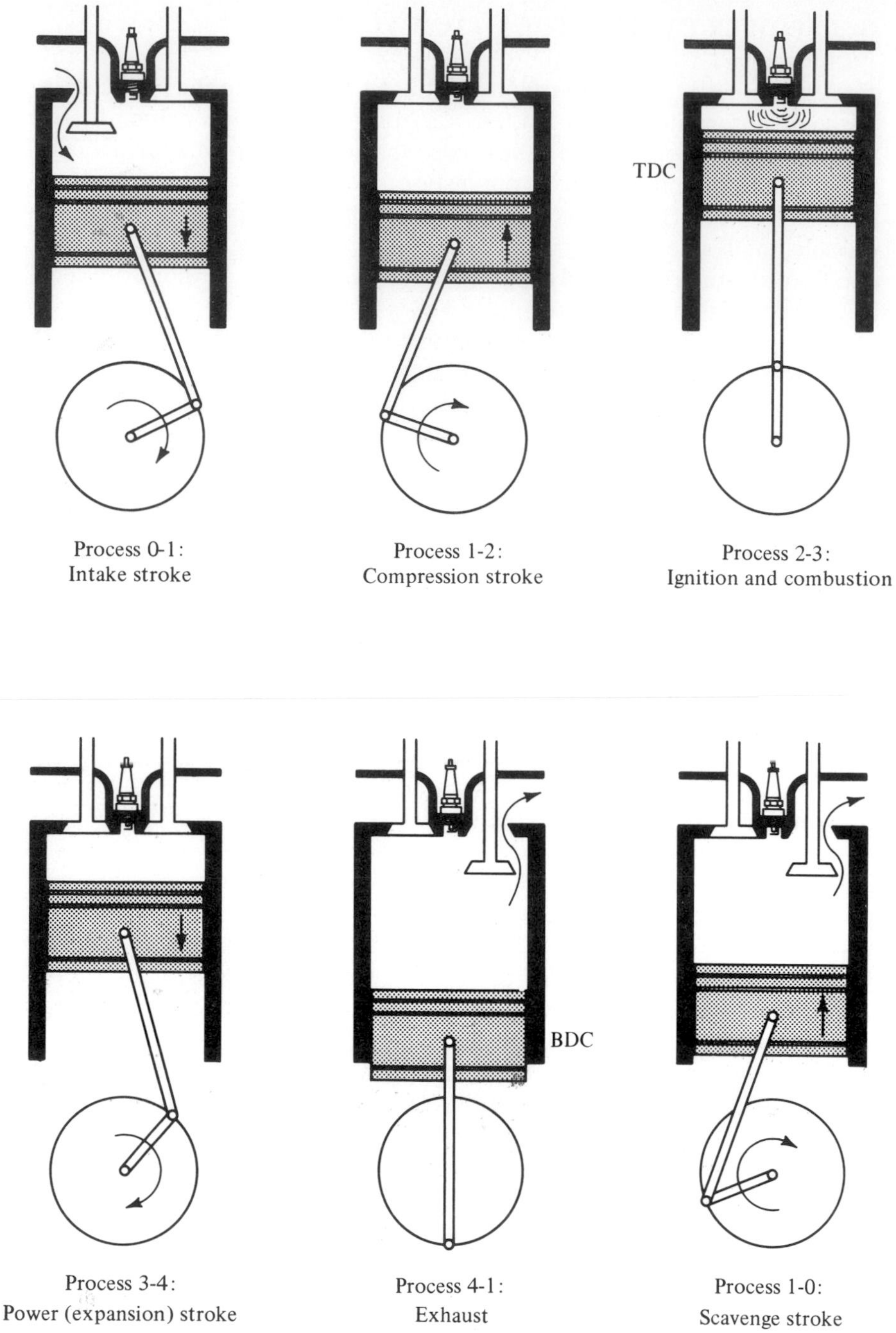

FIGURE 10.57 *Ideal Four-Stroke Otto Engine.*

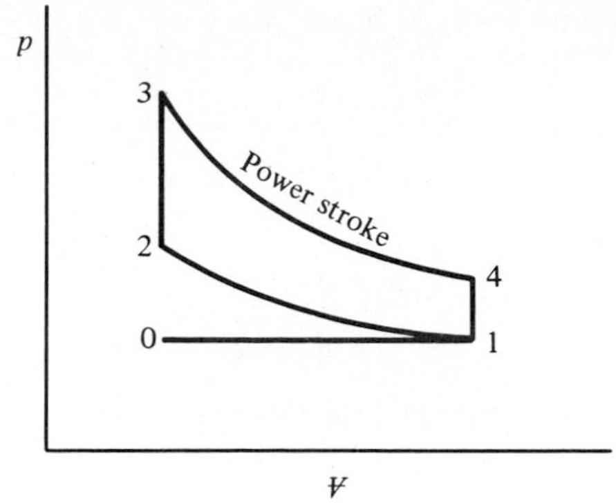

FIGURE 10.58 *p-V Diagram of Ideal Otto Cycle.*

included in Figure 10.59. Note that the Carnot cycle reaches appreciably higher values of specific volume, resulting in lower values of net work output per specific volume change. Further, the maximum pressure in the Carnot cycle reaches much higher values than in the comparable Otto cycle.

For the four processes of the closed system, ideal Otto cycle of Figure 10.59, we obtain the following from the energy equation:

(1–2) Isentropic compression process:

$$q_{in} = 0 \qquad s_2 = s_1$$

$$w_{in} = u_2 - u_1$$

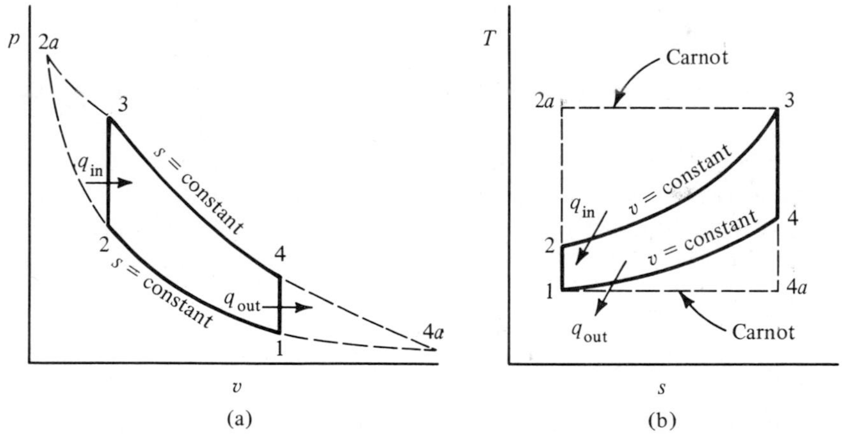

FIGURE 10.59 *p-v and T-s Diagrams of Closed System, Ideal Otto Cycle.*

(2–3) Constant volume (energy input) process:

$$w_{out} = 0$$

$$q_{in} = u_3 - u_2$$

(3–4) Isentropic expansion (power stroke) process:

$$q_{in} = 0$$

$$w_{out} = u_3 - u_4$$

(4–1) Constant volume (energy rejection) process:

$$q_{out} = u_4 - u_1$$

$$w_{out} = 0$$

The net work output of the cycle thus becomes

$$\text{net } w_{out} = w_{out(3-4)} - w_{in(1-2)} = u_3 - u_4 - (u_2 - u_1)$$

The thermal efficiency of the cycle is, from Equation (6.2),

$$\eta = 1 - \frac{q_{out(4-1)}}{q_{in(2-3)}} = 1 - \frac{u_4 - u_1}{u_3 - u_2} \tag{10.15}$$

If we further assume the working fluid to be an ideal gas with constant specific heats (that is, an air-standard cycle), we obtain, for the isentropic compression stroke (1–2), $T_2/T_1 = (\mathcal{V}_1/\mathcal{V}_2)^{\gamma-1}$ and, for the isentropic expansion (power) stroke (3–4), $T_3/T_4 - (\mathcal{V}_4/\mathcal{V}_3)^{\gamma-1}$. Since the piston moves between the same two positions in its strokes, $\mathcal{V}_1 = \mathcal{V}_4$ and $\mathcal{V}_2 = \mathcal{V}_3$. Therefore, $T_2/T_1 = T_3/T_4$ or $T_4 = T_3T_1/T_2$. From Equation (10.15), the thermal efficiency of the air-standard Otto cycle thus becomes

$$\eta = 1 - \frac{c_v(T_4 - T_1)}{c_v(T_3 - T_2)} = 1 - \frac{T_1(T_3/T_2 - 1)}{T_2(T_3/T_2 - 1)} = 1 - \frac{T_1}{T_2}$$

or

$$\eta = 1 - (\mathcal{V}_1/\mathcal{V}_2)^{1-\gamma} \tag{10.16}$$

The ratio between the volume available to the gases at BDC and the volume available at TDC is called the *compression ratio*, $\overline{CR}$. Further, clearance volume C is defined as the volume at TDC, and displacement volume D is the volume swept out by the piston in moving from BDC to TDC. Therefore,

$$\overline{CR} = \frac{D + C}{C} = \frac{\mathcal{V}_1 - \mathcal{V}_2 + \mathcal{V}_2}{\mathcal{V}_2} = \frac{\mathcal{V}_1}{\mathcal{V}_2}$$

Thermal efficiency for the ideal Otto cycle is plotted against compression ratio in Figure 10.60 for $\gamma = 1.4$, $\gamma = 1.35$, and $\gamma = 1.3$. Thermal

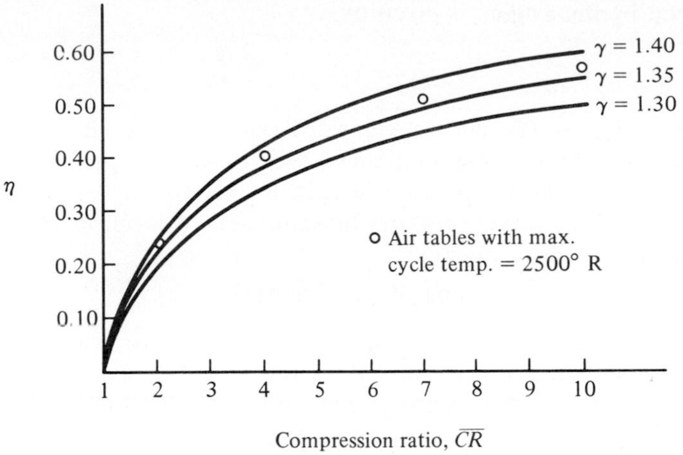

Figure 10.60 *Thermal Efficiency vs. Compression Ratio for Otto Cycle.*

efficiencies calculated by use of air tables (Table A.13) for a maximum cycle temperature of 2500°R have also been included. This figure shows the advantages of operating at as high a compression ratio as possible. However, the p-V diagram of Figure 10.58 also shows that an increase of compression ratio also increases maximum cycle pressure and temperature. At high temperatures, several phenomena may occur in the engine. First, the temperature of the fuel-air mixture may be high enough so that combustion is initiated before the spark is fired. Or, even after the spark is fired and the flame is initiated in the mixture, the temperature and pressure in the gases ahead of the flame may be high enough to cause spontaneous ignition of these gases. In either case, autoignition leads to a rate of pressure increase in the cylinder much greater than usual; the audible sound accompanying this sudden pressure rise is called *knock*. Resultant vibrations of the engine can cause severe damage, so knock is to be avoided. Thus, the maximum available compression ratio that can be used with an Otto cycle engine is limited.

The ability of fuels to resist knock is measured by *octane number*. The addition of tetraethyl lead to gasoline increased the octane numbers of gasoline to over 100, so that compression ratios of up to 10 were used in automobile engines in the 1960s. The use of lead-free fuel in the 1970s, and consequent reduction of octane numbers to the low 90s and high 80s, has necessitated a reduction of compression ratio to around 8.0. This reduction has resulted in decrease in thermal efficiency and a worsening of the fuel economy of the engine.

For each complete cycle of the four-stroke engine just described, there are four strokes and hence two crankshaft revolutions. The power developed by the engine is given by

$$\text{Power} = \text{net}\,(W_{\text{out}})\,\frac{N}{2} \tag{10.17}$$

where net W_{out} is the net engine work output per cycle and N is the crankshaft revolutions per unit time. Mean effective pressure, mep, is defined as that constant pressure which, if exerted on the piston during one power stroke, would produce the same net work output as the actual cycle. In other words,

$$\text{net}\ W_{\text{out}} = (\text{mep})D \tag{10.18}$$

where D is the total engine displacement (which is displacement per cylinder times the number of cylinders). Combining, we obtain the following expression for engine power output of the four-stroke engine:

$$\text{Power} = (\text{mep})\,\frac{N}{2}\,D \tag{10.19}$$

It becomes clear that mep is a measure of engine power; that is, high power outputs are associated with large meps. High maximum cylinder pressure required thicker cylinder walls and, hence, a heavier engine. In an automotive application, where weight is an important factor affecting vehicle acceleration and peformance, tradeoffs must be effected between power output and weight.

The actual performance of a reciprocating, spark-ignited engine differs in many respects from that of the ideal Otto engine just described. Certainly, the compression and power strokes are not isentropic; friction is present between piston and cylinder, and there is heat loss through the cylinder walls to the engine cooling system. In fact, the heat loss from the engine to the cooling fluid, and thence via the radiator to ambient air, may be equivalent to the energy lost to the exhaust. Further, instantaneous combustion at TDC cannot occur. A finite time is required for the flame initiated by the spark to propagate through the entire fuel-air mixture. For this reason, the spark must be advanced to fire before the piston reaches TDC, to ensure complete combustion of the mixture. The pressure in the cylinder at the beginning of the compression stroke is less than atmospheric pressure, since there are pressure drops as the flow crosses the intake valves and is fed into the cylinders. The pressure during the scavenging stroke 1–0 of Figure 10.58 is greater than atmospheric pressure, again because of pressure drops as the flow is pushed out of the cylinders. A p-$\forall$ diagram for the actual engine is shown in Figure 10.61.

The derivation of Otto cycle efficiency was based on the assumption that the working fluid is air for the entire cycle. In actuality, the gaseous products of combustion have properties different from those of air. For example, at the very high cylinder pressures and temperatures involved,

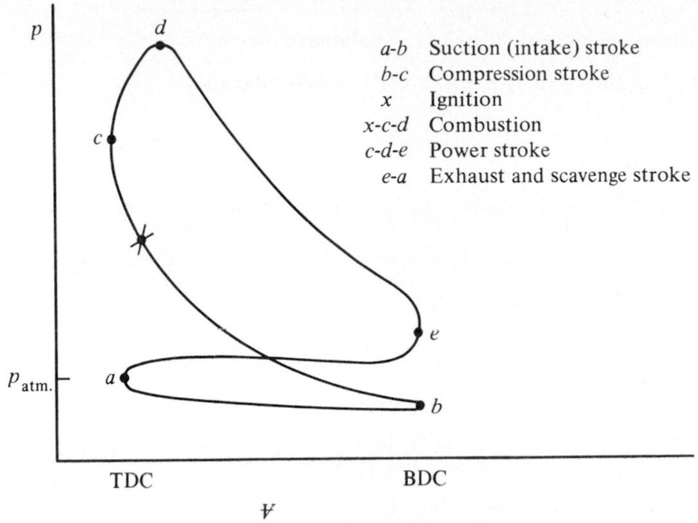

FIGURE 10.61 *Diagram of Actual Otto Cycle.*

some of the energy released during combustion goes into dissociation, or splitting apart of the gaseous molecules. The energy is therefore not available for useful work. When imperfect gas effects, heat losses, friction, and so on, are taken into account, it follows that actual thermal efficiencies of Otto cycle engines are somewhat less than those of Figure 10.60. Maximum actual Otto engine thermal efficiencies are currently about 30 percent.

EXAMPLE 10.21

A four-stroke engine operating on the Otto cycle has a compression ratio of 8.0. The pressure and temperature in the cylinder at the beginning of the compression stroke are 14.7 psia and 100°F, respectively. The ratio of the mass of air to mass of fuel in the cylinders is 16 to 1, with a heat release due to combustion of 20,000 Btu per pound of fuel. Assume the working fluid to have the properties of air, with $R = 53.3$ ft-lbf/lbm°R, $c_v = 0.171$ Btu/lbm°R = constant and $\gamma = 1.4$ = constant. For this engine operating on the ideal Otto cycle, determine the following:

a. cycle thermal efficiency;
b. maximum cylinder pressure and temperature;
c. mean effective pressure (mep); and
d. power output at 2500 rpm for an engine displacement of 350 in^3.

Solution

a. From Equation (10.16), the cycle thermal efficiency is

$$\eta = 1 - (\overline{CR})^{1-\gamma} = 1 - 8^{-0.4} = 0.565$$

b. As shown in Figure 10.59, the compression stroke 1–2 is isentropic, so that

$$p_2/p_1 = (V_1/V_2)^{\gamma} = 8^{1.4} = 18.38 \quad \text{and} \quad p_2 = 270 \text{ psia}$$

From the ideal gas law

$$\frac{T_2}{T_1} = \frac{p_2 V_2}{p_1 V_1} = 18.38/8 = 2.30 \quad \text{and} \quad T_2 = 560°\text{R}(2.30) = 1287°\text{R}$$

For process 2–3, $q_{in} = c_v(T_3 - T_2)$, where $q_{in} = 20{,}000/16 = 1250$ Btu/lbm of air. We will neglect the mass of fuel in comparison to the air mass; hence, we assume constant mass in the cylinder throughout the cycle. Thus, $T_3 - T_2 =$ (1250 Btu/lbm)/0.171 Btu/lbm°R = 7310°R or $T_3 = 8597°\text{R}$ = maximum temperature:

$$\text{Maximum pressure} = p_3 = p_2 \frac{T_3}{T_2} = 270\left(\frac{8597}{1287}\right) = 1804 \text{ psia}$$

For the isentropic expansion stroke, we get

$$\frac{T_4}{T_3} = \left(\frac{V_3}{V_4}\right)^{\gamma-1} = \left(\frac{V_2}{V_1}\right)^{\gamma-1}$$

Hence, $T_4 = 8597(1/8)^{\gamma-1} = 3742°\text{R}$. From the ideal gas law,

$$p_4 = p_3 \frac{T_4 V_3}{T_3 V_4} = p_3 \frac{T_4 V_2}{T_3 V_1} = 1804\left(\frac{3742}{8597}\right)\frac{1}{8} = 98.2 \text{ psia}$$

c. From Equation (10.18),

$$\text{mep} = \frac{\text{net } W_{out}}{D} = \frac{\text{net } w_{out}}{v_1 - v_2}$$

Now

$$v_1 = \frac{RT_1}{p_1} = \frac{(53.3 \text{ ft-lbf/lbm°R})(560°\text{R})}{(14.7 \text{ lbf/in}^2)(144 \text{ in}^2/\text{ft}^2)} = 14.10 \text{ ft}^3/\text{lbm}$$

and $$v_2 = \frac{14.10}{8} = 1.76 \text{ ft}^3/\text{lbm}$$

Using Equation (6.1), we obtain

$$\text{net } w_{out} = \eta q_{in} = 0.565(1250) = 706 \text{ Btu/lbm}$$

$$\text{mep} = \frac{(706 \text{ Btu/lbm})(778 \text{ ft-lbf/Btu})}{(14.10 - 1.76) \text{ ft}^3/\text{lbm}(144 \text{ in}^2/\text{ft}^2)} = 309 \text{ psi}$$

d. From Equation (10.19), the power output of the four-stroke engine is

$$\text{Power} = (\text{mep})D(N/2) = \frac{(309\ \text{lbf/in}^2)(350\ \text{in}^3)2500\ \text{min}^{-1}}{(12\ \text{in/ft})(60\ \text{s/min})^2}$$

$$= 187{,}760\ \text{ft-lbf/s}$$

$$= 341\ \text{hp}$$

The efficiency of an equivalent Carnot engine is

$$\eta_{\text{Carnot}} = 1 - \frac{T_H}{T_L} = 1 - \frac{560}{8597} = 0.935\ (\text{vs. } 0.565)$$

EXAMPLE 10.22

Perform a second law analysis of the engine operating on an ideal Otto cycle as described in Example 10.21.

Solution

Working medium: air
Reference conditions: $T_0 = 520°\text{R}$ (60°F), $p_0 = 14.7$ psia, $u_0 = 0$ Btu/lbm, $s_0 = 0$ Btu/lbm°R, $v_0 = 13.093$ ft³/lbm

State or process	p (psia)	T (°R)	v (ft³/lbm)	u (Btu/lbm)	s (Btu/lbm°R)	a (Btu/lbm)
1	14.7	560	14.10	6.8	0.01775	0.4
Compression						
2	270	1287	1.764	131.2	0.01764	91.2
Energy input						
3	1804	8597	1.764	1381.2	0.34236	1172.3
Expansion						
4	98.2	3742	14.10	551.0	0.34255	375.6
Energy rejection						
1	14.7	560	14.10	6.8	0.01775	0.4

State or process	q_{in} (Btu/lbm)	w_{out} (Btu/lbm)	Δa (Btu/lbm)	i (Btu/lbm)
1				
Compression	0	−124.4	90.8	0
2				
Energy input	1250.0	0	1081.1	
3				
Expansion	0	830.2	−796.7	0
4				
Energy rejection	−544.2	0	−375.2	375.2
1				
Net	705.8	705.8	0	375.2

Values of T, v, q_{in}, and w_{out} were taken directly from Example 10.21. The specific volume at the reference conditions is

$$v_0 = RT/p_0 = \frac{(53.3 \text{ ft-lbf/lbm°R})(520\text{°R})}{(144 \text{ in.}^2/\text{ft}^2)(14.7 \text{ lbf/in.}^2)} = 13.093 \text{ ft}^3/\text{lbm}$$

Values of the internal energy u were calculated from $c_v(T - T_0) = 0.171$ Btu/lbm°R$(T - 520)$. c_p was determined from

$$c_p = c_v + R = 0.171 \text{ Btu/lbm°R} + \frac{53.3 \text{ ft-lbf/lbm°R}}{778 \text{ ft-lbf/Btu}} = 0.2395 \text{ Btu/lbm}$$

Values of entropy were calculated using (6.25). The availability at each stage in the closed Otto cycle was calculated using Equation (7.7), namely,

$$a = u - u_0 + p_0(v - v_0) - T_0(s - s_0)$$

where $p_0 = 14.7(144)$ lbf/ft^2(ft^3/lbm) $= 2.721$ Btu/lbm. Reference temperature T_0 was chosen as 520°R (60°F), which is a typical air intake temperature for engines operating on the Otto cycle. p_0 was taken as 14.7 psia. At the reference condition, u_0 and s_0 were taken as zero, with $v_0 = 13.093$ ft^3/lbm. From the first law for a closed system such as the Otto cycle, we have from (2.5) for the compression process (1–2) $w_{out} = u_1 - u_2 = 6.8$ Btu/lbm $- 131.2$ Btu/lbm $= -124.4$ Btu/lbm. From the second law for a closed system, we get from (10.4) for the compression process, $w_{out} = a_1 - a_2 - i + p_0(v_2 - v_1) = -90.8 - 0 + 2.721(1.746 - 14.10) = -90.8 - 33.6 = -124.5$Btu/lbm. Similarly, for the expansion process (3–4), we have from Equation (2.5) $w_{out} = u_3 - u_4 = 1381.2 - 551.0 = 830.2$ Btu/lbm, while from (10.4) we get $w_{out} = a_3 - a_4 - i + p_0(v_4 - v_1) = 796.7 - 0 + 2.721(14.10 - 1.764) = 796.7 + 33.5 = 830.2$ Btu/lbm.

As mentioned in previous second-law analysis examples, we have omitted the value of irreversibility for the availability input process (2–3) of the cycle.

The first-law distribution of energy in the cycle is:

Work output	705.8 Btu/lbm	56.5%	$(= \eta)$
Energy rejection	544.2	43.5%	
Energy input	1250.0 Btu/lbm	100.0%	

From Equation (10.6) we have the second-law efficiency for a closed cycle such as the Otto cycle

$$\eta_2 = \frac{n_{ct} w_{out}}{a_{in}} = \frac{705.8}{1081.1} = 65.3\%$$

The second-law distribution of pertinent quantities is:

Net work output	705.8 Btu/lbm	65.3%	$(= \eta_2)$
Compression irreversibility	0	0	
Expansion irreversibility	0	0	
Energy rejection irreversibility	375.2	34.7%	
Availability input	1081.0 Btu/lbm	100.0%	

Since we assumed ideal compression, expansion, and energy input in this example, the energy rejection process appears as the likely candidate for improvement of the cycle.

Since, in the Otto engine, a spark ignites a combustible mixture of air and fuel, a fairly constant ratio of air and fuel (about 15 to 1) must be maintained for all load conditions. Thus, the load and speed of an Otto engine are controlled by the use of a valve (throttle) which restricts the quantity of the fuel-air mixture taken into the engine. During throttling, lower pressures at points 1, 2, 3, and 4 of the cycle will result. However, in an ideal cycle, the net work output per unit mass of working fluid and the thermal efficiency remain constant. A cutaway view of an automobile engine based on the Otto cycle is shown in Figure 10.62(a); external views are shown in Figures 10.62(b) and 10.62(c).

With the four-stroke cycle, there is one power stroke for each two crankshaft revolutions. The spark-ignited Otto engine can also operate on a two-stroke cycle, in which there is one power stroke for each crankshaft revolution. As shown in Figure 10.63, toward the end of the compression stroke and at TDC, fresh fuel-air mixture is admitted to the crankcase through port I. Near the end of the power stroke, the exhaust port E is uncovered by the piston, and the products of combustion are allowed to flow out of the cylinder. At the very end of the power stroke, and the beginning of the compression stroke, another port O is uncovered, allowing fresh fuel-air mixture to flow into the cylinder from the crankcase and further scavenge the exhaust gases. The pressure necessary to force the fresh mixture into the cylinder may be supplied either by the compression of crankcase gases during the power stroke or by an external blower. The combining of exhaust and intake strokes with the power and compression strokes makes it possible to achieve the Otto cycle in two piston strokes.

The advantage of the two-stroke Otto cycle is that it provides twice as many power strokes as the four-stroke cycle per cylinder per crankshaft revolution. However, as shown in Figures 10.63(c) and (d), the processes of intake and scavenging are not as efficient as with a four-stroke cycle. It is possible to lose some of the fresh fuel-air mixture out of the exhaust prior to combustion, and it is also possible for an appreciable fraction of the burned gases to remain in the cylinder. For these reasons, the actual power output of the two-stroke cycle, while greater than that of the four-stroke cycle, is certainly not twice as great, as might be predicted from the number of power strokes per revolution. Also, the poor scavenging and intake efficiency of the two-stroke engine lead to a worsening in fuel economy, compared to a four-stroke engine. Finally, with the crankcase of the two-stroke engine used for compressing the

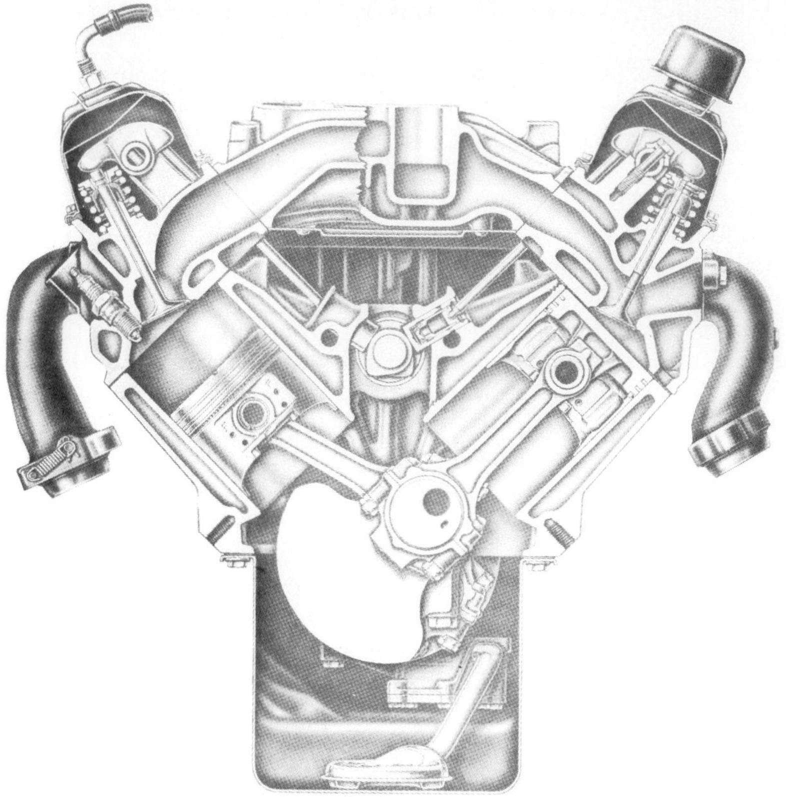

(a) Cross-sectional view

(b) Exterior view

FIGURE 10.62 *Chrysler 318 CID (Cubic Inch Displacement) Automobile Engine. (Art and Photos Courtesy of Chrysler Corporation, Detroit, Michigan.)*

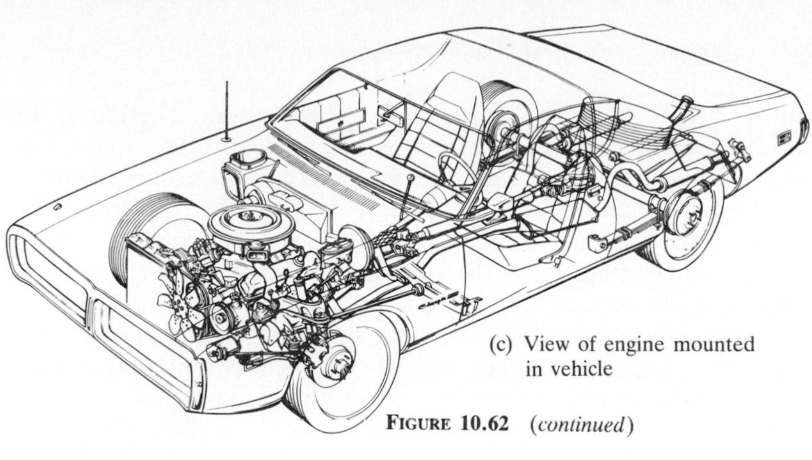

(c) View of engine mounted in vehicle

FIGURE 10.62 (*continued*)

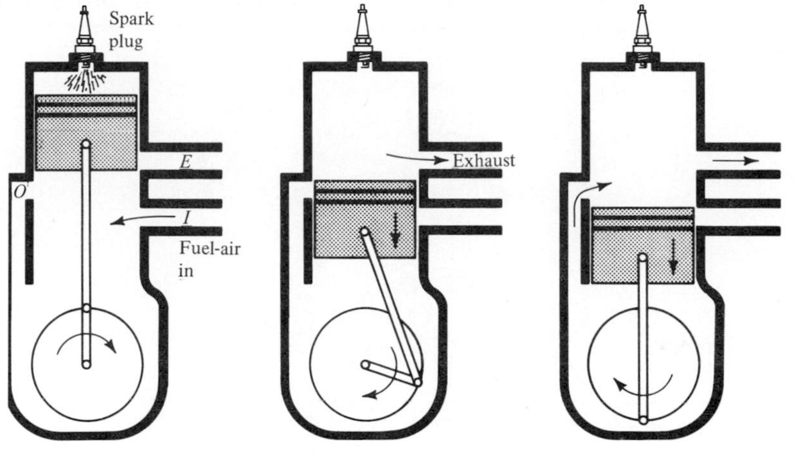

TDC
(a) Ignition and combustion

(b) Power and exhaust

BDC
(c) Intake, exhaust, and scavenge

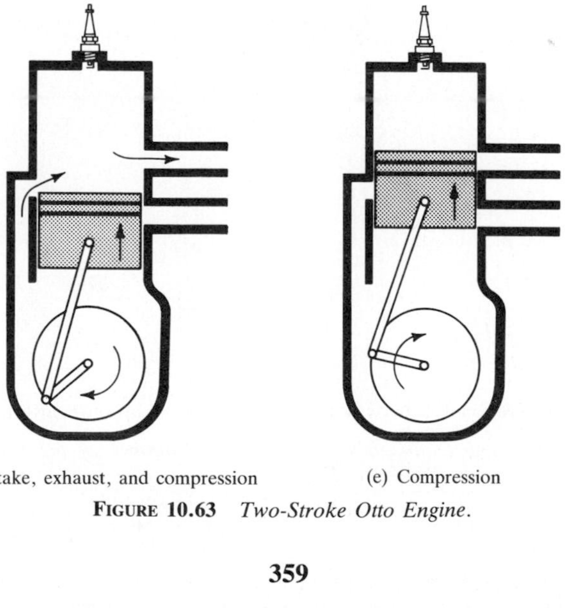

(d) Intake, exhaust, and compression

(e) Compression

FIGURE 10.63 *Two-Stroke Otto Engine.*

incoming charge, it is not available for lubrication. Therefore, the two-stroke engine cannot be lubricated as easily as the four-stroke engine; oil must be mixed with the fuel in the two-stroke engine to achieve adequate lubrication. For all these reasons, the two-stroke Otto engine sees only a limited application, in which fuel economy is not a primary factor. Examples are the outboard motor boat (Figure 10.64) and chain-saw engines.

To this point, we have discussed reciprocating, spark-ignited, internal combustion engines. Another type of spark-ignited, internal combustion engine that operates on the Otto cycle is the *rotary* engine, perfected in the 1950s. In place of the reciprocating action of the piston, the rotary engine substitutes the rotary motion of a triangular-shaped rotor inside a

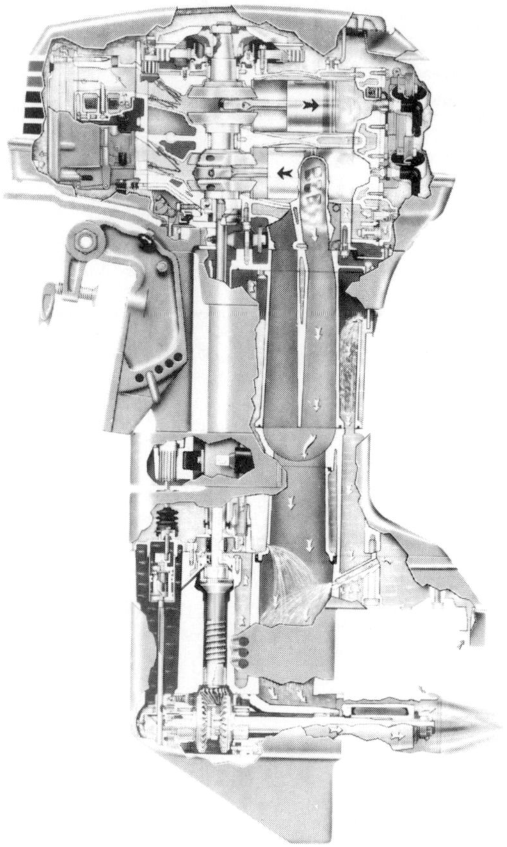

(a) Cross-sectional view of two-cycle Evinrude 50-hp outboard motor

FIGURE 10.64 *Two-Cycle Outboard Motor.* (*Photos Courtesy of the Outboard Marine Corporation, Waukegan, Illinois.*)

(b) Charging process for two-cycle engine

FIGURE 10.64 (*continued*)

housing to compress and expand the working fluid. As shown in Figure 10.65, the rotor divides the housing into three volumes. Let us follow volume A as it passes through a cycle. The air-fuel mixture is admitted to the engine in process 0–1. As the rotor turns, this volume is sealed off and compressed, corresponding to the compression stroke 1–2 of Figure 10.58. When the volume reaches a minimum (process 2–3, corresponding to TDC), the spark is fired and combustion takes place. The hot gases then expand and turn the rotor in the power stroke, 3–4. Finally, in process 4–1–0, the exhaust ports are uncovered to the volume, and the gases are exhausted from the engine. The p-$\mathcal{V}$ diagram is thus exactly the same as that of Figure 10.58. Note that there are three volumes of gas at various stages of the cycle at a given time; in other words, there are three power strokes per rotor revolution. The output shaft of the engine is geared to run at three times the rotor angular velocity, so that there is one power stroke for each output shaft revolution.

The rotary engine has a high power-to-weight ratio, with fewer parts

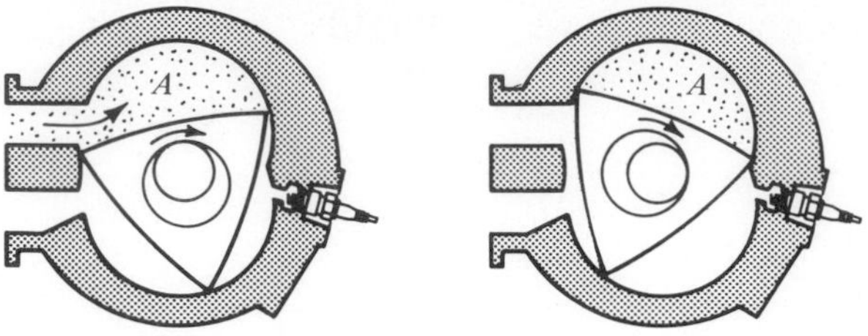

Process 0-1: Intake   Process 1-2: Compression

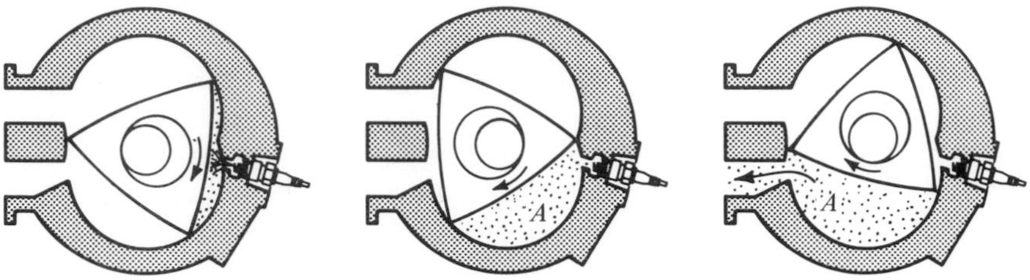

Process 2-3: Ignition and combustion   Process 3-4: Power   Process 4-1-0: Exhaust and scavenge

FIGURE 10.65 *Rotary Spark-Ignited Engine.*

than that of a conventional piston engine. For example, a six-cylinder piston engine, with twelve valves and accompanying hardware to control their motion, can be replaced by a two-rotor rotary engine with no valves (ports are used to control intake and exhaust). Further, the high inertial forces of the reciprocating piston and the accompanying noise and vibration are replaced by the smooth, quiet rotary motion of the engine. It would seem to be an ideal engine for automotive or light-duty aircraft applications. However, several disadvantages and problem areas have arisen. First, there is the problem of achieving a reliable, durable, moving seal between the three corners of the rotor and the housing. Leakage past the seals leads to a reduction of compression pressure and overall decrease in engine thermal efficiency. Perhaps the most basic problem, however, is the shape of the combustion chamber itself (2–3 in Figure 10.65). The long, thin chamber possesses a large surface-to-volume ratio, which causes a relatively large heat loss from the gas volume and thus a reduction in overall thermal efficiency. Further, the flame initiated at the spark plug is quenched, or extinguished, in the vicinity of the cooler wall. This region near the wall where the flame cannot propagate leads to the presence of unburned fuel, ultimately swept out into the engine exhaust.

Thus, the rotary engine with its high surface-to-volume ratio suffers from relatively high emissions of unburned fuel or, as it is generally termed, *unburned hydrocarbons*. Unburned hydrocarbons are contributors to the formation of photochemical smog.

The rotary engine, used as a power plant for automobiles, has been marketed for several years. However, the problems of reducing emissions to meet standards as well as achieving a competitive fuel economy have prevented widespread use of the engine to date. Unless these problems can be solved, the future of the rotary engine is in doubt.

10.6 POWER CYCLES: THE DIESEL CYCLE

In the Otto engine, a premixed charge of fuel and air is compressed in the compression stroke and then ignited with a spark plug near TDC. Limitations on compression ratio and cycle efficiency are imposed by the necessity of preventing preignition. With the Diesel engine, only air is compressed in the compression stroke, with fuel sprayed into the cylinders starting at the very end of the compression stroke. The high-pressure air is hot enough to cause the fuel to burn shortly after it is injected, so that no spark plug is necessary. Such an engine is called a *compression-ignition* engine.

Like the Otto engine, the Diesel can operate on either a two-stroke or a four-stroke cycle. The ideal four-stroke Diesel cycle engine is shown in Figure 10.66(a). The fuel is sprayed into the cylinder and burned while the piston is moving down in the cylinder, thereby achieving constant pressure combustion.

As we did for the Otto cycle, we will approximate the open loop Diesel cycle by a closed system cycle (Figure 10.67). For the four processes of the ideal closed system Diesel cycle of Figure 10.67, we obtain the following from the energy equation (2.5):

(1–2) Isentropic compression process:

$$q_{in} = 0 \qquad s_2 = s_1$$

$$w_{in} = u_2 - u_1$$

(2–3) Constant pressure (energy input) process:

$$q_{in} = u_3 - u_2 + p(v_3 - v_2)$$

$$= h_3 - h_2$$

$$w_{out} = \int_2^3 p\, dv = p(v_3 - v_2)$$

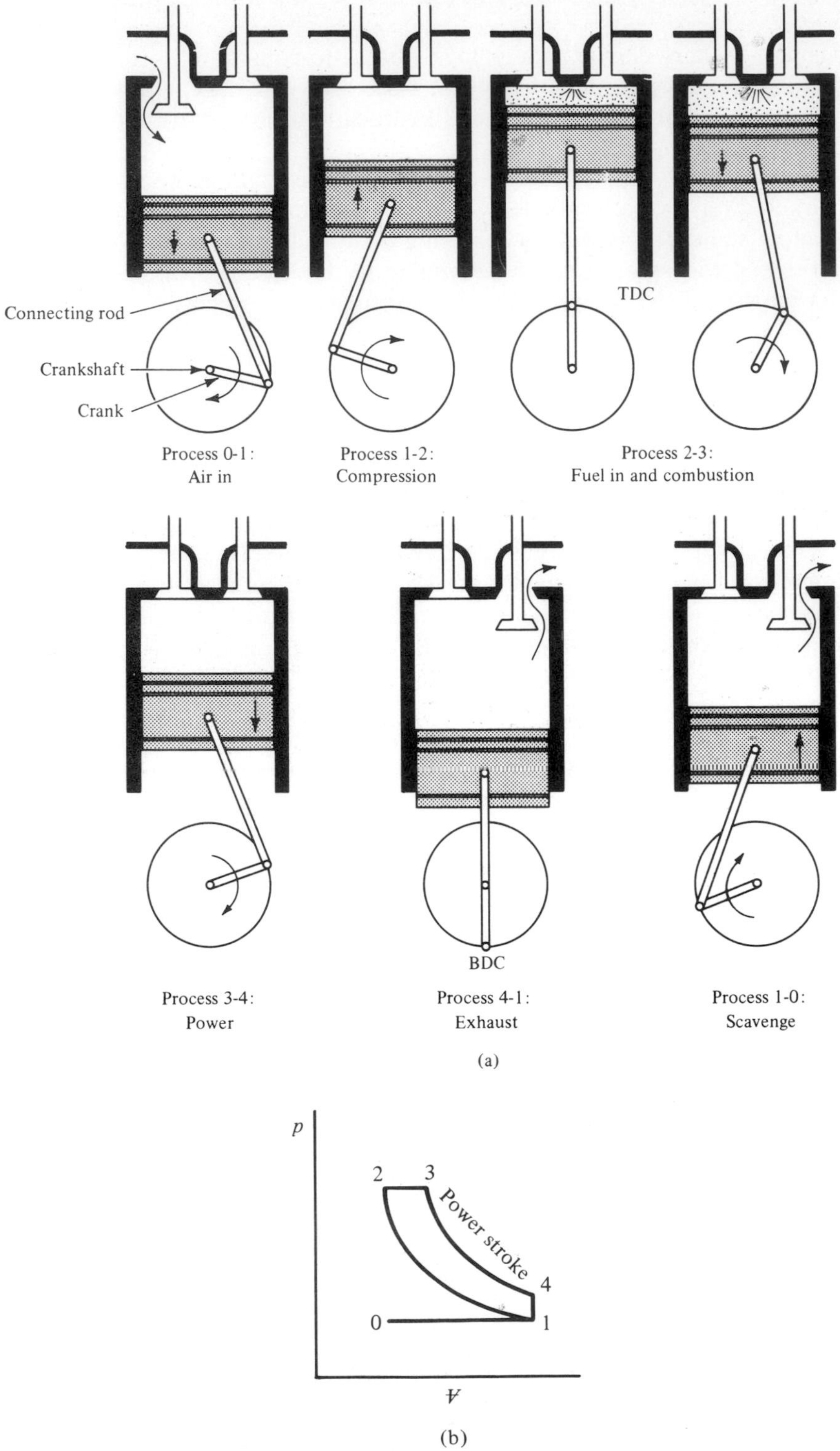

Figure 10.66 *Four-Stroke Diesel Engine.*

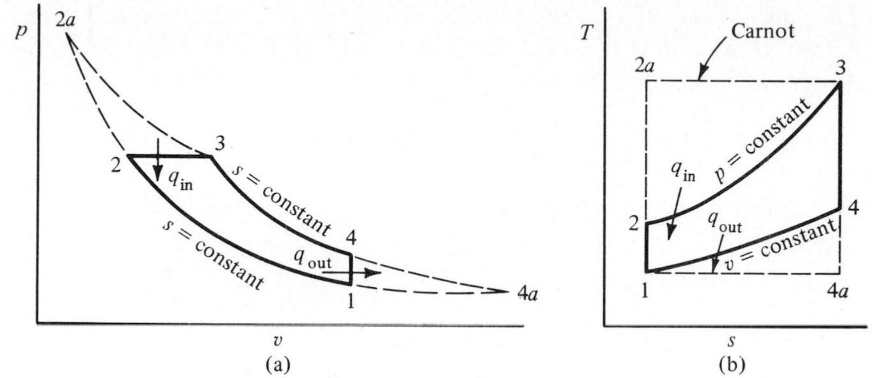

FIGURE 10.67 *p-v and T-s Diagrams of Ideal Diesel Cycle.*

(3–4) Isentropic expansion (power stroke) process:

$$q_{in} = 0 \qquad s_3 = s_4$$

$$w_{out} = u_3 - u_4$$

(4–1) Constant volume (energy rejection) process:

$$q_{out} = u_4 - u_1$$

$$w_{out} = 0$$

The net work output of the cycle is

$$\begin{aligned} \text{net } w_{out} &= w_{out(3-4)} - w_{in(1-2)} + w_{out(2-3)} \\ &= u_3 - u_4 - (u_2 - u_1) + p(v_3 - v_2) \\ &= h_3 - h_2 - (u_4 - u_1) \end{aligned}$$

The thermal efficiency of the cycle is, from (6.2):

$$\eta = 1 - \frac{q_{out(4-1)}}{q_{in(2-3)}} = 1 - \frac{u_4 - u_1}{h_3 - h_2} \qquad \textbf{(10.20)}$$

If we further assume the working fluid to be an ideal gas with constant specific heats, we obtain, for the isentropic compression,

$$\frac{T_2}{T_1} = \left(\frac{\mathcal{V}_1}{\mathcal{V}_2}\right)^{\gamma-1} = \overline{CR}^{\gamma-1}$$

where the ratio $\mathcal{V}_1/\mathcal{V}_2$, as with the Otto cycle, is called the *compression ratio*, $\overline{CR}$. For the constant pressure energy input process, the equation of state gives, with $p_2 = p_3$,

$$\frac{p_2 v_2}{p_3 v_3} = \frac{RT_2}{RT_3} \quad \text{or} \quad \frac{T_3}{T_2} = \frac{\mathcal{V}_3}{\mathcal{V}_2}$$

The ratio $\forall_3/\forall_2$ (that is, the ratio of volume after combustion to volume before combustion) is called the *fuel cutoff ratio*, $\overline{CO}$. The isentropic power stroke yields (since $\forall_4 = \forall_1$)

$$\frac{T_4}{T_3} = \left(\frac{\forall_3}{\forall_4}\right)^{\gamma-1} = \left(\frac{\forall_3}{\forall_1}\right)^{\gamma-1}$$

From Equation (10.20), the thermal efficiency becomes

$$\eta = 1 - \frac{c_v(T_4 - T_1)}{c_p(T_3 - T_2)} = 1 - \frac{T_1}{T_2}\frac{1}{\gamma}\frac{(T_4/T_1 - 1)}{(T_3/T_2 - 1)}$$

The temperature ratio T_4/T_1 can be written as

$$\frac{T_4}{T_1} = \frac{T_4 T_3 T_2}{T_3 T_2 T_1} = \left(\frac{\forall_3}{\forall_1}\right)^{\gamma-1}\left(\frac{\forall_3}{\forall_2}\right)\left(\frac{\forall_1}{\forall_2}\right)^{\gamma-1} = \left(\frac{\forall_3}{\forall_2}\right)^{\gamma}$$

Hence, the thermal efficiency of the air-standard Diesel cycle becomes

$$\eta = 1 - \left(\frac{\forall_1}{\forall_2}\right)^{1-\gamma}\left\{\frac{1}{\gamma}\frac{(\forall_3/\forall_2)^{\gamma} - 1}{\forall_3/\forall_2 - 1}\right\} \qquad \textbf{(10.21)}$$

or, written in terms of $\overline{CR}$ and $\overline{CO}$,

$$\eta = 1 - (\overline{CR})^{1-\gamma}\left\{\frac{1}{\gamma}\frac{(\overline{CO}^{\gamma} - 1)}{(\overline{CO} - 1)}\right\}$$

The expression for the air-standard Otto cycle is, from (10.16),

$$\eta = 1 - (\overline{CR})^{1-\gamma}$$

Since the term in braces { } in Equation (10.21) is always greater than 1, it follows that, for the same compression ratio, Otto cycle efficiency is greater than Diesel efficiency. However, the compression ratio of the Diesel engine is not limited by preignition, with air being compressed only on the compression stroke. In fact, high compression ratios are desirable to cause compression ignition. Compression ratios for present-day Diesel engines are in the range of 12 to 24, with the higher values used for passenger car applications. It follows that a Diesel engine with a compression ratio of 15 to 20 is more efficient than an Otto engine with a compression ratio limited to 8 or 9.

The cutoff ratio is a measure of the time during which fuel is injected; larger cutoff ratios are associated with more fuel injected and greater power. As shown in Equation (10.21), higher cutoff ratios mean lower thermal efficiencies. Figure 10.68 presents the thermal efficiency of a Diesel cycle as a function of compression ratio $\overline{CR}$ for several cutoff ratios $\overline{CO}$. The curve with $\overline{CO} = 1$ corresponds to the Otto cycle. It must be remembered that our analysis has assumed all processes reversible, with no heat losses. The actual Diesel engine has efficiencies somewhat lower than those indicated on Figure 10.68.

As shown in Figure 10.68 it is desirable to burn the fuel as close to TDC as possible, that is, with the cutoff ratio close to 1. In this way, the energy generated by the combustion process will be available throughout the entire expansion stroke to produce power. To achieve this, fuel injection must begin somewhat before TDC depending on the engine rotational speed, so as to ensure complete combustion. It follows that the p-$\forall$ diagram of the compression-ignition engine may look very similar to that of the Otto engine, shown in Figure 10.61. Figures 10.69(a) and (b) show, respectively, a cross-sectional and an exterior view of a truck engine based on the Diesel cycle.

Comparing the Otto engine with a Diesel engine of greater compression ratio, one can see that the Diesel engine possesses a greater overall thermal efficiency and therefore better fuel economy. The higher compression ratio, however, also means greater cylinder pressures. Furthermore, whereas combustion in the Otto engine occurs via a flame-initiated process at a single location, combustion in the Diesel is initiated simultaneously at many points through the gases in the cylinder; greater rates of pressure increase result for the Diesel. Both of these factors

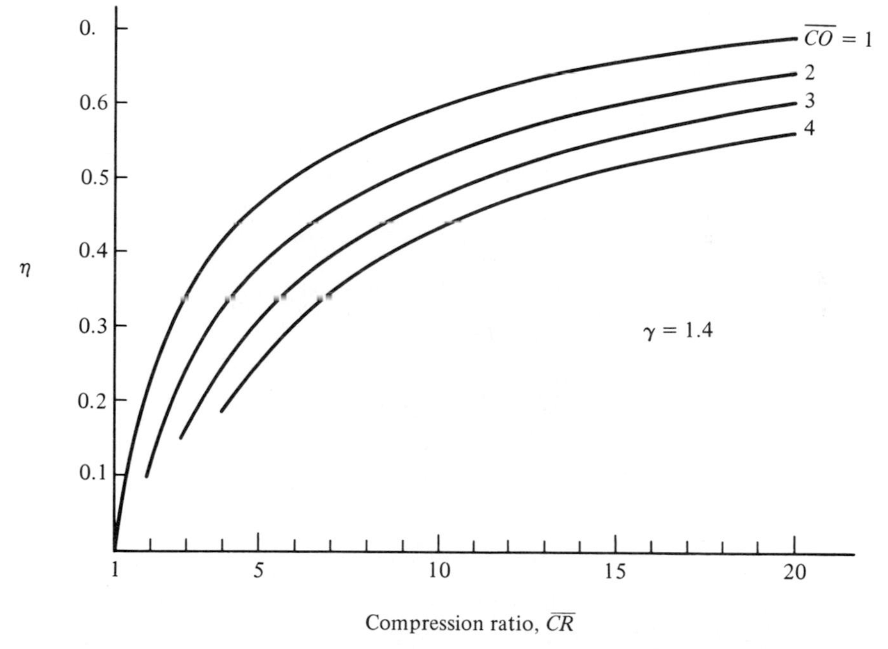

FIGURE 10.68 *Thermal Efficiency of Ideal Diesel Cycle.*

require that the Diesel engine possess thicker, heavier walls than the Otto. Hence, the power-to-weight ratio of a Diesel engine is less than that of a spark-ignited Otto engine. A Diesel-powered vehicle will require a longer time to accelerate from one velocity to another and, in general, will not provide as good an accelerative performance as a vehicle powered by an Otto engine of the same displacement. Furthermore, the Diesel uses a fuel-injection system which must inject a closely controlled amount of fuel into the cylinders against cylinder pressures as high as 50 to 80 atmospheres. Remember that in the Otto engine, the fuel-air mixture is admitted to the cylinders at close to atmospheric pressure.

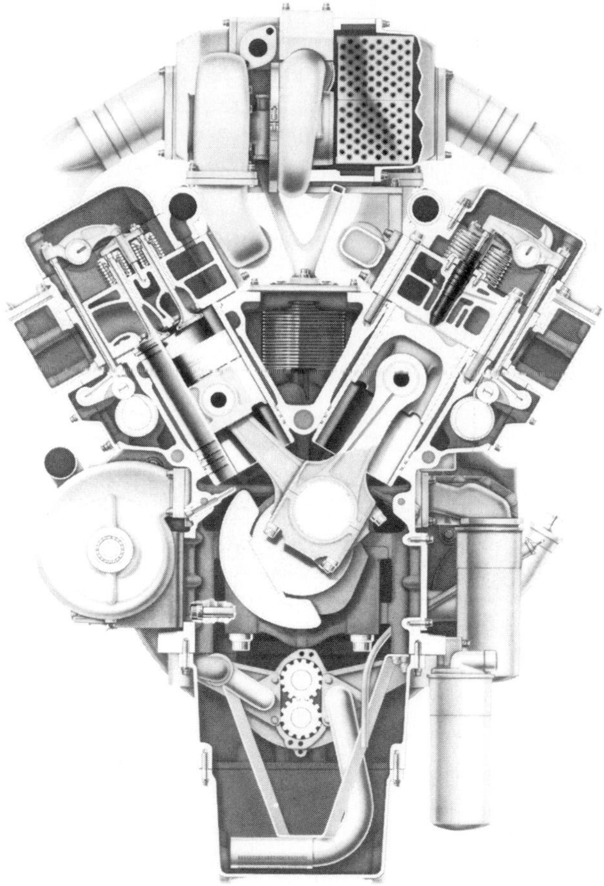

(a) Cross-sectional view

FIGURE 10.69 *Diesel Engine. (Photos Courtesy of Cummins Engine Co., Inc., Columbus, Indiana.)*

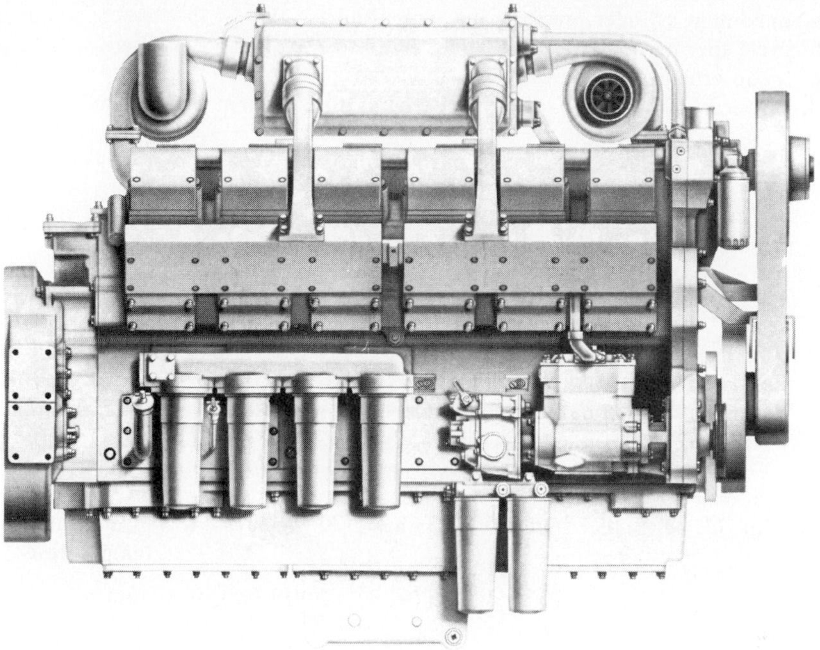

(b) Exterior view

FIGURE 10.69 (*continued*)

Thus, the Diesel engine is considerably more expensive than an Otto engine of the same size. Because of its good fuel economy, the Diesel is used extensively for truck applications, where engine weight is only a small fraction of vehicle total weight and acceleration performance is not an important criterion. Further, Diesel fuel, with no octane requirement, has generally been cheaper to purchase, adding to the overall economy of the Diesel. At present, with economy and the shortage of petroleum becoming increasingly important factors, the Diesel is seeing more and more use as a power plant for the light-duty motor vehicle, in spite of its poorer accelerative characteristics in comparison with the Otto engine.

EXAMPLE 10.23

A four-stroke Diesel engine has a compression ratio of 18. The pressure and temperature at the beginning of the compression stroke are 1 atm and 27°C (300 K), respectively. The ratio of mass of air to mass of fuel in the cylinders is 40 to 1, with a heat release due to combustion of 45 000 kJ/kg of fuel. For the ideal air-standard cycle with $\gamma = 1.40$, $R = 0.287$ kJ/kg K, and $c_p = 1.004$ kJ/kg K, determine

a. maximum cylinder pressure and temperature,
b. cycle thermal efficiency,
c. mean effective pressure (mep), and
d. power output at 2000 rpm for an engine displacement of 5000 cm^3.

Solution

a. The energy input is

$$q_{in} = 45\,000 \text{ kJ/kg}/40 = 1125 \text{ kJ/kg} = c_p(T_3 - T_2)$$

Hence

$$T_3 - T_2 = (1125 \text{ kJ/kg})/(1.004 \text{ kJ/kg K}) = 1120.5 \text{ K}$$

For isentropic compression, we have $p_2/p_1 = (V_1/V_2)^\gamma = 18^{1.4} = 57.2$, so that $p_2 = 57.2$ atm. From the equation of state, we obtain

$$\frac{T_2}{T_1} = \frac{p_2 V_2}{p_1 V_1} = \frac{57.2}{18} = 3.178$$

or $T_2 = 953.3$ K and $T_3 = 953.3 + 1120.5 = 2073.8$ K = maximum temperature

$$p_3 = p_2 = 57.2 \text{ atm} = 57.2(101.3) = 5794 \text{ kPa} = \text{maximum } p$$

The fuel cutoff ratio $\overline{CO} = V_3/V_2 = T_3/T_2 = 2073.8/953.3 = 2.175$. Above, we showed that $T_4/T_1 = (V_3/V_2)^\gamma = \overline{CO}^\gamma = 2.175^{1.4} = 2.968$. Here, $T_4 = 2.968(300 \text{ K}) = 890.4$ K. From the equation of state, $p_4/p_1 = T_4/T_1 = 2.968$; hence, $p_4 = 2.968$ atm.

b. From Equation (10.21), the thermal efficiency of the cycle is

$$\eta = 1 - \overline{CR}^{1-\gamma}\left\{\frac{1}{\gamma}\,\frac{\overline{CO}^\gamma - 1}{\overline{CO} - 1}\right\} = 1 - (18)^{-0.4}\left\{\frac{1}{1.4}\,\frac{2.175^{1.4} - 1}{1.175}\right\} = 62.4\%$$

c. From Equation (10.18), the mean effective pressure is

$$\text{mep} = \frac{\text{net } W_{out}}{D} = \frac{\text{net } w_{out}}{v_1 - v_2} = \frac{\eta q_{in}}{\dfrac{RT_1}{p_1}\left(1 - \dfrac{1}{18}\right)} = \frac{0.624(1125 \text{ kJ/kg})}{\dfrac{(0.287 \text{ kJ/kg K})(300 \text{ K})}{101.3 \text{ kN/m}^2}\,\dfrac{17}{18}}$$

$$= 874.5 \text{ kPa}$$

d. From Equation (10.19), the power output of a four-stroke engine is

$$\text{Power} = (\text{mep})D(N/2) = (874.5 \text{ kN/m}^2)(5000 \times 10^{-6} \text{ m}^3)\left(\frac{2000}{2(60)} \text{ s}^{-1}\right)$$

$$= 72.88 \text{ kW}$$

The efficiency of an equivalent Carnot power cycle is

$$\eta_{\text{Carnot}} = 1 - \frac{T_L}{T_H} = 1 - \frac{300}{2074} = 0.855$$

Comparing the results of Examples 10.21 and 10.23, one can see that the Diesel cycle has greater efficiency but a lower mep and less power output than does the Otto. The Otto engine burns a charge of fuel and air that has been preheated and compressed in the compression stroke. With the Diesel engine, liquid fuel is introduced into the cylinders near the end of the compression stroke. This fuel must be heated and atomized after injection and prior to combustion, events that take time to occur. If too much fuel is injected into the Diesel or if insufficient time is allowed to completely burn the fuel (too high rpm), black smoke will appear in the exhaust. The maximum engine rpm of the Diesel engine is limited by the appearance of smoke, and it is lower than the maximum rpm achievable with an Otto engine.

Combustion in the spark-ignition engine occurs in a homogeneous fuel-air mixture. For a flame to propagate through such a mixture, the air-fuel mass ratio cannot be much greater than 20 to 1. For the Diesel engine, combustion occurs in a heterogeneous mixture, taking place in pockets around the fuel droplets. Whereas the air-fuel ratio in the vicinity of the burning is still less than 20 to 1, the overall air-fuel ratio in the cylinder can be 50 to 1 or 100 to 1. The ability of the Diesel engine to operate at large air-fuel ratios gives it advantages for light load operation, such as idling, where very little fuel and power are required. However, the spark-ignited engine can operate at higher rpm and larger fuel-air ratios than the Diesel engine without smoking. This gives the Otto engine advantages with respect to output power for a given displacement.

In compression-ignition engines, a constant quantity of air is inducted into each cylinder. Hence, the load is controlled by varying the quantity of fuel injected into a constant quantity of air. Thus, the air-fuel ratio can be varied over a large range. At partial loads, the cutoff ratio is less than at maximum load, resulting in slightly higher thermal efficiencies than at maximum load (as shown in Figure 10.68).

10.7 GAS POWER CYCLES WITH REGENERATION: THE STIRLING AND ERICSSON CYCLES

Our examination of the Brayton power cycle showed that the use of regeneration can be an effective means (particularly at low compressor pressure ratios) to increase the thermal efficiency of the cycle and approach the efficiency of the Carnot cycle. In this section, we will discuss attempts that have been made, by use of regeneration, to achieve, or at least come close to, Carnot efficiency in gas power cycles.

One such cycle is the *Stirling cycle*, which consists of two constant temperature processes and two constant volume processes, as shown in Figure 10.70. Heat is added to the working fluid (process 3–4) and is rejected from the working fluid (process 1–2) while the fluid is maintained at constant temperature. In addition, heat must be added during the constant volume compression process (2–3) and removed during the constant volume expansion process (4–1). If the two heat exchanges at constant volume are carried out in a regenerator, it is possible, in the limit, to involve no external heat transfer in these two processes. The Stirling cycle is adaptable for use with reciprocating machinery, and hence we will carry out the analysis for a closed system.

The four processes in an ideal, reversible, closed system Stirling cycle are as follows:

(1–2) Isothermal compression (energy rejection):

$$q_{out} = w_{in} + u_1 - u_2 = T_1(s_1 - s_2)$$

$$w_{in} = \int_2^1 p\,dv$$

(2–3) Constant volume compression process:

$$q_{in} = u_3 - u_2$$

$$w_{out} = 0$$

(3–4) Isothermal expansion (energy input and power stroke):

$$q_{in} = w_{out} + u_4 - u_3 = T_3(s_4 - s_3)$$

$$w_{out} = \int_3^4 p\,dv$$

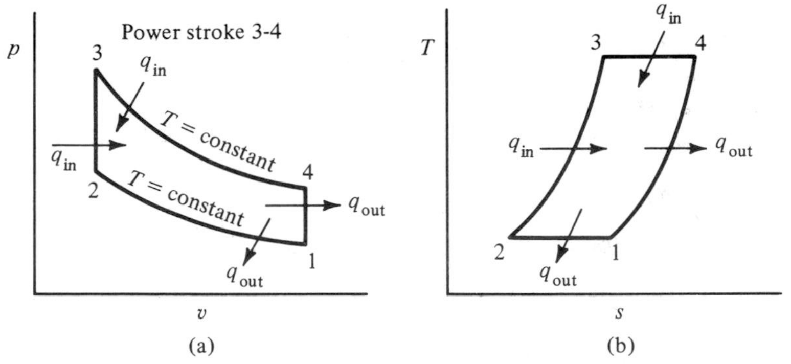

FIGURE 10.70 *Ideal Stirling Cycle.*

(4–1) Constant volume expansion process:

$$q_{out} = u_4 - u_1$$

$$w_{out} = 0$$

The net work output of the cycle is

$$\text{net } w_{out} = w_{out(3-4)} - w_{in(1-2)}$$

In an ideal regenerator, the quantity of heat rejected during process 4–1 is stored in the regenerator and then is restored to the fluid during process 2–3. In an actual regenerator, the heat released by the regenerator will be less than that received. The effectiveness of the regenerator is

$$e = \frac{q_{out(actual)}}{q_{out(4-1)}}$$

Hence the net energy input to the cycle during processes 2–3 and 4–1 is

$$q_{in} = q_{in(2-3)} - q_{out(actual)} = q_{in(2-3)} - eq_{out(4-1)}$$

The thermal efficiency of the cycle is

$$\eta = \frac{\text{net } w_{out}}{q_{in}} = \frac{w_{out(3-4)} - w_{in(1-2)}}{q_{in(2-3)} - eq_{out(4-1)} + q_{in(3-4)}}$$

$$= \frac{w_{out(3-4)} - w_{in(1-2)}}{u_3 - u_2 - e(u_4 - u_1) + w_{out(3-4)} + u_4 - u_3}$$

or

$$\eta = \frac{w_{out(3-4)} - w_{in(1-2)}}{u_4 - u_2 - e(u_4 - u_1) + w_{out(3-4)}} \quad \textbf{(10.22)}$$

For an ideal gas undergoing a reversible isothermal process, the work output is, from Equation (8.4),

$$w_{out(1-2)} = RT_1 \ln v_2/v_1 \quad \text{and} \quad w_{out(3-4)} = RT_3 \ln v_4/v_3$$

With $v_1 = v_4$ and $v_2 = v_3$, $w_{out(3-4)} = RT_3 \ln v_1/v_2$. Thus for an ideal gas with constant specific heats, we obtain, with $T_4 = T_3$, $T_2 = T_1$, $c_p - c_v = R$, and $\gamma = c_p/c_v$, $c_v/R = 1/(\gamma - 1)$, hence

$$\eta = \frac{R(T_3 - T_1) \ln v_1/v_2}{(1-e)c_v(T_3 - T_1) + RT_3 \ln v_1/v_3}$$

$$= \frac{(T_3/T_1 - 1) \ln v_1/v_2}{(1-e)[1/(\gamma-1)](T_3/T_1 - 1) + (T_3/T_1) \ln v_1/v_2} \quad \textbf{(10.23)}$$

The thermal efficiency of a Stirling cycle without regeneration ($e = 0$) is shown in Figure 10.71 as a function of the compression pressure ratio for several temperature ratios T_3/T_1. When comparing the Stirling

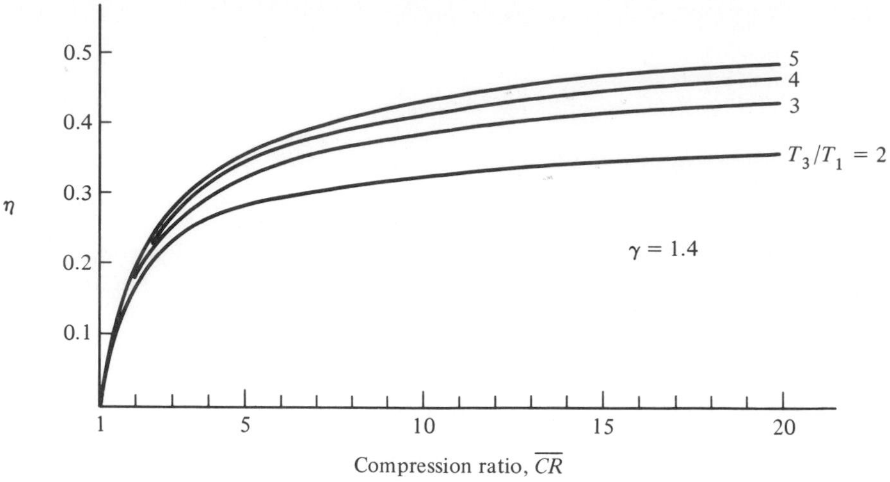

FIGURE 10.71 *Thermal Efficiency of Ideal Stirling Cycle Without Regeneration.*

cycle with the Otto cycle (Figure 10.60) and the Diesel cycle (Figure 10.68), one can see that the thermal efficiency of the Stirling cycle without regeneration is appreciably below that of those two cycles. With increasing regenerator effectiveness, the thermal efficiency of the Stirling cycle approaches that of the Carnot cycle (Figure 10.72). With ideal regeneration ($e = 1$), the thermal efficiency of the Stirling cycle equals

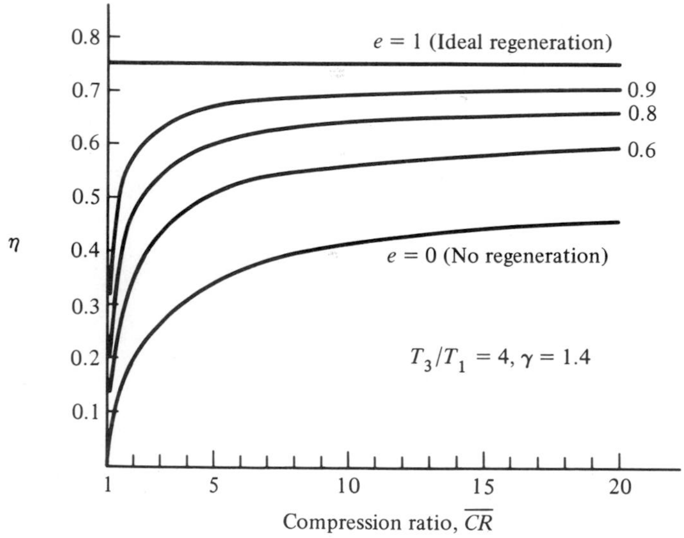

FIGURE 10.72 *Stirling Cycle Efficiency as a Function of Regenerator Effectiveness.*

that of a Carnot cycle operating between the same temperatures. From Equation (10.23), with $e = 1$,

$$\eta = \frac{(T_3/T_1 - 1)\ln v_1/v_2}{(T_3/T_1)\ln v_1/v_2} = 1 - \frac{T_1}{T_3} \tag{10.24}$$

The operation of a practical external combustion Stirling engine is shown in Figure 10.73. There are two pistons in the cylinder. One is a power piston *P*, tightly sealed with the cylinder wall to prevent leakages. The other piston is the displacer piston *D* (not sealed to the cylinder), the sole purpose of which is to move the working fluid around from one volume to another through the regenerator. The tubing labeled "hot" is maintained at a uniform high temperature by an external heat source, for example, combustion of a fuel. The tubing labeled "cold" is maintained at a uniform low temperature, for example, ambient temperature, by being maintained in contact with a heat sink. The regenerator consists of a material to store heat, but it must also prevent direct heat flow from the

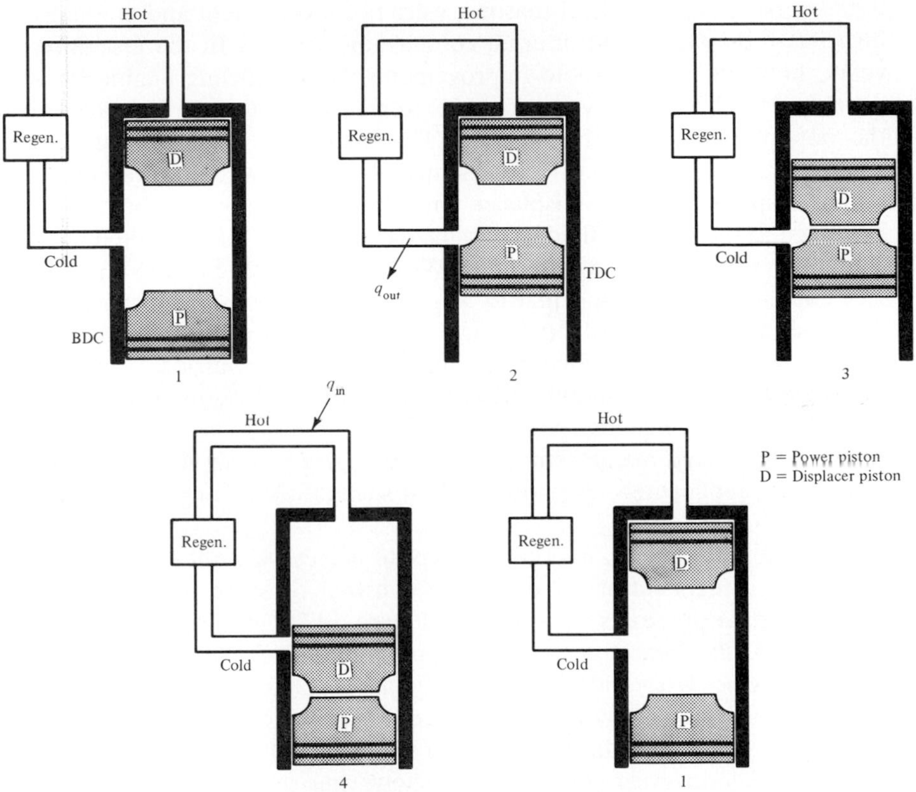

FIGURE 10.73 *Stirling Engine.*

heat source to the heat sink. At 1, the power piston is at BDC, with the displacer at its TDC position. From 1–2, the power piston moves from its BDC to TDC, compressing the working fluid. During this process, the working fluid in the cylinder is in contact with the low-temperature reservoir, so the temperature remains constant. For this isothermal compression, heat must be removed from the working fluid, as shown in 2. From 2–3, the displacer moves downward, pushing the fluid through the regenerator where it picks up sufficient heat to reach T_3 (see Figure 10.70). From 3–4, P and D move down together, increasing the volume available to the working fluid. During this process, the fluid is in contact with the heat source, so its temperature is maintained at $T_4 = T_3$. During the expansion in 3–4, heat must be added to the working fluid from the heat source to maintain constant temperature. Finally, from 4 to 1, D moves upward to its TDC, while P remains at its BDC. The fluid is thereby pushed through the regenerator, where it gives up heat and is cooled to T_1.

Although the Stirling engine was invented in the early 1800s, the initial versions, for practical reasons, were not too efficient and gradually lost out to the steam and internal combustion engines. In the last thirty years, however, considerable improvements in the Stirling engine have been made. These improvements have created a great deal of interest in the potential of the Stirling engine. The improvements have included better seals, an improved drive mechanism to maintain the desired phase relationship between the displacer and power pistons, and the use of hydrogen as the working fluid to minimize losses as the gas is pushed back and forth through the regenerator. With such improvements, thermal efficiencies approximately equal to 50 percent of Carnot efficiency can be achieved over a range of operating conditions. Since the engine is an external combustion engine (heat applied from combustion occurring externally) with a continuous rather than intermittent burning, as in the Otto and Diesel engines, pollutant emissions tend to be low and the engine is less noisy. At present, the Stirling engine is being developed for automotive applications. A photograph of such a Stirling engine is shown in Figure 10.74.

The *Ericsson cycle*, another attempt to achieve Carnot efficiency by the use of a regenerator, involves two constant pressure processes and two isothermal processes, as shown in Figure 10.75. Again, for an ideal regenerator, the heat transfer during process 2–3 is balanced by that during process 4–1, and the efficiency approaches that of a Carnot engine. As with the Brayton cycle, the application of the Ericsson cycle is most appropriate to rotating machinery; we will therefore analyze an ideal closed loop cycle. Figure 10.76 shows the schematic of a power generating system based on the Ericsson cycle.

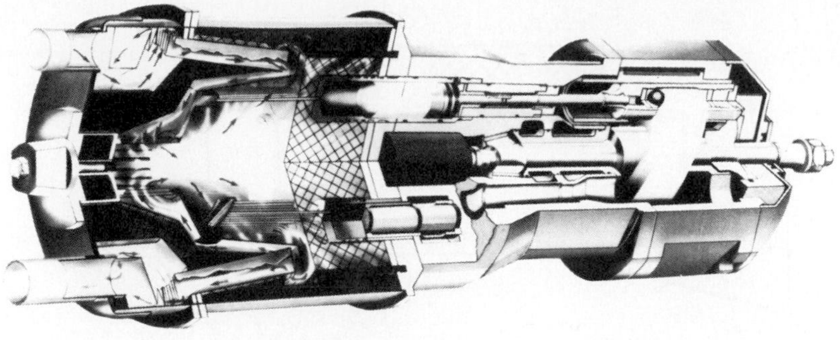

FIGURE 10.74 *Version of Stirling Engine for Automotive Application Being Developed by Ford Motor Co. and Philips. (Photo Courtesy of Jet Propulsion Laboratory, California Institute of Technology.)*

The four reversible processes of an ideal, closed loop Ericsson cycle are as follows:

(1–2) Isothermal compression (energy rejection):

$$q_{out} = w'_{in} - (h_2 - h_1)$$

$$w'_{in} = \int_1^2 v\, dp$$

(2–3) Constant pressure process:

$$q_{in} = h_3 - h_2$$

$$w'_{out} = 0$$

(3–4) Isothermal expansion (energy input and power output):

$$q_{in} = w'_{out} + (h_4 - h_3)$$

$$w'_{out} = \int_4^3 v\, dp$$

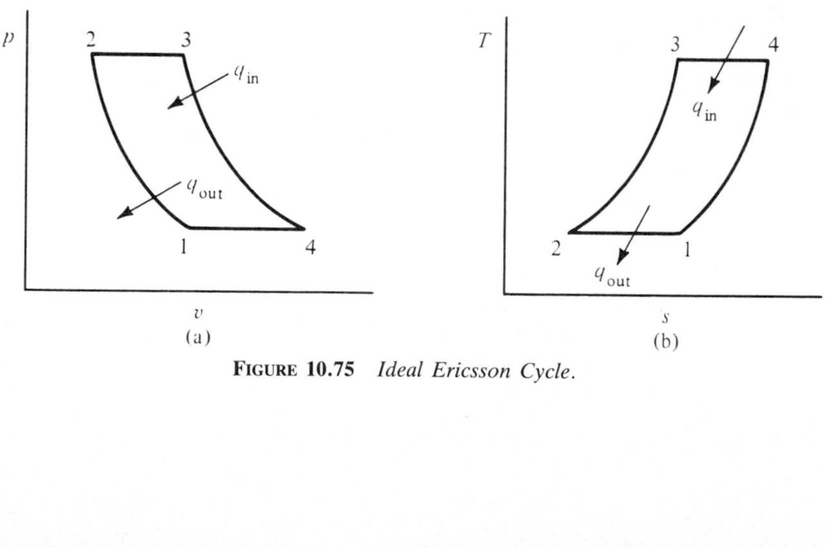

FIGURE 10.75 *Ideal Ericsson Cycle.*

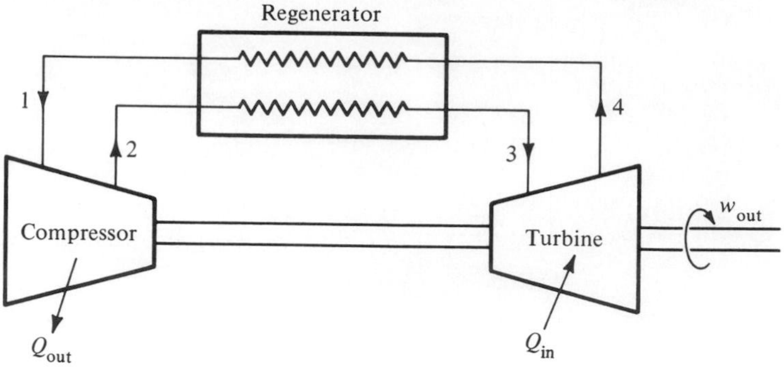

FIGURE 10.76 *Schematic of Closed Loop Ericsson Cycle.*

(4–1) Constant pressure process:

$$q_{out} = h_4 - h_1$$

$$w'_{in} = 0$$

The net work output of the cycle is

$$\text{net } w'_{out} = w'_{out(3-4)} - w'_{in(1-2)}$$

For an ideal gas with constant specific heats, the thermal efficiency of an Ericsson cycle with ideal regeneration is, with $T_4 = T_3$,

$$\eta = \frac{w'_{out(3-4)} - w'_{in(1-2)}}{w'_{out(3-4)}}$$

Now

$$w'_{out(3-4)} = \int_4^3 v\,dp = \int_4^3 \frac{RT}{p}\,dp = RT_3 \ln \frac{p_3}{p_4}$$

With $p_3 = p_2$ and $p_4 = p_1$,

$$w'_{out(3-4)} = RT_3 \ln \frac{p_2}{p_1}$$

$$w'_{in(1-2)} = \int_1^2 v\,dp = RT_1 \ln \frac{p_2}{p_1}$$

Hence

$$\eta = \frac{RT_3 \ln p_2/p_1 - RT_1 \ln p_2/p_1}{RT_3 \ln p_2/p_1} = 1 - \frac{T_1}{T_3} \qquad \textbf{(10.25)}$$

PROBLEMS

10.1 A Rankine cycle, using H_2O as the working fluid, is to operate with turbine inlet conditions of 8000 kPa and 600°C and a condenser temperature of 30°C. Assuming that the turbine and pump are isentropic and that all processes are reversible, determine the thermal efficiency of the cycle. Calculate the rate of thermal energy input to the steam generator if the net power output is 150 megawatts. Repeat the problem for turbine efficiencies of 90 percent and 80 percent.

10.2 A Rankine power plant is to be used for arctic applications with R-22 as the working fluid. At the turbine inlet, the pressure is 400 psia and the temperature is 200°F. The condenser is air cooled and has a temperature of 0°F. Assuming all processes to be ideal, calculate the cycle thermal efficiency.

10.3 In a Rankine power cycle, steam is supplied to the turbine at a pressure of 15,000 kPa and a temperature of 500°C. The pressure in the condenser is maintained at 5 kPa. Assume a turbine efficiency of 85 percent and a pump efficiency of 90 percent. Calculate the thermal efficiency and the second law efficiency of the cycle.

10.4 For the Rankine power plant of Example 10.4, choose a reference temperature of 77°F and calculate the second law efficiency for a system boundary which excludes the furnace.

10.5 In an electric power generating station, a nuclear reactor is used to supply sufficient thermal energy so that steam leaves as saturated vapor at 800 psia at point 5 in Figure 10.77. In the superheater, 5–6, heat is added from combustion of a fossil fuel to bring the temperature up to 1200°F. The efficiency of the high-pressure turbine is 85 percent, and the efficiency of the low-pressure turbine is 100 percent. The condenser pressure is 1.0 psia.

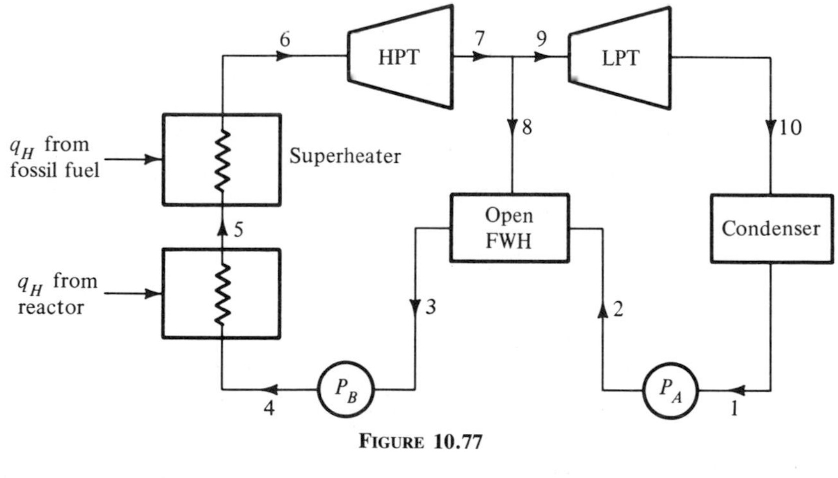

Figure 10.77

One open feedwater heater is to be used, with steam extracted from the high-pressure turbine at 20 psia ($p_7 = p_8 = p_9 = 20$ psia). Pump efficiencies are each 80 percent. The power plant is to provide a net power output of 1000 megawatts. Find the following:

a. the plant thermal efficiency;
b. the horsepower required for each pump;
c. the rate at which heat must be supplied from the reactor and from the fossil fuel;
d. the rate at which heat is rejected from the condenser; and
e. the plant second-law efficiency.

10.6 A Rankine power plant is to be used for an automotive application. The maximum cycle pressure and temperature are to be 7000 kPa and 600°C. The condenser pressure is 100 kPa. Determine the ideal cycle thermal efficiency. Then repeat the problem for an actual engine in which turbine efficiency is 80 percent and pump efficiency is 75 percent. In both cases, water is the working fluid.

10.7 In a steam power plant, the steam pressure and temperature entering the high-pressure turbine (HPT) are 1500 psia, 1200°F. Part of the steam flow leaving the high-pressure turbine is reheated in the boiler to 1000°F and then is flowed to a low-pressure turbine (LPT). The remainder is extracted for open feedwater heating. Exit pressure from the high-pressure turbine is 100 psia. The condenser pressure is 0.95 psia; the high-pressure turbine efficiency, 90 percent; the low pressure-turbine efficiency, 80 percent; and the pump efficiencies, 70 percent. For a net plant output of 500 megawatts, find the rate of heat rejection in the condenser and the horsepower output of each turbine. Also, for an allowable cooling water temperature rise of 8°F, determine the condenser cooling water flow requirements (Figure 10.78).

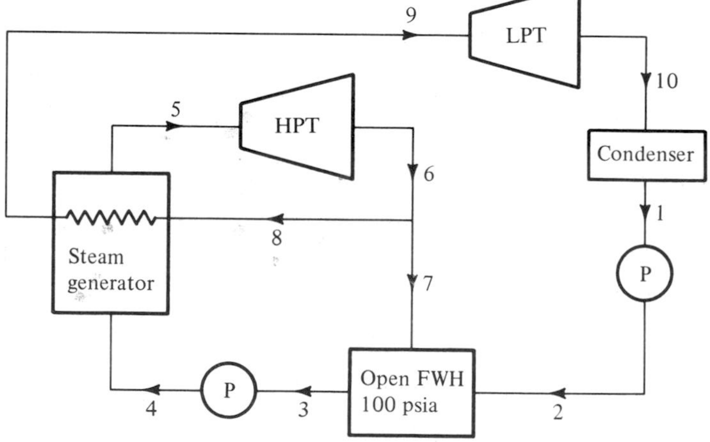

FIGURE 10.78

10.8 The maximum cycle pressure and temperature for a Rankine power plant are 5000 kPa and 500°C. The steam is extracted from the turbine at 400 kPa and is reheated to 500°C; then it is expanded in the turbine to a condenser pressure of 10 kPa. Assuming all processes to be ideal, calculate the mass flow entering the boiler for a net plant output of 400 megawatts. Also, assuming that the heating value of the gaseous fuel used is 46,000 kJ/kg, determine the required fuel flow rate to the plant. Assume that 90 percent of the heat given off from the combustion process goes into the working fluid, and the remainder is lost out the stack.

10.9 For the supercritical steam power plant shown in Figure 10.79, the maximum pressure is 4000 psia, and the maximum temperature is 1200°F. The plant uses one closed feedwater heater, with a drip pump to take the liquid condensate from the feedwater heater and raise it to boiler pressure. The efficiency of the high-pressure turbine is 90 percent; the efficiency of the low-pressure turbine, 85 percent; and the efficiency of each pump, 80 percent. The condenser pressure is 2.0 psia, and the plant net power output is 300 megawatts. Find the plant thermal efficiency, the mass flow of steam at the boiler exit, and the horsepower required of each pump. (p_8 = 100 psia.)

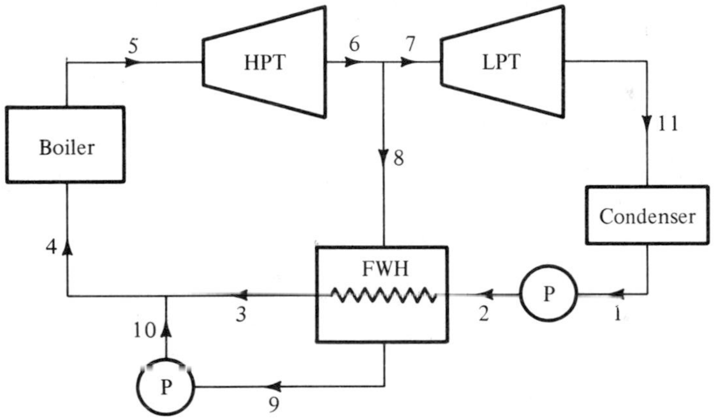

FIGURE 10.79

10.10 In the steam power plant shown in Figure 10.80, two closed feedwater heaters are used to preheat the water before it flows into the boiler. Liquid condensate from the high-pressure heater is drained through a trap to the low-pressure heater; condensate from the low-pressure heater is drained through a trap to the condenser. Assume that the feedwater in each heater is raised to the temperature of the condensing steam and that the pump and turbine are isentropic. Calculate the plant thermal efficiency.

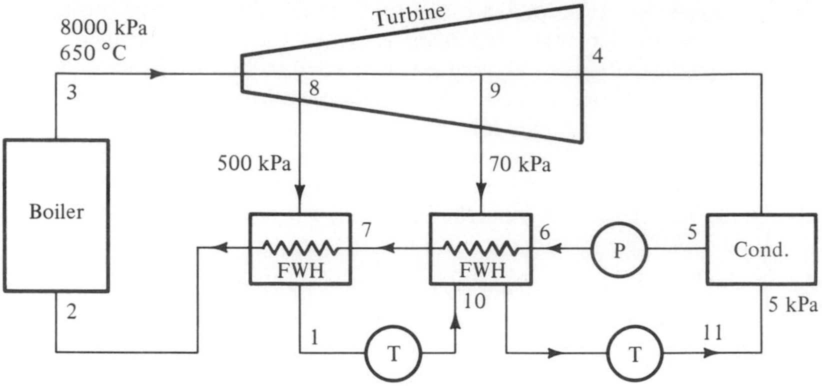

FIGURE 10.80

10.11 A throttle valve is used to control the flow of steam to the turbine in the power plant of Figure 10.81. In addition, steam is extracted from the turbine at 140 psia and reheated to 800°F in the boiler. For the conditions shown, find the plant thermal efficiency. Compare the thermal efficiency with and without the throttle valve.

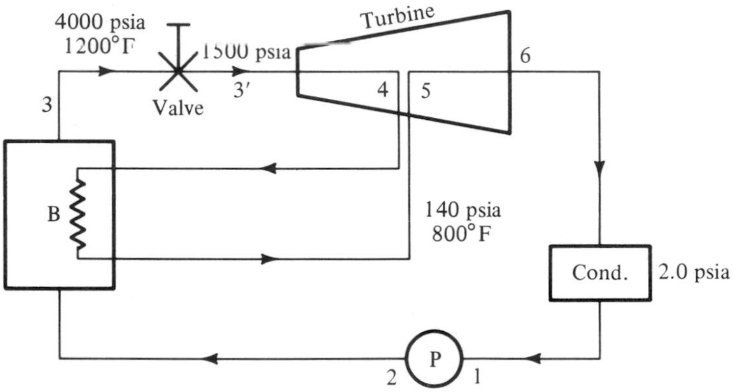

FIGURE 10.81

10.12 The conditions shown in Figure 10.82 indicate pressure and heat losses throughout the cycle. Determine the cycle thermal efficiency. For a net power output of 250 megawatts, find the rate of heat loss in kW from the piping between the boiler and the turbine.

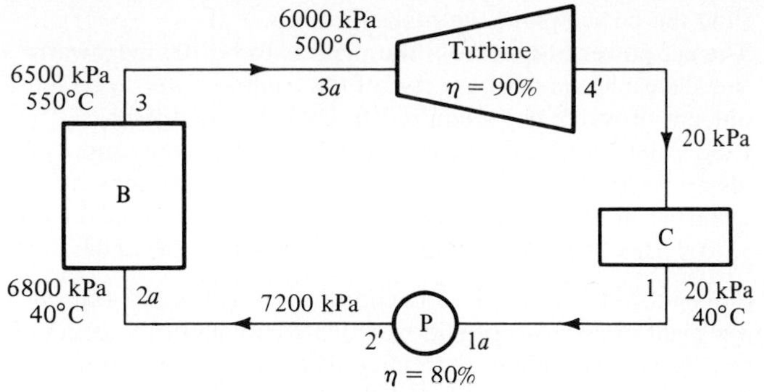

FIGURE 10.82

10.13 A Rankine power plant operates with reheat and one open feedwater heater (Figure 10.83). The steam flow is expanded in the turbine to 2500 kPa and then is reheated in the boiler to 600°C. Some steam is extracted at 100 kPa for open feedwater heating. Assuming that the net plant output is to be 1000 megawatts, find the rate of heat rejection from the condenser, the pump power required, and the total rate of energy input to the boiler.

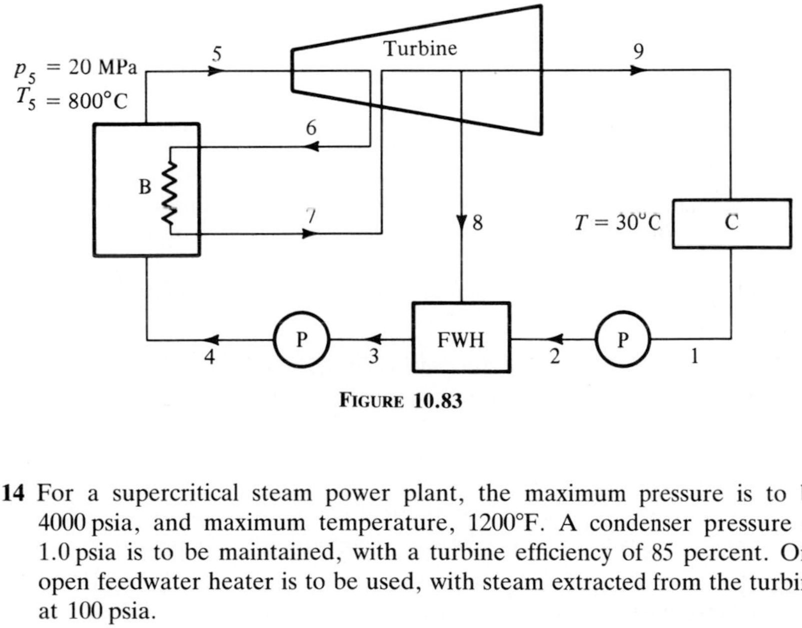

FIGURE 10.83

10.14 For a supercritical steam power plant, the maximum pressure is to be 4000 psia, and maximum temperature, 1200°F. A condenser pressure of 1.0 psia is to be maintained, with a turbine efficiency of 85 percent. One open feedwater heater is to be used, with steam extracted from the turbine at 100 psia.

a. Find the power plant thermal efficiency.

b. The net power output from the plant is to be 700 megawatts. Assuming the allowable temperature rise of the cooling water to be 5.0°F, find the amount of water flow required to cool the condenser.

c. Coal is to be used as the energy source, and a 30-day supply of coal must always be on hand to fuel the plant. How many tons of coal must be available at the plant? Assume that for each pound of coal burned, 11,000 Btus of heat are added to the water in the boiler.

10.15 A cooling tower is to be used in conjunction with a 1000-megawatt nuclear power plant in order to provide adequate cooling of the condenser cooling water. The power plant thermal efficiency is 33 percent. Estimate the cooling water flow required and the flow of makeup water necessary to replace that lost due to evaporation. The temperature rise of the cooling water across the condenser is to be 10°C (Figure 10.84).

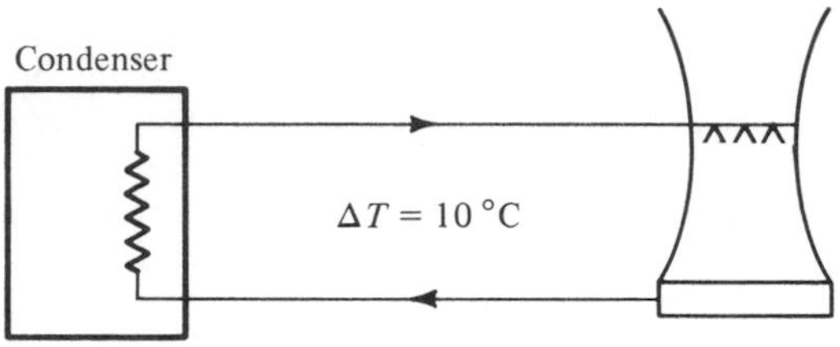

FIGURE 10.84

10.16 A new design is proposed for a Rankine power plant which will involve maximum steam pressures of 4500 psia and maximum temperatures of 1500°F. Design a reheat regenerative system for this power plant such that the thermal efficiency will be 50 percent, with a moisture content at the turbine exhaust not to exceed 8 percent. Assume the turbine efficiency to be 85 percent; the pump efficiency, 80 percent; and all other processes ideal. In your design, attempt to minimize the number of feedwater heaters.

10.17 A reheat-regenerative steam cycle operates under the following conditions: Inlet pressure and temperature to the first turbine stage = 7000 kPa, 550°C, respectively, with the condenser pressure = 15 kPa. The steam extraction pressure is 700 kPa. After partial extraction of the steam for the open feedwater heater, the steam is reheated to 550°C and enters the second turbine stage. Determine the thermal efficiency of the cycle. Take turbine stage efficiency at 87 percent and pump efficiency at 70 percent.

10.18 Exhaust gases from a gas turbine are to be used to supply heat to a steam turbine. The gas turbine utilizes a two-stage compressor with an ideal intercooler. The compressor inlet conditions are 101 kPa and 20°C, with an outlet pressure of 1000 kPa. The inlet temperature to the turbine is 1000°C.

The compressor and turbine efficiencies are each 85 percent. The exhaust gas leaves the steam heat exchanger at 250°C. The steam turbine inlet conditions are 350°C and 3500 kPa. The low pressure in the steam cycle is 10 kPa. The steam turbine efficiency is 80 percent. Calculate the ratio of steam and air mass flow rates and the thermal efficiency of the cycle.

10.19 A gas turbine power plant is used to supply compressed air to various industrial air tools. A single compressor delivers compressed air to the combustion chamber and to the air tools. The following conditions prevail:

	Inlet	Discharge
Compressor pressure, psia	14.5	60
Compressor temperature, °F	68	
Turbine temperature, °F	1000	
Turbine velocity, fps	~0	400

Assume that turbine exhausts to 14.7 psia, assume isentropic compressor and turbine, and assume that all turbine work output is used to drive the compressor. Determine the ratio of air intake and air delivery mass flow rates.

10.20 To accommodate a large load variation (no load to full load), a gas turbine power plant utilizes two separate turbines. The first turbine drives the compressor, while the second turbine provides the power as needed. A control valve located between the two turbines regulates the power output. The following conditions are applicable:

	Inlet	Exit
Compressor (efficiency = 0.75) pressure, kPa	100	500
Compressor temperature, °C	20	
Combustion chamber temperature, °C		600
First turbine (efficiency = 0.85)		
Second turbine (efficiency = 0.80) pressure, kPa		100

Determine the pressure change across the second turbine and the combustion energy input for the following conditions: full power (control valve wide open), no power, and half power. At full power, the power plant delivers 200 kW. What is the air-fuel ratio for a fuel with a heating value of 45,000 kJ/kg?

10.21 An air conditioning system for a bus uses R-22 as the working fluid. The evaporator pressure is 500 kPa, and the condenser outlet pressure is 1500 kPa. The temperature of the R-22 leaving the evaporator is 5°C above saturation. For a compressor efficiency of 80 percent, determine the coefficient of performance of this cycle. For a heat load of 35 kW, determine the flow of refrigerant and compressor power required.

10.22 Determine the performance of R-22 in a refrigeration cycle. Assume that the refrigerant leaves the evaporator at the saturation temperature, as

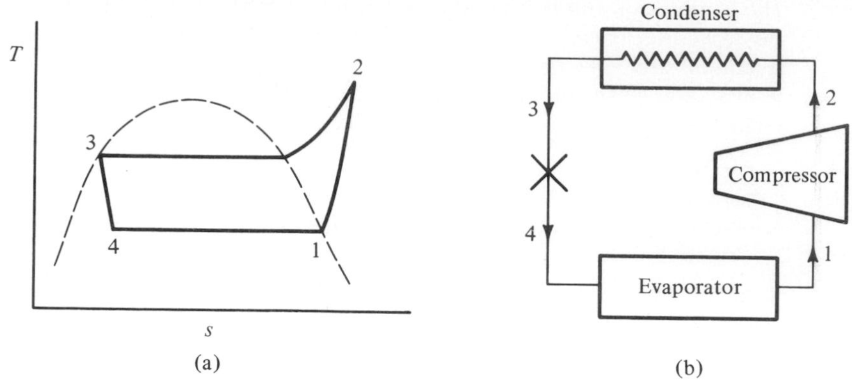

Evaporator $T = 5°F$
$T_3 = 112.76°F$

FIGURE 10.85

shown in Figure 10.85, and determine the compressor horsepower required per ton of refrigeration delivered, for a compressor efficiency of 80 percent.

10.23 For the air conditioning system of Example 10.10, choose 50°F as the temperature of the surroundings and calculate the second-law efficiency of the system.

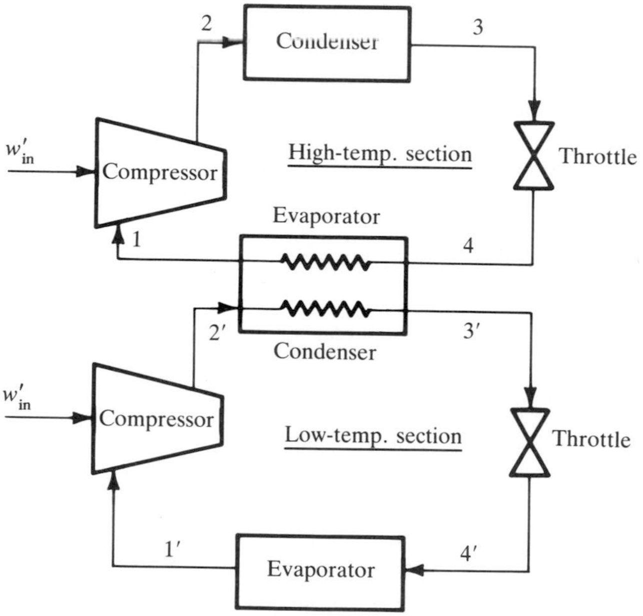

FIGURE 10.86 *Two-Refrigerant Refrigerion System.*

10.24 When a refrigeration cycle operates over a large temperature range, two working media are utilized in separate subsystems (Figure 10.86). In such a system, the high-temperature section (condenser side) uses refrigerant R-12 and the low-temperature section employs refrigerant R-13. The saturation temperature in the evaporator of the R-12 section is 5°F; the saturation temperature in the condenser is 104.35°F. The conditions and thermodynamic properties of R-12 required for the calculations are as follows:

Refrigerant R-12 (Dichlorodifluoromethane)

T (°F)	p (psia)	h_f (Btu/lbm)	h_g (Btu/lbm)	h (Btu/lbm)	s_f (Btu/lbm°R)	s_g (Btu/lbm°R)	s (Btu/lbm°R)
5	26.48	9.60	77.81		0.0216	0.1684	
20	26.48			79.99			0.1730
104.35	140.0	32.10	87.39		0.0649	0.1630	
160.7	140.0			97.61			0.1803

Reference temperature $T_0 = -120$°F, $h_0 = -16.56$ Btu/lbm, $s_0 = -0.0437$ Btu/lbm°R.

The saturation temperature in the evaporator of the R-13 section is −110°F; the saturation temperature in the condenser is 15°F. The conditions and thermodynamic properties of R-13 required for the calculations are as follows:

Refrigerant R-13 (Chlorotrifluoromethane)

T (°F)	p (psia)	h_f (Btu/lbm)	h_g (Btu/lbm)	h (Btu/lbm)	s_f (Btu/lbm°R)	s_g (Btu/lbm°R)	s (Btu/lbm°R)
−110	16.81	−15.527	47.789		−0.03977	0.14129	
−100	16.81			49.15			0.1453
15	223.6	14.065	57.359		0.03070	0.12194	
94.8	223.6			72.64			0.1518

Reference temperature $T_0 = -120$°F, $h_0 = -17.67$ Btu/lbm, $s_0 = -0.0459$ Btu/lbm°R.

Calculate the coefficient of performance and the second-law efficiency of the entire system.

10.25 A refrigeration system utilizes refrigerants R-22 and R-503 in separate subsystems to achieve cryogenic cooling (see Figure 10.86). R-22 is used in the high-temperature section and R-503 is employed in the low-temperature section of the system. The saturation temperature in the evaporator of the R-22 section is −40°F; the saturation temperature in the condenser is 96.27°F. Thermodynamic properties of R-22 required for the calculations are given in Appendix A.

State	T (°F)	p (psia)
1	−40	15.22
2	200	200
3	96.27	200

Reference temperature $T_0 = -150$°F, $h_0 = -25.97$ Btu/lbm, $s_0 = -0.0715$ Btu/lbm°R.

The saturation temperature in the evaporator of the R-503 section is −150°F; the saturation temperature in the condenser is −28.63°F. The conditions and thermodynamic properties of R-503 required for the calculations are as follows:

Refrigerant R-503 (Azeotropic mixture of R-23 and R-13)

T (°F)	p (psia)	h_f (Btu/lbm)	h_g (Btu/lbm)	h (Btu/lbm)	s_f (Btu/lbm°R)	s_g (Btu/lbm°R)	s (Btu/lbm°R)
−150	7.08	−45.088	30.174		0.44334	0.68636	
−28.63	150	−19.438	41.905		0.51158	0.65389	
120	150			66.966			0.70358

Reference temperature $T_0 = -120$°F, $h_0 = -16.56$ Btu/lbm, $s_0 = 0.4433$ Btu/lbm°R.

Calculate the coefficient of performance and second-law efficiency of the entire cooling system.

10.26 A gas turbine unit is to provide peaking power for an electric utility, with a net power output of 10 megawatts (Figure 10.50). The pressure ratio across the compressor is 7.0; the efficiency of the compressor, 80 percent; and the efficiency of the turbine, 92 percent. In order to conserve fuel, a regenerator with an effectiveness of 85 percent is used. The maximum temperature of the cycle is 2200°R; the conditions at the compressor inlet are 20°F and 15 psia. Assume the working fluid to be air and that the air behaves as a perfect gas with constant specific heats ($c_p = 0.24$ Btu/lbm°R, and $\gamma = 1.4$). Also, neglect pressure drops in the combustion chamber and regenerator. Determine the required air flow and fuel flow for a fuel heating value of 18,000 Btu/lbm and the power plant thermal efficiency.

10.27 A gas turbine unit is to provide a power output of 500 kW. The compressor pressure ratio is 6.0, and the maximum cycle temperature is 1100 K; the system incorporates a regenerator as shown in Figure 10.50. The unit uses natural gas, with 50 000 kJ given off for each kilogram of fuel burned. Assuming all components to be ideal, determine the flow of natural gas required and the overall plant thermal efficiency. Assume the working fluid to be air, with constant properties as given in Table B.1.

10.28 Repeat Problem 10.27, assuming a compressor efficiency of 80 percent, a turbine efficiency of 90 percent, a regenerator effectiveness of 80 percent, and a pressure drop of 70 kPa across the regenerator (2–3 and 5–6).

10.29 A closed cycle gas turbine is to be used to supply power in a remote location. The net output of the power plant is to be 100 kW, with nitrogen the working fluid. Treat the nitrogen as a perfect gas with constant specific heat (see Table A.1). The compressor efficiency is 80 percent; the turbine efficiency, 86 percent; and the regenerator effectiveness, 85 percent. At 1 (Figure 10.87), the temperature and pressure are 60°F and 80 psia. The compressor pressure ratio is 2.5, the maximum cycle temperature is 1200°F. Neglect pressure drops due to friction in the cooler, regenerator, and reactor. Determine the following:

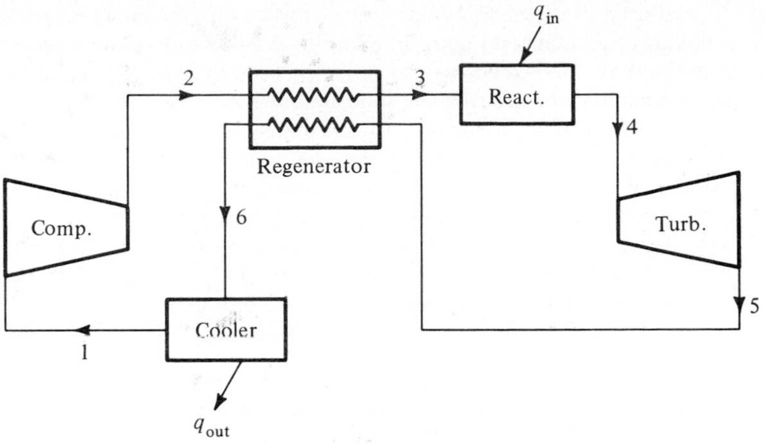

FIGURE 10.87

a. rate of heat rejection in the cooler;
b. turbine horsepower output;
c. compressor horsepower required; and
d. rate of heat addition in the reactor.

10.30 The compressor pressure ratio in a Brayton cycle plant is 6.5, with compressor inlet conditions of 1 atm and 20°C. The maximum cycle temperature is 1200 K, with air the working fluid. In order to account, at least to a certain degree, for the variation in thermodynamic properties of air with temperature, take c_p to be 1.004 kJ/kg K from the inlet to the combustion chamber and 1.18 kJ/kg K from the combustion chamber outlet to the exhaust. Assume all components to be ideal. Calculate the cycle thermal efficiency, and compare it with that calculated on the basis of constant specific heat (c_p = 1.004 kJ/kg K) throughout the entire cycle.

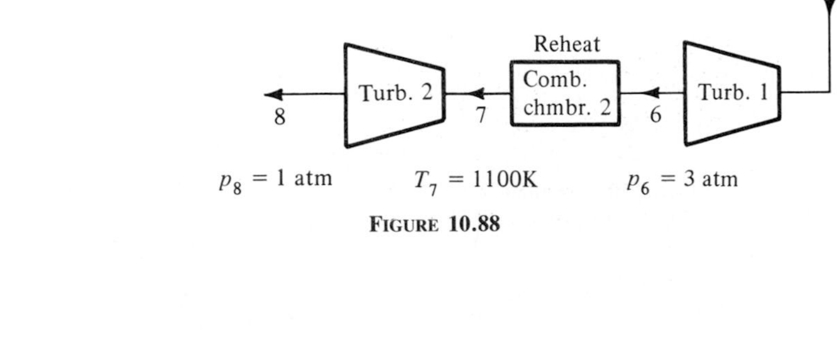

FIGURE 10.88

10.31 A modification of the Brayton cycle involves the use of intercooling and reheating. As shown in Figure 10.88, after the pressure of the working fluid is raised in the first compressor, a cooler is used to remove heat and reduce the temperature of the working fluid before the fluid is compressed in the second compressor. Likewise, after expansion in the first turbine, the working fluid is reheated by a second combustor, and then it is expanded through a second turbine. For the conditions shown in Figure 10.88, determine the cycle thermal efficiency, and compare it with the efficiency of a Brayton cycle operating between the same maximum and minimum pressures and temperatures but with no intercooling or reheating. In both cases, assume that all components are ideal and that the working fluid is air, with constant properties as given in Table B.1.

10.32 Calculate the second-law efficiency of the modified Brayton cycle of Problem 10.31 and its components.

10.33 Incorporate an ideal regenerator into the cycle of Problem 10.31, and calculate the thermal efficiency of the resultant cycle.

10.34 A gas turbine is to be used for automotive application, as shown in Figure 10.89. The high-pressure turbine is used solely to drive the compressor, and the low-pressure turbine is used to supply the useful output power. The net power output from the engine is to be 150 horsepower, with a maximum cycle temperature of 2000°F. Take the efficiency of each turbine to be 90 percent; the compressor efficiency, 80 percent; and the regenerator effectiveness, 55 percent. The working fluid can be taken as air, behaving as a perfect gas with constant specific heat (Table A.1). Calculate the cycle thermal efficiency and the fuel flow required for a fuel heating value of 19,000 Btu/lbm. Also determine the exhaust temperature T_7.

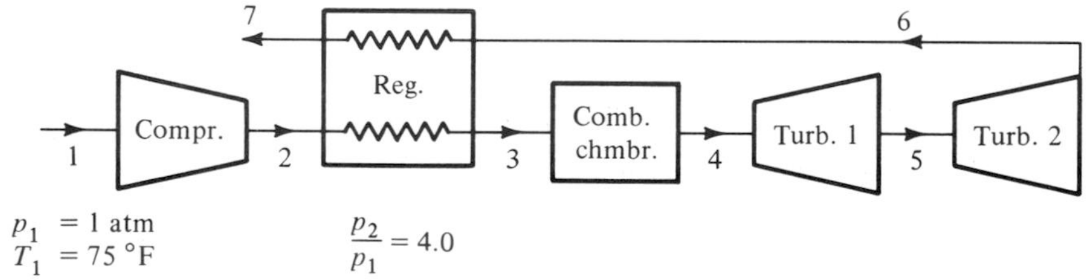

FIGURE 10.89

10.35 Perform a second-law analysis of the automotive gas turbine cycle of Problem 10.34.

10.36 With the development of ceramic turbine blades, it may be possible to achieve turbine inlet temperatures of 3000°F. Repeat Problem 10.34 with this higher temperature.

10.37 Consider the performance of an ideal turbojet engine (Figure 10.47) operating at an altitude of 9000 meters (ambient pressure 30 kPa, ambient temperature −45°C) and a forward speed of 200 m/s. Assume isentropic flow in the diffuser, compressor, turbine, and nozzle; neglect pressure drop in the combustor. The turbine inlet temperature is 1350 K; the compressor pressure ratio, 8:1; and the heating value of the fuel, 50 000 kJ/kg(fuel). Assume the working fluid to be air, behaving as a perfect gas with constant specific heat (Table B.1).

a. Determine specific fuel consumption (thrust per unit mass flow of fuel).

b. The engine thrust is 70 000 N; find the nozzle exit area and the total engine air mass flow. (In your calculations, neglect the mass flow of fuel in comparison to the air mass flow).

10.38 Consider the performance of the turbojet engine of Problem 10.37 under actual conditions. Assume that the actual pressure rise in the diffuser is only 90 percent of that of an isentropic diffuser. Take the compressor efficiency to be 80 percent; the turbine efficiency, 90 percent; and the pressure drop in the combustor, 5 percent of the combustor inlet pressure. Assume that the actual exit velocity achieved in the nozzle is only 95 percent of that of an isentropic nozzle. Find the thrust specific fuel consumption at an altitude of 9000 meters.

10.39 An open-loop, steady-flow Brayton cycle power plant operates at the following conditions:

Compressor inlet: $T_1 = 60°F$, $p_1 = 14.7$ psia
Compressor exit: $T_2 = 450°F$, $p_2 = 75$ psia
Energy input exit: $T_3 = 1200°F$, $p_3 = 75$ psia
Turbine exit: $T_4 = 600°F$, $p_4 = 14.7$ psia

The schematic of the system is shown in Figure 10.41 with the *T*-*s* diagram shown in Figure 10.90. Assume air as the working medium with constant specific heat and perform first- and second-law analyses of the cycle.

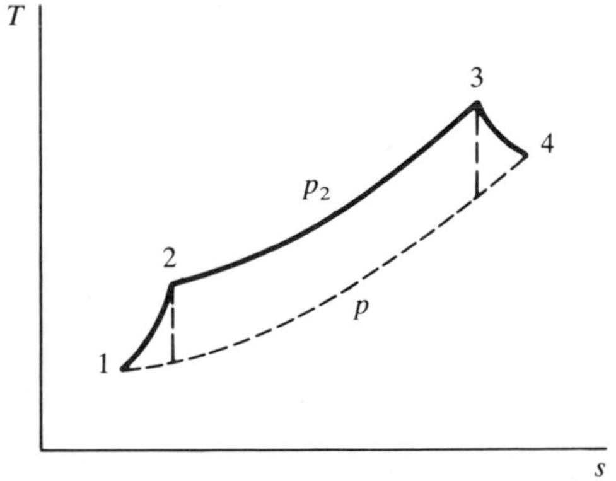

FIGURE 10.90 *Brayton Power Cycle of Problem 10.39.*

10.40 A turbojet engine is to be designed to produce 17,500 pounds of takeoff thrust at sea level. The compressor pressure ratio is 10:1, and the turbine inlet temperature is 1800°F. Ambient conditions are 14.7 psia and 80°F. Assume the properties of air to be the following: $c_p = 0.24$ Btu/lbm°F from the diffuser inlet to the combustor inlet; and $c_p = 0.276$ Btu/lbm°F from the combustor outlet to the nozzle exit. Take all components to be ideal and the heating value of the fuel to be 20,000 Btu/lbm. Find the following:

a. the nozzle exit area required;
b. the fuel flow;
c. the thrust specific fuel consumption; and
d. the turbine horsepower output.

10.41 The exhaust from a Brayton power plant is to be used to supply the energy input for a Rankine steam power plant. For the conditions shown in Figure 10.91, determine the following:

a. the thermal efficiency of the combined power plant;
b. power output from the Rankine plant;
c. the rate of heat rejection to the condenser cooling water; and
d. the second-law efficiency of the combined power plant.

Assume all components to be ideal, with air, the working fluid of the Brayton cycle, behaving as a perfect gas with constant specific heats. Take the air turbine exhaust temperature T_4 at 400°F.

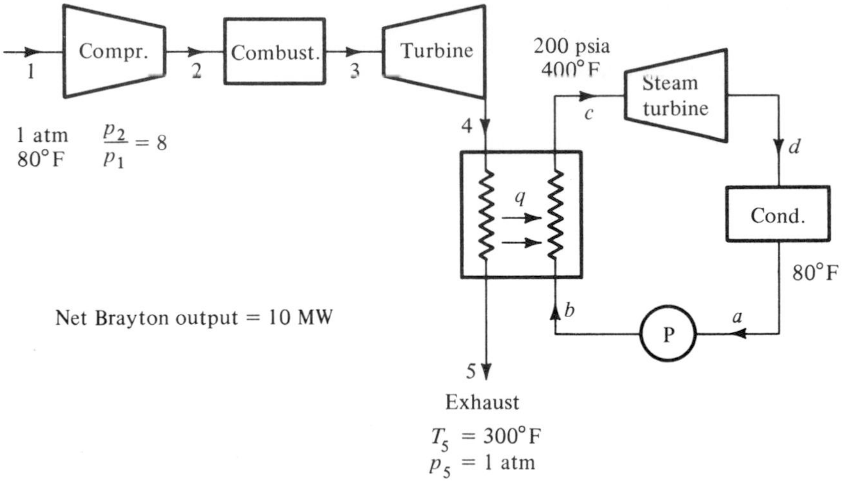

FIGURE 10.91

10.42 Suppose that the exhaust from the Brayton turbine of the previous problem were used to provide heat for a regenerator, as shown in Figure 10.50. For 100 percent effectiveness, determine the efficiency of this power plant for maximum cycle temperature $T_4 = 1100$°F, 1300°F, and 1500°F.

10.43 An air conditioning unit used in an aircraft operates on a reversed Brayton cycle with air as the working fluid. The compressor inlet conditions are 1 atmosphere and 30°C. The compressor pressure ratio is 6 : 1, and the air is delivered to the cabin at 12°C. The system is to deliver 100 kilograms of air per minute to the cabin. Determine the compressor power required, the turbine power output, and the coefficient of performance of the cycle. Assume ideal components.

10.44 Repeat Problem 10.43 with a turbine efficiency of 85 percent and a compressor efficiency of 78 percent.

10.45 Repeat Problem 10.43 with a turbine efficiency of 75 percent and a compressor efficiency of 60 percent.

10.46 Perform a second-law analysis of the air conditioning unit of Problem 10.45.

10.47 A four-stroke engine operating on the Otto cycle has a compression ratio of 7.9. The pressure and temperature in the cylinder at the beginning of the compression stroke are 100 kPa and 26°C. The heating value of the gasoline used is 44 500 kJ/kg(fuel), with the air-fuel ratio 17 : 1 on a mass basis. Assume the working fluid to have the following properties: $\gamma = 1.3$ and mean molecular mass = 28. Determine the following:

a. cycle thermal efficiency;
b. maximum cycle pressure and temperature;
c. mean effective pressure;
d. engine displacement for a power output of 200 kW at 3000 rpm;
e. fuel flow required in kg/min at 3000 rpm;
f. second-law efficiency.

10.48 In order to utilize the waste heat from the Otto engine of Problem 10.47, a Rankine steam cycle is incorporated into a combined power source, as shown in Figure 10.92. Assuming ideal Rankine cycle components, determine the combined cycle efficiency. Take the maximum pressure in the Rankine cycle to be 20 atmospheres, and the maximum temperature, 500°C, with a condenser temperature of 30°C.

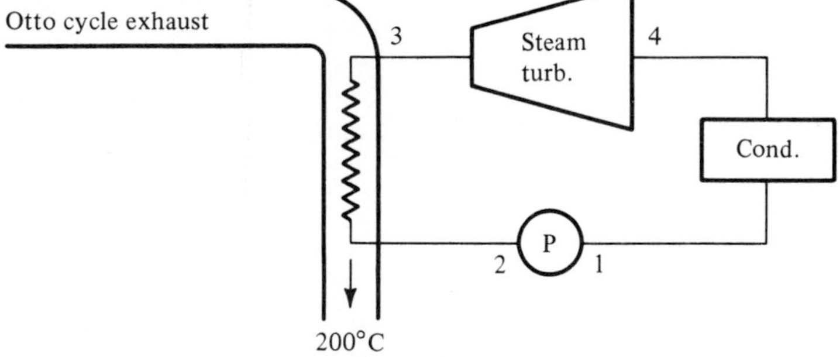

FIGURE 10.92

10.49 A throttled Otto engine has an intake pressure of 6.0 psia at 65°F, with 8:1 compression ratio. For a fuel heating value of 20,000 Btu/lbm and a 16:1 air-fuel ratio, determine the cycle thermal efficiency, the maximum pressure and temperature of the cycle, and the power output at 2800 rpm for an engine displacement of 300 cubic inches. Assume the working fluid to be a perfect gas with constant specific heats, $\gamma = 1.35$, and mean molecular mass = 28.

10.50 A Diesel engine has a compression ratio of 20:1. Intake conditions are 100 kPa and 25°C; the heating value of the fuel used is 46 000 kJ/kg(fuel). For an engine displacement of 4.0 liters and an engine speed of 2000 rpm, plot power output and cycle efficiency vs. air-fuel ratio over the range 15:1 to 40:1. Assume the working fluid to be air, with properties given in Table B.1.

10.51 Perform a second-law analysis of the Diesel cycle of Example 10.23.

10.52 The so-called dual power cycle involves two heat addition processes, one at constant volume and the other at constant pressure (Figure 10.93). Heat addition at constant volume occurs in process (2–3), while process (3–4) involves heat addition at constant pressure. Heat rejection takes place in process (5–1). Show that an expression for the thermal efficiency of the dual power cycle is

$$\eta = 1 - \frac{\dfrac{p_3}{p_2}\left(\dfrac{v_4}{v_3}\right)^{\gamma} - 1}{\left(\dfrac{p_3}{p_2} - 1\right) + \gamma\,\dfrac{p_3}{p_2}\left(\dfrac{v_4}{v_3} - 1\right)}\left(\frac{v_2}{v_1}\right)^{\gamma - 1}$$

When $v_4/v_3 = 1$, the expression for η reduces to that of the Otto cycle; when $p_3/p_2 = 1$, the expression reduces to that of the Diesel cycle.

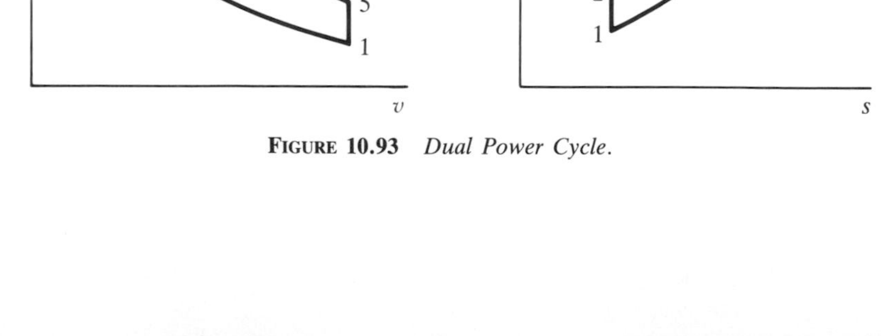

FIGURE 10.93 *Dual Power Cycle.*

10.53 The temperature and pressure at the beginning of compression of a dual power cycle (Figure 10.93) are 65°F and 14.7 psia, respectively. The diameter of the cylinder is six inches and the stroke is five inches. The compression ratio (v_1/v_2) is 6. The pressure at the end of the constant volume process is 600 psia. The constant pressure process occurs during 5 percent of the stroke. Calculate the thermal efficiency and the second-law efficiency of the cycle, assuming the working fluid is air.

10.54 The compression ratio for an ideal Stirling cycle is 8.0, with a minimum pressure of 1 atmosphere, a minimum temperature of 100°F, and a maximum temperature of 2000°F. Determine the thermal efficiency of the cycle without a regenerator. Assume the working fluid to be air, with properties given in Table A.1. Then repeat the problem with an ideal regenerator incorporated into the cycle. Compare your results with an Otto engine of the same compression ratio.

10.55 Repeat Problem 10.54 with hydrogen the working fluid in the Stirling engine.

10.56 Perform a second-law analysis of the Stirling cycle of Problem 10.54.

chapter

11

Combustion Processes

11.1 INTRODUCTION

In virtually all of the thermodynamic processes that we have studied up to this point, a pure substance or a mixture of pure substances has been involved with no change in chemical composition. In a chemical reaction, such as a combustion process, chemical changes do occur, such that the products of the reaction may be entirely different from the initial participants (called *reactants*). Since a large percentage of electric power generating stations and automobile and truck engines use thermal energy generated by a combustion reaction, it is important to be able to analyze combustion processes thermodynamically.

Combustion generally means the oxidation of a fuel accompanied by heat release. Fuels of primary interest are coal (solid), and liquid and gaseous hydrocarbons. Hydrocarbons are compounds of hydrogen and carbon and are generally of the form C_xH_y; for example, methane CH_4, acetylene C_2H_2, propane C_3H_8, benzene C_6H_6, and octane C_8H_{18}. Commonly used fuels, such as gasoline and natural gas, consist of mixtures of hydrocarbons. For example, natural gas is mostly methane, but it also contains ethane, propane, butane, and nitrogen. Gasoline can be approximated by octane (C_8H_{18}), but it contains many other liquid hydrocarbons.

We are interested, then, in chemical reactions between hydrocarbon fuels and oxygen, where the oxygen is usually supplied from air. Specifically, we want to be able to find the products of a combustion reaction, the heat given off by the reaction, and the temperature attainable from the reaction.

11.2 BALANCING A CHEMICAL EQUATION

In order to describe a chemical reaction such as occurs during a combustion process, we utilize a chemical equation, which specifies the quantities of products obtained for given quantities of reactants. The chemical equation expresses the fact that during a reaction, the number of atoms of each element, as well as the total mass of the participants, is preserved.

Since combustion involves oxidation of the carbon atoms in a hydrocarbon fuel to CO_2 and oxidation of the hydrogen atom to H_2O, we would obtain the following chemical equation for the complete combustion of octane (C_8H_{18}) with oxygen:

$$\underbrace{\underset{fuel}{C_8H_{18}} + \underset{oxidizer}{12\tfrac{1}{2}O_2}}_{reactants} \rightarrow \underbrace{8CO_2 + 9H_2O}_{products} + \text{energy release}$$

To obtain this balanced equation (in which the same number of atoms of C, H, and O are present before and after the reaction, though in different molecular combination), we start with one molecule of C_8H_{18}, which upon combustion would produce 8 molecules of CO_2 and 9 molecules of H_2O. To obtain the correct number of molecules of oxygen, we recognize that there are 8 O_2's in 8 CO_2 molecules and 4.5 O_2's in 9 H_2O molecules; thus we require $8 + 4.5 = 12.5$ molecules of O_2.

Since a molecule by itself is too small, a larger quantity of a substance, the *mole*, is used as the standard in chemical equations. A mole represents the molecular mass in mass units. That is, one pound mole is equal to the molecular mass of the substance in pounds, and a kilogram mole is equal to the molecular mass of the substance in kilograms. In each mole of any substance there will be an equal number of molecules present. For example, in each kilogram mole (kgmol), there will be 6.023×10^{26} molecules (Avogadro's number), and in each pound mole (lbmol), there will be 2.732×10^{26} molecules present. Thus, the above chemical equation for the combustion of octane can be considered on a molal basis; in other words, the reactants consist of one mole of octane and $12\frac{1}{2}$ moles of molecular oxygen. The products consist of 8 moles of carbon dioxide and 9 moles of water. Note that the total number of moles (in the above example, $13\frac{1}{2}$ moles of reactants vs. 17 moles of products) is not conserved.[1] However, conservation of mass requires that the same mass be present before and after the reaction.

1. An exception would be a reaction such as

$$\tfrac{1}{2}N_2 + \tfrac{1}{2}O_2 \rightarrow NO$$

TABLE 11.1 *Atomic Masses of Elements in Combustion.**

Carbon	12.011
Hydrogen	1.008
Nitrogen	14.007
Oxygen	15.999

* Adapted from *Handbook of Chemistry and Physics*: Cleveland, CRC Press, 1977.

With the atomic mass of hydrogen equal to 1.008, that of oxygen equal to 15.999, and that of carbon equal to 12.011 (Table 11.1), it follows that one pound mole of octane has a mass of $8(12.011) + 18(1.008) = 114.232$ pounds; likewise, one kilogram mole of octane has a mass of 114.232 kilograms. Returning to the chemical equation for the combustion of octane, note that for each pound mole of octane burned, the reactants consist of 114.232 pounds of octane and $12\frac{1}{2}(31.998) = 399.975$ pounds of oxygen. The products of combustion consist of $8(44.009) = 352.072$ pounds of carbon dioxide and $9(18.015) = 162.135$ pounds of water. The total mass of reactants is 514.207 pounds; the total mass of products is 514.207 pounds. As is evident, mass is conserved. For each kilogram mole of octane burned, there would be 514.207 kilograms of reactants and 514.207 kilograms of products.

We can also write the chemical equation on a mass basis as follows:

$$114.232 \text{ lbm } C_8H_{18} + 399.975 \text{ lbm } O_2 \rightarrow 352.072 \text{ lbm } CO_2 + 162.135 \text{ lbm } H_2O$$

or dividing by 114.232 lbm, we obtain

$$C_8H_{18} + 3.501 \text{ lbm } O_2 \rightarrow 3.082 \text{ lbm } CO_2 + 1.419 \text{ lbm } H_2O$$

In the above example of the combustion of octane, oxygen was supplied to the reaction. In most combustion reactions, however, oxygen is supplied from atmospheric air. The composition of dry atmospheric air on a percent mole basis is

78.09%	molecular nitrogen (N_2)
20.95%	molecular oxygen (O_2)
0.93%	argon (Ar)
0.03%	carbon dioxide (CO_2)
100.00%	

with a resulting molecular mass of 28.964 (see Example 5.1).

For combustion analysis, we will simplify the above and consider air to consist of 79 percent molecular nitrogen and 21 percent molecular oxygen. With this approximation, there will be $0.79/0.21 = 3.76$ moles of nitrogen for each mole of oxygen in the air. The molecular mass of air for this composition is $0.79(28.014) + 0.21(31.998) = 28.851$.

Thus, if octane is burned completely with dry air, the chemical equation of the reaction becomes

$$C_8H_{18} + 12\tfrac{1}{2}(O_2 + 3.76N_2) \rightarrow 8CO_2 + 9H_2O + 47N_2$$

Except at very high temperature (3000°F and above), nitrogen is relatively inactive and does not participate in the combustion reactions. However, it does become one of the constituents of the products of combustion.

For the above reaction, the ratio of mass of air to mass of fuel in the reactants (air-fuel ratio) is

$$A/F = \frac{\text{mass of air}}{\text{mass of fuel}} = 12\tfrac{1}{2}\{31.998 + 3.76(28.014)\}/114.232 = 15.03$$

A value of about 15 for the air-fuel ratio is typical for hydrocarbon fuels.

In the reactions that we have considered, we have assumed that the exact amount of air has been provided in the reaction to completely oxidize the fuel. The amount of air required for complete combustion is called *theoretical* or *stoichiometric* air. If more air is provided than that required to completely combust the fuel, oxygen will appear in the products. *Excess air* is usually expressed in percent of that required for complete combustion. Thus, for example, 20 percent excess air is equivalent to 120 percent theoretical air. If octane is burned with 20 percent excess air, the chemical equation of the combustion becomes

$$C_8H_{18} + 1.2\{12\tfrac{1}{2}(O_2 + 3.76N_2)\} \rightarrow 8CO_2 + 9H_2O + 56.4N_2 + 2\tfrac{1}{2}O_2$$

If less air is provided than that required for complete combustion, the products will contain carbon monoxide (CO), hydrogen (H_2), and perhaps even unburned fuel and carbon. The chemical equation for octane combusted with 80 percent theoretical air is

$$C_8H_{18} + 0.8\{12\tfrac{1}{2}(O_2 + 3.76N_2)\}$$
$$\rightarrow aCO_2 + bCO + cH_2 + dH_2O + eC + 37.6N_2$$

where a, b, c, d, and e, the number of moles of each product, cannot be determined by simple mass or atom balances. We will show later how the composition of the products can be found for combustion processes with insufficient air and also for fuel-rich processes.

EXAMPLE 11.1

Propane (C_3H_8) is burned with a theoretical amount of dry atmospheric air. Determine the mole fraction and the gas constant of the products of combustion.

Solution

The chemical equation for the combustion of propane with theoretical dry air is

$$C_3H_8 + 5O_2 + 18.8N_2 \rightarrow 3CO_2 + 4H_2O + 18.8N_2$$

The total number of moles of products is $3 + 4 + 18.8 = 25.8$. The mole fraction of each of the products therefore becomes

$$\begin{aligned} CO_2&: \quad 3/25.8 = 0.1163 \\ H_2O&: \quad 4/25.8 = 0.1550 \\ N_2&: \quad 18.8/25.8 = \underline{0.7287} \\ & \qquad\qquad\quad 1.0000 \end{aligned}$$

From Equation (5.3), the molecular mass of the products is

$$\begin{aligned} \overline{M} &= \frac{N_A}{N}\overline{M}_A + \frac{N_B}{N}\overline{M}_B + \frac{N_C}{N}\overline{M}_C \\ &= 0.1163(44.009) + 0.1550(18.015) + 0.7287(28.014) = 28.324 \end{aligned}$$

Hence, the gas constant R of the products is, from Equation (5.12),

$$R_u/\overline{M} = 1545.3/28.324 = 54.56 \text{ ft-lbf/lbm°R} \quad (\text{vs. } 53.35 \text{ ft-lbf/lbm°R for air})$$

When atmospheric air is supplied to a combustion process, the air will usually also contain some moisture. The presence of this water vapor must be included in the chemical equation, as illustrated in the following example.

EXAMPLE 11.2

Propane is burned with 20 percent excess atmospheric air having a specific humidity of 0.01 pound mass of water vapor per 1.0 pound mass of dry air. Determine the mole fractions and the gas constant of the products of combustion.

Solution

The chemical equation for the combustion of propane using the theoretical amount of dry air is

$$C_3H_8 + 5O_2 + 18.8N_2 \rightarrow 3CO_2 + 4H_2O + 18.8N_2$$

For 20 percent excess dry air, we have

$$C_3H_8 + (1.2)(5O_2 + 18.8N_2) \rightarrow 3CO_2 + 4H_2O + 1.2(18.8)N_2 + O_2$$

The specific humidity on a mole basis is, with $\overline{M}_{\text{water}} = 18.015$ and $\overline{M}_{\text{air}} = 28.851$,

$$\omega = \{(0.01 \text{ lbm(water)/lbm(dry air)}\}(28.851/18.015)$$
$$= 0.016 \text{ mole(water)/mole(dry air)}$$

For 20 percent excess moist air, we get

$$C_3H_8 + 6O_2 + 22.56N_2 + 28.56 \times 0.016H_2O$$
$$\rightarrow 3CO_2 + (4 + 28.56 \times 0.016)H_2O + 22.56N_2 + O_2$$

The total number of moles of products is $3 + 4.457 + 22.56 + 1 = 31.017$. The mole fractions of the products therefore become

CO_2:	$3/31.017 = 0.0967$
H_2O:	$4.457/31.017 = 0.1438$
N_2:	$22.56/31.017 = 0.7273$
O_2:	$1/31.017 = 0.0322$
	1.0000

The molecular mass of the products is

$$\overline{M} = 0.0967(44.009) + 0.1438(18.015) + 0.7273(28.014) + 0.0322(31.998)$$
$$= 28.251$$

Hence

$$R = R_u/\overline{M} = 1545.3/28.251 = 54.70 \text{ ft-lbf/lbm°R}$$

So far we have discussed the combustion process using a single compound as fuel. Often, however, the fuel is actually a mixture of several hydrocarbons. For example, natural gas, used widely for heating and cooking purposes, is a mixture of methane (CH_4), ethane (C_2H_6), and propane (C_3H_8). To obtain the chemical equation for the combustion of a fuel mixture, we add the chemical equations for each constituent fuel. The chemical equations for the constituents using theoretical air are as follows:

$$\text{Methane:} \quad CH_4 + 2O_2 + 7.52N_2 \rightarrow CO_2 + 2H_2O + 7.52N_2$$
$$\text{Ethane:} \quad C_2H_6 + 3.5O_2 + 13.16N_2 \rightarrow 2CO_2 + 3H_2O + 13.16N_2$$
$$\text{Propane:} \quad C_3H_8 + 5O_2 + 18.80N_2 \rightarrow 3CO_2 + 4H_2O + 18.80N_2$$

A typical composition of natural gas on a mole basis is

70% methane

15% ethane

15% propane

Hence

$$0.70CH_4 + 1.40O_2 + 5.264N_2 \rightarrow 0.70CO_2 + 1.40H_2O + 5.264N_2$$
$$0.15C_2H_6 + 0.525O_2 + 1.974N_2 \rightarrow 0.30CO_2 + 0.45H_2O + 1.974N_2$$
$$0.15C_3H_8 + 0.75O_2 + 2.820N_2 \rightarrow 0.45CO_2 + 0.60H_2O + 2.820N_2$$

Adding, we obtain the chemical equation for the fuel mixture:

$$0.7CH_4 + 0.15C_2H_6 + 0.15C_3H_8 + 2.675O_2 + 10.058N_2 \rightarrow 1.45CO_2 + 2.45H_2O + 10.058N_2$$

The air-fuel ratio on a molal basis is

$$A/F = (2.675 + 10.058)/1 = 12.733 \text{ moles(air)/mole(fuel)}$$

The molecular mass of the fuel is

$$\overline{M}_{\text{fuel}} = 0.7(16.043) + 0.15(30.070) + 0.15(44.097) = 22.355$$

Thus the theoretical air-fuel ratio on a mass basis is, with $\overline{M}_{\text{air}} = 28.851$,

$$A/F = 12.733(28.851)/22.355 = 16.43 \text{ lbm(air)/lbm(fuel)}$$

11.3 FIRST LAW FOR COMBUSTION PROCESSES

The first law states that in a steady-flow process with no work and no changes of potential or kinetic energies (Figure 11.1)

$$Q_{\text{out}} = H_{\text{in}} - H_e$$

We are very much interested in determining the heat liberated from a combustion reaction; however, in order to use the above equation, with the reactants and products consisting of different substances, it is necessary to adopt a common baseline to evaluate the enthalpies for the different substances. As a baseline for enthalpy we will use $h = 0$ for an element in its natural state at 25°C (77°F), 1 atm pressure. Hence, the enthalpy of gaseous nitrogen or oxygen at 25°C, 1 atm is zero, as is the enthalpy of solid carbon (graphite). The enthalpy of compounds at the reference state can be determined experimentally. For example, to

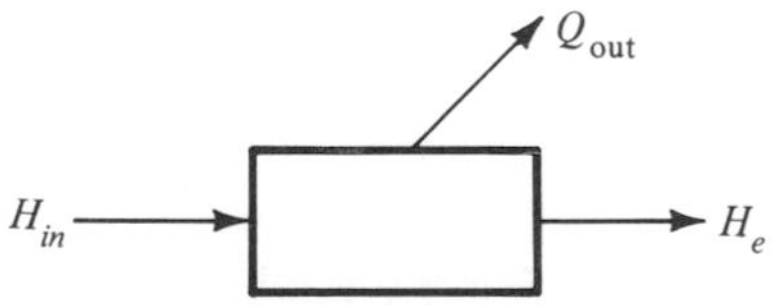

FIGURE 11.1

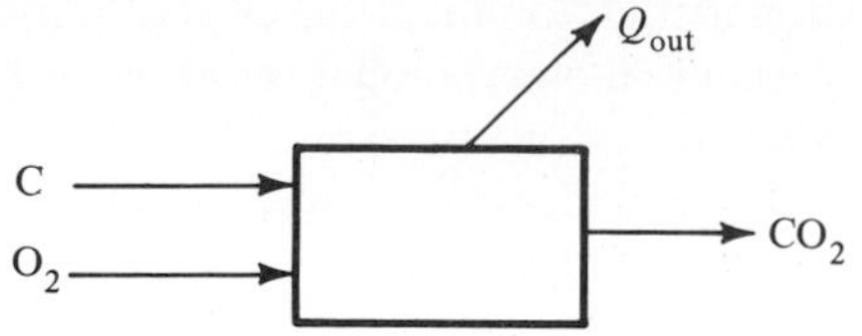

FIGURE 11.2

determine the enthalpy of carbon dioxide at 25°C and 1 atmosphere pressure, react 1 mole of carbon with 1 mole of oxygen in a steady-flow process at 1 atmosphere, cool the products down to 25°C, and carefully measure the heat liberated (Figure 11.2). Applying the first law, we have

$$Q_{out} = H_{reactants} - H_{products} = N_C h'_C + N_{O_2} h'_{O_2} - N_{CO_2} h'_{CO_2}$$

where h' denotes enthalpy per mole and N is the number of moles of the constituent involved in the reaction. At 25°C, $h'_C = h'_{O_2} = 0$ since C and O_2 are elements in their natural state at 25°C. Therefore, $Q_{out} = -N_{CO_2} h'_{CO_2}$. The measured value of heat liberated for this experiment is $Q_{out} = 393\,510$ kJ/kgmol = 169,290 Btu/lbmol. It follows that at 25°C and 1 atmosphere pressure, the enthalpy h' of CO_2 is −169,290 Btu/lbmol or −393 510 kJ/kgmol. The enthalpy of a substance at 25°C and 1 atmosphere is called the *enthalpy of formation* h'_f of that substance, since it is numerically equal to the heat liberated when the substance is formed. Experiments similar to the one mentioned for carbon dioxide have been carried out for the other compounds normally found in combustion reactions. Results are shown in Table 11.2.

TABLE 11.2 *Enthalpies of Formation at 25°C (77°F), 1 atmosphere**

	h'_f (Btu/lbmol)	h'_f (kJ/kgmol)
CO_2	−169,290	−393 510
CO	−47,550	−110 530
H_2O(liquid)	−122,970	−285 830
H_2O(vapor)	−104,030	−241 810
CH_4(methane)	−32,200	−74 850
C_2H_6(ethane)	−36,420	−84 670
C_3H_8(propane)	−44,670	−103 850
C_6H_6(benzene, gas)	+35,680	+82 930
C_8H_{18}(octane, gas)	−89,680	−208 450
C_8H_{18}(octane, liquid)	−107,510	−249 900

* Values adapted from *Handbook of Chemistry and Physics*, Cleveland: CRC Press, 1977, pp. D82–86.

It has been shown that for ideal gases, enthalpy is independent of pressure. Further, except at very high pressures, thermodynamic properties of liquids are insensitive to pressure. Therefore, in our analyses of combustion processes, we will assume that the enthalpies of the various substances are independent of pressure.

With baseline values of enthalpies available, it is now possible to calculate the amount of heat given off in a combustion reaction.

EXAMPLE 11.3

Liquid octane is reacted with 10 percent excess air in a steady-flow combustor at 1 atmosphere pressure. Determine the heat given off per kilogram mole of octane if reactants and products are both maintained at 25°C and water in the products is in liquid form. Then repeat the problem for water vapor in the products.

Solution

For 10 percent excess air,

$$C_8H_{18} + 13.75(O_2 + 3.76N_2) \rightarrow 8CO_2 + 9H_2O + 1.25O_2 + 51.7N_2$$

For steady flow,

$$Q_{out} = H_{reactants} - H_{products}$$

$$H_{products} = 8h'_{f(CO_2)} + 9h'_{f(H_2O)} + 1.25h'_{f(O_2)} + 51.7h'_{f(N_2)}$$

where $h'_{f(N_2)} = h'_{f(O_2)} = 0$ at 25°C since these are elements.

$$H_{products} = 8(-393\,510) + 9(-285\,830) \quad \text{for liquid water in the products}$$

$$= -5\,720\,550 \text{ kJ}$$

$$H_{reactants} = -249\,900 \text{ kJ}$$

since, again, $h'_{f(O_2)} = h'_{f(N_2)} = 0$ at 25°C.

$$Q_{out} = -249\,900 + 5\,720\,550$$

$$= +5\,470\,650 \text{ kJ/kgmol(octane)}$$

with liquid water in the products. With water vapor in the products,

$$Q_{out} = 8(393\,510) + 9(241\,810) - 249\,900 = 5\,074\,470 \text{ kJ/kgmol } C_8H_{18}$$

As this example shows, more heat is given off with liquid water in the products than with water vapor. The difference is due to the latent heat of H_2O; that is, with water vapor in the products, some of the energy that would otherwise be liberated goes into vaporizing the water. The amount of liquid water in the products is determined by the relationship between the partial pressure of the water and the saturation pressure.

The heat given off with reactants and products both at 25°C is independent of the amount of excess air, since the enthalpies of O_2 and N_2 are both zero at 25°C. In other words, the results of Example 11.2 would be the same for theoretical air or any percent excess air. Naturally, if not enough air is provided to completely burn the fuel, less heat will be released.

The heat given off when a fuel is completely oxidized, with reactants and products both at 25°C, is called the *heat of combustion*. Higher heating value (HHV) of a fuel is the heat of combustion with liquid water in the products; lower heating value (LHV) is the heat of combustion with water vapor in the products. Values of HHV and LHV for some gaseous hydrocarbon fuels are given in Tables A.18 and B.18. For example, the HHV of gaseous octane at 25°C is listed as 48 255 kJ/kg. From Example 11.3, the HHV for liquid octane is 5 470 650 kJ/kgmol or 47 892 kJ/kg. For gaseous octane, with more energy in the reactants, HHV = 47 892 + latent heat of octane, where values of latent heat are given in Tables A.19 and B.19. For gaseous octane, HHV = 47 892 + 363 = 48 255 kJ/kg.

If the chemical reaction occurs in a closed volume, we must use the first law for a closed system. With no work term, $Q_{out} = U_1 - U_2$. Since, by definition, $H = U + p\forall$, we have

$$Q_{out} = H_1 - H_2 - (p\forall)_1 + (p\forall)_2$$

We shall assume that since gases occupy much greater volumes than do liquids or solids, moles of liquid and solid can be neglected in the evaluation of the $p\forall$ term. If the gases can be assumed to behave as perfect gases, $Q_{out} = H_1 - H_2 + N_2R_uT_2 - N_1R_uT_1$, where N_1 and N_2 refer to gaseous moles.

EXAMPLE 11.4

Calculate the heat given off (in Btu/lbm) when gaseous propane is reached with theoretical air in a closed container. Assume liquid water in the products, with reactants and products both at 77°F.

Solution

For theoretical air,

$$C_3H_8 + 5(O_2 + 3.76N_2) \rightarrow 3CO_2 + 4H_2O(\text{liquid}) + 18.8N_2$$

$$Q_{out} = H_1 - H_2 + N_2R_uT_2 - N_1R_uT_1$$

$$H_1 - H_2 = H_{reactants} - H_{products}$$

$$H_{reactants} = -44{,}670\ \text{Btu/lbmol} \quad \text{(Table 11.2)}$$

$$H_{\text{products}} = 3(-169{,}290) + 4(-122{,}970) = -999{,}750 \text{ Btu/mol}$$

$$H_1 - H_2 = -44{,}670 + 999{,}750 = 955{,}080 \text{ Btu/lbmol(propane)}$$

$$\begin{aligned} N_2R_uT_2 - N_1R_uT_1 &= (18.8 + 3)(1545 \text{ ft-lbf/mol°R})(537\text{°R})/778 \text{ ft-lbf/Btu} \\ &\quad - (1 + 5 + 18.8)(1545 \text{ ft-lbf/mol°R})(537\text{°R})/778 \text{ ft-lbf/Btu} \\ &= 23{,}250 \text{ Btu/mol} - 26{,}450 \text{ Btu/mol} \end{aligned}$$

Therefore,

$$Q_{\text{out}} = 951{,}880 \text{ Btu/lbmol(propane)}$$

or

$$Q_{\text{out}} = 21{,}585 \text{ Btu/lbm}(C_3H_8)$$

If reactants or products are at a temperature other than the baseline temperature of 25°C, then additional terms must be included to account for the enthalpy or internal energy difference between each substance at the specified temperature and each substance at the reference temperature. Values of such internal energy and enthalpy differences as a function of temperature are given in Tables A.16, A.17, B.16, and B.17.

EXAMPLE 11.5

Gaseous propane at 100°C is reacted with theoretical air at 0°C in a steady-flow combustion process. The products are at 700°C. Calculate the heat given off in kJ/kg of propane.

Solution

For this steady-flow process,

$$Q_{\text{out}} = H_{\text{reactants}} - H_{\text{products}}$$

For theoretical air,

$$C_3H_8 + 5(O_2 + 3.76N_2) \rightarrow 3CO_2 + 4H_2O + 18.8N_2$$

For the reactants,

$$\begin{aligned} H_{\text{reactants}} &= h'_{f(C_3H_8)} + (h'_{100°C} - h'_{25°C})_{C_3H_8} \\ &\quad + 5(h'_{0°C} - h'_{25°C})_{O_2} + 18.8(h'_{0°C} - h'_{25°C})_{N_2} \\ &= -103\,850 \text{ kJ/kgmol} + (140 \text{ kJ/kg})(44.1 \text{ kg/mol}) \\ &\quad + 5 \text{ mol}(-22.9 \text{ kJ/kg})32 \text{ kg/mol} + 18.8 \text{ mol}(-26.0 \text{ kJ/kg})28 \text{ kg/mol} \\ &= -103\,850 + 6174 - 3664 - 13\,686 \\ &= -115\,000 \text{ kJ/kgmol(propane)} \end{aligned}$$

where the enthalpy differences have been found from Table B.17 in kJ/kg.

For the products,

$$H_{\text{products}} = 3h'_{f(CO_2)} + 3(h'_{700°C} - h'_{25°C})_{CO_2}$$
$$+ 4h'_{f(H_2O)} + 4(h'_{700°C} - h'_{25°C})_{H_2O}$$
$$+ 18.8(h'_{700°C} - h'_{25°C})_{N_2}$$
$$= 3[-393\,510 \text{ kJ/mol} + (725.5 \text{ kJ/kg})(44 \text{ kg/mol})]$$
$$+ 4[-241\,810 \text{ kJ/mol} + (1381 \text{ kJ/kg})(18 \text{ kg/mol})]$$
$$+ 18.8[(734.9 \text{ kJ/kg})(28 \text{ kg/mol})]$$
$$= -1\,565\,700 \text{ kJ/kgmol of propane}$$

$$Q_{\text{out}} = -115\,000 + 1\,565\,700$$
$$= 1\,450\,700 \text{ kJ/kgmol(propane)}$$
$$= 32\,900 \text{ kJ/kg(propane)}$$

If the combustion process is caused to proceed adiabatically, that is, if the reaction chamber or steady-flow reaactor is well insulated, then all the energy released from the chemical reaction goes into heating up the products. For this case, we call the temperature of the products the *adiabatic flame temperature*; it represents the maximum temperature achievable for given reactants. For an adiabatic steady-flow reaction, with no work and no changes of potential or kinetic energy, the first law yields

$$H_{\text{reactants}} = H_{\text{products}}$$

For an adiabatic reaction in a closed, rigid, well-insulated container,

$$U_{\text{reactants}} = U_{\text{products}}$$

EXAMPLE 11.6

Calculate the adiabatic flame temperature T_f in steady flow for gaseous propane burned with theoretical air. Fuel and air are at 77°F.

Solution

The chemical equation for the reaction of propane with theoretical air is

$$C_3H_8 + 5O_2 + 18.8N_2 \rightarrow 3CO_2 + 4H_2O + 18.8N_2$$

For this adiabatic process,

$$H_{\text{reactants}} = H_{\text{products}}$$

$$H_{\text{reactants}} = -44{,}670 \text{ Btu}$$

$$H_{\text{products}} = 3h'_{f(CO_2)} + 3(h'_{T_f} - h'_{77°F})_{CO_2} + 4h'_{f(H_2O)} + 4(h'_{T_f} - h'_{77°F})_{H_2O}$$
$$+ 18.8(h'_{T_f} - h'_{77°F})_{N_2}$$

where h'_{T_f}, for example, denotes the enthalpy of the constituent per mole at the adiabatic flame temperature.

Since H_{products} must equal −44,670 Btu, it is now necessary to assume a flame temperature T_f and check for an energy balance. As a first trial, let $T_f = 3000°F$. Using Tables 11.2 and A.17, we have

$$H_{\text{products}} = 3[-169{,}290 \text{ Btu/mol} + (847.6 \text{ Btu/lbm})(44 \text{ lbm/mol})]$$
$$+ 4[-104{,}030 \text{ Btu/mol} + (1640.5 \text{ Btu/lbm})(18 \text{ lbm/mol})]$$
$$+ 18.8(818.7 \text{ Btu/lbm})(28 \text{ lbm/mol})$$
$$= -263{,}000 \text{ Btu}$$

Since the value of H_{products} for the assumed temperature is too negative, as a second trial, assume a higher temperature. Let $T_f = 4000°F$. At this temperature,

$$H_{\text{products}} = 3[-169{,}290 \text{ Btu/mol} + (1177.7 \text{ Btu/lbm})(44 \text{ lbm/mol})]$$
$$+ 4[-104{,}030 \text{ Btu/mol} + (2333.3 \text{ Btu/lbm})(18 \text{ lbm/mol})]$$
$$+ 18.8(1127.8 \text{ Btu/lbm})(28 \text{ lbm/mol})$$
$$= -7000 \text{ Btu}$$

Assuming a linear interpolation between the two assumed temperatures, we have the following:

Assumed T_f	H_{products}
3000°F	−263,000 Btu
T_f	−44,670 Btu
4000°F	−7,000 Btu

If follows that $T_f = 3850°F$.

EXAMPLE 11.7

Liquid octane and theoretical air are contained in a well-insulated rigid vessel at 1 atmosphere pressure at 25°C. The mixture is ignited. Find the final pressure and temperature in the vessel.

Solution

For theoretical air,

$$C_8H_{18} + 12.5(O_2 + 3.76N_2) \rightarrow 8CO_2 + 9H_2O + 47N_2$$

From the first law,

$$Q_{\text{out}} = U_1 - U_2 = H_1 - H_2 - (N_1R_uT_1 - N_2R_uT_2)$$
$$= 0 \quad \text{for an adiabatic process}$$

$$H_1 - H_2 = h'_{f(C_8H_{18})} - 8h'_{f(CO_2)} - 9h'_{f(H_2O)} - 47h'_{f(N_2)} - 8(h'_{T_f} - h'_{77°F})_{CO_2}$$

$$- 9(h'_{T_f} - h'_{77°F})_{H_2O} - 47(h'_{T_f} - h'_{77°F})_{N_2}$$

$$N_1 = 59.5 \quad \text{(count only gaseous moles)}$$

$$N_2 = 64$$

Now, assume T_2 until $Q_{out} = 0$. As a first trial, let $T_2 = 2500°C$. Using Tables 11.2 and B.17, we have

$$H_1 - H_2 = -249\,900 \text{ kJ} + 8(393\,510) \text{ kJ} + 9(241\,810) \text{ kJ} - 47(0)$$

$$- 8(3153.1 \text{ kJ/kg})44 \text{ kg/mol} - 9(6317.5 \text{ kJ/kg})18 \text{ kg/mol}$$

$$- 47(3010.4 \text{ kJ/kg})28 \text{ kg/mol}$$

$$= 5\,074\,470 \text{ kJ} - 6\,095\,012 \text{ kJ}$$

$$= -1\,020\,540 \text{ kJ}$$

$$U_1 - U_2 = -1\,020\,540 \text{ kJ/mol} - 59.5(8.314 \text{ kJ/mol K})298 \text{ K}$$

$$+ 64(8.314 \text{ kJ/mol K})2773 \text{ K}$$

$$= +307\,550 \text{ kJ}$$

Since we wish to arrive at a value of T_2 such that $U_1 - U_2 = 0$, as a second trial, try $T_2 = 3000°C$. In this case,

$$U_1 - U_2 = 5\,074\,470 \text{ kJ} - 8(3860.2 \text{ kJ/kg})44 \text{ kg/mol}$$

$$- 9(7864.1 \text{ kJ/kg})(18 \text{ kg/mol}) - 47(3671.6 \text{ kJ/kg})28 \text{ kg/mol}$$

$$+ 64(8.314 \text{ kJ/mol K})3273 \text{ K} - 59.5(8.314 \text{ kJ/mol K})298 \text{ K}$$

$$= -796\,000 \text{ kJ}$$

Interpolation yields the following:

Assumed T_f	$U_1 - U_2$
2500°C	+307 550 kJ
T_f	0
3000°C	−796 000 kJ

It follows that $T_f = 2640°C$.

For this constant process,

$$p_2 = p_1 \frac{N_2}{N_1} \frac{T_2}{T_1} = (1) \text{ atm} \frac{64}{59.5} \frac{2690 \text{ K}}{298 \text{ K}} = 9.71 \text{ atmospheres}$$

As an alternate procedure for finding T_f, we could calculate the heat given off by the chemical reaction with reactants and products both at 25°C. Then, to find T_f, we equate this heat release with $(U_{T_f} - U_{25°C})_{\text{products}}$. For example, with reactants and products at 25°C, for the combustion of octane in a closed vessel,

$$Q_{\text{out}(25°\text{C})} = (\Delta U)_{25°\text{C}} = (\Delta H)_{25°\text{C}} - (\Delta N)R_u 298$$

$$(\Delta H)_{25°\text{C}} = 8(-393\,510\ \text{kJ}) + 9(-241\,810\ \text{kJ}) + 249\,900\ \text{kJ} = -5\,074\,470\ \text{kJ}$$

$$(\Delta U)_{25°\text{C}} = -5\,074\,470\ \text{kJ} - 4.5(8.314\ \text{kJ/mol K})298\ \text{K} = -5\,085\,620\ \text{kJ}$$

$$= (U_{T_f} - U_{25°\text{C}})_{\text{products}}$$

where internal energies of the product gases are found in Table B.17. For example, as a first trial, let $T_f = 2500°\text{C}$. Then

$$(U_{2500°\text{C}} - U_{25°\text{C}}) = 8(2685.5\ \text{kJ/kg})44\ \text{kg/mol} + 9(5175.2\ \text{kJ/kg})18\ \text{kg/mol}$$
$$+ 47(2275.8\ \text{kJ/kg})28\ \text{kg/mol}$$
$$= 4\,778\,630\ \text{kJ}$$

Since this value is too low, assume a higher T_f. Let $T_f = 3000°\text{C}$.

$$(U_{3000°\text{C}} - U_{25°\text{C}}) = 8(3298.2\ \text{kJ/kg})44\ \text{kg/mol} + 9(6491.0\ \text{kJ/kg})18\ \text{kg/mol}$$
$$+ 47(2788.6\ \text{kJ/kg})28\ \text{kg/mol}$$
$$= 5\,882\,310\ \text{kJ}$$

Interpolating, we obtain the following:

Assumed T_f	$(U_{T_f} - U_{25°\text{C}})_{\text{products}}$
2500°C	4 778 630 kJ
T_f	5 085 620 kJ
3000°C	5 882 310 kJ

Solving, we obtain $T_f = 2640°\text{C}$.

11.4 SECOND LAW FOR COMBUSTION PROCESSES

In this section we will apply the concepts of availability and irreversibility introduced in Chapter 7 to combustion processes and determine changes in availability of fuel–air reactants with respect to products of combustion.

Using Equation (10.3a) we obtain the availability balance equation for a steady-flow, open-loop system in which there is no work:

$$A_{\text{in}} - A_e = I$$

For constant volume combustion processes in a closed system, we get

$$A_1 - A_2 = I$$

where the capital letters indicate total quantities, i.e., Btu's or kJ's.

To evaluate the availability of reactive constituents in combustion processes we take as reference state the products of combustion in gaseous form at a temperature of 77°F (25°C) and one atmosphere. Hence, we can use the lower heating values of the various fuels (Tables A.18 or B.18) as reference values of availability a_0. The reactants and products are gas mixtures and the pressure of each component is the partial pressure in the gas mixture. From Equation (5.1), we have $p_i = (N_i/N)p$, where N_i/N is the mole fraction and p is the total pressure of the mixture. In general, the constituents will not be at the standard reference conditions of $T_0 = 77°\text{F}$ and $p_0 = 1$ atmosphere. Hence, we have to correct the availability values for pressure and temperature.

For processes involving chemical reactions, we need to include the chemical availability a_0 of the fuel in the definition of availability as given in Equation (7.6) for steady-flow systems. Thus, the expression for the availability of each constituent becomes, omitting kinetic and potential energies,

$$a = h - h_0 - T_0(s - s_0) + a_0$$

Further, for an ideal gas, we have from Equation (6.25)

$$s - s_0 = c_p \ln (T/T_0) - R \ln (p/p_0)$$

Thus

$$a - a_0 = h - h_0 - T_0[c_p \ln (T/T_0) - R \ln (p/p_0)]$$

For closed systems, we modify Equation (7.7) to include the chemical availability of the fuel and obtain

$$a = u - u_0 + p_0(v - v_0) - T_0[c_p \ln (T/T_0) - R \ln (p/p_0)] + a_0$$

Detailed aspects of the calculation procedures to determine availability changes in combustion will be shown in the next two examples.

EXAMPLE 11.8

Methane is burned with 20 percent excess dry air in a steady-flow process at a pressure of one atmosphere. Determine the adiabatic flame temperature and the loss of availability due to the combustion. The mixture of methane and air enters the burner at 77°F and 1 atmosphere.

Solution

The chemical equation for the combustion of methane with 20 percent excess air is

$$CH_4 + 2.4O_2 + 9.024N_2 \rightarrow CO_2 + 2H_2O + 9.024N_2 + 0.4O_2$$

The total number of moles of reactants is $1 + 2.4 + 9.024 = 12.424$; the total number of moles of products is $1 + 2.0 + 9.024 + 0.4 = 12.424$. The mole fraction of each reactant and product therefore becomes

CH_4:	$1/12.424 = 0.0805$	CO_2:	$1/12.424 = 0.0805$
O_2:	$2.4/12.424 = 0.1932$	H_2O:	$2/12.424 = 0.1610$
N_2:	$9.024/12.424 = 0.7263$	N_2:	$9.024/12.424 = 0.7263$
	1.0000	O_2:	$0.4/12.424 = 0.0322$
			1.0000

For an adiabatic process, we have $H_{\text{reactants}} = H_{\text{products}}$. From Table 11.2 the value of enthalpy of formation of methane is $-32{,}200$ Btu/lbmol. Hence, for one mole of CH_4, $H_{\text{reactants}} = -32{,}200$ Btu. The enthalpy of the products is

$$H_{\text{products}} = h'_{f(CO_2)} + (h'_{T_f} - h'_{77°F})_{CO_2} + 2h'_{f(H_2O)} + 2(h'_{T_f} - h'_{77°F})_{H_2O}$$
$$+ 9.024(h'_{T_f} - h'_{77°F})_{N_2} + 0.4(h'_{T_f} - h'_{77°F})_{O_2}$$

where h'_{T_f} denotes the enthalpy of the constituents per mole at the adiabatic flame temperature. As a first trial, assume a flame temperature $T_f = 3200°F$. Using Tables 11.2 and A.17, we have

$$H_{\text{products}} = -169{,}290 \text{ Btu/lbmol} + (913.6 \text{ Btu/lbm})(44 \text{ lbm/mol})$$
$$+ 2[-104{,}030 \text{ Btu/lbmol} + (1779.1 \text{ Btu/lbm})(18 \text{ lbm/mol})]$$
$$+ 9.024(880.5 \text{ Btu/lbm})(28 \text{ lbm/mol})$$
$$+ 0.4(813.1 \text{ Btu/lbm})(32 \text{ lbm/mol}) = -40{,}216 \text{ Btu}$$

As a second trial, assume a higher temperature and let $T_f = 3300°F$. At this temperature we get

$$H_{\text{products}} = -169{,}290 + (946.6)44 + 2[-104{,}030 + (1848.3)18]$$
$$+ 9.024(911.4)28 + 0.4(841.6)32 = -28{,}104 \text{ Btu}$$

Interpolating, we have

Assumed T_f	H_{products}
3200°F	$-40{,}216$ Btu
T_f	$-32{,}200$ Btu
3300°F	$-28{,}104$ Btu

Hence, we obtain $T_f = 3266°F$.

In this example, the total pressure equals one atmosphere. Hence, the partial pressure of each constituent will equal the value of its mole fraction. To determine the availability of reactant or product gases, use

$$a - a_0 = h - h_0 - T_0\left(c_p \ln \frac{T}{T_0} - R \ln \frac{p}{p_0}\right)$$

where $T_0 = 537°R$. Values of R are given in Table A.1; values of c_p for product gases are averaged over the range from 77°F to 3266°F. For example, for CO_2,

$$c_p = \frac{h - h_0}{T - T_0} = \frac{935.4 \text{ Btu/lbm}}{3266°\text{F} - 77°\text{F}} = 0.293 \text{ Btu/lbm°R}$$

and

$$\begin{aligned} a - a_0 &= 935.4 \text{ Btu/lbm} - 537°\text{R}\Big(0.293 \text{ Btu/lbm°R} \ln \frac{3726}{537} \\ &\quad - \frac{35.11 \text{ ft-lbf/lbm°R}}{778 \text{ ft-lbf/Btu}} \ln 0.0805\Big) \\ &= 569.5 \text{ Btu/lbm} \end{aligned}$$

Using the same calculation procedure, we obtain the results shown in the following table.

Reactant	$RT_0 \ln p/p_0$ (Btu/lbm)	$a - a_0$ (Btu/lbm)	a_0 (Btu/lbm)	$A - A_0$ (Btu/lbm)
CH_4	−167.5	167.5	21,500	341,320
O_2	−54.8	−54.8	0	−4,210
N_2	−12.2	−12.2	0	−3,080

Total = 334,030 Btu/mol methane

Product	$h - h_0$ (Btu/lbm)	c_p (Btu/lbm°R)	$c_p \ln T/T_0$ (Btu/lbm°R)	$R \ln p/p_0$ (Btu/lbm°R)	$a - a_0$ (Btu/lbm)	$A - A_0$ (Btu/mol)
CO_2	935.4	0.293	0.568	−0.1136	569.5	25,060
H_2O	1824.8	0.572	1.108	−0.2014	1121.7	40,380
N_2	900.9	0.283	0.548	−0.0227	594.3	150,160
O_2	831.9	0.261	0.506	−0.2134	445.6	5,704

Total = 221,304 Btu/mol methane

Hence, the decrease in availability due to adiabatic combustion of methane with 20 percent excess air is

$$334{,}030 - 221{,}304 = 112{,}726 \text{ Btu/mol } CH_4$$

In other words, 34 percent of the entering availability is lost in the conversion of chemical to thermal energy.

The adiabatic flame temperature is the temperature achieved when combustion takes place without heat loss from the burner. From the results of the second-law analysis, it is seen that the combustion process incurs a large loss in availability. In this case, the availability contained in the fuel was reduced by 34 percent due to combustion, while energy was conserved.

We see that in a furnace (e.g., Example 10.4) an appreciable portion of the availability of the fuel is lost in the process of converting chemical to thermal energy (combustion). Additional availability losses occur in the heat exchanger of the furnace.

EXAMPLE 11.9

Calculate the adiabatic flame temperature and the change in availability when gaseous propane is burned with theoretical air in a well-insulated rigid vessel. The fuel and air are at 25°C and atmospheric pressure.

Solution

The chemical equation for the reaction of propane with theoretical air is

$$C_3H_8 + 5O_2 + 18.8N_2 \rightarrow 3CO_2 + 4H_2O + 18.8N_2$$

The number of moles of reactants is 24.8; the number of moles of products is 25.8. The mole fraction of each reactant and product becomes

C_3H_8:	1/24.8	0.0403	CO_2:	3/25.8	0.1163
O_2:	5/24.8	0.2016	H_2O:	4/25.8	0.1550
N_2:	18.8/24.8	0.7581	N_2:	18.8/25.8	0.7287
		1.0000			1.0000

For this adiabatic process, we have

$$U_1 - U_2 = H_1 - H_2 - (N_1R_uT_1 - N_2R_uT_2) = 0$$

Now,

$$\begin{aligned} H_1 - H_2 = h'_{f(C_3H_8)} - 3h'_{f(CO_2)} - 4h'_{f(H_2O)} - 18.8h'_{f(N_2)} \\ - 3(h'_{T_f} - h'_{25°C})_{CO_2} - 4(h'_{T_f} - h'_{25°C})_{H_2O} \\ - 18.8(h'_{T_f} - h'_{25°C})_{N_2} \end{aligned}$$

As a first trial, assume $T_2 = 2500°C$. Using Tables 11.2 and B.17, we have

$$\begin{aligned} H_1 - H_2 &= -103{,}850 \text{ kJ} + 3(393{,}510) \text{ kJ} + 4(241{,}810) \text{ kJ} - 18.8(0) \\ &\quad - 3(3153.1 \text{ kJ/kg})44 \text{ kg/mol} - 4(6317.5 \text{ kJ/kg})18 \text{ kg/mol} \\ &\quad - 18.8(3010.4 \text{ kJ/kg})28 \text{ kg/mol} \\ &= 2{,}043{,}920 - 2{,}455{,}744 = -411{,}824 \text{ kJ} \end{aligned}$$

Hence,

$$U_1 - U_2 = -411{,}824 \text{ kJ} - 61{,}444 \text{ kJ} + 25.8(8.314 \text{ kJ/mol K})2773 \text{ K}$$
$$= 121{,}544 \text{ kJ}$$

As a second trial, try $T_2 = 3000°C$. In this case, we get

$$H_1 - H_2 = 2{,}043{,}920 - 3(3860.2)44 - 4(7864.1)18 - 18.8(3671.6)28$$
$$= 2{,}043{,}920 - 3{,}008{,}491 = -964{,}571 \text{ kJ}$$

Hence,

$$U_1 - U_2 = -964{,}571 - 24.8(8.314 \text{ kJ/mol K})298 \text{ K} + 25.8(8.314)3273$$
$$= -323{,}953 \text{ kJ}$$

Interpolation yields

$$T_2 = 2636°C = 2909 \text{ K}$$

For this constant volume process, we obtain the final pressure from

$$p_2 = p_1 \frac{N_2}{N_1}\frac{T_2}{T_1} = (1)\left(\frac{25.8}{24.8}\right)\left(\frac{2909}{298}\right) = 10.155 \text{ atm}$$

To obtain the availability of the constituent gases at their partial pressures, we use

$$a - a_0 = u - u_0 + p_0(v - v_0) - T_0(s - s_0)$$

where

$$s - s_0 = c_p \ln \frac{T}{T_0} - R \ln \frac{p}{p_0}$$

For the reactants at 25°C, $u - u_0 = 0$ and $T = T_0$; so $a - a_0 = p_0(v - v_0) + RT_0 \ln p/p_0$.

For example, for C_3H_8,

$$v = \frac{RT}{p} = \frac{(0.1885 \text{ kJ/kg K})298 \text{ K}}{0.0403 \times 101.3 \text{ kN/m}^2} = 13.76 \text{ m}^3/\text{kg}$$

$$v_0 = \frac{RT_0}{p_0} = \frac{(0.1885 \text{ kJ/kg K})(298 \text{ K})}{(101.3 \text{ kN/m}^2)} = 0.555 \text{ m}^3/\text{kg}$$

$$p_0(v - v_0) = (101.3 \text{ kN/m}^2)(13.20 \text{ m}^3/\text{kg}) = 1337 \text{ kJ/kg}$$

$$RT_0 \ln \frac{p}{p_0} = (0.1885 \text{ kJ/kg K})(298 \text{ K}) \ln 0.0403$$
$$= -180.4 \text{ kJ/kg}$$

$$a - a_0 = 1337 \text{ kJ/kg} - 180.4 \text{ kJ/kg} = 1157 \text{ kJ/kg}$$

For propane, $a_0 = 46\,350$ kJ/kg (LHV); therefore, $a = 47\,507$ kJ/kg or

$$A_{C_3H_8} = (47\,507 \text{ kJ/kg})(44.1 \text{ kg/mol})$$
$$= 2\,095\,060 \text{ kJ/mol } C_3H_8$$

For the other reactants,

Reactant	R (kJ/kg K)	p/p_0	v (m^3/kg)	v_0 (m^3/kg)	$p_0(v - v_0)$ (kJ/kg)
O_2	0.2598	0.2016	3.791	0.764	307
N_2	0.2968	0.7581	1.152	0.873	28.3

Reactant	$RT_0 \ln p/p_0$ (kJ/kg)	$a - a_0$ (kJ/kg)
O_2	124	183
N_2	24.5	3.8

Summing for the reactants, we get

$$\begin{aligned} A &= 2\,095\,060 \text{ kJ/mol } C_3H_8 \\ &\quad + 5(32.0)183 \text{ kJ/mol } C_3H_8 \\ &\quad + 18.8(28.0)3.8 \\ &= 2\,126\,340 \text{ kJ/mol } C_3H_8 \end{aligned}$$

For the products at 2909 K we have the following:

$$a - a_0 = u - u_0 + p_0(v - v_0) - c_p T_0 \ln \frac{T}{T_0} + RT_0 \ln \frac{p}{p_0}$$

Values of $u - u_0$ are provided in Appendix Table B.16. Values of c_p for the range 25°C to 2636°C are calculated from

$$c_p = \frac{h - h_0}{T - T_0}$$

Product	R (kJ/kg K)	p/p_0	v (m^3/kg)	v_0 (m^3/kg)	$u - u_0$ (kJ/kg)	$p_0(v - v_0)$ (kJ/kg)
CO_2	0.1889	1.1811	4.593	0.556	2852.2	408.9
H_2O	0.4615	1.5741	8.419	1.358	5369.9	715.3
N_2	0.2968	7.4002	1.152	0.873	2415.3	28.3

Product	$h - h_0$ (kJ/kg)	c_p (kJ/kg K)	$T_0 c_p \ln T/T_0$ (kJ/kg)	$T_0 R \ln p/p_0$ (kJ/kg)	$a - a_0$ (kJ/kg)
CO_2	3346.0	1.2815	870.2	9.37	2400.3
H_2O	6739.4	2.5812	1752.5	62.39	4395.1
N_2	3190.8	1.2221	829.9	177.03	1790.7

with values of $h - h_0$ provided in Appendix Table B.17. For the products,

$$\begin{aligned} A &= (2400.3 \text{ kJ/kg})(3 \times 44.0) \text{ kg/mol propane} \\ &\quad + (4395.1 \text{ kJ/kg})(4 \times 18.0) \text{ kg/mol propane} \\ &\quad + (1790.7 \text{ kJ/kg})(18.8 \times 28.0) \text{ kg/mol propane} \\ &= 1\,575\,920 \text{ kJ/mol propane} \end{aligned}$$

The decrease in availability due to adiabatic combustion of propane with theoretical air in a closed system is

$$\begin{aligned} A_{\text{reactants}} - A_{\text{products}} &= 2\,126\,340 - 1\,575\,920 \\ &= 550\,420 \text{ kJ/mol } C_3H_8 \end{aligned}$$

In other words, $(550\,420/2\,126\,340) \times 100$ or 26 percent of the initial availability is lost in the conversion of chemical to thermal energy.

11.5 CHEMICAL EQUILIBRIUM

Combustion temperatures are often high enough that dissociation of the product molecules must be accounted for. By *dissociation*, we mean the breakup of molecules into their base constituents. For example, at high temperatures, water molecules, at least to some extent, will dissociate into hydrogen and oxygen; carbon dioxide will dissociate into carbon monoxide and oxygen; and even molecular hydrogen will dissociate into atomic hydrogen. It is clear that dissociation will affect the composition of the products of combustion. Also, since energy is required for dissociation, the temperature achieved in an actual combustion reaction with dissociation will be less than that calculated on the basis of no dissociation. In this section, we will develop methods to handle combustion reactions with dissociation.

Let us consider a familiar example. We start with water at 1 atmosphere and room temperature, heat it to 500 K, and let the products reach equilibrium. At 500 K, H_2O is quite stable, and there is very little tendency toward dissociation into hydrogen and oxygen. Starting with 1 mole of H_2O, we can write the chemical equation as follows:

$$H_2O \rightarrow aH_2O + bH_2 + cO_2 \qquad \textbf{(11.1)}$$

where $a \gg b$ and $a \gg c$. In fact, at 500 K and 1 atmosphere, b and c are both less than 10^{-15} moles and can be neglected.

Now raise the temperature of the product mixture to 2500 K and let the H_2O, O_2, and H_2O reach a new equilibrium. Now b and c will be much greater percentages of a.

In the mixture of H_2, O_2, and H_2O, H_2O is dissociating into H_2 and O_2, but also H_2 and O_2 are recombining to form H_2O. These reactions are occurring as follows:

$$\left.\begin{aligned} H_2 + \tfrac{1}{2}O_2 &\rightarrow H_2O \\ H_2O &\rightarrow H_2 + \tfrac{1}{2}O_2 \end{aligned}\right\} \tag{11.2}$$

or, as it is usually written,

$$H_2 + \tfrac{1}{2}O_2 \rightleftharpoons H_2O$$

At equilibrium, the rate of formation of H_2O is the same as the rate of dissociation of H_2O, so the number of moles (a, b, and c) of H_2O, H_2, and O_2 remains constant. Note that the coefficients of H_2, O_2, and H_2O in Equation (11.2) are not the same as a, b, and c of Equation (11.1). Equation (11.2) expresses the fact that for each mole of H_2O formed, 1 mole of H_2 and $\frac{1}{2}$ mole of O_2 are used up. Likewise, for each mole of H_2O that dissociates, 1 mole of H_2 and $\frac{1}{2}$ mole of O_2 are formed. Whatever the values of a, b, and c at equilibrium, the equilibrium reactions proceed according to Equation (11.2).

As the temperature of the product mixture is increased, the equilibrium value of a will decrease, b and c will increase, and other species will start to appear in appreciable quantities. At higher temperatures, Equation (11.1) becomes

$$H_2O \rightarrow aH_2O + bH_2 + cO_2 + dOH + eO + fH \tag{11.3}$$

In the product mixture, equilibrium reactions are occurring according to the following equations:

$$\left.\begin{aligned} H_2 + \tfrac{1}{2}O_2 &\rightleftharpoons H_2O \\ H_2O &\rightleftharpoons \tfrac{1}{2}H_2 + OH \\ H_2 &\rightleftharpoons 2H \\ O_2 &\rightleftharpoons 2O \end{aligned}\right\} \tag{11.4}$$

Again, at equilibrium, the rate at which each reaction proceeds to the right is balanced by the rate at which it proceeds to the left so that at equilibrium, there is no variation of product composition, and a, b, c, d, e, and f reach steady-state values.

In order to be able to determine product compositions with dissociation, it is necessary to have a criterion for chemical equilibrium. Such a criterion for equilibrium in a mixture at a given pressure and temperature is provided by the thermodynamic property called *Gibbs function*. Gibbs function G is defined by

$$G = H - TS \tag{11.5}$$

In the differential form,

$$dG = dH - T\,dS - S\,dT$$

But, from Equation (6.21),

$$T\,dS = dH - V\,dp$$

so that

$$dG = V\,dp - S\,dT \tag{11.6}$$

For a system that has reached pressure and temperature equilibrium, $dp = dT = 0$, so that our criterion for chemical equilibrium is

$$dG = 0 \tag{11.7}$$

To apply this criterion, let us consider a general case in which species A, B, C, and D are in chemical equilibrium at temperature T and pressure p, and a moles of A, b moles of B, c moles of C, and d moles of D. In this system, reactions are occurring according to

$$\nu_A A + \nu_B B \rightleftharpoons \nu_C C + \nu_D D \tag{11.8}$$

From our previous example, referring to Equation (11.2), A is H_2, $\nu_A = 1$; B is O_2, $\nu_B = \frac{1}{2}$; and C is H_2O, $\nu_C = 1$.

If the reaction described by Equation (11.8) occurs to an infinitesimal extent to the right (or left), then our criterion of equilibrium requires that $dG = 0$. In other words, for each infinitesimal reaction,

$$(\nu_C g'_C + \nu_D g'_D - \nu_A g'_A - \nu_B g'_B)\,d\varepsilon = 0 \tag{11.9a}$$

where $d\varepsilon$ is a measure of the extent of the reaction, such that $\nu_A\,d\varepsilon$ represents, for example, da, the infinitesimal change in the number of moles of A as a result of the equilibrium reaction; and g'_A denotes the Gibbs function per mole of constituent A. A positve value of $d\varepsilon$ indicates that the reaction in Equation (11.8) proceeds toward the right; a negative value that it proceeds toward the left. Since $d\varepsilon$ can have both positive and negative values, the term in brackets in Equation (11.9a) must be zero:

$$\nu_C g'_C + \nu_D g'_D - \nu_A g'_A - \nu_B g'_B = 0 \tag{11.9b}$$

In order to evaluate Gibbs function, let $g' = g'_{ref} + (g' - g'_{ref})$, where g'_{ref} = Gibbs function per mole at temperature T and 1 atmosphere pressure. For a given constituent, the term $g' - g'_{\text{ref}}$ represents the difference in Gibbs function per mole for that constituent at temperature T between 1 atmosphere pressure and the partial pressure of that constituent in the mixture. To evaluate this difference, use Equation (11.6):

$$dG = \forall\, dp - S\, dT = \forall\, dp$$

for a given temperature T ($dT = 0$). Now we shall assume that all constituents behave as ideal gases, so that $p\forall = NR_uT$. Therefore,

$$dg' = \forall \frac{dp}{N} = R_uT \frac{dp}{p}$$

Integrating between 1 atmosphere and a partial pressure of p atmospheres, we obtain

$$\int dg' = R_uT \int_{p=1\text{ atm}}^{p=p\text{ atm}} dp/p$$

or $g' - g'_{\text{ref}} = R_uT \ln p$ where, again, p is expressed in atmospheres. Applying this result to each of the constituents in Equation (11.8), we obtain

$$\nu_C g'_C = \nu_C g'_{C(\text{ref})} + \nu_C R_u T \ln p_C$$

$$\nu_D g'_D = \nu_D g'_{D(\text{ref})} + \nu_D R_u T \ln p_D$$

$$\nu_A g'_A = \nu_A g'_{A(\text{ref})} + \nu_A R_u T \ln p_A$$

$$\nu_B g'_B = \nu_B g'_{B(\text{ref})} + \nu_B R_u T \ln p_B$$

where p_A, for example, represents the partial pressure of A in atmospheres. Combining yields for our criterion for equilibrium (11.9b):

$$\nu_C g'_{C(\text{ref})} + \nu_D g'_{D(\text{ref})} - \nu_A g'_{A(\text{ref})} - \nu_B g'_{B(\text{ref})} = R_uT \ln \frac{p_A^{\nu_A} p_B^{\nu_B}}{p_C^{\nu_C} p_D^{\nu_D}} \qquad \textbf{(11.10)}$$

The terms on the left side of Equation (11.10) represent values of Gibbs function at 1 atmosphere and temperatre T. By its definition, Gibbs function is a thermodynamic property, so that values of the left side can be determined as a function of temperature. For convenience, Equation (11.10) can be rearranged as follows:

$$\ln \frac{p_A^{\nu_A} p_B^{\nu_B}}{p_C^{\nu_C} p_D^{\nu_D}} = \frac{\nu_C g'_{C(\text{ref})} + \nu_D g'_{D(\text{ref})} - \nu_A g'_{A(\text{ref})} - \nu_B g'_{B(\text{ref})}}{R_uT} \qquad \textbf{(11.11)}$$

The term $(p_C^{\nu_C} p_D^{\nu_D})/(p_A^{\nu_A} p_B^{\nu_B})$ is defined as the equilibrium constant K_p, so that

$$\ln K_p = -\frac{\nu_C g'_{C(\text{ref})} + \nu_D g'_{D(\text{ref})} - \nu_A g'_{A(\text{ref})} - \nu_B g'_{B(\text{ref})}}{R_u T} \tag{11.12}$$

It follows that K_p is a function of temperature only. Values of K_p for various equilibrium reactions are given in Tables A.20 and B.20. The following examples indicate the utilization of the equilibrium constant for obtaining reaction products with dissociation.

EXAMPLE 11.10

A mixture of 50 percent CO_2 and 50 percent O_2 by volume is heated in a steady-flow process at 1 atmosphere to 3000°C. Determine the equilibrium composition of the products.

Solution

Taking atomic balances for C and O (see Figure 11.3), we have

$$\text{C:}\quad 1 = a + c$$

$$\text{O:}\quad 4 = 2a + 2b + c$$

To solve for three unknowns a, b, and c, we need one more equation, which is provided by the equilibrium constant K_p for the reaction

$$CO_2 \rightleftharpoons CO + \tfrac{1}{2}O_2$$

where

$$K_p = \frac{p_{CO}\, p_{O_2}^{1/2}}{p_{CO_2}}$$

At 3000°C, from Table B.20, $\ln K_p = -0.206$ or $K_p = 0.8138$. From Equation (5.1), in the mixture of products, the partial pressure of CO_2 is

$$p_{CO_2} = \left(\frac{a}{a+b+c}\right) p_e$$

where p_e = total pressure of product gases. Likewise,

$$p_{O_2} = \left(\frac{b}{a+b+c}\right) p_e \quad \text{and} \quad p_{CO} = \left(\frac{c}{a+b+c}\right) p_e$$

Combining yields

$$0.8138 = \frac{c\sqrt{b}\sqrt{p_e}}{a\sqrt{a+b+c}}$$

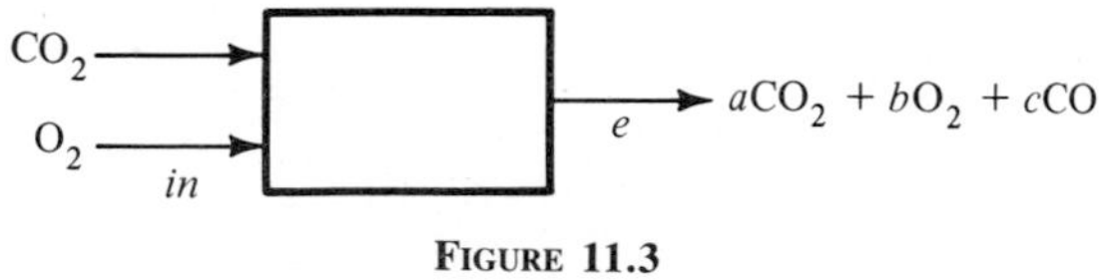

FIGURE 11.3

where $p_e = 1$ atmosphere. Using the atomic balance equations, we obtain $a = 1 - c$, $b = 1 + (c/2)$, so that

$$0.8138 = \frac{c\sqrt{1 + (c/2)}}{(1 - c)\sqrt{2 + (c/2)}}$$

Solving the resultant cubic equation, we have

$$c = 0.521 \qquad a = 0.479 \qquad b = 1.261$$

EXAMPLE 11.11

One mole of CO_2 is contained in a closed, rigid vessel at 300 K and 1 atmosphere. The contents are heated to 3000°C. Find the equilibrium contents and the final pressure (see Figure 11.4).

Solution

Assuming all gases ideal, we have

$$\frac{p_2}{p_1} = \frac{N_2}{N_1}\frac{T_2}{T_1}$$

since $\forall$ = constant.

$$p_2 = p_1(a + b + c)\,\frac{3273}{300} = 10.91(a + b + c)$$

where $p_1 = 1$ atmosphere. Taking atomic balances, we have

$$\text{C:} \quad 1 = a + c$$

$$\text{O:} \quad 2 = 2a + 2b + c$$

or

$$a = 1 - c\,, \quad b = c/2$$

From Example 11.8, for $CO_2 \rightleftharpoons CO + \frac{1}{2}O_2$, $K_p = 0.8138$ at 3000°C, where

$$K_p = \frac{p_{CO}\,p_{O_2}^{1/2}}{p_{CO_2}} = \frac{\left[\dfrac{c}{a + b + c}\,10.91(a + b + c)\right]\left[\dfrac{b}{a + b + c}\,10.91(a + b + c)\right]^{1/2}}{\left[\dfrac{a}{a + b + c}\,10.91(a + b + c)\right]}$$

$$0.8138 = \frac{c\sqrt{10.91b}}{a}$$

$$= \frac{c\sqrt{5.455c}}{1 - c}$$

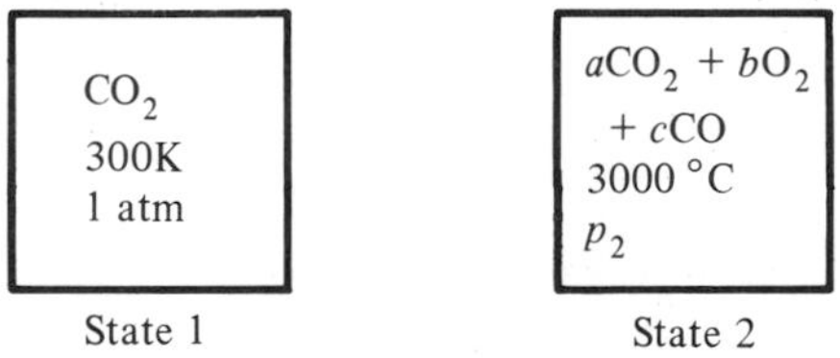

FIGURE 11.4

Solving, we have

$$c = 0.366 \qquad a = 0.634 \qquad b = 0.183$$

$$p_2 = 10.91(0.634 + 0.183 + 0.366) = 12.91 \text{ atmospheres}$$

The equilibrium constant can also be used to determine equilibrium compositions at lower temperatures. In Section 11.2, we noted that for combustion processes with insufficient air, product composition could not be determined by simple atomic balances. The following example will illustrate the use of thc cquilibrium constant to solve such problems.

EXAMPLE 11.12

Propane is burned with 80 percent theoretical air in a steady-flow, constant pressure process. Determine the products at 1000°C. Assume the products to contain CO_2, CO, H_2O, H_2, and N_2.

Solution

The chemical equation for this process is

$$C_3H_8 + 4(O_2 + 3.76N_2) \rightarrow aCO_2 + bCO + cH_2O + dH_2 + 15.04N_2$$

Three equations are provided by atomic balances:

$$\text{C:} \quad 3 = a + b$$

$$\text{H:} \quad 8 = 2c + 2d$$

$$\text{O:} \quad 8 = 2a + b + c$$

or

$$b = 3 - a, \quad c = 5 - a, \quad d = a - 1$$

We must have one more equation to solve for a, b, c, and d. The reaction in the products between CO_2, CO, H_2O, and H_2 is called the *water-gas reaction*:

$$CO_2 + H_2 \rightleftharpoons CO + H_2O$$

where

$$K_p = \frac{p_{CO}p_{H_2O}}{p_{CO_2}p_{H_2}} = \frac{\left(\dfrac{bp_e}{a+b+c+d+15.04}\right)\left(\dfrac{cp_e}{a+b+c+d+15.04}\right)}{\left(\dfrac{ap_e}{a+b+c+d+15.04}\right)\left(\dfrac{dp_e}{a+b+c+d+15.04}\right)}$$

$$= \frac{bc}{ad}$$

Note that for the water-gas reaction, there is no change in the number of moles, so p_e cancels. From Table B.20, ln $K_p = 0.496$, or $K_p = 1.642$. Therefore

$$\frac{bc}{ad} = \frac{(3-a)(5-a)}{a(a-1)} = 1.642$$

Solving, we obtain

$$a = 1.968 \qquad b = 1.032 \qquad c = 3.032 \qquad d = 0.968$$

More than one equilibrium or dissociation reaction may be involved in a given situation. For example, if H_2O were heated to a high temperature, for example, 4000 K, at atmospheric pressure, four equilibrium constants would be involved. The chemical equation would appear as follows:

$$H_2O \rightarrow aH_2 + bO_2 + cH_2O + dH + eO + fOH$$

Taking atomic balances, we have

$$\text{H:} \quad 2 = 2a + 2c + d + f$$

$$\text{O:} \quad 1 = 2b + c + e + f$$

The following equilibrium reactions must be considered:

$$H_2 \rightleftharpoons 2H \qquad K_{p_1} = \frac{p_H^2}{p_{H_2}} = \frac{d^2/(a+b+c+d+e+f)}{a} p_e$$

$$O_2 \rightleftharpoons 2O \qquad K_{p_2} = \frac{p_O^2}{p_{O_2}} = \frac{e^2/(a+b+c+d+e+f)}{b} p_e$$

$$H_2O \rightleftharpoons H_2 + \tfrac{1}{2}O_2 \qquad K_{p_3} = \frac{p_{H_2} p_{O_2}^{1/2}}{p_{H_2O}} = \frac{a\sqrt{b}}{c\sqrt{a+b+c+d+e+f}} \sqrt{p_e}$$

$$H_2O \rightleftharpoons \tfrac{1}{2}H_2 + OH \qquad K_{p_4} = \frac{p_{H_2}^{1/2} p_{OH}}{p_{H_2O}} = \frac{f\sqrt{a}}{c\sqrt{a+b+c+d+e+f}} \sqrt{p_e}$$

The solution of this set of six equations with six unknowns is clearly beyond the realm of hand calculation and should be performed on a high-speed computer.

In the calculation of adiabatic flame temperature, it is often necessary to take dissociation into account. In Example 11.6, we calculated the adiabatic flame temperature of propane with theoretical air, neglecting dissociation, the result being 3850°F. From Tables A.20 and B.20, at this temperature all the equilibrium constants have relative low values, the largest being that for $CO_2 \rightleftharpoons CO + \frac{1}{2}O_2$ ($K_p = 0.02$). Thus, even though there will be a small amount of CO in the exhaust, the equilibrium constant is of a low enough value that it does not have a significant effect on the calculation of flame temperature.

Some fuel-oxidizer combinations exhibit high enough combustion temperatures that dissociation must be accounted for. An example would be the combustion of hydrogen with a theoretical amount of air. In this case, the relationships to determine the adiabatic combustion temperature are as follows:

The chemical equation for the combustion reaction will be

$$H_2 + \tfrac{1}{2}\{O_2 + 3.76N_2\} \rightarrow aH_2O + bH_2 + cO_2 + dOH + eH$$
$$+ fO + gN + hNO + iN_2$$

Here, we have 10 unknowns, namely, a, b, c, d, e, f, g, h, and i, and the combustion temperature T_2. There are three atomic balance equations and six dissociation equilibrium reactions:

$$H_2 \rightleftharpoons 2H$$
$$O_2 \rightleftharpoons 2O$$
$$N_2 \rightleftharpoons 2N$$
$$H_2O \rightleftharpoons H_2 + \tfrac{1}{2}O_2$$
$$H_2O \rightleftharpoons \tfrac{1}{2}H_2 + OH$$
$$\tfrac{1}{2}N_2 + \tfrac{1}{2}O_2 \rightleftharpoons NO$$

The tenth equation required is the energy equation, which for a steady-flow process becomes

$$H_{\text{reactants}} = H_{\text{products}}$$

Again, it is obvious that this type of problem must be solved on a high-speed computer.

11.6 POLLUTANT FORMATION IN COMBUSTION PROCESSES

As a result of the combustion processes occurring in power plants, car and truck engines, space heaters, forest fires, and so on, a considerable amount of air pollutants are emitted into the atmosphere. Some of the principal pollutants include carbon monoxide, sulfur oxides, and nitrogen oxides. Even the relatively small concentrations of these gases in the atmosphere have the potential for severely deleterious effects to both human health and plant and crop growth.

Sulfur oxides, SO_x, generated by combustion processes consist mainly of SO_2 (sulfur dioxide), along with a much smaller amount of SO_3 (sulfur trioxide). Sulfur is present in coal, residual fuel oil, and, to a much lesser extent, natural gas and the lighter fractions of crude oil, such as kerosene and gasoline. During the combustion process, sulfur in the fuel is oxidized to SO_x.

EXAMPLE 11.13

A 1000-megawatt power plant with a thermal efficiency of 40 percent uses coal of 2 percent sulfur content by weight. In the combustion process, 10 percent excess air is used. Take the coal to be, on a weight basis, 75 percent carbon and 2 percent sulfur with the remainder ash and other noncombustibles. The coal has a heating value of 11,000 Btu/lbm. Write the chemical equation for the process, and calculate the number of pounds of SO_2 given off per day.

Solution

In one pound of coal, there are 0.75 pound of carbon and 0.02 pound of sulfur. Since the atomic masses of C and S are, respectively, 12.011 and 32.064, these masses correspond to $0.75/12.011 = 0.06244$ mole of carbon and $0.02/32.064 = 0.000624$ mole of sulfur. Therefore, we can write

$$0.06244\text{C} + 0.000624\text{S} + 1.1(0.06244)(\text{O}_2 + 3.76\text{N}_2)$$
$$\rightarrow 0.06244\text{CO}_2 + 0.000624\text{SO}_2 + 0.25825\text{N}_2 + 0.00562\text{O}_2$$

In other words, for each pound of coal burned, there is produced $(0.000624)64.062 = 0.04$ pound of SO_2. (The molecular mass of SO_2 is 64.062.)

For a 1000-megawatt output, the rate of thermal energy input must be $1000/0.40 = 2500$ megawatts, which is equivalent to

$$(3.412 \times 10^6)2500 = 8.530 \times 10^9 \text{ Btu/h}$$

For a heating value of 11,000 Btu/lbm, there are required

$$\frac{8.530 \times 10^9 \text{ Btu/h}}{11{,}000 \text{ Btu/lb}} = 7.755 \times 10^5 \text{ lbm(coal)/h}$$

For this amount of coal, $0.04(7.755 \times 10^5)$ or 31,000 pounds of SO_2 are emitted into the atmosphere per hour, or 744,400 pounds of SO_2 per day.

The large amount of SO_2 put into the air by the combustion of coal in power plants has necessitated, at least in some areas, the switch either to low-sulfur coal (less than 1 percent sulfur) or to fuel oil or natural gas. The shortage of the latter two fuels and the cost of transporting low-sulfur coal over relatively long distances have aggravated this problem.

Carbon monoxide is present in the exhaust from the combustion of a hydrocarbon fuel if the air-fuel ratio is less than theoretical (stoichiometric), in other words, if there is not a sufficient number of oxygen atoms present to complete the oxidation of carbon to carbon dioxide. The calculation of equilibrium exhaust CO concentration was shown in Example 11.12.

At the very high combustion temperatures that occur in rocket motor combustion chambers, internal combustion engines, and some other combustion devices, it is possible for the CO_2 molecules in the exhaust to dissociate and yield CO, even though the air-fuel mixture ratio may be stoichiometric or lean.

EXAMPLE 11.14

Coal is reacted with a stoichiometric amount of air in a furnace at 1 atmosphere pressure. Calculate the equilibrium exhaust concentration of CO at combustion temperatures of 2000 K and 3000 K. Treat the coal as pure carbon, and assume that the exhaust consists of CO, CO_2, O_2, and N_2.

Solution

For stoichiometric air, the chemical equation is given by

$$C + O_2 + 3.76N_2 \rightarrow aCO + bCO_2 + cO_2 + 3.76N_2$$

Taking atomic balances, we obtain

$$C:\quad 1 = a + b$$

$$O:\quad 2 = a + 2b + 2c$$

or

$$b = 1 - a \qquad c = \tfrac{1}{2}a$$

For the third equation, use the equilibrium reaction $CO_2 \rightleftharpoons CO + \frac{1}{2}O_2$ with $K_p = (p_{CO} p_{O_2}^{1/2})/p_{CO_2}$. Hence

$$p_{CO} = \frac{a}{a + b + c + 3.76}\,(1\text{ atm})$$

$$p_{O_2} = \frac{c}{a + b + c + 3.76}\,(1\text{ atm})$$

$$p_{CO_2} = \frac{b}{a + b + c + 3.76}\,(1\text{ atm})$$

Combining, we get

$$K_p = \frac{a\sqrt{c}}{b\sqrt{a + b + c + 3.76}} = \frac{a\sqrt{a/2}}{(1 - a)\sqrt{4.76 + a/2}}$$

At 2000 K, from Table B.20, $\ln K_p = -6.641$, or $K_p = 0.001306$. At 3000 K, $\ln K_p = -1.117$, or $K_p = 0.3272$. Hence at 2000 K,

$$0.001306 = \frac{a\sqrt{a/2}}{(1 - a)\sqrt{4.76 + a/2}}$$

Solving by trial and error, we obtain $a = 0.025$, $b = 0.975$, and $c = 0.0125$. Thus, the chemical equation for the combustion of coal becomes

$$C + O_2 + 3.76N_2 \rightarrow 0.025CO + 0.975CO_2 + 0.0125O_2 + 3.76N_2$$

(Note that for complete combustion we would obtain $a = 0$, $b = 1$, $c = 0$.) The mole percent of CO in the exhaust is

$$\frac{0.025(100)}{0.025 + 0.975 + 0.0125 + 3.76} = 0.524\%$$

At 3000 K,

$$0.3273 = \frac{a\sqrt{a/2}}{(1-a)\sqrt{4.76 + a/2}}$$

Again, solving by trial and error, we obtain $a = 0.58$, $b = 0.42$, and $c = 0.29$. The mole percent of CO in the exhaust is

$$\frac{0.58(100)}{0.58 + 0.42 + 0.29 + 3.76} = 11.5\%$$

In power generating stations, air-fuel ratios are generally set lean of stoichiometric to ensure burnup of the fuel. Further, in these continuous burning systems, temperatures are usually maintained below 2000 K. For these reasons, the contribution of CO from stationary power plants is small. Temperatures achieved in the internal combustion engine, however, are higher. Also, in the conventional automobile engine, the air-fuel ratio cannot always be maintained lean. Rich operation is required to provide power during starting (choking) and during periods of acceleration. For these reasons, the automobile has been the prime source of CO. The effect of air-fuel ratio on exhaust CO concentration for automobile engines is shown in Figure 11.5.

It must be remembered that all calculations made to determine the composition of the products of combustion assume that the products are in thermodynamic equilibrium. The time required for the products to reach equilibrium is a function of the rate of the chemical reactions that are occurring. Such time-dependent calculations cannot be handled by the use of the principles of equilibrium thermodynamics, and thus they will

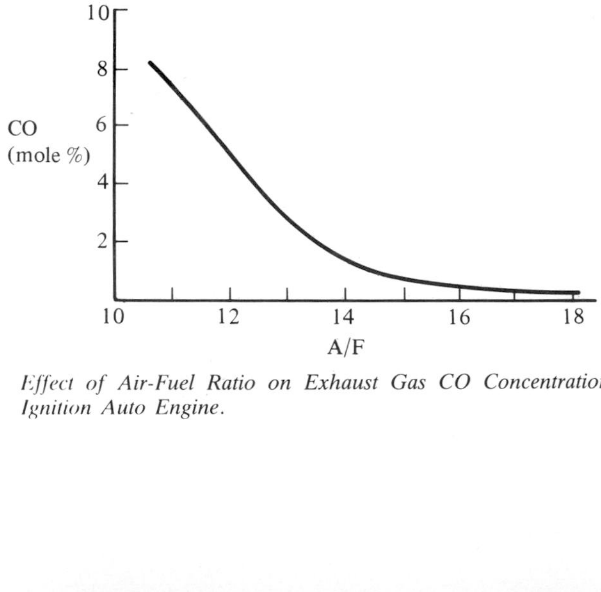

FIGURE 11.5 *Effect of Air-Fuel Ratio on Exhaust Gas CO Concentration for a Spark Ignition Auto Engine.*

not be covered in the text. It is sufficient to note that, at least with the automobile engine, the times during which the gases are exposed to high temperatures are short enough that significant departures from equilibrium concentrations do occur.

Harmful emissions from combustion processes also include the oxides of nitrogen, NO_x. As we have seen, if the combustion temperature is not high, the nitrogen in the air will not enter into the combustion reaction. However, at sufficiently high temperatures, nitrogen and oxygen react to produce NO in the exhaust. After contact with outside air, NO is converted, at least to some extent, to NO_2. The following example illustrates the effect of temperature on NO_x formation.

EXAMPLE 11.15

Air is heated in a steady-flow process at 1 atmosphere to high temperatures. Calculate the mole percent NO in the exhaust for temperatures of 1000°C, 2000°C, 3000°C, and 4000°C.

Solution

The chemical equation can be written as

$$O_2 + 3.76N_2 \rightarrow aO_2 = bN_2 + cNO$$

Taking atomic balances, we obtain

$$\text{O:} \qquad 2 = 2a + c$$

$$\text{N:} \quad 7.52 = 2b + c$$

or

$$a = 1 - \tfrac{1}{2}c \qquad b = 3.76 - \tfrac{1}{2}c$$

The third equation required for solution is that involving the equilibrium constant for the reaction $\frac{1}{2}N_2 + \frac{1}{2}O_2 \rightleftharpoons NO$. Hence

$$K_p = \frac{p_{NO}}{p_{N_2}^{1/2}\, p_{O_2}^{1/2}} = \frac{c}{\sqrt{ab}}$$

Notice that for this equilibrium reaction with the same number of moles on each side of the equation, the result is independent of the total pressure. Further substitution gives

$$K_p = \frac{c}{\sqrt{(1 - c/2)(3.76 - c/2)}}$$

or

$$K_p^2 = \frac{c^2}{3.76 - 2.38c + 0.25c^2}$$

From Table B.20, the values of K_p are: at 1000°C, $K_p = 0.000893$; at 2000°C, $K_p = 0.0384$; at 3000°C, $K_p = 0.1650$; and at 4000°C, $K_p = 0.3542$. Solving the above quadratic equation for c, we obtain at 1000°C, $c = 0.00173$; mole % NO = 0.036%

at 2000°C, $c = 0.0728$; mole % NO = 1.53%

at 3000°C, $c = 0.290$; mole % NO = 6.09%

at 4000°C, $c = 0.561$; mole % NO = 11.8%

Another variable that is important in the formation of NO_x is the air-fuel ratio (A/F). Since the presence of both oxygen and nitrogen atoms is required, peak NO_x exhaust concentrations occur when the air-fuel ratio is lean of stoichiometric, with excess oxygen available. Too lean an air-fuel ratio would lead to lower combustion temperatures because of dilution, and hence a reduction in NO_x. A typical curve of NO_x vs. air-fuel ratio is shown in Figure 11.6.

Again, care must be exercised in using equilibrium thermodynamics calculations for processes that may not have time to reach equilibrium. Calculations such as those of Example 11.15 may be useful in indicating significant variables that affect NO_x levels; however, for nonequilibrium processes, large differences may exist between actual data and these calculations.

The automobile and stationary power plant are both major sources of NO_x. Unfortunately, attempts to achieve good economy by running combustion systems at high temperatures and slightly lean of stoichiometric have led to increases in NO_x emission. It is to be hoped that emission-control techniques will be perfected which will allow low NO_x emission and minimum energy consumption.

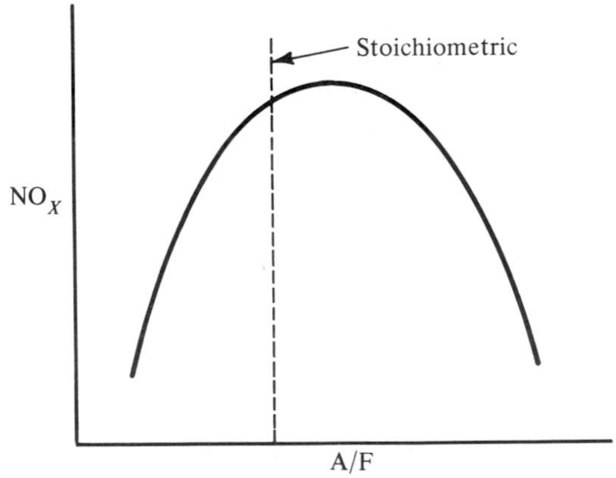

FIGURE 11.6 *Effect of Air-Fuel Ratio on Exhaust Gas NO_x Concentration.*

PROBLEMS

11.1 Acetylene is reacted with theoretical air. Neglecting dissociation, determine the air-fuel ratio on a mass basis and the products of combustion. Then repeat the problem for 10 percent excess air.

11.2 A mixture of 90 percent octane and 10 percent ethanol by weight is reacted with a stoichiometric amount of air. Neglecting dissociation, determine the air-fuel ratio on a mass basis and the products of combustion.

11.3 Ethane is reacted with excess air. A dry analysis (that is, excluding water) of the products reveals 3 percent oxygen by volume. Determine the air-fuel ratio used in the combustion process on a volume and on a mass basis. Neglect dissociation.

11.4 A hydrocarbon fuel is burned with air. The products of combustion have the following dry volumetric composition: 10 percent CO_2, 2 percent O_2, 1 percent CO, and 87 percent N_2. Determine the composition of the fuel, that is, the mass of hydrogen to the mass of carbon.

11.5 A certain coal contains 4 percent sulfur, 4 percent hydrogen, 76 percent carbon, 6 percent oxygen, and 10 percent ash (noncombustible), all on a weight basis. The coal is burned in a furnace at an air-fuel ratio of 15:1 on a mass basis. Determine the composition of the stack gases on a volume basis. Neglect dissociation.

11.6 Octane is burned with 8 percent excess air having a specific humidity of 0.005. Determine the composition of the products on a volume and on a mass basis.

11.7 Liquid methanol at 25°C is reacted with stoichiometric air at 25°C in a steady-flow process. The temperature of the products is 600°C. Calculate the heat given off in kJ/kg of fuel. (See Table B.18.)

11.8 Methane at 77°F is burned with 10 percent excess air of 50 percent relative humidity in a steady-flow process at 1 atmosphere. Determine the heat given off in Btu/lbm of methane if the product temperature is 1000°F and the air inlet temperature is 77°F.

11.9 A mixture of gaseous propane and 10 percent excess air is contained in a closed vessel of 200 cm^3 volume at 25°C and 1 atmosphere. The mixture is caused to react, with the resultant temperature measured to be 500°C. Calculate the final pressure in the vessel and the heat given off. Neglect dissociation.

11.10 Natural gas of mole composition 85 percent methane and 15 percent ethane is burned in a steady-flow process at 1 atmosphere with 10 percent excess air. Determine the adiabatic combustion temperature for a reactant temperature of 25°C.

11.11 Determine the availability change due to the adiabatic combustion of natural gas described in Problem 11.10.

11.12 Butane gas is burned with 150 percent theoretical air in a steady-flow reactor. The inlet air and butane are both at 77°F and the products leave the reactor at 1200°F. Find the air-fuel ratio on a mass basis and the heat given off in Btu/lbm of butane.

11.13 At the end of the compression stroke in an Otto cycle engine, just prior to ignition, the pressure is 125 psia, and the temperature is 500°F. The mixture, which has an air-fuel ratio of 18:1 by mass, is then sparked. Neglecting dissociation and losses from the cylinder, determine the pressure and the temperature after the constant volume combustion. Take the fuel to be gaseous octane.

11.14 Hydrogen peroxide (H_2O_2) gas at 77°F is caused to decompose in a steady-flow process, the products of the decomposition being water and oxygen. The heat of formation of H_2O_2 is 1680 Btu/lbm. Determine the adiabatic decomposition temperature.

11.15 Propane at 25°C is reacted with air at 300°C in a steady-flow adiabatic process at 1 atmosphere. Neglecting dissociation, determine the resultant product temperature. Would you expect a higher or a lower temperature if the air were initially at 25°C?

11.16 A mixture of 20 percent carbon monoxide and 80 percent air by volume is heated in a closed vessel from 100°F to 1000°F. Determine the final composition and the heat given off.

11.17 The gases emitted by an Otto cycle engine contain, by volume, 2 percent oxygen, 2 percent carbon monoxide, 8 percent carbon dioxide, 10 percent water vapor, and the remainder nitrogen. This mixture at 570°F is reacted with O_2 over a catalyst to remove all the CO. Determine the heat given off during the reaction per pound of tailpipe gases. The engine has a displacement of 350 in.^{3} and a speed of 2000 rpm. Find the increase in temperature over the catalyst bed. List all assumptions.

11.18 A coal contains 82 percent carbon, 6 percent oxygen, 10 percent ash, and 2 percent hydrogen by weight. The coal is burned in a steady-flow process at an air-fuel ratio of 16:1. Neglecting dissociation, determine the adiabatic combustion temperature. Assume the temperature of the reactants to be 25°C.

11.19 Determine the availability change due to the adiabatic combustion of coal described in Problem 11.18.

11.20 Liquid octane and theoretical air at 77°F are contained in a rigid vessel having a 300-in.^{3} volume. The mixture is sparked. Neglecting dissociation, determine the adiabatic combustion temperature.

11.21 Determine the availability change due to the adiabatic combustion of octane described in Problem 11.20.

11.22 A mixture of 10 moles of argon, 10 moles of nitrogen, and 2 moles of oxygen is heated in a rigid container from 1 atmosphere, 300 K, to 4000 K.

Products at the final state consist of Ar, O_2, N_2, and NO. Find the mole percent at the final state.

11.23 Water vapor is heated in a steady-flow process from 1 atmosphere and 300°C to a temperature at which the water is 10 percent dissociated. Find the final temperature, and determine the heat added per kilogram of H_2O.

11.24 One mole of carbon monoxide and 1 mole of water vapor are heated in a closed container from 1 atmosphere, 200°F, to 1000°F. Determine the volumetric concentrations of CO, H_2, H_2O, and CO_2 at the final state. Also, find the final pressure in the container and the heat added.

11.25 Propane is to be burned in a gas turbine combustor with sufficient excess air so that the temperature on the turbine blades does not exceed 1800°F. At the combustor inlet, the air temperature is 260°F; fuel enters the combustor at 100°F as a gas. Determine the required percent excess air so that the temperature of the combustion gases at the combustor outlet will be 1800°F.

11.26 Determine the availability change due to the adiabatic combustion of propane described in Problem 11.25.

11.27 Molecular oxygen at 200°C, 1 atmosphere, is heated to 4000°C in a constant pressure, steady-flow process. Determine the mass percentages of atomic and molecular oxygen at the final state. Then repeat the problem for pressures of 5 atmospheres and 10 atmospheres.

11.28 Hydrogen is burned with 150 percent excess air in a constant volume process. Determine the adiabatic combustion temperature for an initial pressure of 500 kPa and initial temperature of 500°C.

11.29 Methane is burned with 90 percent theoretical air in a steady-flow process at 1 atmosphere, the products consisting of CO_2, CO, H_2O, H_2, and N_2. Determine the heat given off for a reactant temperature of 50°C and a product temperature of 1000°C.

11.30 Methane is reacted with air in a steady-flow process at 1 atmosphere. Calculate the volumetric concentration (in ppm) of CO in the exhaust at a combustion temperature of 3000°C. The exhaust consists of N_2, O_2, CO, CO_2, H_2O, and H_2. Perform the above calculation for 90 percent, 100 percent, and 110 percent theoretical air.

11.31 Coal containing 2 percent sulfur by mass is reacted with stoichiometric air in a furnace at 1 atmosphere pressure. Calculate the exhaust concentrations of NO and SO_2 at a combustion temperature of 2000°C. Treat the coal as pure carbon (and sulfur), and assume that the exhaust consists of CO, CO_2, O_2, N_2, NO, and SO_2. Express your answer in ppm by volume.

11.32 A mixture of 20 percent oxygen (O_2) and 80 percent nitrogen (N_2) by volume is heated to 2000°C in a steady-flow process at 3 atmospheres. Determine the mole percent NO in the exhaust. Assume that the products consist of N_2, O_2, and NO.

11.33 An analysis of No. 6 fuel oil reveals, on a mass basis, 2.3 percent sulfur, 9.7 percent hydrogen, and 88 percent carbon. This fuel is burned with 20 percent excess air in a 500-megawatt power plant of 40 percent thermal efficiency. Determine the pounds of SO_2 put out by the plant per day. The heating value of the oil is 18,300 Btu/lbm.

11.34 The volumetric analysis of a gaseous fuel derived from coal reveals 45 percent hydrogen, 37 percent methane, 6 percent carbon monoxide, 4 percent nitrogen, 5 percent ethylene (C_2H_4), and 3 percent carbon dioxide. For this fuel, determine the higher heating value (HHV).

11.35 One method of eliminating unburned hydrocarbon, carbon monoxide, and oxides of nitrogen from combustion exhausts is to burn hydrogen with oxygen. Determine the adiabatic combustion temperature and the products of combustion for a steady-flow mixture of 20 parts oxygen and 1 part hydrogen by mass. Assume the products to consist of H_2O, O_2, and H_2. The reactants are gases at 25°C and 1 atmosphere.

11.36 The effect of water vapor injection on a combustion process is to be determined. Consider the combustion of propane gas in steady flow with 50 percent excess air at 300°F. Water vapor at 300°F is injected into this mixture until it constitutes 15 percent by volume of the entire reactant mixture. Determine the adiabatic combustion temperature and exhaust composition, assuming products to consist of N_2, O_2, H_2O, CO_2, and H_2. Then repeat the problem for no water injection. The pressure is one atmosphere.

11.37 A rigid vessel having a 30-cm^3 internal volume is filled with molecular nitrogen to an absolute pressure of 15 kPa, with an initial temperature of −100°C. The nitrogen is heated to 3000°C. Determine the resultant composition and pressure.

11.38 Air is heated in a steady-flow process at 10 atmospheres to high temperatures. Calculate the mole percent NO in the exhaust for a temperature of 1000°C.

chapter

12

Thermodynamics of Some New Energy Conversion Systems

12.1 INTRODUCTION

The current annual energy consumption rate of the United States is approximately 2×10^{12} W. The total energy use in the United States is divided approximately along the following lines (1):

Transportation	25%
Electric utilities	25%
Industry	30%
Residential and commercial[1]	20%

This energy demand is supplied almost entirely by the combustion of fossil fuels, with about 19 percent by coal, 47 percent by liquid petroleum, and 26 percent by natural gas. The remaining 8 percent is provided by hydroelectric and nuclear plants. Energy consumption in the United States had been increasing at the rate of approximately 4.2 percent per year until 1973. Since that time, due to higher prices and shortages of energy, rates of consumption have decreased so that energy consumption for 1982 in the United States was 10 percent less than for 1979. The supply of fossil fuels in the United States is not unlimited. Shortages of petroleum products have forced the United States to import,

---

1. Residential and commercial needs include air heating and conditioning, water heating, and cooking.

at present, up to 50 percent of its petroleum needs. The future availability of oil from the North Slope of Alaska and from currently undeveloped offshore deposits may help to reduce this dependency on imports; however, the fact remains that only finite reserves of fossil fuels are available. In addition, there are great concerns about the effect of fossil fuels on the environment. Thus, alternate energy sources must be developed.

The first alternate source that comes to mind is *solar energy*. The total rate of impingement of solar energy on the United States is about 600 times the rate of energy consumption. This source of free energy is virtually limitless. The costs involved in collecting, storing, and using the solar energy are, however, substantial. Solar energy is comparatively low in intensity per unit area, which makes collection for a large central power station an expensive project, but also makes solar energy atractive for heating of buildings and residences.

Solar collectors are currently available which occupy part of the roof area and collect solar thermal energy. This collected energy is stored and used as needed, for both space and water heating. A typical system is shown in Figure 12.1. The use of solar energy for air conditioning will be described in Section 12.5. In spite of the relatively low operating cost of such solar heating and cooling systems, the high initial construction and installation costs have prevented their large-scale use. However, current increases in prices (and lack of availability) of gas, oil, and electricity have made solar heating and cooling a much more attractive proposition; it will be in such systems that solar energy will first see widespread use.

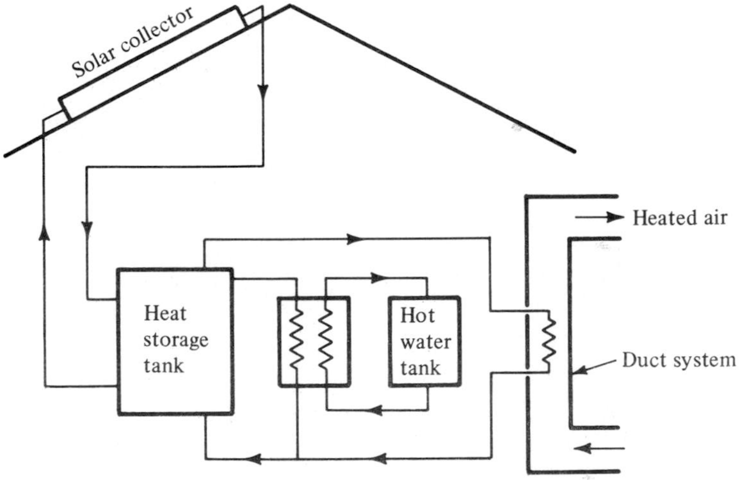

FIGURE 12.1 *Domestic Solar Energy System.*

Another possible way to utilize solar thermal energy is with the solar tower concept. As shown in Figure 12.2, the sun's energy is focused on a central collector by a large number of reflecting mirrors; these mirrors are controllable so that as the position of the sun changes, maximum utilization of solar energy is always obtained. It is estimated that temperatures of 500°C to 1000°C can be achieved at the collector; the thermal energy could be transported to and used in a conventional Rankine cycle power plant located at the base of the tower. A 1-square-mile array of mirrors would be required to provide a 100-megawatt output, with the tower located in the southwestern United States.

Solar photovoltaic cells, which convert sunlight directly into electricity, have been used for many years to provide power for spacecraft. The cost of such cells has, to date, prevented their use for earth applications. However, with the cost of conventional power sources increasing, the use of solar cells may appear attractive in the future, with greater effort put into attempts to reduce manufacturing costs.

A large amount of solar energy is stored naturally in the upper layers of the *oceans*. Semitropical and tropical waters provide temperature differences of up to 20°C between the warmed surface waters and the cooler waters at great depth. It has been proposed to operate a Rankine cycle power plant using this temperature difference. Although the thermal efficiency would be small, the large amounts of available heat might render such a system practical. This system will be discussed in detail in Section 12.2.

A small fraction of incident solar energy is converted to wind energy, also available for providing useful work. A windmill farm with total capacity of 670 megawatts in 1987 is available near San Francisco for the Pacific Gas and Electric Company. In some areas, the windmill may, in the future, be an important source of power.

Geothermal energy, another large alternate source of power, is actually the natural heat of the earth. The temperatures in the earth increase with depth, however, only the geothermal energy that exists in the upper 10 kilometers of the earth's surface can be extracted economically by drilling. Also, most of the available geothermal energy is far too diffuse to be recovered; however, when the heat is concentrated and stored in hot rocks, hot water, or steam, it is possible to utilize this energy for the production of electric power. For example, the Geysers plant in California has a generating capacity of 1000 megawatts from geothermal deposits. Methods of utilizing geothermal energy will be discussed in Section 12.3.

Another way to utilize more efficiently the available fuels to produce electric power is by development of more efficient *energy conversion systems*. For example, fuel cells are devices that convert chemical energy

(a)

(b)

Figure 12.2 *Solar Tower Concept for Electric Power Plant. (Photos Courtesy of U.S. Department of Energy.) (a) The World's First Solar Electric Power Plant Is Scheduled for Operation near Barstow, California, in the Mojave Desert. The Facility Will Generate Enough Power to Maintain a City of 10,000. (b) This Is an Aerial View of the Solar Thermal Test Facility at Sandia Laboratories. Heliostat Arrays, Each Containing 25 Four-foot-square Mirrors, Are Shown on the Opposite Side of the 20-story-high "Power Tower."*

directly into electrical energy. The fuel cell is not a heat engine and therefore is not subject to the Carnot limitation on maximum cycle efficiency. The operation of fuel cells will be described in Section 12.4.

The purpose of this chapter is to present the thermodynamics of some of these new energy conversion systems. None of these systems is in use to any great extent today. However, energy needs seem to dictate that some, if not all, of these systems will be providing useful energy by the year 2000.

12.2 ELECTRIC POWER FROM OCEAN THERMAL GRADIENTS

The oceans collect large amounts of solar thermal energy. In the tropical oceans, solar-heated surface water in combination with colder water coming from the poles along the ocean floor yields temperature differences of approximately 35°F to 40°F, with surface layers between 75°F and 85°F and lower layers at 40°F to 45°F. In the Caribbean Sea, for example, within a few miles from land, ocean currents with the above temperature differences occur within 2000 to 3000 feet of each other. Similar temperature gradients are found in other tropical seas within 20 degree latitude of the equator.

The first attempt at recovering ocean thermal energy was by Claude in the 1920s (2). His system, an open Rankine cycle, involved the vaporization to steam of the warmed surface water in a vacuum chamber. The steam flowed through a turbine, producing useful work, and then discharged through a condenser back to the ocean. The condenser tubes were cooled with cold water brought up from over a mile down in the ocean. The Claude system produced 20 to 30 kilowatts, and while demonstrating the feasibility of extracting useful energy from ocean temperature gradients, it was abandoned. One of the problems with such a system was that of maintaining low pressures throughout the system with the use of water as the working medium (the saturation pressure of water at 80°F is only 0.5 psia). Also, the large specific volume of steam at these low pressures requires a very large turbine.

A more practical design has been proposed by Anderson (2, 3). In his system, a working fluid with much higher saturation pressure at ocean temperature, such as ammonia, propane, or Freon-22, is used in a closed Rankine cycle. The working fluid is heated to its saturation temperature and vaporized in the boiler, with heat supplied from the warm surface water. The vapor is expanded through a turbine to produce useful work and then is converted back to liquid in the condenser, with cold water from the ocean depths serving as the heat sink for the condenser. Some scientists have proposed to float the power plant near the ocean surface, as close as possible to land. An artist's conception of one such system is shown in Figure 12.3(a); a model of another is shown in Figure 12.3(b). The following example illustrates the calculation of performance achievable with such a power plant.

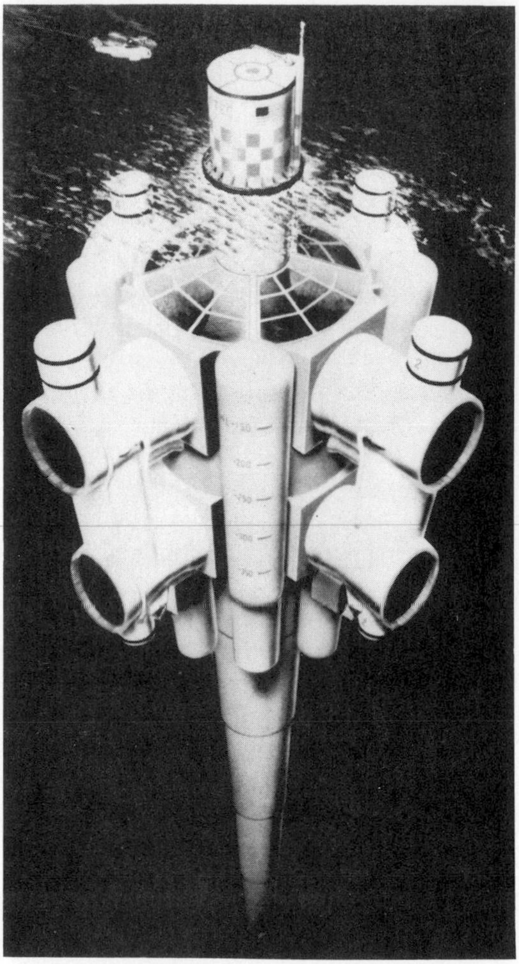

(a)

FIGURE 12.3 *Ocean Thermal Power. (Photos Courtesy of U.S. Department of Energy.) (a) The Ocean Thermal Energy Conversion (OTEC) System Proposed by Lockheed's Ocean Systems Division Will Generate Enough Power to Maintain a City of 100,000. The 300,000-ton Structure Is 1600 Feet Long and 250 Feet in Diameter. The Platform on the Top Consists of Crew Quarters and Maintenance Facilities, Surrounded by Turbine Generators and Pumps. (b) The OTEC System Proposed by TRW Will Use a Rankine-style Cycle Heat Engine with Ammonia as the Working Fluid to Generate Enough Power to Maintain a City of 60,000. The Platform is 340 Feet Across and Seventeen Stories High; a 50-foot-wide Fiberglass-reinforced Plastic Pipe Will Stretch 4000 Feet Down to Suck up Cold Water.*

(b)

FIGURE 12.3 *(continued)*

EXAMPLE 12.1

An ocean thermal energy plant is to be constructed in a location where ocean surface temperatures are 80°F and temperatures of the colder water at depth are 40°F. Assume that flows of ocean water are provided so that the R-22 working fluid is raised to a saturation temperature of 70°F in the boiler and is condensed at 50°F. The pump efficiency is 80 percent; the turbine efficiency, 90 percent. For a net power output of 100 megawatts, determine the following:

a. maximum and minimum cycle pressures;
b. flow of R-22 required;
c. plant thermal efficiency; and

d. cold ocean water flow required (assume the condenser to be an ideal heat exchanger, with the coolant water outlet temperature equal to 50°F). See Figure 12.4.

Solution

a. From Table A.10, the saturation pressure of R-22 at 70°F (the maximum cycle pressure) is 136.12 psia. The saturation pressure at 50°F (the minimum cycle pressure) is 98.727 psia.

b. At point 3, with saturated vapor at 70°F,

$$h_3 = 110.41 \text{ Btu/lbm} \quad \text{and} \quad s_3 = 0.21456 \text{ Btu/lbm°R}$$

For an isentropic turbine,

$$s_3 = s_{4s} = 0.21456 \text{ Btu/lbm°R}$$

$$= s_f + x_{4s}s_{fg} = 0.05190 + x_{4s}0.16613$$

$$x_{4s} = 0.979$$

$$h_{4s} = h_f + x_4 h_{fg} = 24.27 + 0.979(84.68)$$

$$= 107.17 \text{ Btu/lbm}$$

Isentropic turbine work

$$w_s' = h_3 - h_{4s} = 3.24 \text{ Btu/lbm}$$

Actual turbine work

$$= \eta w_s' = 0.9(3.24) = 2.916 \text{ Btu/lbm}$$

Isentropic pump work required

$$= v(p_2 - p_1)$$

$$= (0.01281 \text{ ft}^3/\text{lbm})(136.12 - 98.727) \text{ lbf/in}^2(144/778) \frac{\text{in}^2/\text{ft}^2}{\text{ft-lbf/Btu}}$$

$$= 0.0887 \text{ Btu/lbm}$$

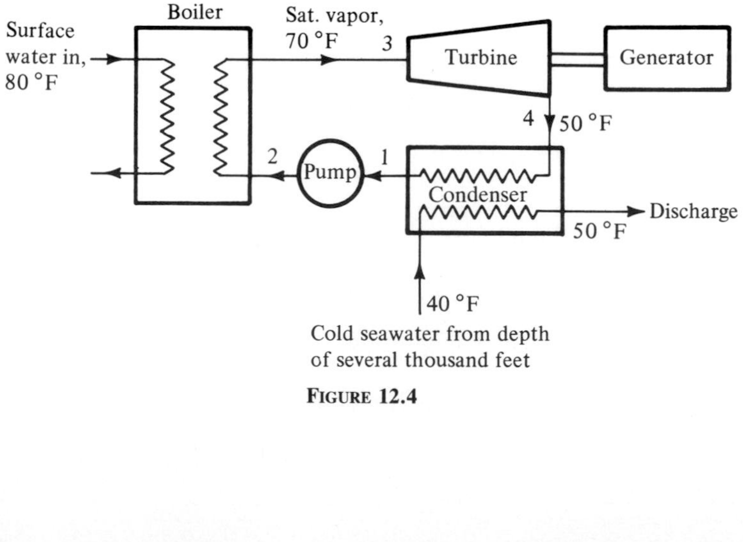

FIGURE 12.4

Actual pump work required

$$= 0.0887/\eta_p = 0.0887/0.80 = 0.111 \text{ Btu/lbm}$$

Net plant work output

$$= 2.916 \text{ Btu/lbm} - 0.111 \text{ Btu/lbm}$$

$$= 2.805 \text{ Btu/lbm}$$

For a 100-megawatt net power output,

$$\dot{m}_{\text{R-22}} = \frac{100 \times 3.41 \times 10^6 \text{ Btu/h}}{2.805 \text{ Btu/lbm}}$$

$$= 1.216 \times 10^8 \text{ lbm/h}$$

c. For the boiler,

$$q_{\text{in}} = h_3 - h_2 \quad \text{where } h_2 = h_1 + \text{pump work}$$

$$h_1 = h_f \text{ (at 50°F)} = 24.27 \text{ Btu/lbm}$$

so that

$$h_2 = 24.27 + 0.11 = 24.38 \text{ Btu/lbm}$$

$$q_{\text{in}} = 110.41 - 24.38 = 86.03 \text{ Btu/lbm}$$

$$\eta = \frac{\text{net work out}}{q_{\text{in}}} = \frac{2.805 \text{ Btu/lbm}}{86.03 \text{ Btu/lbm}}$$

$$= 3.26\%$$

d. For the cycle,

$$q_{\text{out}} = h_4 - h_1 \quad \text{where } h_4 = h_3 - \text{turbine work}$$

$$h_4 = 110.41 - 2.916 = 107.49 \text{ Btu/lbm}$$

$$q_{\text{out}} = 107.49 - 24.27 = 83.22 \text{ Btu/lbm}$$

The rate of heat rejection from the condenser is

$$(\dot{m}_{\text{R-22}})q_{\text{out}} = (1.216 \times 10^8 \text{ lbm/h})(83.22 \text{ Btu/lbm})$$

or

$$\dot{q}_{\text{out}} = 1.012 \times 10^{10} \text{ Btu/h}$$

The cold water flow required is

$$\dot{m}_{\text{cold water}} = \frac{1.012 \times 10^{10} \text{ Btu/h}}{c_p \, \Delta T}$$

$$= \frac{1.012 \times 10^{10} \text{ Btu/h}}{(1.0 \text{ Btu/lbm°F})(50°\text{F} - 40°\text{F})}$$

$$= 1.01 \times 10^9 \text{ lbm/h}$$

Note: This flow must be pumped up to the power plant from a depth of several thousand feet.

12.3 GEOTHERMAL ENERGY

Geothermal energy represents heat stored in underground deposits of superheated steam, hot water, or hot rock. Geothermal deposits that have current economic potential occur where elevated temperatures (50°C to 400°C) are present in permeable rock layers at depths that can be reached by drilling (5 to 10 kilometers) (4). The pores in the permeable layer act as a reservoir for storage of hot water and steam. If the rock above the storage layer is also permeable, the heated water rises to the surface, with the energy dissipated in a geyser or hot spring. If, however, the rock down to the storage layer has very low permeability,

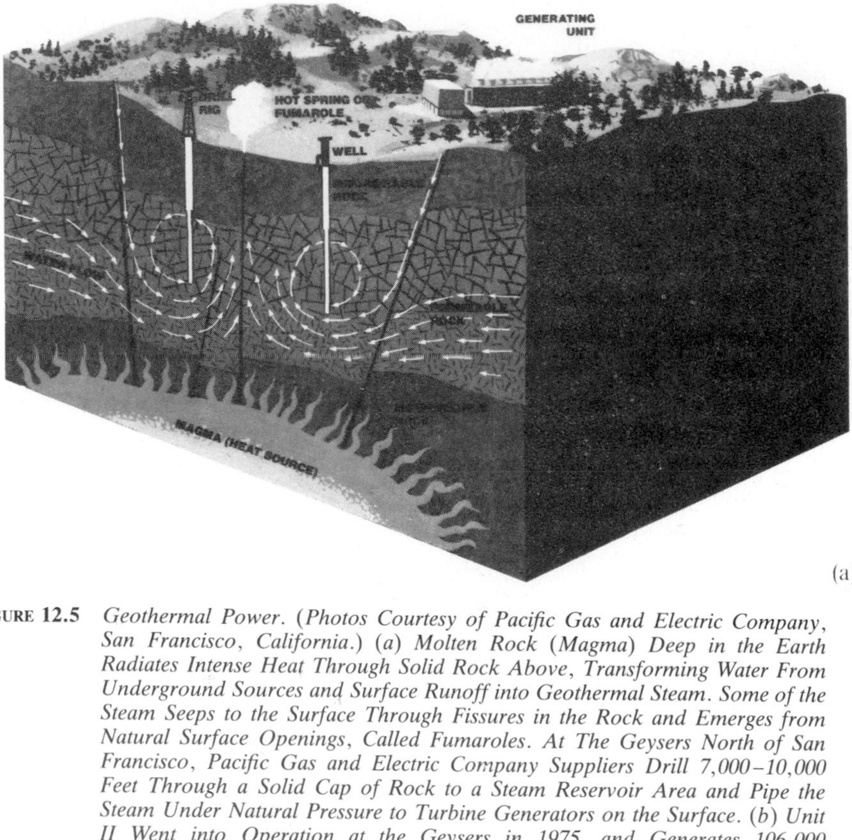

(a)

FIGURE 12.5 *Geothermal Power. (Photos Courtesy of Pacific Gas and Electric Company, San Francisco, California.) (a) Molten Rock (Magma) Deep in the Earth Radiates Intense Heat Through Solid Rock Above, Transforming Water From Underground Sources and Surface Runoff into Geothermal Steam. Some of the Steam Seeps to the Surface Through Fissures in the Rock and Emerges from Natural Surface Openings, Called Fumaroles. At The Geysers North of San Francisco, Pacific Gas and Electric Company Suppliers Drill 7,000–10,000 Feet Through a Solid Cap of Rock to a Steam Reservoir Area and Pipe the Steam Under Natural Pressure to Turbine Generators on the Surface. (b) Unit II Went into Operation at the Geysers in 1975, and Generates 106,000 Kilowatts. The Entire Facility Generates 1000 Megawatts, Enough Power to Serve One Million Residential Customers. The Geysers Is the World's Largest Geothermal Field, a Distinction Held Previously by the 390,000-Kilowatt Facility at Larderello, Italy. Planned Expansion Could Raise Capacity to 2 Million Kilowatts.*

(b)

FIGURE 12.5 (*continued*)

the thermal energy in the hot water or steam is retained naturally and can be tapped with a well (Figure 12.5).

There are two types of geothermal systems: vapor-dominated and hot-water. With the vapor-dominated system, earth heating is adequate to provide superheated steam at the wellhead, with as much as 50°C superheat at pressures of 10 atmospheres. Such steam can then be piped directly to a turbine to produce electrical power. This type of system exists at The Geysers, in California, with a net output of 1000 megawatts.

By far the most prevalent type of geothermal deposit discovered to date is the hot-water system; these systems appear to be about 20 times more common than vapor-dominated systems. However, there are technical difficulties associated with extracting the energy from such hot-water wells. The water may contain a high concentration of dissolved solids, and utilization of such corrosive hot brine solutions in a turbine may not be feasible. Further, conventional turbines have corrosion problems with handling liquid-vapor mixtures of appreciable liquid content. With these restrictions, we will discuss two conversion systems for handling the output from hot-water geothermal wells, the flashed steam system and the binary cycle (5).

With the flashed steam system, the wellhead product is fed into a flash separator, where the mixture is throttled to a lower pressure, increasing the vapor fraction. The vapor is then removed in a separator and used to drive a turbine; the liquid brine solution is discarded. The following example illustrates the conversion efficiency of such a system.

EXAMPLE 12.2

In order to convert the geothermal energy of a hot-water well to useful electrical energy, a flashed steam system is to be used. Assume a reservoir temperature of 250°C, with saturated liquid in the reservoir. The wellhead pressure is 700 kPa, with the water flowed isentropically from reservoir to wellhead. The resultant liquid-vapor mixture undergoes an isenthalpic expansion in the separator to 270 kPa, with the vapor fraction used to drive a turbine. The turbine exhaust is flowed into a condenser, so that the turbine exhaust pressure is maintained at 15 kPa. Determine the conversion efficiency of this process, that is, the ratio of actual turbine power output to the total power available from the mixture at the wellhead (Figure 12.6). Assume the turbine to be isentropic.

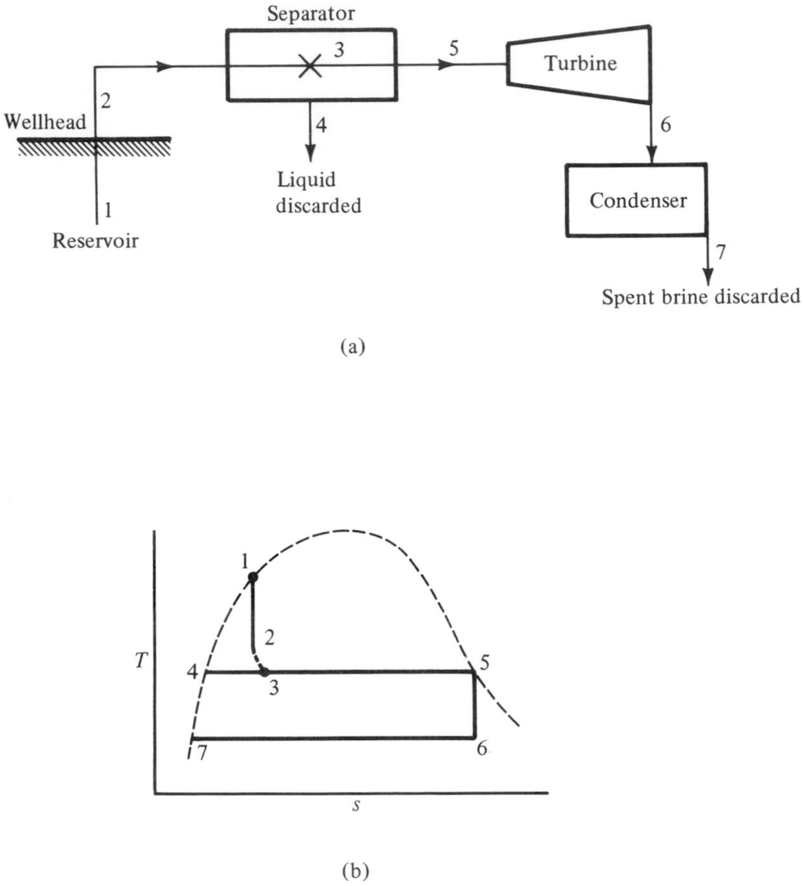

FIGURE 12.6 *Flashed Steam System.*

Solution

In the reservoir, at 1, $T_1 = 250°\text{C}$. From Table B.2,

$$h_1 = h_f = 1085.8\ \text{kJ/kg} \quad \text{and} \quad s_1 = s_f = 2.7935\ \text{kJ/kg K}$$

For the isentropic expansion 1–2, $s_1 = s_2 = 2.7935\ \text{kJ/kg K}$. For $p_2 = 700\ \text{kPa}$, from Table B.2, with $T_2 = 165°\text{C}$,

$$s_f = 1.9923\ \text{kJ/kg K} \quad \text{and} \quad s_{fg} = 4.7125\ \text{kJ/kg K}$$

Therefore

$$x_2 = \frac{2.7935 - 1.9923}{4.7125} = 0.17$$

$$h_2 = h_f + x_2 h_{fg} \quad \text{where } h_f \text{ and } h_{fg} \text{ are found in Table B.2}$$

$$h_2 = 697.25 + 0.17(2064.8)$$

$$= 1048.3\ \text{kJ/kg}$$

For the throttling process, $h_2 = h_3 = 1048.3\ \text{kJ/kg}$. At 3, with $p_3 = 270\ \text{kPa}$,

$$h_3 = 1048.3 = h_f + x_3 h_{fg}$$

$$= 546 + x_3(2173.6)$$

$$x_3 = 0.231$$

At this point, the liquid fraction is discarded and the vapor fraction, 23.1 percent of the total mass flow, is expanded through the turbine. At 5, $p_5 = 270\ \text{kPa}$, $h_5 = h_g = 2719.9\ \text{kJ/kg}$, and $s_5 = s_g = 7.0261\ \text{kJ/kg K}$. For the isentropic turbine expansion, $s_5 = s_6 = 7.0261\ \text{kJ/kg K}$.

From Table B.2, $s_{f6} = 0.7549\ \text{kJ/kg K}$ and $s_{fg6} = 7.2544\ \text{kJ/kg K}$. Therefore,

$$x_6 = \frac{7.0261 - 0.7549}{7.2544} = 0.864$$

and

$$h_6 = h_f + x_6 h_{fg}$$

$$= 225.97 + 0.864(2373.2) = 2276.4\ \text{kJ/kg}$$

$$\text{Total power output} = \dot{m}_5(h_5 - h_6)$$

$$\text{Total power available} = \dot{m}_2(h_2 - h_7)$$

$$\text{Conversion efficiency} = \frac{\dot{m}_5}{\dot{m}_2}\,\frac{h_5 - h_6}{h_2 - h_7}$$

$$= 0.231\,\frac{2719.9 - 2276.4}{1048.3 - 225.97}$$

$$= 0.125$$

In other words, only 12.5 percent of the available power is converted into useful power. It must be noted that the spent brine is not completely lost but can be pumped back into the well to replenish it.

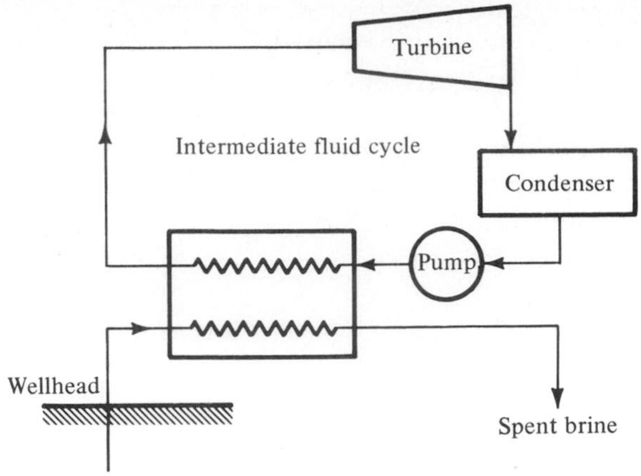

FIGURE 12.7 *Binary Cycle.*

Another possible system to use with hot-water deposits is the binary cycle. Here, the problems with turbine blade corrosion are eliminated with the use of an intermediate fluid, such as Freon, ammonia, or propane, as shown in Figure 12.7. Since the temperature of the liquid-vapor mixture in the heat exchanger must be high to enable the required heat transfer, there is a considerable amount of energy lost in the spent brine. Therefore, overall conversion efficiency is again low.

Naturally, it would be desirable to expand the entire wellhead product through a turbine (see Figure 12.8) to produce power and thereby maximize conversion efficiency. This system awaits development of turbine materials adequate to withstand the very corrosive brine solutions with high liquid content.

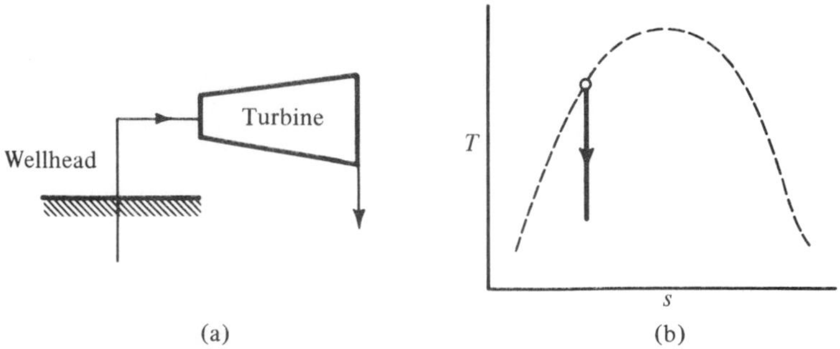

FIGURE 12.8 *Total Expansion of Wellhead Product.*

12.4 FUEL CELLS

The maximum thermal efficiency of a cyclic heat engine operating between two reservoirs is that of a Carnot engine. Thus, a limitation is imposed on the efficiency of a heat engine cycle. In the fuel cell, however, chemical energy is converted directly into electrical energy. Not being a heat engine, the fuel cell is not subject to the Carnot limitation; therefore, the potential exists for greater energy conversion efficiencies.

A simple hydrogen-oxygen fuel cell is depicted schematically in Figure 12.9. The fuel (hydrogen) is fed into the left chamber, while the oxidizer (oxygen) is flowed into thc right chamber. The anode and cathode are separated by an acid electrolyte. For the fuel cell under consideration, the reactions that occur at anode and cathode are the following.[2]

$$\left.\begin{aligned} \text{Anode:}\quad & 2H_2 \rightarrow 4H^+ + 4e^- \\ \text{Cathode:}\quad & 4H^+ + 4e^- + O_2 \rightarrow 2H_2O \end{aligned}\right\} \qquad \textbf{(12.1)}$$

The electrodes are made of a material (such as graphite) that must have good electrical conductivity (to minimize the internal resistance of the cell) and high porosity (to allow for contact between the electrolyte and fuel and electrolyte and oxidizer). The electrodes must also be coated with a catalyst to promote the reactions in Equation (12.1).

At the anode, hydrogen is split into hydrogen ions and electrons; the electrons are allowed to pass through an external load resistance to do useful work, with the hydrogen ions going into the electrolyte. At the

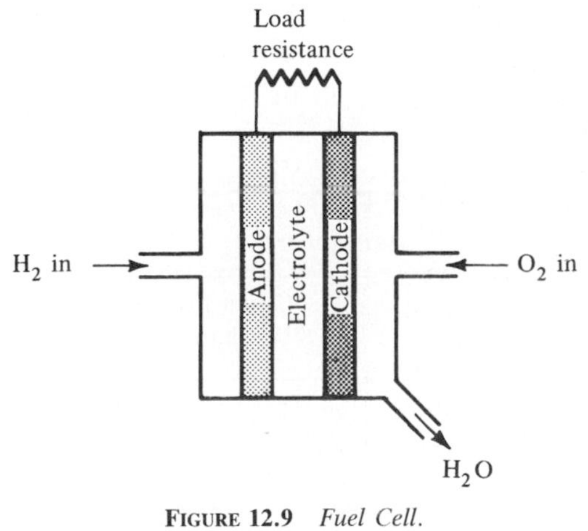

FIGURE 12.9 *Fuel Cell.*

2. H^+ is a positive hydrogen ion; e^- is a free electron.

cathode, the electrons that have passed through the external circuit combine with the hydrogen ions from the electrolyte and oxygen to yield water, which is drawn off from the fuel cell.

For each gram mole (gmol) of hydrogen that is reacted (6.02×10^{23} molecules of hydrogen), 12.04×10^{23} electrons pass through the external circuit [see Equation (12.1), anode reaction]. Since each electron has a charge of 1.602×10^{-19} coulombs, 6.02×10^{23} electrons have a total charge of 96,490 coulombs, called a *faraday*. In our example, for each gram mole of hydrogen reacted, 2 faradays or 192,980 coulombs of electricity are provided.

For the hydrogen-oxygen fuel cell, the overall reaction is

$$2H_2 + O_2 \rightarrow 2H_2O$$

Whereas in Chapter 11 we calculated the heat liberated from such a chemical reaction, we now ask, How much useful electrical work can be derived from this reaction in a fuel cell?

For a steady-flow system, with no changes of potential or kinetic energies of the working fluid, the first law, Equation (3.15), yields

$$q_{\text{in}} = w'_{\text{out}} + h_e - h_{\text{in}}$$

Let us now assume that the fuel cell reaction occurs at constant temperature. The work output of the cell will be maximum for a reversible process. From the second law, $q_{\text{in}} = T(s_e - s_{\text{in}})$ for a reversible isothermal process. Therefore, $w'_{\text{max}} = T(s_e - s_{\text{in}}) + h_{\text{in}} - h_e$. But, from Chapter 11, $h - Ts = g$, with $g =$ Gibbs free energy, so that

$$w'_{\text{max}} = g_{\text{in}} - g_e \tag{12.2}$$

Gibbs free energy for several compounds is given in Table 12.1. (Gibbs free energy for elements in their natural state at 25°C, 1 atmosphere is taken as 0.)

TABLE 12.1 *Gibbs Free Energy at 25°C, 1 Atmosphere.*

Compound	Gibbs Free Energy ($kJ/kgmol \times 10^{-6}$)
H_2O (liquid)	−0.237
H_2O (vapor)	−0.229
CO_2	−0.394
CH_4	−0.0508
CO	−0.137
H_2	0
O_2	0

EXAMPLE 12.3

Determine the output voltage of an H_2-O_2 fuel cell operating at 25°C and 1 atmosphere. Assume that the cell operates reversibly, with H_2O (liquid) a product of the cell reaction. Find the heat liberated and the cell efficiency, the latter defined as

$$\eta_{\text{cell}} = \frac{w'_{\text{out}}}{h_e - h_{\text{in}}}$$

Solution

From the electrode reactions (12.1), for each mole of H_2O produced, 192 980 coulombs of electricity are provided. Since coulombs are ampere-seconds, it follows that

$$\text{Work done} = (\text{voltage})(\text{ampere-seconds}) = \text{joules}$$

where work done, for this reversible cell, is given by Equation (12.2). Therefore

$$\text{Voltage} = \frac{(237\,000 \text{ joules/gmol})}{192\,980 \text{ coulombs/gmol}} = 1.23 \text{ volts}$$

From the first law, $q_{\text{in}} = w'_{\text{out}} + h_e - h_{\text{in}}$. For liquid water, the enthalpy of formation is given in Table 11.2 as $h_f = -285\,830$ kJ/kgmol. It follows that $h_e - h_{\text{in}} = -285\,830$ kJ/kgmol of H_2 consumed. Therefore,

$$q_{\text{in}} = 237\,000 - 285\,830 = -48\,830 \text{ kJ/kgmol } H_2$$

$$\eta_{\text{cell}} = \frac{237\,000}{285\,830} = 82.9\%$$

The above calculations have been performed for an ideal, reversible fuel cell. The actual conversion efficiency of an H_2-O_2 fuel cell is less than 82.9 percent due to various losses. For example, there is an internal resistance of the fuel cell, leading to a drop in output voltage as the cell is operated. Some of the fuel and oxidizer may react directly, leading to a heat release but producing no useful work. A certain activation energy is required to promote the desired electrode reactions, energy that would otherwise be available for useful work.

In spite of the above losses, the fuel cell has the potential for achieving greater efficiencies than the conventional steam power plant in the conversion of chemical energy of a fuel to useful electrical energy. Although the output of a fuel cell is only of the order of 1 volt, many cells can be stacked together to produce the desired voltage and power characteristics (6). Such a power plant would be easily transportable and would be pollutant free.

In order to develop fuel cells for large, central station power plants,

FIGURE 12.10 *High Temperature Fuel Cells. Molten Carbonate Fuel Cells Are Being Developed for Commercial Use. These Will Provide an Electrical Efficiency of About 50 Percent. (Photo Courtesy of U.S. Department of Energy.)*

many companies are investigating the use of high temperature fuel cells. For this application, coal would first be burned in the presence of steam to produce hydrogen gas, the hydrogen used as fuel for the cell. A molten carbonate is used as the electrolyte, with the fuel cell operating at temperatures close to 1000 K.

Electrode reactions are

$$\left.\begin{array}{rl}\text{Anode:} & H_2 + CO_3^{=} \rightarrow H_2O + CO_2 + 2e^- \\ \text{Cathode:} & CO_2 + \tfrac{1}{2}O_2 + 2e^- \rightarrow CO_3^{=}\end{array}\right\} \qquad \textbf{(12.3)}$$

A futuristic view of a central power station using fuel cells is shown in Figure 12.10.

12.5 SOLAR AIR CONDITIONING SYSTEMS

With the rising cost of conventional fuel, the use of solar energy for heating and cooling will become more competitive with use of the other energy sources. Solar radiation is received at the earth's atmosphere at a rate of 10,130 Btu/ft^2-day (which equals 1.353 kW/m^2). However, as it passes through the atmosphere, the amount of solar radiation is reduced by scattering and absorption. The amount of radiation that reaches the ground can vary from almost none under heavy cloud cover to about 85 percent of that reaching the earth's atmosphere. Furthermore, there is a considerable seasonal variation of solar radiation at a point on the earth's surface due to the variation in the angle of tilt between the earth's axis and its orbit around the sun. The combination of these factors results in an annual average solar incidence in the continental United States of about 1400 Btu/ft^2-day (7).

In general, the utilization of solar energy is a capital-intensive process; that is, large investments in equipment are required to save operating costs (fuel purchases). Hence, the underlying economic problem in the utilization of solar energy is the balancing of the annual cost of extra investment (principal and interest) against annual fuel savings.

Since solar energy is a form of thermal energy, it can be used directly, upon collection in suitable solar collectors, to heat residential and commercial space and water. However, if it is used for air conditioning or refrigeration purposes, a special refrigeration cycle (the absorption cycle) is most adapable to solar energy input.

Recall that in the conventional vapor compression cycle, which was discussed in Section 10.3, the refrigerant is compressed in its vapor phase with an accompanying large rise in temperature by utilizing electric energy input to the compressor. On the other hand, in an absorption refrigeration cycle, the refrigerant is compressed in the liquid phase with little change in temperature. The increase in temperature of the refrigerant in the cycle is accomplished by the addition of heat, separate from the compression process. Solar energy can be used as the source of the heat required in the absorption cycle. The energy requirements for the pump are small (since the compression process takes place in the liquid phase) and are usually supplied by electric energy or fossil fuel.

A simple vapor compression refrigeration system is illustrated in Figure 12.11. The refrigerant enters the condenser as vapor at high pressure and is condensed there to a liquid. The liquid refrigerant enters the expansion valve, where the refrigerant pressure is reduced to its low value. The low-pressure refrigerant is then vaporized in the evaporator.

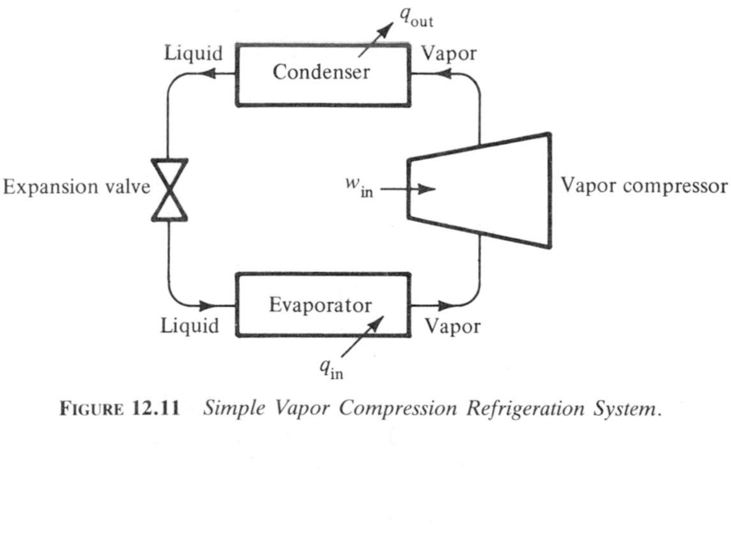

FIGURE 12.11 *Simple Vapor Compression Refrigeration System.*

Finally, the refrigerant vapor is raised to the high pressure in the vapor compressor to complete the cycle.

In an absorption refrigeration cycle, the vapor compressor is replaced by an absorber, a liquid pump, a generator, and a liquid return line, as illustrated in Figure 12.12. Further, both a refrigerant working fluid and a liquid solvent carrier are utilized. The function of the solvent carrier is to transport dissolved refrigerant, in liquid form, from the low- to the high-pressure region in the cycle. In the absorber, the low-pressure refrigerant vapor is absorbed into solution. Since the absorption process is exothermic, some heat must be removed to keep the temperature constant. The strong refrigerant solution is raised to the high pressure by the pump. In the generator, heat is added to the solution, thereby driving refrigerant vapor out of solution, since at a given pressure the amount of refrigerant dissolved in the solvent decreases with increase in temperature. Thus, high-pressure refrigerant vapor leaves the generator to enter the condenser, while a weak refrigerant solution is returned to the absorber. The solvent, which travels through the absorber, pump, and generator, serves to transport the refrigerant, in liquid phase, from the low pressure in the evaporator to the high pressure in the condenser. Examples of fluid combinations used in absorption systems are ammonia as refrigerant with water as solvent, and water as refrigerant with lithium bromide as solvent.

As in a vapor compression cycle, a coefficient of performance

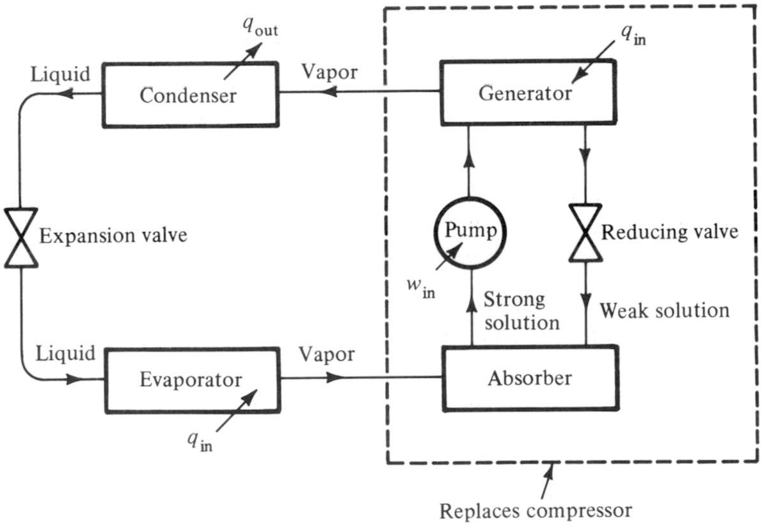

FIGURE 12.12 *Simple Absorption Refrigeration System.*

(COP) is used to evaluate the effectiveness of an aborption cycle and is given by

$$\text{COP} = \frac{\text{useful energy output}}{\text{required energy input}}$$

For refrigeration or air conditioning operation, the useful energy output is that of the evaporator. The required energy input in an absorption cycle consists of the heat input to the generator and the work input to the pump.

Typical COPs for water–ammonia absorption systems range from 0.4 to 0.5; for lithium bromide–water systems, COPs are slightly higher, in the range of 0.6 to 0.7. The lithium bromide–water system has the advantage for solar air conditioning of being able to operate at lower generator temperatures than the water–ammonia system. Note that the COPs are smaller by a factor of 5 to 10 than those of a typical vapor compression mechanical refrigeration cycle. However, the energy input in the absorption system can be provided by the sun; there is not the relatively expensive shaft work of the mechanical system.

As an example of the design of a solar air conditioning system, let the cooling requirement be 5 tons for a system with a COP of 0.6. Assuming that the air conditioning unit operates 8 hours per day, it would require a total cooling capacity per day of:

$$5 \text{ tons} \times 12{,}000 \text{ Btu/h/ton} \times 8 \text{ hr} = 480{,}000 \text{ Btu}$$

If the pump work is neglected, the energy input to the system consists of solar heat input to the generator. This would amount to:

$$\dot{q}_{\text{in}} = \frac{480{,}000 \text{ Btu}}{0.6} = 800{,}000 \text{ Btu/day}$$

Using an average insolation during the summer months of 1800 Btu/ft^2-day, with a collector efficiency of 50 percent, we find that the required collector area is

$$\frac{800{,}000 \text{ Btu/day}/1800 \text{ Btu/ft}^2\text{-day}}{0.5} = 889 \text{ ft}^2$$

In actual absorption cycle installations, a heat exchanger is added to the system, as shown in Figure 12.13. In this heat exchanger, between the generator and the absorber, the strong solution flowing from the absorber to the generator is heated, at the same time cooling the weak solution returning to the absorber.

Since solar energy must be collected and stored for a number of hours prior to its usage, due to the occurrence of clouds and for nighttime operation, installations based on solar energy supply will contain solar collectors, energy storage units, and an auxiliary energy source. Solar

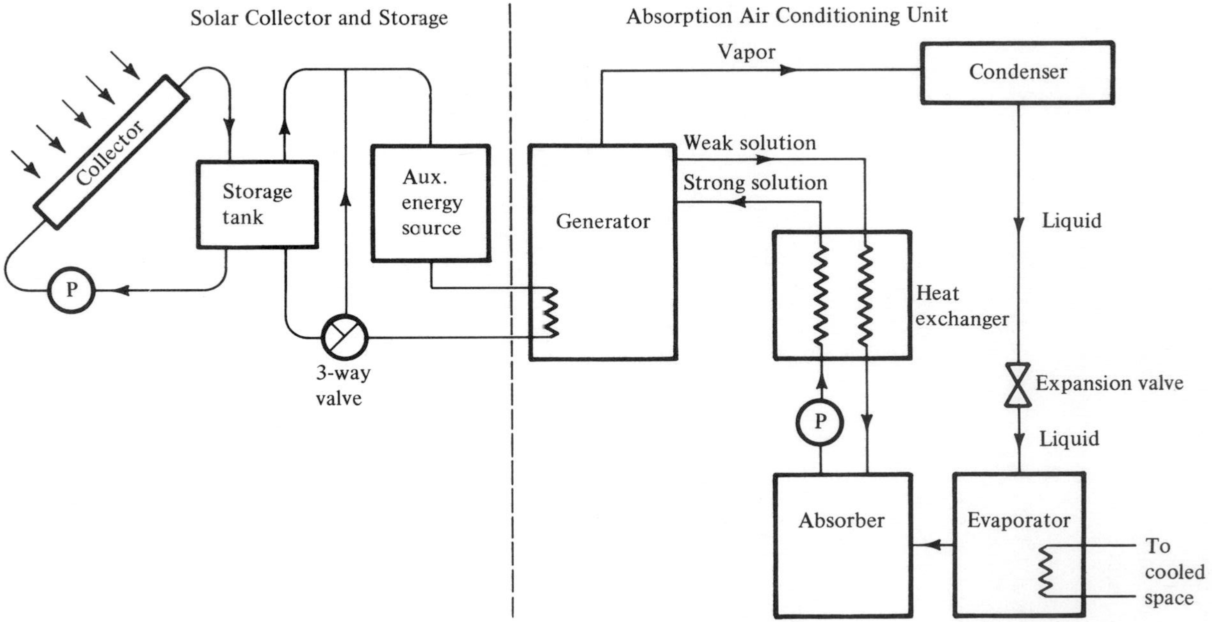

Figure 12.13 *Solar Absorption Air Conditioning System.*

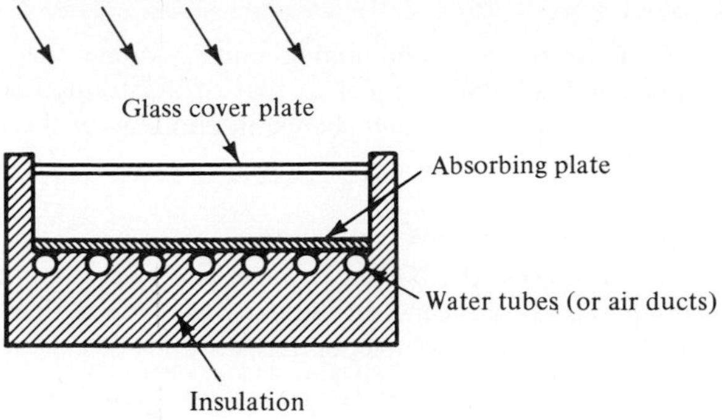

FIGURE 12.14 *Flat Plate Solar Collector.*

collectors usually consist of flat black plates, which absorb the solar energy, and air ducts or water tubes, which remove the energy from the collector plates (Figure 12.14). The energy storage units accumulate the solar energy when it is available. For the air duct collector, pebble bed storage units are used, while for the water tube collector, an insulated water tank is used. A schematic diagram of a typical solar air conditioning system is shown in Figure 12.13.

Since solar collectors must be oriented toward the sun within narrow limits for efficient energy collection, it is difficult to fit solar heating and cooling systems to many existing buildings; exceptions are low-rise buildings with flat roofs, such as shopping centers and schools. For new residential houses and other buildings, the solar collectors must be considered as the initial design stage and must be integrated into the envelopes of the building at nearly optimum orientations.

EXAMPLE 12.4

A solar absorption air conditioning system uses water as refrigerant and lithium bromide (LiBr) as absorbent. The condenser and evaporator operate at a pressure of 10 kPa and 1 kPa, respectively. The refrigerant (water) and the concentrated absorbent solution (weak refrigerant solution) leave the generator at the same temperature. The concentrated absorbent solution leaves the generator in a saturated state with a mass concentration of 0.42 kg H_2O/kg solution. The energy required to boil off the refrigerant in the generator is supplied by heating coils using hot water from solar collector panels. Since lithium bromide is a hygroscopic salt of low volatility, the vapor leaving the generator will not contain any lithium bromide vapor. The dilute absorbent solution (strong refrigerant solution) leaves the absorber in a saturated state and a mass concentration of 0.46 kg H_2O/kg solution. A heat exchanger between the dilute and concentrated absorbents reduces the exit temperature of the concentrated absorbent to 43.5°C (Figure 12.15).

Find the COP of the air conditioning unit. Assume that the value of enthalpy of a subcooled solution is equal to that of a saturated solution at the same temperature and neglect the small change in enthalpy of the dilute absorbent across the pump.

Properties of Water–Lithium Bromide Solutions

Unlike pure substances, the boiling temperature of a binary solution, at a given pressure, depends on the concentration (x) of the solution. For the two given concentrations, the saturation temperatures are

$$\text{for } x = 0.46 \qquad T_f(°\text{C}) = 27.2 + 1.136\,T_f(\text{refrig})$$

$$\text{for } x = 0.42 \qquad T_f(°\text{C}) = 34.0 + 1.168\,T_f(\text{refrig})$$

where T_f(refrig) is the saturation temperature of the refrigerant (water) at a given pressure.

The saturation enthalpy of a binary solution will also be a function of concentration and temperature. Thus

$$\text{for } x = 0.46 \qquad h_f(\text{kJ/kg}) = 60 + 2.11(T_f - 25°\text{C})$$

$$\text{for } x = 0.42 \qquad h_f(\text{kJ/kg}) = 77 + 1.98(T_f - 25°\text{C})$$

Solution

The total mass flowing through the absorber is (see Figure 12.15),

$$m_1(\text{kg refrig}) + m_7(\text{kg sol}) = m_2(\text{kg sol})$$

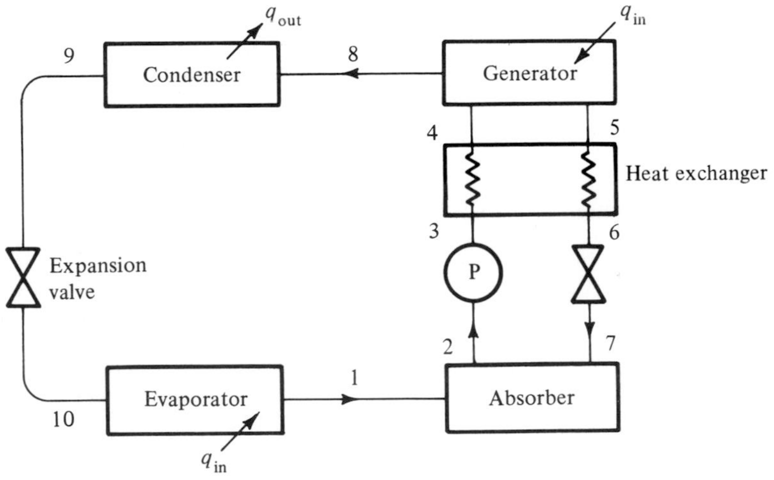

FIGURE 12.15

The mass of water flowing through the absorber is

$$0.46m_2(\text{kg H}_2\text{O/kg sol})(\text{kg sol}) = 0.42m_7 + m_1$$

Combining the two equations, we get

$$m_2 = 14.5 \text{ kg dilute absorbent solution/kg refrig}$$

$$m_7 = 13.5 \text{ kg concentrated absorbent solution/kg refrig}$$

The energy balance (first law) for the heat exchanger is

$$13.5(h_5 - h_6) = 14.5(h_4 - h_3)$$

From Table B.2a we obtain the saturation temperatures of the refrigerant (water) at a pressure of 1 kPa and 10 kPa, namely 6.98°C and 45.83°C, respectively. We thus calculate

	T_f(sol)(°C)	
p(kPa)	$x = 0.46$	$x = 0.42$
1	35.1	42.2
10	79.3	87.5

or $T_2 = 35.1$°C, $T_3 = T_2 = 35.1$°C, $T_5 = 87.5$°C, and $T_6 = 43.5$°C. From the expressions for the saturation enthalpy we calculate that $h_2 = 60 + 2.11(35.1 - 25) = 81.3$ kJ/kg solution, $h_5 = 200.8$ kJ/kg, and $h_6 = 113.6$ kJ/kg. With h_2 approximately equal to h_3, we have for the heat exchanger

$$13.5(200.8 - 113.6) = 14.5(h_4 - 81.3)$$

or $h_4 = 162.5$ kJ/kg solution. From the relation for the saturation enthalpy at $x = 0.46$, $162.5 = 60 + 2.11\ (T_4 - 25)$, we get $T_4 = 73.6$°C. The energy balance for the generator is

$$h_8 + 13.5h_5 = 14.5h_4 + q_{\text{in}} \ (\text{kJ/kg refrig})$$

The enthalpy of the water vapor entering the condenser h_g at a temperature of 87.5°C and a pressure of 10 kPa is obtained from Table B.3 as 2664.3 kJ/kg refrig; hence

$$2664.3 + 13.5(200.8) = 14.5(162.5) + q_{\text{in}} \ (\text{kJ/kg refrig})$$

or

$$q_{\text{in}} = 3018.9 \text{ kJ/kg refrig}$$

This is the energy input required to boil off the refrigerant in the generator.

Since the process in the expansion valve is isenthalpic, $h_9 = h_{10} = h_f$ (at 10 kPa). The cooling effect (q_{in}) of the evaporator is h_g (at 1 kPa) $- h_f$ (at 10 kPa) $= h_1 - h_{10} = 2514.4 - 191.8 = 2322.6$ kJ/kg refrig. Hence the coefficient of performance is

$$\text{COP} = 2322.6/3018.9 = 0.769$$

EXAMPLE 12.5

A solar-absorption air conditioning system utilizes ammonia as refrigerant and water as absorbent. The evaporator operates at a temperature of 44°F and a corresponding pressure of 80 psia. The condenser temperature is maintained at 96°F with a corresponding saturation pressure of 200 psia. The temperature of the refrigerant and the weak refrigerant solution leaving the generator is 180°F. This value corresponds to the saturation temperature of an ammonia–water solution with a concentration of 0.46 lbm ammonia/lbm solution. The strong refrigerant solution leaves the absorber saturated at a temperature of 98°F with a concentration of 0.54 lbm ammonia/lbm solution. Use of a heat exchanger between the two solutions lowers the temperature of the weak refrigerant solution to 115°F (Figure 12.15). Unlike the lithium bromide, the absorbent water is volatile. A characteristic of the ammonia–water system is that when liquid and vapor phases of the mixture exist in equilibrium, the concentration of the two components will not be the same in the two phases, but will show a far greater concentration of ammonia in the vapor phase. Therefore, assume that the vapor leaving the generator is pure ammonia vapor.

Determine the coefficient of performance of the unit and the heat that must be removed from the absorber to allow continuous absorption of refrigerant. Assume the value of enthalpy of a subcooled solution is equal to that of a saturated solution at the same temperature.

Properties of Ammonia

p(psia)	T(°F)	h_f(Btu/lbm)	h_g(Btu/lbm)	h(Btu/lbm)
80	44	92	624	
200	96	151	633	
200	180			691

Properties of Ammonia–Water Solutions

	$x = 0.46$		$x = 0.54$	
T_f(°F)	p_f(psia)	h_f(Btu/lbm)	p_f(psia)	h_f(Btu/lbm)
0	6	−64	10	−61
50	22	−14	33	−11
100	60	42	83	45
150	130	100	180	103
200	250	164	340	168

where x is in lbm ammonia/lbm solution.

Solution

The mass flowing through the absorber is (referring to Figure 12.15),

$$m_1(\text{lbm refrig}) + m_7(\text{lbm sol}) = m_2(\text{lbm sol})$$

The ammonia mass flowing through the absorber is

$$0.54m_2(\text{lbm ammonia/lbm solution})(\text{lbm solution}) = 0.46m_7 + m_1$$

Combining the two equations, we obtain

$$m_2 = 6.75 \text{ lbm dilute absorbent solution/lbm refrig}$$

$$m_7 = 5.75 \text{ lbm concentrated absorbent solution/lbm refrig}$$

The energy balance in the heat exchanger is

$$5.75(h_5 - h_6) = 6.75(h_4 - h_3)$$

From the table of properties of ammonia–water solutions we get, by interpolation,

at 180°F, 200 psia, and $x = 0.46$	$h_5 = 138$ Btu/lbm solution
at 98°F, 80 psia, and $x = 0.54$	$h_2 = 43$ Btu/lbm solution
at 115°F and $x = 0.46$	$h_6 = 59$ Btu/lbm solution

With $h_2 = h_3$ (that is, neglecting the small change in enthalpy across the pump) we get

$$5.75(138 - 59) = 6.75(h_4 - 43)$$

or $h_4 = 110$ Btu/lbm solution. For $x = 0.54$, we obtain $T_4 = 155°\text{F}$, by interpolation.

The enthalpy of the ammonia leaving the generator at a temperature of 180°F is $h_8 = 691$ Btu/lbm ammonia. The energy balance for the generator is

$$691 + 5.75(h_5) = q_{in} + 6.75(h_4)$$

$$691 + 5.75(138) = q_{in} + 6.75(110)$$

or $q_{in} = 742$ Btu/lbm refrig.

Since the process in the expansion valve is isenthalpic, $h_9 = h_{10}$ and the cooling effect (q_{in}) of the evaporator is h_g (at 80 psia) $- h_f$ (at 200 psia) $= h_1 - h_9 = 624 - 151 = 473$ Btu/lbm refrig. Hence the coefficient of performance is

$$\text{COP} = 473/742 = 0.64$$

The cooling requirement of the absorber is found from the energy balance for steady flow

$$q_{out} = 6.75h_2 = h_1 + 5.75h_7$$

With $h_6 = h_7$ across the reducing valve, we get

$$q_{out} = 624 + 5.75(59) - 6.75(43) = 673 \text{ Btu/lbm refrig}$$

PROBLEMS

12.1 An OTEC power plant, using ammonia as the working fluid, is to operate between an evaporator temperature of 75°F and a condenser temperature of 45°F. Hot surface water is available at 80°F; cold water from the ocean floor is available at 40°F. Take ΔT for hot water and cold water to be 3°F. Determine the cycle thermal efficiency (do not include hot water and cold water pumping requirements in this calculation). Take a turbine efficiency of 90 percent, pump efficiency of 80 percent. Determine the flows of hot water and cold water required for evaporator and condenser.

Properties of Ammonia

T (°F)	p_{sat} (psia)	v_f (ft^3/lbm)	h_f (Btu/lbm)	h_g (Btu/lbm)	s_f (Btu/lbm°R)	s_g (Btu/lbm°R)
45	80.96	0.02548	92.3	624.1	0.1996	1.2535
75	140.5	0.02650	126.2	629.9	0.2643	1.2065

12.2 An OTEC electric power station is to use propane as the working fluid in a closed Rankine cycle, with propane properties given below. The warmed water at the ocean surface is at 80°F; the evaporator temperature is 70°F, the condenser temperature 50°F, turbine efficiency 90 percent, pump efficiency 80 percent. The temperature rise of the cold sea water across the condenser is 6°F; the temperature drop of the warm surface sea water across the evaporator is 5°F. Find the flows of cold and warm water required, the cycle thermal efficiency and flow of propane for a net plant output of 20 megawatts.

Properties of Propane

T (°F)	p_{sat} (psi)	v_f (ft^3/lbm)	h_f (Btu/lbm)	h_g (Btu/lbm)	s_f (Btu/lbm°R)	s_g (Btu/lbm°R)
50	92.2	0.03101	51.8	205.0	0.1108	0.4114
70	124.6	0.03202	64.2	209.9	0.1343	0.4094

12.3 A geothermal energy conversion system uses a flash separation, in which the mixture at the wellhead is separated into liquid and vapor. The liquid portion is discarded, and the pure vapor is expanded through a turbine to produce useful power, as shown in Figure 12.16. Assume the flow is isentropic from the well reservoir to the wellhead. The wellhead pressure is 100 psi; the turbine exhaust pressure = condenser pressure = 2.0 psia. The turbine efficiency = 85 percent; assume saturated liquid in the well reservoir at 400°F. Find:

a. Conversion efficiency = $\dfrac{\text{Turbine power}}{\dot{m}_2(h_2 - h_6)}$

b. Mass flow for a turbine power output of 10 Mwatts.

c. Rate of heat rejection in condenser.

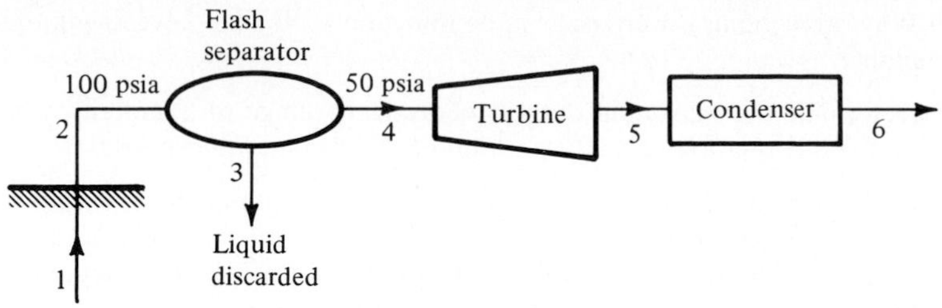

FIGURE 12.16

12.4 A binary system is to be used to recover the geothermal energy of a hot water well. Assume a reservoir temperature of 250°C, with saturated liquid in the reservoir. The wellhead pressure is 700 kPa, with the water flowed isentropically from reservoir to wellhead. The pressure just downstream of the valve shown is 1 atmosphere, with the steam exhausting from the heat exchanger at 1 atmosphere, 50°C.

The fluid in the Rankine cycle is R-22, with $T_3 = 80°C$, saturated vapor at 3; $T_1 = T_4 = 25°C$. The efficiency of the turbine is 70 percent; the efficiency of pump is 70 percent. Determine the conversion efficiency of this system. What flow of hot water is necessary for a power output of 100 kilowatts? (See Figure 12.17.)

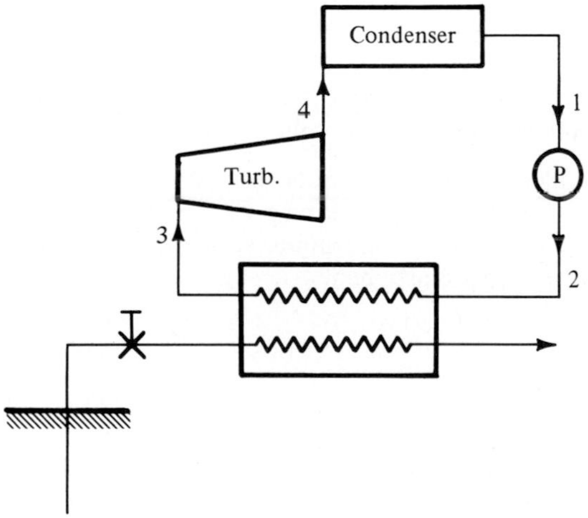

FIGURE 12.17

12.5 Why are cooling towers used in conjunction with The Geysers geothermal plant?

12.6 Determine the maximum output theoretical voltage of a molten carbonate fuel cell [Equation (12.3)] operating at 25°C, one atmosphere.

12.7 A H_2-O_2 fuel cell operating steadily at 25°C, one atmosphere, is to produce 5 amperes for one hour. Determine the flows of hydrogen and oxygen required if the cell voltage is 0.8 volt. Compute the cell efficiency, based on liquid water in the products.

12.8 100 tons of air conditioning are to be provided by a solar absorption unit at a location where the solar incidence in summer months is 30 MJ/m^2 per day. Determine the area of the flat plate solar collector that must be provided for a collection efficiency of 50 percent. Assume air conditioning operates, at most, 12 hours per day.

12.9 Calculate the quantity of heat rejected in the absorber and the condenser of the water–lithium bromide absorption system of Example 12.4.

12.10 Determine the COP of the water–lithium bromide absorption system of Example 12.4 if no heat exchanger were used.

12.11 Determine the COP of the ammonia–water absorption system of Example 12.5, if the generator temperature is raised to 200°F. The enthalpy of superheated ammonia at a temperature of 200°F and a pressure of 250 psia is 699 Btu/lbm, while that of saturated liquid ammonia at 250 psia is 168 Btu/lbm.

12.12 Determine the solar collector area required for a domestic solar water heater, if the home requires 75 gallons of water per day at a temperature of 130°F. Water is supplied at 60°F. Assume a 40 percent collector efficiency and heat losses amounting to 15 percent of the solar energy collected. The average *insolation* (impingement of solar energy) on the collector is 2000 Btu/ft^2-day. Note that collector efficiency is defined as the ratio of energy collected by the solar collector and insolation on the collector.

12.13 A solar collector has an area of 50 m^2 and collects 200 W/m^2. The collected energy is transferred to a storage tank containing four cubic meters of water. What will be the maximum temperature increase in the water per day, if no losses are assumed?

REFERENCES

1. Krenz, J. H. *Energy Conversion and Utilization*, 2nd ed. (Boston, Mass.: Allyn and Bacon, Inc., 1984).
2. Dugger, G. L., et al. "Floating Ocean Thermal Power Plants and Potential Products." *Journal of Hydronautics* 9 (October 1975): 129–41.
3. Proceedings, Third Workshop on OTEC, May 1975, ERDA.

4. Muffler, L., and White, D. "Geothermal Energy." *Perspectives on Energy*. (New York: Oxford University Press, 1975), pp. 352–59.
5. Austin, A., and Landberg, A. "Electric Power Generation from Geothermal Hot Water Deposits." *Mechanical Engineering*, December 1975, pp. 18–25.
6. "Commercial Fuel Cell." *Mechanical Engineering*, February 1974, p. 50.
7. Duffie, J. A., and Beckman, W. A. "Solar Heating and Cooling." *Science* 191 (1976): 143–49.
8. Culp, A. W. *Principles of Energy Conversion* (New York: McGraw-Hill Book Co., 1979).

chapter

13

Fundamentals of Engineering Heat Transfer

13.1 INTRODUCTION

In Section 2.3 we defined heat as thermal energy in transfer and distinguished three physical mechanisms by which heat can flow from a system to its surroundings: conduction, convection, and radiation. In general, thermal energy will flow, in accordance with the second law of thermodynamics, from a region of higher to a region of lower temperature. In conduction and convection, contact between regions is required. In conduction, the mechanism of transfer is the interaction between adjacent molecules; in convection, the mechanism of conduction is enhanced by external movement of one of the media. In radiation, thermal energy is emitted from a body by means of electromagnetic waves, is transmitted through vacuum or space containing a material medium, and is then absorbed by another body. Within a solid medium, heat is generally transferred only by conduction. In gaseous and liquid media, heat is transferred by conduction, but is considerably enhanced by the movement of the fluid particles (convection). This convection may be due to a forced motion or may be naturally induced by density differences within the fluid. In transparent solids, liquids, and gases, thermal radiation can be another mode of heat transfer. Thus, all three forms of heat transfer are often involved, with one or two modes predominating.

Throughout the text there has been some discussion of the other form of energy transfer, namely work. A number of ways of obtaining work from a system or delivering it to a system were discussed. Examples mentioned were the cylinder-piston arrangement in an automotive engine or the rotating blades of a gas turbine. However, little has been said so far on how to effect the transfer of heat in such applications as the boiler or condenser in a vapor power cycle, or the regenerator of a Brayton gas turbine cycle. It is the purpose of this chapter to discuss the transfer of heat in some detail.

As an illustrative example of conduction and convection heat transfer, consider the case of a cold wind blowing past a warm wall (Figure 13.1). Measurements show that the wind velocity is usually uniform everywhere except in the immediate vicinity of the wall. In this region adjacent to the wall, the wind velocity decreases rapidly to zero as the wall is approached. This region or film is called the *boundary layer*. The temperature of the air, in a fashion similar to the velocity, also approaches the value of the wall temperature. If on the other side of the wall there is a warm room, heat will be transferred from the still air of the warm room through the wall. There is a small variation of temperature within the wall, where heat is transferred by conduction. On the other hand, in the outside air where the energy is transferred by convection, the rapid change in temperature occurs in a thin layer in the moving air adjacent to the wall; the main body of the outside air remains unaffected. A similar situation occurs inside the room, except that the mechanism causing the motion of the air is the difference in density of the air in contact with the wall and the warmer ambient air within the room. The cooler, more dense air in contact with the inside wall moves downward due to gravity. Such self-induced motion is called natural (also free) convection. The convection occurring on the outside wall is called forced convection, since it is caused by an external force other than gravity.

The basic equation giving a quantitative description of the relation between work, heat, and other forms of energy is the energy equation (first law of thermodynamics). To determine the amount of work, such as performed by the movement of a piston in a cylinder, we also required an

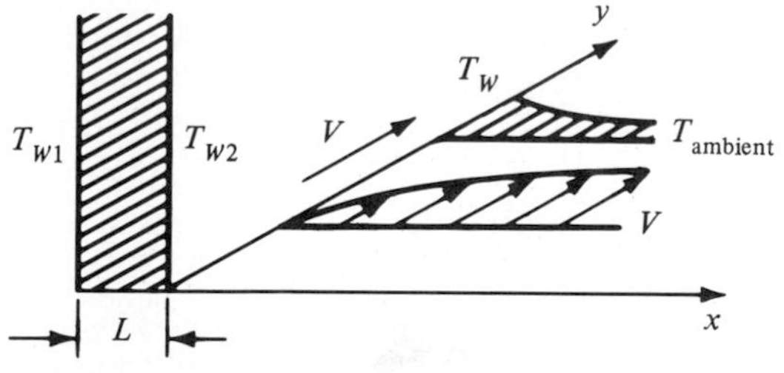

FIGURE 13.1 *Cold Wind along Warm Wall.*

additional equation in the form of (2.3). Just as in the case of work, an additional equation will be required to fully quantify heat. The form of the descriptive equation will, of course, be different for the three types of heat transfer mechanisms.

The amount of heat transferred by conduction through a solid wall will depend on the material of the wall, the surface area of the wall, and the temperature difference on the two sides of the wall. Hence, to predict the quantity of heat transferred through a given wall, we require a relationship between heat quantity and temperature. Such a relationship is the Fourier law of heat conduction and is given in the case of a flat wall by

$$\dot{Q} = -kA_s(T_{w2} - T_{w1})/L$$

where T_{w1} and T_{w2} are the temperatures of the two faces of the wall, A is the surface area of the wall normal to heat flow, and L is the wall thickness. It states that the energy transferred by conduction is proportional to surface area and temperature gradient. The negative sign is introduced to account for the fact that energy is conducted in the direction of decreasing temperature. The constant of proportionality k is called the *thermal conductivity*. The value of k will be different for different substances. For a given substance, it will be a function of the temperature of the substance, some increasing with temperature (e.g., water, air), others decreasing with temperature (e.g., oil, most metals). Metals have the largest values of thermal conductivity. The best thermal conductors are generally the best electrical conductors. The thermal conductivity of fluids is very low, so that the heat transferred in totally still liquids and gases is very small in comparison with the augmentation in heat transfer due to the convective motion of the fluid. For example, if air is confined in very small spaces and hence is restrained from circulating, as in the pores of corkboard or wool clothing, convection heat transfer is inhibited. Since air with its low thermal conductivity is a poor conductor, such materials can be used as heat insulators.

The amount of heat transferred by convection from a solid surface to a moving fluid (or vice versa) can be calculated from a knowledge of the fluid temperature gradient at the solid surface. This follows from the fact that, since the velocity of the fluid is zero at the surface, the heat transfer at the wall is by conduction only. Thus, the surface heat transfer rate is proportional to the product of thermal conductivity and temperature gradient. However, this gradient is a function of the strength of convection in the flow stream away from the wall. For the simple case of a flat wall, it is relatively easy to obtain an analytical solution for the temperature distribution in the film, and hence the temperature gradient at the wall. In most cases, however, measurement of temperature distribution would have to be made. Since the film is relatively thin, such measurements would be difficult and time consuming. It is much simpler to measure the ambient and wall temperatures for a known heat flow. The relationship between ambient and wall temperatures and heat transfer rate in convection is called Newton's law of cooling.

$$\dot{Q} = hA_s(T_w - T_f)$$

where T_w is the wall temperature, T_f is the ambient temperature of the fluid outside the film, and A_s is the surface area. The constant of proportionality h is called the convective heat transfer coefficient. This constant, usually determined experimentally, will in general be a function of the configuration and size of the body, the temperature of the fluid and of the body's surface, the velocity of the fluid, and physical properties of the fluid (particularly the thermal conductivity k).

The amount of heat transferred by radiation between two bodies at different surface temperatures is T_1, and T_2 will be proportional to the difference of the fourth power of the absolute temperatures of each body:

$$\dot{Q} = F(\epsilon, G)A_s(T_1^4 - T_2^4)$$

where the constant $F(\epsilon, G)$ is called the *radiation factor*, which is a function of the emissivity ϵ of the surfaces and the radiation shape factor G (also called *geometric view factor*), and A_s is the surface area of the body. The emissivity ϵ, a function of the character of the surface, expresses the fraction of energy radiated from a given surface as compared to that of an ideal radiator at the same temperature, while the geometric view factor indicates the portion of the radiation emitted by one body which falls upon the other body.

In actual physical situations involving the transfer of energy as heat, different modes of heat transfer usually occur at the same time, either in series such as convection and conduction in a refrigeration condenser or in parallel such as convection and radiation from a bare, high-temperature metal pipe to room air. In the sections which follow, each mode of heat transfer will be described separately. In the next chapter, combined modes of heat transfer will be discussed in various applications of heat transfer equipment.

We will consider steady-state transfer phenomena only, in which the heat transfer rates and temperature differences remain constant with time. Thus, heat transfer during start-up or shutdown of a power plant or during change from one operation condition to another will not be considered here.

13.2 CONDUCTION HEAT TRANSFER

In heat transfer by conduction, thermal energy is transferred between two bodies in direct contact (or regions of one body) by molecular interaction. In gases, liquids, and most solids, molecules at higher energy levels (as indicated by a higher temperature) impart some of their energy to adjacent molecules at lower energy levels. In pure metals, the mechanism is that of diffusion of free electrons.

The descriptive equation relating the transfer rate and temperature gradient in a body is given by Fourier's law. Its one-dimensional form is

$$\dot{Q} = -kA_s \, dT/dx \tag{13.1}$$

The constant of proportionality k, the thermal conductivity, is in general a function of the temperature of the substance. As an example, Figure 13.2 shows the temperature dependence of k for copper. Over a small temperature range, the value of k can be assumed to be constant. For larger temperature ranges, an average value can be used.

As an example of the use of Equation 13.1, consider a simple flat wall of thickness t with each face at a different temperature as shown in Figure 13.3. Assume the value of the thermal conductivity k to be constant. We can integrate (13.1) to obtain the expression for the temperature variation in the wall, namely,

$$T = -\frac{\dot{Q}}{A_s k} x + c$$

Using the conditions $T = T_1$ at $x = 0$ and $T = T_2$ at $x = t$, we get a linear temperature distribution within the wall

$$T = T_1 - (T_1 - T_2)x/t \tag{13.2}$$

and a heat transfer rate which is proportional to the temperature difference

$$\dot{Q} = \frac{A_s(T_1 - T_2)k}{t} \tag{13.3}$$

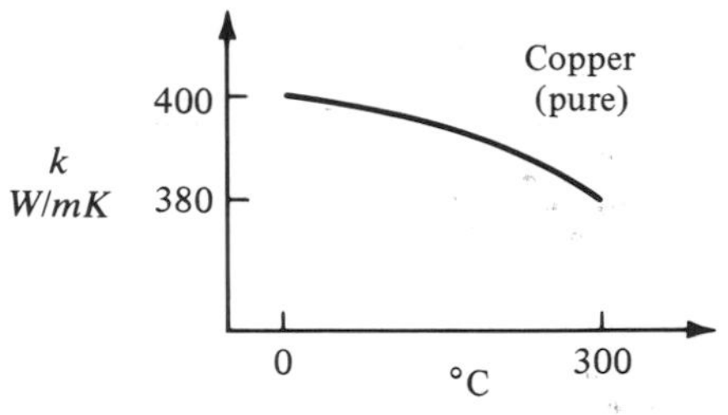

FIGURE 13.2 *Thermal Conductivity of Copper.*

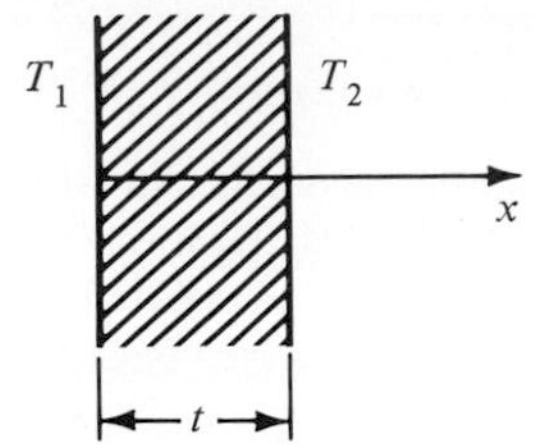

FIGURE 13.3 *Conduction through a Flat Wall.*

Usually walls consist of several layers of different materials. For example, the walls of a dwelling typically would consist of a layer of brick, a layer of cement blocks, and a layer of plaster or dry wall. The result obtained for a single-layer wall can be used to determine the temperature distribution within the composite wall by recognizing that, at steady-state condition, the heat transfer rate in each layer must be equal. For the two-layer wall shown in Figure 13.4, we have $\dot{Q}_a = \dot{Q}_b = \dot{Q}$. Using (13.3), we get

$$\dot{Q}_a = (k_a A_s / t_a)(T_1 - T_2)$$

and

$$\dot{Q}_b = (k_b A_s / t_b)(T_2 - T_3)$$

Thus $T_1 - T_2 = \dot{Q} t_a / k_a A_s$ and $T_2 - T_3 = \dot{Q} t_b / k_b A_s$. Adding, we get an expression for the heat flow rate

$$\dot{Q} = \frac{T_1 - T_3}{(t_a / k_a A_s) + (t_b / k_b A_s)} = (T_1 - T_3)/(R_a + R_b) \qquad \textbf{(13.4)}$$

where the quantity t/kA_s is called the thermal resistance R. This relation indicates that the total thermal resistance (R) of a composite wall is the sum of the individual resistances of each layer (R_a and R_b). The temperature distribution within each layer of the wall is linear, but each distribution has a different slope.

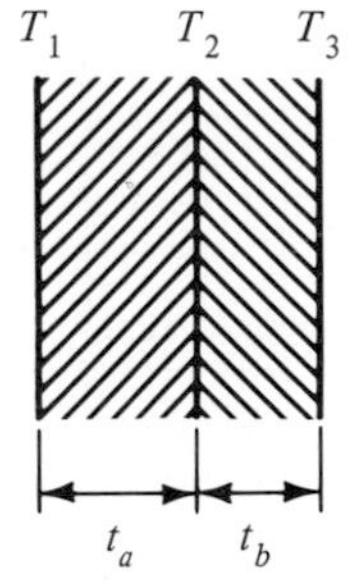

FIGURE 13.4 *Conduction in a Two-Layer Wall.*

EXAMPLE 13.1

Determine the heat transfer rate through a one-inch-thick wall made of carbon steel having a thermal conductivity of 35 Btu/ft-h-°R. The inner surface of the wall is covered with a $\frac{1}{8}$-in.-thick scale of thermal conductivity of 0.5 Btu/ft-h-°R. The temperature of the outside steel surface is 430°F and that of the inner surface of the scale is 350°F.

Solution

From (13.4), we have with $t_a = 1$ in, $k_a = 35$ Btu/ft-h-°R, and $t_b = \frac{1}{8}$ in, $k_b =$ 0.5 Btu/ft-h-°R,

$$\dot{Q}/A_s = \frac{T_1 - T_3}{t_a/k_a + t_b/k_b} = \frac{430°\text{F} - 350°\text{F}}{\dfrac{1.0\text{ ft}}{12(35\text{ Btu/ft-h-°R})} + \dfrac{0.125\text{ ft}}{12(0.5\text{ Btu/ft-h-°R})}}$$

$$= 3446\text{ Btu/ft}^2\text{-h}$$

To determine the temperature of the inner surface of the steel wall, use

$$T_1 - T_2 = \frac{\dot{Q}}{A}\frac{t_a}{k_a} = 3446\left(\frac{1/12}{35}\right) = 8.2°\text{F} \quad \text{or} \quad T_2 = 421.8°\text{F}$$

In deriving the results for a composite wall, we assumed that there is perfect contact between the various layers and that the temperature of the contacting faces of two adjacent layers is the same. However, if the surface is rough, perfect contact is not possible and a thin air layer forms between two solid layers of the wall. Since air is a poor heat conductor ($k = 0.013$ Btu/ft-h-°R, or 0.023 W/m K), even a thin air film may lead to a considerable reduction in heat flow through the wall.

When metal bolts extend through a wall (as shown in Figure 13.5), there is an additional path of heat flow. In this case, we get

$$\dot{Q}_a = \frac{k_a A_a}{t}(T_1 - T_2) \quad \text{and} \quad \dot{Q}_b = \frac{k_b A_b}{t}(T_1 - T_2)$$

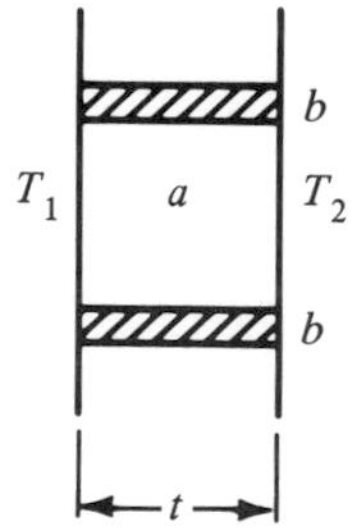

FIGURE 13.5 *Metal Bolts in a Wall.*

Thus the total heat transfer rate is

$$\dot{Q} = \dot{Q}_a + \dot{Q}_b = [k_a A_a + k_b A_b](T_1 - T_2)/t \qquad \textbf{(13.5)}$$

EXAMPLE 13.2

Metal bolts of thermal conductivity 60 W/m K extend through an insulating wall with conductivity of 0.06 W/m K. The area of the bolts is 0.2 percent of the wall area. Calculate the effect of the bolts on the heat loss through the wall.

Solution

From (13.5),

$$\dot{Q} = [0.06A_a + 60A_b](T_1 - T_2)/t$$

With $A_b/A_a = 0.002$,

$$\dot{Q} = [0.06A_a + 60(0.002)A_a](T_1 - T_2)/t = 0.18\,\frac{A_a}{t}\,(T_1 - T_2)$$

Without metal bolts,

$$\dot{Q} = Q_a = k_a\,\frac{A_a}{t}\,(T_1 - T_2) = 0.06\,\frac{A_a}{t}\,(T_1 - T_2)$$

The presence of the bolts increases the heat loss by a factor of 3.

Another application of (13.1) is the heat conduction through the wall of a circular tube. The inner and outer surfaces of the tube are held at constant, but different temperatures (Figure 13.6). From (13.1) we get

$$\dot{Q} = -k2\pi RL\,\frac{dT}{dR}$$

where L is the length of the tube. Rearranging, we have

$$dT = -\frac{\dot{Q}}{2\pi kL}\,\frac{dR}{R}$$

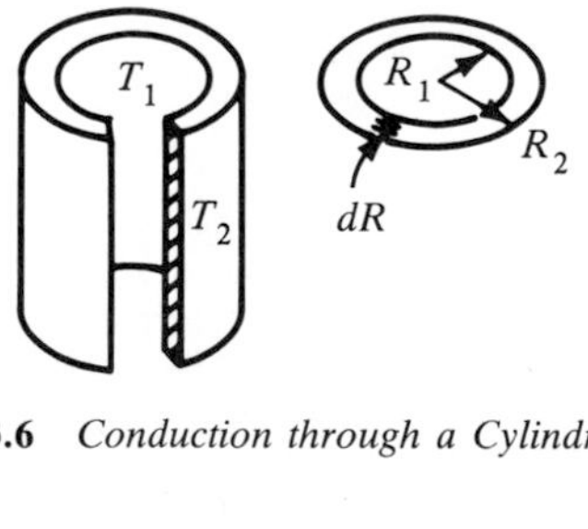

FIGURE 13.6 *Conduction through a Cylindrical Wall.*

Integrating, we get

$$T = -\frac{\dot{Q}}{2\pi kL} \ln R + c$$

Using the temperature conditions that $T = T_1$ at $R = R_1$ and $T = T_2$ at $R = R_2$, we obtain a nonlinear temperature distribution within the cylinder wall

$$T = T_1 - \frac{T_1 - T_2}{\ln R_2/R_1} \ln R/R_1$$

and a heat transfer rate of

$$\dot{Q} = 2\pi L(T_1 - T_2)k/(\ln R_2/R_1) \qquad \textbf{(13.6)}$$

Similar to composite plane walls, piping installations also consist of composite walls, for example, pipes with insulation on the outside and scale deposits on the inside. For steady-state conditions of heat flow through the pipe wall *a* of Figure 13.6 covered by an insulating material *b* (from R_2 to R_3), $\dot{Q}_a = \dot{Q}_b = \dot{Q}$:

$$\dot{Q}_a = 2\pi L(T_1 - T_2)k_a/\ln (R_2/R_1)$$

and

$$\dot{Q}_b = 2\pi L(T_2 - T_3)k_b/\ln (R_3/R_2)$$

Thus

$$T_1 - T_2 = \frac{\dot{Q}}{2\pi Lk_a} \ln (R_2/R_1) \quad \text{and} \quad T_2 - T_3 = \frac{\dot{Q}}{2\pi Lk_b} \ln (R_3/R_2)$$

Adding, we get an expression for the heat flow rate through the composite cylindrical wall:

$$\dot{Q} = \frac{2\pi L(T_1 - T_3)}{(1/k_a) \ln (R_2/R_1) + (1/k_b) \ln (R_3/R_2)} \qquad \textbf{(13.7)}$$

EXAMPLE 13.3

A stainless steel tube with one inch ID and 1.25 inch OD ($k = 12.5$ Btu/ft-h-°R) is covered with a one-inch layer of insulation ($k = 0.15$ Btu/ft-h-°R). The inside wall temperature of the tube is 800°F, while the outside of the insulation is kept at 80°F. Calculate the heat loss per foot of tube length.

Solution

From (13.7),

$$\dot{Q}/L = \frac{2\pi(T_1 - T_3)}{\ln (R_2/R_1)/k_a + \ln (R_3/R_2)/k_b} = \frac{2\pi(800 - 80)}{\ln \frac{0.625}{0.50}\Big/ 12.5 + \ln \frac{1.625}{0.625}\Big/ 0.15}$$

$$= 708.2 \text{ Btu/ft-h}$$

EXAMPLE 13.4

If the insulation of Example 13.3 is placed on the inside of the steel tube, what must the insulation thickness be for the same heat loss?

Solution

From (13.7),

$$\dot{Q}/L = \frac{2\pi(800-80)}{\ln \frac{0.625}{0.50}\Big/ 12.5 + \ln \frac{0.50}{R}\Big/ 0.15} = 708.2 \text{ Btu/ft-h}$$

Hence,

$$\ln \frac{0.50}{R}\Big/ 0.15 = 6.389 - 0.01785 = 6.371 \quad \text{and} \quad R = 0.192 \text{ in.}$$

The inside thickness of the inside insulation becomes

$$t = 0.50 - 0.192 = 0.308 \text{ in.}$$

Thus, less thickness of insulation having the same value of thermal conductivity k is required when placed on the inside of the tube. Disadvantages of such use lie in the manufacture of such a tube and in the inspection for integrity of inside insulation.

13.3 CONVECTION HEAT TRANSFER

The process of heat transfer by convection involves the transfer of heat from a liquid or gas to a solid through direct contact. While the heat is actually transferred at the solid surface by conduction, it is augmented by the movement of the fluid particle (convection) in the liquid or gas; the combined phenomena are termed *convection heat transfer*. Convection heat transfer depends on many factors, mainly on the type of fluid flow, the physical properties of the flow, and the size and shape of the surface of the solid.

A study of fluid flow shows that there are two types of flow: *laminar* and *turbulent*. In laminar flow, the particles of the fluid move in layers (lamina) parallel to the walls of the passage or body; in turbulent flow, the particles move in somewhat random, but generally directed fashion. The magnitude of the velocity of a turbulent flow at a given location might vary as indicated in Figure 13.7. In addition, the direction of the velocity would also fluctuate. Since these fluctuations are rapid, the flow velocity at a given location can be averaged with respect to time. The mixing motion of the fluid particles in a turbulent flow causes additional

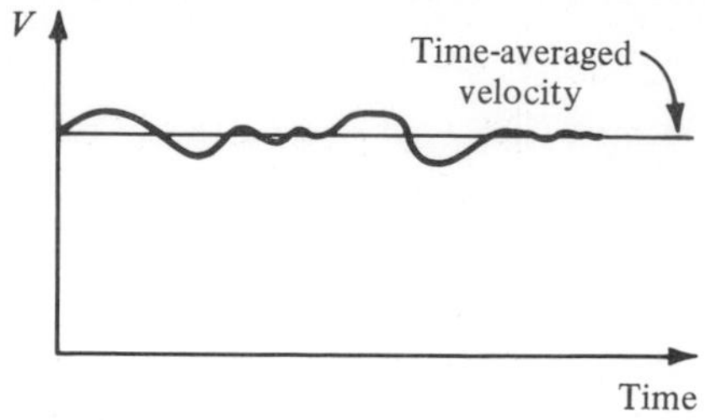

FIGURE 13.7 *Velocity Fluctuation in Turbulent Flow.*

energy dissipation. Thus, in laminar flow in a pipe, the frictional loss is proportional to the average velocity ($\bar{V}$) of the flow, whereas in turbulent flow, the frictional loss is proportional to $\bar{V}^2$.

In laminar flow, heat is transferred in a direction normal to the wall mainly by conduction, and the rate is proportional to the thermal conductivity of the fluid. In turbulent flow, this mode of heat transfer occurs only in the laminar sublayer, which exists in the immediate vicinity of the wall. In the main turbulent flow field, the heat is transferred by an intensive intermixing of the particles. The determining factor in this case is the thermal resistance of the thin sublayer, with the thermal resistance of the fluid in the turbulent region having only secondary effect.

There are two types of convection—*free* and *forced* convection. In free convection, the motion of the fluid is due solely to the difference in densities of the warmer and colder particles of the fluid. In forced convection, the motion of the fluid occurs due to the influence of an external agent, such as a pump, a fan, or the wind. Forced convection may be accompanied by free convection. The effect of the free convection increases with increase in difference in temperature of different fluid particles; the effect decreases, however, with increase in the velocity of the forced flow, becoming negligible at high flow velocities.

The basic equation utilized for convective heat transfer is Newton's law of heat transfer,

$$\dot{Q} = hA_s(T_w - T_f) \qquad \textbf{(13.8)}$$

which states that the convective heat transfer rate is proportional to the temperature difference between wall and ambient fluid and the surface area A_s. For a given heat transfer situation, we need to establish an applicable value of the convection heat transfer coefficient h. Table 13.1 presents typical ranges of the heat transfer coefficient h for the several types of convection.

TABLE 13.1 *Typical Ranges of Heat Transfer Coefficient h*

Mechanism	h (Btu/ft^2-h-°R)	h (W/m^2 K)
Forced convection, air	5–1,000	25–5,000
water	50–2,500	250–15,000
oil	10–400	50–2,000
Free convection, air	0.5–10	2–50
water	100–200	500–1,000
Boiling water	200–10,000	1,000–50,000
Condensing water (filmwise)	1,000–5,000	5,000–20,000
water (dropwise)	5,000–25,000	20,000–150,000

13.3a HEAT TRANSFER BY FORCED CONVECTION

As was indicated earlier, the geometric configuration of the solid surface is one of the factors affecting the convection heat transfer rate. The number of possible useful configurations is obviously large. However, three flow configurations are most frequently encountered in actual applications, namely flow within circular tubes, flow about circular tubes, and flow along flat walls. These cases will now be discussed in some detail.

1. Flow within Circular Tubes

Convective heat transfer depends on the nature of the flow, i.e., laminar or turbulent. In general, laminar flow occurs at low velocities and turbulent at high velocities. The criterion that establishes the nature of the flow is the nondimensional parameter Reynolds number Re where $\mathrm{Re} = \bar{V}L/\nu$, L is a characteristic length of the flow, and ν is a physical property of the fluid, kinematic viscosity. For flow in a circular tube, the characteristic length is the tube diameter. For Reynolds numbers less than a critical Reynolds number $\mathrm{Re}_{\mathrm{crit}}$ flow is laminar; for Reynolds numbers greater than this critical Reynolds number, the flow is turbulent. For example, for pipe flow, the critical Reynolds number can be taken as 2200. Thus, for water flowing in a one inch (25.4 mm) tube at 50°F, flow will be laminar for velocities less than

$$\bar{V} = \frac{\nu\,\mathrm{Re}_{\mathrm{crit}}}{D} = \frac{(1.41 \times 10^{-5}\ \mathrm{ft^2/s})(2200)}{(\frac{1}{12}\ \mathrm{ft})} = 0.37\ \mathrm{ft/s} \quad \text{or} \quad \bar{V} = 0.11\ \mathrm{m/s}$$

and flow will be turbulent for velocities greater than 0.37 ft/s (0.11 m/s). Values of ν are found in Appendix D.

When the flow within circular tubes is laminar and Re < 2000, an experimentally based relation for the heat transfer coefficient is given by

$$\frac{hD}{k_f} = 0.17\,\mathrm{Re}_f^{0.33}\,\mathrm{Pr}_f^{0.4}\left(\frac{\mathrm{Pr}_f}{\mathrm{Pr}_w}\right)^{0.25} \tag{13.9}$$

where Pr, called Prandtl number, $c_p\mu/k$ represents a dimensionless ratio of physical properties of the fluid. The property μ is dynamic viscosity and is equal to $\mu = \nu\rho$. The subscript f refers to the properties at fluid temperature, while w refers to fluid properties at wall temperature. The $\mathrm{Pr}_f/\mathrm{Pr}_w$ term takes into account the effect of the direction of heat flow (heating or cooling). The relation (13.9) is based on measurements of heat transfer involving a large number of fluids flowing in forced convection within circular tubes.

In turbulent flow, heat transfer within the fluid is caused mainly by mixing. This mixing causes the temperature to remain practically constant over the entire cross section with the exception of a thin film at the wall of the tube. In this film, the temperature changes sharply to reach the temperature of the tube wall. An analysis of experimental data on forced convection through smooth circular tubes shows that for fully developed turbulent flow (Re > 10,000) the heat transfer coefficient is given by

$$\frac{hD}{k_f} = 0.023\,\mathrm{Re}_f^{0.8}\,\mathrm{Pr}_f^{0.4}\left(\frac{\mathrm{Pr}_f}{\mathrm{Pr}_w}\right)^{0.25} \tag{13.10}$$

In the transition region (Re = 2000 to 10,000), the heat transfer coefficient increases sharply to the value for turbulent flow. Figure 13.8 summarizes the relations for the three flow regimes.

Figure 13.9 illustrates the behavior of the forced convection heat transfer coefficient h for water flowing in a circular tube as a function of the product of average flow velocity $\bar{V}$ and tube diameter D.

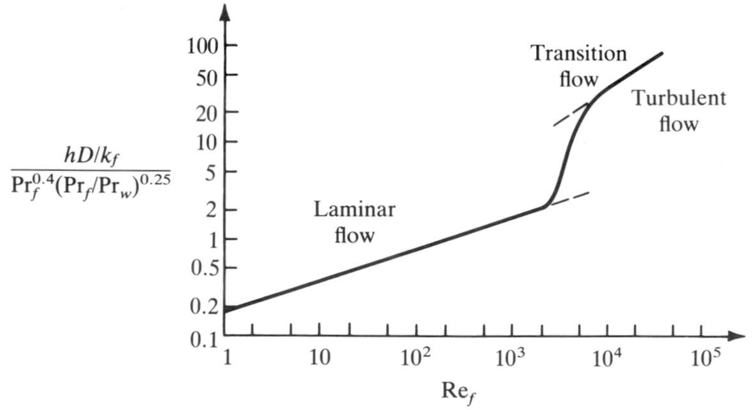

FIGURE 13.8 *Forced Convection Heat Transfer Coefficients Inside Circular Tubes.*

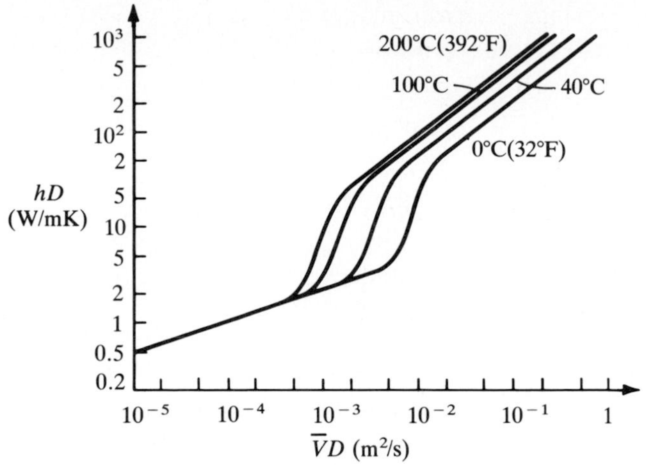

FIGURE 13.9 *Forced Convection Heat Transfer Coefficients for Saturated Water Inside Circular Tubes.*

For turbulent flow, the expression for heat transfer coefficient (13.10) is also applicable to noncircular tubes, provided an equivalent diameter is used in the calculation. The equivalent diameter is defined as

$$D_{eq} = 4A/P \tag{13.11}$$

where A is the cross-sectional area of the pipe and P is the wetted perimeter. Values of physical properties are found in Appendix D.

EXAMPLE 13.5

Hot air at an average temperature of 100°C flows at 5 m/s through a 3-m-long tube with an inside diameter of 60 mm. Determine the amount of heat transferred to the wall if the temperature of the tube wall is kept at 20°C.

Solution

The Reynolds number of the flow is

$$\text{Re} = \frac{\bar{V}D}{\nu} = \frac{(5\ \text{m/s})(0.06\ \text{m})}{2.30 \times 10^{-5}\ \text{m}^2/\text{s}} = 13{,}040$$

hence turbulent flow occurs in the tube. From (13.10), we get

$$\frac{hD}{k_f} = 0.023\ \text{Re}^{0.8}\ \text{Pr}_f^{0.4}\ (\text{Pr}_f/\text{Pr}_w)^{0.25}$$

For air, Pr is constant over a large temperature range with a value of 0.71 at 20°C as well as 100°C. Thus

$$\frac{hD}{k_f} = 0.023(13{,}040)^{0.8}0.71^{0.4} = 39.3$$

and
$$h = \frac{0.0307 \text{ W/m K}}{0.06 \text{ m}} 39.3 = 20.1 \text{ W/m}^2 \text{ K}$$

The amount of heat transferred to the wall is

$$\dot{Q} = h(T_f - T_w)A_s = (20.1 \text{ W/m}^2 \text{ K})(100 \text{ K} - 20 \text{ K})\pi(0.06)3 \text{ m}^2 = 909 \text{ W}$$

EXAMPLE 13.6

Water flows inside a two-inch-diameter tube, 15 ft long with an average velocity of 2.5 ft/s. Find the rate of heat transferred for an average water temperature of 120°F and a tube wall temperature of 160°F.

Solution

At $T_f = 120°\text{F}$, $k_f = 0.371$ Btu/ft-h-°R, $\nu_f = 0.639 \times 10^{-5}$ ft^2/s, $\text{Pr}_f = 4.34$; at $T_w =$ 160°F, $\text{Pr}_w = 3.65$. Hence

$$\text{Re} = \frac{\bar{V}D}{\nu} = \frac{2.5(2)}{0.639 \times 10^{-5}(12)} = 6.52 \times 10^4$$

hence turbulent flow occurs in the tube and, from (13.10),

$$\frac{hD}{R_f} = 0.023(6.52 \times 10^4)^{0.8}(4.34)^{0.4}\left(\frac{4.34}{3.65}\right)^{0.25} = 307$$

so

$$h = \frac{(0.371 \text{ Btu/ft-h-°R})307}{2/12 \text{ ft}} = 683 \text{ Btu/ft}^2\text{-h-°R}$$

$$\dot{Q} = 683 \text{ Btu/ft}^2\text{-h-°R}(160 - 120)\text{°R}\,\pi(2/12)15 \text{ ft}^2 = 214{,}600 \text{ Btu/h}$$

EXAMPLE 13.7

Oil flows at a rate of 1.67 kg/s through a 10-mm-diameter tube 2 m long. The tube wall is kept at 80°C. The oil enters at 20°C. Calculate the heat supplied to the tube per second and the outlet oil temperature. Use the following data for the oil: $k_f = 0.144$ W/m K, $\nu_f = 0.00089$ m^2/s, $\rho_f = 894$ kg/m^3, $k_w = 0.141$ W/m K, $\nu_w =$ 0.000039 m^2/s, $\rho_w = 851$ kg/m^3, $c_{pf} = 1850$ J/kg K, and $c_{pw} = 2080$ J/kg K.

Solution

$$V = \frac{\dot{m}}{\rho A} = \frac{1.67}{894\left(\frac{\pi}{4} 0.01^2\right)} = 23.8 \text{ m/s}$$

So

$$\text{Re} = \frac{\bar{V}D}{\nu} = \frac{23.8(0.01)}{890 \times 10^{-6}} = 267$$

hence laminar flow exists in the tube. From (13.9),

$$\frac{hD}{k_f} = 0.17 \,\text{Re}_f^{0.33}\, \text{Pr}_f^{0.4}\, (\text{Pr}_f/\text{Pr}_w)^{0.25}$$

$$\text{Pr}_f = \frac{c_{pf}\mu_f}{k_f} = \frac{(1850\ \text{J/kg K})(0.00089\ \text{m}^2/\text{s} \times 894\ \text{kg/m}^3)}{(0.144\ \text{W/m K})} = 10{,}220$$

$$\text{Pr}_w = \frac{c_{pw}\mu_w}{k_w} = \frac{(2080\ \text{J/kg K})(0.000039\ \text{m}^2/\text{s} \times 851\ \text{kg/m}^3)}{(0.141\ \text{W/m K})} = 490$$

Thus

$$\frac{hD}{k_f} = 0.17(267)^{0.33}10{,}220^{0.4}\left(\frac{10{,}220}{490}\right)^{0.25} = 92.2$$

and

$$h = \frac{(0.144\ \text{W/m K})92.2}{0.01\ \text{m}} = 1328\ \text{W/m}^2\ \text{K}$$

Hence,

$$\begin{aligned}\dot{Q} &= hA\,\Delta T\\ &= (1328\ \text{W/m}^2\ \text{K})(\pi \times 0.01\ \text{m} \times 2\ \text{m})(80°\text{C} - 20°\text{C})\\ &= 5000\ \text{W}\end{aligned}$$

Also,

$$\dot{Q} = \dot{m}C_p(T_2 - T_1)$$

so

$$T_2 - T_1 = \frac{5000\ \text{W}}{(1.67\ \text{kg/s})(1850\ \text{J/kg K})} = 1.6°\text{C} \quad \text{and} \quad T_2 = 21.6°\text{C}$$

EXAMPLE 13.8

Cool air at an average temperature of 40°F flows in a rectangular duct (0.8 in. by 1.6 in.), 30 ft long, at a velocity of 10 fps. Determine the amount of heat transferred if the duct wall temperature is 68°F.

Solution

From (13.11),

$$D_{\text{eq}} = 4\,\frac{0.8(1.6)}{2(0.8 + 1.6)} = 1.067\ \text{in.}$$

The thermal conductivity at 40°F is 0.0139 Btu/ft-h-°R and the kinematic viscosity is 1.38×10^{-4} ft²/s. The Reynolds number of the flow is

$$\text{Re} = \frac{\bar{V}D_{\text{eq}}}{\nu_f} = \frac{10(1.067)}{1.38 \times 10^{-4}(12)} = 6443$$

hence transitional flow exists in the duct. From Figure 13.8,

$$\frac{hD}{k_f}\Big/\text{Pr}_f^{0.4}\left(\frac{\text{Pr}_f}{\text{Pr}_w}\right)^{0.25} = 17.7$$

For air, the Prandtl number Pr is constant between temperature values 40°F and 70°F and equal to 0.71. Hence

$$\frac{hD_{\text{eq}}}{k_f} = 17.7(0.71)^{0.4} = 15.4$$

and $$h = \frac{k_f}{D_{eq}} 15.4 = \frac{0.0139(12)}{1.067} 15.4 = 2.41 \text{ Btu/ft}^2\text{-h-°R}$$

$$\dot{Q} = h(T_f - T_w)A_s = 2.41(40 - 68) \frac{2(0.8 + 1.6)}{4} 30 = -67.5 \text{ Btu/h}$$

---

2. *Perpendicular Flow about Circular Tubes*

In general, the heat transfer coefficient will vary locally around the circumference of the cylinder. However, usually we are interested only in the heat transfer coefficient averaged over the entire cylinder surface. Based on experimental data, the average heat transfer coefficient for a single cylinder with flow normal to the cylinder can be represented by a relation of the form

$$\frac{hD}{k_f} = C \text{ Re}_f^n \text{ Pr}_f^{0.38} (\text{Pr}_f/\text{Pr}_w)^{0.25} \tag{13.12}$$

where the coefficient C and exponent n vary with the Reynolds number of the approach flow as follows:

Re_f	C	n
0.1–10	0.72	0.27
10–1,000	0.59	0.47
1,000–200,000	0.21	0.62

EXAMPLE 13.9

Determine the heat transfer coefficient for air flowing across a single tube 20 mm in diameter if the fluid temperature $T_f = 30°\text{C}$ and the stream velocity $V = 5$ m/s.

Solution

At a fluid temperature of $T_f = 30°\text{C}$, the thermal conductivity $k_f = 0.026$ W/m K, the kinematic viscosity $\nu_f = 1.60 \times 10^{-5}$ m^2/s, and Pr = 0.71. The Reynolds number of the flow

$$\text{Re}_f = \frac{VD}{\nu_f} = \frac{5(0.02)}{1.60 \times 10^{-5}} = 6250$$

For air, Pr does not vary much with moderate temperature change; hence from (13.12)

$$\frac{hD}{k_f} = 0.21(6250)^{0.62}(0.71)^{0.38} = 41.6$$

and $$h = \frac{(0.026 \text{ W/m K})}{0.02 \text{ m}} 41.6 = 54.1 \text{ W/m}^2 \text{ K}$$

EXAMPLE 13.10

Determine the heat transfer coefficient for water flowing across a single tube 20 mm in diameter if the temperature of the fluid $T_f = 20°C$ and $T_w = 40°C$ and the fluid velocity $V = 0.5$ m/s.

Solution

At a temperature $T_f = 20°C$, the thermal conductivity $k_f = 0.60$ W/m K, the kinematic viscosity $\nu_f = 1 \times 10^{-6}\ m^2/s$ and $Re_f = VD/\nu_f = 0.5(0.02)/10^{-6} =$ 10,000, $Pr_f = 7.02$, and $Pr_w = 4.31$. From (13.12),

$$\frac{hD}{k_f} = 0.21\ Re_f^{0.62}\ Pr_f^{0.38}\ (Pr_f/Pr_w)^{0.25}$$

$$= 0.21(10{,}000)^{0.62}(7.02)^{0.38}(7.02/4.31)^{0.25} = 150$$

and

$$h = \frac{k_f}{D}\,150 = \frac{0.60}{0.02}\,150 = 4500\ \text{W/m}^2\ \text{K}$$

In actual heat transfer applications, tubes are arranged in banks, most frequently staggered or in-line (Figure 13.10). Based on test data, the following relations for the heat transfer coefficient may be given for

Staggered banks of tubes:
$$\frac{hD}{k} = 0.37\ Re^{0.60}\ Pr_f^{0.33}\ (Pr_f/Pr_w)^{0.25} \tag{13.13}$$

In-line banks of tubes:
$$\frac{hD}{k} = 0.22\ Re^{0.65}\ Pr_f^{0.33}\ (Pr_f/Pr_w)^{0.25} \tag{13.14}$$

Both relations are applicable in the Reynolds number range of 200 to 200,000. In these relations, the mean fluid temperature, the velocity in the narrowest cross section of a row, and the tube diameter are used. Such usage of reference quantities makes the heat transfer coefficient virtually independent of relative spacing in the transverse and longitudinal directions.

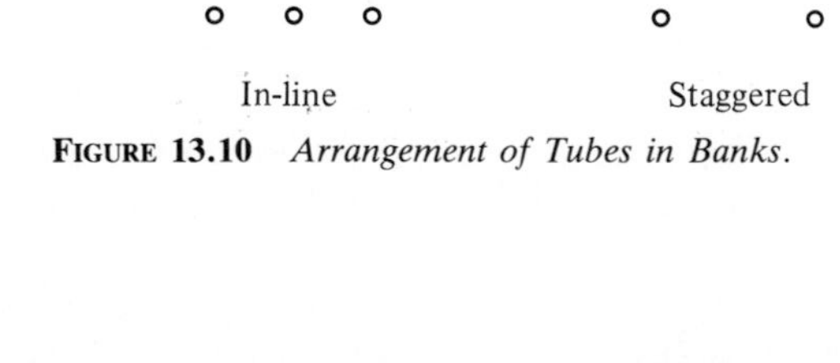

FIGURE 13.10 *Arrangement of Tubes in Banks.*

EXAMPLE 13.11

Air at atmospheric pressure and a temperature of 68°F enters a tube bank with an average velocity of 15 fps. The twenty-five tubes, each 5 ft long, have an OD of 1.25 in. and are arranged in line five rows deep (Figure 13.11). The centerline spacing in the longitudinal direction is 2 in. The tube walls are maintained at a temperature of 120°F. Determine the heat transfer coefficient for the air.

Solution

At the first step, determine the physical properties at the inlet temperature (T_1) of 68°F: $\rho = 0.074$ lbm/ft^3, $c_p = 0.24$ Btu/lbm°R, $k = 0.0144$ Btu/ft-h-°R, $\mathrm{Pr}_f = 0.71$, $\nu_f = 1.62 \times 10^{-4}$ ft^2/s, $\mathrm{Pr}_w = 0.71$. From continuity we get

$$\dot{m} = \rho V_1 A_1 = \rho V_{\max} A_{\min}$$

and

$$V_{\max} = V_1 A_1 / A_{\min} = 15\,\frac{2(5)}{(2-1.25)5} = 40 \text{ fps}$$

Hence

$$\mathrm{Re} = \frac{V_{\max} D}{\nu} = \frac{40(1.25)}{1.62 \times 10^{-4}(12)} = 25{,}600$$

From (13.14),

$$\frac{hD}{k} = 0.22(25{,}600)^{0.65}(0.71)^{0.33} = 144$$

and

$$h = \frac{(0.0144 \text{ Btu/ft-h-°R})144}{(1.25/12)\text{ ft}} = 19.9 \text{ Btu/ft}^2\text{-h-°R}$$

A good approximation to the exit temperature T_2 can be obtained by equating the heat transferred from the tube walls to the air to the rate of increase of the enthalpy of the air as it passes through the tube

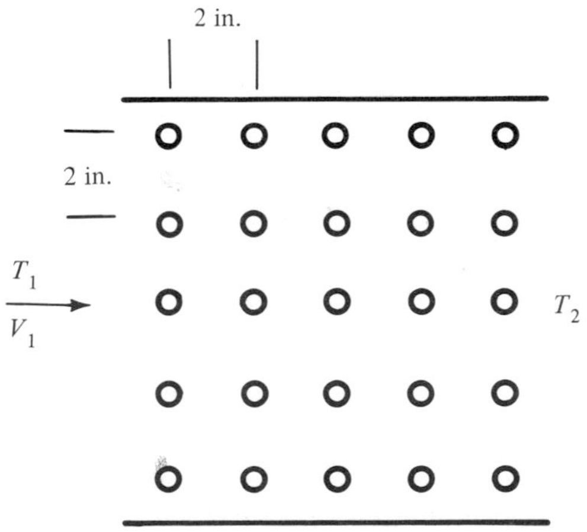

FIGURE 13.11 *Arrangement of Tube Bank in Example 13.11.*

$$\dot{Q} = hA_s(T_w - T_f) = (19.9\ \text{Btu/ft}^2\text{-h-°R})\left(\pi \times \frac{1.25}{12}\ \text{ft} \times 5\ \text{ft} \times 25\right)(120 - T_f)$$

$$= 814(120 - T_f)\ \text{Btu/h}$$

Also,

$$\dot{Q} = \dot{m}c_p(T_2 - T_1)$$

where

$$\dot{m} = \rho AV = (0.074\ \text{lbm/ft}^3)\left(5 \times \frac{2}{12}\ \text{ft} \times 5\ \text{ft}\right)(15\ \text{ft/s})$$

$$= 4.625\ \text{lbm/s}$$

Thus,

$$\dot{Q} = (4.625\ \text{lbm/s})(0.24\ \text{Btu/lbm°F})(T_2 - T_1)$$

$$= 1.11(T_2 - T_1)\ \text{Btu/s}$$

Approximate $T_f = (T_1 + T_2)/2$ and equate:

$$814\left(120 - \frac{T_1 + T_2}{2}\right) = 1.11(T_2 - T_1)3600$$

$$T_2 = 78.0°\text{F}$$

---

3. *Flow along Flat Walls*

Along a flat plate, the fluid flows parallel to the surface of the plate and its velocity is constant everywhere except in the region near the surface of the plate (boundary layer), where due to frictional effects the velocity rapidly decreases to zero at the wall. Experiments have shown that the thickness of the boundary layer gradually increases in the direction of flow. The flow in the boundary layer may be laminar or turbulent, except in the immediate vicinity of the wall, where the flow is always laminar.

For a laminar boundary layer flow along a flat plate, the heat transfer coefficient averaged over the flat plate is given by

$$\frac{hL}{k_f} = 0.76\ \text{Re}_f^{0.5}\ \text{Pr}_f^{0.4}\left(\frac{\text{Pr}_f}{\text{Pr}_w}\right)^{0.25} \qquad \textbf{(13.15)}$$

while for turbulent boundary layer flow, the average coefficient is

$$\frac{hL}{k_f} = 0.036\ \text{Re}_f^{0.8}\ \text{Pr}_f^{0.4}\left(\frac{\text{Pr}_f}{\text{Pr}_w}\right)^{0.25} \qquad \textbf{(13.16)}$$

Figure 13.12 presents a summary of the relations for the heat transfer coefficients for a flat wall.

EXAMPLE 13.12

Air flows along a smooth wall one meter wide (b) and 1.2 m long (L) at a velocity

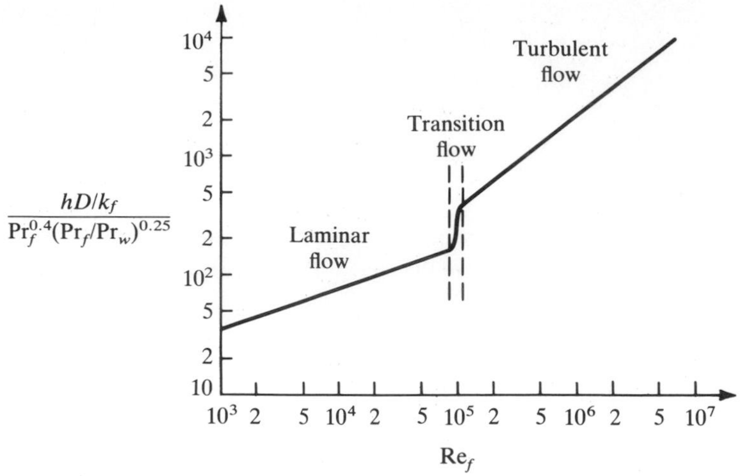

FIGURE 13.12 *Forced Convection Heat Transfer Coefficients for Flat Wall.*

of 9 m/s. Determine the heat transfer coefficient and the total heat quantity transferred if $T_w = 80°\text{C}$ and $T_f = 20°\text{C}$.

Solution

At $T_f = 20°\text{C}$, $k_f = 0.0251$ W/m K, $\nu_f = 1.51 \times 10^{-5}$ m²/s, and $\mathrm{Pr}_f = 0.71$. Hence

$$\mathrm{Re}_L = \frac{VL}{\nu_f} = \frac{9(1.2)}{1.51 \times 10^{-5}} = 7.15 \times 10^5$$

hence turbulent flow will occur. From (13.16),

$$\frac{hL}{k_f} = 0.036\,\mathrm{Re}_f^{0.8}\,\mathrm{Pr}_f^{0.4} = 0.036(7.15 \times 10^5)^{0.8}(0.71)^{0.4} = 1514$$

and

$$h = \frac{k_f}{L}\,1514 = \frac{0.0251}{1.2}\,(1514) = 31.7\ \mathrm{W/m^2\,K}$$

Thus

$$\dot{Q} = hA_s(T_w - T_f) = (31.7)(1 \times 1.2)(60) = 2282\ \mathrm{W}$$

13.3b HEAT TRANSFER BY FREE CONVECTION

Free convection is the flow of fluid which occurs solely due to the difference in density within a fluid. Take, for example, a room in which the air is at constant temperature throughout the entire volume. When a heated object is brought into the room, there will be an exchange of heat between the object and the room. Air in contact with the object is heated and becomes less dense than the air in the surroundings. The difference in density of heated and cold air results in a buoyant force which causes the heated air to rise. The rising hot air is replaced by fresh cold air, which in turn is heated up and also rises. If the object introduced into the room is colder than the room air, the air in contact with the object will cool, becoming more dense, and will move downward. Free convection is thus totally determined by the exchange of heat which is proportional to the surface area of the object, the difference in temperature of object surface and fluid, and the orientation of the heated surface.

The results of experimental measurements on tubes, plates, and spheres can be presented in the form

$$\frac{hL}{k_m} = C\,\mathrm{Ra}^n \qquad \textbf{(13.17)}$$

where the product Ra equals $(g\beta/\nu_m^2\,\mathrm{Pr}/_m/L^3)(T_w - T_f)$ and is called Rayleigh number. β is the coefficient of thermal expansion with units of 1/°C (1/°R) and describes the change of density of a fluid with temperature at a given pressure and is thus proportional to the buoyancy force. The coefficient C and exponent n vary with Rayleigh number as follows:

Ra	C	n
<0.001	0.50	0
0.001–500	1.18	$\frac{1}{8}$
$500–2\times10^7$	0.54	$\frac{1}{4}$
$2\times10^7–10^{13}$	0.135	$\frac{1}{3}$

The linear dimension L is taken as the diameter D for tubes and spheres, and as the height H for vertical plates. The mean temperature T_m at which the physical properties of the fluid are evaluated is $T_m = \frac{1}{2}(T_w + T_f)$, where T_w is the wall temperature and T_f is the fluid temperature.

For horizontal plates the reference dimension is the smaller side of the plate, with constant C and exponent n given in the following table:

	C		
Ra	Upward facing	Downward facing	n
<0.001	0.65	0.35	0
0.001–500	1.53	0.83	$\frac{1}{8}$
$500–2\times10^7$	0.70	0.38	$\frac{1}{4}$
$2\times10^7–10^{13}$	0.176	0.095	$\frac{1}{3}$

EXAMPLE 13.13

Determine the heat loss from a bare vertical steam pipe having a 4-in. diameter and a 10-ft length. The pipe wall temperature is 350°F and the surrounding air temperature is 85°F.

Solution

The mean air temperature is

$$T_m = \tfrac{1}{2}(T_w - T_f) = \tfrac{1}{2}(350 + 85) = 217.5°\text{F}$$

At 217.5°F, $k_m = 0.0178$ Btu/ft-h-°R, $\nu_m = 2.51 \times 10^{-4}$ ft²/s, $\text{Pr}_m = 0.71$, $\beta_m = 0.00148$/°R, and hence

$$\begin{aligned}\text{Ra} &= \frac{g\beta_m}{\nu_m^2}\,\text{Pr}_m\,L^3(T_w - T_f)\\ &= \frac{(32.2\ \text{ft/s}^2)(0.00148/°\text{R})}{(2.51\times10^{-4}\ \text{ft}^2/\text{s})^2}\,(0.71)\left(\frac{4}{12}\ \text{ft}\right)^3 265°\text{F}\\ &= 5.27\times10^6\end{aligned}$$

Thus,

$$C = 0.54\,, \qquad n = 0.25\,, \quad \text{and} \quad \frac{hD}{k} = 0.54(5.27\times10^6)^{0.25} = 25.9$$

so

$$h = \frac{k}{D}(25.9) = \frac{(0.0178\ \text{Btu/ft-hr-°F})25.9}{(4/12\ \text{ft})} = 1.38\ \text{Btu/ft}^2\text{-hr-°F}$$

The heat loss is

$$\begin{aligned}\dot{Q} &= hA_s(T_w - T_f)\\ &= (1.38\ \text{Btu/ft}^2\text{-hr-°F})\left(\pi\times\frac{4}{12}\ \text{ft}\times 10\ \text{ft}\right)(265°\text{F})\\ &= 3838\ \text{Btu/hr}\end{aligned}$$

13.3c CONVECTION HEAT TRANSFER WITH PHASE CHANGE

When convection heat transfer is accompanied by a change in phase of the fluid, i.e. by evaporation or by condensation, higher heat transfer rates than during the usual single-phase convection processes are attained. Evaporation will take place within a liquid when the solid surface in contact with the liquid is kept at a temperature higher than the saturation temperature of the liquid, or when the pressure of the liquid is suddenly reduced. Condensation of a saturated vapor will take place when the vapor comes in contact with a solid surface kept at a temperature lower than the saturation temperature of the vapor. In the case of evaporation, the absorption of latent heat of vaporization results in the generation of vapor bubbles or a vapor layer attached to the solid heating surface. Subsequently, vapor bubbles detach themselves. The movement of the vapor bubbles in the liquid increases the convective circulation and hence the heat transfer rates. Because of the generation of bubbles which accompanies evaporation, the process is also called *boiling*. In the case of condensation, the removal of latent heat from the vapor results in higher heat transfer rates. However, when the liquid condensate forms a continuous film on the solid surface, the heat transfer rates are reduced.

1. *Heat Transfer during Evaporation*

As in convective processes, the quantity of heat transferred during evaporation is expressed in terms of a heat transfer coefficient h defined as

$$\dot{Q} = h(T_w - T_s)A_s$$

However, the heat transfer coefficient is itself a function of the difference between heating surface and saturation temperature. The variation of h for water in a container boiling at atmospheric pressure is shown in Figure 13.13. At low temperature difference [less than 5°F (3°C)] the coefficient is small and is determined only by free convection occurring in a single-phase liquid. In this region, the heat is transferred by liquid which was in contact with the heated surface, was superheated there, and rose to the liquid-vapor interface, where it changed into vapor. In the

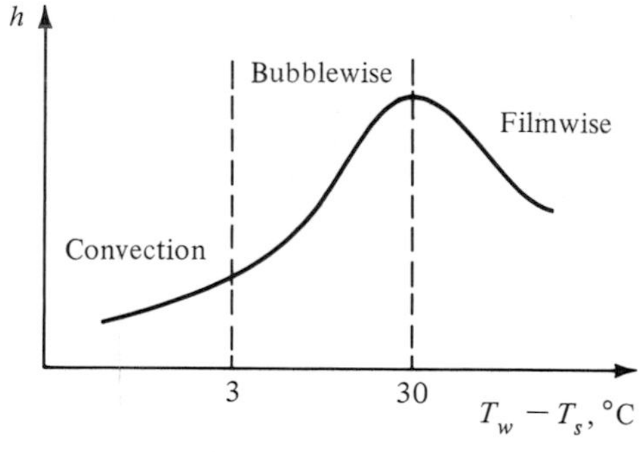

FIGURE 13.13 *Boiling Heat Transfer Coefficients for a Horizontal Heating Wire (1 mm diameter) in Water at Atmospheric Pressure.*

region of moderate temperature differences [from 5°F to about 90°F (3°C to about 30°C)], the heat transfer is determined by increased convection of the liquid resulting from the growth and movement of vapor bubbles. These vapor bubbles form at discrete locations on the heating surface. The heat transfer coefficient increases rapidly with increase in temperature difference, reaching a maximum value at about 90°F (30°C) temperature difference. This mode of boiling is called *nucleate* or *ordinary boiling*.

After reaching the maximum value, the heat transfer coefficient decreases sharply with increase in temperature difference ΔT. In the nucleate boiling regime, the number of starting points or nuclei increased with increase in ΔT; at the point of maximum 'h' value, the many vapor bubbles coalesce to form a continuous vapor film which separates the liquid from the heating surface. This mode is therefore called *film boiling*. Vapor bubbles continue to separate from the vapor film, with new film forming on the freed surface. In general, the maximum value of h will depend on the physical properties of the boiling liquid.

In the free convection evaporative region, the heat transfer coefficient h is determined from equation (13.17) with the physical properties of the liquid evaluated at the mean temperature $T_m = \frac{1}{2}(T_w + T_s)$. In the bubblewise evaporative region, h is given by the relation

$$h = C\mu_l\sqrt{\frac{g(\rho_l - \rho_v)}{\sigma}}\;\frac{c_{pl}^3(T_w - T_s)^2}{h_{fg}^2(\text{Pr}_l)^4} \tag{13.18}$$

where the constant C depends on the roughness and wettability of the boiling surface. Its value must be obtained for each surface-fluid combination; for example, for water and copper, steel, or platinum, the value of the nondimensional constant is 4.5×10^5. The surface tension (σ) with units of lbf/ft (N/m) is proportional to the force tending to retain a vapor bubble on the heating surface during evaporation. The buoyancy force (which is proportional to the density difference between liquid and vapor) tends to separate the vapor bubble from the heating surface. Relationship (13.18) has been determined from experimental measurements of heat transfer on a variety of liquids over a range of saturation temperatures. The subscript l refers to the liquid phase (earlier in the thermodynamic part of the text the subscript f was used) and the subscript v refers to the vapor phase (earlier the subscript g was used). T_s is the saturation temperature, while T_w is the wall temperature. All thermal properties are taken at the saturation temperature. In the filmwise evaporative region, an experimentally determined relation for h for laminar motion within the film is

$$h = 3.52k_v\left[\frac{g\rho_v(\rho_l - \rho_v)}{\mu_v^2 D}\,\text{Pr}_v\left(\frac{h_{fg}}{c_{pv}(T_w - T_s)} + 0.4\right)\right]^{1/4} \tag{13.19}$$

Figure 13.14 shows calculated heat transfer coefficients for horizontal heating tubes in water as a function of the difference between wall and saturation temperatures over a range of saturation temperatures. At the lower temperature differences, h is determined by free convection only and hence is a function of the diameter of the heating tube. In the bubblewise region, the heat transfer coefficient is independent of heating tube size. Figure 13.15 shows the evaporative heat transfer coefficients for horizontal heating tubes in refrigerant R-22.

EXAMPLE 13.14

A copper tube 1 m long, 20 mm in diameter with a surface temperature of 160°C is placed into a container of saturated water at a temperature of 150°C. Determine the resulting heat transfer and evaporation rates.

Solution

From Figure 13.14 at 150°C and $T_w - T_s = 160 - 150 = 10$°C,

$$h = 32{,}000 \text{ W/m}^2 \text{ K}$$

hence

$$\dot{Q} = h(T_w - T_s)\pi DL = 32{,}000(10)\pi(0.020)1 = 20{,}100 \text{ W} = 20.1 \text{ kW}$$

Now

$$\dot{Q} = \dot{m}h_{fg}$$

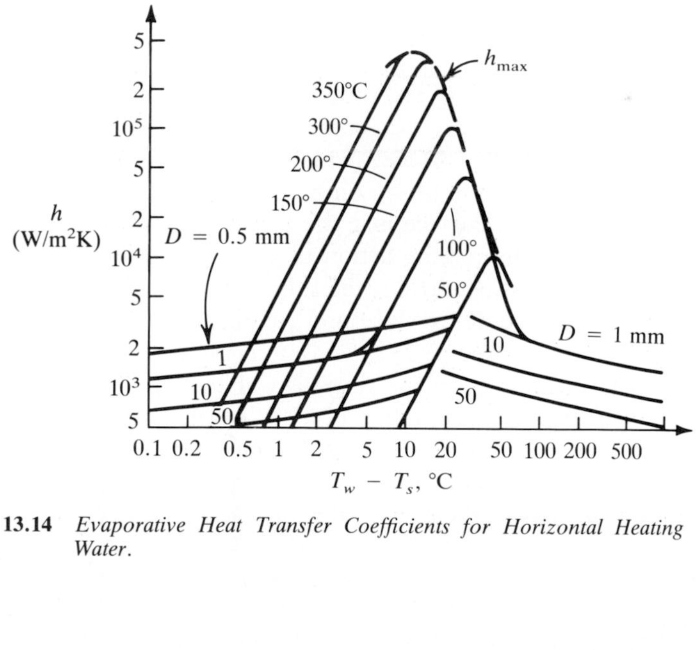

FIGURE 13.14 *Evaporative Heat Transfer Coefficients for Horizontal Heating Tubes in Water.*

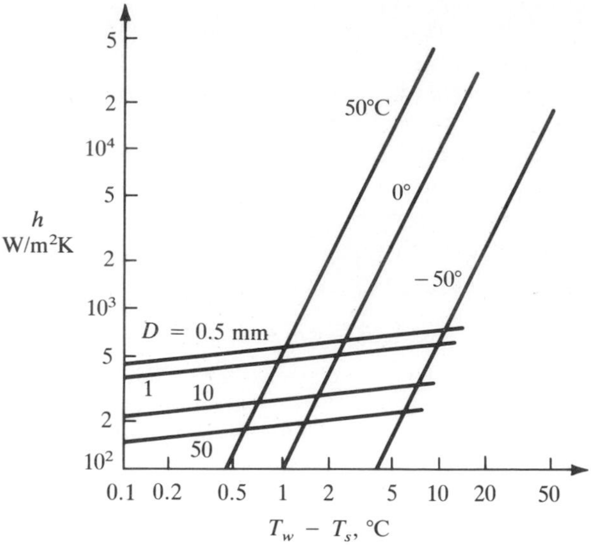

FIGURE 13.15 *Evaporative Heat Transfer Coefficients for Horizontal Heating Tubes in Refrigerant R-22.*

From Table B.2 at 150°C,

$$h_{fg} = 2113.2 \text{ kJ/kg}$$

$$\dot{m} = \frac{20.1}{2113.2} = 0.00951 \text{ kg/s} = 34.2 \text{ kg/h}$$

---

2. *Heat Transfer during Condensation*

As in evaporation, the configuration accompanying the phenomenon plays an important role in determining the magnitude of the heat transfer rate. In condensation, two configurations of the generated liquid occur on the heat transfer surface, namely, liquid film and liquid drops. Filmwise condensation is the most common and easiest to achieve. If the condensate on the cooler surface is in the form of drops which individually run off the surface under the action of gravity, the heat transfer rate will be greater than that occurring when the condensate forms a continuous film covering the cooler surface and running off as a film.

First consider the filmwise condensation process. The rate of heat flow will depend primarily on the thickness of the condensate film, with thickness depending on the relative rates of condensation of vapor and rate of condensate removal. On a vertical or inclined surface, the removal of the condensate is by the action of gravity. The film thickness will increase downward from the top of the surface. When the character of the flow within the condensate film is laminar (which is the case in most applications), the heat transfer across the film will be largely by conduction; hence $\dot{Q}/A = -k(T_s - T_w)/L$ and, with increasing condensate thickness, a decrease in heat transfer rate occurs.

The overall heat transfer coefficient h is the integrated value of the local coefficients h_x. Thus, the heat transfer by condensation is

$$\dot{Q} = hA(T_s - T_w) = \dot{m}h_{fg} \tag{13.20}$$

For laminar flow in the condensate film on a vertical plate or tube, experimental measurements on a variety of substances give the expression for the overall heat transfer coefficient:

$$h = 0.95[L(T_s - T_w)]^{-0.22}\left(\frac{g}{\nu^2}\right)^{0.26}(\mu h_{fg})^{0.22}k^{0.78}\left(\frac{\mathrm{Pr}_s}{\mathrm{Pr}_w}\right)^{0.25} \tag{13.21}$$

where L is the vertical height of wall or vertical tube, or diameter of horizontal tube. For mixed flow in the condensate film (i.e., laminar flow in the upper portion of the wall and turbulent flow in the lower portion),

$$h = \frac{\mu_l h_{fg}}{L(T_s - T_w)}\left[89.44 + 0.24\,\mathrm{Pr}_s^{0.5}\left\{\left(\frac{g}{\nu}\right)^{1/3}\frac{k[L(T_s - T_w)]}{\mu_l h_{fg}} - 2300\right\}\right]^{4/3} \tag{13.22}$$

Figure 13.16 gives calculated values of h for condensing water over a range of saturation temperatures, including values in the mixed region. Turbulent flow in the lower portion of the condensate film will be reached only when the vertical dimension of the plate is large or when the difference between wall and saturation temperatures is large.

EXAMPLE 13.15

A vertical wall 3 ft high and 10 ft long is kept at a temperature of 60°F. Saturated steam at $T_s = 80$°F condenses on the wall. Find the mass flow rate of condensate.

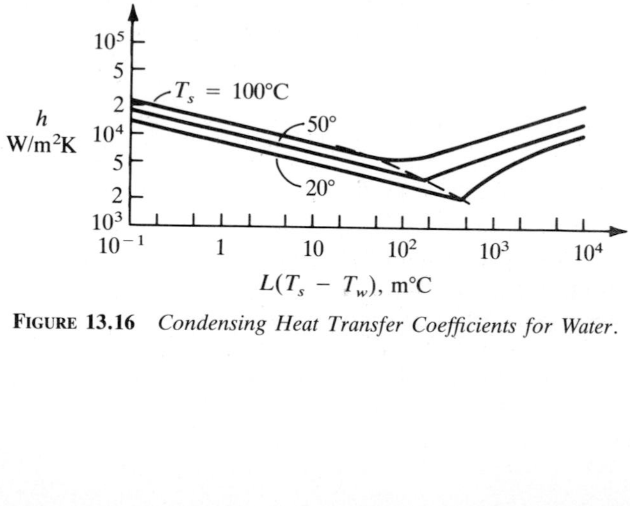

FIGURE 13.16 *Condensing Heat Transfer Coefficients for Water.*

Solution

From Figure 13.16 at $L(T_s - T_w) = 3(20) = 60$ ft-°F $= 60(0.3048/1.8) = 10.16$ m°C and $T_s = 80°\text{F} = 26.7°\text{C}$,

$$h = 5300 \text{ W/m}^2 \text{ K} = 5300(0.1761) = 930 \text{ Btu/ft}^2\text{-h-°R}$$

From Table A.2 at $T_s = 80°\text{F}$,

$$h_{fg} = 1048.4 \text{ Btu/lbm}$$

$$\dot{Q} = h(T_s - T_w)A = 930 \text{ Btu/ft}^2\text{-h-°F}(80 - 60)°\text{F}(3 \text{ ft} \times 10 \text{ ft}) = 558{,}000 \text{ Btu/h}$$

Now

$$\dot{Q} = \dot{m}h_{fg}$$

hence

$$\dot{m} = 558{,}000/1048.4 = 532.2 \text{ lbm/h}$$

For condensation inside horizontal tubes, we get for laminar condensate flow

$$\frac{hD}{k_l} = 13.8 \, \text{Re}_v^{0.2} \left(\frac{\mu h_{fg}}{k_l(T_s - T_w)} \right)^{1/6} (\text{Pr}_l)^{1/6} \left(\frac{\nu_v}{\nu_l} \frac{\rho_v}{\rho_l} \right)^{0.1} \tag{13.23}$$

while for turbulent condensate flow

$$\frac{hD}{k_l} = 0.026 \left(\text{Re}_l + \text{Re}_v \frac{\nu_v}{\nu_l} \right)^{0.8} \text{Pr}_l^{0.33} \tag{13.24}$$

When a condensate surface is coated or contaminated with a substance that prevents the condensate film from totally wetting the surface, the vapor condenses in drops which periodically run off the surface. Since much of the surface is not covered by a condensate film with its insulating effect, the heat transfer coefficient can be several times as large as for filmwise condensation. However, dropwise condensation is difficult to maintain over long time periods; thus heat transfer equipment is usually designed for filmwise condensation.

13.4 RADIATION HEAT TRANSFER

In heat transfer by radiation, thermal energy is transferred by electromagnetic waves between bodies not in contact with each other. The transfer process takes place as follows: Some of the internal energy of a body is continuously transformed into radiant energy. The intensity of radiation increases with increasing temperature, since the internal energy of a substance also increases with temperature. The radiant energy is transmitted from the body by electromagnetic waves until it reaches another body, where a fraction of the radiant energy is absorbed and retransformed into internal thermal energy. The process of radiation heat transfer therefore involves conversion of internal thermal energy of one body into electromagnetic energy and its eventual reconversion into internal energy of another body (or bodies). Electromagnetic waves are propagated in vacuum at the speed of light, namely 186,000 miles per second (9.82×10^8 fps) or 3×10^8 m/s.

The ideal maximum radiant energy emitted by a body is proportional to the fourth power of its absolute temperature, namely

$$\dot{Q}_{\text{ideal}} = \sigma_0 A_s T^4$$

where the constant of proportionality σ_0 is called the Stefan-Boltzmann constant, which equals 0.171×10^{-8} Btu/ft^2-h-°R or 5.67×10^{-8} W/m K^4. The radiant energy emitted by a real body equals

$$\dot{Q} = \epsilon\sigma_0 A_s T^4$$

where the factor ϵ is called emissivity. Its value can vary from zero to unity. The value of emissivity and hence the magnitude of radiation of a substance will depend on the nature of the substance, the character of the emitting surface, and, with some gases, on the thickness of the gas layer and its pressure. For most solids and liquids, only a thin surface layer participates in the process of thermal radiation. In gases, all particles of the gas emit energy, and the process of thermal radiation from gases is therefore volumetric in nature.

The radiant energy reaching a body will be partly absorbed by the body and partly reflected, and part will pass through the body. When the incident energy is reflected in its entirety, the body is called a *white body*. When the incident energy is entirely absorbed by a body, the body is called a *black body*. When the incident radiant energy passes in its entirety through the body, the body is called a *transparent body*. Most solids and liquids are opaque (nontransparent) to radiant energy. Gases can be transparent (e.g., air) or almost opaque (e.g., water vapor, carbon dioxide). For solid bodies, the reflection and absorption of incident radiant energy depend largely on the finish of the surface. The reflectivity of a polished surface can be several times larger than that of rough surfaces, irrespective of their color. The absorptivity of a body can be increased by coating its surface with a dark, rough layer.

Next, let us examine the net radiation heat exchange between two solid bodies. Each body will emit energy according to its temperature, the emissivity of its surface, and its surface area. However, not all the radiation emitted by one body will reach the other body (Figure 13.17).

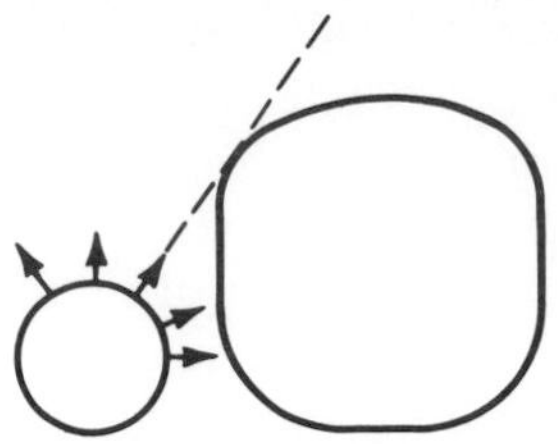

FIGURE 13.17 *Radiation Near Body.*

This is accounted for by the radiation shape factor (geometric view factor). The deviation of each surface from an ideal radiator (black body) as well as reflections between the bodies are accounted for by use of the emissivity factor. The resulting net radiation heat transfer between two bodies is

$$\dot{Q}_{12} = F_{12}\sigma_0 A_1(T_1^4 - T_2^4) \tag{13.25}$$

where the radiation factor F_{12} is given by

$$F_{12} = \frac{1}{1/\epsilon_1 + 1/G_{12} + (1/\epsilon_2 - 1)A_1/A_2 - 1} \tag{13.26}$$

where the geometric view factor G_{12} (also called angle factor, radiation shape factor, configuration factor) is a function of the relative geometric configuration of the two bodies. The view factor indicates the portion of radiation emitted by one body which falls on the other. As an example, for a circular cylinder near a plane wall (Figure 13.18), only rays emanating from one-half of the cylinder reach the plane wall, hence $G_{12} = 1/2$. On the other hand, the view factor for two parallel planes would be unity, since all the radiation leaving one plane reaches the other.

When the surface area of one body is much smaller than that of the other, the area ratio A_1/A_2 approaches zero and hence from (13.26)

$$F_{12} = \frac{1}{1/\epsilon_1 + 1/G_{12} - 1}$$

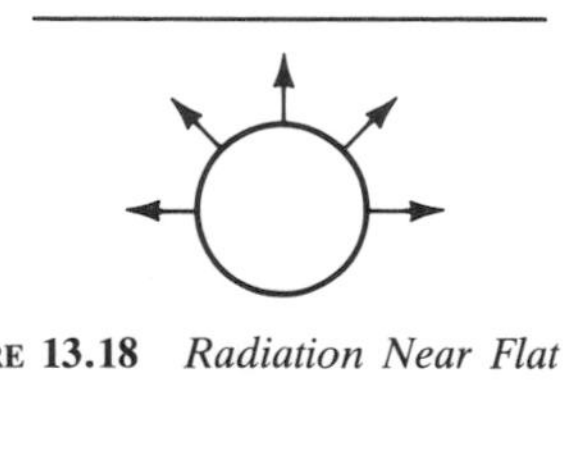

FIGURE 13.18 *Radiation Near Flat Wall.*

For example, for a small body within a large enclosure (room), $G_{12} = 1$; hence $F_{12} = \epsilon_1$ and

$$\dot{Q}_{12} = \epsilon_1 \sigma_0 A_1 (T_1^4 - T_2^4) \tag{13.27}$$

EXAMPLE 13.16

Two large plates are kept a small distance apart at 1200°C and 750°C, respectively. The emissivities of the plates are 0.8 and 0.6, respectively. Determine the net radiation heat transfer between the plates.

Solution

The view factor G_{12} is unity. From (13.26), the radiation factor is

$$F_{12} = \frac{1}{1/\epsilon_1 + 1/G_{12} + (1/\epsilon_2 - 1)A_1/A_2 - 1} = \frac{1}{1/0.8 + 1 + (1/0.6 - 1) - 1} = 0.5217$$

Hence from (13.25),

$$\dot{Q}_{12}/A = 0.5217(5.67)10^{-8}(1473^4 - 1023^4) = 106.87 \text{ kW/m}^2$$

To intensify radiation heat transfer, the temperature of the radiating body can be raised or the emissivity of the surface improved. On the other hand, to diminish radiation heat transfer, the temperature of the emitting body can be reduced or the emissivity of the surface lowered. When is it not possible to reduce the temperature or lower its emissivity, special thin-walled radiation shields can be placed between radiating surfaces to reduce the radiation heat transfer. Radiation shields are discussed in further detail in Chapter 14.

So far we have discussed radiation heat transfer when the medium between two bodies exchanging radiant energy is transparent or when a vacuum exists between the bodies. This assumption is valid for monatomic and diatomic gases, nitrogen, oxygen, hydrogen, in particular. However, polyatomic gases, carbon dioxide, water vapor, sulfur dioxide, and the oxides of nitrogen, in particular, possess considerable emissive power and absorptivity. Thus, the products of combustion are emitting-absorbing gases. Unlike most solids, which emit and absorb radiation over a large spectrum of wavelengths, these gases emit and absorb within certain limits of wavelengths. Outside these bands, these gases are transparent. Further, unlike opaque solids, where emission and absorption of thermal energy occur in the surface layers, the entire gas volume partakes of the emission and absorption phenomenon. The emissivity of gas is a function of its temperature, the mean thickness of the gas layer, and the partial pressure of the gas. Figure 13.19 gives the emissivity of

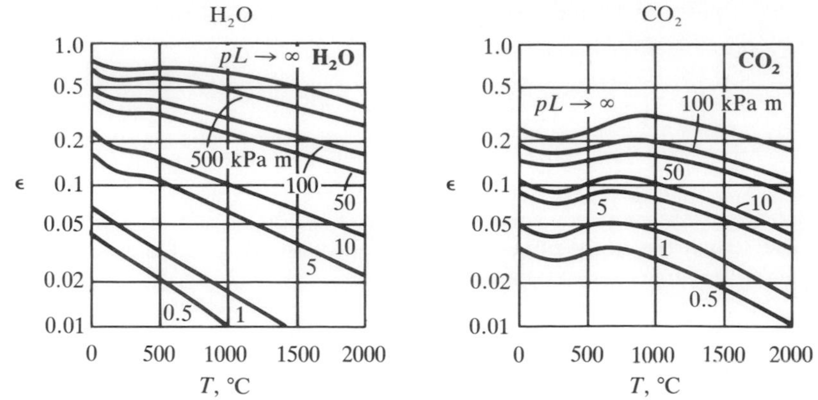

FIGURE 13.19 *Emissivity of Water Vapor and Carbon Dioxide (Total Pressure = 1 atm).*

carbon dioxide and water vapor as function of its temperature and the parameter pL, where p is the partial pressure of the gas and L is the thickness of the gas layer. The mean thickness L can be approximated by the value of four times the ratio of gas volume to bounding surface area. Hence, for a long cylinder, L equals the cylinder diameter; for a sphere, L equals two-thirds the sphere diameter; and for a cube, L equals two-thirds the cube edge length.

When a gas is confined within a solid shell with surface temperature different from that of the radiating gas, the radiation emitted by the shell will be partially absorbed by the gas. The emission of gas to the surface is $\sigma_0 \epsilon T_s^4$, while the absorption by the gas of radiation from the surface is $\sigma_0 \alpha T_w^4$, where α is the absorptivity of the gas for radiation from a black surface. The value of α can be approximated by the value of emissivity ϵ of the gas at the wall temperature. To account for the nonideal character of the wall, an effective wall emissivity ϵ_e is used ($\epsilon_e \sim 0.9$). Thus, the net radiation heat transfer between gas and its bounding surface becomes

$$\dot{Q} = \sigma_0 \epsilon_e A_w [\epsilon T_g^4 - \alpha T_w^4] \qquad \textbf{(13.28)}$$

EXAMPLE 13.17

A 5-ft-diameter duct transports exhaust gas at 2200°F and atmospheric pressure. The volumetric composition of the exhaust gas is 12 percent CO_2, 16 percent water vapor, and 72 percent N_2. The duct walls have an effective emissivity of 0.9 and are kept at a temperature of 600°F. Determine the net radiation heat transfer.

Solution

The partial pressures are obtained from (5.1)

$$p_{CO_2} = 0.12(14.7) = 1.764 \text{ psia}$$

$$P_{H_2O} = 0.16(14.7) = 2.352 \text{ psia}$$

With $L = 5$ ft,

$$pL_{CO_2} = 1.764(5) = 8.82 \text{ psia-ft} = 8.82(6.8948)0.3048 = 18.54 \text{ kPa} \cdot \text{m}$$

$$pL_{H_2O} = 2.352(5) = 11.76 \text{ psia-ft} = 11.76(2.1013) = 24.71 \text{ kPa} \cdot \text{m}$$

From Figure 13.19 at $T = 2200°\text{F} = 1204°\text{C}$,

$$\epsilon_{CO_2} = 0.11 \quad \text{and} \quad \epsilon_{H_2O} = 0.14$$

Hence

$$\epsilon = (0.11 + 0.14) - 0.25$$

At $T_w = 600°\text{F} = 316°\text{C}$

$$\epsilon_{CO_2} = 0.11 \quad \text{and} \quad \epsilon_{H_2O} = 0.26$$

Hence

$$\alpha = \epsilon - 0.11 + 0.26 = 0.37$$

Hence from (13.28)

$$\dot{Q}/L = 0.171 \times 10^{-8}(0.9)\pi(5)[0.25(2660)^4 - 0.37(1060)^4]$$

$$= 2.4175 \times 10^{-8}[1.2516 \times 10^{13} - 4.67 \times 10^{11}] = 291{,}300 \text{ Btu/h-ft}$$

REFERENCES

ARPACI, V. S., and LARSEN, P. S., *Convection Heat Transfer*. Englewood Cliffs, N.J., Prentice Hall, 1984.

HOLMAN, J. P., *Heat Transfer*. New York, McGraw-Hill Book Co., Inc., 1981.

INCROPERA, F. P., and DEWITT, D. P., *Fundamentals of Heat Transfer*. New York, John Wiley & Sons, 1981.

JANNA, W. S., *Engineering Heat Transfer*. Boston, PWS Engineering, 1986.

KREITH, F., and BOHN, M., *Principles of Heat Transfer*. New York, Harper & Row, 1986.

OZISIK, M., *Heat Conduction*. New York, John Wiley & Sons, 1980.

SIEGEL, R., and HOWELL, J. R., *Thermal Reduction Heat Transfer*. New York, McGraw-Hill Book Co., Inc., 1980.

PROBLEMS

13.1 The heat transfer rate through a $\frac{1}{2}$-in. thick plane wall is 20 Btu/ft^2-h. Find the temperature difference and temperature gradient across the wall if the wall material is (a) stainless steel, and (b) alumina.

13.2 Find the heat loss from a wall (height = 4 m, width = 5 m, thickness = 200 mm) made of brick (thermal conductivity $k = 0.7$ W/m°C). The wall surface temperatures are kept at 110°C and 40°C, respectively.

13.3 Calculate the thermal conductivity of a 50-mm thick, plane wall material, if the difference in wall surface temperatures is 20°C and the heat transfer rate is 150 W/m^2.

13.4 The walls of a rectangular tank (surface area = 15 ft^2) are covered with two layers of insulating material. The steel wall has a thermal conductivity of 25 Btu/ft-h-°R. The first insulation layer is 2 in. thick and has a thermal conductivity of 0.08 Btu/ft-h-°R, while the second insulation layer is $\frac{1}{2}$ in. thick with a thermal conductivity of 0.35 Btu/ft-h-°R. The temperature of the inside surface of the tank is 500°F. The outside insulation surface is kept at a temperature of 100°F. Determine the heat transfer rate through the wall and the temperatures at the insulation interfaces.

13.5 A steam pipe (I.D. = 150 mm, O.D. = 160 mm) has a 100-mm thick insulation ($k = 0.10$ W/m°C). The temperature of the steam is 500°C, while that of the outside insulation surface is 50°C. Determine the heat transfer per unit length of pipe and the temperature at the interface.

13.6 Hot water at 80°F flows through a one inch inner diameter tube with a wall thickness of 0.25 in. The convective heat transfer coefficient on the inside of the wall is 20 Btu/ft^2-h-°F, the free convection heat transfer coefficient on the outside of the pipe is 0.5 Btu/ft^2-h-°F. For an ambient air temperature of 40°F, calculate the heat loss per foot of tube. The thermal conductivity of the wall material is 20 Btu/ft-h-°F.

13.7 It is desired to reduce the heat loss from the pipe of Problem 13.6 by adding insulation of thermal conductivity 0.08 Btu/ft-h-°F. Determine the insulation thickness required as a function of heat loss.

13.8 A thin flat plate, 6 ft long and 5 ft wide, is exposed to air flowing parallel to its surface. The velocity and temperature of the approach flow are 10 ft/s and 70°F, respectively. The plate temperature is 150°F. Determine the heat transfer rate from the plate.

13.9 Calculate the heat transfer coefficient for oil flowing at a velocity of 0.5 m/s in an 8-mm-diameter tube. The average oil temperature is 75°C, while that of the wall is 20°C. ($\nu_f = 4.2 \times 10^{-5}$ m/s, $k_f = 0.14$ W/m-K, $\text{Pr}_f = 550$, $\text{Pr}_w = 13000$.)

13.10 Water flows in a 5-mm-diameter tube at a velocity of 0.5 m/s. The tube wall temperature is 50°C. Determine the length of tube for the following water temperature conditions: inlet = 10°C, outlet = 25°C.

13.11 A one-inch round copper bus bar is cooled by cross flow of air. The velocity and temperature of the cross flow are 3 ft/s and 70°F, respectively. Calculate the heat transfer coefficient for a bus bar temperature of 175°F.

13.12 An air heater consists of a bank of 40 mm diameter, in-line tubes with transverse and longitudinal pitches of 2.5 times the tube diameter. There are five rows of tubes, each row consisting of six tubes. Air enters the heater at a temperature of 20°C and leaves at 75°C. The outside wall temperature of the tubes is 175°C. Determine the length of the tubes for a heat transfer rate of 100 kW and an air velocity of 10 m/s.

13.13 The in-line arrangement of Problem 13.10 is replaced with a staggered tube arrangement. The air velocity in the narrowest section of the tube bank is increased to 15 m/s, all other conditions remaining the same. Determine the length of the tubes.

13.14 A horizontal cylinder is cooled by free convection of 30°C ambient air. The diameter and temperature of the cylinder are 200 mm and 200°C, respectively. Determine the convective heat transfer rate.

13.15 Determine the heat transfer coefficient of a horizontal plate (5 ft × 10 ft) with its heated surface facing upward. The surface temperature of the plate is kept at 200°F, while that of the ambient air is 60°F.

13.16 What will be the heat transfer coefficient if the heated surface of Problem 13.13 faces downward, all other conditions remaining unchanged?

13.17 Saturated steam condenses in a filmwise manner on a vertical tube of 2 m height. The pressure of the steam is 5 kPa. The surface temperature of the tube is 25°C. Find the overall heat transfer coefficient, assuming laminar flow in the condensate film.

13.18 What must be the temperature difference for filmwise condensation of saturated steam ($p = 14.7$ psia) on a horizontal $1\frac{1}{2}$-in.-diameter tube if the heat transfer rate is to be maintained at 20,000 Btu/ft^2-h? Also, find the value of the heat transfer coefficient.

13.19 Water at a pressure of 500 kPa boils on the surface of an electric immersion heater having a diameter of 40 mm and a length of 0.75 m. The inside surface of the heater tube is heated electrically, consuming 7 kW of power. Determine the outside surface temperature of the tube.

13.20 Determine the net heat transfer rate by radiation between two infinitely large parallel plates which are maintained at temperatures of 1000°C and 750°C, respectively. The emissivity of the 1000°C plate is 0.85, while that of the 750°C plate is 0.6.

13.21 The plates of Problem 13.20 are of finite dimensions so that the view factor is 0.7. Estimate the net heat transfer rate.

13.22 Exhaust gas, having a volume composition of 13 percent CO_2, 12 percent H_2O, and 75 percent N_2, flows through a 6-ft-diameter duct. The temperature and pressure of the exhaust gas are 2200°F and one atmosphere, respectively. The duct wall has an emissivity of 0.75 and a temperature of 540°F. Estimate the net radiation heat transfer rate.

13.23 Air at a temperature of 70°F and one atmosphere pressure flows through a one-inch-diameter duct at a rate of 5 lbm/h. Calculate the mean heat transfer coefficient.

13.24 Water at 100°F flows through a $2\frac{1}{2}$-in.-diameter tube at a Reynolds number of 90,000. Find the average heat transfer coefficient.

13.25 A one-foot-long plate is immersed in a flowing stream of oil at 60°F, with $k = 0.08$ Btu/ft-h-°R. The temperature of the plate is held at 250°F. The Reynolds number of the stream is 10,000, with Pr_f and $\text{Pr}_w = 9000$ and 200, respectively. Calculate the heat transfer rate per unit width.

13.26 A 0.5-m-diameter circular flue exhausts gases containing 10 percent CO_2 and 5 percent water vapor by volume. The total pressure of the exhaust gases is 99 kPa. The gas temperature at the flue inlet is 850°C and at the outlet 700°C. The flue walls have an emissivity of 0.85 and a temperature of 400°C. Calculate the net radiation heat transfer. What would be the net radiation heat transfer if the exhaust gases contained no carbon dioxide or water vapor?

13.27 In Problem 13.25, the flue has a rectangular cross section with dimensions 0.25 and 1.5 m. Determine the net radiation heat transfer.

chapter

14

Heat Transfer Applications

14.1 COMBINED MODES OF HEAT TRANSFER

In the preceding chapter, the three basic modes of heat transfer, conduction, convection, and radiation, were discussed separately. Actually, these modes of heat transfer often occur simultaneously and usually are coupled and influence each other. Examples of combined convection and conduction occur in gas turbine regenerators, steam turbine condensers, and refrigeration evaporators. Examples of combined convection and radiation occur in the cooling of semiconductor devices by use of a heat sink.

1. Convection-Conduction Heat Transfer

In the tubes of a gas turbine regenerative heat exchanger there occur the following heat transfer modes: convection inside the tubes, conduction through the tube wall, and convection outside the tubes. Each mode of heat transfer is coupled to the others.

Consider the case shown in Figure 14.1. Here, hot fluid at a temperature T_{f1} flows inside the tube of length L and cold fluid at a temperature T_{f2} flows on the outside. The tube is made of solid material with a thermal conductivity k. The wall temperatures of the tube are unknown and designated T_{w1} and T_{w2}. The convection heat transfer coefficient on the hot fluid side is h_1 and that of the cold fluid side is h_2. Under steady-state conditions, the rate of heat transferred from the hot fluid to the wall is the same as that transferred through the solid wall and that transferred from the wall to the cold fluid. Hence, we write, from

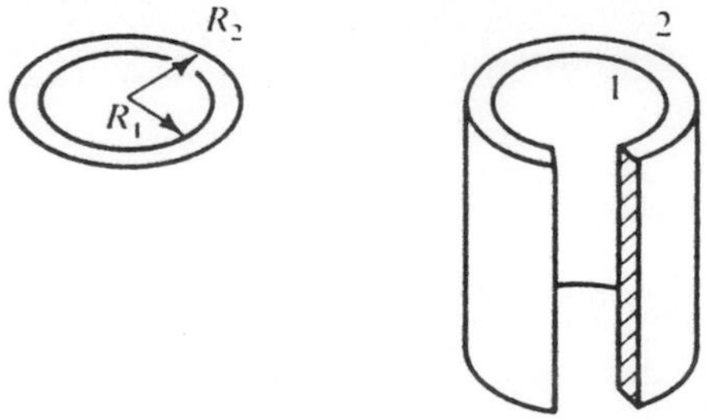

FIGURE 14.1 *Convection-Conduction in a Cylindrical Tube.*

(13.9) for inside convection

$$\dot{Q}/L = h_1(T_{f1} - T_{w1})\pi D_1$$

from (13.6) for conduction in the tube wall,

$$\dot{Q}/L = 2\pi(T_{w1} - T_{w2})k/\ln(D_2/D_1) \tag{14.1}$$

from (13.9) for outside convection:

$$\dot{Q}/L = h_2(T_{w2} - T_{f2})\pi D_2$$

Rewriting the above equations in terms of the three temperature differences and adding these differences (thereby eliminating the two unknown wall temperatures), we obtain

$$T_{f1} - T_{f2} = (\dot{Q}/L\pi)\left(\frac{1}{h_1 D_1} + \frac{1}{2k}\ln\frac{D_2}{D_1} + \frac{1}{h_2 D_2}\right)$$

or

$$\dot{Q} = UA_1(T_{f1} - T_{f2}) \tag{14.2}$$

U is called the overall heat transfer coefficient (expressed in units of Btu/ft^2-h-°R or W/m^2 K) and is based on the inside surface A_1. Rewriting as thermal resistance, we get

$$\frac{1}{U} = \frac{1}{h_1} + \frac{D_1}{2k}\ln\frac{D_2}{D_1} + \frac{1}{h_2}\frac{D_1}{D_2} \tag{14.3}$$

which states that the overall thermal resistance is the sum of the individual resistances $1/h_1$, $(D_1/2k)\ln(D_2/D_1)$, and $(1/h_2)(D_1/D_2)$. If the pipe consisted of several layers (e.g., metal tube and insulation) instead of the single layer used above, the resistance of the additional layers would be added to (14.3) to obtain the overall coefficient. For the case of a thin tube of large diameter or a flat plate, (14.3) reduces to [with $(D_1/2k)\ln(D_2/D_1) = (D_2 - D_1)/2k + \cdots$]

$$\frac{1}{U} = \frac{1}{h_1} + \frac{t}{k} + \frac{1}{h_2} \tag{14.4}$$

TABLE 14.1 *Typical Ranges of Overall Heat Transfer Coefficient U*

Fluids	U (Btu/ft^2-h-°R)	U (W/m^2 K)
Gas to gas	2–10	10–50
Gas to water	5–10	30–60
Water to water	100–400	600–2000
Condensing steam to water	200–1000	1000–6000
Condensing R-12 to water	50–200	300–1000

If one of the convective heat transfer coefficients is much smaller than the other, the smaller one will control the quantity of heat transferred. Suppose, for example, the heat transfer coefficient of a liquid in a tube to be $h_1 = 2000$ W/m^2 K and the coefficient of air outside the tube to be $h_2 = 50$ W/m^2 K; then from (14.4),

$$\frac{1}{U} = \frac{1}{2000} + \frac{1}{50} = 0.0005 + 0.02 = 0.0205$$

Hence $U = 48.78$ W/m^2 K versus 50 W/m^2 K, when the larger coefficient is neglected.

Table 14.1 presents typical ranges of the overall heat transfer coefficient U for several media combinations.

Sometimes it is permissible to neglect the conduction resistance in calculating the overall heat transfer coefficient U, depending on the relative values of thermal resistance, as illustrated in the next example.

EXAMPLE 14.1

A hot water heater consists of several plane walls. On the water side the heat transfer coefficient h_2 equals 1000 Btu/ft-h-°R. The metal wall has a thickness of 1/8 in with a thermal conductivity of 20 Btu/ft-h-°R (steel). Examine the effect of neglecting conduction resistance if the water is heated by

a. hot air with $h_1 = 8$ Btu/ft^2-h-°R

b. condensing steam with $h_1 = 2100$ Btu/ft^2-h-°R, and

c. condensing steam, but the metal wall has a thermal conductivity of 200 Btu/ft-h-°R (copper).

Solution

a. From (14.4),

$$\frac{1}{U} = \frac{1}{h_1} + \frac{t}{k} + \frac{1}{h_2} = \frac{1}{8} + \frac{1}{8(12)20} + \frac{1}{1000}$$

$$U = 7.90 \text{ Btu/ft}^2\text{-h-°R}$$

When the conduction resistance is neglected, $U = 7.94$ Btu/ft^2-h-°R, a negligible change.

b.

$$\frac{1}{U} = \frac{1}{2100} + \frac{1}{8(12)20} + \frac{1}{1000}$$

$$U = 501 \text{ Btu/ft}^2\text{-h-°R}$$

When the conduction resistance is neglected,

$$U = 677 \text{ Btu/ft}^2\text{-h-°R}$$

a 35 percent change.

c.

$$\frac{1}{U} = \frac{1}{200} + \frac{1}{8(12)20} + \frac{1}{1000}$$

$$U = 153.4 \text{ Btu/ft}^2\text{-h-°R}$$

When the conduction resistance is neglected, $U = 166.7$ Btu/ft^2-h-°R, an 8.6 percent change.

The unknown wall temperatures T_{w1} and T_{w2} can be determined by substituting the expression for $\dot{Q}$ given in (14.2) into expressions for the various temperature differences obtainable from (14.1). Solving the equations, we get

$$T_{w1} = T_{f1} + \frac{U}{h_1}(T_{f2} - T_{f1}) \tag{14.5}$$

and

$$T_{w2} = T_{f2} - \frac{U}{h_2}\frac{D_1}{D_2}(T_{f2} - T_{f1}) \tag{14.6}$$

To evaluate the component thermal resistances (specifically h_1 and h_2) of U for a given flow configuration, it is often necessary to obtain an estimate of values of the usually unspecified wall temperatures T_{w1} and T_{w2}. These heat transfer coefficients are often strongly wall temperature dependent, as is the case for boiling and condensing. As we have seen in Section 13.3, in forced convection, the heat transfer coefficient h is mainly (aside from tube size and fluid velocity) a function of fluid temperature through the physical properties of the fluid, but has a lesser dependence on wall temperature through the Prandtl number Pr, which is a function of temperature. In free convection, the dependence of h on wall temperature is much stronger, since the physical properties of the fluid are evaluated at an average fluid-wall temperature. In addition the Rayleigh number is directly proportional to the difference between wall and fluid temperatures. In convection with phase change, the heat transfer coefficient h is strongly dependent on wall temperature, since it is proportional to a power of the difference between saturation and wall temperatures.

In these instances, to calculate a value of U, a wall temperature is assumed and an initial estimate of U is obtained. A better estimate of T_w can then be obtained from equations (14.5) and (14.6). By further iteration, a satisfactory value of U is calculated. The following example will illustrate the procedure.

EXAMPLE 14.2

Steam at a temperature of 50°C condenses on the outside of a brass tube having an O.D. of 25 mm and an I.D. of 22.5 mm. Cooling water flows within the tube at a velocity of 1 m/s and an average temperature of 30°C. Determine the overall heat transfer coefficient. The conductivity of brass is $k = 113$ W/m K.

Solution

To obtain h_1 (water), use (13.10):

$$\frac{h_1 D}{k_f} = 0.023\left(\frac{VD}{\nu_f}\right)^{0.8}(\mathrm{Pr}_f)^{0.4}\left(\frac{\mathrm{Pr}_f}{\mathrm{Pr}_w}\right)^{0.25}$$

At $T_f = 30$°C,

$$\nu_f = 0.830 \times 10^{-6}\ \mathrm{m^2/s}\,, \qquad k_f = 0.615\ \mathrm{W/m\,K}\,, \quad \text{and} \quad \mathrm{Pr}_f = 5.67$$

As a first estimate, assume $T_{w1} = T_{w2} = 45$°C; hence $\mathrm{Pr}_w = 3.98$, and

$$\frac{h_1 D}{k_f} = 176.8\,, \qquad h_1 = 4830\ \mathrm{W/m^2\,K}$$

From (13.21),

$$h_2[D(T_s - T_w)]^{0.22} = 10{,}643$$

Hence h_2 (steam) is seen to be very sensitive to wall temperature. For $T_{w2} = 45$°C, h_2 becomes 16,817 W/m^2 K.

From (14.3),

$$\frac{1}{U} = \frac{1}{h_1} + \frac{D_1}{2k}\ln\frac{D_2}{D_1} + \frac{1}{h_2}\frac{D_1}{D_2} = \frac{1}{4830} + \frac{0.0225}{2(113)}\ln\frac{25}{22.5} + \frac{1}{16{,}817}\,\frac{22.5}{25.0}$$

$$U = 3690\ \mathrm{W/m^2\,K}$$

From (14.5),

$$T_{w1} = T_f + \frac{U}{h_1}(T_s - T_f) = 45.28°\mathrm{C}$$

Hence h_1 remains unchanged at 4830 W/m^2 K. From (14.6),

$$T_{w2} = T_s - \frac{U}{h_2}\frac{D_1}{D_2}(T_s - T_f) = 50 - \frac{3690}{16{,}817}\,\frac{22.5}{25.0}\,20 = 46.05°\mathrm{C}$$

hence

$$h_2 = \frac{10{,}643}{[0.025(50 - 46.05)]^{0.22}} = 17{,}700\ \text{W/m}^2\ \text{K} \quad \text{and} \quad U = 3730\ \text{W/m}^2\ \text{K}$$

For a second estimate we get

$$T_{w1} = 30 + \frac{3730}{4830}\,20 = 45.4°\text{C} \quad \text{and} \quad T_{w2} = 50 - \frac{3730}{17{,}700}\,\frac{22.5}{25.0}\,20 = 46.20°\text{C}$$

$$h_2 = 17{,}860\ \text{W/m}^2\ \text{K} \quad \text{and} \quad U = 3656\ \text{W/m}^2\ \text{K}$$

In some heat transfer processes, it may be necessary to increase the quantity of heat transferred; in others, to decrease it. In conduction, the thermal resistance of a wall may be reduced by decreasing its thickness and by using a material of greater thermal conductivity. Heat transfer by convection may be increased by stirring or by increasing the flow velocity. In boiling, an increase can be obtained by stirring and cleaning the heating surface; in condensation, by rearranging the cooling surface, cleaning it, and by increasing the flow velocity of the vapor. In radiation, the heat transfer process can be increased by increasing the emissivity and the temperature of the emitting surface.

One way of increasing the heat transfer between a solid surface and a fluid is to increase the surface area in contact with the fluid. This increase may be accomplished by the use of fins or spines of different geometric configurations (called *extended surfaces*), as illustrated in Figures 14.2 through 14.5.

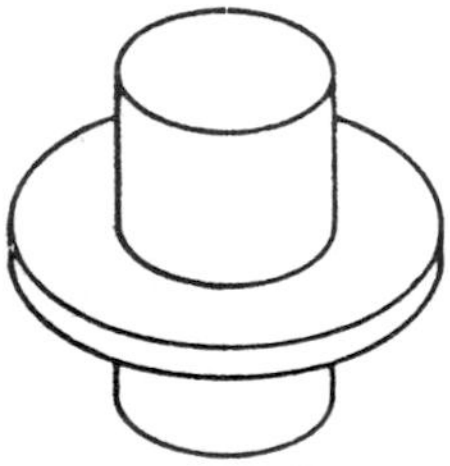

FIGURE 14.2 *Circular Fin as Extended Surface.*

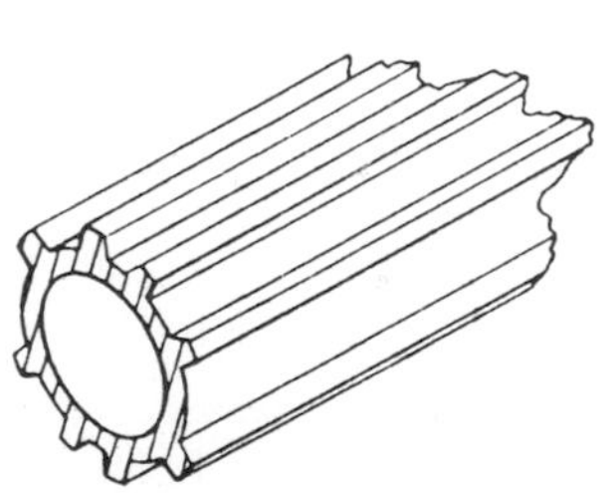

FIGURE 14.3 *Straight Fins as Extended Surfaces.*

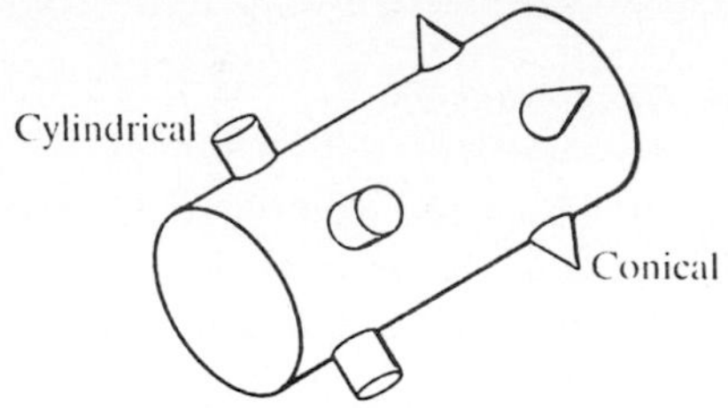

FIGURE 14.4 *Spines as Extended Surfaces.*

FIGURE 14.5 *Heat Sink (Extended Surface) Used with Semiconductors (Courtesy of Aavid Engineering, Laconia, New Hampshire).*

2. *Convection-Radiation Heat Transfer*

In the preceding section we discussed combined heat transfer in which the different modes of heat transfer occur in sequence, that is, in series. In this section, we will discuss combined heat transfer in which different modes occur simultaneously, that is, in parallel. Such parallel heat transfer occurs, for example, when a high-temperature steam pipe is exposed to atmospheric air. Heat losses will be due to convection and radiation. Convection heat transfer can be either free or forced. Also, because of the high pipe wall temperature, radiation may make a significant contribution to the total heat loss. Another example of combined convection and radiation occurs in the cooling of semiconductor devices using blackened heat sinks. This example will be discussed in detail in Section 14.3.

EXAMPLE 14.3

A bare horizontal steam pipe made of steel has a diameter of 25 mm and a surface temperature of 250°C. It passes through still air at a temperature of 25°C. The emissivity of the pipe is 0.95. Determine the total heat loss from the pipe.

Solution

The mean temperature

$$T_m = \tfrac{1}{2}(T_w + T_f) = \tfrac{1}{2}(250 + 25) = 137.5°\text{C}$$

At 137.5°C,

$$k_m = 0.0331 \text{ W/m K}, \qquad \nu_m = 2.73 \times 10^{-5} \text{ m}^2/\text{s}$$

$$\text{Pr}_m = 0.71, \quad \text{and} \quad \frac{g\beta}{\nu^2} = 3.25 \times 10^7 \text{ 1/K m}^3$$

hence

$$\text{Ra} = \frac{g\beta}{\upsilon^2}\text{Pr}_m D^3(T_w - T_f) = 8.11 \times 10^4$$

Thus

$$C = 0.54 \quad \text{and} \quad n = 0.25$$

From (13.17),

$$\frac{hD}{k_m} = 0.54(8.11 \times 10^4)^{0.25} = 9.11 \quad \text{and} \quad h = \frac{0.0331}{0.025}\,9.11 = 12.1 \text{ W/m}^2 \text{ K}$$

The heat loss by free convection is, from (13.8),

$$\dot{Q}/A = h(T_w - T_f) = 12.1(225) = 2720 \text{ W/m}^2$$

The heat loss by radiation is, from (13.28),

$$\dot{Q}/A = 5.67 \times 10^{-8}(0.95)(T_w^4 - T_f^4) = 5.67 \times 10^{-8}(0.95)(523^4 - 298^4)$$

$$= 3605 \text{ W/m}^2$$

The total heat loss is 2720 + 3605 = 6325 W/m^2. Thus, radiation and free convection are of comparable importance in determining the heat loss of the hot pipe.

EXAMPLE 14.4

In Example 14.3, instead of still air, the air flow is forced at right angles to the pipe with a velocity of 20 m/s.

Solution

At 25°C, $k_f = 0.0255$ W/m K, $\nu_f = 15.6 \times 10^{-6}$ m^2/s, and $\mathrm{Pr}_f = 0.71$; at 250°C, $\mathrm{Pr}_w = 0.71$. The Reynolds number is $\mathrm{Re} = [20(0.025)/15.6 \times 10^{-6}] = 32{,}050$. From (13.10),

$$\frac{hD}{k_f} = 0.023\,\mathrm{Re}_f^{0.8}\,\mathrm{Pr}_f^{0.4}\left(\frac{\mathrm{Pr}_f}{\mathrm{Pr}_w}\right)^{0.25} = 0.023(32{,}050)^{0.8}(0.71)^{0.4} = 80.7$$

and

$$h = \frac{(0.0255)80.7}{0.025} = 82.3\ \mathrm{W/m^2\,K}$$

hence

$$\dot{Q}/A = h(T_w - T_f) = 82.3(225) = 18{,}520\ \mathrm{W/m^2}$$

The total heat loss is $\dot{Q}/A = 18{,}520 + 3605 = 22{,}125$ W/m^2. Thus, for forced convection and radiation at a pipe temperature of 250°C, radiation accounts for a smaller portion of the heat loss. At a pipe temperature of 500°C, however, the radiation heat loss would be equivalent to the convection heat loss.

14.2 HEAT EXCHANGERS

A heat exchanger is a device that transfers heat between two fluids. Open-type exchangers achieve the transfer by complete mixing of the two fluid streams. Closed-type exchangers achieve the transfer through a solid wall which separates the two fluids. The open and closed feedwater heaters of a steam power plant, described in Chapter 10, are examples of the two types. Other examples of a closed heat exchanger are the condenser and evaporator of the vapor compression evaporation plant (Section 10.3). In both, the temperature of one fluid remains constant during its passage through the device, while that of the other fluid varies. In a general heat exchanger, however, both fluid temperatures will vary. An example is the regenerator used in gas turbines to improve its cycle efficiency (Section 10.4). The thermodynamic performance of regenerators, evaporators, and condensers was discussed in Chapter 10. In this section we will establish their heat transfer area requirements. This information will allow trade-off between heat exchanger size (initial cost) and cycle efficiency improvement (plant operating cost).

The temperature distribution in a single pass condenser is shown in Figure 14.6. The temperature difference between the two fluids decreases as the coolant proceeds along the heat exchanger. Usually the inlet temperatures are specified and it is desired to obtain the heat transfer area for a specified amount of heat to be transferred. Since the temperature of the coolant varies within the heat exchanger, the rate of heat exchange between condensing fluid and coolant will vary locally and can be expressed in terms of temperature difference ΔT and an overall heat transfer coefficient

$$d\dot{Q} = U \, \Delta T \, dA_x$$

The total amount of heat transferred through the entire heat exchanger is obtained by integration:

$$\dot{Q} = \int_0^A U \, \Delta T \, dA_x \tag{14.7}$$

Under steady-state conditions, the incremental amount of heat transferred must be equal to that acquired by the coolant during an increase in temperature dT_{2x}, namely,

$$d\dot{Q} = \dot{m}_2 c_{p2} \, dT_{2x}$$

The resulting differential equation for T_{2x} is

$$\frac{dT_{2x}}{T_1 - T_{2x}} = \frac{U \, dA_x}{\dot{m}_2 c_{p2}}$$

If the coefficient U is assumed to remain constant, this equation can easily be integrated between limits T_{2i} and T_{2e} and 0 and A to obtain

$$\ln \frac{T_1 - T_{2e}}{T_1 - T_{2i}} = -\frac{UA}{\dot{m}_2 c_{p2}}$$

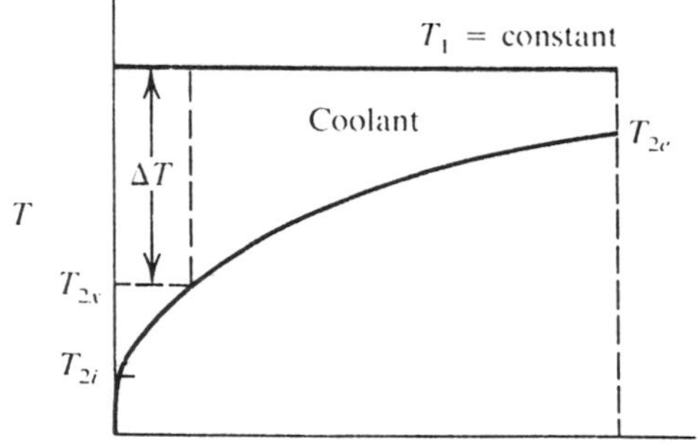

FIGURE 14.6 *Temperature Distribution in a Condenser.*

or
$$T_{2e} - T_1 = (T_{2i} - T_1)e^{-UA/\dot{m}_2 c_{p2}} \tag{14.8}$$

At any area A_x within the condenser, the temperature of the coolant T_{2x} can be shown to be

$$T_{2x} - T_1 = (T_{2i} - T_1)e^{-UA_x/\dot{m}_2 c_{p2}}$$

This relation indicates an exponential change in coolant temperature. If the overall difference between inlet and exit temperatures of the coolant is small, the temperature increases almost linearly.

From (14.8) the coolant temperature at exit can be calculated if the total heat exchanger area A is known. We can estimate the overall heat transfer coefficient U, using techniques described in Section 14.1. However, in designing a heat exchanger, the inlet and exit temperatures are usually specified, and it is required to determine the heat transfer area A for a given quantity of heat to be transferred. To obtain such a relation, proceed as follows. For constant U, we get, from (14.7),

$$\dot{Q} = UA \int_0^1 \Delta T \, d(A_x/A)$$

Letting

$$\int_0^1 \Delta T \, d(A_x/A) = \Delta T_m$$

with ΔT_m called log mean temperature difference, we have

$$\dot{Q} = UA \, \Delta T_m \tag{14.9}$$

From (14.7) and $\dot{Q} = \dot{m}_2 c_{p2}(T_{2e} - T_{2i})$, we get

$$\ln \frac{T_1 - T_{2e}}{T_1 - T_{2i}} = -\frac{UA}{\dot{m}_2 c_{p2}} = -\frac{(T_{2e} - T_{2i})UA}{\dot{Q}} = -\frac{T_{2e} - T_{2i}}{\Delta T_m}$$

or
$$\frac{\Delta T_m}{T_{2e} - T_{2i}} = -\frac{1}{\ln \dfrac{T_1 - T_{2e}}{T_1 - T_{2i}}} \tag{14.10}$$

In designing condensers, it is required to find the necessary heat transfer area when inlet and exit temperatures of coolant, condensing temperature, and flow rate of coolant or quantity of heat transferred are specified. On the other hand, in the operation of a given condenser, that is, for known heat transfer area and inlet temperatures, it is required to find the exit temperature of the coolant, when its flow rate is changed from design condition. The following example will illustrate the calculations.

EXAMPLE 14.5

The condenser of an automobile air conditioner is designed to remove 50,000 Btu/h when the outside ambient air temperature is 95°F. The saturation temperature of the R-12 refrigerant is 150°F with $h_{fg} = 46.9$ Btu/lbm.

a. Calculate the condenser surface area and refrigerant flow rate if the temperature of the coolant air rises 10°F across the condenser. The overall heat transfer coefficient has been determined to be 35 Btu/ft^2-h-°R. Assume the R-12 enters and leaves the condenser in a saturated condition.

b. Determine the exit temperature of the air if its flow rate is decreased to one-half of its design value due to operation of the automobile at a lower speed. Assume that inlet conditions remain unchanged, but U decreases to 25 Btu/ft^2-h-°R.

c. Calculate the condenser performance when the ambient air temperature increases to 110°F.

Solution

a. From (14.9), $\dot{Q} = UA\,\Delta T_m$ and from (14.10)

$$\Delta T_m = -(T_{2e} - T_{2i})/\ln\frac{T_1 - T_{2e}}{T_1 - T_{2i}} = -10/\ln\frac{150-105}{150-95} = 49.83°\text{F}$$

Note that the arithmetic mean temperature difference in this case is $\frac{1}{2}$(150–105 + 150 − 95) = 50°F and could be used in the calculation here.

$$A = \frac{\dot{Q}}{U\,\Delta T_m} = \frac{50{,}000 \text{ Btu/h}}{(35 \text{ Btu/ft}^2\text{-h-°F})(49.83°\text{F})} = 28.7 \text{ ft}^2$$

Now $\dot{Q} = \dot{m}_1 h_{fg}$; hence $\dot{m}_1$ = (50,000 Btu/h)/(46.9 Btu/lbm) = 0.296 lbm/s refrigerant flow.

b. To calculate design air flow, use $\dot{Q} = \dot{m}_2 c_{p2}(T_{2e} - T_{2i})$ where c_{p2} = 0.24 Btu/lbm°F. Hence $\dot{m}_2$ = (50,000 Btu/h)/[(0.24 Btu/lbm°F)(10°F)] = 20,833 lbm/h = 5.78 lbm/s = design air flow. To calculate the exit air temperature, use (14.8):

$$\begin{aligned} T_{2e} &= T_1 - (T_1 - T_{2i})e^{-UA/\dot{m}_2 c_{p2}} \\ &= 150°\text{F} - (150°\text{F} - 95°\text{F})e - 25(28.7)/10417(0.24) \\ &= 108.7°\text{F} \end{aligned}$$

In this case,

$$\begin{aligned} \dot{Q} &= \dot{m}c_p(T_{2e} - T_{2i}) \\ &= \left(\frac{20{,}833}{2} \text{ lbm/h}\right)(0.24 \text{ Btu/lbm°F})(13.7°\text{F}) \\ &= 34{,}250 \text{ Btu/h} \end{aligned}$$

c. The design air flow is 20,833 lbm/h. So

$$\dot{Q} = (20{,}833)(0.24)(T_{2e} - 110)$$

Also, from (14.9), $\dot{Q} = UA\,\Delta T_m$ or

$$\dot{Q} = 35(28.7)\,\frac{150 - 110 - 150 - T_{2e}}{\ln \dfrac{150 - 110}{150 - T_{2e}}}$$

$$= (20{,}833)(0.24)(T_{2e} - 110)$$

So $T_{2e} = 117.3$°F and $\dot{Q} = 36{,}500$ Btu/hr. Hence refrigerant flow $\dot{m}_1$ must be decreased so that the refrigerant will leave the condenser as saturated liquid. The lower mass flow is $\dot{m}_1 = 36{,}500/46.9 = 778.3$ lbm/h $= 0.216$ lbm/s refrigerant flow. Otherwise, it would leave the condenser as wet vapor.

EXAMPLE 14.6

A steam condenser consists of brass tubes as specified in Example 14.2. The cooling water enters the tubes at 25°C and leaves them at 35°C. Determine the heat transfer area and cooling water flow for a steam flow of 10 kg/s. Steam condenses at 50°C.

Solution

The average fluid temperature is $T_f = \frac{1}{2}(25 + 35) = 30$°C; hence $c_{p2} =$ 4.179 kJ/kg K. From Example 14.2,

$$h_1 = 4830\ \text{W/m}^2\,\text{K}, \qquad h_2 = 17{,}860\ \text{W/m}^2\,\text{K}, \quad \text{and} \quad U = 3656\ \text{W/m}^2\,\text{K}$$

At 50°C,

$$h_{fg} = 2382.9\ \text{kJ/kg}$$

Now

$$\dot{Q} = \dot{m}_1 h_{fg} = 10(2382.9)\ \text{kW} = 23{,}829\ \text{kW}$$

Also

$$\dot{Q} = \dot{m}_2 c_{p2}(T_{2e} - T_{2i}) = \dot{m}_2 4.179(35 - 25)\ \text{kW}$$

$$\dot{m}_2 = \frac{10(2382.9)\ \text{kW}}{(4.179\ \text{kJ/kg K})10\ \text{K}} = 570\ \text{kg/s}$$

From (4.9),

$$\dot{Q} = UA\,\Delta T_m$$

where from (14.10),

$$\Delta T_m = \frac{(50 - 25) - (50 - 35)}{\ln \dfrac{50 - 25}{50 - 35}} = 19.6\text{°C}$$

$$A = 333\ \text{m}^2$$

In a fashion similar to that for a condenser, we can obtain the corresponding relation for an evaporator (Figure 14.7):

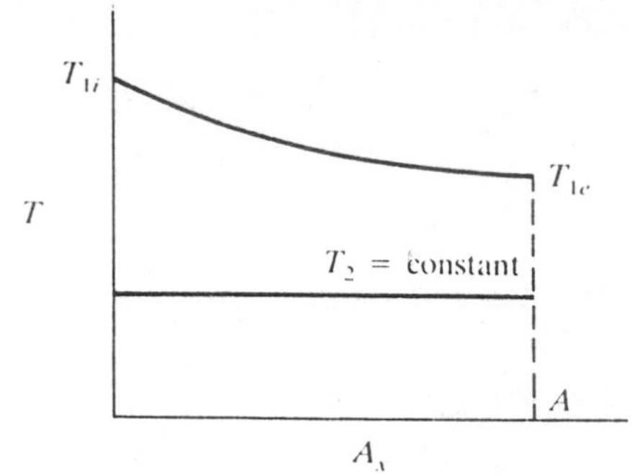

FIGURE 14.7 *Temperature Distribution in an Evaporator.*

$$\frac{T_{1e} - T_2}{T_{1i} - T_2} = e^{-UA/\dot{m}_1 c_{p1}} \tag{14.11}$$

$$\frac{T_{1x} - T_2}{T_{1i} - T_2} = e^{-UA_x/\dot{m}_1 c_{p1}}$$

and

$$\frac{\Delta T_m}{T_{1i} - T_{1e}} = -1/\ln \frac{T_{1e} - T_2}{T_{1i} - T_2} \tag{14.12}$$

In the special case of a condenser and evaporator, the temperature of only one fluid varies during the passage through the heat exchanger. In the general case, the temperatures of both fluids vary during their passage through the heat exchanger. The heat transferred and temperatures attained at exits and within the heat exchanger will depend on the flow arrangement within the exchanger. The basic flow arrangements utilized the parallel flow and cross flow (Figure 14.8). In addition, multipass flow arrangements are also used (Figure 14.9). A photograph of a shell-and-tube heat exchanger is shown in Figure 14.10.

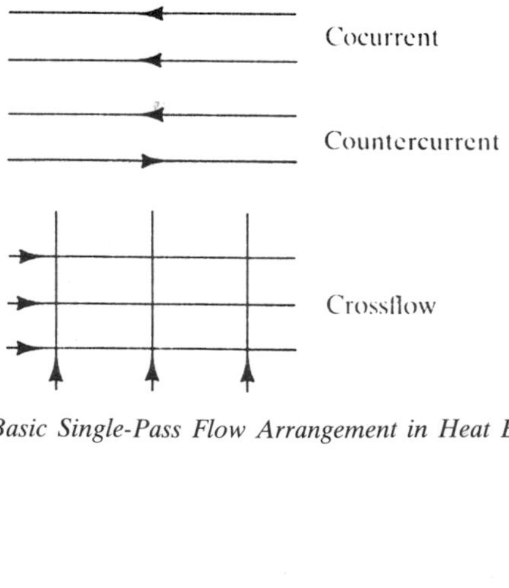

FIGURE 14.8 *Basic Single-Pass Flow Arrangement in Heat Exchangers.*

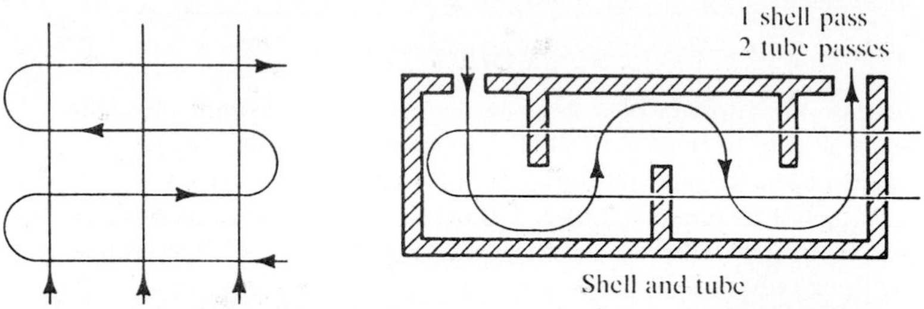

FIGURE 14.9 *Examples of Multipass Flow Arrangements in Heat Exchangers.*

FIGURE 14.10 *Shell and Tube Heat Exchanger (Courtesy of Heat Transfer Division, ITT-Standard, Buffalo, New York).*

The concept of log mean temperature difference presented in (14.9) for condensers and evaporators, namely $\dot{Q} = UA\,\Delta T_m$, can also be used for other types of heat exchangers. Using an analysis similar to that for the condenser, we can obtain analytic solutions for the temperature relations for a countercurrent, cocurrent, and a tube-and-shell exchanger. A numerical solution is required for the cross-flow exchanger. As an example, the expression for the log mean temperature difference for a countercurrent exchanger is

$$\Delta T_m = \frac{T_{1i} - T_{2e} - T_{1e} + T_{2i}}{\ln \dfrac{T_{1i} - T_{2e}}{T_{1e} - T_{2i}}}$$

For a cocurrent one,

$$\Delta T_m = \frac{T_{1i} - T_{2i} - T_{1e} + T_{2e}}{\ln \dfrac{T_{1i} - T_{2i}}{T_{1e} - T_{2e}}}$$

The expression for the tube-and-shell exchanger is more complex.

EXAMPLE 14.7

In a counterflow double pipe heat exchanger, water flowing through the inner pipe at 20°C is to be heated to 60°C with hot oil ($c_p = 2.0$ kJ/kg K) flowing in the reverse direction through the outer annulus of the pipe. The water flow rate is 2.0 kg/s, the oil flow rate is 2.0 kg/s, the oil enters at 150°C, and the overall heat transfer coefficient between oil and water is 600 W/m^2 °C. Determine the heat transfer area required.

Solution

For water at 40°C, $c_p = 4.175$ kJ/kg K. Therefore,

$$\dot{Q} = (2.0\ \text{kg/s})(4.175\ \text{kJ/kg K})40°\text{C} = 334\ \text{kW}$$

For the oil,

$$\dot{m}c_p\,\Delta T_{\text{oil}} = (2.0\ \text{kg/s})(2.0\ \text{kJ/kg K})\,\Delta T_{\text{oil}} = \dot{Q}$$

So $\Delta T_{\text{oil}} = 83.5$°C and $T_{e_{\text{oil}}} = 66.5$°C. At the side of the exchanger where the water enters, the temperature difference between the two fluids is $66.5 - 20 =$ 46.5°C (oil exits at this side of the exchanger). At the side of the exchanger where the water exits, the temperature difference is $150 - 60 = 90$°C. So

$$\Delta T_m = \frac{90 - 46.5}{\ln \dfrac{90}{46.5}} = 65.8°\text{C}$$

$$\begin{aligned} A &= \frac{\dot{Q}}{U\,\Delta T_m} \\ &= \frac{334\ \text{kW}}{(600\ \text{W/m}^2\ °\text{C})(0.001\ \text{kW/W})(65.8°\text{C})} \\ &= 8.46\ \text{m}^2 \end{aligned}$$

Calculated results of various temperature relations are presented in Figures 14.11 through 14.18 for the four commonly used flow arrangements, namely, a single-pass countercurrent, an unmixed cross-flow, a cocurrent, and a 1-2 shell-and-tube (one shell pass, two tube passes) heat exchanger. The log mean temperature difference versus the temperasture drop in one fluid is shown as a function of the parameter

$$R = \frac{\dot{m}_2 c_{p2}}{\dot{m}_1 c_{p1}} = \frac{T_{1i} - T_{1e}}{T_{2e} - T_{2i}}$$

Alternatively, the temperature drop in one fluid versus the drop in the second fluid is shown as a function of the overall heat transfer parameter $UA/\dot{m}_1 c_{p1}$. The utility of these figures will be illustrated by the following examples.

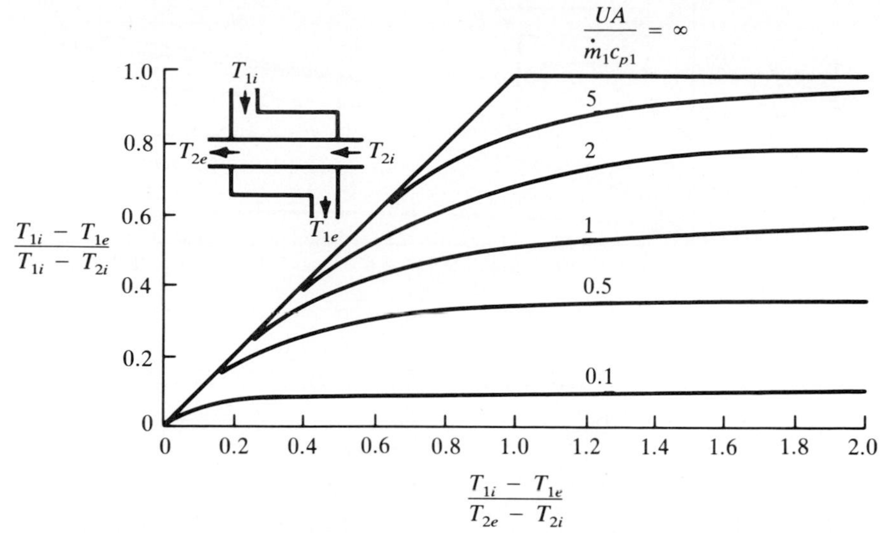

FIGURE 14.11 *Temperature Ratios for a Countercurrent Heat Exchanger.*

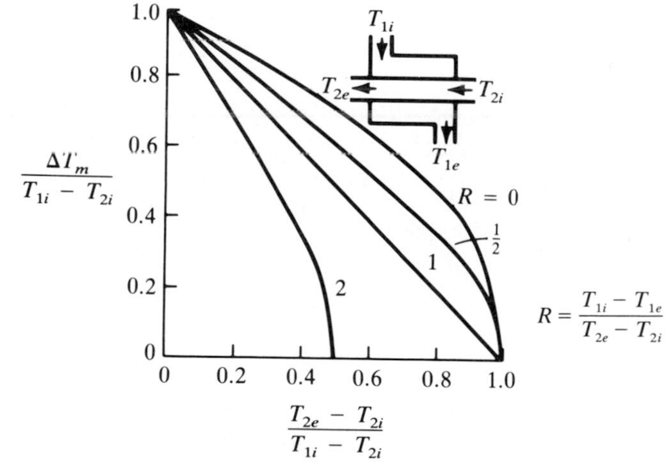

FIGURE 14.12 *Temperature Ratios for a Countercurrent Heat Exchanger.*

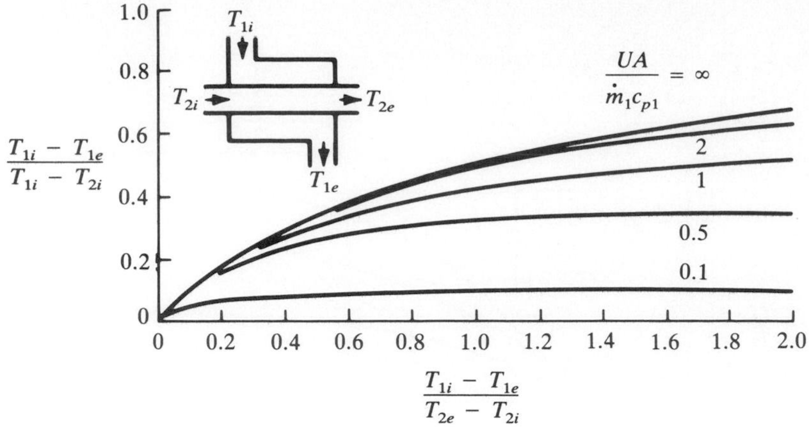

FIGURE 14.13 *Temperature Ratios for a Cocurrent Heat Exchanger.*

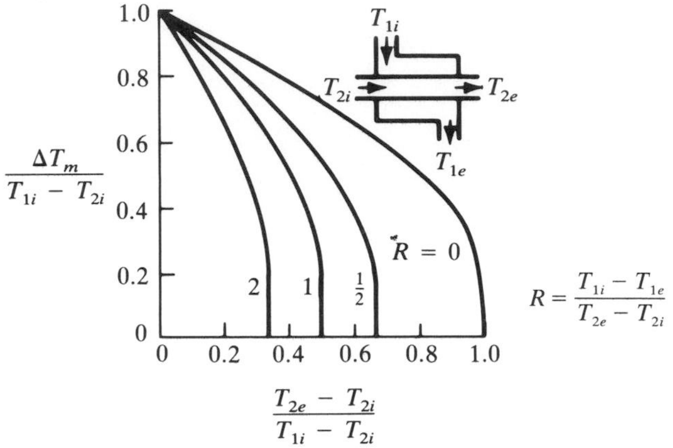

FIGURE 14.14 *Temperature Ratios for a Cocurrent Heat Exchanger.*

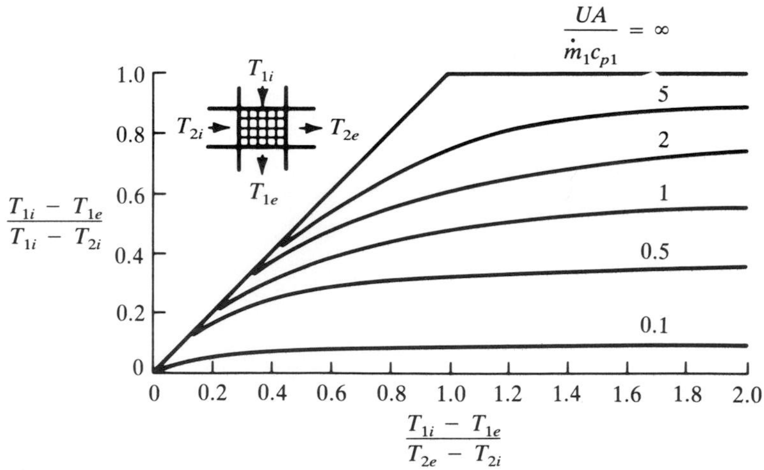

FIGURE 14.15 *Temperature Ratios for an Unmixed Cross Flow Heat Exchanger.*

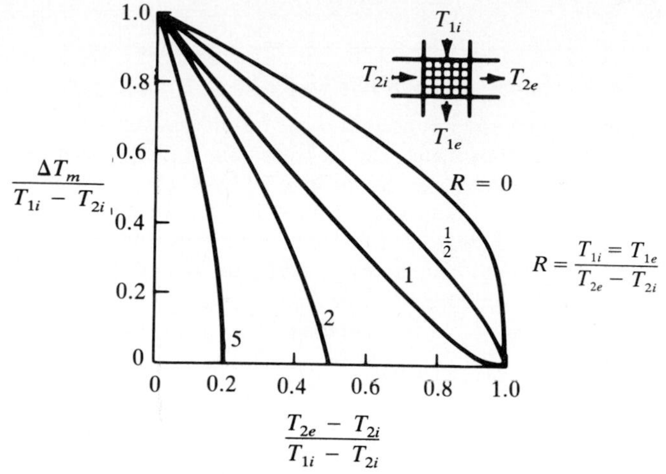

FIGURE 14.16 *Temperature Ratios for an Unmixed Cross Flow Heat Exchanger.*

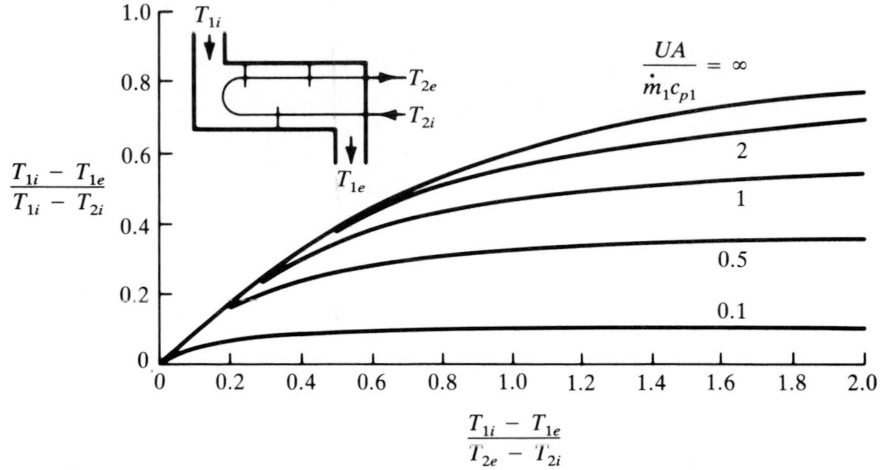

FIGURE 14.17 *Temperature Ratios for a 1-2 Shell and Tube Heat Exchanger.*

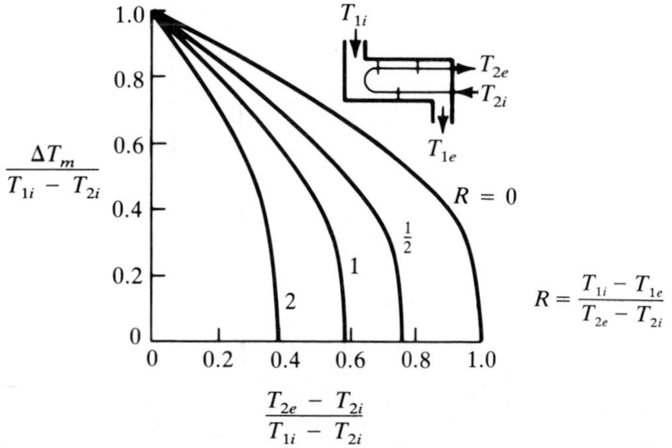

FIGURE 14.18 *Temperature Ratios for a 1-2 Shell and Tube Heat Exchanger.*

EXAMPLE 14.8

A countercurrent regenerator is used in a gas turbine power plant to preheat the air before it enters the combustor. The air leaves the compressor at a temperature of 350°C (T_{2i}). Exhaust gas leaves the turbine at 700°C (T_{1i}). The mass flow rates of air and gas are 5 kg/s. Take the c_p of air and gas to be equal to 1.05 kJ/kg K. Determine the required heat transfer area as a function of the regenerator effectiveness, if the overall heat transfer coefficient U is 75 W/m^2 K.

Solution

The effectiveness of a heat exchanger is given by (10.12), which in the nomenclature of this chapter becomes

$$e = \frac{T_{2e} - T_{2i}}{T_{1i} - T_{2i}} = \frac{T_{2e} - 350}{700 - 350}$$

For example, for $e = 0.8$, air is heated to a temperature T_{2e} of 630°C. Now

$$R = \frac{\dot{m}_2 c_{p2}}{\dot{m}_1 c_{p1}} = \frac{5(1.05)}{5(1.05)} = 1 = \frac{T_{1i} - T_{1e}}{T_{2e} - T_{2i}} = \frac{700 - T_{1e}}{630 - 350}$$

hence the exit temperature of the exhaust gas T_{1e} becomes 420°C. The rate of heat exchange is

$$\dot{Q} = \dot{m}_1 c_{p1}(T_{1i} - T_{1e}) = 5(1.050)(700 - 420) = 1470 \text{ W}$$

Now from Figure 14.12 at $R = 1$ and $e = 0.8$, $\Delta T_m/(T_{1i} - T_{2i}) = 0.2$; thus

$$\Delta T_m = (700 - 350)(0.2) = 70\text{°C}$$

Hence from (14.9)

$$A = \frac{\dot{Q}}{U\,\Delta T_m} = \frac{1{,}470{,}000}{75(70)} = 280 \text{ m}^2$$

Similarly, we calculate required areas for other values of e.

e	T_{2e}	T_{1e}	$\dot{Q}$ (kW)	ΔT_m	A (m^2)
1.0	700	350	1838	0	∞
0.9	665	385	1654	35	630
0.8	630	420	1470	70	280
0.7	595	455	1286	105	163
0.6	560	490	1103	140	105
0.5	525	525	919	175	70
0.4	490	560	735	210	47
0.3	455	595	551	245	30
0.2	420	630	368	280	18
0.1	385	665	184	315	8
0	350	700	0	350	0

EXAMPLE 14.9

Determine the heat transfer area requirements of Example 14.8 if a 1-2 shell and tube, an unmixed cross flow, and a cocurrent heat exchanger are used, respectively.

Solution

For each value of effectiveness e, the values of temperatures will be the same as in Example 14.8. The mean temperature difference ΔT_m, however, will be different for each type.

				Cross flow		Shell & tube		Cocurrent	
e	T_{2e}	T_{1e}	$\dot{Q}$ (kW)	ΔT_m	A (m^2)	ΔT_m	A (m^2)	ΔT_m	A (m^2)
1.0	700	350	1838	0	∞				
0.9	665	385	1654	8	2864				
0.8	630	420	1470	35	560				
0.7	595	455	1286	74	231				
0.6	560	490	1103	117	125				
0.5865	555	495	1078	120	120	0	∞		
0.5	525	525	919	158	78	142	86	0	∞
0.4	490	560	735	199	49	193	51	173	57
0.3	455	595	551	239	31	237	31	230	32
0.2	420	630	368	274	18	275	18	275	18
0.1	385	665	184	313	8	313	8	313	8
0	350	700	0	350	0	350	0	350	0

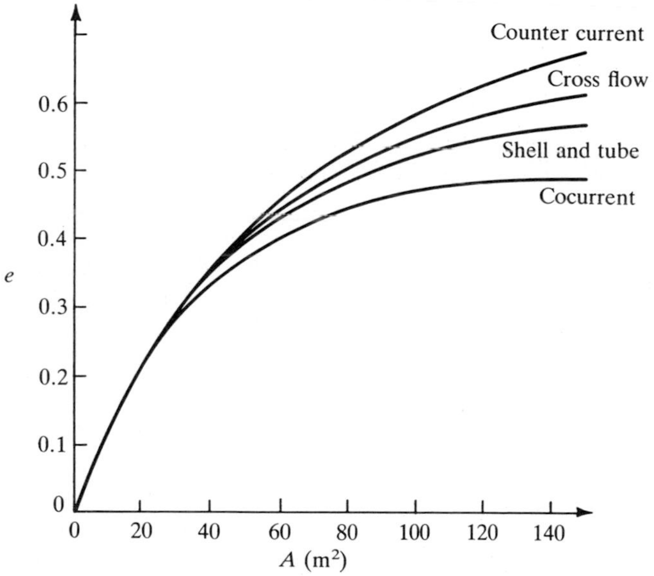

FIGURE 14.19 *Effectiveness of Heat Exchangers.*

From the table and Figure 14.19 we see that for a desired value of effectiveness, the heat transfer area requirement is smallest for a countercurrent heat exchanger, and largest for the cocurrent type, bracketing the area requirements for the other two types. For the countercurrent and cross-flow types, the maximum achievable effectiveness is unity, while for the cocurrent and shell-tube types the maximum achievable e is less than 0.6.

In obtaining the results given in Figures 14.11 through 14.18, it was assumed that the overall heat transfer coefficient U is constant, that the outer shell of the heat exchanger is perfectly insulated, and that changes in kinetic energy are negligible. In general, the overall heat transfer coefficient U is not constant within a heat exchanger. However, usually it is adequate to base the calculation on an average U value calculated at the midpoint of the heat exchanger. If more accurate calculations are required, the heat exchanger is divided, for calculation purposes, into a number of small sections. For each section, Equation (14.9) is used. The total heat transferred is obtained by summing the incremental heat transfer of each section.

In addition to the four flow arrangements described, other flow arrangements have been utilized in heat exchangers. Examples are cross flow with one fluid mixed, the other unmixed; cross flow with both fluids mixed; and multiple shell and tube passes in shell and tube exchangers. The performance of all variations in flow arrangements falls between two limits, those of the countercurrent and cocurrent arrangements.

14.3 COOLING OF ELECTRONIC COMPONENTS

Modern electronic systems and their electronic components, such as semiconductors, printed circuit boards (PCBs), and microwave devices, have been shrinking in size. At the same time, their power has been increasing. The temperature at which these devices operate must be controlled; otherwise electronic malfunction can occur. Heat is generated in electronic components by the flow of electric current. It must be dissipated by providing a good heat flow path from the components to the surroundings. For small devices, air is used as the coolant, utilizing free convection and radiative transfer or, alternately, forced convection. For larger devices, liquid forced convection or even evaporative cooling is employed. For these methods, the temperature of the electronic components will always be higher than that of the ambient surroundings. Thus, when the temperature of the surrounding air is too high for safe operation of special electronic equipment, refrigeration systems must be employed for cooling.

Free air convection requires no additional equipment and hence is the least complex method of cooling electronic components. Fins are used to increase the heat transfer area. Forced air convection requires the addition of a fan. Forced liquid convection is more complex and requires a pump, a coolant liquid, a coolant reservoir, and a control system.

The methods outlined in Chapter 13 can be used to evaluate the several means of cooling electronic components. Heat generated by specific semiconductor devices is transferred by conduction to a "heat sink" with fins which are used to increase the convective heat transfer area. Since the surface heat sinks are black, cooling by radiation is significant when accompanied by free convection. On the other hand, radiative cooling becomes small when accompanied by forced convection. Electronic components, such as microprocessors (chips), resistors, and power transistors, are mounted on printed circuit boards. In this case, radiation can be neglected, since the board surface is bright with low emissivity. The circuit boards are usually cooled by use of free convection. If free convection cooling is insufficient, forced convection is provided.

1. *Free convection and radiation*

The relationships developed in Section 13.3b for free convection heat transfer from different surfaces are directly applicable here. To utilize the relationship for radiative heat transfer given in Section 13.4, we require additional information on the view factor of the fins. As seen from Equation (13.25), the rate of radiative heat transfer from a heat sink will depend on the temperature of the heat sink, the ambient temperature, the emissivity of the heat sink surface, and the radiation interference between adjacent fins and fin base. This interference effectively reduces the total surface area of the fins available for transfer of heat. The radiation factor F as given in Equation (13.26) includes the combined effect of geometric view factor G (fin interference) and emissivity of heat sink and environment. The view factor expresses the fraction of thermal radiation leaving a diffuse surface of the heat sink, passing through a nonparticipating medium (air) to the environment. The heat transfer from a trough between two parallel rectangular fins (Figure 14.20) will be

$$\dot{Q} = [A_1F_{13} + A_2F_{23} + A_1F_{14} + A_2F_{24}]\sigma_0(T_s^4 - T_a^4)$$
$$= [2hLF_{13} + wLF_{23} + 2hLF_{14} + wLF_{24}]\sigma_0(T_s^4 - T_a^4)$$

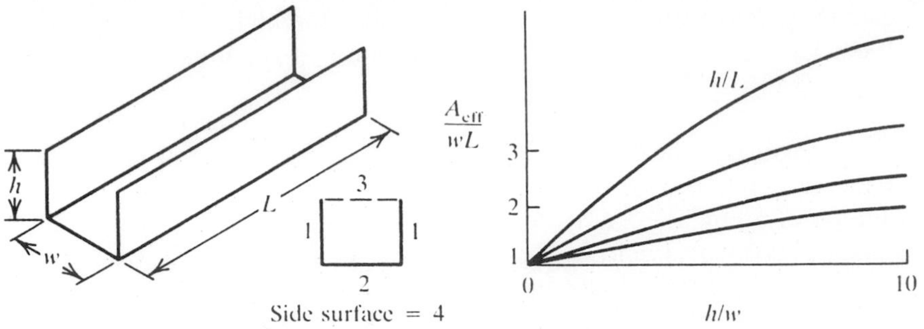

FIGURE 14.20 *Rectangular Fin Extended Surfaces.*

where T_s is the temperature of the heat sink surface and T_a is the temperature of ambient air. As a good approximation, we can take the emissivities of the black heat sink surface and the surrounding as unity. In this case, we see from equation (13.26) that the radiation factor will equal the view factor. Hence,

$$\dot{Q} = [2(h/w)(G_{13} + G_{14}) + (G_{23} + G_{24})]wL\sigma_0(T_s^4 - T_a^4)$$
$$= A_{\text{eff}}\sigma_0(T_s^4 - T_a^4) \qquad \textbf{(14.13)}$$

Based on values of geometric view factors on rectangular surfaces available in the literature, values of effective area A_{eff} were calculated for a range of geometric variables and are presented in Figure 14.21.

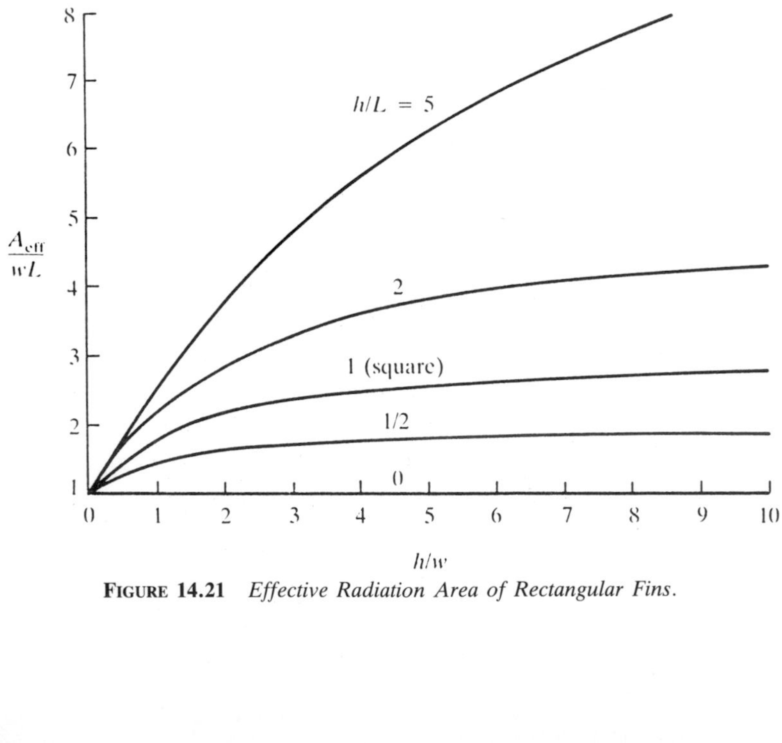

FIGURE 14.21 *Effective Radiation Area of Rectangular Fins.*

2. *Forced convection*

The relationships developed in Section 13.3 for forced convection heat transfer on circular cylinders in cross flow and on flat walls in parallel flow can be used directly to establish cooling requirements for electronic components.

EXAMPLE 14.10

Calculate the heat transferred from the heat sink shown in Figure 14.22 when its fins are at 200°F and the surroundings are at 70°F. The physical dimensions of the heat sink are shown in the figure.

Solution

First, calculate the heat transferred by free convection. In actual use, the heat sink will be mounted with its fins in a vertical position. The vertical length is 1 in., with a total surface area of 12 fins × 2 sides (1 in. × 0.9 in.) + 2 sides (1.5 in. × 1 in.) + 8(0.3 in. × 1 in.) = 27.0 in^2. For free convection, we get from (13.17) $hL/k_m = C\,\text{Ra}^n$. Next, calculate the Rayleigh number Ra, using the following properties of air at the mean air temperature $T_m = \frac{1}{2}(T_w + T_f) = \frac{1}{2}(200 + 70) =$ 135°F:

$$k_m = 0.0159\ \text{Btu/ft-h-°R}\,, \qquad \nu_m = 2.00 \times 10^{-4}\ \text{ft}^2\text{/s}$$

$$\text{Pr}_m = 0.71\,, \qquad \frac{g\beta}{\nu^2} = 1.36 \times 10^6\ 1/\text{°R-ft}$$

Hence

$$\text{Ra} = 1.36 \times 10^6(0.71)(\tfrac{1}{12})^3(200 - 70) = 0.73 \times 10^5$$

hence,

$$C = 0.54\,, \qquad n = 0.25$$

and, from (13.17),

$$\frac{h(1)}{(0.0159)(12)} = 0.54(0.73 \times 10^5)^{0.25} = 8.88$$

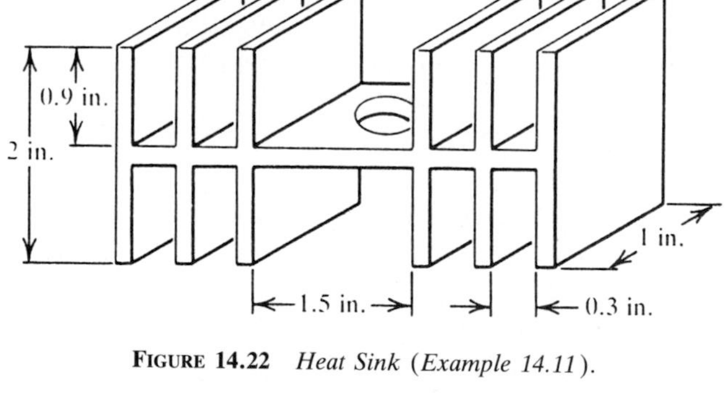

Figure 14.22 *Heat Sink (Example 14.11).*

and
$$h = \frac{0.0159(12)}{1} 8.88 = 1.69 \text{ Btu/ft}^2\text{-h-°R}$$

$$\dot{Q} = hA_s(T_w - T_f) = (1.69) \frac{27}{144} (200 - 70) = 41.2 \text{ Btu/h}$$

Next, calculate the heat transferred by radiation. The effective radiation surface area is obtained as follows:

8 troughs: $h/w = 0.9/0.3 = 3$, $h/L = 0.9/1 = 0.9$

From Figure 14.21,

$$A_{\text{eff}} = 2.0wL$$

2 troughs: $h/w = 0.9/1.5 = 0.6$, $h/L = 0.9/1 = 0.9$

From Figure 14.21,

$$A_{\text{eff}} = 1.4wL$$

2 ends: $A = 2(1) = 2.0 \text{ in}^2$

Total area:

$$8(2.0)(0.3 \times 1) + 2(1.4)(1.5 \times 1) + 2(2) = 13.2 \text{ in}^2$$

Thus

$$\dot{Q} = \sigma(T_s^4 - T_a^4)A_{\text{eff}} = 0.171 \times 10^{-8}(660^4 - 530^4) \frac{13.0}{144} = 17.1 \text{ Btu/h}$$

Thus the total heat transferred from the heat sink by free convection and radiation is $41.2 + 17.1 = 58.3$ Btu/h (17.21 W).

EXAMPLE 14.11

Calculate the heat transferred from a power transistor on a circuit board when the card is cooled by forced air. In general, electronic components are mounted on one side of the PCB (printed circuit board), while the connections between components are made with printed circuitry on the opposite side of the PCB. The chassis box (Figure 14.23) measures 2 in. × 6 in. × 10 in. Each PCB measures 4 by 9 in. The geometric dimensions of the transistor case designated as TO-5 are diameter = $\frac{5}{16}$ in. and height = $\frac{1}{4}$ in. The cross-sectional area of the box is 2 in. by 6 in. A fan provides an air flow of 50 cfm. Assume a 100°F temperature difference between forced air stream and transistor case.

Solution

The cross-sectional area of each duct between adjacent PCBs is $\frac{1}{2}$ in. by 4 in.; hence the air flow in this duct is

$$\frac{0.5 \text{ in.} \times 4 \text{ in.}}{2 \text{ in.} \times 6 \text{ in.}} (50 \text{ cfm}) = 8.33 \text{ cfm}$$

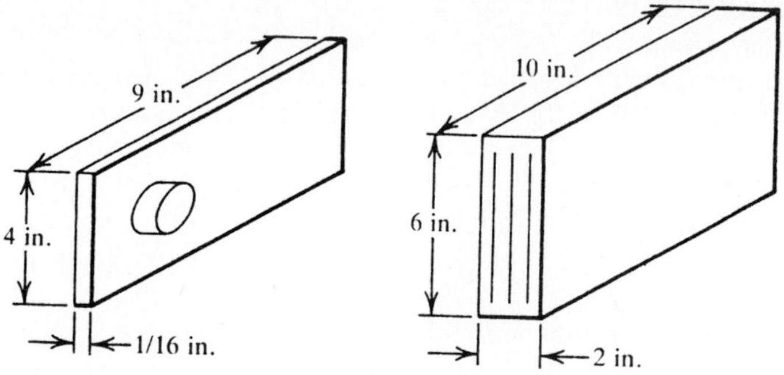

FIGURE 14.23 *Heat Transfer in Electronic Chassis Box (Example 14.11).*

The average velocity $\bar{V}$ in the duct is

$$\frac{(8.33\ \text{ft}^3/\text{m})144\ \text{in.}^2/\text{ft}^2}{(60\ \text{s/m})(\frac{1}{2}\ \text{in.} \times 4\ \text{in.})} = 10\ \text{fps}$$

Obtain an approximation to the heat transfer from a stubby cylinder by using the relationship for the heat transfer per unit length for a long cylinder (i.e., neglect end effects). From (13.12), we have

$$\frac{hD}{k_f} = C\,\text{Re}^n\,\text{Pr}_f^{0.38}\,(\text{Pr}_f/\text{Pr}_w)^{0.25}$$

The Reynolds number Re is

$$\frac{\bar{V}D}{\nu} = \frac{10(5/16)}{(1.63 \times 10^{-4})12} = 1600$$

For Re = 1600,

$$C = 0.21\,, \qquad n = 0.62$$

$$k_f = 0.0149\ \text{Btu/ft-h-°R}\,, \qquad \text{Pr}_f = 0.71$$

Thus

$$\frac{hD}{k} = 0.21(1600)^{0.62}(0.71)^{0.38} = 17.9 \quad \text{and} \quad h = \frac{0.0149(12)}{0.3125}\,17.9$$

$$= 10.2\ \text{Btu/ft}^2\text{-h-°R}$$

$$\dot{Q} = h(T_w - T_f)A_s = h(T_w - T_f)\pi DL = 10.2(100)\pi\left(\frac{5}{16}\,\frac{1}{12}\right)\left(\frac{0.25}{12}\right)$$

$$= 1.74\ \text{Btu/h}$$

14.4 INSULATION

In conventional insulation techniques, an extra layer of material is added to reduce the heat transfer by conduction. Thus the essential characteristic of a good insulation material is a low thermal conductivity, usually less than 0.02 Btu/ft-h-°R (0.1 W/m K). Since air possesses a low thermal conductivity, many good insulation materials contain air voids or fibrous layering. Examples of conduction insulation have been given in Section 13.2.

In addition to conventional insulation techniques, radiation heat transfer can also be utilized to effect a reduction in heat loss in low-temperature storage devices. In storing liquified natural gas at a low temperature, a vacuum spacing can be used to serve as an insulation layer in addition to conventional insulation. The small gap is evacuated to eliminate heat loss by convection. In this application, the different modes of heat transfer occur in series.

To "insulate" against radiative heat transfer, thin-walled radiation shields can be placed near a high-temperature source to reflect some of the energy radiated. As a simple example of the use of radiation shields, consider two radiating parallel planes with emissivities ϵ_1 and ϵ_2 (Figure 14.24). Here the view factor G_{12} is equal to unity and A_1 equals A_2. Hence, the radiation factor is

$$F_{12} = \frac{1}{1/\epsilon_1 + 1/\epsilon_2 - 1}$$

Therefore, the heat transfer between the two planes is

$$\dot{Q}_{12} = \frac{\sigma_0 A_1}{1/\epsilon_1 + 1/\epsilon_2 - 1}(T_1^4 - T_2^4)$$

Now introduce a shield with emissivity ϵ_s between the two radiating surfaces. Upon reaching thermal equilibrium, the heat transferred to the shield from surface 1 must equal that reaching it from surface 2; that is,

$$\dot{Q}_{1s} = \dot{Q}_{2s}$$

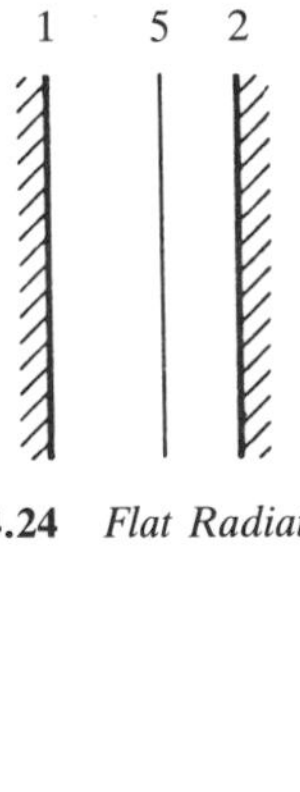

FIGURE 14.24 *Flat Radiation Shield.*

Specifically

$$\dot{Q}_{1s} = \frac{\sigma_0 A}{1/\epsilon_1 + 1/\epsilon_s - 1}(T_1^4 - T_s^4) \quad \text{and} \quad \dot{Q}_{s2} = \frac{\sigma A}{1/\epsilon_2 + 1/\epsilon_s - 1}(T_s^4 - T_2^4)$$

Hence equating the two expressions and solving for T_s, we have

$$T_s^4 = \frac{1/\epsilon_s + 1/\epsilon_2 - 1}{1/\epsilon_1 + 1/\epsilon_2 + 2/\epsilon_s - 2} T_1^4 + \frac{1/\epsilon_1 + 1/\epsilon_s - 1}{1/\epsilon_1 + 1/\epsilon_2 + 2/\epsilon_s - 2} T_2^4$$

$$\dot{Q}_{1s} = \sigma_0 A(T_1^4 - T_2^4)/(1/\epsilon_1 + 1/\epsilon_2 + 2/\epsilon_s - 2)$$

Thus the ratio of the two heat transfer rates becomes

$$\frac{\dot{Q}_{1s}}{\dot{Q}_{12}} = \frac{1/\epsilon_1 + 1/\epsilon_2 - 1}{1/\epsilon_1 + 1/\epsilon_2 + 2/\epsilon_s - 2} \tag{14.14}$$

The reduction in radiation heat transfer from the surface is a function of the relative emissivities of the plane walls and the shield surface. For example, for $\epsilon_1 = \epsilon_2 = \epsilon_s = \epsilon$,

$$\frac{\dot{Q}_{1s}}{\dot{Q}_{12}} = \frac{2/\epsilon - 1}{4/\epsilon - 2} = \frac{1}{2}$$

(that is, independent of the value of ϵ). The use of one shield reduces the heat transferred by half. It can also be shown that two shields will reduce the heat transferred to one-third, three shields to one-quarter, and n shields to $1/(n+1)$. Greater reductions can be obtained by using a shield with low emissivity (a highly polished surface); for example, let

$$\epsilon_1 = \epsilon_2 = \epsilon = 0.8 \quad \text{and} \quad \epsilon_s = 0.1\epsilon$$

Thus, the reduction in radiation heat transfer becomes, from (14.14),

$$\frac{\dot{Q}_{1s}}{\dot{Q}_{12}} = \frac{2/\epsilon - 1}{2/\epsilon + 20/\epsilon - 2} = 0.059$$

From the point of view of thermal radiation, the shield distance from the plane is of no importance. However, if the effects of convection in the medium between the plane walls are included, the use of a large number of shields results in small distances between them, thereby also reducing the convection heat transfer. Unlike flat shields, cylindrical and spherical shields are most effective when they are placed near the body at the higher temperature.

EXAMPLE 14.12

A wood-burning stove utilizes a hollow lining with radiation shields to reduce its outside temperature (Figure 14.25). Determine the number of shields required to

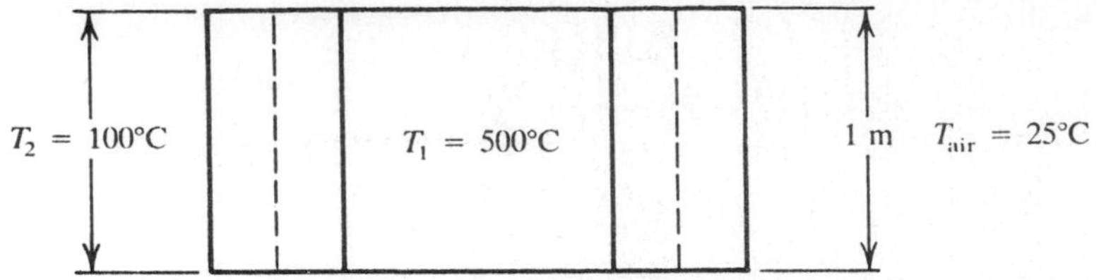

FIGURE 14.25 *Radiation Shields for Stove.*

keep the outside of the stove at 100°C. Assume the emissivity of all surfaces to be 0.87. The temperatue of the still atmospheric air is 25°C and the temperature in the stove is 500°C.

Solution

For n shields within the hollow lining, the radiation heat exchange between surfaces 1 and 2 becomes

$$\dot{Q}_{12}/A_1 = \frac{1}{n+1}\frac{\sigma_0}{2/\epsilon - 1}(T_1^4 - T_2^4) = \frac{1}{n+1}\frac{5.67\times 10^{-8}}{1.299}(773^4 - 373^4)$$
$$= \frac{14{,}740}{n+1}\ \text{W/m}^2$$

Free convection:

The mean film temperature is

$$T_m = \tfrac{1}{2}(100 + 25) = 62.5°\text{C}$$

At 62.5°C,

$$k_m = 0.0279\ \text{W/m K}, \qquad \text{Pr}_m = 0.71, \quad \text{and} \quad \frac{g\beta}{\nu^2} = 78.2\times 10^6$$

hence

$$\text{Ra} = 78.2\times 10^6(0.71)(75)(1)^3 = 4.17\times 10^9$$

Thus

$$C = 0.135 \quad \text{and} \quad n = \tfrac{1}{3}$$

From (13.17),

$$\frac{hD}{k} = 0.135(4.17\times 10^9)^{1/3} = 217.2 \quad \text{and} \quad h = \frac{0.0279}{1}217.2 = 6.06\ \text{W/m}^2\,\text{K}$$

$$\frac{\dot{Q}}{A} = h(T_2 - T_a) = 6.06(100 - 25) = 454.5\ \text{W/m}^2$$

Radiation into atmosphere:

From (13.27),

$$\dot{Q}/A = \epsilon\sigma_0(T_2^4 - T_a^4) = (0.87)5.67\times 10^{-8}(373^4 - 298^4) = 566\ \text{W/m}^2$$

Hence,

$$\text{Total loss from surface } 2 = 454.5 + 566 = 1020.5\ \text{W/m}^2$$

Therefore,

$$\frac{14{,}740}{n+1} = 1020.5$$

and $n = 13$ shields are required.

REFERENCES

ELLISON, G. N., *Thermal Computations for Electronic Equipment*. New York, Van Nostrand Reinhold, 1984.

KAYS, W. M., and LONDON, A. L., *Compact Heat Exchangers*. New York, McGraw-Hill Book Company, 1984.

KRAUS, A. D., and BAR-COHEN, A., *Thermal Analysis and Control of Electronic Equipment*. New York, McGraw-Hill Book Company, 1982.

TABOREK, J., and HEWITT, G. F., *Heat Exchangers: Theory and Practice*. New York, McGraw-Hill Book Company, 1982.

PROBLEMS

14.1 An air heater consists of five 2 in. × 12 in. fins placed between two heated parallel planes. The resulting ducts through which the air passes have a 1 in. by 2 in. cross section. Air enters at a rate of 100 lbm/h at atmospheric pressure and 68°F. Calculate the air temperature leaving the heater if the heated plates are kept at 250°F.

14.2 A gas-to-gas heat exchanger consists of two concentric pipes of 50-ft length. The cold gas, which flows in the concentric annulus, is heated from a temperature of 69°F to 150°F. The hot gas, which flows in the inner pipe, is cooled from 250°F to 160°F. It is intended to increase the capability of the heat exchanger by lengthening the pipes so that the cold gas is heated to a temperature of 200°F. The mass flow of the two gases is to remain unchanged. Calculate the required length of the redesigned heat exchanger.

14.3 Oil enters a cooler at a temperature of 70°C and leaves the cooler at 30°C. Water is used as coolant and enters the cooler at a temperature of 20°C. Find the water outlet temperature if the oil and water mass flow rates are 10,000 kg/h and 20,000 kg/h, respectively. $c_p = 2.0$ kJ/kg K.

14.4 If the water flow rate of Problem 14.3 is decreased to 10,000 kg/h, what will be the water outlet temperature?

14.5 In a tubular steam-to-water heat exchanger, saturated steam condenses at a pressure of 350 kPa on the external surface of the tube. The water flows inside the tube, changing its temperature from 20°C to 90°C. Determine the log mean temperature difference of the heat exchanger.

14.6 Calculate the steam flow rate of the heat exchanger of Problem 14.5 if the water flow rate is 8000 kg/h.

14.7 What is the effect of raising the steam pressure of Problem 14.5 on the log mean temperature difference?

14.8 The heat sink of Example 14.10 is to be cooled by a forced air stream flowing parallel to the rectangular fins. Determine the heat transferred if the air stream velocity is 10 ft/s.

14.9 The heat transferred from the heat sink of Example 4.10 is to be increased to 30 W. Calculate which method would be more effective (i.e., require less fin material):

a. increasing the length of the fins from one inch to the required dimension, or

b. increasing the number of fins, leaving the space between them at 0.3 in. (7.62 mm).

14.10 Calculate the effect on heat transfer rate of the power transistor of Example 14.11 if the spacing between the PCBs is reduced to 0.30 in., maintaining the total air flow through the chassis at 50 cfm.

14.11 A cross flow heat exchanger is used to preheat air entering at 20°C by means of 800°C turbine exhaust gas. Determine the exit temperature of the air for an air flow rate of 0.75 kg/s and hot gas flow of 0.50 kg/s. ($UA = 0.375$ kW/k; c_p for gas same as air.)

14.12 An exchanger is to heat 1.5 lbm/s water from 60°F to 160°F by using saturated steam condensing on the outside of 1.5-in., $\frac{3}{32}$-in.-thick, 10-ft-long brass tubes ($h = 2500$ Btu/ft^2-h). Determine the required heat transfer area for a shell-and-tube exchanger and the required number of tubes. The water velocity in the tubes is not to exceed 15 fps. ($T_{\text{steam}} = 212$°F; k for brass = 65 Btu/ft-h-°R.)

14.13 A 1-in. diameter pipe carries a flow of liquid nitrogen at 140°R. The pipe is concentric with a 2-in. diameter pipe, and the space between the two pipes is evacuated. Assume the emissivity of the highly polished pipe surfaces to be 0.03. Determine the rate of heat addition to the liquid nitrogen per foot of pipe length. Neglect thermal resistance at the inside surface of the inner pipe, neglect the thermal resistance of the pipe walls, and assume the pipe is in a room of temperature 70°C.

14.14 Repeat the calculation of Problem 14.13 for a 4-in. diameter outer pipe.

14.15 Develop an expression for the steady-state heat flow through a spherical container with wall thickness t. The inner wall temperature is T_1, the outer wall temperature is T_2, and the thermal conductivity of the wall material is k. Start with let R_1 = radius of inner wall of sphere.

14.16 A cryogenic fluid at −100°F is to be stored in a 10-ft diameter spherical container. The container is covered with a 2-in. thick layer of insulation ($k = 0.05$ Btu/ft-hr-°F). Determine the heat flow to the fluid if the sphere is in a room with an ambient temperature of 80°F. Neglect radiative heat transfer from the outer surface of the tank.

14.17 Determine the effect of a 10-mm thickness of insulation ($k = 0.1$ W/m K) on the heat loss of the steam pipe of Examples 14.3 and 14.4, respectively.

appendix

A

Thermodynamic Properties (English Units)

LIST OF TABLES

USING THE TABLES OF THERMODYNAMIC PROPERTIES OF SUBSTANCES

In the tables of Appendices A and B, the thermodynamic properties v, u, h, and s of four substances (water, carbon dioxide, refrigerant R-22, and air) are presented as a function of temperature (T) and pressure (p). In most tables in which the thermodynamic properties of substances are presented, the increments of the independent variables (typically temperature and pressure) are usually small. Hence, linear interpolation can be used to obtain the thermodynamic properties of the substance for values of the independent variables not given in the tables. The tables given in these appendices, however, are of necessity in abbreviated form, and hence interpolation must be used with special care to obtain accurate values of thermodynamic properties.

1. *Saturation Tables*

In the saturation tables, the increment of temperature is 10°F (50°C). In the other tables, the increment of temperature is 100°F (50°C), with varying increments in the pressure. Hence, in the *saturation* tables, linear interpolation can be used for all properties; thus we have

$$\frac{X - X_1}{X_2 - X_1} = \frac{T - T_1}{T_2 - T_1}$$

where X is a thermodynamic property, the subscripts 1 and 2 signify the endpoints of the interval for which properties are given in the tables, and T is the corresponding temperature. For example, to find the saturated vapor enthalpy of water at 93°F, we determine the following from Table A.2:

°F	h_g (Btu/lbm)
90	1100.8
100	1105.1

Hence

$$\frac{h_g - 1100.8}{1105.1 - 1100.8} = \frac{93 - 90}{100 - 90}$$

$$h_g = 1100.8 + 0.3(4.3) = 1100.8 + 1.29 = 1102.1 \text{ Btu/lbm}$$

2. *Superheat Tables*

In the *superheat* tables, where the increments of the two independent variables are larger, the interpolation rules given in the following paragraphs should be used.

For the *specific volume* at a given value of pressure, use linear interpolation to obtain

$$\frac{v - v_1}{v_2 - v_1} = \frac{T - T_1}{T_2 - T_1}$$

For interpolation at a given value of temperature, one can obtain better accuracy by using linear interpolation of the reciprocal of pressure instead of pressure. Hence, at a given temperature, the specific volume is given by

$$\frac{v - v_1}{v_2 - v_1} = \frac{1/p - 1/p_1}{1/p_2 - 1/p_1}$$

Since for a gas, v is approximately proportional to T/p, the specific volume is proportional to temperature and inversely proportional to pressure. As an example, we obtain from Table A.3 the following values of specific volume of water in ft^3/lbm for the temperatures and pressures indicated:

p (psia)	1000°F	1100°F	1200°F
10	86.91	92.87	98.84
15	57.926		
20	43.435		

Thus at a pressure of 10 psia and a temperature of 1100°F, we obtain, by linear interpolation between temperatures of 1000° and 1200°F,

$$\frac{v - 86.91}{98.84 - 86.91} = \frac{1100 - 1000}{1200 - 1000}$$

or $v = 86.91 + 5.96 = 92.87 \text{ ft}^3/\text{lbm}$, which is identical to the value given in the table at a temperature of 1100°F. At a temperature of 1000°F, we obtain, at a pressure of 15 psia,

$$\frac{v - 86.91}{98.84 - 86.91} = \frac{1/15 - 1/10}{1/20 - 1/10}$$

or $v = 86.91 - 43.47(0.6666) = 57.930$ psia, as compared to 57.926 psia given in the table for a pressure of 15 psia. If we had used linear interpolation in the pressure p, we would have gotten a value of 65.17 psia. Thus the use of the reciprocal of the pressure should be used in such interpolations.

For determining the values of *internal energy* and *enthalpy* from the tables, use linear interpolation at given values of pressure or temperature. As an example, we obtain from Table A.3 the following values of internal energy of water for the temperatures and pressures indicated:

p (psia)	1000°F	1100°F	1200°F
10	1373.8	1414.7	1456.6
15	1373.7		
20	1373.5		

At a pressure of 10 psia and a temperature of 1100°F, we obtain, by linear interpolation between temperatures of 1000° and 1200°F,

$$\frac{u - u_1}{u_2 - u_1} = \frac{T - T_1}{T_2 - T_1}$$

$$u = 1373.8 + (1456.6 - 1373.8)\,\frac{1100 - 1000}{200}$$

$$= 1373.8 + 41.4 = 1415.2 \text{ Btu/lbm (vs. 1414.7 given in the table)}$$

The values of *entropy* at a given pressure are interpolated by use of the logarithm of the temperature ratio:

$$\frac{s - s_1}{s_2 - s_1} = \frac{\ln T/T_1}{\ln T_2/T_1}$$

and similarly at a given temperature by the pressure ratio:

$$\frac{s - s_1}{s_2 - s_1} = \frac{\ln p/p_1}{\ln p_2/p_1}$$

For a gas, the change in entropy is proportional to $\ln T_2/T_1$ and $\ln p_2/p_1$. (Recall that for an ideal gas, the change in entropy is given by $s_2 - s_1 = c_p \ln T_2/T_1 - R \ln p_2/p_1$.) Thus, for example, at a given pressure,

$$s - s_1 = c_p \ln T/T_1 \quad \text{and} \quad s_2 - s_1 = c_p \ln T_2/T_1$$

Hence,

$$s - s_1 = \frac{s_2 - s_1}{\ln T_2/T_1} \ln T/T_1$$

$$\frac{s - s_1}{s_2 - s_1} = \frac{\ln T/T_1}{\ln T_2/T_1}$$

As an example, Table A.13 gives the following values of entropy for dry air in Btu/lbm°R for the temperatures and pressures indicated:

p (psia)	600°F	700°F	800°F
100	1.0070	1.0298	1.0510
150	0.9792		
300	0.9311		

For a pressure of 100 psia and a temperature of 700°F, we obtain, by logarithmic interpolation between temperatures of 600 and 800°F,

$$s = s_1 + (s_2 - s_1)\frac{\ln T/T_1}{\ln T_2/T_1}$$

$$= 1.0070 + (1.0510 - 1.0070)\frac{\ln 1159.67/1059.67}{\ln 1259.67/1059.67}$$

$$= 1.0070 + 0.0440\frac{0.09017}{0.17289} = 1.0070 + 0.0229$$

$$= 1.0299 \text{ Btu/lbm°R} \quad \text{(vs. 1.0298 given in the table)}$$

For a temperature of 600°F and a pressure of 150 psia, we obtain, by logarithmic interpolation between pressures of 100 and 300 psia,

$$s = s_1 + (s_2 - s_1)\frac{\ln p/p_1}{\ln p_2/p_1}$$

$$= 1.0070 + (0.9311 - 1.0070)\frac{\ln 150/100}{\ln 300/100}$$

$$= 1.0070 - 0.0759(0.4055)/1.0986 = 1.0070 - 0.0280$$

$$= 0.9790 \text{ Btu/lbm°R} \quad \text{(vs. 0.9792 given in the table)}$$

TABLE A.1 *Properties of Various Gases.*

Gas	Chemical Formula	Molecular Mass	Gas Constant R (ft-lbf/lbm°R)	c_p (Btu/lbm°R)	c_v (Btu/lbm°R)	$\gamma = \frac{c_p}{c_v}$
Acetylene	C_2H_2	26.038	59.35	0.409	0.333	1.23
Air		28.967	53.35	0.240	0.171	1.40
Ammonia	NH_3	17.031	90.74	0.500	0.384	1.30
Argon	Ar	39.948	38.68	0.124	0.074	1.67
Butane	C_4H_{10}	58.124	26.59	0.400	0.367	1.09
Carbon dioxide	CO_2	44.009	35.11	0.202	0.156	1.30
Carbon monoxide	CO	28.010	55.17	0.249	0.178	1.40
Ethane	C_2H_6	30.070	51.39	0.422	0.357	1.18
Ethylene	C_2H_4	28.054	55.08	0.411	0.340	1.21
Hydrogen	H_2	2.016	766.53	3.43	2.44	1.40
Methane	CH_4	16.043	96.32	0.532	0.403	1.32
Nitrogen	N_2	28.014	55.16	0.248	0.177	1.40
Oxygen	O_2	31.998	48.29	0.219	0.156	1.40
Propane	C_3H_8	44.097	35.04	0.404	0.360	1.12
Refrigerant R-12	CCl_2F_2	120.92	12.78	0.140	0.124	1.13
Refrigerant R-22	$CHClF_2$	86.476	17.87	0.150	0.128	1.17
Water vapor	H_2O	18.015	85.78	0.446	0.336	1.33

NOTE: c_p and c_v are for ideal gases at 77°F.

TABLE A.2 *Thermodynamic Properties of Water—Saturated Steam.*

Temp. (°F)	Abs. Press. (psia)	Specific Volume (ft^3/lbm) Sat. Liq.	Evap.	Sat. Vap	Internal Energy (Btu/lbm)* Sat. Liq.	Evap.	Sat. Vap.
T	p	v_f	v_{fg}	v_g	u_f	u_{fg}	u_g
32.018	0.08865	0.016022	3302.4	3302.4	0.000	1021.3	1021.3
40	0.12163	0.016019	2445.8	2445.8	8.027	1015.9	1024.0
50	0.17796	0.016023	1704.8	1704.8	18.053	1009.2	1027.3
60	0.25611	0.016033	1207.6	1207.6	28.059	1002.4	1030.5
70	0.36292	0.016050	868.3	868.4	38.051	995.7	1033.8
80	0.50683	0.016072	633.3	633.3	48.036	989.0	1037.0
90	0.69813	0.016099	468.1	468.1	58.016	982.3	1040.3
100	0.94924	0.016130	350.4	350.4	67.996	975.6	1043.5
110	1.2750	0.016165	265.37	265.39	77.98	968.7	1046.7
120	1.6927	0.016204	203.25	203.26	87.97	962.0	1049.9
130	2.2230	0.016247	157.32	157.33	97.95	955.1	1053.1
140	2.8892	0.016293	122.98	123.00	107.94	948.3	1056.2
150	3.7184	0.016343	97.05	97.07	117.94	941.4	1059.3
160	4.7414	0.016395	77.27	77.29	127.95	934.4	1062.4
170	5.9926	0.016451	62.04	62.06	137.95	927.4	1065.4
180	7.5110	0.016510	50.21	50.22	147.98	920.4	1068.4
190	9.340	0.016572	40.941	40.957	158.01	913.3	1071.3
200	11.526	0.016637	33.622	33.639	168.05	906.2	1074.3
210	14.123	0.016705	27.799	27.816	178.11	898.9	1077.0
212	14.696	0.016719	26.782	26.799	180.12	897.5	1077.6
220	17.186	0.016775	23.131	23.148	188.18	891.6	1079.8
230	20.779	0.016849	19.364	19.381	198.27	884.3	1082.6
240	24.968	0.016926	16.304	16.321	208.37	876.8	1085.2
250	29.825	0.017006	13.802	13.819	218.50	869.2	1087.7
260	35.427	0.017089	11.745	11.762	228.65	861.6	1090.3
270	41.856	0.017175	10.042	10.060	238.82	853.9	1092.7
280	49.200	0.017264	8.627	8.644	249.01	846.1	1095.1
290	57.550	0.01736	7.4430	7.4603	259.2	838.2	1097.4
300	67.005	0.01745	6.4483	6.4658	269.5	830.0	1099.5
310	77.667	0.01755	5.6085	5.6260	279.7	821.9	1101.6
320	89.643	0.01766	4.8961	4.9138	290.1	813.6	1103.7
330	103.045	0.01776	4.2892	4.3069	300.5	805.1	1105.6
340	117.992	0.01787	3.7699	3.7878	310.9	796.5	1107.4
350	134.604	0.01799	3.3238	3.3418	321.4	787.7	1109.1
360	153.010	0.01811	2.9392	2.9573	331.8	778.9	1110.7

* The internal energy and the entropy of saturated water are taken as zero at the triple point (32.018°F).

Temp. (°F)	Enthalpy (Btu/lbm) Sat. Liq.	Evap.	Sat. Vap.	Entropy (Btu/lbm°R)* Sat. Liq.	Evap.	Sat. Vap.
T	h_f	h_{fg}	h_g	s_f	s_{fg}	s_g
32.018	0.0003	1075.5	1075.5	0.0000	2.1872	2.1872
40	8.027	1071.0	1079.0	0.0162	2.1432	2.1594
50	18.054	1065.3	1083.4	0.0361	2.0901	2.1262
60	28.060	1059.7	1087.7	0.0555	2.0391	2.0946
70	38.052	1054.0	1092.1	0.0745	1.9900	2.0645
80	48.037	1048.4	1096.4	0.0932	1.9426	2.0359
90	58.018	1042.7	1100.8	0.1115	1.8970	2.0086
100	67.999	1037.1	1105.1	0.1295	1.8530	1.9825
110	77.98	1031.4	1109.3	0.1472	1.8105	1.9577
120	87.97	1025.6	1113.6	0.1646	1.7693	1.9339
130	97.96	1019.8	1117.8	0.1817	1.7295	1.9112
140	107.95	1014.0	1122.0	0.1985	1.6910	1.8895
150	117.95	1008.2	1126.1	0.2150	1.6536	1.8686
160	127.96	1002.2	1130.2	0.2313	1.6174	1.8487
170	137.97	996.2	1134.2	0.2473	1.5822	1.8295
180	148.00	990.2	1138.2	0.2631	1.5480	1.8111
190	158.04	984.1	1142.1	0.2787	1.5148	1.7934
200	168.09	977.9	1146.0	0.2940	1.4824	1.7764
210	178.15	971.6	1149.7	0.3091	1.4509	1.7600
212	180.17	970.3	1150.5	0.3121	1.4447	1.7568
220	188.23	965.2	1153.4	0.3241	1.4201	1.7442
230	198.33	958.7	1157.1	0.3388	1.3902	1.7290
240	208.45	952.1	1160.6	0.3533	1.3609	1.7142
250	218.59	945.4	1164.0	0.3677	1.3323	1.7000
260	228.76	938.6	1167.4	0.3819	1.3043	1.6862
270	238.95	931.7	1170.6	0.3960	1.2769	1.6729
280	249.17	924.6	1173.8	0.4098	1.2501	1.6599
290	259.4	917.4	1176.8	0.4236	1.2238	1.6473
300	269.7	910.0	1179.7	0.4372	1.1979	1.6351
310	280.0	902.5	1182.5	0.4506	1.1726	1.6232
320	290.4	894.8	1185.2	0.4640	1.1477	1.6116
330	300.8	886.9	1187.7	0.4772	1.1231	1.6003
340	311.3	878.8	1190.1	0.4902	1.0990	1.5892
350	321.8	870.6	1192.3	0.5032	1.0752	1.5784
360	332.2	862.1	1194.4	0.5161	1.0517	1.5678

TABLE A.2 (Continued) *Thermodynamic Properties of Water—Saturated Steam.*

Temp. (°F)	Abs. press. (psia)	Specific Volume (ft³/lbm) Sat. Liq.	Evap.	Sat. Vap.	Internal Energy (Btu/lbm) Sat. Liq.	Evap.	Sat. Vap.
T	p	v_f	v_{fg}	v_g	u_f	u_{fg}	u_g
370	173.339	0.01823	2.6062	2.6244	342.3	769.8	1112.1
380	195.729	0.01836	2.3170	2.3353	352.9	760.5	1113.4
390	220.321	0.01850	2.0649	2.0833	363.5	751.1	1114.7
400	247.259	0.01864	1.8444	1.8630	374.2	741.6	1115.8
410	276.694	0.01878	1.6510	1.6697	385.0	731.6	1116.6
420	308.780	0.01894	1.4808	1.4997	395.8	721.6	1117.4
430	343.674	0.01909	1.3306	1.3496	406.7	711.4	1118.1
440	381.54	0.01926	1.19761	1.21687	417.6	700.8	1118.5
450	422.55	0.01943	1.07962	1.09905	428.7	690.1	1118.8
460	466.87	0.01961	0.97463	0.99424	439.8	679.1	1118.9
470	514.67	0.01980	0.88095	0.90076	451.0	667.8	1118.8
480	566.15	0.02000	0.79716	0.81717	462.4	656.1	1118.5
490	621.48	0.02021	0.72203	0.74224	473.8	644.2	1117.9
500	680.86	0.02043	0.65448	0.67492	485.3	631.8	1117.2
510	744.47	0.02067	0.59362	0.61429	497.1	619.1	1116.2
520	812.53	0.02091	0.53864	0.55956	508.9	606.0	1114.9
530	885.23	0.02118	0.48886	0.51004	520.8	592.5	1113.3
540	962.79	0.02146	0.44367	0.46513	533.0	578.5	1111.4
550	1045.43	0.02176	0.40256	0.42432	545.3	563.8	1109.1
560	1133.38	0.02207	0.36507	0.38714	557.8	548.7	1106.5
570	1226.88	0.02242	0.33079	0.35321	570.5	532.9	1103.4
580	1326.17	0.02279	0.29937	0.32216	583.5	516.4	1099.9
590	1431.5	0.02319	0.27049	0.29368	596.8	499.1	1095.9
600	1543.2	0.02364	0.24384	0.26747	610.3	481.0	1091.3
610	1661.6	0.02412	0.21915	0.24327	624.4	461.7	1086.1
620	1786.9	0.02466	0.19615	0.22081	638.7	441.4	1080.2
630	1919.5	0.02526	0.17459	0.19986	653.7	419.5	1073.2
640	2059.9	0.02595	0.15427	0.18021	669.2	395.8	1065.0
650	2208.4	0.02674	0.13499	0.16173	685.5	369.8	1055.3
660	2365.7	0.02768	0.11663	0.14431	702.8	341.0	1043.8
670	2532.2	0.02884	0.09871	0.12755	722.2	307.7	1030.0
680	2708.6	0.03037	0.08080	0.11117	743.3	269.5	1012.8
690	2895.7	0.03256	0.06203	0.09459	767.1	222.9	989.9
700	3094.3	0.03662	0.03857	0.07519	801.4	150.7	952.1
705.47	3208.2	0.05078	0	0.05078	875.9	0	875.9

critical point →

Temp. (°F)	Enthalpy (Btu/lbm)			Entropy (Btu/lbm°R)		
	Sat. Liq.	Evap.	Sat. Vap.	Sat. Liq.	Evap.	Sat. Vap.
T	h_f	h_{fg}	h_g	s_f	s_{fg}	s_g
370	342.9	853.4	1196.3	0.5289	1.0286	1.5575
380	353.6	844.5	1198.0	0.5416	1.0057	1.5473
390	364.3	835.3	1199.6	0.5542	0.9831	1.5372
400	375.1	825.9	1201.0	0.5667	0.9607	1.5274
410	386.0	816.2	1202.1	0.5791	0.9385	1.5176
420	396.9	806.2	1203.1	0.5915	0.9165	1.5080
430	407.9	796.0	1203.9	0.6038	0.8946	1.4985
440	419.0	785.4	1204.4	0.6161	0.8729	1.4890
450	430.2	774.5	1204.7	0.6283	0.8514	1.4797
460	441.5	763.2	1204.8	0.6405	0.8299	1.4704
470	452.9	751.6	1204.6	0.6527	0.8084	1.4611
480	464.5	739.6	1204.1	0.6648	0.7871	1.4518
490	476.1	727.2	1203.3	0.6769	0.7657	1.4426
500	487.9	714.3	1202.2	0.6890	0.7443	1.4333
510	499.9	700.9	1200.8	0.7012	0.7228	1.4240
520	512.0	687.0	1199.0	0.7133	0.7013	1.4146
530	524.3	672.6	1196.9	0.7255	0.6796	1.4051
540	536.8	657.5	1194.3	0.7378	0.6577	1.3954
550	549.5	641.8	1191.2	0.7501	0.6356	1.3856
560	562.4	625.3	1187.7	0.7625	0.6132	1.3757
570	575.6	608.0	1183.6	0.7750	0.5905	1.3654
580	589.1	589.9	1179.0	0.7876	0.5673	1.3550
590	602.9	570.8	1173.7	0.8004	0.5437	1.3442
600	617.1	550.6	1167.7	0.8134	0.5196	1.3330
610	631.8	529.2	1160.9	0.8267	0.4947	1.3214
620	646.9	506.3	1153.2	0.8403	0.4689	1.3092
630	662.7	481.6	1144.2	0.8542	0.4419	1.2962
640	679.1	454.6	1133.7	0.8686	0.4134	1.2821
650	696.4	425.0	1121.4	0.8837	0.3830	1.2667
660	714.9	392.1	1107.0	0.8995	0.3502	1.2498
670	735.8	354.0	1089.8	0.9174	0.3133	1.2307
680	758.5	310.1	1068.5	0.9365	0.2720	1.2086
690	784.5	256.1	1040.6	0.9583	0.2227	1.1810
700	822.4	172.7	995.2	0.9901	0.1490	1.1390
705.47	906.0	0	906.0	1.0612	0	1.0612

TABLE A.2(a) *Thermodynamic Properties of Water—Saturated Steam.*

Abs. Press. (psia)	Temp. (°F)	Specific Volume (ft³/lbm) Sat. Liq.	Evap.	Sat. Vap.	Internal Energy (Btu/lbm) Sat. Liq.	Evap.	Sat. Vap.
p	T	v_f	v_{fg}	v_g	u_f	u_{fg}	u_g
15.0	213.03	0.016726	26.274	26.290	181.16	896.7	1077.9
14.696	212.00	0.016719	26.782	26.799	180.12	897.5	1077.6
14.5	211.32	0.016714	27.121	27.138	179.44	898.0	1077.4
14.0	209.56	0.016702	28.027	28.043	177.67	899.2	1076.9
13.5	207.75	0.016689	28.997	29.014	175.84	900.6	1076.4
13.0	205.88	0.016676	30.040	30.057	173.96	901.9	1075.9
12.5	203.95	0.016663	31.163	31.180	172.02	903.3	1075.3
12.0	201.96	0.016650	32.377	32.394	170.02	905.3	1075.3
11.5	199.89	0.016636	33.693	33.710	167.94	906.3	1074.2
11.0	197.75	0.016622	35.125	35.142	165.79	907.8	1073.6
10.5	195.52	0.016607	36.689	36.705	163.55	909.4	1072.9
10.0	193.21	0.016592	38.404	38.420	161.23	911.1	1072.3
9.5	190.80	0.016577	40.293	40.310	158.81	912.8	1071.6
9.0	188.27	0.016561	42.385	42.402	156.28	914.5	1070.8
8.5	185.63	0.016545	44.716	44.733	153.63	916.4	1070.0
8.0	182.86	0.016527	47.328	47.345	150.84	918.4	1069.2
7.5	179.93	0.016510	50.277	50.294	147.91	920.5	1068.4
7.0	176.84	0.016491	53.634	53.650	144.81	922.6	1067.4
6.5	173.56	0.016472	57.490	57.506	141.52	925.0	1066.5
6.0	170.05	0.016451	61.967	61.984	138.01	927.4	1065.4
5.5	166.29	0.016430	67.232	67.249	134.24	930.1	1064.3
5.0	162.24	0.016407	73.515	73.532	130.18	932.9	1063.1
4.5	157.81	0.016384	81.185	81.201	125.81	935.9	1061.7
4.0	152.96	0.016358	90.63	90.64	120.90	939.3	1060.2
3.5	147.56	0.016330	102.80	102.82	115.49	943.1	1058.6
3.0	141.47	0.016300	118.71	118.73	109.41	947.3	1056.7
2.5	134.41	0.016267	141.08	141.10	102.35	952.1	1054.5
2.0	126.07	0.016230	173.74	173.76	94.03	957.8	1051.8
1.5	115.62	0.016187	228.66	228.68	83.59	955.0	1048.6
1.0	101.74	0.016136	333.59	333.60	69.73	974.4	1044.1
0.50	79.59	0.016071	641.5	641.5	47.62	989.3	1036.9
0.25	59.32	0.016032	1235.5	1235.5	27.38	1002.9	1030.3
0.10	35.02	0.016020	2945.5	2945.5	3.03	1019.3	1022.3

Abs. Press. (psia)	Enthalpy (Btu/lbm) Sat. Liq.	Evap.	Sat. Vap.	Entropy (Btu/lbm°R) Sat. Liq.	Evap.	Sat. Vap.
p	h_f	h_{fg}	h_g	s_f	s_{fg}	s_g
15.0	181.21	969.7	1150.9	0.3137	1.4415	1.7552
14.696	180.17	970.3	1150.5	0.3121	1.4447	1.7568
14.5	179.48	970.7	1150.2	0.3111	1.4468	1.7579
14.0	177.71	971.9	1149.6	0.3085	1.4522	1.7607
13.5	175.89	973.0	1148.9	0.3058	1.4579	1.7636
13.0	174.00	974.2	1148.2	0.3029	1.4638	1.7667
12.5	172.06	975.4	1147.5	0.3000	1.4698	1.7699
12.0	170.05	976.6	1146.7	0.2970	1.4762	1.7731
11.5	167.98	977.9	1145.9	0.2938	1.4827	1.7766
11.0	165.82	979.3	1145.1	0.2906	1.4896	1.7802
10.5	163.59	980.7	1144.2	0.2872	1.4968	1.7839
10.0	161.26	982.1	1143.3	0.2836	1.5043	1.7879
9.5	158.84	983.6	1142.4	0.2799	1.5121	1.7920
9.0	156.30	985.1	1141.4	0.2760	1.5204	1.7964
8.5	153.65	986.8	1140.4	0.2719	1.5292	1.8011
8.0	150.87	988.5	1139.3	0.2676	1.5384	1.8060
7.5	147.93	990.2	1138.2	0.2630	1.5482	1.8112
7.0	144.83	992.1	1136.9	0.2581	1.5587	1.8168
6.5	141.54	994.1	1135.6	0.2530	1.5699	1.8229
6.0	138.03	996.2	1134.2	0.2474	1.5820	1.8294
5.5	134.26	998.5	1132.7	0.2414	1.5951	1.8365
5.0	130.20	1000.9	1131.1	0.2349	1.6094	1.8443
4.5	125.77	1003.5	1129.3	0.2277	1.6253	1.8530
4.0	120.92	1006.4	1127.3	0.2199	1.6428	1.8626
3.5	115.51	1009.6	1125.1	0.2110	1.6627	1.8737
3.0	109.42	1013.2	1122.6	0.2009	1.6854	1.8864
2.5	102.36	1017.3	1119.7	0.1891	1.7124	1.9015
2.0	94.03	1022.1	1116.2	0.1750	1.7450	1.9200
1.5	83.59	1028.1	1111.7	0.1570	1.7872	1.9442
1.0	69.73	1036.1	1105.8	0.1326	1.8455	1.9781
0.50	47.62	1048.6	1096.3	0.0925	1.9446	2.0370
0.25	27.38	1060.1	1087.4	0.0542	2.0425	2.0967
0.10	3.03	1073.8	1076.8	0.0061	2.1705	2.1766

TABLE A.3 *Thermodynamic Properties of Water—Superheated Steam.*

Specific Volume, v (ft^3/lbm)

Abs. Press. (psia)	Sat. Temp. (°F)	Sat. Spec. Vol.	Temperature (°F)			
p	T_g	v_g	100	200	300	400
0.2	53.16	1526.3	1666.1	1964.3	2262.3	2560.3
0.5	79.59	641.5	665.9	785.5	904.8	1024.0
1	101.74	333.6		392.5	452.3	511.9
2	126.07	173.76		196.03	225.99	255.87
5	162.24	73.63		78.14	90.24	102.24
10	193.21	38.42		38.84	44.98	51.03
15	213.03	26.290			29.899	33.963
20	227.96	20.087			22.356	25.428
40	267.25	10.497			11.036	12.624
60	292.71	7.174			7.257	8.354
80	312.04	5.471				6.218
100	327.82	4.431				4.935
120	341.27	3.7275				4.0786
140	353.04	3.2190				3.4661
160	363.55	2.8336				3.0060
180	373.08	2.5312				2.6474
200	381.80	2.2873				2.3598
250	400.97	1.8432				
300	417.35	1.5427				
350	431:73	1.3255				
400	444.60	1.1610				
500	467.01	0.9276				
600	486.20	0.7697				
700	503.08	0.6556				
800	518.21	0.5690				
900	531.95	0.5009				
1000	544.58	0.4460				
1500	596.20	0.2772				
2000	635.80	0.1883				
2500	668.11	0.1307				
3000	695.33	0.0850				
3500						
4000						
5000						
6000						
8000						
10,000						
12,500						
15,000						

Specific Volume, v (ft^3/lbm)

Abs. Press. (psia)	Temperature (°F)					
p	500	600	700	800	900	1000
0.2	2858.1	3156.0	3453.9	3751.9	4049.7	4347.5
0.5	1143.2	1262.3	1381.5	1500.8	1619.9	1739.0
1	571.5	631.1	690.7	750.4	809.9	869.5
2	285.70	315.52	345.32	375.12	404.91	434.71
5	114.21	126.15	138.08	150.01	161.94	173.86
10	57.04	63.03	69.00	74.98	80.94	86.91
15	37.985	41.986	45.978	49.964	53.946	57.926
20	28.457	31.466	34.465	37.458	40.447	43.435
40	14.165	15.685	17.195	18.699	20.199	21.697
60	9.400	10.425	11.438	12.446	13.450	14.452
80	7.018	7.794	8.560	9.319	10.075	10.829
100	5.588	6.216	6.833	7.443	8.050	8.655
120	4.6341	5.1637	5.6813	6.1928	6.7006	7.2060
140	3.9526	4.4119	4.8588	5.2995	5.7364	6.1709
160	3.4413	3.8480	4.2420	4.6295	5.0132	5.3945
180	3.0433	3.4093	3.7621	4.1084	4.4508	4.7907
200	2.7247	3.0583	3.3783	3.6915	4.0008	4.3077
250	2.1504	2.4262	2.6872	2.9410	3.1909	3.4382
300	1.7665	2.0044	2.2263	2.4407	2.6509	2.8585
350	1.5483	1.7028	1.8970	2.0832	2.2652	2.4445
400	1.2841	1.4763	1.6499	1.8151	1.9759	2.1339
500	0.9919	1.1584	1.3037	1.4397	1.5708	1.6992
600	0.7944	0.9456	1.0726	1.1892	1.3008	1.4093
700		0.7928	0.9072	1.0102	1.1078	1.2023
800		0.6774	0.7828	0.8759	0.9631	1.0470
900		0.5869	0.6858	0.7713	0.8504	0.9262
1000		0.5137	0.6080	0.6875	0.7603	0.8295
1500		0.2820	0.3717	0.4350	0.4894	0.5394
2000			0.2488	0.3072	0.3534	0.3942
2500			0.1681	0.2293	0.2712	0.3068
3000			0.0982	0.1759	0.2161	0.2484
3500			0.0307	0.1364	0.1764	0.2066
4000			0.0287	0.1052	0.1463	0.1752
5000			0.0268	0.0591	0.1038	0.1312
6000			0.0256	0.0397	0.0757	0.1020
8000			0.0242	0.0306	0.0465	0.0671
10,000			0.0233	0.0276	0.0362	0.0495
12,500			0.0224	0.0256	0.0309	0.0390
15,000			0.0218	0.0244	0.0288	0.0337

TABLE A.3 (Continued) *Thermodynamic Properties of Water—Superheated Steam.*

Specific Volume, v (ft^3/lbm)

Abs. Press. (psia) p	Temperature (°F) 1100	1200	1300	1400	1500	1600
0.2	4645.3	4943.0	5240.8	5538.6	5836.4	6135.7
0.5	1858.1	1977.2	2096.3	2215.4	2334.5	2454.3
1	929.1	988.6	1048.2	1107.7	1167.3	1227.1
2	464.50	494.29	524.08	553.86	583.65	613.57
5	185.78	197.70	209.62	221.53	233.45	245.43
10	92.87	98.84	104.80	110.76	116.72	122.71
15	61.905	65.882	69.858	73.833	77.807	81.809
20	46.420	49.405	52.388	55.370	58.352	61.357
40	23.194	24.689	26.183	27.676	29.168	30.678
60	15.452	16.450	17.448	18.445	19.441	20.450
80	11.581	12.331	13.081	13.829	14.577	15.325
100	9.258	9.860	10.460	11.060	11.659	12.258
120	7.7096	8.2119	8.7130	9.2134	9.7130	10.213
140	6.6036	7.0349	7.4652	7.8946	8.3233	8.752
160	5.7741	6.1522	6.5293	6.9055	7.2811	7.656
180	5.1289	5.4657	5.8014	6.1363	6.4704	6.804
200	4.6128	4.9165	5.2191	5.5209	5.8219	6.123
250	3.6837	3.9278	4.1709	4.4131	4.6546	4.896
300	3.0643	3.2688	3.4721	3.6746	3.8764	4.078
350	2.6219	2.7980	2.9730	3.1471	3.3205	3.493
400	2.2901	2.4450	2.5987	2.7515	2.9037	3.055
500	1.8256	1.9507	2.0746	2.1977	2.3200	2.442
600	1.5160	1.6211	1.7252	1.8284	1.9309	2.033
700	1.2948	1.3858	1.4757	1.5647	1.6530	1.7405
800	1.1289	1.2093	1.2885	1.3669	1.4446	1.5214
900	0.9998	1.0720	1.1430	1.2131	1.2825	1.3509
1000	0.8966	0.9622	1.0266	1.0901	1.1529	1.2146
1500	0.5869	0.6327	0.6773	0.7210	0.7639	0.8056
2000	0.4320	0.4680	0.5027	0.5365	0.5695	0.6011
2500	0.3390	0.3692	0.3980	0.4259	0.4529	0.4784
3000	0.2770	0.3033	0.3282	0.3522	0.3753	0.3966
3500	0.2326	0.2563	0.2784	0.2995	0.3198	0.3381
4000	0.1994	0.2210	0.2410	0.2601	0.2783	0.2943
5000	0.1529	0.1718	0.1890	0.2050	0.2203	0.2348
6000	0.1221	0.1391	0.1544	0.1684	0.1817	0.19420
8000	0.0845	0.0989	0.1115	0.1230	0.1338	0.14372
10,000	0.0633	0.0757	0.0865	0.0963	0.1054	0.11378
12,500	0.0486	0.0583	0.0673	0.0756	0.0832	0.09029
15,000	0.0405	0.0479	0.0552	0.0624	0.0690	0.07507

Internal Energy, u (Btu/lbm)

Abs. Press. (psia)	Sat. Temp. (°F)	Sat. Int. Ener.	Temperature (°F)			
p	T_g	u_g	100	200	300	400
0.2	53.16	1028.2	1043.9	1077.8	1112.2	1147.1
0.5	79.59	1036.9	1043.8	1077.6	1112.1	1147.1
1	101.74	1044.1		1077.6	1112.0	1147.1
2	126.07	1051.9		1077.2	1111.9	1147.0
5	162.24	1063.0		1076.3	1111.3	1146.7
10	193.21	1072.2		1074.7	1110.5	1146.2
15	213.03	1077.9			1109.5	1145.6
20	227.96	1082.0			1108.7	1145.1
40	267.25	1092.1			1104.9	1143.0
60	292.71	1097.9			1101.0	1140.7
80	312.04	1102.1				1138.4
100	327.82	1105.2				1136.1
120	341.27	1107.6				1133.5
140	353.04	1109.6				1131.0
160	363.55	1111.2				1128.4
180	373.08	1112.6				1125.6
200	381.80	1113.6				1122.8
250	400.97	1115.8				
300	417.35	1117.3				
350	431.73	1118.2				
400	444.60	1118.7				
500	467.01	1118.9				
600	486.20	1118.2				
700	503.08	1116.9				
800	518.21	1115.2				
900	531.95	1113.0				
1000	544.58	1110.4				
1500	596.20	1093.2				
2000	635.80	1068.6				
2500	668.11	1032.8				
3000	695.33	973.1				
3500						
4000						
5000						
6000						
8000						
10,000						
12,500						
15,000						

TABLE A.3 (Continued) *Thermodynamic Properties of Water—Superheated Steam.*

Internal Energy, *u* (Btu/lbm)

Abs. Press. (psia)	Temperature (°F)					
p	500	600	700	800	900	1000
0.2	1182.9	1219.4	1256.7	1294.8	1333.9	1373.9
0.5	1182.8	1219.4	1256.7	1294.8	1333.9	1373.9
1	1182.8	1219.3	1256.7	1294.8	1333.9	1373.9
2	1182.8	1219.3	1256.6	1294.8	1333.9	1373.9
5	1182.5	1219.2	1256.5	1294.8	1333.9	1373.8
10	1182.2	1218.9	1256.3	1294.6	1333.7	1373.8
15	1181.9	1218.7	1256.2	1294.5	1333.6	1373.7
20	1181.6	1218.4	1255.9	1294.3	1333.5	1373.5
40	1180.2	1217.5	1255.2	1293.7	1333.0	1373.1
60	1178.8	1216.6	1254.5	1293.1	1331.8	1372.7
80	1177.4	1215.5	1253.8	1292.5	1331.9	1372.3
100	1175.9	1214.6	1253.1	1292.0	1331.4	1371.8
120	1174.5	1213.5	1252.2	1291.3	1331.0	1371.4
140	1172.9	1212.5	1251.5	1290.7	1330.5	1370.9
160	1171.4	1211.5	1250.8	1290.1	1330.0	1370.6
180	1169.8	1210.4	1250.0	1289.5	1329.6	1370.1
200	1168.2	1209.4	1249.3	1288.9	1328.9	1369.7
250	1164.0	1206.8	1247.3	1287.3	1327.7	1368.5
300	1159.6	1203.9	1245.3	1285.8	1326.4	1367.5
350	1151.2	1201.1	1243.3	1284.3	1325.1	1366.4
400	1150.1	1198.1	1241.3	1282.6	1323.8	1365.3
500	1139.4	1191.9	1237.1	1279.5	1321.3	1363.1
600	1127.7	1185.3	1232.7	1276.3	1318.6	1360.9
700		1178.3	1228.1	1272.8	1315.9	1358.7
800		1170.8	1223.4	1269.4	1313.2	1356.4
900		1162.9	1218.5	1265.9	1310.6	1354.2
1000		1154.2	1213.4	1262.4	1307.8	1351.9
1500		1098.0	1184.7	1243.3	1293.4	1340.4
2000			1148.8	1221.7	1277.9	1328.2
2500			1098.9	1197.3	1261.2	1315.6
3000			1006.0	1169.3	1243.2	1302.3
3500			759.5	1136.3	1224.0	1288.4
4000			741.8	1096.4	1203.3	1273.9
5000			721.2	988.2	1156.9	1243.2
6000			707.7	901.0	1104.8	1210.3
8000			688.5	833.8	1005.5	1141.7
10,000			674.4	803.3	944.3	1081.0
12,500			669.8	779.4	900.4	1025.0
15,000			649.5	761.8	871.0	987.1

Internal Energy, u (Btu/lbm)

Abs. Press (psia)	Temperature (°F)					
p	1100	1200	1300	1400	1500	1600
0.2	1414.9	1456.7	1499.4	1543.0	1587.5	1632.6
0.5	1414.9	1456.7	1499.4	1543.0	1587.5	1632.6
1	1414.9	1456.7	1499.4	1543.0	1587.5	1632.6
2	1414.9	1456.7	1499.4	1543.0	1587.5	1632.6
5	1414.8	1456.7	1499.3	1543.0	1587.5	1632.6
10	1414.7	1456.6	1499.3	1542.9	1587.4	1632.5
15	1414.6	1456.5	1499.3	1542.9	1587.4	1632.5
20	1414.5	1456.5	1499.2	1542.9	1587.3	1632.4
40	1414.1	1456.1	1498.9	1542.6	1587.1	1632.2
60	1413.7	1455.8	1498.7	1542.3	1586.9	1631.9
80	1413.5	1455.5	1498.3	1542.1	1586.7	1631.7
100	1413.1	1455.1	1498.0	1541.8	1586.5	1631.4
120	1412.7	1454.7	1497.8	1541.6	1586.3	1631.5
140	1412.3	1454.4	1497.5	1541.4	1586.1	1631.3
160	1411.9	1454.1	1497.2	1541.1	1585.8	1631.1
180	1411.6	1453.8	1497.0	1540.9	1585.7	1630.9
200	1411.2	1453.4	1496.6	1540.7	1585.4	1630.7
250	1410.2	1452.7	1495.9	1540.0	1584.9	1630.2
300	1409.3	1451.8	1495.2	1539.4	1584.4	1629.6
350	1408.4	1451.1	1494.5	1538.8	1583.8	1629.2
400	1407.4	1450.2	1493.8	1538.2	1583.3	1628.7
500	1405.5	1448.6	1492.4	1537.0	1582.2	1627.7
600	1403.6	1447.0	1491.1	1535.8	1581.2	1626.8
700	1401.7	1445.3	1489.5	1534.5	1580.2	1625.7
800	1399.8	1443.7	1488.2	1533.3	1579.0	1624.8
900	1397.9	1442.1	1486.7	1532.1	1578.0	1623.8
1000	1396.0	1440.3	1485.3	1530.8	1577.0	1621.3
1500	1386.3	1432.1	1478.2	1524.7	1571.7	1617.9
2000	1376.3	1423.7	1471.0	1518.4	1566.3	1612.9
2500	1366.1	1415.1	1463.7	1512.2	1560.6	1608.0
3000	1355.6	1406.4	1456.3	1505.9	1555.5	1603.1
3500	1344.9	1397.6	1448.9	1499.6	1550.1	1597.7
4000	1333.7	1388.6	1441.4	1493.2	1544.6	1593.4
5000	1310.6	1370.1	1426.0	1480.3	1533.6	1582.7
6000	1286.7	1351.5	1410.6	1467.2	1522.5	1571.7
8000	1237.1	1313.2	1379.4	1441.0	1500.0	1553.1
10,000	1188.2	1275.2	1348.5	1414.9	1477.8	1534.5
12,500	1135.5	1230.5	1311.5	1383.3	1450.6	1511.2
15,000	1094.4	1193.0	1277.1	1353.2	1424.4	1488.6

TABLE A.3 (Continued) *Thermodynamic Properties of Water—Superheated Steam.*

Enthalpy, h (Btu/lbm)

Abs. Press. (psia)	Sat. Temp. (°F)	Sat. Enth.	Temperature (°F)			
p	T_g	h_g	100	200	300	400
0.2	53.16	1084.7	1105.6	1150.5	1195.9	1241.9
0.5	79.59	1096.3	1105.4	1150.3	1195.8	1241.9
1	101.74	1105.8		1150.2	1195.7	1241.8
2	126.07	1116.2		1149.8	1195.5	1241.7
5	162.24	1131.1		1148.6	1194.8	1241.3
10	193.21	1143.3		1146.6	1193.7	1240.6
15	213.03	1150.9			1192.5	1239.9
20	227.96	1156.3			1191.4	1239.2
40	267.25	1169.8			1186.6	1236.4
60	292.71	1177.6			1181.6	1233.5
80	312.04	1183.1				1230.5
100	327.82	1187.2				1227.4
120	341.27	1190.4				1224.1
140	353.04	1193.0				1220.8
160	363.55	1195.1				1217.4
180	373.08	1196.9				1213.8
200	381.80	1198.3				1210.1
250	400.97	1201.1				
300	417.35	1202.9				
350	431.73	1204.0				
400	444.60	1204.6				
500	467.01	1204.7				
600	486.20	1203.7				
700	503.08	1201.8				
800	518.21	1199.4				
900	531.95	1196.4				
1000	544.58	1192.9				
1500	596.20	1170.1				
2000	635.80	1138.3				
2500	668.11	1093.3				
3000	695.33	1020.3				
3500						
4000						
5000						
6000						
8000						
10,000						
12,500						
15,000						

Enthalpy, h (Btu/lbm)

Abs. Press. (psia)	Temperature (°F)					
p	500	600	700	800	900	1000
0.2	1288.7	1336.2	1384.5	1433.7	1483.8	1534.8
0.5	1288.6	1336.2	1384.5	1433.7	1483.8	1534.8
1	1288.6	1336.1	1384.5	1433.7	1483.8	1534.8
2	1288.5	1336.1	1384.4	1433.7	1483.8	1534.8
5	1288.2	1335.9	1384.3	1433.6	1483.7	1534.7
10	1287.8	1335.5	1384.0	1433.4	1483.5	1534.6
15	1287.3	1335.2	1383.8	1433.2	1483.4	1534.5
20	1286.9	1334.9	1383.5	1432.9	1483.2	1534.3
40	1285.0	1333.6	1382.5	1432.1	1482.5	1533.7
60	1283.2	1332.3	1381.5	1431.3	1481.1	1533.2
80	1281.3	1330.9	1380.5	1430.5	1481.1	1532.6
100	1279.3	1329.6	1379.5	1429.7	1480.4	1532.0
120	1277.4	1328.2	1378.4	1428.8	1479.8	1531.4
140	1275.3	1326.8	1377.4	1428.0	1479.1	1530.8
160	1273.3	1325.4	1376.4	1427.2	1478.4	1530.3
180	1271.2	1324.0	1375.3	1426.3	1477.7	1529.7
200	1269.0	1322.6	1374.3	1425.5	1477.0	1529.1
250	1263.5	1319.0	1371.6	1423.4	1475.3	1527.6
300	1257.7	1315.2	1368.9	1421.3	1473.6	1526.2
350	1251.5	1311.4	1366.2	1419.2	1471.8	1524.7
400	1245.1	1307.4	1363.4	1417.0	1470.1	1523.3
500	1231.2	1299.1	1357.7	1412.7	1466.6	1520.3
600	1215.9	1290.3	1351.8	1408.3	1463.0	1517.4
700		1281.0	1345.6	1403.7	1459.4	1514.4
800		1271.1	1339.3	1399.1	1455.8	1511.4
900		1260.6	1332.7	1394.4	1452.2	1508.5
1000		1249.3	1325.9	1389.6	1448.5	1505.4
1500		1176.3	1287.9	1364.0	1429.2	1490.1
2000			1240.9	1335.4	1408.7	1474.1
2500			1176.7	1303.4	1386.7	1457.5
3000			1060.5	1267.0	1363.2	1440.2
3500			779.4	1224.6	1338.2	1422.2
4000			763.0	1174.3	1311.6	1403.6
5000			746.0	1042.9	1252.9	1364.6
6000			736.1	945.1	1188.8	1323.6
8000			724.3	879.1	1074.3	1241.0
10,000			717.5	854.5	1011.3	1172.6
12,500			712.6	838.6	971.9	1115.2
15,000			710.0	829.5	950.9	1080.6

TABLE A.3 (Continued) *Thermodynamic Properties of Water—Superheated Steam.*

Enthalpy, h (Btu/lbm)

Abs. Press (psia)	Temperature (°F)					
p	1100	1200	1300	1400	1500	1600
0.2	1586.8	1639.6	1693.4	1748.0	1803.5	1859.7
0.5	1586.8	1639.6	1693.4	1748.0	1803.5	1859.7
1	1586.8	1639.6	1693.4	1748.0	1803.5	1859.7
2	1586.8	1639.6	1693.4	1748.0	1803.5	1859.7
5	1586.7	1639.6	1693.3	1748.0	1803.5	1859.7
10	1586.6	1639.5	1693.3	1747.9	1803.4	1859.6
15	1586.5	1639.4	1693.2	1747.8	1803.4	1859.6
20	1586.3	1639.3	1693.1	1747.8	1803.3	1859.5
40	1585.8	1638.8	1692.7	1747.5	1803.0	1859.3
60	1585.3	1638.4	1692.4	1747.1	1802.8	1859.0
80	1584.9	1638.0	1692.0	1746.8	1802.5	1858.8
100	1584.4	1637.6	1691.6	1746.5	1802.2	1858.5
120	1583.9	1637.1	1691.3	1746.2	1802.0	1858.3
140	1583.4	1636.7	1690.9	1745.9	1801.7	1858.0
160	1582.9	1636.3	1690.5	1745.6	1801.4	1857.8
180	1582.4	1635.9	1690.2	1745.3	1801.2	1857.5
200	1581.9	1635.4	1689.8	1745.0	1800.9	1857.3
250	1580.6	1634.4	1688.9	1744.2	1800.2	1856.7
300	1579.4	1633.3	1688.0	1743.4	1799.6	1856.0
350	1578.2	1632.3	1687.1	1742.6	1798.9	1855.4
400	1576.9	1631.2	1686.2	1741.9	1798.2	1854.8
500	1574.4	1629.1	1684.4	1740.3	1796.9	1853.6
600	1571.9	1627.0	1682.6	1738.8	1795.6	1852.5
700	1569.4	1624.8	1680.7	1737.2	1794.3	1851.2
800	1566.9	1622.7	1678.9	1735.7	1792.9	1850.0
900	1564.4	1620.6	1677.1	1734.1	1791.6	1848.8
1000	1561.9	1618.4	1675.3	1732.5	1790.3	1846.1
1500	1549.2	1607.7	1666.2	1724.8	1783.7	1841.5
2000	1536.2	1596.9	1657.0	1717.0	1777.1	1835.4
2500	1522.9	1585.9	1647.8	1709.2	1770.4	1829.3
3000	1509.4	1574.8	1638.5	1701.4	1763.8	1823.3
3500	1495.5	1563.6	1629.2	1693.6	1757.2	1816.7
4000	1481.3	1552.2	1619.8	1685.7	1750.6	1811.2
5000	1452.1	1529.1	1600.9	1670.0	1737.4	1798.2
6000	1422.3	1505.9	1582.0	1654.2	1724.2	1787.3
8000	1362.2	1459.6	1544.5	1623.1	1698.1	1765.9
10,000	1305.3	1415.3	1508.6	1593.1	1672.8	1745.0
12,500	1247.9	1365.4	1467.2	1558.2	1643.1	1720.1
15,000	1206.8	1326.0	1430.3	1526.4	1615.9	1697.0

Entropy, s (Btu/lbm°R)

Abs. Press. (psia)	Sat. Temp. (°F)	Sat. Entropy	Temperature (°F)			
p	T_g	s_g	100	200	300	400
0.2	53.16	2.1160	2.1550	2.2287	2.2928	2.3497
0.5	79.59	2.0370	2.0537	2.1276	2.1917	2.2487
1	101.74	1.9781		2.0509	2.1152	2.1722
2	126.07	1.9200		1.9740	2.0385	2.0957
5	162.24	1.8443		1.8716	1.9369	1.9943
10	193.21	1.7879		1.7928	1.8593	1.9173
15	213.03	1.7552			1.8134	1.8720
20	227.96	1.7320			1.7805	1.8397
40	267.25	1.6765			1.6992	1.7608
60	292.71	1.6440			1.6492	1.7134
80	312.04	1.6208				1.6790
100	327.82	1.6027				1.6516
120	341.27	1.5879				1.6286
140	353.04	1.5752				1.6085
160	363.55	1.5641				1.5906
180	373.08	1.5543				1.5743
200	381.80	1.5454				1.5593
250	400.97	1.5264				
300	417.35	1.5105				
350	431.73	1.4968				
400	444.60	1.4847				
500	467.01	1.4639				
600	486.20	1.4461				
700	503.08	1.4304				
800	518.21	1.4163				
900	531.95	1.4032				
1000	544.58	1.3910				
1500	596.20	1.3373				
2000	635.80	1.2881				
2500	668.11	1.2345				
3000	695.33	1.1619				
3500						
4000						
5000						
6000						
8000						
10,000						
12,500						
15,000						

TABLE A.3 (Continued) *Thermodynamic Properties of Water—Superheated Steam.*

Entropy, *s* (Btu/lbm°R)

Abs. Press. (psia)	Temperature (°F)					
p	500	600	700	800	900	1000
0.2	2.4011	2.4482	2.4918	2.5326	2.5643	2.6006
0.5	2.3001	2.3472	2.3908	2.4316	2.4699	2.5062
1	2.2237	2.2708	2.3144	2.3552	2.3935	2.4298
2	2.1472	2.1943	2.2380	2.2787	2.3170	2.3532
5	2.0460	2.0932	2.1369	2.1776	2.2159	2.2521
10	1.9692	2.0166	2.0603	2.1011	2.1394	2.1757
15	1.9242	1.9717	2.0155	2.0563	2.0946	2.1309
20	1.8921	1.9397	1.9836	2.0244	2.0628	2.0991
40	1.8143	1.8624	1.9065	1.9476	1.9860	2.0224
60	1.7681	1.8168	1.8612	1.9024	1.9410	1.9774
80	1.7349	1.7842	1.8289	1.8702	1.9089	1.9454
100	1.7088	1.7586	1.8036	1.8451	1.8839	1.9205
120	1.6872	1.7376	1.7829	1.8246	1.8635	1.9001
140	1.6686	1.7196	1.7652	1.8071	1.8461	1.8828
160	1.6522	1.7039	1.7499	1.7919	1.8310	1.8678
180	1.6376	1.6900	1.7362	1.7784	1.8176	1.8545
200	1.6242	1.6773	1.7239	1.7663	1.8057	1.8426
250	1.5951	1.6502	1.6976	1.7405	1.7801	1.8173
300	1.5703	1.6274	1.6758	1.7192	1.7591	1.7964
350	1.5483	1.6077	1.6571	1.7009	1.7411	1.7787
400	1.5282	1.5901	1.6406	1.6850	1.7255	1.7632
500	1.4921	1.5595	1.6123	1.6578	1.6990	1.7371
600	1.4590	1.5329	1.5884	1.6351	1.6769	1.7155
700		1.5090	1.5673	1.6154	1.6580	1.6970
800		1.4869	1.5484	1.5980	1.6413	1.6807
900		1.4659	1.5311	1.5822	1.6263	1.6662
1000		1.4457	1.5149	1.5677	1.6126	1.6530
1500		1.3431	1.4443	1.5073	1.5572	1.6004
2000			1.3794	1.4578	1.5138	1.5603
2500			1.3076	1.4129	1.4766	1.5269
3000			1.1966	1.3692	1.4429	1.4976
3500			0.9508	1.3242	1.4112	1.4709
4000			0.9343	1.2754	1.3807	1.4461
5000			0.9153	1.1593	1.3207	1.4001
6000			0.9026	1.0746	1.2615	1.3574
8000			0.8844	1.0122	1.1613	1.2798
10,000			0.8710	0.9842	1.1039	1.2185
12,500			0.8576	0.9618	1.0637	1.1653
15,000			0.8466	0.9455	1.0382	1.1302

Entropy, s (Btu/lbm°R)

Abs. Press. (psia)	Temperature (°F)					
p	1100	1200	1300	1400	1500	1600
0.2	2.6350	2.6679	2.6993	2.7295	2.7586	2.7929
0.5	2.5406	2.5735	2.6049	2.6351	2.6642	2.6919
1	2.4642	2.4971	2.5285	2.5588	2.5878	2.6155
2	2.3876	2.4204	2.4519	2.4821	2.5112	2.5391
5	2.2866	2.3194	2.3509	2.3811	2.4101	2.4381
10	2.2101	2.2430	2.2744	2.3046	2.3337	2.3616
15	2.1653	2.1982	2.2297	2.2599	2.2890	2.3169
20	2.1336	2.1665	2.1979	2.2282	2.2572	2.2852
40	2.0569	2.0899	2.1214	2.1516	2.1807	2.2087
60	2.0120	2.0450	2.0765	2.1068	2.1359	2.1639
80	1.9800	2.0131	2.0446	2.0750	2.1041	2.1321
100	1.9552	1.9883	2.0199	2.0502	2.0794	2.1074
120	1.9349	1.9680	1.9996	2.0300	2.0592	2.0872
140	1.9176	1.9508	1.9825	2.0129	2.0421	2.0701
160	1.9027	1.9359	1.9676	1.9980	2.0273	2.0553
180	1.8894	1.9227	1.9545	1.9849	2.0142	2.0422
200	1.8776	1.9109	1.9427	1.9732	2.0025	2.0305
250	1.8524	1.8858	1.9177	1.9482	1.9776	2.0057
300	1.8317	1.8652	1.8972	1.9278	1.9572	1.9853
350	1.8141	1.8477	1.8798	1.9105	1.9400	1.9681
400	1.7988	1.8325	1.8647	1.8955	1.9250	1.9531
500	1.7730	1.8069	1.8393	1.8702	1.8998	1.9280
600	1.7517	1.7859	1.8184	1.8494	1.8792	1.9074
700	1.7335	1.7679	1.8006	1.8318	1.8617	1.8899
800	1.7175	1.7522	1.7851	1.8164	1.8464	1.8747
900	1.7033	1.7382	1.7713	1.8028	1.8329	1.8613
1000	1.6905	1.7256	1.7589	1.7905	1.8207	1.8492
1500	1.6395	1.6759	1.7101	1.7425	1.7734	1.8020
2000	1.6014	1.6391	1.6743	1.7075	1.7389	1.7678
2500	1.5703	1.6094	1.6456	1.6796	1.7116	1.7409
3000	1.5434	1.5841	1.6214	1.6561	1.6888	1.7184
3500	1.5194	1.5618	1.6002	1.6358	1.6691	1.6989
4000	1.4976	1.5417	1.5812	1.6177	1.6516	1.6818
5000	1.4582	1.5061	1.5481	1.5863	1.6216	1.6523
6000	1.4229	1.4748	1.5194	1.5593	1.5960	1.6279
8000	1.3603	1.4208	1.4705	1.5140	1.5533	1.5876
10,000	1.3065	1.3749	1.4295	1.4763	1.5180	1.5545
12,500	1.2534	1.3264	1.3860	1.4363	1.4808	1.5197
15,000	1.2139	1.2880	1.3491	1.4022	1.4491	1.4900

TABLE A.4 *Thermodynamic Properties of Water—Compressed Water.*

Abs. Press. (psia)	Temperature (°F)			
p	32	100	200	300
	Specific Volume, v (ft³/lbm)			
0.2	0.01602			
1	0.01602	0.01613		
10	0.01602	0.01613		
100	0.01602	0.01613	0.01663	0.01745
500	0.01599	0.01611	0.01661	0.01742
1000	0.01597	0.01608	0.01658	0.01738
2000	0.01591	0.01603	0.01653	0.01731
3000	0.01586	0.01599	0.01648	0.01724
4000	0.0158	0.0159	0.0164	0.0172
	Internal Energy, u (Btu/lbm)			
0.2	0.00			
1	0.00	68.00		
10	0.00	67.99		
100	0.01	67.96	167.99	269.45
500	0.04	67.83	167.65	268.90
1000	0.06	67.65	167.26	268.22
2000	0.12	67.33	166.48	266.91
3000	0.17	67.00	165.73	265.65
4000	0.22	66.73	165.1	264.44
	Enthalpy, h (Btu/lbm)			
0.2	0.00			
1	0.00	68.00		
10	0.03	68.02		
100	0.31	68.26	168.29	269.77
500	1.52	69.32	169.19	270.51
1000	3.02	70.63	170.33	271.44
2000	6.01	73.26	172.60	273.32
3000	8.97	75.88	174.88	275.22
4000	11.92	78.5	177.2	277.1
	Entropy, s (Btu/lbm°R)			
0.2	0.0000			
1	0.0000	0.1295		
10	0.0000	0.1295		
100	0.0000	0.1295	0.2939	0.4371
500	0.0000	0.1292	0.2934	0.4364
1000	0.0001	0.1289	0.2928	0.4355
2000	0.0002	0.1283	0.2916	0.4337
3000	0.0002	0.1277	0.2904	0.4320
4000	0.0002	0.1271	0.2893	0.4304

Abs. Press. (psia)	Temperature (°F)			Sat. Temp. (°F)	Sat. Liq.
p	400	500	600	T_f	
		Specific Volume, v (ft³/lbm)			v_f
0.2				53.16	0.01603
1				101.74	0.01614
10				193.21	0.01659
100				327.82	0.01774
500	0.01861			467.01	0.01975
1000	0.01855	0.02036		544.58	0.02159
2000	0.01844	0.02014	0.02332	635.80	0.02565
3000	0.01833	0.01995	0.02276	695.33	0.03428
4000	0.0182	0.0198	0.0223		
		Internal Energy, u (Btu/lbm)			u_f
0.2				53.16	21.22
1				101.74	69.73
10				193.21	161.23
100				327.82	298.21
500	373.66			467.01	447.69
1000	372.53	484.02		544.58	538.55
2000	370.37	480.08	605.85	635.80	662.62
3000	368.29	476.44	603.26	695.33	782.81
4000	366.3	473.0	590.4		
		Enthalpy, h (Btu/lbm)			h_f
0.2				53.16	21.22
1				101.74	69.73
10				193.21	161.26
100				327.82	298.54
500	375.38			467.01	449.52
1000	375.96	487.79		544.58	542.55
2000	377.19	487.53	614.48	635.80	672.11
3000	378.47	487.52	610.08	695.33	801.84
4000	379.8	487.7	606.9		
		Entropy, s (Btu/lbm°R)			s_f
0.2				53.16	0.0422
1				101.74	0.1326
10				193.21	0.2836
100				327.82	0.4743
500	0.5660			467.01	0.6490
1000	0.5647	0.6876		544.58	0.7434
2000	0.5621	0.6834	0.8091	635.80	0.8625
3000	0.5597	0.6796	0.8009	695.33	0.9728
4000	0.5573	0.6760	0.7940		

TABLE A.5 *Thermodynamic Properties of Water—Saturated Ice-Vapor.*

Temp.(°F)	Abs. Press. (psia)	Specific Volume (ft^3/lbm) Sat. Ice	Subl.	Sat. Vap.	Internal Energy (Btu/lbm) Sat. Ice	Subl.	Sat. Vap.
T	p	v_i	v_{ig}	v_g	u_i	u_{ig}	u_g
32.018	0.08865	0.01747	3302	3302	−143.35	1164.6	1021.3
30	0.0808	0.01747	3609	3609	−144.35	1165.4	1021.0
20	0.0505	0.01745	5658	5658	−149.31	1167.0	1017.7
10	0.0309	0.01744	9050	9050	−154.17	1168.6	1014.4
0	0.0185	0.01742	14770	14770	−158.93	1170.0	1011.1
−10	0.0108	0.01741	24670	24670	−163.59	1171.4	1007.8
−20	0.0062	0.01739	42200	42200	−168.16	1172.7	1004.5
−30	0.0035	0.01738	74100	74100	−172.63	1173.8	1001.2
−40	0.0019	0.01737	133900	133900	−177.00	1174.9	997.9

Temp.(°F)	Enthalpy (Btu/lbm) Sat. Ice	Subl.	Sat. Vap.	Entropy (Btu/lbm°R) Sat. Ice	Subl.	Sat. Vap.
T	h_i	h_{ig}	h_g	s_i	s_{ig}	s_g
32.018	−143.35	1218.8	1075.5	−0.2916	2.4788	2.1872
30	−144.35	1219.3	1074.9	−0.2936	2.4897	2.1961
20	−149.31	1219.9	1070.6	−0.3038	2.5425	2.2387
10	−154.17	1220.4	1066.2	−0.3141	2.5977	2.2836
0	−158.93	1220.7	1061.8	−0.3241	2.6546	2.3305
−10	−163.59	1221.0	1057.4	−0.3346	2.7143	2.3797
−20	−168.16	1221.2	1053.0	−0.3448	2.7764	2.4316
−30	−172.63	1221.2	1048.6	−0.3551	2.8411	2.4860
−40	−177.00	1221.2	1044.2	−0.3654	2.9087	2.5433

TABLE A.6 *Thermodynamic Properties of Carbon Dioxide—Saturated Liquid-Vapor.*

Temp. (°F)	Abs. Press. (psia)	Specific Volume (ft³/lbm) Sat. Liq.	Evap.	Sat. Vap.	Internal Energy (Btu/lbm) Sat. Liq.	Evap.	Sat. Vap.
T	p	v_f	v_{fg}	v_g	u_f	u_{fg}	u_g
−69.9	75.146	0.01360	1.1434	1.1570	−13.9	133.8	119.9
−60	94.75	0.01384	0.9112	0.9250	−9.3	129.8	120.5
−50	118.27	0.01409	0.7359	0.7500	−4.9	125.8	120.9
−40	145.87	0.01437	0.5969	0.6113	−0.4	121.8	121.4
−30	178.07	0.01465	0.4882	0.5028	4.2	117.6	121.8
−20	215.02	0.01498	0.4015	0.4165	8.6	113.5	122.1
−10	257.58	0.01533	0.3315	0.3468	13.3	109.1	122.4
0	305.76	0.01571	0.2793	0.2905	17.9	104.6	122.5
10	360.5	0.01614	0.2274	0.2435	21.8	100.8	122.6
20	421.8	0.01662	0.1882	0.2048	28.3	94.2	122.5
30	490.8	0.01719	0.1551	0.1723	34.2	88.1	122.3
40	567.3	0.01786	0.1263	0.1442	39.9	81.8	121.7
50	652.9	0.01870	0.1020	0.1207	46.2	73.8	120.0
60	747.4	0.01970	0.0798	0.0995	53.0	65.4	118.4
70	852.8	0.02113	0.0590	0.0801	60.9	55.6	116.6
80	969.3	0.02370	0.0363	0.0600	69.7	38.5	108.2
87.87	1070.0	0.03423	0	0.03423	90.3	0	90.3

Temp. (°F)	Enthalpy (Btu/lbm)* Sat. Liq.	Evap.	Sat. Vap.	Entropy (Btu/lbm°R)* Sat. Liq.	Evap.	Sat. Vap.
T	h_f	h_{fg}	h_g	s_f	s_{fg}	s_g
−69.9	−13.7	149.7	136.0	−0.0333	0.3839	0.3506
−60	−9.1	145.8	136.7	−0.0221	0.3650	0.3429
−50	−4.6	141.9	137.3	−0.0109	0.3463	0.3354
−40	0.0	137.9	137.9	0.0000	0.3285	0.3285
−30	4.7	133.7	138.4	0.0106	0.3113	0.3219
−20	9.2	129.5	138.7	0.0210	0.2945	0.3155
−10	14.0	124.9	138.9	0.0314	0.2778	0.3092
0	18.8	120.1	138.9	0.0419	0.2611	0.3030
10	24.0	114.8	138.8	0.0525	0.2445	0.2970
20	29.6	108.9	138.5	0.0636	0.2273	0.2909
30	35.8	102.1	137.9	0.0754	0.2095	0.2849
40	41.8	95.0	136.8	0.0872	0.1903	0.2775
50	48.5	86.1	134.6	0.0999	0.1699	0.2698
60	55.7	76.5	132.2	0.1136	0.1470	0.2606
70	63.8	63.9	127.7	0.1283	0.1203	0.2486
80	74.0	45.0	119.0	0.1469	0.0836	0.2304
87.87	97.1	0	97.1	0.1880	0	0.1880

* The enthalpy and entropy of saturated liquid carbon dioxide are taken as zero at a temperature of −40°F.

TABLE A.7 *Thermodynamic Properties of Carbon Dioxide—Superheated CO_2.*

Abs. Press. (psia)	Sat. Temp. (°F)	Sat. Vapor	Temperature (°F)				
p	T_g		0	100	200	300	400
		v_g	Specific Volume, v (ft³/lbm)				
0.2	−182.8	315.0	560.20	682.28	804.20	926.12	1048.04
0.5	−170.9	140.2	224.08	272.91	321.68	370.44	419.21
1	−159.8	76.0	111.95	136.40	160.81	185.20	209.59
2	−147.6	37.2	55.98	68.20	80.40	92.60	104.80
5	−131.9	16.0	22.33	27.24	32.14	37.02	41.91
10	−118.3	8.66	11.14	13.60	16.06	18.54	20.95
15	−109.4	5.78	7.406	9.057	10.70	12.33	13.96
20	−102.5	4.26	5.554	6.793	8.023	9.247	10.47
50	−79.7	1.68	2.166	2.681	3.188	3.686	4.180
100	−57.6	0.880	1.042	1.315	1.579	1.833	2.084
150	−38.7	0.596	0.6678	0.8583	1.042	1.216	1.385
200	−23.9	0.450	0.4813	0.6318	0.7742	0.9073	1.036
250	−11.6	0.357	0.3693	0.4951	0.6134	0.7220	0.8260
300	−1.1	0.296	0.2930	0.4036	0.5062	0.5985	0.6862
350	8.2	0.251		0.3382	0.4296	0.5103	0.5862
400	16.6	0.217		0.2890	0.3722	0.4442	0.5116
500	31.3	0.167		0.2196	0.2915	0.3516	0.4068
600	44.0	0.134		0.1725	0.2374	0.2898	0.3370
800	65.1	0.0904		0.1142	0.1696	0.2127	0.2498
1000	82.5	0.055		0.0790	0.1296	0.1665	0.1976
2000					0.0517	0.0741	0.0941
3000					0.0325	0.0466	0.0604
		u_g	Internal Energy, u (Btu/lbm)				
0.2	−182.8	112.1	134.9	150.1	166.8	184.5	203.3
0.5	−170.9	112.5	134.8	150.1	166.8	184.5	203.3
1	−159.8	112.8	134.6	150.1	166.8	184.5	203.3
2	−147.6	113.9	134.5	150.1	166.7	184.5	203.3
5	−131.9	115.6	134.4	150.0	166.7	184.5	203.3
10	−118.3	116.0	134.4	149.9	166.7	184.5	203.3
15	−109.4	117.5	134.2	149.9	166.6	184.5	203.3
20	−102.5	118.8	134.0	149.8	166.5	184.5	203.3
50	−79.7	120.3	133.2	149.3	166.2	184.3	203.0
100	−57.6	120.5	131.3	148.5	165.8	183.9	202.6
150	−38.7	121.5	129.5	147.7	165.3	183.3	202.3
200	−23.9	121.3	127.8	146.6	164.8	182.8	201.9
250	−11.6	122.3	125.3	145.7	164.5	182.5	201.6
300	−1.1	122.5	122.7	144.7	163.5	181.9	201.0
350	8.2	122.5		143.7	162.9	181.4	200.7
400	16.6	122.5		142.4	162.3	181.1	200.4
500	31.3	122.3		140.2	161.0	180.3	199.6
600	44.0	121.3		137.8	159.6	179.2	198.8
800	65.1	116.8		132.2	156.6	177.4	197.1
1000	82.5	106.4		123.9	153.0	175.1	195.5
2000					129.6	163.9	185.4
3000					112.2	154.5	178.5

Abs. Press. (psia)	Temperature (°F)					
p	500	600	700	800	900	1000
			Specific Volume, v (ft^3/lbm)			
0.2	1169.94	1291.85	1413.76	1535.68	1657.59	1779.50
0.5	467.98	516.74	565.50	614.27	663.04	711.80
1	233.98	258.37	282.75	307.13	331.52	355.90
2	116.99	129.18	141.37	153.57	165.76	177.95
5	46.70	51.67	56.55	61.43	66.30	71.18
10	23.39	25.83	28.27	30.71	33.15	35.59
15	15.59	17.22	18.85	20.48	22.10	23.73
20	11.69	12.92	14.14	15.36	16.58	17.80
50	4.672	5.162	5.652	6.140	6.629	7.117
100	2.332	2.578	2.824	3.068	3.313	3.558
150	1.552	1.717	1.881	2.045	2.208	2.371
200	1.162	1.286	1.410	1.533	1.656	1.778
250	0.9277	1.028	1.127	1.226	1.324	1.422
300	0.7717	0.8556	0.9385	1.021	1.103	1.185
350	0.6602	0.7334	0.8037	0.8745	0.9448	1.015
400	0.5767	0.6402	0.7028	0.7648	0.8266	0.8885
500	0.4597	0.5110	0.5613	0.6111	0.6608	0.7105
600	0.3817	0.4248	0.4670	0.5086	0.5502	0.5919
800	0.2843	0.3172	0.3490	0.3805	0.4119	0.4436
1000	0.2259	0.2526	0.2782	0.3035	0.3289	0.3546
2000	0.1099	0.1237	0.1366	0.1494	0.1626	0.1767
3000	0.0715	0.0809	0.0897	0.0984	0.1076	0.1175
			Internal Energy, u (Btu/lbm)			
0.2	223.1	243.8	265.2	287.2	309.9	333.1
0.5	223.1	243.8	265.2	287.2	309.9	333.1
1	223.1	243.8	265.2	287.2	309.9	333.1
2	223.1	243.8	265.2	287.2	309.9	333.1
5	223.1	243.8	265.2	287.2	309.9	333.1
10	231.1	243.8	265.2	287.2	309.9	333.1
15	223.1	243.8	265.2	287.2	309.9	333.1
20	223.1	243.6	265.2	287.2	309.8	333.1
50	222.9	242.6	265.0	287.1	309.8	333.0
100	222.5	242.4	264.8	287.0	309.8	333.0
150	222.2	242.1	264.7	286.8	309.7	333.0
200	221.9	242.0	264.6	286.8	309.6	333.0
250	221.7	241.7	264.5	286.7	309.5	332.9
300	221.3	241.5	264.3	286.5	309.4	332.8
350	221.0	241.2	264.1	286.5	309.3	332.8
400	220.8	241.1	264.0	286.3	309.1	332.6
500	220.2	240.6	263.7	286.1	309.0	332.5
600	219.5	240.1	263.2	285.8	308.8	332.4
800	218.3	239.3	262.6	285.4	308.4	331.9
1000	217.0	238.6	262.0	284.8	307.1	331.6
2000	209.0	233.9	258.6	282.8	306.5	329.8
3000	203.9	231.9	257.4	281.8	305.6	328.7

TABLE A.7 (Continued) *Thermodynamic Properties of Carbon Dioxide—Superheated CO_2.*

Abs. Press. (psia)	Sat. Temp. (°F)	Sat. Vapor	Temperature (°F)				
p	T_g		0	100	200	300	400
		h_g	Enthalpy, h (Btu/lbm)				
0.2	−182.8	123.8	155.6	175.4	196.6	218.8	242.1
0.5	−170.9	125.5	155.5	175.4	196.6	218.8	242.1
1	−159.8	126.9	155.3	175.3	196.6	218.8	242.1
2	−147.6	127.7	155.2	175.3	196.5	218.8	242.1
5	−131.9	130.4	155.1	175.2	196.4	218.8	242.1
10	−118.3	132.0	155.0	175.1	196.4	218.8	242.1
15	−109.4	133.5	154.8	175.0	196.3	218.8	242.1
20	−102.5	134.6	154.6	174.9	196.2	218.8	242.1
50	−79.7	135.8	153.2	174.1	195.7	218.4	241.7
100	−57.6	136.8	150.6	172.8	195.0	217.8	241.2
150	−38.7	138.0	148.0	171.5	194.2	217.1	240.7
200	−23.9	138.6	145.4	170.0	193.5	216.4	240.2
250	−11.6	138.8	142.4	168.6	192.9	215.9	239.8
300	−1.1	138.9	139.0	167.1	191.6	215.1	239.1
350	8.2	138.8		165.6	190.7	214.5	238.7
400	16.6	138.6		163.8	189.9	214.0	238.3
500	31.3	137.8		160.5	188.0	212.8	237.2
600	44.0	136.2		157.0	186.0	211.4	236.2
800	65.1	130.2		149.1	181.7	208.9	234.1
1000	82.5	116.6		138.5	177.0	205.9	232.1
2000					148.7	191.3	220.2
3000					130.2	180.4	212.0
		s_g	Entropy, s (Btu/lbm°R)				
0.2	−182.8	0.577	0.662	0.702	0.736	0.768	0.798
0.5	−170.9	0.543	0.620	0.660	0.694	0.727	0.756
1	−159.8	0.515	0.589	0.629	0.663	0.696	0.725
2	−147.6	0.488	0.558	0.598	0.632	0.665	0.694
5	−131.9	0.453	0.517	0.556	0.590	0.623	0.652
10	−118.3	0.431	0.485	0.525	0.559	0.592	0.621
15	−109.4	0.412	0.467	0.507	0.541	0.574	0.603
20	−102.5	0.403	0.454	0.494	0.528	0.561	0.590
50	−79.7	0.366	0.409	0.451	0.486	0.518	0.548
100	−57.6	0.341	0.371	0.417	0.454	0.485	0.515
150	−38.7	0.328	0.352	0.398	0.436	0.466	0.497
200	−23.9	0.318	0.335	0.383	0.421	0.452	0.484
250	−11.6	0.310	0.319	0.370	0.410	0.442	0.474
300	−1.1	0.304	0.306	0.360	0.401	0.434	0.465
350	8.2	0.298		0.350	0.390	0.427	0.458
400	16.6	0.293		0.342	0.385	0.420	0.451
500	31.3	0.284		0.328	0.373	0.410	0.440
600	44.0	0.274		0.316	0.363	0.400	0.431
800	65.1	0.255		0.289	0.344	0.387	0.417
1000	82.5	0.226		0.263	0.329	0.370	0.404
2000					0.268	0.325	0.363
3000					0.227	0.290	0.335

Abs. Press. (psia)	Temperature (°F)					
p	500	600	700	800	900	1000
			Enthalpy, h (Btu/lbm)			
0.2	266.4	291.6	317.5	344.0	371.2	399.0
0.5	266.4	291.6	317.5	344.0	371.2	399.0
1	266.4	291.6	317.5	344.0	371.2	399.0
2	266.4	291.6	317.5	344.0	371.2	399.0
5	266.4	291.6	317.5	344.0	371.2	399.0
10	266.4	291.6	317.5	344.0	371.2	399.0
15	266.4	291.6	317.5	344.0	371.2	399.0
20	266.4	291.4	317.5	344.0	371.2	399.0
50	266.1	290.4	317.3	343.9	371.1	398.9
100	265.7	290.1	317.1	343.8	371.1	398.9
150	265.3	289.8	316.9	343.6	371.0	398.8
200	264.9	289.6	316.8	343.5	370.9	398.8
250	264.6	289.3	316.6	343.4	370.8	398.7
300	264.1	289.0	316.4	343.2	370.6	398.6
350	263.8	288.7	316.2	343.1	370.5	398.5
400	263.5	288.5	316.0	342.9	370.3	398.4
500	262.7	287.9	315.6	342.6	370.1	398.2
600	261.9	287.3	315.1	342.3	369.9	398.0
800	260.4	286.3	314.3	341.7	369.4	397.6
1000	258.8	285.3	313.5	341.0	368.0	397.2
2000	249.7	279.7	309.2	338.1	366.7	395.2
3000	243.6	276.8	307.2	336.4	365.3	393.9
			Entropy, s (Btu/lbm°R)			
0.2	0.824	0.848	0.872	0.895	0.917	0.936
0.5	0.782	0.806	0.830	0.853	0.875	0.894
1	0.751	0.775	0.799	0.822	0.844	0.863
2	0.720	0.744	0.768	0.791	0.813	0.832
5	0.678	0.702	0.726	0.749	0.771	0.790
10	0.647	0.671	0.695	0.718	0.740	0.759
15	0.629	0.653	0.677	0.700	0.722	0.741
20	0.616	0.640	0.664	0.683	0.709	0.728
50	0.574	0.598	0.622	0.645	0.668	0.687
100	0.542	0.566	0.590	0.613	0.627	0.654
150	0.524	0.548	0.572	0.595	0.619	0.636
200	0.511	0.535	0.559	0.582	0.606	0.623
250	0.501	0.525	0.549	0.572	0.596	0.613
300	0.493	0.517	0.541	0.564	0.588	0.605
350	0.486	0.510	0.534	0.557	0.581	0.598
400	0.480	0.504	0.528	0.550	0.574	0.592
500	0.469	0.494	0.518	0.540	0.564	0.582
600	0.461	0.486	0.510	0.532	0.556	0.574
800	0.448	0.473	0.497	0.519	0.543	0.561
1000	0.436	0.462	0.486	0.508	0.532	0.550
2000	0.396	0.427	0.452	0.475	0.499	0.517
3000	0.372	0.405	0.431	0.455	0.479	0.497

TABLE A.8 *Thermodynamic Properties of Carbon Dioxide—Compressed Liquid.*

Abs. Press. (psia)	Temperature (°F)				Sat. Temp. (°F)	Sat. Liq.
p	25	50	75	100	T_f	
			Specific Volume, v (ft³/lbm)			v_f
500	0.01683				31.22	0.01719
1000	0.01641	0.01790	0.02123		82.40	0.02691
2000	0.01602	0.01687	0.01843	0.02119		
3000	0.01565	0.01639	0.01767	0.01981		
5000	0.01492	0.01542	0.01614	0.01704		
8000	0.01428	0.01467	0.01523	0.01579		
			Internal Energy, u (Btu/lbm)			u_f
500	30.1				31.22	34.9
1000	27.9	41.9	57.0		82.40	76.0
2000	24.3	36.7	51.0	66.5		
3000	21.3	33.0	46.3	59.5		
5000	17.7	27.5	39.1	51.4		
8000	12.9	22.4	32.8	43.5		
			Enthalpy, h (Btu/lbm)			h_f
500	31.7				31.22	36.5
1000	30.9	45.2	60.9		82.40	81.0
2000	30.2	42.9	57.8	74.3		
3000	30.0	42.1	56.1	70.5		
5000	31.5	41.8	54.0	67.2		
8000	34.0	44.1	55.3	66.9		
			Entropy, s (Btu/lbm°R)			s_f
500	0.0687				31.22	0.0768
1000	0.0635	0.0922	0.1190		82.40	0.1594
2000	0.0559	0.0815	0.1089	0.1542		
3000	0.0490	0.0719	0.0974	0.1249		
5000	0.0392	0.0614	0.0840	0.1049		
8000	0.0285	0.0497	0.0705	0.0901		

TABLE A.9 *Thermodynamic Properties of Carbon Dioxide—Saturated Solid-Vapor.*

Temp. (°F)	Abs. Press. (psia)	Specific Volume (ft^3/lbm) Sat. Solid	Subl.	Sat. Vapor	Internal Energy (Btu/lbm) Sat. Solid	Subl.	Sat. Vapor
T	p	v_i	v_{ig}	v_g	u_i	u_{ig}	u_g
−69.9	75.146	0.01059	1.146	1.157	−97.9	217.9	120.0
−70	75.14	0.01058	1.147	1.158	−98.1	218.0	119.9
−80	51.00	0.01047	1.675	1.685	−102.4	222.3	119.9
−90	34.70	0.01039	2.490	2.500	−106.6	225.8	119.2
−100	22.00	0.01032	3.820	3.830	−110.0	228.8	118.8
−109.4	14.696	0.01025	5.730	5.740	−112.9	230.7	117.8
−110	14.40	0.01024	5.890	5.900	−113.1	230.7	117.6
−120	9.22	0.01018	9.070	9.080	−116.0	232.6	116.6
−130	5.407	0.01012	14.68	14.69	−118.7	234.4	115.7
−140	3.190	0.01007	24.19	24.20	−121.4	236.1	114.7
−150	1.795	0.01004	43.04	43.05	−123.9	237.7	113.8
−160	0.985	0.01000	76.25	76.26	−126.4	239.4	113.0
−170	0.525	0.009965	137.9	137.9	−128.9	241.1	112.2
−180	0.259	0.009928	265.0	265.0	−131.5	243.0	111.5
−190	0.120	0.009892	544.9	544.9	−134.1	244.9	110.8
−200	0.048	0.009856	1222.3	1222.3	−137.1	247.3	110.2

Temp. (°F)	Enthalpy (Btu/lbm) Sat. Solid	Subl.	Sat. Vap.	Entropy (Btu/lbm°R) Sat. Solid	Subl.	Sat. Vapor
T	h_i	h_{ig}	h_g	s_i	s_{ig}	s_g
−69.9	−97.8	233.8	136.0	−0.2493	0.5999	0.3506
−70	−98.0	234.0	136.0	−0.2494	0.6002	0.3508
−80	−102.3	238.1	135.8	−0.2610	0.6275	0.3665
−90	−106.5	241.8	135.3	−0.2720	0.6542	0.3822
−100	−110.0	244.4	134.4	−0.2815	0.6796	0.3981
−109.4	−112.9	246.3	133.4	−0.2898	0.7032	0.4134
−110	−113.1	246.4	133.3	−0.2904	0.7049	0.4145
−120	−116.0	248.1	132.1	−0.2986	0.7304	0.4318
−130	−118.7	248.4	131.0	−0.3068	0.7568	0.4500
−140	−121.4	250.7	129.0	−0.3153	0.7849	0.4696
−150	−123.9	252.0	128.1	−0.3200	0.8165	0.4965
−160	−126.4	253.3	126.9	−0.3291	0.8426	0.5135
−170	−128.9	254.5	125.6	−0.3385	0.8719	0.5334
−180	−131.5	255.7	124.2	−0.3480	0.9123	0.5643
−190	−134.1	257.0	122.9	−0.3570	0.9501	0.5931
−200	−137.1	258.2	121.1	−0.3667	0.9913	0.6246

TABLE A.10 *Thermodynamic Properties of R-22 (Chlorodifluoromethane)— Saturated Liquid-Vapor.*

Temp. (°F)	Abs. Press. (psia)	Specific Volume (ft³/lbm) Sat. Liq.	Evap.	Sat. Vap.	Internal Energy (Btu/lbm) Sat. Liq.	Evap.	Sat. Vap.
T	p	v_f	v_{fg}	v_g	u_f	u_{fg}	u_g
−150	0.2716	0.01018	141.22	141.23	−25.98	106.40	80.42
−140	0.4469	0.01027	88.522	88.532	−23.73	105.09	81.36
−130	0.7106	0.01037	57.346	57.356	−21.46	103.77	83.31
−120	1.0954	0.01046	38.270	38.280	−10.19	102.45	83.26
−110	1.6417	0.01056	26.231	26.242	−16.89	101.11	84.22
−100	2.3989	0.01066	18.422	18.433	−14.57	99.76	85.19
−90	3.4229	0.01077	13.224	13.235	−12.22	98.38	86.16
−80	4.7822	0.01088	9.6840	9.6949	−9.85	96.98	87.13
−70	6.5522	0.01100	7.2208	7.2318	−7.49	95.59	88.10
−60	8.8180	0.01111	5.4733	5.4844	−5.01	94.08	89.07
−50	11.674	0.01124	4.2112	4.2224	−2.54	92.56	90.02
−40	15.222	0.01136	3.2843	3.2957	−0.03	91.01	90.97
−30	19.773	0.01150	2.5934	2.6049	2.51	89.40	91.91
−20	24.845	0.01163	2.0710	2.0826	5.13	87.82	92.95
−10	31.162	0.01178	1.6707	1.6825	7.68	86.07	93.75
0	38.657	0.01193	1.3604	1.3723	10.32	84.33	94.65
10	47.464	0.01209	1.1169	1.1290	13.00	82.53	95.53
20	57.727	0.01226	0.92405	0.93631	15.71	80.67	96.38
30	69.591	0.01243	0.76956	0.78208	18.45	78.76	97.21
40	83.206	0.01262	0.64491	0.65753	21.23	76.79	98.02
50	98.727	0.01281	0.54325	0.55606	24.04	74.75	98.79
60	116.31	0.01302	0.45970	0.47272	26.80	72.65	99.54
70	136.12	0.01325	0.39048	0.40373	29.78	70.46	100.24
80	158.33	0.01349	0.33272	0.34621	32.71	68.20	100.91
90	183.09	0.01375	0.28414	0.29789	35.69	65.84	101.53
100	210.60	0.01404	0.24298	0.25702	38.72	63.37	102.09
110	241.04	0.01435	0.20787	0.22222	41.81	60.78	102.59
120	274.60	0.01469	0.17768	0.19238	44.96	58.05	103.01
130	311.50	0.01508	0.15153	0.16661	48.19	55.14	103.33
140	351.94	0.01552	0.12866	0.14418	51.52	52.02	103.54
150	396.19	0.01602	0.10846	0.12448	54.97	48.63	103.60
160	444.53	0.01663	0.09038	0.10701	58.58	44.88	103.46
170	497.26	0.01737	0.07391	0.09128	62.42	40.62	103.04
180	554.78	0.01833	0.05846	0.07679	66.62	35.57	102.19
190	617.59	0.01973	0.04311	0.06284	71.46	29.09	100.55
200	686.36	0.02244	0.02500	0.04744	78.30	19.13	97.43
204.81	721.91	0.03053	0	0.03053	87.25	0	87.25

Temp. (°F)	Enthalpy (Btu/lbm)* Sat. Liq.	Evap.	Sat. Vapor	Entropy (Btu/lbm°R)* Sat. Liq.	Evap.	Sat. Vapor
T	h_f	h_{fg}	h_g	s_f	s_{fg}	s_g
−150	−25.97	113.49	87.52	−0.07147	0.36648	0.29501
−140	−23.72	112.40	88.68	−0.06432	0.35161	0.28729
−130	−21.46	111.31	89.85	−0.05736	0.33763	0.28027
−120	−19.19	110.21	91.02	−0.05055	0.32443	0.27388
−110	−16.89	109.09	92.20	−0.04389	0.31194	0.26805
−100	−14.56	107.93	93.37	−0.03734	0.30008	0.26274
−90	−12.22	106.76	94.54	−0.03091	0.28878	0.25787
−80	−9.84	105.55	95.71	−0.02457	0.27799	0.25342
−70	−7.43	104.30	96.87	−0.01832	0.26764	0.24932
−60	−4.99	103.00	98.01	−0.01214	0.25770	0.24556
−50	−2.51	101.65	99.14	−0.00604	0.24813	0.24209
−40	0	100.26	100.26	0.00000	0.23888	0.23888
−30	2.55	98.80	101.35	0.00598	0.22993	0.23591
−20	5.13	97.29	102.42	0.01189	0.22126	0.23315
−10	7.75	95.70	103.45	0.01776	0.21282	0.23058
0	10.41	94.06	104.47	0.02357	0.20460	0.22817
10	13.10	92.34	105.44	0.02932	0.19660	0.22592
20	15.84	90.54	106.38	0.03503	0.18876	0.22379
30	18.61	88.67	107.28	0.04070	0.18108	0.22178
40	21.42	86.72	108.14	0.04632	0.17354	0.21986
50	24.27	84.68	108.95	0.05190	0.16613	0.21803
60	27.17	82.54	109.71	0.05745	0.15882	0.21627
70	30.12	80.29	110.41	0.06296	0.15160	0.21456
80	33.11	77.94	111.05	0.06846	0.14442	0.21288
90	36.16	75.46	111.62	0.07394	0.13728	0.21122
100	39.27	72.84	112.11	0.07942	0.13014	0.20956
110	42.45	70.05	112.50	0.08491	0.12296	0.20787
120	45.70	67.08	112.78	0.09042	0.11571	0.20613
130	49.06	63.88	112.94	0.09598	0.10833	0.20431
140	52.53	60.40	112.93	0.10163	0.10072	0.20235
150	56.14	56.59	112.73	0.10739	0.09281	0.20020
160	59.95	52.31	112.26	0.11334	0.08442	0.19776
170	64.02	47.40	111.42	0.11959	0.07531	0.19490
180	68.50	41.57	110.07	0.12635	0.06498	0.19133
190	73.71	34.02	107.73	0.13409	0.05237	0.18646
200	80.86	21.99	102.85	0.14460	0.03334	0.17794
204.81	91.33	0	91.33	0.16016	0	0.16016

* The enthalpy and entropy of saturated liquid R-22 are taken as zero at a temperature of −40°F.

TABLE A.11 *Thermodynamic Properties of R-22 (Chlorodifluoromethane)—Superheated $CHClF_2$.*

Abs. Press. (psia)	Sat. Temp. (°F)	Sat. Vapor	Temperature (°F)					
p	T_g		−100	0	100	200	300	400
		v_g	Specific Volume, v (ft³/lbm)					
0.2	−155.80	188.29	223.01	285.14	347.22	409.32	471.37	533.42
0.5	−137.64	79.698	89.095	114.00	138.85	163.70	188.55	213.37
1	−122.16	41.678	44.458	56.949	69.397	81.83	94.27	106.68
2	−104.87	21.831	22.139	28.426	34.668	40.893	47.12	53.34
5	−78.62	9.3011		11.311	13.831	16.333	18.83	21.34
10	−55.58	4.8778		5.6060	6.8855	8.1464	9.399	10.65
15	−40.57	3.3412		3.7037	4.5701	5.4174	6.256	7.10
20	−29.12	2.5527		2.7521	3.4122	4.0529	4.685	5.325
40	1.63	1.3285			1.6749	2.0060	2.3281	2.657
60	22.03	0.90222			1.0952	1.3235	1.5424	1.767
80	37.76	0.68318			0.80477	0.98209	1.1495	1.323
100	50.77	0.54908			0.63003	0.77712	0.91372	1.046
120	61.95	0.45822			0.51309	0.64036	0.75651	0.8678
140	71.83	0.39243			0.42911	0.54258	0.64419	0.7409
160	80.71	0.34249			0.36568	0.46914	0.55993	0.6447
180	88.81	0.30323			0.31587	0.41194	0.49437	0.5715
200	96.27	0.27150			0.27553	0.36609	0.44190	0.51218
250	112.76	0.21351				0.28325	0.34740	0.40549
300	126.98	0.17400				0.22759	0.28431	0.33436
350	139.54	0.14514				0.18738	0.23917	0.28356
400	150.82	0.12297				0.15674	0.20524	0.24546
500	170.50	0.09053				0.11220	0.15757	0.19212
		u_g	Internal Energy, u (Btu/lbm)					
0.2	−155.80	79.88	85.37	96.49	109.20	123.42	139.07	156.01
0.5	−137.64	81.58	85.34	96.47	109.19	123.42	139.06	156.00
1	−122.16	83.05	85.30	96.45	109.18	123.40	139.04	155.98
2	−104.87	84.72	85.22	96.41	109.15	123.38	139.02	155.95
5	−78.62	87.27		96.30	109.07	123.33	138.96	155.87
10	−55.58	89.49		96.04	108.93	123.24	138.86	155.78
15	−40.57	90.92		95.81	108.80	123.15	138.80	155.69
20	−29.12	92.00		95.57	108.66	123.06	138.74	155.60
40	1.63	94.79			108.09	122.69	138.49	155.28
60	22.03	96.55			107.50	122.32	138.23	155.02
80	37.76	97.84			106.89	121.94	137.96	154.86
100	50.77	98.85			106.25	121.55	137.70	154.77
120	61.95	99.68			105.59	121.16	137.43	154.58
140	71.83	100.37			104.89	120.75	137.15	154.41
160	80.71	100.96			104.16	120.34	136.88	154.24
180	88.81	101.46			103.38	119.92	136.60	154.07
200	96.27	101.89			102.55	119.49	136.31	153.82
250	112.76	102.71				118.36	135.59	153.29
300	126.98	103.24				117.15	134.84	152.75
350	139.54	103.54				115.85	134.06	152.20
400	150.82	103.60				114.42	133.26	151.64
500	170.50	103.01				111.05	131.55	150.48

Abs. Press. (psia)	Sat. Temp. (°F)	Sat. Vapor	Temperature (°F)					
p	T_g		−100	0	100	200	300	400
		h_g	Enthalpy, h (Btu/lbm)					
0.2	−155.80	86.85	93.62	107.04	122.05	138.57	156.52	175.75
0.5	−137.64	88.96	93.59	107.02	122.04	138.56	156.51	175.74
1	−122.16	90.77	93.53	106.99	122.02	138.54	156.49	175.72
2	−104.87	92.80	93.42	106.93	121.98	138.51	156.46	175.69
5	−78.62	95.87		106.74	121.87	138.44	156.38	175.61
10	−55.58	98.52		106.41	121.67	138.31	156.26	175.49
15	−40.57	100.19		106.09	121.48	138.18	156.16	175.39
20	−29.12	101.44		105.76	121.28	138.06	156.08	175.31
40	1.63	104.63			120.49	137.54	155.72	174.95
60	22.03	106.57			119.66	137.01	155.35	174.64
80	37.76	107.95			118.80	136.48	154.98	174.45
100	50.77	109.01			117.91	135.93	154.61	174.13
120	61.95	109.85			116.98	135.38	154.23	173.85
140	71.83	110.54			116.01	134.81	153.84	173.60
160	80.71	111.10			114.99	134.23	153.46	173.33
180	88.81	111.56			113.90	133.64	153.06	173.10
200	96.27	111.93			112.75	133.03	152.67	172.78
250	112.76	112.59				131.46	151.66	172.05
300	126.98	112.90				129.78	150.62	171.31
350	139.54	112.94				127.98	149.55	170.57
400	150.82	112.70				126.02	148.45	169.81
500	170.50	111.38				121.43	146.13	168.26
		s_g	Entropy, s (Btu/lbm°R)					
0.2	−155.80	0.29985	0.32028	0.35311	0.38261	0.40982	0.43515	0.45889
0.5	−137.64	0.28557	0.29917	0.33204	0.36155	0.38878	0.41411	0.43785
1	−122.16	0.27521	0.28314	0.31607	0.34561	0.37274	0.39807	0.42193
2	−104.87	0.26527	0.26700	0.30005	0.32964	0.35679	0.38212	0.40590
5	−78.62	0.25283		0.27872	0.30845	0.33566	0.36099	0.38477
10	−55.58	0.24399		0.26230	0.29229	0.31961	0.34494	0.36872
15	−40.57	0.23906		0.25248	0.28273	0.31016	0.33553	0.35931
20	−29.12	0.23566		0.24535	0.27588	0.30342	0.32884	0.35262
40	1.63	0.22780			0.25893	0.28694	0.31259	0.33637
60	22.03	0.22337			0.24855	0.27706	0.30293	0.32679
80	37.76	0.22029			0.24083	0.26987	0.29598	0.31998
100	50.77	0.21790			0.23454	0.26415	0.29050	0.31464
120	61.95	0.21593			0.22912	0.25936	0.28595	0.31024
140	71.83	0.21425			0.22428	0.25520	0.28205	0.30648
160	80.71	0.21276			0.21984	0.25149	0.27862	0.30320
180	88.81	0.21142			0.21566	0.24813	0.27554	0.30026
200	96.27	0.21018			0.21165	0.24503	0.27274	0.29760
250	112.76	0.20740				0.23814	0.26665	0.29186
300	126.98	0.20487				0.23204	0.26146	0.28705
350	139.54	0.20244				0.22641	0.25688	0.28286
400	150.82	0.20001				0.22104	0.25273	0.27914
500	170.50	0.19474				0.21034	0.24532	0.27267

TABLE A.12 *Thermodynamic Properties of Dry Air—Saturated Liquid-Vapor.*

Vapor Temp. (°F)	Abs. Press. (psia)	Liq. Temp. (°F)	Specific Volume (ft³/lbm) Sat. Liq.	Evap.	Sat. Vap.	Internal Energy (Btu/lbm) Sat. Liq.	Evap.	Sat. Vap.
T	p	T	v_f	v_{fg}	v_g	u_f	u_{fg}	u_g
−340	1.65	−345.6	0.01696	29.88	29.90	−5.08	83.97	78.89
−330	4.09	−335.4	0.01738	12.57	12.59	−0.89	81.81	80.92
−328	4.82	−333.4	0.01747	10.71	10.73	−0.02	81.35	81.33
−320	8.99	−325.2	0.01788	5.976	5.994	3.43	79.39	82.82
−310	17.24	−315.0	0.01842	3.133	3.151	7.83	77.06	84.89
−300	30.74	−304.8	0.01902	1.830	1.849	12.42	73.90	86.32
−290	51.07	−294.5	0.01965	1.124	1.144	17.14	70.31	87.45
−280	80.11	−284.2	0.02045	0.7166	0.7388	21.96	66.30	88.26
−270	119.94	−273.8	0.02139	0.4724	0.4938	26.92	61.61	88.53
−260	172.56	−263.4	0.02247	0.3157	0.3382	32.06	56.53	88.59
−250	240.46	−252.9	0.02388	0.2111	0.2350	32.51	50.54	88.05
−240	325.57	−242.3	0.02587	0.1371	0.1630	43.34	42.98	86.32
−230	430.99	−231.5	0.02955	0.0792	0.1087	51.40	31.92	83.32
−221.1	546.65	−221.1	0.05006	0	0.05006	71.72	0	71.72

Vapor Temp. (°F)	Enthalpy (Btu/lbm)* Sat. Liq.	Evap.	Sat. Vap.	Entropy (Btu/lbm°R)* Sat. Liq.	Evap.	Sat. Vap.
T	h_f	h_{fg}	h_g	s_f	s_{fg}	s_g
−340	−5.08	93.07	87.99	−0.0423	0.8104	0.7681
−330	−0.88	91.32	90.44	−0.0068	0.7295	0.7227
−328	0	90.91	90.91	0	0.7151	0.7151
−320	3.44	89.35	92.79	0.0262	0.6585	0.6847
−310	7.89	87.05	94.94	0.0574	0.5951	0.6525
−300	12.53	84.31	96.84	0.0874	0.5370	0.6244
−290	17.33	80.93	98.26	0.1162	0.4830	0.5992
−280	22.26	76.95	99.21	0.1444	0.4317	0.5761
−270	27.39	72.10	99.49	0.1714	0.3833	0.5547
−260	32.78	66.61	99.39	0.1980	0.3361	0.5341
−250	38.57	59.94	98.51	0.2253	0.2918	0.5171
−240	44.90	51.24	96.14	0.2544	0.2300	0.4844
−230	53.76	38.23	91.99	0.2909	0.1706	0.4615
−221.1	76.78	0	76.78	0.3681	0	0.3681

* The enthalpy and entropy of saturated liquid are taken as zero at a temperature of −333.4 F. At the corresponding saturation pressure (4.82 psia), the temperature of the saturated vapor is −328°F.

TABLE A.13 *Thermodynamic Properties of Dry Air—Superheated Air.*

Abs. Press. (psia)	Sat. Vap. Temp. (°F)	Sat. Vapor	Temperature (°F)			
p	T_g		−300	−200	−100	0
		Specific Volume, v (ft^3/lbm)				
		v_g				
0.2	−358.7	191.8	295.52	480.93	666.16	851.37
1	−345.2	45.50	59.00	96.13	133.21	170.27
5	−327.6	10.08	11.71	19.19	26.62	34.04
10	−318.2	5.274	5.807	9.574	13.30	17.02
14.7	−312.4	3.571	3.928	6.503	9.045	11.58
50	−290.4	1.165		1.872	2.640	3.394
100	−274.7	0.5874		0.9154	1.311	1.693
150	−264.1	0.3905		0.5962	0.8675	1.126
300	−242.8	0.1831		0.2558	0.4207	0.5577
500	−224.0	0.06841		0.1197	0.2420	0.3304
700					0.1683	0.2346
1000					0.1130	0.1627
1500					0.07198	0.1077
3000					0.04088	0.05683
5000					0.02961	0.03811
7000					0.02539	0.03236
10,000					0.02223	0.02805
		Internal Energy, u (Btu/lbm)				
		u_g				
0.2	−358.7	76.84	87.40	104.47	121.56	138.65
1	−345.2	78.57	87.35	104.44	121.55	138.64
5	−327.6	81.67	87.24	104.39	121.52	138.62
10	−318.2	83.41	87.05	104.32	121.47	138.57
14.7	−312.4	84.69	86.93	104.24	121.41	138.54
50	−290.4	84.96		103.71	121.08	138.08
100	−274.7	88.49		102.77	120.51	137.96
150	−264.1	88.59		101.98	120.05	137.58
300	−242.8	86.78		98.49	118.47	136.46
500	−224.0	75.11		93.83	116.37	134.97
700					113.55	133.32
1000					109.31	130.85
1500					103.27	127.56
3000					92.70	120.55
5000					84.10	113.96
7000					81.19	108.56
10,000					76.83	100.45

Abs. Press. (psia)	Temperature (°F)					
p	100	200	300	400	500	600
			Specific Volume, v (ft^3/lbm)			
0.2	1036.59	1221.80	1407.01	1592.22	1777.43	1962.63
1	207.32	244.36	281.41	318.45	355.50	392.54
5	41.46	48.88	56.29	63.70	71.10	78.52
10	20.73	24.44	28.15	31.85	35.56	39.26
14.7	14.10	16.64	19.15	21.68	24.20	26.72
50	4.193	4.889	5.634	5.377	7.120	7.863
100	2.070	2.446	2.819	3.191	3.569	3.935
150	1.379	1.631	1.881	2.130	2.379	2.627
300	0.6885	0.8171	0.9439	1.060	1.195	1.320
500	0.4123	0.4915	0.5691	0.6457	0.7218	0.7974
700	0.2945	0.3521	0.4081	0.4634	0.5180	0.5723
1000	0.2062	0.2475	0.2874	0.3266	0.3652	0.4035
1500	0.1381	0.1665	0.1938	0.2205	0.2465	0.2724
3000	0.06355	0.08634	0.1003	0.1137	0.1270	0.1399
5000	0.04731	0.05676	0.06573	0.07427	0.08276	0.09088
7000	0.03815	0.04513	0.05154	0.05758	0.06372	0.06951
10,000	0.03128	0.03641	0.04089	0.04506	0.04944	0.05349
			Internal Energy, u (Btu/lbm)			
0.2	155.79	172.99	190.33	207.85	225.62	243.65
1	155.78	172.98	190.33	207.85	225.61	243.65
5	155.76	172.96	190.31	207.83	225.61	243.64
10	155.73	172.93	190.29	207.82	225.59	243.63
14.7	155.72	172.90	190.28	207.79	225.57	243.62
50	155.51	172.76	190.15	207.69	225.46	243.52
100	155.21	172.52	189.94	207.54	225.24	243.42
150	154.98	172.33	189.78	207.39	225.20	243.30
300	154.14	171.49	189.18	206.86	224.73	242.87
500	153.01	170.69	188.38	206.16	224.11	242.29
700	151.82	169.78	187.65	205.56	223.61	241.88
1000	150.04	168.41	186.56	204.65	222.87	241.26
1500	147.70	166.59	185.03	203.34	221.75	240.28
3000	140.90	160.99	180.55	199.45	218.40	237.34
5000	133.13	154.61	175.11	194.59	214.06	233.15
7000	128.33	149.85	170.73	190.69	210.44	229.87
10,000	121.12	142.72	164.16	184.83	205.00	224.95

TABLE A.13 (Continued) *Thermodynamic Properties of Dry Air—Superheated Air.*

Abs. Press. (psia)	Temperature (°F)					
p	700	800	900	1000	1100	1200
	Specific Volume, *v* (ft³/lbm)					
0.2	2147.84	2333.05	2518.26	2703.47	2888.64	3073.85
1	429.58	466.62	503.66	542.70	577.64	614.67
5	85.93	93.34	100.75	108.16	115.54	122.95
10	42.96	46.67	50.38	54.08	57.78	61.48
14.7	29.24	31.76	34.28	36.81	39.32	41.84
50	8.604	9.346	10.09	10.83	11.57	12.31
100	4.308	4.678	5.050	5.416	5.791	6.162
150	2.875	3.123	3.371	3.619	3.865	4.112
300	1.445	1.569	1.694	1.818	1.941	2.066
500	0.8726	0.9477	1.023	1.097	1.172	1.247
700	0.6263	0.6801	0.7340	0.7872	0.8407	0.8943
1000	0.4416	0.4794	0.5173	0.5549	0.5922	0.6297
1500	0.2980	0.3235	0.3488	0.3741	0.3991	0.4242
3000	0.1528	0.1656	0.1782	0.1908	0.2172	0.2206
5000	0.09903	0.1072	0.1150	0.1230	0.1307	0.1386
7000	0.07534	0.08124	0.08682	0.09260	0.09811	0.1037
10,000	0.05758	0.06177	0.06568	0.06980	0.07367	0.07760
	Internal Energy, *u* (Btu/lbm)					
0.2	261.99	280.65	299.64	318.93	338.63	358.57
1	261.99	280.65	299.64	318.93	338.63	358.57
5	261.98	280.64	299.62	318.91	338.62	358.55
10	261.97	280.64	299.61	318.91	338.60	358.54
14.7	261.96	280.63	299.60	318.89	338.59	358.53
50	261.87	280.53	299.48	318.79	338.49	358.44
100	261.76	280.47	299.45	318.72	338.47	358.42
150	261.68	280.36	299.36	318.68	338.40	358.34
300	261.29	280.00	298.99	318.39	338.15	358.08
500	260.77	279.52	298.49	318.00	337.82	357.74
700	260.41	279.26	298.26	317.76	337.60	357.55
1000	259.88	278.86	297.92	317.41	337.27	357.27
1500	259.02	277.98	297.28	316.78	336.72	356.79
3000	256.40	275.21	294.95	314.59	334.45	354.55
5000	252.65	272.18	291.70	311.66	331.60	352.04
7000	249.64	269.33	289.04	309.11	329.19	349.79
10,000	245.13	265.06	285.04	305.28	325.57	346.42

Abs. Press (psia)	Temperature (°F)					
p	1300	1400	1500	1600	1700	1800
	Specific Volume, v (ft^3/lbm)					
0.2	3258.44	3444.26	3629.47	3814.68	3999.89	4185.10
1	651.71	688.74	725.78	762.96	800.00	837.05
5	130.36	137.76	145.17	152.17	160.01	167.42
10	65.18	68.89	72.59	76.31	80.01	83.72
14.7	44.36	46.88	49.40	51.93	54.45	56.97
50	13.05	13.79	14.53	15.28	16.02	16.76
100	6.532	6.903	7.274	7.646	8.016	8.387
150	4.359	4.606	4.854	5.102	5.349	5.596
300	2.189	2.312	2.437	2.561	2.685	2.808
500	1.321	1.395	1.470	1.544	1.619	1.693
700	0.9473	1.000	1.054	1.107	1.160	1.213
1000	0.6671	0.7044	0.7417	0.7791	0.8164	0.8535
1500	0.4491	0.4736	0.4990	0.5240	0.5489	0.5737
3000	0.2330	0.2451	0.2586	0.2707	0.2836	0.2959
5000	0.1462	0.1542	0.1619	0.1695	0.1771	0.1847
7000	0.1092	0.1150	0.1205	0.1260	0.1314	0.1369
10,000	0.08145	0.08560	0.08949	0.09330	0.09720	0.1010
	Internal Energy, u (Btu/lbm)					
0.2	378.73	399.17	419.89	440.83	462.01	483.45
1	378.72	399.17	419.89	440.82	462.01	483.44
5	378.70	399.16	419.88	440.82	462.00	483.43
10	378.70	399.14	419.87	440.80	461.99	483.42
14.7	378.68	399.13	419.86	440.79	461.97	483.41
50	378.62	399.07	419.79	440.72	461.91	483.34
100	378.58	399.02	419.75	440.68	461.88	483.32
150	378.49	398.97	419.70	440.62	461.84	483.28
300	378.28	398.81	419.53	440.50	461.71	483.18
500	378.00	398.60	419.30	440.33	461.53	483.04
700	377.84	398.41	419.15	440.17	461.40	482.93
1000	377.59	398.13	418.93	439.93	461.21	482.76
1500	377.15	397.86	418.52	439.59	460.90	482.48
3000	375.18	396.00	416.63	437.93	459.38	480.67
5000	372.44	393.20	414.38	435.34	456.95	478.19
7000	370.28	391.07	412.34	433.39	455.03	476.38
10,000	367.05	387.87	409.29	430.46	452.16	473.66

TABLE A.13 (Continued) *Thermodynamic Properties of Dry Air—Superheated Air.*

Abs. Press. (psia)	Sat. Vap. Temp. (°F)	Sat. Vapor	Temperature (°F)			
p	T_g		−300	−200	−100	0
		Enthalpy, h (Btu/lbm)				
		h_g				
0.2	−358.7	83.94	98.34	122.27	146.21	170.16
1	−345.2	86.99	98.27	122.23	146.20	170.15
5	−327.6	91.00	98.08	122.14	146.15	170.12
10	−318.2	93.17	97.80	122.04	146.08	170.07
14.7	−312.4	94.41	97.61	121.92	146.01	170.03
50	−290.4	98.25		121.03	145.51	169.71
100	−247.7	99.36		119.71	144.77	169.29
150	−264.1	99.43		118.53	144.13	168.83
300	−242.8	96.94		112.69	141.83	167.42
500	−224.0	81.44		104.90	138.76	165.54
700					135.35	163.71
1000					130.22	160.96
1500					123.25	157.45
3000					115.39	152.10
5000					111.50	149.22
7000					114.08	150.48
10,000					117.97	152.36
		Entropy, s (Btu/lbm°R)				
		s_g				
0.2	−358.7	0.8756	0.9772	1.0936	1.1718	1.2305
1	−345.2	0.7906	0.8666	0.9832	1.0614	1.1201
5	−327.6	0.6970	0.7552	0.8726	0.9509	1.0097
10	−318.2	0.6774	0.7937	0.8248	0.9033	0.9622
14.7	−312.4	0.6581	0.6771	0.7982	0.8769	0.9357
50	−290.4	0.5987		0.7112	0.7915	0.8509
100	−274.7	0.5633		0.6609	0.7428	0.8026
150	−264.1	0.5412		0.6301	0.7138	0.7744
300	−242.8	0.4939		0.5618	0.6583	0.7228
500	−224.0	0.4301		0.5115	0.6174	0.6848
700					0.5868	0.6583
1000					0.5544	0.6303
1500					0.5051	0.5960
3000					0.4666	0.5509
5000					0.4382	0.5177
7000					0.4126	0.4919
10,000					0.3854	0.4645

Abs. Press. (psia)	Temperature (°F)					
p	100	200	300	400	500	600
			Enthalpy, h (Btu/lbm)			
0.2	194.15	218.21	242.40	266.78	291.40	316.29
1	194.14	218.20	242.40	266.78	291.40	316.29
5	194.12	218.19	242.39	266.77	291.39	316.29
10	194.09	218.16	242.37	266.76	291.39	316.28
14.7	194.06	218.15	242.36	266.75	291.38	316.28
50	193.84	218.00	242.28	266.69	291.34	316.27
100	193.52	217.78	242.11	266.59	291.28	316.24
150	193.26	217.60	241.99	266.51	291.24	316.22
300	192.36	216.85	241.58	265.58	291.07	316.15
500	191.16	216.17	341.04	265.90	290.89	316.07
700	189.97	215.39	240.51	265.59	290.71	316.01
1000	188.20	214.21	239.74	265.09	290.45	315.93
1500	186.01	212.81	238.82	264.55	290.18	315.89
3000	176.18	208.92	236.23	262.57	288.90	315.01
5000	176.90	207.13	235.93	263.31	290.63	317.24
7000	177.75	208.31	237.49	265.28	292.98	319.91
10,000	179.00	210.10	239.83	268.21	296.49	323.93
			Entropy, s (Btu/lbm°R)			
0.2	1.2778	1.3173	1.3514	1.3816	1.4087	1.4333
1	1.1674	1.2069	1.2411	1.2712	1.2984	1.3230
5	1.0570	1.0966	1.1307	1.1609	1.1880	1.2126
10	1.0095	1.0491	1.0832	1.1134	1.1405	1.1683
14.7	0.9831	1.0227	1.0568	1.0870	1.1142	1.1387
50	0.8986	0.9390	0.9726	1.0028	1.0299	1.0546
100	0.8507	0.8912	0.9249	0.9552	0.9823	1.0070
150	0.8225	0.8632	0.8969	0.9273	0.9545	0.9792
300	0.7725	0.8136	0.8482	0.8788	0.9063	0.9311
500	0.7357	0.7770	0.8123	0.8431	0.8707	0.8956
700	0.7106	0.7526	0.7882	0.8193	0.8471	0.8721
1000	0.6840	0.7267	0.7627	0.7941	0.8220	0.8471
1500	0.6524	0.6964	0.7331	0.7652	0.7932	0.8185
3000	0.6049	0.6454	0.6706	0.7159	0.7437	0.7690
5000	0.5699	0.6120	0.6478	0.6795	0.7073	0.7326
7000	0.5436	0.5852	0.6209	0.6526	0.6805	0.7059
10,000	0.5157	0.5568	0.5923	0.6241	0.6520	0.6776

TABLE A.13 (Continued) *Thermodynamic Properties of Dry Air—Superheated Air.*

Abs. Press. (psia)	Temperature (°F)					
p	700	800	900	1000	1100	1200
			Enthalpy, h (Btu/lbm)			
0.2	341.48	367.00	392.84	418.99	445.52	472.31
1	341.48	367.00	392.84	418.99	445.52	472.31
5	341.48	367.00	392.84	418.99	445.52	472.31
10	341.48	367.00	392.84	418.99	445.52	472.31
14.7	341.48	367.00	392.84	418.99	445.52	472.31
50	341.48	367.00	392.84	418.99	445.52	472.38
100	341.48	367.04	392.84	418.99	445.65	472.45
150	341.48	367.05	392.87	419.13	445.68	472.48
300	341.51	367.10	393.03	419.32	445.90	472.77
500	341.51	367.21	393.14	419.50	446.26	473.12
700	341.54	367.36	393.34	419.32	446.50	473.39
1000	341.60	367.57	393.65	420.09	446.86	473.80
1500	341.74	367.78	394.10	420.62	447.50	474.54
3000	341.23	367.14	393.88	420.51	449.80	477.05
5000	344.28	371.37	398.10	425.47	452.53	480.28
7000	347.23	374.56	401.50	429.06	456.28	484.12
10,000	351.68	379.37	406.58	434.44	461.90	490.02
			Entropy, s (Btu/lbm°R)			
0.2	1.4561	1.4772	1.4956	1.5119	1.5331	1.5497
1	1.3457	1.3668	1.3852	1.4015	1.4228	1.4394
5	1.2353	1.2565	1.2748	1.2912	1.3124	1.3291
10	1.1879	1.2090	1.2274	1.2438	1.2649	1.2815
14.7	1.1615	1.1826	1.2010	1.2174	1.2385	1.2551
50	1.0774	1.0985	1.1170	1.1333	1.1323	1.1489
100	1.0298	1.0510	1.0694	1.0857	1.1069	1.1235
150	1.0020	1.0232	1.0416	1.0580	1.0790	1.0957
300	0.9540	0.9753	0.9938	1.0102	1.0313	1.0480
500	0.9187	0.9400	0.9585	0.9749	0.9962	1.0129
700	0.8952	0.9166	0.9352	0.9516	0.9730	0.9896
1000	0.8703	0.8918	0.9104	0.9269	0.9484	0.9650
1500	0.8419	0.8635	0.8822	0.8988	0.9201	0.9368
3000	0.7924	0.8141	0.8329	0.8495	0.8709	0.8876
5000	0.7560	0.7777	0.7965	0.8131	0.8346	0.8514
7000	0.7294	0.7512	0.7701	0.7868	0:8086	0.8254
10,000	0.7012	0.7231	0.7422	0.7590	0.7808	0.7978

Abs. Press (psia)	Temperature (°F)					
	1300	1400	1500	1600	1700	1800
			Enthalpy, *h* (Btu/lbm)			
0.2	499.32	526.62	554.20	582.01	610.05	638.34
1	499.32	526.62	554.20	582.01	610.05	638.34
5	499.32	526.62	554.20	582.01	610.05	638.34
10	499.32	526.62	554.20	582.01	610.05	638.34
14.7	499.32	526.62	554.20	582.01	610.05	638.34
50	499.38	526.68	554.26	582.07	610.12	638.40
100	499.45	526.76	554.36	582.17	610.26	638.54
150	499.48	526.82	554.43	582.24	610.32	638.61
300	499.80	527.16	554.82	582.67	610.77	639.07
500	500.23	527.67	555.31	583.19	611.30	639.68
700	500.55	527.94	555.68	583.56	611.66	640.06
1000	501.04	528.48	556.18	584.10	612.28	640.70
1500	501.81	529.32	557.03	585.04	613.26	641.74
3000	504.51	532.05	560.20	588.22	616.81	644.95
5000	507.71	535.87	564.18	592.17	620.81	649.08
7000	511.73	540.03	568.43	696.60	625.24	653.71
10,000	517.77	546.27	574.89	603.11	632.03	660.56
			Entropy, *s* (Btu/lbm°R)			
0.2	1.5655	1.5806	1.5951	1.6089	1.6223	1.6350
1	1.4552	1.4703	1.4848	1.4987	1.5120	1.5247
5	1.3449	1.3600	1.3744	1.3883	1.4016	1.4144
10	1.3050	1.1805	1.3268	1.3407	1.3540	1.3668
14.7	1.2709	1.2860	1.3005	1.3144	1.3277	1.3404
50	1.1648	1.1799	1.1943	1.2082	1.2216	1.2343
100	1.1394	1.1545	1.1690	1.1828	1.1962	1.2089
150	1.1115	1.1266	1.1411	1.1549	1.1682	1.1810
300	1.0638	1.0789	1.0934	1.1073	1.1206	1.1333
500	1.0287	1.0438	1.0583	1.0722	1.0855	1.0982
700	1.0054	1.0206	1.0351	1.0490	1.0623	1.0750
1000	0.9808	0.9960	1.0105	1.0244	1.0377	1.0504
1500	0.9443	0.9678	0.9823	0.9962	1.0095	1.0222
3000	0.8952	0.9188	0.9333	0.9473	0.9606	0.9733
5000	0.8590	0.8827	0.8972	0.9112	0.9245	0.9373
7000	0.8331	0.8569	0.8715	0.8856	0.8990	0.9119
10,000	0.8057	0.8295	0.8442	0.8584	0.8719	0.8849

TABLE A.14 *Thermodynamic Properties of Dry Air—Compressed Liquid.*

Abs. Press. (psia)	Temperature (°F)		Temperature (°F)	
p	−325	−300	−275	−250
	Specific Volume, v (ft^3/lbm)			
100	0.01790	0.01931		
150	0.01788	0.01928	0.02127	
300	0.01784	0.01921	0.02113	0.02420
500	0.01779	0.01912	0.02094	0.02408
700	0.01773	0.01903	0.02076	0.02356
1000	0.01767	0.01892	0.02054	0.02302
3000	0.01727	0.01829	0.01950	0.02102
5000	0.01686	0.01775	0.01871	0.01982
7000	0.01658	0.01739	0.01820	0.01913
	Internal Energy, u (Btu/lbm)			
100	3.35	14.47		
150	3.31	14.39	26.27	
300	3.11	14.12	25.78	38.73
500	2.94	13.77	25.18	37.27
700	2.76	13.47	24.64	36.50
1000	2.46	12.97	23.87	35.20
3000	0.29	9.99	19.80	29.59
5000	−0.81	9.32	17.64	26.82
7000	−2.48	6.51	15.31	24.15
	Enthalpy, h (Btu/lbm)			
100	3.68	14.83		
150	3.81	14.93	26.86	
300	4.10	15.19	26.95	40.07
500	4.59	15.54	27.12	38.50
700	5.06	15.94	27.33	39.55
1000	5.66	16.47	27.67	39.46
3000	9.88	20.14	30.63	41.26
5000	14.79	25.75	34.95	45.16
7000	19.00	28.91	38.89	48.93
	Entropy, s (Btu/lbm°R)			
100	0.0264	0.1009		
150	0.0257	0.0996	0.1678	
300	0.0243	0.0983	0.1652	0.2317
500	0.0226	0.0954	0.1616	0.2255
700	0.0207	0.0931	0.1582	0.2199
1000	0.0183	0.0901	0.1540	0.2136
3000	0.0043	0.0728	0.1324	0.1858
5000	−0.0059	0.0565	0.1141	0.1655
7000	−0.0173	0.0450	0.1018	0.1525

Abs. Press. (psia) p	-225	-200	Sat. Temp. (°F) T_f	Sat. Liq.
		Specific Volume, v (ft³/lbm)		v_f
100			−286.6	0.02028
150			−268.5	0.02193
300			−246.4	0.02500
500	0.03232		−224.5	0.03436
700	0.02850	0.06820		
1000	0.02649	0.03666		
3000	0.02257	0.02469		
5000	0.02085	0.02216		
7000	0.01997	0.02097		
		Internal Energy, u (Btu/lbm)		u_f
100			−286.6	20.72
150			−268.5	29.48
300			−246.4	40.64
500	57.72		−224.5	58.98
700	51.55	82.24		
1000	48.13	67.04		
3000	39.71	50.39		
5000	36.14	45.40		
7000	33.07	41.70		
		Enthalpy, h (Btu/lbm)		h_f
100			−286.6	21.10
150			−268.5	30.09
300			−246.4	42.39
500	60.71		−224.5	62.16
700	55.24	91.07		
1000	53.03	73.82		
3000	52.24	64.10		
5000	55.43	65.90		
7000	58.94	68.86		
		Entropy, s (Btu/lbm°R)		s_f
100			−286.6	0.1380
150			−268.5	0.1850
300			−246.4	0.2426
500	0.3201		−224.5	0.3255
700	0.2921	0.4354		
1000	0.2752	0.3590		
3000	0.2360	0.2841		
5000	0.2125	0.2552		
7000	0.1984	0.2393		

TABLE A.15 *Specific Heat Capacities for Ideal Gases (Btu/lbm°R).*

Temp. T(°F)	Water Vapor			Carbon Dioxide		
	c_p	c_v	c_p/c_v	c_p	c_v	c_p/c_v
−100	0.442	0.331	1.333	0.174	0.129	1.346
0	0.443	0.333	1.331	0.191	0.145	1.311
100	0.446	0.336	1.328	0.205	0.160	1.283
200	0.451	0.341	1.323	0.217	0.172	1.262
300	0.457	0.347	1.318	0.229	0.184	1.246
400	0.464	0.354	1.312	0.239	0.193	1.233
500	0.472	0.361	1.305	0.247	0.202	1.223
600	0.480	0.370	1.298	0.256	0.210	1.214
700	0.489	0.378	1.291	0.263	0.217	1.208
800	0.497	0.387	1.285	0.269	0.224	1.202
900	0.506	0.396	1.278	0.275	0.230	1.196
1000	0.515	0.405	1.272	0.280	0.235	1.192
1100	0.525	0.414	1.243	0.285	0.240	1.188
1200	0.534	0.424	1.260	0.289	0.244	1.185
1300	0.543	0.433	1.255	0.293	0.248	1.182
1400	0.552	0.442	1.249	0.297	0.251	1.179
1500	0.561	0.451	1.244	0.300	0.255	1.177
1600	0.570	0.460	1.240	0.303	0.258	1.175
1700	0.579	0.469	1.235	0.306	0.261	1.173
1800	0.587	0.477	1.231	0.308	0.263	1.172
1900	0.596	0.486	1.227	0.310	0.265	1.170
2000	0.604	0.494	1.223	0.312	0.267	1.169
2100	0.612	0.502	1.220	0.314	0.269	1.168
2200	0.619	0.509	1.216	0.316	0.271	1.166
2300	0.627	0.516	1.213	0.318	0.273	1.165
2400	0.634	0.524	1.211	0.320	0.274	1.164
2500	0.640	0.530	1.208	0.322	0.277	1.163

Temp. T(°F)	Refrigerant R-22 c_p	c_v	c_p/c_v	Dry Air c_p	c_v	c_p/c_v
−100	0.126	0.103	1.223	0.239	0.171	1.402
0	0.142	0.119	1.193	0.240	0.171	1.401
100	0.158	0.135	1.170	0.240	0.172	1.400
200	0.173	0.150	1.153	0.241	0.173	1.397
300	0.186	0.163	1.141	0.243	0.174	1.394
400	0.198	0.175	1.131	0.245	0.176	1.389
500	0.208	0.185	1.124	0.247	0.179	1.383
600	0.216	0.193	1.119	0.250	0.181	1.378
700	0.224	0.201	1.114	0.254	0.185	1.371
800	0.230	0.207	1.111	0.257	0.188	1.364
900	0.237	0.214	1.107	0.260	0.191	1.358
1000	0.242	0.219	1.105	0.263	0.194	1.353
1100	0.247	0.224	1.103	0.266	0.197	1.347
1200	0.251	0.228	1.101	0.269	0.200	1.342
1300	0.254	0.231	1.099	0.271	0.203	1.338
1400	0.258	0.235	1.098	0.274	0.205	1.334
1500	0.260	0.238	1.097	0.276	0.208	1.330
1600	0.263	0.240	1.096	0.279	0.210	1.326
1700	0.265	0.242	1.095	0.281	0.212	1.323
1800	0.267	0.244	1.094	0.283	0.214	1.320
1900	0.270	0.247	1.093	0.284	0.216	1.318
2000	0.271	0.248	1.093	0.286	0.217	1.316
2100	0.273	0.250	1.092	0.287	0.219	1.313
2200	0.274	0.251	1.091	0.289	0.220	1.312
2300	0.277	0.254	1.091	0.290	0.221	1.310
2400	0.279	0.256	1.090	0.290	0.222	1.309
2500	0.281	0.258	1.089	0.291	0.223	1.308

TABLE A.15 (Continued) *Specific Heat Capacities for Ideal Gases (Btu/lbm°R).*

Temp.	Oxygen, O_2			Nitrogen, N_2			Hydrogen, H_2		
T(°F)	c_p	c_v	c_p/c_v	c_p		c_p/c_v	c_p	c_v	c_p/c_v
−100	0.217	0.155	1.400	0.248	0.177	1.400	3.229	2.244	1.439
0	0.218	0.156	1.399	0.248	0.177	1.400	3.395	2.400	1.415
100	0.220	0.158	1.394	0.248	0.178	1.399	3.426	2.441	1.404
200	0.223	0.161	1.387	0.249	0.178	1.398	3.451	2.466	1.399
300	0.226	0.164	1.378	0.250	0.179	1.396	3.461	2.476	1.398
400	0.230	0.168	1.369	0.251	0.180	1.394	3.465	2.480	1.397
500	0.234	0.172	1.360	0.253	0.182	1.390	3.469	2.484	1.397
600	0.239	0.176	1.352	0.255	0.184	1.385	3.474	2.489	1.396
700	0.243	0.181	1.344	0.258	0.187	1.379	3.480	2.495	1.395
800	0.246	0.184	1.337	0.261	0.190	1.373	3.486	2.501	1.394
900	0.250	0.187	1.331	0.263	0.193	1.367	3.499	2.514	1.392
1000	0.252	0.190	1.326	0.269	0.198	1.359	3.513	2.528	1.390
1100	0.255	0.193	1.322	0.272	0.201	1.353	3.530	2.545	1.387
1200	0.257	0.195	1.318	0.275	0.204	1.348	3.549	2.564	1.384
1300	0.259	0.197	1.314	0.278	0.207	1.343	3.570	2.585	1.381
1400	0.261	0.199	1.311	0.280	0.209	1.338	3.592	2.607	1.378
1500	0.263	0.201	1.309	0.283	0.212	1.334	3.618	2.633	1.374
1600	0.265	0.203	1.306	0.285	0.214	1.331	3.644	2.659	1.370
1700	0.266	0.204	1.304	0.287	0.217	1.327	3.671	2.686	1.367
1800	0.268	0.206	1.302	0.290	0.219	1.324	3.702	2.717	1.363
1900	0.269	0.207	1.300	0.292	0.221	1.321	3.729	2.744	1.359
2000	0.270	0.208	1.298	0.293	0.222	1.319	3.758	2.773	1.355
2100	0.271	0.209	1.297	0.295	0.224	1.316	3.787	2.802	1.352
2200	0.272	0.210	1.295	0.296	0.226	1.314	3.816	2.830	1.348
2300	0.274	0.212	1.293	0.298	0.227	1.312	3.844	2.859	1.345
2400	0.275	0.213	1.292	0.299	0.228	1.311	3.872	2.887	1.341
2500	0.276	0.214	1.290	0.300	0.229	1.309	3.898	2.913	1.338

TABLE A.16 *Ideal Gas Internal Energies of Various Substances (Btu/lbm).*

Temp. T(°F)	Methane, CH_4	Acetylene, C_2H_2	Ethylene, C_2H_4	Ethane, C_2H_6	Propane, C_3H_8	Octane, C_8H_{18}
−100	−68.4	−51.3	−49.1	−53.7	−53.0	
0	−29.5	−22.2	−21.4	−25.7	−24.9	
77	0	0	0	0	0	0
100	10.3	8.4	7.1	8.3	9.0	7.8
200	54.1	44.4	40.7	47.8	48.7	45.8
300	101.8	82.2	79.0	92.6	94.6	88.3
400	154.1	123.2	122.3	143.1	146.5	136.1
500	211.7	164.1	169.5	198.3	203.5	189.2
600	272.7	205.0	220.8	260.4	265.5	247.5
700	340.6	257.1	275.8	326.5	332.2	311.1
800	410.2	304.3	307.1	396.7	403.4	380.0
900	487.9	354.1	395.3	471.0	478.9	454.1
1000	566.9	404.2	459.5	549.1	557.6	533.5
1100	651.8	458.4	526.2	628.4	638.6	616.1
1200	739.3	509.1	600.0	715.9	723.5	702.3
1300	830.8	563.2	666.4	803.7	811.3	793.3
1400	925.9	618.3	740.1	894.7	902.0	
1500	1023.4	674.3	816.4	988.3	995.3	
1600	1125.0	730.2	893.3	1085.2	1090.9	
1700	1227.6	789.4	972.6	1183.7	1188.6	
1800	1334.9	848.6	1053.5	1284.5	1287.9	
1900	1443.1	907.2	1135.4	1387.5	1378.7	
2000	1554.4	963.2	1218.4	1492.7	1491.3	
2100	1669.7	1024.8	1302.1	1599.0	1595.9	
2200	1784.8	1086.9	1387.2	1707.4	1699.8	
2300	1903.1	1194.6	1473.4	1817.2	1804.2	
2400	2019.9	1212.9	1560.5	1928.2	1909.6	
2500	2149.8	1276.8	1648.6	2041.3	2016.0	

NOTE: Internal energy is taken as zero at 77°F.

Temp. T(°F)	Dry Air	Oxygen, O_2	Nitrogen, N_2	Water Vapor, H_2O	Carbon Dioxide, CO_2	Carbon Monoxide, CO	Hydrogen, H_2
−100	−30.0	−27.6	−31.1	−58.8	−25.4	−31.4	−415.5
0	−13.2	−12.0	−13.6	−25.7	−11.3	−13.7	−183.4
77	0	0	0	0	0	0	0
100	4.1	3.5	4.1	7.8	4.0	4.1	58.0
200	21.4	19.4	21.9	41.5	20.7	21.9	203.5
300	38.7	35.7	39.7	75.9	38.4	39.8	550.7
400	56.2	52.3	57.6	110.9	57.2	57.9	798.0
500	73.8	69.5	75.8	146.8	77.0	76.3	1046.6
600	92.0	86.9	94.3	183.2	97.7	95.0	1295.3
700	110.4	104.8	113.0	220.4	119.1	113.9	1544.3
800	129.0	123.1	132.0	258.4	141.0	133.4	1794.2
900	148.1	141.7	151.3	297.8	163.7	153.0	2045.9
1000	167.3	160.6	170.9	337.8	187.0	173.5	2297.1
1100	186.8	179.8	190.7	378.6	210.8	193.4	2550.9
1200	206.8	199.2	211.0	420.5	235.0	214.0	2806.2
1300	226.9	218.9	231.5	463.2	259.7	235.1	3063.7
1400	247.4	238.8	252.4	506.9	284.7	256.2	3323.4
1500	268.0	258.9	273.5	551.2	310.0	277.8	3585.4
1600	289.0	279.1	294.9	596.5	335.7	299.5	3849.4
1700	310.3	299.5	316.4	643.1	361.5	321.6	4116.9
1800	331.5	319.8	338.2	691.1	388.0	343.5	4386.8
1900	352.9	340.5	360.2	740.1	414.2	365.9	4627.0
2000	375.2	361.3	382.4	789.3	441.2	388.3	4935.8
2100	396.8	382.2	404.8	840.8	467.7	411.2	5214.5
2200	418.2	403.2	427.3	891.6	494.8	434.1	5496.5
2300	440.3	427.3	449.9	943.0	521.9	457.1	5781.1
2400	462.7	445.5	472.7	995.2	549.3	480.2	6068.3
2500	485.2	466.8	495.6	1047.9	576.8	503.3	6358.0
3000	598.8	574.6	611.5	1318.3	715.7	620.6	7847.8
4000	832.3	797.9	849.7	1900.9	1000.7	861.0	11,002
5000	1071.7	1030.4	1092.9	2561.7	1291.4	1105.9	14,339
6000	1315.4	1270.8	1339.5	3156.1	1585.9	1353.9	17,827
7000	1562.2	1517.6	1588.3	1810.9	1821.5	1604.1	21,445
8000	1811.5	1769.2	1839.0	4477.3	2183.4	1856.2	25,175
9000	2063.0	2023.9	2091.7	5152.5	2485.7	2109.5	29,017
10,000	2317.0	2281.3	2347.0	5836.1	2790.8	2364.7	32,960

TABLE A.17 *Ideal Gas Enthalpies of Various Substances (Btu/lbm).*

Temp. T(°F)	Methane, CH_4	Ethane, CH_6	Acetylene, C_2H_2	Ethylene, C_2H_4	Propane, C_3H_8	Octane, C_8H_{18}
−100	−90.3	−65.4	−64.8	−61.6	−61.0	
0	−39.1	−30.8	−28.1	−26.8	−28.4	
77	0	0	0	0	0	0
100	13.1	9.8	10.1	8.7	10.0	8.2
200	69.6	55.9	53.7	49.4	54.2	47.9
300	129.4	107.3	99.2	94.8	104.6	92.2
400	194.1	164.4	147.8	145.2	161.0	141.7
500	264.0	226.2	196.3	199.5	222.5	196.5
600	337.4	294.9	244.8	257.8	289.0	256.6
700	417.7	367.6	304.6	319.9	360.2	321.9
800	499.7	444.4	359.4	385.3	435.9	392.6
900	589.7	525.3	416.8	453.6	515.9	468.4
1000	681.1	610.0	474.6	524.8	599.1	549.5
1100	778.4	699.0	536.4	598.6	684.7	633.9
1200	878.3	790.0	594.7	679.5	774.1	721.8
1300	982.2	884.4	656.4	753.0	866.4	814.6
1400	1089.6	982.0	719.2	833.8	961.6	
1500	1199.5	1082.2	782.8	917.1	1059.4	
1600	1313.5	1185.8	846.3	1001.1	1159.5	
1700	1428.5	1290.9	913.1	1087.5	1261.7	
1800	1548.2	1398.3	980.0	1175.5	1365.5	
1900	1668.7	1507.9	1046.2	1264.5	1470.8	
2000	1792.4	1619.7	1109.8	1354.5	1577.9	
2100	1920.1	1732.6	1179.1	1445.3	1687.0	
2200	2047.6	1847.6	1248.8	1537.5	1795.4	
2300	2176.6	1964.0	1319.1	1630.8	1904.3	
2400	2307.5	2081.6	1390.1	1725.0	2014.2	
2500	2449.8	2201.3	1461.6	1820.2	2125.1	

NOTE: Enthalpy is taken as zero at 77°F.

Temp. T(°F)	Dry Air	Oxygen, O_2	Nitrogen, N_2	Water Vapor, H_2O	Carbon Dioxide, CO_2	Carbon Monoxide, CO	Hydrogen, H_2
−100	−42.1	−38.6	−43.9	−78.3	−33.4	−44.0	−589.8
0	−18.5	−16.8	−19.1	−34.2	−14.7	−19.2	−259.2
77	0	0	0	0	0	0	0
100	5.7	5.0	5.7	10.3	5.1	5.7	80.7
200	29.8	27.1	30.6	55.1	26.3	30.6	424.7
300	54.0	49.6	55.5	100.5	48.5	55.6	770.4
400	78.4	72.4	80.5	146.5	71.8	80.8	1116.6
500	103.0	95.8	105.8	193.4	96.1	106.3	1463.3
600	127.9	119.4	131.3	240.8	121.3	132.0	1810.5
700	153.1	143.5	157.1	289.1	147.2	158.0	2158.0
800	178.6	168.0	183.2	338.3	173.7	184.6	2506.4
900	204.5	192.8	209.6	388.5	200.9	211.3	2856.6
1000	230.6	217.9	236.3	439.5	228.7	238.3	3206.3
1100	257.0	243.3	263.2	491.4	257.0	265.9	3558.6
1200	283.8	268.9	290.6	544.3	285.7	293.6	3912.4
1300	310.8	294.8	318.2	598.0	314.9	321.8	4268.4
1400	338.1	320.9	346.1	652.7	344.4	350.0	4626.6
1500	365.6	347.2	374.3	708.1	374.3	378.7	4987.1
1600	393.4	373.6	402.8	764.4	404.5	407.4	5349.6
1700	421.6	400.2	431.4	822.0	434.8	436.6	5715.6
1800	449.6	426.7	460.3	881.0	465.8	465.6	6084.0
1900	477.9	453.7	489.4	941.0	496.5	495.1	6455.7
2000	507.1	480.7	518.7	1001.3	528.0	524.6	6830.0
2100	535.5	507.8	548.2	1063.9	559.0	554.6	7207.2
2200	563.8	535.0	577.8	1125.7	590.6	584.6	7587.7
2300	592.8	565.3	607.5	1188.1	622.3	614.7	7970.8
2400	622.0	589.7	637.4	1251.3	654.2	644.9	8356.5
2500	651.4	617.2	667.4	1315.1	686.2	675.1	8744.7
3000	799.2	756.0	818.7	1640.5	847.6	827.8	10,727
4000	1101.3	1041.4	1127.8	2333.3	1177.7	1139.1	14,866
5000	1409.2	1335.9	1441.9	3104.4	1513.5	1454.9	19,188
6000	1721.5	1638.4	1759.3	3809.0	1853.2	1773.8	22,035
7000	2036.8	1947.2	2079.0	4574.0	2133.9	2094.9	28,264
8000	2354.7	2260.9	2400.6	5350.7	2540.9	2417.9	32,979
9000	2674.7	2577.7	2724.2	6136.0	2888.4	2577.7	37,806
10,000	2997.3	2897.1	3050.4	6929.9	3238.6	3068.2	42,734

TABLE A.17 (Continued) *Ideal Gas Enthalpies of Various Substances (Btu/lbm).*

Temp. T(°F)	Hydroxyl, OH	Nitric Oxide, NO	Atomic Hydrogen, H	Atomic Nitrogen, N	Atomic Oxygen, O
0	−32.7	−18.3	−378.1	−27.0	−25.3
77	0	0	0	0	0
1000	384.5	228.0	4547.4	324.4	290.4
2000	821.4	501.3	9472.8	675.8	599.8
3000	1291.1	788.2	14,398	1027.2	908.2
4000	1785.2	1081.6	19,311	1378.8	1216.6
5000	2295.5	1378.6	24,249	1731.9	1525.7
6000	2817.0	1678.1	29,175	2089.4	1836.9
7000	3347.2	1979.3	34,100	2455.5	2151.4
8000	3884.1	2282.1	39,026	2834.6	2469.7
9000	4426.7	2586.1	43,951	3230.7	2792.1
10,000	4974.3	2891.4	48,877	3645.7	3118.6

TABLE A.18 *Heating Values of Some Hydrocarbon Fuels at Constant Pressure.*

Substance	Chemical Formula	Molecular Mass	HHV [Btu/lbm(fuel)]	LHV [Btu/lbm(fuel)]
Methane	CH_4	16.04	23,860	21,500
Acetylene	C_2H_2	26.04	21,470	20,730
Ethylene	C_2H_4	28.05	21,630	20,280
Ethane	C_2H_6	30.07	22,320	20,430
Propane	C_3H_8	44.10	21,660	19,940
Butane	C_4H_{10}	58.12	21,290	19,665
Octane	C_8H_{18}	114.23	20,760	19,265
Methanol	CH_4O	32.04	10,260	9,075
Ethanol	C_2H_6O	46.07	13,160	11,925

NOTE: Combustion reactants and products are at 77°F. Fuels supplied are in gaseous phase.

TABLE A.19 *Latent Heat of Vaporization of Several Hydrocarbon Fuels.*

Substance	h_{fg} (Btu/lbm)
Propane	147.1
Butane	155.8
Octane	156.1
Methanol	502.2
Ethanol	395.4

NOTE: Vaporization is at atmospheric pressure.

TABLE A.20 *Equilibrium Constant* (ln K_p) *for Various Reactions.*

Temp. T(°F)	$H_2O \rightleftharpoons H_2 + \frac{1}{2}O_2$	$H_2O \rightleftharpoons \frac{1}{2}H_2 + OH$	$CO_2 \rightleftharpoons CO + \frac{1}{2}O_2$	$CO_2 + H_2 \rightleftharpoons CO + H_2O$
77	−92.214	−106.234	−103.768	−11.554
1000	−30.091	−34.052	−31.460	−1.369
2000	−15.138	−16.770	−14.437	0.701
3000	−8.764	−9.437	−7.319	1.445
4000	−5.229	−5.383	−3.434	1.795
5000	−2.981	−2.815	−0.997	1.984
6000	−1.424	−1.043	0.667	2.091
7000	−0.281	0.252	1.874	2.155
8000	0.596	1.241	2.788	2.192
9000	1.293	2.020	3.498	2.206
10,000	1.859	2.696	4.109	2.210

Temp. T(°F)	$\frac{1}{2}N_2 + \frac{1}{2}O_2 \rightleftharpoons NO$	$H_2 \rightleftharpoons 2H$	$O_2 \rightleftharpoons 2O$	$N_2 \rightleftharpoons 2N$
77	−34.933	−163.999	−186.988	−367.493
1000	−11.886	−52.229	−59.295	−125.808
2000	−6.434	−25.414	−28.817	−68.343
3000	−4.132	−13.953	−15.878	−43.992
4000	−2.864	−7.570	−8.704	−30.510
5000	−2.063	−3.496	−4.144	−21.947
6000	−1.513	−0.670	−0.992	−16.011
7000	−1.115	1.409	1.323	−11.658
8000	−0.815	2.998	3.090	−8.322
9000	−0.580	4.258	4.482	−5.673
10,000	−0.379	5.347	5.696	−3.368

NOTE: Better accuracy in the value of the equilibrium constant is obtained by use of linear interpolation of the reciprocal of temperature, similar to that described for the specific volume of superheated substances.

TABLE A.21 *Availability and Entropy of Water*

(a) Saturated Steam

Abs Press (psia)	Temp °F	Availability (Btu/lbm) Sat. Liq.	Evap.	Sat. Vap.	Entropy (Btu/lbm°R) Sat. Liq.	Evap.	Sat. Vap.
p	T	a_f	a_{fg}	a_g	s_f	s_{fg}	s_g
1	101.74	0.5	45.0	45.5	0.1326	1.8455	1.9781
5	162.24	6.0	136.7	142.7	0.2349	1.6094	1.8443
10	193.21	11.3	173.8	185.1	0.2872	1.5007	1.7879
14.7	212.00	14.5	194.5	209.0	0.3121	1.4447	1.7568
30	250.34	23.1	230.3	253.4	0.3682	1.3313	1.6995
50	281.02	31.4	254.0	285.4	0.4112	1.2474	1.6586
100	327.82	45.8	282.7	328.5	0.4743	1.1284	1.6027
200	381.80	65.4	305.0	370.4	0.5438	1.0016	1.5454
300	417.35	80.1	313.6	393.7	0.5882	0.9223	1.5105
400	442.60	92.3	317.0	409.3	0.6217	0.8630	1.4847
500	467.01	103.0	317.6	420.6	0.6490	0.8149	1.4639
600	486.20	112.6	316.5	429.1	0.6723	0.7738	1.4461
700	503.08	121.5	314.1	435.6	0.6928	0.7376	1.4304
800	518.21	129.9	310.9	440.8	0.7111	0.7052	1.4163
900	531.95	137.8	307.0	444.8	0.7279	0.6753	1.4032
1000	544.58	145.4	302.5	447.9	0.7434	0.6476	1.3910
1500	596.20	179.5	274.4	453.9	0.8085	0.5288	1.3373
2000	635.80	210.9	237.7	448.6	0.8625	0.4256	1.2881
2500	668.11	242.9	189.4	432.3	0.9139	0.3206	1.2345
3000	695.33	281.4	116.9	398.3	0.9728	0.1891	1.1619
3208.2	705.47	338.1	0	338.1	1.0612	0	1.0612

The reference conditions of the surroundings are taken as $T_0 = 537$°R and $p_0 =$ 14.7 psia; hence $h_0 = 45.09$ Btu/lbm, $s_0 = 0.0876$ Btu/lmb°R. Availability is taken as $a = h - h_0 - T_0 (s - s_0)$. In this form, availability is also called *exergy*.

(*b*) *Superheated Steam and Compressed Water.*

Abs. Press. (psia) p	Sat. Temp. (°F) T_g	Sat. Liq. a_f	Sat. Vapor a_g	Availability (Btu/lbm) Temperature (°F) 100	200	300	400	500	600
0.5	79.59	0.1	4.4	4.5	9.7	20.8	36.3	55.4	77.7
1.0	101.74	0.5	45.5	0.4	50.8	61.8	77.3	96.4	118.6
5	162.24	6.0	142.7	0.4	145.2	156.6	172.4	191.6	213.8
10	193.21	11.3	185.2	0.4	185.8	197.2	213.0	232.3	254.6
50	281.02	31.4	285.4	0.6	12.3	288.2	305.2	325.4	348.2
100	327.82	45.8	328.5	0.7	12.3	37.0	342.5	363.6	378.2
500	467.01	103.0	420.6	1.9	13.6	38.1	73.4	431.9	463.6
1000	544.58	145.4	447.9	3.4	15.1	39.5	74.7	120.5	474.9
2000	635.80	210.9	448.6	6.3	18.0	42.4	77.3	122.5	182.0
3000	695.33	281.4	398.3	9.3	20.9	45.2	79.9	124.5	182.0
4000				12.2	23.8	47.9	82.5	126.7	182.5
5000				15.5	26.8	50.9	85.2	129.0	183.4
6000				18.1	29.5	53.6	87.8	131.2	184.6
7000				20.9	32.4	56.4	90.4	133.5	186.0
8000				23.9	35.3	59.2	93.1	135.8	187.7
9000				26.3	38.2	62.0	95.8	138.3	189.4
10000				29.7	41.1	64.8	98.5	140.6	191.2
15000				44.0	55.3	78.8	111.7	152.7	201.1

p (psia)	Temperature (°F) 700	800	900	1000	1100	1200	1300	1400	1500
0.5	102.6	130.0	159.0	191.1	224.1	259.4	297.7	354.6	374.4
1	143.6	167.5	199.7	232.5	265.8	301.0	336.2	376.2	415.9
5	238.7	266.2	295.7	327.3	360.8	376.0	432.8	471.3	511.2
10	279.6	307.1	336.6	368.2	401.7	437.0	473.9	512.3	552.2
50	373.5	401.2	430.0	462.6	496.3	531.6	568.5	607.1	647.0
100	412.9	440.8	470.7	502.7	536.4	571.8	608.9	647.5	687.5
500	493.9	524.4	556.2	589.4	624.3	660.8	698.7	738.0	778.7
1000	514.4	549.7	584.5	619.7	656.1	693.7	732.7	773.0	814.5
2000	502.1	554.5	597.8	638.2	678.2	718.7	759.9	802.0	845.3
3000	419.9	533.7	590.3	638.0	682.6	726.1	769.8	814.0	858.9
4000	263.2	491.4	572.1	629.0	679.1	726.3	772.7	818.0	865.7
5000	256.4	422.3	545.6	614.7	671.0	722.3	771.5	820.1	868.6
6000	253.4	370.9	513.3	596.6	660.2	715.9	768.0	818.8	869.1
7000	251.9	348.0	479.5	576.4	647.5	707.7	762.9	816.0	868.1
8000	251.3	337.5	452.6	555.7	633.7	698.6	756.8	812.0	865.9
9000	251.3	331.6	433.6	536.5	619.8	688.9	750.0	807.4	863.0
10000	251.7	327.9	410.2	520.2	605.7	678.9	742.9	802.3	859.6
15000	257.3	323.7	395.4	475.6	556.9	636.3	707.8	775.4	839.7

TABLE A.22 *Availability and Entropy of R-22 (Chlorodifluoromethane)*

(*a*) *Saturated Refrigerant R-22*

Abs. Press. (psia)	Temp. °F	Availability (Btu/lbm) Sat. Liq.	Evap.	Sat. Vap.	Entropy (Btu/lbm°R) Sat. Liq.	Evap.	Sat. Vap.
p	T	a_f	a_{fg}	a_g	s_f	s_{fg}	s_g
0.5	−137.65	25.63	−58.77	−33.14	−0.06291	0.34874	0.28583
1	−122.16	23.75	−49.79	−26.04	−0.05190	0.32701	0.27511
2	−104.87	22.16	−41.39	−19.23	−0.04060	0.30593	0.26533
5	−78.62	20.13	−30.27	−10.14	−0.02394	0.27693	0.25299
10	−55.59	18.74	−21.90	−3.16	−0.00969	0.25383	0.24414
20	−29.12	17.58	−13.58	4.00	0.00657	0.22906	0.23563
40	1.63	16.83	−5.75	11.08	0.02472	0.20299	0.22771
60	22.03	16.66	−2.55	14.11	0.03617	0.18721	0.22338
80	37.76	16.71	1.34	18.05	0.04520	0.17504	0.22024
100	50.77	16.86	4.42	20.28	0.05245	0.16540	0.21785
150	76.38	17.44	6.72	24.16	0.06626	0.14729	0.21355
200	96.27	18.17	8.74	26.91	0.07723	0.13300	0.21023
250	112.76	19.00	9.99	28.99	0.08656	0.12080	0.20736
300	126.98	19.83	10.69	30.52	0.09431	0.11056	0.20487
350	139.54	20.73	11.05	31.78	0.10163	0.10072	0.20235
400	150.82	21.61	11.10	32.71	0.10797	0.09200	0.19997
450	161.08	22.50	10.92	33.42	0.11395	0.08355	0.19750
500	170.50	23.53	10.45	33.98	0.12024	0.07434	0.19458
550	179.20	24.46	9.93	34.39	0.12564	0.06609	0.19173
600	187.29	25.69	8.70	34.39	0.13242	0.05518	0.18760
650	194.83	26.88	7.30	34.18	0.13873	0.04429	0.18302
700	201.88	29.45	3.79	33.24	0.14779	0.02676	0.17475
721.9	204.81	30.85	0	30.85	0.16016	0	0.16016

The reference conditions of the surroundings are taken as $T_0 = 490°R$ and $p_0 = 14.7$ psia; hence $h_0 = 110.59$ Btu/lbm, $s_0 = 0.26242$ Btu/lbm°R. Availability is taken as $a = h - h_0 - T_0\,(s - s_0)$. In this form, availability is also called *exergy*.

(*b*) *Superheated Refrigerant R-22*

Abs. Press (psia) p	Sat. Temp. (°F) T_g	Sat. Vapor Avail. a_g	Availability (Btu/lbm) Temperature (°F) −100	0	100	200	300	400
0.2	−155.80	−42.08	−45.32	−47.99	−47.43	−44.24	−38.71	−31.11
0.5	−137.64	−33.14	−35.01	−37.68	−37.12	−33.94	−28.41	−20.81
1	−122.16	−26.04	−27.21	−29.89	−29.33	−26.10	−20.57	−13.03
2	−104.87	−19.23	−19.41	−22.10	−21.55	−18.32	−12.78	−5.20
5	−78.62	−10.02		−11.83	−11.27	−8.04	−2.51	5.07
10	−55.58	−3.16		−4.12	−3.55	−0.30	5.24	12.82
15	−40.57	1.05		0.37	0.94	4.20	9.75	17.33
20	−29.12	4.00		3.53	4.10	7.38	12.95	20.52
40	1.63	11.08			14.94	17.48	20.55	28.13
60	22.03	14.11			15.87	19.25	24.92	32.51
80	37.76	18.05			18.79	22.24	27.95	35.68
100	50.77	20.28			20.98	29.40	30.26	37.95
120	61.95	22.04			22.71	26.29	32.11	39.83
140	71.83	23.56			24.11	27.76	33.64	41.22
160	80.71	24.85			25.26	29.00	34.93	42.76
180	88.81	25.96			26.23	30.05	36.04	43.97
200	96.27	26.91			27.04	30.97	37.02	44.95
250	112.76	28.99				32.77	39.00	51.94
300	126.98	30.52				34.08	40.50	48.66
350	139.54	31.78				35.04	41.68	49.96
400	150.82	32.71				35.71	42.61	55.44
450	161.08	33.42				36.15	43.35	51.91
500	170.50	33.98				36.36	43.92	52.15

appendix

B

Thermodynamic Properties (SI Units)

LIST OF TABLES

TABLE B.1 *Properties of Various Gases.*

Gas	Chemical Formula	Molecular Mass	Gas Constant, R (kJ/kg K)	c_p (kJ/kg K)	c_v (kJ/kg K)	$\gamma = c_p/c_v$
Acetylene	C_2H_2	26.038	0.3193	1.711	1.393	1.23
Air		28.967	0.2870	1.004	0.715	1.40
Ammonia	NH_3	17.031	0.4882	2.092	1.607	1.30
Argon	Ar	39.948	0.2081	0.519	0.310	1.67
Butane	C_4H_{10}	58.124	0.1430	1.674	1.536	1.09
Carbon dioxide	CO_2	44.009	0.1889	0.845	0.653	1.30
Carbon monoxide	CO	28.010	0.2968	1.042	0.745	1.40
Ethane	C_2H_6	30.070	0.2765	1.766	1.494	1.18
Ethylene	C_2H_4	28.054	0.2964	1.720	1.423	1.21
Hydrogen	H_2	2.016	4.124	14.35	10.21	1.40
Methane	CH_4	16.043	0.5182	2.223	1.686	1.32
Nitrogen	N_2	28.014	0.2968	1.038	0.741	1.40
Oxygen	O_2	31.998	0.2598	0.916	0.653	1.40
Propane	C_3H_8	44.097	0.1885	1.690	1.506	1.12
Refrigerant R-12	CCl_2F_2	120.92	0.0688	0.586	0.519	1.13
Refrigerant R-22	$CHClF_2$	86.476	0.0961	0.628	0.536	1.17
Water vapor	H_2O	18.015	0.4615	1.866	1.406	1.33

NOTE: c_p and c_v are for ideal gases at 25°C.

TABLE B.2 *Thermodynamic Properties of Water—Saturated Steam.*

Temp. (°C)	Abs. Press. (kPa)	Specific Volume (m^3/kg)			Internal Energy (kJ/kg)*		
		Sat. Liq.	Evap.	Sat. Vapor	Sat. Liq.	Evap.	Sat. Vap.
T	p	v_f	v_{fg}	v_g	u_f	u_{fg}	u_g
0.01	0.6112	0.0010002	206.16	206.16	0.000	2375.5	2375.5
5	0.8718	0.0010000	147.16	147.16	21.006	2361.4	2382.4
10	1.2270	0.0010003	106.43	106.43	41.993	2347.4	2389.4
15	1.7039	0.0010008	77.98	77.98	62.940	2333.2	2396.1
20	2.3366	0.0010017	57.84	57.84	83.859	2319.1	2403.0
25	3.1660	0.0010029	43.40	43.40	104.77	2305.0	2409.8
30	4.2415	0.0010043	32.93	32.93	125.66	2291.0	2416.6
35	5.6216	0.0010060	25.25	25.25	146.55	2276.8	2423.4
40	7.3748	0.0010078	19.55	19.55	167.44	2262.8	2430.2
45	9.5823	0.0010099	15.28	15.28	188.34	2248.5	2436.9
50	12.335	0.0010121	12.05	12.05	209.25	2234.3	2443.6
55	15.741	0.0010145	9.578	9.579	230.15	2220.1	2450.2
60	19.920	0.0010171	7.678	7.679	251.07	2205.7	2456.7
65	25.009	0.0010199	6.201	6.202	271.99	2191.3	2463.3
70	31.162	0.0010228	5.045	5.046	292.94	2176.7	2469.7
75	38.549	0.0010259	4.133	4.134	313.90	2162.1	2476.0
80	47.360	0.0010292	3.408	3.409	334.87	2147.5	2482.3
85	57.806	0.0010326	2.828	2.829	355.86	2132.6	2488.5
90	70.106	0.0010362	2.360	2.361	376.87	2117.7	2494.6
95	84.523	0.0010399	1.981	1.982	397.90	2123.7	2500.6
100	101.325	0.0010437	1.672	1.673	418.95	2087.5	2506.5
105	120.80	0.0010477	1.418	1.419	440.04	2072.2	2512.3
110	143.27	0.0010518	1.209	1.210	461.17	2077.9	2517.9
115	169.08	0.0010562	1.035	1.036	482.32	2041.2	2523.5
120	198.54	0.0010607	0.8904	0.8915	503.51	2025.5	2529.0
125	232.13	0.0010652	0.7691	0.7702	524.74	2009.5	2534.2
130	270.13	0.0010700	0.6670	0.6681	546.02	1993.4	2539.4
135	313.11	0.0010749	0.5817	0.5818	567.34	1977.1	2544.4
140	361.38	0.0010800	0.5074	0.5085	588.71	1960.6	2549.3
145	415.56	0.0010853	0.4449	0.4460	610.15	1943.8	2554.0
150	476.00	0.0010906	0.3913	0.3924	631.63	1927.0	2558.6
155	543.38	0.0010962	0.3453	0.3464	653.18	1909.8	2563.0
160	618.07	0.0011022	0.3057	0.3068	674.73	1874.7	2549.5
165	700.82	0.0011082	0.2713	0.2724	696.47	1874.6	2571.1
170	792.02	0.0011145	0.2415	0.2426	718.24	1856.7	2575.0
175	892.50	0.0011209	0.2154	0.2165	740.07	1838.5	2578.6
180	1002.66	0.0011274	0.1927	0.1938	761.99	1820.0	2582.0
185	1123.3	0.0011344	0.1728	0.1739	783.99	1801.1	2585.1

* The internal energy and the entropy of saturated water are taken as zero at the triple point (0.01°C).

Temp. (°C)	Enthalpy (kJ/kg) Sat. Liq.	Evap.	Sat. Vap.	Entropy (kJ/kg K)* Sat. Liq.	Evap.	Sat. Vap.
T	h_f	h_{fg}	h_g	s_f	s_{fg}	s_g
0.01	0.0007	2501.6	2501.6	0.0000	9.1575	9.1575
5	21.006	2489.7	2510.7	0.0762	8.9507	9.0269
10	41.994	2477.9	2519.9	0.1510	8.7510	8.9020
15	62.942	2466.0	2529.0	0.2244	8.5582	8.7827
20	83.861	2454.3	2538.2	0.2963	8.3721	8.6684
25	104.77	2442.5	2547.2	0.3672	8.1923	8.5591
30	125.66	2430.7	2556.4	0.4365	8.0181	8.4546
35	146.56	2418.8	2565.3	0.5049	7.8494	8.3543
40	167.45	2406.9	2574.4	0.5721	7.6862	8.2583
45	188.35	2394.9	2583.3	0.6383	7.5278	8.1661
50	209.26	2382.9	2592.2	0.7035	7.3741	8.0776
55	230.17	2370.8	2601.0	0.7677	7.2249	7.9926
60	251.09	2358.6	2609.7	0.8310	7.0798	7.9108
65	272.02	2346.3	2618.4	0.8933	6.9389	7.8322
70	292.97	2334.0	2626.9	0.9548	6.8017	7.7565
75	313.94	2321.5	2635.4	1.0154	6.6681	7.6835
80	334.92	2308.8	2643.8	1.0753	6.5379	7.6132
85	355.92	2296.5	2652.0	1.1343	6.4111	7.5454
90	376.94	2283.2	2660.1	1.1925	6.2874	7.4799
95	397.99	2270.2	2668.1	1.2501	6.1665	7.4166
100	419.06	2256.9	2676.0	1.3069	6.0485	7.3554
105	440.17	2243.6	2683.7	1.3630	5.9332	7.2962
110	461.32	2230.0	2691.3	1.4185	5.8203	7.2388
115	482.50	2216.2	2698.7	1.4733	5.7099	7.1832
120	503.72	2202.2	2706.0	1.5276	5.6017	7.1293
125	524.99	2188.0	2713.0	1.5813	5.4956	7.0769
130	546.31	2173.6	2719.9	1.6344	5.3917	7.0261
135	567.68	2158.9	2726.6	1.6869	5.2897	6.9766
140	589.10	2144.0	2733.1	1.7390	5.1894	6.9284
145	610.60	2128.7	2739.3	1.7906	5.0909	6.8815
150	632.15	2113.2	2745.4	1.8416	4.9942	6.8358
155	653.78	2097.4	2751.2	1.8923	4.8888	6.7811
160	675.47	2081.3	2756.7	1.9425	4.8050	6.7475
165	697.25	2064.8	2762.0	1.9923	4.7125	6.7048
170	719.12	2047.9	2767.1	2.0416	4.6214	6.6630
175	741.07	2030.7	2771.8	2.0906	4.5315	6.6221
180	763.12	2013.1	2776.3	2.1393	4.4426	6.5819
185	785.26	1995.2	2780.4	2.1876	4.3548	6.5424

TABLE B.2 (Continued) *Thermodynamic Properties of Water—Saturated Steam.*

Temp. (°C)	Abs. Press. (kPa)	Specific Volume (m^3/kg) Sat. Liq.	Specific Volume Evap.	Specific Volume Sat. Vap.	Internal Energy (kJ/kg) Sat. Liq.	Internal Energy Evap.	Internal Energy Sat. Vap.
T	p	v_f	v_{fg}	v_g	u_f	u_{fg}	u_g
190	1255.1	0.0011415	0.1552	0.1563	806.09	1782.0	2588.1
195	1399.4	0.0011489	0.1397	0.1408	828.27	1762.5	2590.8
200	1554.9	0.0011565	0.1261	0.1272	850.57	1742.5	2593.1
205	1724.4	0.0011644	0.1183	0.1150	872.98	1722.5	2595.5
210	1907.7	0.0011726	0.1030	0.1042	895.50	1701.9	2597 4
215	2106.1	0.0011811	0.09345	0.09463	918.14	1680.9	2599.0
220	2319.8	0.0011900	0.08485	0.08604	940.91	1659.4	2600.3
225	2550.2	0.0011992	0.07715	0.07835	963.83	1637.6	2601.4
230	2797.6	0.0012087	0.07024	0.07145	987.22	1614.9	2602.1
235	3063.3	0.0012187	0.06403	0.06525	1010.1	1592.4	2602.4
240	3347.8	0.0012291	0.05842	0.05965	1033.5	1569.0	2602.5
245	3652.4	0.0012399	0.05337	0.05461	1057.1	1545.1	2602.1
250	3977.6	0.0012513	0.04879	0.05004	1080.8	1520.5	2601.4
255	4324.8	0.0012632	0.04464	0.04590	1104.7	1495.5	2600.2
260	4694.4	0.0012756	0.04085	0.04213	1137.9	1460.7	2598.6
265	5087.9	0.0012887	0.03742	0.03871	1153.3	1443.2	2596.5
270	5505.8	0.0013025	0.03429	0.03559	1178.0	1415.9	2593.9
275	5949.8	0.0013170	0.03142	0.03274	1203.1	1387.6	2590.7
280	6420.2	0.0013324	0.02880	0.03013	1228.2	1383.9	2587.0
285	6918.8	0.0013487	0.02638	0.02773	1253.9	1328.8	2582.6
290	7446.1	0.0013659	0.02417	0.02554	1279.8	1297.6	2577.4
295	8003.9	0.0013844	0.02213	0.02351	1306.2	1265.4	2571.6
300	8592.7	0.0014041	0.02025	0.02165	1332.9	1232.0	2565.0
305	9214.5	0.0014252	0.01850	0.01993	1360.3	1197.2	2557.5
310	9869.8	0.0014480	0.01688	0.01833	1388.1	1161.0	2549.1
315	10,561	0.0014726	0.01539	0.01686	1416.5	1123.0	2539.5
320	11,288	0.0014995	0.01398	0.01548	1445.7	1083.3	2529.0
325	12,056	0.0015289	0.01266	0.01419	1475.6	1041.4	2516.9
330	12,863	0.0015615	0.01143	0.01299	1506.4	996.7	2503.1
335	13,712	0.0015978	0.01025	0.01185	1538.4	948.8	2487.2
340	14,605	0.0016387	0.009142	0.010781	1571.6	897.2	2468.7
345	15,546	0.0016858	0.008077	0.009763	1606.3	840.8	2447.1
350	16,535	0.0017411	0.007058	0.008799	1643.1	779.1	2422.2
355	17,577	0.0018085	0.006051	0.007859	1684.8	707.5	2392.3
360	18,675	0.0018959	0.001896	0.006940	1728.8	627.0	2355.8
365	19,833	0.0020160	0.003996	0.006012	1778.0	530.7	2308.8
370	21,054	0.0022136	0.002759	0.004973	1843.6	394.5	2238.1
374.15	22,120	0.003170	0	0.003170	2037.3	0	2037.3

Temp. (°C)	Enthalpy (kJ/kg) Sat. Liq.	Evap.	Sat. Vap.	Entropy (kJ/kg K) Sat. Liq.	Evap.	Sat. Vap.
T	h_f	h_{fg}	h_g	s_f	s_{fg}	s_g
190	807.52	1976.7	2784.3	2.2356	4.2680	6.5036
195	829.88	1957.9	2787.8	2.2833	4.1821	6.4654
200	852.37	1938.6	2790.9	2.3307	4.0971	6.4278
205	874.99	1918.8	2793.8	2.3778	4.0128	6.3906
210	897.74	1898.5	2796.2	2.4247	3.9292	6.3539
215	920.63	1877.6	2798.3	2.4713	3.8463	6.3176
220	943.67	1856.2	2799.9	2.5178	3.7639	6.2817
225	966.89	1834.3	2801.2	2.5641	3.6820	6.2461
230	990.26	1811.7	2802.0	2.6102	3.6005	6.2107
235	1013.8	1788.5	2802.3	2.6562	3.5194	6.1756
240	1037.6	1764.6	2802.2	2.7020	3.4386	6.1406
245	1061.6	1740.0	2801.6	2.7478	3.3309	6.1057
250	1085.8	1714.6	2800.4	2.7935	3.2773	6.0708
255	1110.2	1688.5	2798.7	2.8392	3.1967	6.0359
260	1134.9	1661.5	2796.4	2.8848	3.1162	6.0010
265	1159.9	1633.6	2793.5	2.9306	3.0352	5.9658
270	1185.2	1604.6	2789.9	2.9763	2.9541	5.9304
275	1210.9	1574.7	2785.5	3.0223	2.8724	5.8947
280	1236.8	1543.6	2780.4	3.0683	2.7903	5.8586
285	1263.2	1511.3	2774.5	3.1146	2.7074	5.8220
290	1290.0	1477.6	2767.6	3.1611	2.6237	5.7848
295	1317.3	1442.6	2759.8	3.2079	2.5390	5.7469
300	1345.0	1406.0	2751.0	3.2552	2.4529	5.7081
305	1373.4	1367.7	2741.1	3.3029	2.3656	5.6685
310	1402.4	1327.6	2730.0	3.3512	2.2766	5.6278
315	1432.1	1285.5	2717.6	3.4002	2.1856	5.5858
320	1462.6	1241.1	2703.7	3.4500	2.0923	5.5423
325	1494.0	1194.0	2688.0	3.5008	1.9961	5.4969
330	1526.5	1143.6	2670.2	3.5528	1.8962	5.4490
335	1560.3	1089.5	2649.7	3.6063	1.7916	5.3979
340	1595.5	1030.7	2626.2	3.6616	1.6811	5.3427
345	1632.5	996.4	2598.9	3.7193	1.5635	5.2828
350	1671.8	895.7	2567.7	3.7800	1.4377	5.2177
355	1716.6	813.8	2530.4	3.8489	1.2953	5.1442
360	1764.2	721.3	2485.4	3.9210	1.1390	5.0600
365	1818.0	610.0	2428.0	4.0021	0.9558	4.9579
370	1890.2	452.6	2342.8	4.1108	0.7036	4.8144
374.15	2107.4	0	2107.4	4.4429	0	4.4429

TABLE B.2(a) *Thermodynamic Properties of Water—Saturated Steam.*

Abs. Press. (kPa)	Temp. (°C)	Specific Volume (m^3/kg) Sat. Liq.	Evap.	Sat. Vap.	Internal Energy (kJ/kg) Sat. Liq.	Evap.	Sat. Vap.
p	T	v_f	v_{fg}	v_g	u_f	u_{fg}	u_g
105	101.00	0.0010445	1.617	1.618	423.17	2084.5	2507.7
101.325	100.00	0.0010437	1.672	1.673	418.95	2087.5	2506.5
100	99.63	0.0010434	1.693	1.694	417.41	2088.6	2506.0
95	98.20	0.0010423	1.776	1.777	411.39	2093.0	2504.4
90	96.71	0.0010412	1.868	1.869	405.12	2097.6	2502.7
85	95.15	0.0010400	1.971	1.972	398.54	2102.2	2500.8
80	93.51	0.0010387	2.086	2.087	391.64	2107.2	2498.8
75	91.79	0.0010375	2.216	2.217	384.37	2112.4	2496.7
70	89.96	0.0010361	2.364	2.365	376.70	2117.9	2494.6
65	88.02	0.0010347	2.534	2.535	368.55	2123.6	2492.1
60	85.95	0.0010333	2.731	2.732	359.87	2129.8	2489.7
55	83.74	0.0010317	2.963	2.964	350.55	2136.3	2486.9
50	81.35	0.0010301	3.239	3.240	340.51	2143.5	2484.0
45	78.74	0.0010284	3.575	3.576	329.59	2151.2	2480.8
40	75.89	0.0010265	3.992	3.993	317.61	2159.6	2477.2
35	72.70	0.0010245	4.528	4.529	304.26	2168.7	2473.0
30	69.12	0.0010223	5.228	5.229	289.27	2179.3	2468.5
25	64.99	0.0010199	6.203	6.204	271.96	2191.2	2463.2
20	60.09	0.0010172	7.649	7.650	251.43	2205.5	2456.9
15	54.00	0.0010140	10.02	10.02	225.95	2222.9	2448.9
10	45.83	0.0010102	14.67	14.67	191.82	2246.3	2438.1
7.5	40.32	0.0010079	19.24	19.24	168.76	2261.8	2430.6
5.0	32.90	0.0010052	28.19	28.19	137.76	2282.9	2420.7
2.5	21.10	0.0010020	54.26	54.26	88.45	2316.2	2404.6
1.0	6.98	0.0010001	129.20	129.20	29.34	2355.9	2385.2

Abs. Press. (kPa)	Enthalpy (kJ/kg) Sat. Liq.	Evap.	Sat. Vap.	Entropy (kJ/kg K) Sat. Liq.	Evap.	Sat. Vap.
p	h_f	h_{fg}	h_g	s_f	s_{fg}	s_g
105	423.28	2254.3	2677.6	1.3182	6.0252	7.3434
101.325	419.06	2256.9	2676.0	1.3069	6.0485	7.3554
100	417.51	2257.9	2675.4	1.3027	6.0571	7.3598
95	411.49	2261.7	2673.2	1.2865	6.0906	7.3771
90	405.21	2265.6	2670.9	1.2696	6.1258	7.3954
85	398.63	2269.8	2668.4	1.2518	6.1629	7.4147
80	391.72	2274.0	2665.8	1.2330	6.2022	7.4352
75	384.45	2278.6	2663.0	1.2131	6.2439	7.4570
70	376.77	2283.3	2660.1	1.1921	6.2883	7.4804
65	368.62	2288.3	2656.9	1.1696	6.3359	7.5055
60	359.93	2293.6	2653.6	1.1454	6.3873	7.5327
55	350.61	2299.3	2649.9	1.1194	6.4429	7.5623
50	340.56	2305.4	2646.0	1.0912	6.5035	7.5947
45	329.64	2312.0	2641.7	1.0603	6.5704	7.6307
40	317.65	2319.2	2636.9	1.0261	6.6448	7.6709
35	304.30	2327.2	2631.5	0.9877	6.7291	7.7168
30	289.30	2336.1	2625.4	0.9441	6.8254	7.7695
25	271.99	2346.4	2618.3	0.8932	6.9391	7.8323
20	251.45	2358.4	2609.9	0.8321	7.0773	7.9094
15	225.97	2373.2	2599.2	0.7549	7.2544	8.0093
10	191.83	2392.9	2584.8	0.6493	7.5018	8.1511
7.5	168.77	2406.2	2574.9	0.5763	7.6760	8.2523
5.0	137.77	2423.8	2561.6	0.4763	7.9197	8.3960
2.5	88.45	2451.7	2540.2	0.3119	8.3321	8.6440
1.0	29.34	2485.0	2514.4	0.1060	8.8707	8.9767

TABLE B.3 *Thermodynamic Properties of Water—Superheated Steam.*

Specific Volume, v (m^3/kg)

Abs. Press. (kPa)	Sat. Temp. (°C)	Sat. Spec. Vol.	Temperature (°C)				
p	T_g	v_g	50	100	150	200	250
1	6.98	129.2	149.14	172.19	195.29	218.35	241.44
5	32.90	28.19	29.78	34.42	39.04	43.66	48.28
10	45.83	14.67	14.87	17.20	19.51	21.83	24.14
20	60.09	7.650		8.585	9.748	10.907	12.064
40	75.89	3.993		4.279	4.866	5.448	6.028
60	85.95	2.732		2.844	3.238	3.628	4.016
80	93.51	2.087		2.126	2.425	2.718	3.010
100	99.63	1.694		1.696	1.936	2.172	2.406
200	120.23	0.8854			0.9595	1.0804	1.1989
300	133.54	0.6056			0.6337	0.7164	0.7964
400	143.62	0.4622			0.4707	0.5343	0.5952
500	151.84	0.3747				0.4250	0.4744
600	158.84	0.3155				0.3520	0.3939
700	164.96	0.2727				0.2999	0.3364
800	170.41	0.2403				0.2608	0.2932
900	175.36	0.2148				0.2303	0.2596
1000	179.88	0.1943				0.2059	0.2326
1500	198.29	0.1317				0.1324	0.1520
2000	212.37	0.09954					0.1114
2500	223.94	0.07991					0.08699
3000	233.84	0.06663					0.07055
3500	242.54	0.05703					0.05869
4000	250.33	0.04975					
4500	257.41	0.04404					
5000	263.91	0.03943					
6000	275.55	0.03244					
7000	285.79	0.02737					
8000	294.97	0.02353					
9000	303.31	0.02050					
10,000	310.96	0.01804					
20,000	365.70	0.005877					
30,000							
40,000							
50,000							
60,000							
70,000							
80,000							
90,000							
100,000							

Specific Volume, v (m^3/kg)

Abs. Press. (kPa)	Temperature (°C)					
p	300	350	400	450	500	550
1	264.51	287.59	310.67	333.74	356.82	379.89
5	52.90	57.52	62.13	66.74	71.36	75.98
10	26.45	28.76	31.06	33.37	35.68	37.99
20	13.219	14.374	15.529	16.684	17.838	18.992
40	6.607	7.185	7.763	8.340	8.918	9.495
60	4.402	4.788	5.174	5.559	5.944	6.329
80	3.300	3.590	3.879	4.168	4.457	4.746
100	2.639	2.871	3.102	3.334	3.565	3.797
200	1.3162	1.4328	1.5492	1.6653	1.7812	1.8971
300	0.8753	0.9535	1.0314	1.1090	1.1865	1.2639
400	0.6549	0.7139	0.7725	0.8309	0.8892	0.9474
500	0.5226	0.5701	0.6172	0.6640	0.7108	0.7574
600	0.4344	0.4742	0.5136	0.5528	0.5918	0.6308
700	0.3714	0.4057	0.4396	0.4733	0.5069	0.5403
800	0.3241	0.3543	0.3842	0.4137	0.4432	0.4725
900	0.2874	0.3144	0.3410	0.3674	0.3936	0.4197
1000	0.2580	0.2824	0.3065	0.3303	0.3540	0.3775
1500	0.1697	0.1865	0.2029	0.2191	0.2350	0.2509
2000	0.1255	0.1386	0.1511	0.1634	0.1756	0.1876
2500	0.09893	0.10975	0.12004	0.13004	0.13987	0.14958
3000	0.08116	0.09053	0.09931	0.10779	0.11608	0.12426
3500	0.06842	0.07678	0.08449	0.09189	0.09909	0.10617
4000	0.05883	0.06645	0.07338	0.07996	0.08634	0.09260
4500	0.05134	0.05840	0.06472	0.07068	0.07643	0.08204
5000	0.04530	0.05194	0.05779	0.06325	0.06849	0.07360
6000	0.03614	0.04222	0.04738	0.05210	0.05659	0.06094
7000	0.02946	0.03523	0.03992	0.04413	0.04809	0.05189
8000	0.02426	0.02995	0.03431	0.03814	0.04170	0.04510
9000		0.02579	0.02993	0.03348	0.03674	0.03982
10,000		0.02242	0.02641	0.02974	0.03276	0.03560
20,000			0.009947	0.01271	0.01477	0.01655
30,000			0.002831	0.006735	0.008681	0.01017
40,000			0.001909	0.003675	0.005616	0.006982
50,000			0.001729	0.002492	0.003882	0.005113
60,000			0.001632	0.002084	0.002952	0.003947
70,000			0.001567	0.001890	0.002467	0.003222
80,000			0.001518	0.001772	0.002188	0.002764
90,000			0.001479	0.001691	0.002013	0.002458
100,000			0.001446	0.001629	0.001893	0.002246

TABLE B.3 (Continued) *Thermodynamic Properties of Water—Superheated Steam.*
Specific Volume, v (m^3/kg)

Abs. Press. (kPa)	Temperature (°C)					
p	600	650	700	750	800	850
1	402.97	426.05	449.12	472.20	495.27	518.35
5	80.59	85.21	89.82	94.44	99.05	103.67
10	40.29	42.60	44.91	47.22	49.52	51.84
20	20.146	21.300	22.455	23.609	24.762	25.918
40	10.072	10.649	11.227	11.804	12.381	12.959
60	6.714	7.099	7.484	7.869	8.254	8.639
80	5.035	5.324	5.613	5.901	6.190	6.479
100	4.028	4.259	4.490	4.721	4.952	5.183
200	2.0129	2.1286	2.2442	2.3598	2.4754	2.591
300	1.3412	1.4185	1.4957	1.5728	1.6499	1.7271
400	1.0054	1.0634	1.1214	1.1793	1.2372	1.2950
500	0.8039	0.8504	0.8968	0.9432	0.9896	1.0358
600	0.6696	0.7084	0.7471	0.7858	0.8245	0.8631
700	0.5737	0.6070	0.6402	0.6734	0.7066	0.7397
800	0.5017	0.5309	0.5600	0.5891	0.6181	0.6471
900	0.4458	0.4717	0.4976	0.5235	0.5493	0.5751
1000	0.4010	0.4244	0.4477	0.4710	0.4943	0.5174
1500	0.2667	0.2824	0.2980	0.3136	0.3292	0.3447
2000	0.1995	0.2114	0.2232	0.2349	0.2467	0.2583
2500	0.15921	0.16876	0.17826	0.18772	0.19714	0.2065
3000	0.13234	0.14036	0.14832	0.15624	0.16412	0.1720
3500	0.11315	0.12007	0.12694	0.13376	0.14054	0.1473
4000	0.09876	0.10486	0.11090	0.11689	0.12285	0.1288
4500	0.08757	0.09303	0.09843	0.10378	0.10910	0.1144
5000	0.07862	0.08356	0.08845	0.09329	0.09809	0.1028
6000	0.06518	0.06936	0.07348	0.07755	0.08159	0.08554
7000	0.05559	0.05922	0.06279	0.06631	0.06980	0.07321
8000	0.04839	0.05161	0.05477	0.05788	0.06096	0.06398
9000	0.04280	0.04570	0.04853	0.05133	0.05408	0.05678
10,000	0.03832	0.04096	0.04355	0.04608	0.04858	0.05103
20,000	0.01816	0.01967	0.02111	0.02250	0.02385	0.02515
30,000	0.01144	0.01258	0.01365	0.01465	0.01562	0.01655
40,000	0.008088	0.009053	0.009930	0.01075	0.01152	0.01227
50,000	0.006111	0.006960	0.007720	0.008420	0.009076	0.009698
60,000	0.004835	0.005596	0.006269	0.006885	0.007460	0.008003
70,000	0.003972	0.004652	0.005257	0.005808	0.006321	0.006809
80,000	0.003379	0.003974	0.004519	0.005017	0.005481	0.005915
90,000	0.002967	0.003476	0.003964	0.004419	0.004841	0.005237
100,000	0.002668	0.003106	0.003536	0.003952	0.004341	0.004700

Internal Energy, u (kJ/kg)

Abs. Press. (kPa)	Sat. Temp. (°C)	Sat. Int. Energy	Temperature (°C)				
p	T_g	u_g	50	100	150	200	250
1	6.98	2385.2	2445.5	2516.4	2588.4	2661.7	2736.3
5	32.90	2420.7	2444.8	2516.0	2588.2	2661.6	2736.2
10	45.83	2438.1	2444.0	2515.5	2588.0	2661.3	2736.0
20	60.09	2456.9		2514.6	2587.3	2661.1	2735.8
40	75.89	2477.2		2512.6	2586.3	2660.3	2735.4
60	85.95	2489.7		2510.7	2585.1	2659.6	2734.8
80	93.51	2498.8		2508.7	2583.8	2658.9	2734.4
100	99.63	2506.0		2506.6	2582.7	2658.2	2733.9
200	120.23	2529.2			2576.6	2654.4	2731.4
300	133.54	2543.0			2570.3	2650.6	2729.0
400	143.62	2552.7			2563.7	2646.7	2726.4
500	151.84	2560.2				2642.6	2723.9
600	158.84	2566.2				2638.5	2721.3
700	164.96	2571.1				2634.3	2718.5
800	170.41	2575.3				2630.0	2715.8
900	175.36	2578.8				2625.4	2713.2
1000	179.88	2581.9				2620.9	2710.4
1500	198.29	2592.4				2596.1	2695.5
2000	212.37	2598.1					2679.6
2500	223.94	2601.1					2662.0
3000	233.84	2602.4					2643.2
3500	242.54	2602.4					2622.7
4000	250.33	2601.3					
4500	257.41	2599.5					
5000	263.91	2597.1					
6000	275.55	2590.4					
7000	285.79	2581.9					
8000	294.97	2571.7					
9000	303.31	2560.1					
10,000	310.96	2547.3					
20,000	365.70	2300.9					
30,000							
40,000							
50,000							
60,000							
70,000							
80,000							
90,000							
100,000							

TABLE B.3 (Continued) *Thermodynamic Properties of Water—Superheated Steam.*

Internal Energy, *u* (kJ/kg)

Abs. Press. (kPa)	Temperature (°C)					
p	300	350	400	450	500	550
1	2812.3	2889.9	2969.0	3049.9	3132.4	3216.6
5	2812.2	2889.8	2969.0	3049.9	3132.4	3216.6
10	2812.1	2889.7	2969.0	3049.8	3132.3	3216.6
20	2812.0	2889.6	2968.8	3049.7	3132.2	3216.6
40	2811.6	2889.4	2968.6	3049.5	3132.1	3216.4
60	2811.3	2889.1	2968.4	3049.4	3132.0	3216.3
80	2811.0	2888.8	2968.2	3049.2	3131.8	3216.1
100	2810.6	2888.5	2968.0	3049.0	3131.6	3215.9
200	2808.9	2887.2	2966.9	3048.0	3130.8	3215.3
300	2807.1	2885.9	2965.8	3047.1	3130.1	3214.5
400	2805.2	2884.4	2964.6	3046.1	3129.2	3213.8
500	2803.5	2883.1	2963.5	3045.2	3128.4	3213.1
600	2801.7	2881.7	2962.4	3044.3	3127.6	3212.4
700	2799.8	2880.3	2961.3	3043.4	3126.8	3211.7
800	2798.0	2879.0	2960.1	3042.4	3125.9	3211.0
900	2796.0	2877.5	2959.1	3041.4	3125.2	3210.4
1000	2794.1	2876.1	2957.9	3040.5	3124.3	3209.6
1500	2784.4	2869.4	2952.3	3035.7	3120.3	3206.1
2000	2774.0	2861.4	2946.5	3031.0	3116.1	3202.4
2500	2763.1	2853.8	2940.6	3026.2	3112.0	3199.0
3000	2751.6	2845.9	2934.6	3021.2	3108.0	3195.3
3500	2739.5	2837.8	2928.5	3016.4	3103.8	3191.8
4000	2726.7	2829.3	2922.2	3011.4	3099.6	3188.2
4500	2713.2	2820.5	2915.9	3006.3	3095.4	3184.6
5000	2699.0	2811.5	2909.4	3001.3	3091.3	3181.0
6000	2668.2	2792.5	2895.8	2990.9	3082.7	3173.7
7000	2633.2	2772.1	2881.3	2980.2	3074.0	3166.4
8000	2592.7	2750.3	2867.1	2969.2	3065.2	3158.9
9000		2726.9	2851.8	2957.9	3056.1	3151.4
10,000		2701.6	2835.8	2946.2	3047.0	3143.8
20,000			2621.6	2810.1	2945.7	3063.1
30,000			2076.9	2623.6	2824.6	2972.3
40,000			1857.7	2368.6	2682.2	2872.3
50,000			1791.3	2168.6	2528.9	2765.5
60,000			1749.4	2062.1	2393.5	2659.4
70,000			1718.1	1997.6	2294.4	2565.2
80,000			1692.8	1952.3	2222.4	2486.9
90,000			1671.5	1917.1	2168.7	2421.8
100,000			1653.0	1888.3	2126.8	2369.2

Internal Energy, u (kJ/kg)

Abs. Press (kPa)	Temperature (°C)					
p	600	650	700	750	800	850
1	3302.6	3390.3	3479.7	3570.8	3663.4	3758.6
5	3302.6	3390.3	3479.7	3570.8	3663.4	3758.6
10	3302.6	3390.3	3479.7	3570.8	3663.4	3758.6
20	3302.5	3390.2	3479.6	3570.7	3663.4	3758.5
40	3302.4	3390.1	3479.5	3570.6	3663.4	3758.4
60	3302.3	3390.1	3479.5	3570.6	3663.3	3758.4
80	3302.2	3389.9	3479.4	3570.5	3663.2	3758.3
100	3302.0	3389.8	3479.2	3570.4	3663.1	3758.2
200	3301.4	3389.3	3478.8	3569.9	3662.7	3758.0
300	3300.8	3388.7	3478.3	3569.6	3662.3	3757.7
400	3300.1	3388.1	3477.8	3569.1	3662.0	3757.4
500	3299.6	3387.6	3477.4	3568.7	3661.6	3757.1
600	3298.9	3387.1	3476.8	3568.3	3661.2	3756.7
700	3298.3	3386.5	3476.4	3567.8	3660.9	3756.4
800	3297.7	3386.0	3475.9	3567.4	3660.5	3756.1
900	3297.0	3385.5	3475.5	3567.0	3660.1	3755.8
1000	3296.4	3384.9	3475.0	3566.6	3659.8	3755.5
1500	3293.3	3382.1	3472.6	3564.5	3657.9	3754.1
2000	3290.2	3379.3	3470.1	3562.4	3656.0	3752.5
2500	3287.1	3376.7	3467.8	3560.2	3654.2	3751.0
3000	3284.0	3373.9	3465.3	3558.1	3652.3	3749.3
3500	3280.9	3371.2	3462.9	3555.9	3650.5	3747.7
4000	3277.8	3368.5	3460.5	3553.8	3648.6	3746.1
4500	3274.5	3365.7	3458.1	3551.8	3646.8	3744.5
5000	3271.4	3362.9	3455.7	3549.7	3644.9	3743.4
6000	3265.1	3357.3	3450.8	3545.4	3641.2	3740.2
7000	3258.8	3351.9	3445.9	3541.1	3637.4	3737.0
8000	3252.4	3346.3	3441.0	3536.9	3633.6	3733.0
9000	3245.9	3340.7	3436.2	3532.5	3630.0	3730.6
10,000	3239.5	3335.1	3431.3	3528.3	3626.2	3727.4
20,000	3172.3	3307.9	3381.6	3485.0	3588.3	3695.3
30,000	3099.8	3217.6	3330.2	3440.8	3549.9	3662.5
40,000	3022.9	3154.9	3277.6	3396.6	3510.9	3629.2
50,000	2942.8	3090.9	3224.2	3349.9	3471.5	3596.4
60,000	2861.5	3026.6	3170.9	3304.3	3432.0	3563.1
70,000	2834.9	2963.4	3118.3	3259.2	3392.8	3529.5
80,000	2710.0	2971.1	3124.8	3264.4	3360.3	3496.9
90,000	2646.5	2845.4	3017.8	3172.4	3316.7	3464.2
100,000	2590.7	2794.7	2970.8	3130.8	3280.2	3432.5

TABLE B.3 (Continued) *Thermodynamic Properties of Water—Superheated Steam.*

Enthalpy, h (kJ/kg)

Abs. Press. (kPa)	Sat. Temp. (°C)	Sat. Enth.	Temperature (°C)				
p	T_g	h_g	50	100	150	200	250
1	6.98	2514.4	2594.6	2688.6	2783.7	2880.1	2977.7
5	32.90	2561.6	2593.7	2688.1	2783.4	2879.9	2977.6
10	45.83	2584.8	2592.7	2687.5	2783.1	2879.6	2977.4
20	60.09	2609.9		2686.3	2782.3	2879.2	2977.1
40	75.89	2636.9		2683.8	2780.9	2878.2	2976.5
60	85.95	2653.6		2681.3	2779.4	2877.3	2975.8
80	93.51	2665.8		2678.8	2777.8	2876.3	2975.2
100	99.63	2675.4		2676.2	2776.3	2875.4	2974.5
200	120.23	2706.3			2768.5	2870.5	2971.2
300	133.54	2724.7			2760.4	2865.5	2967.9
400	143.62	2737.6			2752.0	2860.4	2964.5
500	151.84	2747.5				2855.1	2961.1
600	158.84	2755.5				2849.7	2957.6
700	164.96	2762.0				2844.2	2954.0
800	170.41	2767.5				2838.6	2950.4
900	175.36	2772.1				2832.7	2946.8
1000	179.88	2776.2				2826.8	2943.0
1500	198.29	2789.9				2794.7	2923.5
2000	212.37	2797.2					2902.4
2500	223.94	2800.9					2879.5
3000	233.84	2802.3					2854.8
3500	242.54	2802.0					2828.1
4000	250.33	2800.3					
4500	257.41	2797.7					
5000	263.91	2794.2					
6000	275.55	2785.0					
7000	285.79	2773.5					
8000	294.97	2759.9					
9000	303.31	2744.6					
10,000	310.96	2727.7					
20,000	365.70	2418.4					
30,000							
40,000							
50,000							
60,000							
70,000							
80,000							
90,000							
100,000							

Enthalpy, h (kJ/kg)

Abs. Press. (kPa) p	Temperature (°C) 300	350	400	450	500	550
1	3076.8	3177.5	3279.7	3383.6	3489.2	3596.5
5	3076.7	3177.4	3279.7	3383.6	3489.2	3596.5
10	3076.6	3177.3	3279.6	3383.5	3489.1	3596.5
20	3076.4	3177.1	3279.4	3383.4	3489.0	3596.4
40	3075.9	3176.8	3279.1	3383.1	3488.8	3596.2
60	3075.4	3176.4	3278.8	3382.9	3488.6	3596.0
80	3075.0	3176.0	3278.5	3382.6	3488.4	3595.8
100	3074.5	3175.6	3278.2	3382.4	3488.1	3595.6
200	3072.1	3173.8	3276.7	3381.1	3487.0	3594.7
300	3069.7	3171.9	3275.2	3379.8	3486.0	3593.7
400	3067.2	3170.0	3273.6	3378.5	3484.9	3592.8
500	3064.8	3168.1	3272.1	3377.2	3483.8	3591.8
600	3062.3	3166.2	3270.6	3376.0	3482.7	3590.9
700	3059.8	3164.3	3269.0	3374.7	3481.6	3589.9
800	3057.3	3162.4	3267.5	3373.4	3480.5	3589.0
900	3054.7	3160.5	3266.0	3372.1	3479.4	3588.1
1000	3052.1	3158.5	3264.4	3370.8	3478.3	3587.1
1500	3038.9	3148.7	3256.6	3364.3	3472.8	3582.4
2000	3025.0	3138.6	3248.7	3357.8	3467.3	3577.6
2500	3010.4	3128.2	3240.7	3351.3	3461.7	3572.9
3000	2995.1	3117.5	3232.5	3344.6	3456.2	3568.1
3500	2979.0	3106.5	3224.2	3338.0	3450.6	3563.4
4000	2962.0	3095.1	3215.7	3331.2	3445.0	3558.6
4500	2944.2	3083.3	3207.1	3324.4	3439.3	3553.8
5000	2925.5	3071.2	3198.3	3317.5	3433.7	3549.0
6000	2885.0	3045.8	3180.1	3303.5	3422.2	3539.3
7000	2839.4	3018.7	3161.2	3289.1	3410.6	3529.6
8000	2786.8	2989.9	3141.6	3274.3	3398.8	3519.7
9000		2959.0	3121.2	3259.2	3386.8	3509.8
10,000		2925.8	3099.9	3243.6	3374.6	3499.8
20,000			2820.5	3064.3	3241.1	3394.1
30,000			2161.8	2825.6	3085.0	3277.4
40,000			1934.1	2515.6	2906.8	3151.6
50,000			1877.7	2293.2	2723.0	3021.1
60,000			1847.3	2187.1	2570.6	2896.2
70,000			1827.8	2129.9	2467.1	2790.7
80,000			1814.2	2094.1	2397.4	2708.0
90,000			1804.6	2069.3	2349.9	2643.0
100,000			1797.6	2051.2	2316.1	2593.8

TABLE B.3 (Continued) *Thermodynamic Properties of Water—Superheated Steam*

Enthalpy, h (kJ/kg)

Abs. Press (k Pa)	Temperature (°C)					
p	600	650	700	750	800	850
1	3705.6	3816.4	3928.9	4043.0	4158.7	4276.9
5	3705.6	3816.3	3928.8	4043.0	4158.7	4276.9
10	3705.5	3816.3	3928.8	4042.9	4158.7	4276.9
20	3705.4	3816.2	3928.7	4042.9	4158.7	4276.9
40	3705.3	3816.1	3928.6	4042.8	4158.6	4276.8
60	3705.1	3816.0	3928.5	4042.7	4158.5	4276.7
80	3705.0	3815.8	3928.4	4042.6	4158.4	4276.6
100	3704.8	3815.7	3928.2	4042.5	4158.3	4276.5
200	3704.0	3815.0	3927.6	4041.9	4157.8	4276.2
300	3703.2	3814.2	3927.0	4041.4	4157.3	4275.8
400	3702.3	3813.5	3926.4	4040.8	4156.9	4275.4
500	3701.5	3812.8	3925.8	4040.3	4156.4	4275.0
600	3700.7	3812.1	3925.1	4039.8	4155.9	4274.6
700	3699.9	3811.4	3924.5	4039.2	4155.5	4274.2
800	3699.1	3810.7	3923.9	4038.7	4155.0	4273.8
900	3698.2	3810.0	3923.3	4038.1	4154.5	4273.4
1000	3697.4	3809.3	3922.7	4037.6	4154.1	4273.0
1500	3693.3	3805.7	3919.6	4034.9	4151.7	4271.1
2000	3689.2	3802.1	3916.5	4032.2	4149.4	4269.1
2500	3685.1	3798.6	3913.4	4029.5	4147.0	4267.2
3000	3681.0	3795.0	3910.3	4026.8	4144.7	4265.2
3500	3676.9	3791.4	3907.2	4024.1	4142.4	4263.2
4000	3672.8	3787.9	3904.1	4021.4	4140.0	4261.3
4500	3668.6	3784.3	3901.0	4018.8	4137.7	4259.3
5000	3664.5	3780.7	3897.9	4016.1	4135.3	4257.4
6000	3656.2	3773.5	3891.7	4010.7	4130.7	4253.4
7000	3647.9	3766.4	3885.4	4005.3	4126.0	4249.5
8000	3639.5	3759.2	3879.2	3999.9	4121.3	4245.6
9000	3631.1	3752.0	3873.0	3994.5	4116.7	4241.6
10,000	3622.7	3744.7	3866.8	3989.1	4112.0	4237.7
20,000	3535.5	3671.1	3803.8	3935.0	4065.3	4198.3
30,000	3443.0	3595.0	3739.7	3880.3	4018.5	4159.0
40,000	3346.4	3517.0	3674.8	3825.5	3971.7	4120.0
50,000	3248.3	3438.9	3610.2	3770.9	3925.3	4081.3
60,000	3151.6	3362.4	3547.0	3717.4	3879.6	4043.3
70,000	3060.4	3289.0	3486.3	3665.8	3835.3	4006.1
80,000	2980.3	3220.3	3428.7	3616.7	3792.8	3970.1
90,000	2913.5	3158.2	3374.6	3570.1	3752.4	3935.5
100,000	2857.5	3105.3	3324.4	3526.0	3714.3	3902.5

Entropy, s (kJ/kg K)

Abs. Press (kPa)	Sat. Temp. (°C)	Sat. Entropy	Temperature (°C)					
p	T_g	s_g	50	100	150	200	250
1	6.98	8.9767	9.2430	9.5136	9.7527	9.9679	10.1641
5	32.90	8.3960	8.4981	8.7698	9.0094	9.2248	9.4211
10	45.83	8.1511	8.1757	8.4486	8.6888	8.9045	9.1010
20	60.09	7.9094		8.1261	8.3676	8.5839	8.7806
40	75.89	7.6709		7.8009	8.0450	8.2625	8.4598
60	85.95	7.5327		7.6085	7.8551	8.0738	8.2718
80	93.51	7.4352		7.4703	7.7195	7.9395	8.1381
100	99.63	7.3598		7.3618	7.6137	7.8349	8.0342
200	120.23	7.1268			7.2794	7.5072	7.7096
300	133.54	6.9909			7.0771	7.3119	7.5176
400	143.62	6.8943			6.9285	7.1708	7.3800
500	151.84	6.8192				7.0592	7.2721
600	158.84	6.7575				6.9662	7.1829
700	164.96	6.7052				6.8859	7.1066
800	170.41	6.6596				6.8148	7.0397
900	175.36	6.6192				6.7508	6.9800
1000	179.88	6.5828				6.6922	6.9259
1500	198.29	6.4406				6.4508	6.7099
2000	212.37	6.3366					6.5454
2500	223.94	6.2536					6.4077
3000	233.84	6.1837					6.2857
3500	242.54	6.1228					6.1732
4000	250.33	6.0685					
4500	257.41	6.0191					
5000	263.91	5.9735					
6000	275.55	5.8908					
7000	285.79	5.8162					
8000	294.97	5.7471					
9000	303.31	5.6820					
10,000	310.96	5.6198					
20,000	365.70	4.9412					
30,000							
40,000							
50,000							
60,000							
70,000							
80,000							
90,000							
100,000							

TABLE B.3 (Continued) *Thermodynamic Properties of Water—Superheated Steam.*

Entropy, s (kJ/kg K)

Abs. Press. (k Pa)	Temperature (°C)					
p	300	350	400	450	500	550
1	10.3450	10.5133	10.6711	10.8200	10.9612	11.0957
5	9.6021	9.7704	9.9283	10.0772	10.2184	10.3529
10	9.2820	9.4504	9.6083	9.7572	9.8984	10.0329
20	8.9618	9.1303	9.2882	9.4372	9.5784	9.7130
40	8.6413	8.8100	8.9680	9.1170	9.2583	9.3929
60	8.4536	8.6224	8.7806	8.9296	9.0710	9.2056
80	8.3202	8.4892	8.6475	8.7966	8.9380	9.0727
100	8.2166	8.3858	8.5442	8.6934	8.8348	8.9656
200	7.8937	8.0638	8.2226	8.3722	8.5139	8.6487
300	7.7034	7.8744	8.0338	8.1838	8.3257	8.4608
400	7.5675	7.7395	7.8994	8.0497	8.1919	8.3271
500	7.4614	7.6343	7.7948	7.9454	8.0879	8.2233
600	7.3740	7.5479	7.7090	7.8600	8.0027	8.1383
700	7.2997	7.4745	7.6362	7.7875	7.9305	8.0663
800	7.2348	7.4107	7.5729	7.7246	7.8678	8.0038
900	7.1771	7.3540	7.5169	7.6689	7.8124	7.9486
1000	7.1251	7.3031	7.4665	7.6190	7.7627	7.8991
1500	6.9207	7.1044	7.2709	7.4253	7.5703	7.7077
2000	6.7696	6.9596	7.1296	7.2859	7.4323	7.5706
2500	6.6470	6.8442	7.0178	7.1763	7.3240	7.4633
3000	6.5422	6.7471	6.9246	7.0854	7.2345	7.3748
3500	6.4491	6.6626	6.8443	7.0074	7.1580	7.2993
4000	6.3642	6.5870	6.7733	6.9388	7.0909	7.2333
4500	6.2852	6.5182	6.7093	6.8774	7.0311	7.1746
5000	6.2105	6.4545	6.6508	6.8217	6.9770	7.1215
6000	6.0692	6.3386	6.5462	6.7230	6.8818	7.0285
7000	5.9327	6.2333	6.4536	6.6368	6.7993	6.9485
8000	5.7942	6.1349	6.3694	6.5597	6.7262	7.8778
9000		6.0408	6.2915	6.4894	6.6600	6.8143
10,000		5.9489	6.2182	6.4243	6.5994	6.7564
20,000			5.5585	5.9089	6.1456	6.3374
30,000			4.4896	5.4495	5.7972	6.0386
40,000			4.1190	4.9511	5.4762	5.7835
50,000			4.0083	4.6026	5.1782	5.5525
60,000			3.9383	4.4246	4.9374	5.3463
70,000			3.8855	4.3182	4.7688	5.1748
80,000			3.8425	4.2434	4.6488	5.0382
90,000			3.8059	4.1852	4.5602	4.9276
100,000			3.7738	4.1373	4.4913	4.8393

Enthalpy, s (kJ/kg)

Abs. Press. (k Pa)	Temperature (°C)					
p	600	650	700	750	800	850
1	11.2243	11.3476	11.4663	11.5807	11.6911	11.7977
5	10.4815	10.6049	10.7235	10.8379	10.9483	11.0549
10	10.1616	10.2849	10.4036	10.5180	10.6284	10.7351
20	9.8416	9.9650	10.0836	10.1980	10.3085	10.4152
40	9.5216	9.6450	9.7636	9.8780	9.9885	10.0953
60	9.3343	9.4577	9.5764	9.6908	9.8013	9.9081
80	9.2014	9.3248	9.4436	9.5580	9.6685	9.7754
100	9.0982	9.2217	9.3405	9.4549	9.5654	9.6724
200	8.7776	8.9012	9.0201	9.1346	9.2452	9.3527
300	8.5898	8.7135	8.8325	8.9471	9.0577	9.1647
400	8.4563	8.5802	8.6992	8.8139	8.9246	9.0322
500	8.3526	8.4766	8.5957	8.7105	8.8213	8.9290
600	8.2678	8.3919	8.5111	8.6259	8.7368	8.8444
700	8.1959	8.3201	8.4395	8.5544	8.6653	8.7730
800	8.1336	8.2579	8.3773	8.4923	8.6033	8.7111
900	8.0785	8.2030	8.3225	8.4376	8.5486	8.6565
1000	8.0292	8.1537	8.2734	8.3885	8.4997	8.6076
1500	7.8385	7.9636	8.0838	8.1993	8.3108	8.4191
2000	7.7022	7.8279	7.9485	8.0645	8.1763	8.2849
2500	7.5956	7.7220	7.8431	7.9595	8.0716	8.1806
3000	7.5079	7.6349	7.7564	7.8733	7.9857	8.0950
3500	7.4332	7.5607	7.6828	7.8000	7.9128	8.0225
4000	7.3680	7.4961	7.6187	7.7363	7.8495	7.9595
4500	7.3100	7.4388	7.5619	7.6799	7.7934	7.9037
5000	7.2578	7.3872	7.5108	7.6292	7.7431	7.8537
6000	7.1664	7.2971	7.4217	7.5409	7.6554	7.7668
7000	7.0880	7.2200	7.3456	7.4657	7.5808	7.6928
8000	7.0191	7.1523	7.2790	7.3999	7.5158	7.6284
9000	6.9574	7.0919	7.2196	7.3414	7.4579	7.5712
10,000	6.9013	7.0373	7.1660	7.2886	7.4058	7.5198
20,000	6.5043	6.6554	6.7953	6.9267	7.0511	7.1715
30,000	6.2340	6.4033	6.5560	6.6970	6.8288	6.9562
40,000	6.0135	6.2035	6.3701	6.5210	6.6606	6.7950
50,000	5.8207	6.0331	6.2138	6.3749	6.5222	6.6638
60,000	5.6477	5.8827	6.0775	6.2483	6.4031	6.5517
70,000	5.4931	5.7480	5.9562	6.1361	6.2979	6.4529
80,000	5.3595	5.6270	5.8470	6.0354	6.2034	6.3644
90,000	5.2468	5.5195	5.7479	5.9439	6.1179	6.2842
100,000	5.1505	5.4267	5.6579	5.8600	6.0397	6.2106

TABLE B.4 *Thermodynamic Properties of Water—Compressed Water.*

Abs. Press. (kPa)	Temperature (°C)			
p	0	50	100	150
	Specific Volume, v (m^3/kg)			
1	0.0010002			
50	0.0010002	0.0010121		
100	0.0010002	0.0010121		
500	0.0010000	0.0010119	0.0010435	0.0010908
1000	0.0009997	0.0010117	0.0010432	0.0010904
5000	0.0009977	0.0010099	0.0010412	0.0010877
10,000	0.0009953	0.0010077	0.0010386	0.0010843
20,000	0.0009904	0.0010034	0.0010337	0.0010779
30,000	0.0009857	0.0009993	0.0010289	0.0010718
50,000	0.0009767	0.0009914	0.0010200	0.0010605
	Internal Energy, u (kJ/kg)			
1	0.0			
50	0.0	209.2		
100	0.0	209.2		
500	0.0	209.2	418.9	631.7
1000	0.0	209.1	418.7	631.4
5000	0.1	208.5	417.5	629.6
10,000	0.1	207.7	416.1	627.3
20,000	0.3	206.3	413.3	622.9
30,000	0.4	205.0	410.7	618.7
50,000	0.5	202.3	405.8	611.1
	Enthalpy, h (kJ/kg)			
1	0.0			
50	0.0	209.3		
100	0.1	209.3		
500	0.5	209.7	419.4	632.2
1000	1.0	210.1	419.7	632.5
5000	5.1	213.5	422.7	635.0
10,000	10.1	217.8	426.5	638.1
20,000	20.1	226.4	434.0	644.5
30,000	30.0	235.0	441.6	650.9
50,000	49.3	251.9	456.8	664.1
	Entropy, s (kJ/kg K)			
1	−0.0002			
50	−0.0002	0.7035		
100	−0.0001	0.7035		
500	−0.0001	0.7033	1.3066	1.8416
1000	−0.0001	0.7030	1.3062	1.8410
5000	0.0002	0.7012	1.3030	1.8366
10,000	0.0005	0.6989	1.2992	1.8312
20,000	0.0008	0.6943	1.2916	1.8207
30,000	0.0008	0.6897	1.2843	1.8105
50,000	−0.0002	0.6807	1.2701	1.7912

	Temperature (°C)			Sat. Temp. (°C)	Sat. Liq.
p	200	250	300	T_f	
		Specific Volume, v (m^3/kg)			v_f
1				6.98	0.0010001
50				81.35	0.0010301
100				99.63	0.0010434
500				151.84	0.0010928
1000				179.88	0.0011274
5000	0.0011530	0.0012494		263.91	0.0012858
10,000	0.0011480	0.0012406	0.0013979	310.96	0.0014526
20,000	0.0011387	0.0012247	0.0013606	365.70	0.0020370
30,000	0.0011301	0.0012107	0.0013316		
50,000	0.0011144	0.0011866	0.0012874		
		Internal Energy, u (kJ/kg)			u_f
1				6.98	29.3
50				81.35	340.5
100				99.63	417.4
500				151.84	639.6
1000				179.88	761.5
5000	848.0	1079.6		263.91	1148.1
10,000	844.4	1073.4	1329.4	310.96	1393.5
20,000	837.6	1062.2	1307.1	365.70	1785.8
30,000	831.3	1052.1	1288.8		
50,000	819.7	1034.3	1259.3		
		Enthalpy, h (kJ/kg)			h_f
1				6.98	29.3
50				81.35	340.6
100				99.63	417.5
500				151.84	640.1
1000				179.88	762.6
5000	853.8	1085.8		263.91	1154.5
10,000	855.9	1085.8	1343.4	310.96	1408.0
20,000	860.4	1086.7	1334.3	365.70	1826.5
30,000	865.2	1088.4	1328.7		
50,000	875.4	1093.6	1323.7		
		Entropy, s (kJ/kg K)			s_f
1				6.98	0.1060
50				81.35	1.0912
100				99.63	1.3027
500				151.84	1.8604
1000				179.88	2.1382
5000	2.3252	2.7910		263.91	2.9206
10,000	2.3176	2.7792	3.2488	310.96	3.3605
20,000	2.3030	2.7574	3.2088	365.70	4.0149
30,000	2.2891	2.7374	3.1576		
50,000	2.2632	2.7015	3.1213		

TABLE B.5 *Thermodynamic Properties of Water—Saturated Ice-Vapor.*

Temp. (°C)	Abs. Press. (kPa)	Specific Volume (m^3/kg) Sat. Ice	Subl.	Sat. Vapor	Internal Energy (kJ/kg) Sat. Ice	Subl.	Sat. Vapor
T	p	v_i	v_{ig}	v_g	u_i	u_{ig}	u_g
0.01	0.611	0.001091	206.16	206.16	−333.4	2708.9	2375.5
−5	0.404	0.001090	309.0	309.0	−345.0	2712.2	2367.2
−10	0.261	0.001090	465.1	465.1	−354.1	2715.4	2361.3
−15	0.165	0.001088	719.2	719.2	−364.2	2718.6	2354.4
−20	0.104	0.001087	1124	1124	−374.0	2721.5	2347.5
−25	0.064	0.001086	1810	1810	−383.7	2724.3	2340.6
−30	0.038	0.001086	2903	2903	−393.2	2726.8	2333.6
−35	0.023	0.001085	4932	4932	−402.6	2729.2	2326.6
−40	0.013	0.001084	8353	8353	−411.7	2731.4	2319.7

Temp. (°C)	Enthalpy (kJ/kg) Sat. Ice	Subl.	Sat. Vapor	Entropy (kJ/kg K) Sat. Ice	Subl.	Sat. Vapor
T	h_i	h_{ig}	h_g	s_i	s_{ig}	s_g
0.01	−333.4	2835.0	2501.6	−1.2171	10.3745	9.1574
−5	−345.0	2835.9	2490.9	−1.255	10.528	9.273
−10	−354.1	2837.0	2482.9	−1.309	10.754	9.445
−15	−364.2	2837.9	2473.7	−1.334	10.945	9.611
−20	−374.0	2838.3	2464.3	−1.371	11.114	9.743
−25	−383.7	2838.9	2455.2	−1.409	11.662	10.253
−30	−393.2	2839.1	2445.9	−1.446	11.571	10.125
−35	−402.6	2839.1	2436.5	−1.490	11.869	10.379
−40	−411.7	2839.1	2427.4	−1.525	12.120	10.595

TABLE B.6 *Thermodynamic Properties of Carbon Dioxide—Saturated Liquid-Vapor.*

Temp. (°C)	Abs. Press. (kPa)	Specific Volume (m^3/kg) Sat. Liq.	Evap.	Sat. Vapor	Internal Energy (kJ/kg) Sat. Liq.	Evap.	Sat. Vapor
T	p	v_f	v_{fg}	v_g	u_f	u_{fg}	u_g
−56.6	517.9	0.0008490	0.07163	0.07248	−31.2	311.4	280.2
−55	554.8	0.0008535	0.06695	0.06780	−28.5	308.9	280.4
−50	683.1	0.0008676	0.04678	0.05546	−19.4	300.4	281.0
−45	832.5	0.0008824	0.04487	0.04575	−10.2	292.1	281.9
−40	1005.1	0.0008980	0.03725	0.03815	−0.9	283.7	282.8
−35	1202.9	0.0009146	0.03108	0.03199	8.6	275.1	283.7
−30	1428.1	0.0009322	0.02607	0.02700	18.2	266.3	284.5
−25	1682.7	0.0009511	0.02194	0.02289	27.9	257.3	285.2
−20	1969.1	0.0009715	0.01851	0.01948	37.8	247.5	285.3
−15	2289.6	0.0009938	0.01563	0.01662	47.9	237.2	285.1
−10	2646.6	0.001018	0.01320	0.01422	58.2	226.5	284.7
−5	3043.1	0.001046	0.01111	0.01216	68.8	214.7	283.5
0	3481.7	0.001077	0.009313	0.01039	80.0	201.9	281.9
5	3965.7	0.001113	0.007676	0.008789	91.6	186.4	278.0
10	4498.8	0.001157	0.006273	0.007430	103.9	169.9	273.8
15	5085.0	0.001212	0.005019	0.006231	117.1	152.2	269.3
20	5728.9	0.001286	0.003879	0.005165	131.7	131.0	262.7
25	6435.6	0.001401	0.002724	0.004125	150.7	102.7	253.4
30	7211.1	0.001687	0.001257	0.002944	179.0	52.9	231.9
31.04	7383.4	0.002137	0	0.002137	207.2	0	207.2

Temp. (°C)	Enthalpy (kJ/kg)* Sat. Liq.	Evap.	Sat. Vapor	Entropy (kJ/kg K)* Sat. Liq.	Evap.	Sat. Vapor
T	h_f	h_{fg}	h_g	s_f	s_{fg}	s_g
−56.6	−30.8	348.5	317.7	−0.131	1.609	1.478
−55	−28.0	346.0	318.0	−0.122	1.585	1.463
−50	−18.8	337.7	318.9	−0.081	1.512	1.431
−45	−9.5	329.5	320.0	−0.041	1.444	1.403
−40	0	321.1	321.1	0	1.377	1.377
−35	9.7	312.5	322.2	0.039	1.313	1.352
−30	19.5	303.6	323.1	0.079	1.249	1.328
−25	29.5	294.2	323.7	0.119	1.185	1.304
−20	39.7	284.0	323.7	0.158	1.122	1.280
−15	50.2	273.0	323.2	0.198	1.058	1.256
−10	60.9	261.4	322.3	0.238	0.993	1.231
−5	72.0	248.5	320.5	0.278	0.927	1.205
0	83.7	234.4	318.1	0.320	0.858	1.178
5	96.0	216.9	312.9	0.364	0.779	1.143
10	109.1	198.1	307.2	0.407	0.700	1.107
15	123.3	177.7	301.0	0.454	0.617	1.071
20	139.1	153.2	292.3	0.506	0.522	1.028
25	159.7	120.2	279.9	0.573	0.403	0.976
30	191.2	61.9	253.1	0.682	0.204	0.886
31.04	223.0	0	223.0	0.780	0	0.780

* The enthalpy and the entropy of saturated liquid carbon dioxide are taken as zero at a temperature of −40°C.

TABLE B.7 *Thermodynamic Properties of Carbon Dioxide—Superheated CO_2.*

Abs. Press. (kPa)	Sat. Temp. (°C)	Sat. Vapor	Temperature (°C)				
p	T_g		−50	0	50	100	150
			Specific Volume, v (m³/kg)				
		v_g					
1	−122.36	28.676	42.157	51.603	61.049	70.495	79.941
5	−109.46	6.175	8.431	10.321	12.210	14.099	15.988
10	−103.42	3.202	4.216	5.160	6.105	7.050	7.994
50	−87.00	0.6944	0.8400	1.028	1.219	1.408	1.598
100	−78.84	0.3583	0.4102	0.5126	0.6083	0.7033	0.7982
200	−69.89	0.1857	0.2021	0.2545	0.3028	0.3508	0.3986
300	−64.47	0.1240	0.1327	0.1685	0.2011	0.2333	0.2653
500	−56.33	0.07503	0.07709	0.09967	0.1197	0.1393	0.1587
1000	−40.00	0.03817		0.04796	0.05865	0.06883	0.07876
1500	−28.59	0.02569		0.03066	0.03827	0.04533	0.05211
2000	−19.52	0.01916		0.02192	0.02807	0.03357	0.03879
3000	−5.54	0.01235		0.01296	0.01783	0.02181	0.02548
4000	5.34	0.008688			0.01267	0.01592	0.01882
5000	14.30	0.006396			0.009528	0.01239	0.01483
6000	21.97	0.004749			0.007391	0.01003	0.01217
7000	28.58	0.003272			0.005814	0.008336	0.01027
8000					0.004564	0.007069	0.008847
9000					0.003508	0.006083	0.007746
10,000					0.002588	0.005297	0.006867
			Internal Energy, u (kJ/kg)				
		u_g					
1	−122.36	257.5	294.9	325.0	357.7	392.8	430.3
5	−109.46	263.3	294.7	324.8	357.5	392.7	430.2
10	−103.42	265.9	294.5	324.7	357.4	392.7	430.2
50	−87.00	272.8	294.2	324.4	357.3	392.6	430.1
100	−87.00	275.9	293.8	324.1	357.1	392.5	430.0
200	−69.89	278.7	291.9	323.4	356.6	392.0	429.6
300	−64.47	279.8	289.9	322.6	356.1	391.6	429.3
500	−56.33	280.2	286.0	320.9	355.0	390.8	428.6
1000	−40.00	282.9		316.5	353.1	388.6	426.8
1500	−28.59	284.8		311.7	349.1	386.4	425.1
2000	−19.52	285.4		306.6	346.0	384.2	423.3
3000	−5.54	283.7		294.2	339.4	379.5	419.8
4000	5.34	277.8			332.2	374.7	416.1
5000	14.30	270.1			324.2	369.7	412.4
6000	21.97	259.4			315.1	364.4	408.6
7000	28.58	236.4			304.4	359.0	404.7
8000					291.2	353.2	400.8
9000					273.8	347.4	396.9
10,000					249.3	341.2	392.8

Abs. Press. (kPa)	Temperature (°C)						
p	200	250	300	350	400	450	500
	Specific Volume, v (m^3/kg)						
1	89.387	98.833	108.279	117.725	127.171	136.617	146.062
5	17.877	19.767	21.656	23.545	25.434	27.323	29.212
10	8.939	9.883	10.828	11.773	12.717	13.662	14.606
50	1.788	1.977	2.166	2.355	2.543	2.732	2.921
100	0.8930	0.9878	1.0824	1.1770	1.2716	1.3661	1.4607
200	0.4460	0.4936	0.5410	0.5884	0.6358	0.6831	0.7304
300	0.2971	0.3288	0.3606	0.3922	0.4237	0.4554	0.4869
500	0.1779	0.1971	0.2162	0.2352	0.2543	0.2732	0.2922
1000	0.08857	0.09827	0.1079	0.1175	0.1271	0.1366	0.1462
1500	0.05877	0.6533	0.07183	0.07828	0.08470	0.09111	0.09748
2000	0.04388	0.04885	0.05379	0.05865	0.06351	0.06834	0.07314
3000	0.02899	0.03239	0.03574	0.03905	0.04232	0.04558	0.04880
4000	0.02154	0.02417	0.02673	0.02925	0.03174	0.03420	0.03664
5000	0.01708	0.01924	0.02133	0.02337	0.02538	0.02737	0.02934
6000	0.01411	0.01595	0.01772	0.01945	0.02115	0.02282	0.02448
7000	0.01199	0.01361	0.01515	0.01666	0.01813	0.01958	0.02100
8000	0.01041	0.01185	0.01323	0.01467	0.01586	0.01714	0.01840
9000	0.009174	0.01049	0.01173	0.01293	0.01410	0.01525	0.01638
10,000	0.0008192	0.009399	0.01054	0.01163	0.01270	0.01374	0.01476
	Internal Energy, u (kJ/kg)						
1	469.7	510.7	553.6	597.9	643.5	690.3	738.3
5	469.7	510.9	553.6	597.9	643.5	690.3	738.3
10	469.6	510.8	553.6	597.9	643.5	690.3	738.3
50	469.5	510.7	553.5	597.9	643.5	690.3	738.3
100	469.4	510.6	553.5	597.8	643.3	690.2	738.2
200	469.2	510.4	553.3	597.6	643.2	690.1	738.0
300	468.9	510.2	553.0	597.4	643.1	690.0	737.9
500	468.4	509.7	552.7	597.1	642.8	689.7	737.7
1000	466.8	508.5	551.7	596.2	642.0	689.1	737.1
1500	465.4	507.3	550.7	595.3	641.3	688.3	736.5
2000	463.9	506.1	549.6	594.5	640.5	687.7	735.9
3000	461.1	503.7	547.6	592.7	639.0	686.4	734.8
4000	458.1	501.3	545.5	590.9	637.4	685.0	733.5
5000	455.2	498.8	543.5	589.3	636.0	683.7	732.4
6000	452.2	496.4	541.5	587.5	634.4	682.4	731.2
7000	449.3	493.9	539.5	585.7	632.9	681.0	730.1
8000	446.2	491.6	537.4	583.1	631.4	679.8	728.9
9000	443.2	489.1	535.3	582.3	630.0	678.5	727.7
10,000	440.2	486.7	533.3	580.6	628.4	677.1	726.6

TABLE B.7 (Continued) *Thermodynamic Properties of Carbon Dioxide—Superheated CO_2.*

Abs. Press. (kPa)	Sat. Temp. (°C)	Sat. Vapor	Temperature (°C)				
p	T_g		−50	0	50	100	150
			Enthalpy h (kJ/kg)				
		h_g					
1	−122.36	286.2	337.1	376.6	418.7	463.3	510.2
5	−109.46	294.2	336.9	376.5	418.6	463.2	510.2
10	−103.42	297.9	336.7	376.3	418.5	463.2	510.1
50	−87.00	307.5	336.2	375.8	418.2	463.0	510.0
100	−78.84	311.7	334.8	375.4	417.9	462.8	509.8
200	−69.89	315.8	332.3	374.3	417.2	462.2	509.3
300	−64.47	317.0	329.7	373.1	416.4	461.6	508.9
500	−56.33	317.7	324.5	370.7	414.8	460.4	507.9
1000	−40.00	321.1		364.5	410.7	457.4	505.6
1500	−28.59	323.3		357.7	406.5	454.4	503.3
2000	−19.52	323.7		350.4	402.1	451.3	500.9
3000	−5.54	320.7		333.1	392.9	444.9	496.2
4000	5.34	312.6			382.9	438.4	491.4
5000	14.30	302.1			371.8	431.6	486.5
6000	21.97	287.9			359.4	424.6	481.6
7000	28.58	259.3			345.1	417.4	476.6
8000					327.7	409.8	471.6
9000					305.4	402.1	466.6
10,000					275.2	394.1	461.5
			Entropy, s (kJ/kg K)				
		s_g					
1	−122.36	2.461	2.721	2.880	3.023	3.152	3.270
5	−109.46	2.199	2.417	2.576	2.719	2.848	2.966
10	−103.42	2.094	2.286	2.445	2.588	2.717	2.835
50	−87.00	1.845	1.982	2.141	2.284	2.413	2.531
100	−78.84	1.737	1.851	2.010	2.153	2.282	2.400
200	−69.89	1.629	1.701	1.876	2.020	2.150	2.268
300	−64.47	1.563	1.615	1.796	1.942	2.072	2.191
500	−56.33	1.483	1.690	1.694	1.842	1.973	2.093
1000	−40.00	1.377		1.546	1.702	1.836	1.957
1500	−28.59	1.321		1.452	1.616	1.754	1.877
2000	−19.52	1.278		1.378	1.552	1.693	1.818
3000	−5.54	1.208		1.251	1.454	1.604	1.733
4000	5.34	1.141			1.377	1.537	1.670
5000	14.30	1.076			1.309	1.481	1.619
6000	21.97	1.010			1.244	1.433	1.576
7000	28.58	0.918			1.180	1.389	1.538
8000					1.110	1.348	1.504
9000					1.028	1.310	1.472
10,000					0.926	1.273	1.443

Abs. Press. (k Pa)	Temperature (°C)						
p	200	250	300	350	400	450	500
	Enthalpy, h (kJ/kg)						
1	559.1	609.7	661.9	715.6	770.7	826.9	884.4
5	559.1	609.7	661.9	715.6	770.7	826.9	884.4
10	559.0	609.6	661.9	715.6	770.7	826.9	884.4
50	558.9	609.5	661.8	715.6	770.6	826.9	884.3
100	558.7	609.4	661.7	715.5	770.5	826.8	884.2
200	558.4	609.1	661.5	715.3	770.4	826.7	884.1
300	558.0	608.8	661.2	715.1	770.2	826.6	884.0
500	557.3	608.2	660.8	714.7	769.9	826.3	883.8
1000	555.4	606.8	659.6	713.7	769.1	825.7	883.3
1500	553.6	605.3	658.4	712.7	768.3	825.0	882.7
2000	551.7	603.8	657.2	711.8	767.5	824.4	882.2
3000	548.1	600.9	654.8	709.8	766.0	823.1	881.2
4000	544.3	598.0	652.4	707.9	764.4	821.8	880.1
5000	540.6	595.0	650.1	706.1	762.9	820.5	879.1
6000	536.9	592.1	647.8	704.2	761.3	819.3	878.1
7000	533.2	589.2	645.5	702.3	759.8	818.1	877.1
8000	529.5	586.4	643.2	700.5	758.3	816.9	876.1
9000	525.8	583.5	640.9	698.7	756.9	815.7	875.1
10,000	522.1	580.7	638.7	696.9	755.4	814.5	874.2
	Entropy, s (kJ/kg K)						
1	3.379	3.481	3.577	3.667	3.752	3.832	3.908
5	3.075	3.177	3.272	3.363	3.448	3.528	3.604
10	2.944	3.046	3.142	3.232	3.317	3.397	3.473
50	2.640	2.742	2.838	2.928	3.013	3.093	3.169
100	2.509	2.611	2.707	2.797	2.882	2.962	3.038
200	2.378	2.480	2.575	2.665	2.750	2.831	2.907
300	2.300	2.402	2.499	2.589	2.674	2.754	2.830
500	2.203	2.305	2.401	2.492	2.577	2.657	2.734
1000	2.069	2.172	2.268	2.359	2.445	2.525	2.602
1500	1.989	2.094	2.190	2.281	2.367	2.448	2.525
2000	1.932	2.037	2.134	2.225	2.321	2.393	2.470
3000	1.849	1.955	2.054	2.146	2.232	2.314	2.392
4000	1.789	1.896	1.996	2.088	2.176	2.258	2.336
5000	1.740	1.850	1.950	2.044	2.131	2.214	2.292
6000	1.700	1.811	1.912	2.007	2.095	2.178	2.257
7000	1.664	1.777	1.880	1.975	2.063	2.147	2.226
8000	1.633	1.747	1.851	1.947	2.036	2.120	2.199
9000	1.604	1.720	1.825	1.922	2.012	2.096	2.176
10,000	1.578	1.696	1.802	1.899	1.990	2.075	2.155

TABLE B.8 *Thermodynamic Properties of Carbon Dioxide—Compressed Liquid.*

Abs. Press. (kPa)	Temperature (°C)				Sat. Temp. (°C)	Sat. Liq.
p	0	10	20	30	T_f	
			Specific Volume, v (m^3/kg)			v_f
5000	0.001059	0.001147			14.30	0.001203
10,000	0.001022	0.001086	0.001170	0.001290		
20,000	0.0009745	0.001014	0.001062	0.001120		
30,000	0.0009430	0.0009740	0.001009	0.001048		
50,000	0.0009020	0.0009248	0.0009499	0.0009765		
80,000	0.0008590	0.0008775	0.0008965	0.0009159		
100,000	0.0008375	0.0008539	0.0008706	0.0008876		
			Internal Energy, u (kJ/kg)			u_f
5000	77.0	102.4			14.30	115.1
10,000	69.3	91.1	115.6	146.1		
20,000	57.9	78.0	99.3	122.1		
30,000	49.5	68.2	87.6	108.4		
50,000	38.1	54.7	71.9	90.3		
80,000	27.4	42.6	58.8	75.9		
100,000	22.0	36.9	52.6	69.3		
			Enthalpy, h (kJ/kg)			h_f
5000	82.3	108.1			14.30	121.1
10,000	79.5	102.0	127.3	159.0		
20,000	77.4	98.3	120.5	144.5		
30,000	77.8	97.4	117.9	139.8		
50,000	83.2	100.9	119.4	139.1		
80,000	96.1	112.8	130.5	149.2		
100,000	105.7	122.3	139.7	158.1		
			Entropy, s (kJ/kg K)			s_f
5000	0.310	0.401			14.30	0.446
10,000	0.280	0.362	0.448	0.562		
20,000	0.235	0.308	0.382	0.463		
30,000	0.202	0.270	0.339	0.409		
50,000	0.154	0.218	0.282	0.345		
80,000	0.101	0.163	0.223	0.282		
100,000	0.072	0.133	0.192	0.250		

TABLE B.9 *Thermodynamic Properties of Carbon Dioxide—Saturated Solid-Vapor.*

Temp. (°C)	Abs. Press. (kPa)	Specific Volume (m^3/kg) Sat. Solid	Subl.	Sat. Vapor	Internal Energy (kJ/kg) Sat. Solid	Subl.	Sat. Vapor
T	p	v_i	v_{ig}	v_g	u_i	u_{ig}	u_g
−56.6	517.9	0.0006609	0.07182	0.07248	−230.2	510.4	280.2
−60	409.7	0.0006570	0.09056	0.09122	−235.0	514.4	280.0
−65	287.0	0.0006519	0.1289	0.1295	−241.7	521.5	279.8
−70	198.1	0.0006472	0.1850	0.1857	−248.3	527.3	279.0
−75	134.5	0.0006431	0.2691	0.2697	−254.9	532.3	277.4
−80	89.62	0.0006390	0.3970	0.3976	−261.3	536.8	275.5
−85	58.47	0.0006353	0.5971	0.5977	−267.4	541.1	273.7
−90	37.27	0.0006321	0.9167	0.9173	−273.6	545.2	271.6
−95	23.15	0.0006293	1.4395	1.4401	−279.6	549.1	269.5
−100	13.97	0.0006270	2.326	2.327	−285.5	553.0	267.5
−105	8.17	0.0006250	3.872	3.873	−291.3	556.6	265.3
−110	4.620	0.0006231	6.648	6.649	−296.9	560.1	263.2
−115	2.514	0.0006211	11.88	11.88	−302.5	563.6	261.1
−120	1.311	0.0006188	22.10	22.10	−308.0	566.8	258.8
−125	0.652	0.0006169	42.99	42.99	−313.4	570.5	257.1
−130	0.308	0.0006150	87.80	87.80	−318.8	574.1	255.3

Temp. (°C)	Enthalpy (kJ/kg) Sat. Solid	Subl.	Sat. Vapor	Entropy (kJ/kg K) Sat. Solid	Subl.	Sat. Vapor
T	h_i	h_{ig}	h_g	s_i	s_{ig}	s_g
−56.6	−229.9	547.6	317.7	−1.051	2.529	1.478
−60	−234.7	552.1	317.4	−1.083	2.592	1.509
−65	−241.5	558.5	317.0	−1.114	2.683	1.569
−70	−248.2	564.0	315.8	−1.147	2.776	1.629
−75	−254.8	568.5	313.7	−1.179	2.869	1.690
−80	−261.2	572.3	311.1	−1.211	2.962	1.751
−85	−267.4	576.0	308.6	−1.244	3.062	1.818
−90	−273.6	579.4	305.8	−1.277	3.163	1.886
−95	−279.6	582.4	302.8	−1.310	3.268	1.958
−100	−285.5	585.5	300.0	−1.344	3.381	2.037
−105	−291.3	588.2	296.9	−1.378	3.498	2.120
−110	−296.9	590.5	293.9	−1.412	3.621	2.209
−115	−302.5	593.5	291.0	−1.447	3.753	2.306
−120	−308.0	595.8	287.8	−1.483	3.891	2.408
−125	−313.4	598.5	285.1	−1.519	4.040	2.521
−130	−318.8	601.1	282.3	−1.555	4.197	2.642

TABLE B.10 *Thermodynamic Properties of R-22 (Chlorodifluoromethane) Saturated Liquid-Vapor.*

Temp. (°C)	Abs. Press. (kPa)	Specific Volume (m^3/kg) Sat. Liq.	Evap.	Sat. Vapor	Internal Energy (kJ/kg) Sat. Liq.	Evap.	Sat. Vapor
T	p	v_f	v_{fg}	v_g	u_f	u_{fg}	u_g
−100	2.0750	0.0006366	8.0083	8.0089	−59.37	246.86	187.49
−95	3.2323	0.0006418	5.2845	5.2851	−54.66	244.12	189.46
−90	4.8994	0.0006470	3.5804	3.5810	−49.92	241.36	191.36
−85	7.2412	0.0006525	2.4847	2.4854	−45.16	238.59	193.43
−80	10.461	0.0006581	1.7626	1.7633	−40.36	235.80	195.44
−75	14.794	0.0006638	1.2757	1.2764	−35.52	233.00	197.48
−70	20.523	0.0006697	0.94033	0.94100	−30.62	230.13	199.51
−65	27.965	0.0006758	0.69876	0.70552	−25.68	227.21	201.53
−60	37.480	0.0006821	0.53649	0.53717	−20.68	224.25	203.57
−55	49.474	0.0006885	0.41416	0.41485	−15.62	221.22	205.60
−50	63.139	0.0006952	0.32387	0.32457	−10.50	218.52	208.02
−45	82.701	0.0007022	0.25630	0.25700	−5.32	214.94	209.62
−40	104.943	0.0007093	0.20505	0.20576	−0.07	211.68	211.61
−35	131.669	0.0007168	0.16569	0.16569	5.24	208.33	213.57
−30	163.470	0.0007245	0.13513	0.13585	10.60	204.91	215.51
−25	200.968	0.0007325	0.11113	0.11186	16.01	233.45	217.44
−20	244.814	0.0007409	0.092106	0.092847	21.55	197.77	219.32
−15	295.686	0.0007496	0.076878	0.077628	27.11	194.07	221.18
−10	354.284	0.0007587	0.064583	0.065342	32.74	190.25	222.99
−5	421.330	0.0007683	0.054573	0.055341	35.52	189.24	224.76
0	497.567	0.0007783	0.046359	0.047137	44.20	182.30	226.50
5	583.756	0.0007889	0.039568	0.040357	50.02	178.15	228.17
10	680.673	0.0008000	0.033915	0.034715	55.92	173.87	229.79
15	789.117	0.0008118	0.029177	0.029989	61.88	169.48	231.36
20	909.899	0.0008243	0.025180	0.026004	67.92	164.92	232.84
25	1043.856	0.0008376	0.021787	0.022625	74.04	160.22	234.26
30	1191.842	0.0008519	0.018891	0.019743	80.23	155.36	235.59
35	1354.741	0.0008673	0.016400	0.017267	86.52	150.30	236.82
40	1533.466	0.0008839	0.018859	0.015137	92.90	145.04	237.94
45	1728.969	0.0009020	0.012384	0.013286	99.40	139.53	238.93
50	1942.254	0.0009219	0.010751	0.011672	106.04	133.73	239.77
55	2174.382	0.0009440	0.009440	0.010257	112.81	127.62	240.43
60	2426.496	0.0009687	0.0080321	0.0090008	119.83	121.01	240.84
65	2699.843	0.0009970	0.0068907	0.0078877	127.04	113.93	240.97
70	2995.810	0.0010298	0.0058593	0.0068891	134.53	106.23	240.76
75	3316.03	0.0010691	0.0049144	0.0059835	142.43	97.62	240.05
80	3662.29	0.0011181	0.0040307	0.0051488	150.92	87.70	238.62
85	4036.81	0.0011832	0.0031751	0.0043583	160.31	75.79	236.10
90	4442.50	0.0012822	0.0022823	0.0035645	171.50	59.90	231.40
95	4883.49	0.0015205	0.0010311	0.0025516	188.92	29.92	218.84
96.01	4977.39	0.0019056	0	0.0019056	203.09	0	203.09

Temp. (°C)	Enthalpy (kJ/kg)* Sat. Liq.	Enthalpy (kJ/kg)* Evap.	Enthalpy (kJ/kg)* Sat. Vapor	Entropy (kJ/kg K)* Sat. Liq.	Entropy (kJ/kg K)* Evap.	Entropy (kJ/kg K)* Sat. Vapor
T	h_f	h_{fg}	h_g	s_f	s_{fg}	s_g
−100	−59.37	263.48	204.11	−0.29317	1.52159	1.22842
−95	−54.66	261.20	206.54	−0.26426	1.46397	1.19971
−90	−49.92	258.91	208.98	−0.24016	1.41356	1.17340
−85	−45.16	256.59	211.43	−0.21447	1.36369	1.14922
−80	−40.35	254.25	213.89	−0.18928	1.31624	1.12696
−75	−35.51	251.87	216.36	−0.16452	1.27098	1.10646
−70	−30.61	249.42	218.82	−0.14012	1.22771	1.08759
−65	−25.66	246.92	221.26	−0.11611	1.18621	1.07010
−60	−20.65	244.35	223.70	−0.09234	1.14629	1.05395
−55	−15.59	241.70	226.12	−0.06891	1.10792	1.03901
−50	−10.46	238.96	228.51	−0.04569	1.07081	1.02512
−45	−5.26	236.13	230.87	−0.02276	1.03495	1.01219
−40	0.00	233.20	233.20	0.00000	1.00014	1.00014
−35	5.33	230.15	235.48	0.02251	0.96638	0.98889
−30	10.72	227.00	237.72	0.04485	0.93354	0.97839
−25	16.19	223.72	239.92	0.06699	0.90152	0.96851
−20	21.73	220.33	242.05	0.08895	0.87032	0.95927
−15	27.33	216.79	244.13	0.11075	0.83977	0.95052
−10	33.01	213.13	246.14	0.13234	0.80990	0.94224
−5	38.76	209.32	248.08	0.15380	0.78057	0.93437
0	44.59	205.36	249.95	0.17178	0.75178	0.92688
5	50.48	201.24	251.73	0.19627	0.72346	0.91973
10	56.46	196.96	253.42	0.21727	0.69559	0.91286
15	62.52	192.49	255.02	0.23819	0.66802	0.90621
20	68.67	187.84	256.50	0.25899	0.64074	0.89973
25	74.91	182.97	257.88	0.27970	0.61367	0.89337
30	81.25	177.87	259.12	0.30037	0.58676	0.88713
35	87.69	172.52	260.22	0.32104	0.55982	0.88086
40	94.26	166.89	261.15	0.34167	0.53291	0.87458
45	100.96	160.94	261.90	0.36233	0.50585	0.86818
50	107.83	154.62	262.44	0.38313	0.47844	0.86157
55	114.86	147.86	262.73	0.40409	0.45058	0.85467
60	122.18	140.50	262.68	0.42547	0.42171	0.84718
65	129.73	132.54	262.27	0.44714	0.39200	0.83914
70	137.62	123.77	261.40	0.46944	0.36071	0.83015
75	145.98	113.90	259.89	0.49267	0.32714	0.81981
80	155.01	102.47	257.48	0.51735	0.29016	0.80751
85	165.09	88.60	253.69	0.54446	0.24736	0.79182
90	177.04	70.04	247.24	0.57664	0.19288	0.76952
95	196.35	34.96	231.30	0.62731	0.09493	0.72224
96.01	212.57	0	212.57	0.67090	0	0.67090

* The enthalpy and the entropy of saturated liquid R-22 are taken as zero at a temperature of −40°C.

TABLE B.11 *Thermodynamic Properties of R-22 (Chlorodifluoromethane)—Superheated* $CHClF_2$.

Abs. Press. (kPa)	Sat. Temp. (°C) T_g	Sat. Vapor	Temperature (°C) −50	0	50	100	150	200
			Specific Volume, v (m^3/kg)					
		v_g						
1	−107.61	15.885	21.455	26.263	31.070	35.876	40.685	45.492
5	−89.79	3.519	4.283	5.248	6.211	7.176	8.137	9.098
10	−80.64	1.8411	2.137	2.621	3.104	3.588	4.069	4.549
20	−70.41	0.9646	1.064	1.308	1.550	1.794	2.034	2.275
40	−58.86	0.5059	0.527	0.651	0.773	0.897	1.017	1.138
60	−51.37	0.3468	0.349	0.432	0.514	0.596	0.673	0.759
80	−45.68	0.2652		0.323	0.385	0.446	0.510	0.570
100	−41.03	0.2152		0.257	0.307	0.356	0.407	0.455
200	−25.12	0.1123		0.1260	0.1519	0.1770	0.2020	0.2260
300	−14.61	0.0766		0.0822	0.1001	0.1172	0.1331	0.1497
400	−6.52	0.0582		0.0598	0.0739	0.0871	0.0994	0.1122
500	0.15	0.0469			0.0586	0.0694	0.0796	0.0895
600	5.88	0.0393			0.0482	0.0573	0.0660	0.0744
700	10.93	0.0338			0.0408	0.0488	0.0564	0.0637
800	15.47	0.0296			0.0347	0.0418	0.0489	0.0555
900	19.61	0.0263			0.0307	0.0373	0.0434	0.0493
1000	23.42	0.0237			0.0272	0.0333	0.0388	0.0442
1500	39.10	0.0155			0.0168	0.0213	0.0253	0.0294
2000	51.28	0.0113				0.0153	0.0184	0.0214
2500	61.38	0.0087				0.0116	0.0144	0.0169
3000	70.07	0.0069				0.0091	0.0117	0.0138
			Internal Energy, u (kJ/kg)					
		u_g						
1	−107.61	184.53	208.93	233.47	260.97	290.87	323.67	358.81
5	−89.79	191.51	208.88	233.44	260.95	290.86	323.66	358.80
10	−80.64	195.18	208.78	233.39	260.92	290.84	323.64	358.79
20	−70.41	199.32	208.56	233.24	260.82	290.82	323.63	358.77
40	−58.86	204.01	208.09	232.94	260.59	290.79	323.62	358.75
60	−51.37	207.04	207.64	232.66	260.40	290.73	323.58	358.72
80	−45.68	209.33		232.38	260.21	290.66	323.52	358.65
100	−41.03	211.19		232.11	260.02	290.52	323.43	358.56
200	−25.12	217.40		230.84	259.29	290.07	323.13	358.41
300	−14.61	221.31		229.38	258.36	289.42	322.77	358.11
400	−6.52	224.22		227.99	257.48	288.76	322.30	357.91
500	0.15	226.55			256.66	288.25	321.90	357.60
600	5.88	228.45			255.67	287.67	321.45	357.23
700	10.93	230.07			254.80	287.02	320.94	356.75
800	15.47	231.48			253.97	286.75	320.60	356.43
900	19.61	232.72			252.84	285.86	320.05	355.93
1000	23.42	233.75			251.61	285.04	319.61	355.51
1500	39.10	237.74			245.91	281.80	317.28	353.30
2000	51.28	239.94				278.10	315.07	352.20
2500	61.38	240.85				274.17	312.35	350.21
3000	70.07	240.68				269.41	309.39	348.33

Abs. Press. (kPa)	Sat. Temp. (°C) T_g	Sat. Vapor	Temperature (°C) −50	0	50	100	150	200
			Enthalpy, h (kJ/kg)					
		h_g						
1	−107.61	200.41	230.38	259.73	292.04	326.75	364.35	404.30
5	−89.79	209.10	230.29	259.68	292.00	326.74	364.34	404.29
10	−80.64	213.59	230.15	259.60	291.96	326.72	364.33	404.28
20	−70.41	218.61	229.84	259.40	291.82	326.70	364.32	404.27
40	−58.86	224.25	229.17	258.98	291.51	326.67	364.31	404.26
60	−51.37	227.85	228.56	258.59	291.25	326.51	364.30	404.25
80	−45.68	230.55		258.22	291.01	326.34	364.28	404.23
100	−41.03	232.71		257.81	290.81	326.12	364.08	404.05
200	−25.12	239.86		256.04	289.67	325.47	363.53	403.60
300	−14.61	244.29		254.04	288.39	324.58	362.72	403.02
400	−6.52	247.50		251.91	287.04	323.60	362.08	402.62
500	0.15	250.00			285.96	322.95	361.70	402.37
600	5.88	252.03			284.59	322.05	361.05	401.87
700	10.93	253.73			283.36	321.18	360.42	401.32
800	15.47	255.16			281.73	320.19	359.72	400.82
900	19.61	256.39			280.47	319.43	359.11	400.31
1000	23.42	257.45			278.81	318.34	358.41	399.66
1500	39.10	260.99			271.11	313.75	355.23	397.40
2000	51.28	262.54				308.70	351.87	395.00
2500	61.38	262.60				303.17	348.35	392.46
3000	70.07	261.38				296.71	344.49	389.73
			Entropy, s (kJ/kg K)					
		s_g						
1	−107.61	1.27520	1.43157	1.55006	1.65851	1.75784	1.85237	1.94160
5	−89.79	1.17223	1.27655	1.39517	1.50370	1.60310	1.69763	1.78686
10	−80.64	1.12969	1.20937	1.32828	1.43689	1.53645	1.63098	1.72049
20	−70.41	1.08905	1.14178	1.26115	1.36998	1.46981	1.56434	1.65417
40	−58.86	1.05047	1.07252	1.19285	1.30205	1.40317	1.49770	1.58763
60	−51.37	1.02880	1.03197	1.15310	1.26268	1.36402	1.45855	1.54858
80	−45.68	1.01390		1.12447	1.23451	1.33601	1.43054	1.52067
100	−41.03	1.00240		1.10201	1.21251	1.31426	1.40924	1.49902
200	−25.12	0.96876		1.03090	1.14387	1.25603	1.35131	1.44089
300	−14.61	0.94985		0.98699	1.10230	1.20627	1.30160	1.39123
400	−6.52	0.93671		0.95345	1.07131	1.17637	1.27173	1.36163
500	0.15	0.92667			1.04749	1.15389	1.25125	1.34573
600	5.88	0.91851			1.02686	1.13460	1.23263	1.32711
700	10.93	0.91160			1.00967	1.11845	1.21698	1.31146
800	15.47	0.90558			0.99155	1.10218	1.20155	1.29603
900	19.61	0.90022			0.97826	1.09031	1.19035	1.28483
1000	23.42	0.89537			0.96455	1.07831	1.17906	1.27354
1500	39.10	0.87571			0.90752	1.03033	1.13463	1.22911
2000	51.28	0.85985				0.99257	1.10115	1.19750
2500	61.38	0.84625				0.96324	1.07357	1.17206
3000	70.07	0.83002				0.92890	1.04919	1.15023

TABLE B.12 *Thermodynamic Properties of Dry Air—Saturated Liquid-Vapor.*

Vapor Temp. (°C)	Abs. Press. (kPa)	Liq. Temp. (°C)	Specific Volume (m^3/kg) Sat. Liq.	Evap.	Sat. Vapor	Internal Energy (kJ/kg) Sat. Liq.	Evap.	Sat. Vapor
T_g	p	T_f	v_f	v_{fg}	v_g	u_f	u_{fg}	u_g
-215	1.98	-218.2	0.001028	7.51	7.51	-31.20	211.62	180.42
-210	5.84	-213.1	0.001048	2.993	2.994	-18.71	202.27	183.56
-205	15.15	-208.0	0.001068	1.302	1.303	-9.11	195.87	186.76
-200	33.26	-203.0	0.001091	0.6368	0.6379	-0.04	190.32	190.28
-195	65.64	-197.9	0.001119	0.3401	0.3412	8.95	184.84	193.79
-190	118.86	-192.8	0.001148	0.1966	0.1967	18.22	179.23	197.45
-185	200.65	-187.7	0.001180	0.1190	0.1202	27.81	172.75	200.56
-180	319.62	-182.6	0.001216	0.07592	0.07714	37.60	165.32	202.92
-175	485.08	-177.4	0.001260	0.05019	0.05145	47.68	157.10	204.78
-170	707.16	-172.2	0.001310	0.03415	0.03546	57.95	147.45	205.40
-165	996.31	-167.0	0.001368	0.02367	0.02503	68.50	137.25	205.75
-160	1363.5	-161.8	0.001432	0.01655	0.01798	79.55	125.93	205.48
-155	1820.3	-156.5	0.001520	0.01150	0.01302	91.20	112.84	204.04
-150	2378.8	-151.2	0.001646	0.007719	0.009365	104.31	96.10	200.41
-145	3052.2	-145.8	0.001860	0.004604	0.006464	121.91	70.20	192.11
-140.6	3769.0	-140.6	0.003125	0	0.003125	166.81	0	166.81

Vapor Temp. (°C)	Enthalpy (kJ/kg)* Sat. Liq.	Evap.	Sat. Vapor	Entropy (kJ/kg K)* Sat. Liq.	Evap.	Sat. Vapor
T_g	h_f	h_{fg}	h_g	s_f	s_{fg}	s_g
-215	-31.20	226.49	195.29	-0.4163	3.9490	3.5327
-210	-18.70	219.74	201.04	-0.2693	3.5978	3.3285
-205	-9.09	215.59	206.50	-0.1290	3.2836	3.1546
-200	0	211.50	211.50	0	2.9941	2.9941
-195	9.02	207.17	216.19	0.1222	2.7330	2.8552
-190	18.36	202.47	220.83	0.2395	2.4842	2.7237
-185	28.05	196.59	224.64	0.3536	2.2566	2.6102
-180	37.99	189.59	227.58	0.4623	2.0554	2.5177
-175	48.29	181.45	229.74	0.5697	1.8609	2.4306
-170	58.87	171.60	230.47	0.6725	1.6827	2.3552
-165	69.86	160.83	230.69	0.7731	1.5041	2.2772
-160	81.50	148.50	230.00	0.8737	1.3230	2.1967
-155	93.97	133.77	227.74	0.9783	1.1354	2.1137
-150	108.23	114.46	222.69	1.0907	0.9300	2.0207
-145	127.59	84.25	211.84	1.2369	0.6783	1.9152
-140.6	178.59	0	178.59	1.5411	0	1.5411

* The enthalpy and the entropy of saturated liquid air are taken as zero at a temperature of -203.0°C. At the corresponding saturation pressure (33.26 kPa), the temperature of the saturated vapor is -200°C.

TABLE B.13 *Thermodynamic Properties of Dry Air—Superheated Air.*

Abs. Press. (kPa)	Sat. Temp. (°C)	Sat. Vapor	Temperature (°C)				
p	T_g		−200	−150	−100	−50	0
			Specific Volume, v (m^3/kg)				
		v_g					
1	−218.3	15.797	21.009	35.369	49.729	64.089	78.449
5	−211.2	3.551	4.202	7.074	9.946	12.818	15.690
10	−207.5	1.874	2.101	3.537	4.973	6.409	7.845
50	−197.5	0.4257		0.7044	0.9931	1.281	1.569
100	−191.5	0.2245		0.3499	0.4954	0.6399	0.7841
200	−185.0	0.1178		0.1729	0.2467	0.3195	0.3918
500	−174.6	0.04940		0.06592	0.09714	0.1272	0.1565
1000	−164.9	0.02489		0.03114	0.04781	0.06311	0.07802
2500	−149.0	0.008763			0.01885	0.02469	0.03097
5000					0.007803	0.01193	0.01534
7500					0.004873	0.007776	0.01016
10,000					0.003347	0.005693	0.007613
25,000						0.002498	0.003277
50,000						0.001746	0.002081
75,000						0.001514	0.001726
100,000						0.001389	0.001548
			Internal Energy, u (kJ/kg)				
		u_g					
1	−218.3	175.89	193.08	228.89	264.68	300.49	336.27
5	−211.2	181.74	192.87	228.83	264.65	300.47	336.27
10	−207.5	184.85	192.66	228.77	264.62	300.44	336.27
50	−197.1	193.01		228.38	264.44	300.30	336.16
100	−191.5	197.09		227.94	264.22	300.16	336.07
200	−185.0	200.99		226.98	263.76	299.83	335.85
500	−174.6	205.19		226.52	262.52	298.88	335.15
1000	−164.9	205.45		218.87	259.90	297.29	334.04
2500	−149.0	198.47			250.09	292.47	330.65
5000					238.97	284.29	324.90
7500					220.25	276.07	319.06
10,000					203.98	267.58	313.56
25,000						229.07	284.51
50,000						200.16	256.31
75,000						186.12	240.83
100,000						177.30	230.81

Abs. Press. (kPa)	Temperature (°C)						
p	50	100	150	200	250	300	350
			Specific Volume, v (m^3/kg)				
1	92.809	107.169	121.530	135.890	150.250	164.610	178.970
5	18.562	21.434	24.306	27.178	30.050	32.922	35.794
10	9.281	10.717	12.153	13.589	15.025	16.461	17.897
50	1.856	2.144	2.431	2.718	3.006	3.293	3.580
100	0.9280	1.072	1.216	1.359	1.503	1.647	1.791
200	0.4640	0.5360	0.6080	0.6799	0.7518	0.8237	0.8956
500	0.1855	0.2145	0.2434	0.2722	0.3010	0.3298	0.3586
1000	0.09271	0.1073	0.1218	0.1363	0.1508	0.1652	0.1796
2500	0.03705	0.04302	0.04892	0.05479	0.06063	0.06646	0.07227
5000	0.01853	0.02162	0.02465	0.02764	0.03061	0.03356	0.03649
7500	0.01239	0.01453	0.01661	0.01858	0.02056	0.02256	0.02469
10,000	0.009340	0.01097	0.01255	0.01410	0.01563	0.01713	0.01863
25,000	0.004021	0.004726	0.005402	0.006057	0.006699	0.007330	0.007953
50,000	0.002429	0.002776	0.003118	0.003452	0.003780	0.004102	0.004420
75,000	0.002171	0.002171	0.002396	0.002619	0.002840	0.003057	0.003272
100,000	0.001710	0.001874	0.002041	0.002208	0.002374	0.002539	0.002702
			Internal Energy, u (kJ/kg)				
1	372.20	408.26	444.58	481.26	518.39	556.04	594.24
5	372.20	408.26	444.58	481.26	518.39	556.04	594.24
10	372.20	408.26	444.58	481.26	518.39	556.04	594.24
50	372.16	408.18	444.54	481.23	518.34	556.00	594.21
100	372.03	408.11	444.44	481.21	518.32	555.94	594.11
200	371.84	407.98	444.34	481.07	518.22	555.87	594.07
500	371.34	407.53	443.96	480.75	517.94	555.63	593.86
1000	370.45	406.83	443.40	480.22	517.43	555.20	593.50
2500	367.81	404.93	441.54	478.62	516.05	553.90	592.29
5000	363.41	401.07	438.48	475.96	513.66	551.77	590.37
7500	358.67	397.05	435.01	473.10	511.75	550.01	587.65
10,000	354.74	394.06	432.55	470.76	509.01	547.66	586.56
25,000	331.93	375.25	416.50	456.74	496.52	536.24	576.11
50,000	306.52	379.92	396.82	439.22	480.76	521.93	562.97
75,000	291.03	338.00	382.78	426.19	468.64	510.69	552.45
100,000	280.75	327.72	372.61	416.30	459.24	501.69	543.96

TABLE B.13 (Continued) *Thermodynamic Properties of Dry Air—Superheated Air.*

Abs. Press. (k Pa)	Temperature (°C)						
p	400	450	500	550	600	650	700
			Specific Volume, v (m^3/kg)				
1	193.330	207.690	221.915	236.266	250.617	264.969	279.320
5	38.666	41.538	44.383	47.253	50.123	52.994	55.864
10	19.335	20.769	22.192	23.627	25.062	26.497	27.932
50	3.868	4.155	4.442	4.729	5.017	5.304	5.591
100	1.934	2.078	2.221	2.365	2.509	2.652	2.796
200	0.9674	1.039	1.111	1.183	1.255	1.327	1.399
500	0.3874	0.4162	0.4449	0.4737	0.5024	0.5312	0.5599
1000	0.1941	0.2085	0.2229	0.2373	0.2517	0.2661	0.2804
2500	0.07807	0.08386	0.08964	0.09543	0.1012	0.1070	0.1127
5000	0.03942	0.04234	0.04525	0.04816	0.05106	0.05396	0.05686
7500	0.02649	0.02845	0.03046	0.03241	0.03436	0.03630	0.03824
10,000	0.02012	0.02160	0.02307	0.02454	0.02601	0.02747	0.02893
25,000	0.008570	0.009002	0.009785	0.01038	0.01098	0.01157	0.01217
50,000	0.004735	0.005047	0.005357	0.005665	0.005972	0.006276	0.006581
75,000	0.003484	0.003693	0.003906	0.004113	0.004319	0.004523	0.004727
100,000	0.002863	0.003022	0.003180	0.003337	0.003492	0.003647	0.003800
			Internal Energy, u (kJ/kg)				
1	633.03	672.42	712.55	753.12	794.27	835.98	878.16
5	633.03	672.42	712.55	753.12	794.27	835.98	878.16
10	633.03	672.42	712.55	753.12	794.27	835.98	878.16
50	633.01	672.36	712.36	752.94	794.04	835.75	877.93
100	632.96	672.32	712.33	752.91	794.01	835.74	877.88
200	632.88	672.27	712.27	752.83	793.94	835.61	877.78
500	632.66	672.05	712.08	752.66	793.84	835.52	877.73
1000	632.26	671.70	711.76	752.34	793.51	835.22	877.48
2500	631.22	670.74	710.84	751.50	792.73	834.43	876.83
5000	629.45	669.08	709.30	750.05	791.38	833.11	875.48
7500	628.13	667.91	707.77	748.62	790.00	831.92	874.18
10,000	625.90	665.83	706.19	747.24	788.67	830.68	872.78
25,000	616.26	661.30	699.05	740.55	782.29	824.71	867.31
50,000	604.09	645.46	687.13	729.16	771.56	814.45	857.63
75,000	593.75	636.26	679.48	721.55	764.31	807.50	850.96
100,000	586.22	628.62	671.87	713.93	757.09	800.48	844.28

Abs. Press. (kPa)	Temperature (°C)					
p	750	800	850	900	950	1000
			Specific Volume, v (m^3/kg)			
1	293.671	308.023	322.374	336.725	351.077	365.428
5	58.734	61.605	64.475	67.345	70.215	73.086
10	29.367	30.802	32.237	33.673	35.108	36.543
50	5.878	6.166	6.453	6.740	7.028	7.315
100	2.940	3.083	3.227	3.371	3.514	3.658
200	1.470	1.542	1.614	1.686	1.758	1.829
500	0.5887	0.6174	0.6462	0.6749	0.7037	0.7324
1000	0.2948	0.3092	0.3236	0.3380	0.3523	0.3667
2500	0.1185	0.1243	0.1300	0.1358	0.1416	0.1473
5000	0.05975	0.06265	0.06554	0.06843	0.07131	0.07420
7500	0.04018	0.04211	0.04404	0.04598	0.04790	0.04983
10,000	0.03039	0.03184	0.03329	0.03475	0.03620	0.03765
25,000	0.01276	0.01335	0.01394	0.01453	0.01512	0.01570
50,000	0.006884	0.007187	0.007488	0.007789	0.008089	0.008388
75,000	0.004930	0.005133	0.005335	0.005536	0.005738	0.005938
100,000	0.003953	0.004106	0.004258	0.004410	0.004562	0.004713
			Internal Energy, u (kJ/kg)			
1	920.91	964.06	1007.6	1051.6	1096.0	1140.7
5	920.91	964.06	1007.6	1051.6	1096.0	1140.7
10	920.91	964.06	1007.6	1051.6	1096.0	1140.7
50	920.68	963.78	1007.5	1051.4	1095.8	1140.4
100	920.58	963.76	1007.4	1051.3	1095.7	1140.3
200	920.48	963.66	1007.3	1051.2	1095.6	1140.2
500	920.43	963.58	1007.2	1051.2	1095.5	1140.2
1000	920.28	963.38	1006.9	1050.9	1095.4	1140.0
2500	919.53	962.60	1006.5	1050.3	1094.6	1139.5
5000	918.33	961.53	1003.3	1049.3	1093.8	1138.5
7500	917.15	960.45	1004.2	1048.4	1092.8	1137.5
10,000	916.08	959.48	1003.3	1047.3	1091.9	1136.6
25,000	910.58	954.13	998.1	1042.4	1087.0	1132.1
50,000	901.28	945.13	989.5	1034.2	1079.1	1124.3
75,000	894.93	939.00	983.6	1028.4	1073.5	1118.9
100,000	888.48	932.78	977.6	1022.5	1067.8	1113.4

TABLE B.13 (Continued) *Thermodynamic Properties of Dry Air—Superheated Air.*

Abs. Press. (kPa) p	Sat. Temp. (°C) T_g	Sat. Vapor	Temperature (°C) −200	−150	−100	−50	0
			Enthalpy, h (kJ/kg)				
		h_g					
1	−218.3	191.69	214.09	264.26	314.41	364.58	414.72
5	−211.2	199.49	213.88	264.20	314.38	364.56	414.72
10	−207.5	203.59	213.67	264.14	314.35	364.53	414.72
50	−197.1	214.29		263.60	314.09	364.35	414.61
100	−191.5	219.54		262.93	313.76	364.15	414.48
200	−185.0	224.55		261.56	313.10	363.73	414.21
500	−174.6	229.89		259.48	311.09	362.48	413.40
1000	−164.9	230.34		250.01	307.71	360.40	412.06
2500	−149.0	220.38			297.21	354.19	408.07
5000					277.98	343.94	401.60
7500					256.80	334.39	395.26
10,000					237.45	324.51	389.69
25,000						291.52	366.43
50,000						287.46	360.36
75,000						299.67	370.28
100,000						316.20	385.61
			Entropy, s (kJ/kg K)				
		s_g					
1	−218.3	3.6907	3.9741	4.4973	4.8391	5.0935	5.3092
5	−211.2	3.3847	3.5119	4.0351	4.3769	4.6312	4.8469
10	−207.5	3.2447	3.3088	3.8353	4.1776	4.4320	4.6350
50	−197.1	2.9087		3.3700	3.7142	3.9715	4.1747
100	−191.5	2.7487		3.1671	3.5138	3.7717	3.9752
200	−185.0	2.6112		2.9359	3.3121	3.5712	3.7753
500	−174.6	2.4278		2.6735	3.0406	3.3037	3.5097
1000	−164.9	2.2957		2.4322	2.8276	3.0975	3.3065
2500	−149.0	1.9973			2.4983	2.8130	3.0310
5000					2.2392	2.5782	2.8117
7500					2.0287	2.4260	2.6728
10,000					1.8606	2.3101	2.5744
25,000						1.9251	2.2288
50,000						1.6802	1.9753
75,000						1.5539	1.8397
100,000						1.4659	1.7467

Abs. Press. (kPa)	Temperature (°C)						
p	50	100	150	200	250	300	350
				Enthalpy, h (kJ/kg)			
1	465.01	515.43	566.11	617.15	668.64	720.65	773.21
5	465.01	515.43	566.11	617.15	668.64	720.65	773.21
10	465.01	515.43	566.11	617.15	668.64	720.65	773.21
50	464.92	515.38	566.09	617.15	668.64	720.65	773.21
100	464.83	515.31	566.04	617.11	668.62	720.64	773.21
200	464.64	515.18	565.94	617.05	668.58	720.61	773.19
500	464.09	514.78	565.66	616.85	668.44	720.53	773.16
1000	463.16	514.13	565.20	616.52	668.23	720.40	773.10
2500	460.43	512.21	563.84	615.59	667.62	720.05	772.96
5000	456.06	509.17	561.73	614.16	666.71	719.57	772.82
7500	451.59	506.02	559.58	612.45	665.95	719.21	772.80
10,000	448.14	503.76	558.05	611.76	665.31	718.96	772.86
25,000	432.45	493.40	551.55	608.16	663.99	719.49	774.93
50,000	427.97	491.72	552.71	611.82	669.76	727.02	783.97
75,000	437.05	500.82	562.48	622.61	681.64	739.96	797.85
100,000	451.75	515.12	576.71	637.10	696.58	755.59	814.16
				Entropy, s (kJ/kg K)			
1	5.4654	5.6106	5.7380	5.8520	5.9727	6.0676	6.1555
5	5.0031	5.1483	5.2757	5.3897	5.5105	5.6054	5.6933
10	4.8041	4.9677	5.0766	5.1906	5.3114	5.4063	5.4942
50	4.3438	4.4890	4.6165	4.7306	4.8340	4.9289	5.0169
100	4.1445	4.2897	4.4173	4.5314	4.6348	4.7298	4.8177
200	3.9449	4.0903	4.2179	4.3320	4.4356	4.5306	4.6185
500	3.6801	3.8259	3.9539	4.0682	4.1719	4.2669	4.3550
1000	3.4783	3.6249	3.7534	3.8680	3.9719	4.0671	4.1553
2500	3.2071	3.3561	3.4859	3.5915	3.7060	3.8018	3.8903
5000	2.9949	3.1477	3.2799	3.3970	3.5026	3.5991	3.6882
7500	2.8624	3.0191	3.1538	3.2807	3.3861	3.4826	3.5717
10,000	2.7711	2.9312	3.0678	3.1878	3.2953	3.3933	3.4834
25,000	2.4511	2.6266	2.7728	2.8993	3.0115	3.1128	3.2055
50,000	2.2029	2.3864	2.5398	2.6719	2.7883	2.8929	2.9881
75,000	2.0643	2.2479	2.4030	2.5373	2.6560	2.7624	2.8593
100,000	1.9692	2.1516	2.3066	2.4415	2.5611	2.6687	2.7667

TABLE B.13 (Continued) *Thermodynamic Properties of Dry Air—Superheated Air.*

Abs. Press. (k Pa)	Temperature (°C)						
p	400	450	500	550	600	650	700
			Enthalpy, h (kJ/kg)				
1	826.36	880.11	934.46	989.39	1044.89	1100.95	1157.5
5	826.36	880.11	934.46	989.39	1044.89	1100.95	1157.5
10	826.36	880.11	934.46	989.39	1044.89	1100.95	1157.5
50	826.36	880.11	934.46	989.39	1044.89	1100.95	1157.5
100	826.36	880.11	934.47	989.41	1044.91	1100.97	1157.5
200	826.36	880.12	934.49	989.43	1044.94	1101.01	1157.6
500	826.36	880.15	934.54	989.51	1045.04	1101.12	1157.7
1000	826.36	880.20	934.63	989.64	1045.21	1101.32	1157.9
2500	826.39	880.39	934.95	990.07	1045.73	1101.93	1158.6
5000	826.55	880.78	935.55	990.85	1046.68	1103.02	1159.7
7500	826.80	881.28	936.25	991.72	1047.70	1104.17	1161.0
10,000	827.13	881.83	936.89	992.64	1048.77	1105.38	1162.5
25,000	830.51	886.35	943.67	990.05	1056.79	1113.96	1171.6
50,000	840.84	897.81	954.97	1012.41	1070.16	1128.25	1186.7
75,000	855.05	913.23	972.43	1030.02	1088.23	1146.72	1205.5
100,000	872.52	930.82	989.88	1047.63	1106.29	1165.18	1224.3
			Entropy, s (kJ/kg K)				
1	6.2375	6.3145	6.3872	6.4560	6.5215	6.5839	6.6436
5	5.7753	5.8523	5.9250	5.9938	6.0593	6.1217	6.1814
10	5.5762	5.6532	5.7259	5.7947	5.8602	5.9226	5.9823
50	5.0989	5.1759	5.2486	5.3174	5.3829	5.4453	5.5050
100	4.8998	4.9768	5.0495	5.1183	5.1838	5.2462	5.3059
200	4.7006	4.7776	4.8503	4.9192	4.9846	5.0470	5.1068
500	4.4371	4.5142	4.5869	4.6577	4.7213	4.7837	4.8434
1000	4.2375	4.3146	4.3874	4.4564	4.5219	4.5843	4.6441
2500	3.9727	4.0501	4.1231	4.1921	4.2578	4.3203	4.3802
5000	3.7711	3.8488	3.9221	3.9914	4.0572	4.1199	4.1799
7500	3.6546	3.7323	3.8044	3.8739	3.9399	4.0028	4.0629
10,000	3.5672	3.6456	3.7193	3.7891	3.8553	3.9183	3.9785
25,000	3.2913	3.3713	3.4464	3.5172	3.5844	3.6482	3.7091
50,000	3.0759	3.1576	3.2340	3.3060	3.3741	3.4387	3.5004
75,000	2.9483	3.0310	3.1083	3.1810	3.2496	3.3147	3.3769
100,000	2.8568	2.9403	3.0184	3.0917	3.1608	3.9834	3.2888

Abs. Press. (kPa)	Temperature (°C)					
p	750	800	850	900	950	1000
			Enthalpy, h (kJ/kg)			
1	1214.6	1272.1	1330.0	1388.3	1447.1	1506.1
5	1214.6	1272.1	1330.0	1388.3	1447.1	1506.1
10	1214.6	1272.1	1330.0	1388.3	1447.1	1506.1
50	1214.6	1272.1	1330.0	1388.3	1447.1	1506.1
100	1214.6	1272.1	1330.1	1388.4	1447.1	1506.1
200	1214.7	1272.2	1330.1	1388.4	1447.2	1506.2
500	1214.8	1272.3	1330.3	1388.6	1447.3	1506.4
1000	1215.1	1272.6	1330.9	1388.9	1447.7	1506.7
2500	1215.8	1273.4	1331.5	1389.8	1448.6	1507.7
5000	1217.1	1274.8	1333.0	1391.4	1450.3	1509.5
7500	1248.5	1276.3	1334.5	1393.2	1452.1	1511.3
10,000	1220.0	1277.9	1336.2	1394.8	1453.9	1513.1
25,000	1229.6	1287.9	1346.6	1405.6	1465.0	1524.6
50,000	1245.5	1304.5	1363.9	1423.6	1483.5	1543.7
75,000	1264.7	1324.0	1383.7	1443.6	1503.8	1564.2
100,000	1283.8	1343.4	1403.4	1463.5	1524.0	1584.7
			Entropy, s (kJ/kg K)			
1	6.7007	6.7557	6.8084	6.8593	6.9083	6.9557
5	6.2385	6.2935	6.3462	6.3971	6.4461	6.9435
10	6.0394	6.0944	6.1471	6.1980	6.2470	6.2944
50	5.5621	5.6171	5.6698	5.7207	5.7697	5.8171
100	5.3630	5.4180	5.4707	5.5216	5.5706	5.6180
200	5.1639	5.2189	5.2716	5.3225	5.3715	5.4189
500	4.9006	4.9556	5.0083	5.0592	5.1078	5.1556
1000	4.7013	4.7563	4.8090	4.8600	4.9089	4.9564
2500	4.4374	4.4925	4.5453	4.5963	4.6453	4.6928
5000	4.2372	4.2924	4.3453	4.3964	4.4455	4.4930
7500	4.1204	4.1756	4.2286	4.2797	4.3289	4.3665
10,000	4.0361	4.0914	4.1445	4.1957	4.2450	4.2926
25,000	3.7672	3.8231	3.8766	3.9282	3.9778	4.0257
50,000	3.5592	3.6157	3.6698	3.7218	3.7719	3.8202
75,000	3.4361	3.4929	3.5472	3.5996	3.6499	3.6984
100,000	3.3483	3.4053	3.4599	3.5124	3.5629	3.6116

TABLE B.14 *Thermodynamic Properties of Dry Air—Compressed Liquid.*

Abs. Press. (kPa)	Temperature (°C)			
p	−200	−190	−180	−170
		Specific Volume, v (m^3/kg)		
500	0.001108	0.001167	0.001241	
1000	0.001107	0.001166	0.001238	0.001334
2500	0.001103	0.001161	0.001231	0.001321
5000	0.001098	0.001153	0.001219	0.001301
7500	0.001093	0.001146	0.001208	0.001284
10,000	0.001088	0.001139	0.001198	0.001269
25,000	0.001061	0.001105	0.001151	0.001202
50,000	0.001027	0.001062	0.001097	0.001133
		Internal Energy, u (kJ/kg)		
500	4.98	23.36	42.55	
1000	4.72	23.05	42.13	62.32
2500	4.21	22.26	41.07	60.66
5000	3.32	21.04	39.33	58.17
7500	2.43	19.86	37.72	56.04
10,000	1.70	18.72	36.32	54.07
25,000	−2.81	13.30	29.35	45.37
50,000	−8.02	6.83	21.68	33.49
		Enthalpy, h (kJ/kg)		
500	5.53	23.92	43.17	
1000	5.83	24.22	43.37	63.65
2500	6.97	25.16	44.15	63.96
5000	8.81	26.80	45.42	64.67
7500	10.63	28.45	46.78	65.67
10,000	12.58	30.11	48.30	66.76
25,000	23.72	40.92	58.12	75.42
50,000	43.33	59.93	76.53	90.14
		Entropy, s (kJ/kg K)		
500	0.0725	0.3006	0.5158	
1000	0.0695	0.2969	0.5112	0.7140
2500	0.0618	0.2879	0.4999	0.6981
5000	0.0492	0.2727	0.4806	0.6732
7500	0.0371	0.2582	0.4629	0.6513
10,000	0.0249	0.2444	0.4461	0.6311
25,000	−0.0304	0.1716	0.3630	0.5363
50,000	−0.0959	0.0821	0.2601	0.4264

Abs. Press. (k Pa)	Temperature (°C)			Sat. Temp. (°C	Sat. Liq.
p	−160	−150	−140	T_f	
		Specific Volume, v (m^3/kg)			v_f
500				−176.99	0.001266
1000				−164.94	0.001396
2500	0.001448			−150.14	0.001682
5000	0.001409	0.001570	0.001944		
7500	0.001379	0.001506	0.001704		
10,000	0.001364	0.001460	0.001605		
25,000	0.001257	0.001318	0.001385		
50,000	0.001170	0.001209	0.001249		
		Internal Energy, u (kJ/kg)			u_f
500				−176.99	48.47
1000				−164.94	68.63
2500	81.70			−150.14	107.40
5000	77.81	99.64	131.38		
7500	74.75	94.56	117.57		
10,000	72.01	90.83	111.26		
25,000	61.42	77.56	94.04		
50,000	51.59	66.44	81.02		
		Enthalpy, h (kJ/kg)			h_f
500				−176.99	49.10
1000				−164.94	70.00
2500	85.32			−150.14	111.60
5000	84.85	107.49	141.10		
7500	85.09	105.85	130.35		
10,000	85.65	105.43	127.31		
25,000	92.84	110.51	128.66		
50,000	110.09	126.89	143.47		
		Entropy, s (kJ/kg K)			s_f
500				−176.99	0.5780
1000				−164.94	0.7750
2500	0.7680			−150.14	1.1170
5000	0.8586	1.0510	1.3156		
7500	0.8294	1.0069	1.2003		
10,000	0.8046	0.9729	1.1473		
25,000	0.6955	0.8460	0.9897		
50,000	0.5803	0.7245	0.8563		

TABLE B.15 *Specific Heat Capacities for Ideal Gases (kJ/kg K).*

Temp.	Water Vapor			Carbon Dioxide		
T (°C)	c_p	c_v	c_p/c_v	c_p	c_v	c_p/c_v
−50	1.851	1.388	1.333	0.761	0.572	1.330
0	1.852	1.390	1.332	0.817	0.628	1.301
50	1.869	1.408	1.328	0.869	0.680	1.278
100	1.890	1.428	1.323	0.916	0.727	1.260
150	1.913	1.452	1.318	0.958	0.769	1.246
200	1.939	1.478	1.312	0.995	0.806	1.234
250	1.967	1.506	1.307	1.029	0.840	1.225
300	1.997	1.536	1.301	1.060	0.871	1.217
350	2.029	1.567	1.295	1.088	0.899	1.210
400	2.061	1.600	1.288	1.113	0.924	1.204
450	2.095	1.633	1.283	1.137	0.948	1.199
500	2.129	1.668	1.277	1.158	0.969	1.195
550	2.164	1.702	1.271	1.177	0.989	1.191
600	2.198	1.737	1.266	1.195	1.006	1.188
650	2.233	1.771	1.261	1.212	1.023	1.185
700	2.266	1.805	1.256	1.227	1.038	1.182
750	2.299	1.838	1.251	1.240	1.051	1.180
800	2.331	1.869	1.247	1.253	1.064	1.178
850	2.369	1.908	1.242	1.264	1.075	1.176
900	2.407	1.946	1.237	1.275	1.086	1.174
950	2.440	1.979	1.233	1.285	1.096	1.172
1000	2.473	2.011	1.229	1.294	1.105	1.171
1050	2.504	2.043	1.226	1.302	1.113	1.170
1100	2.535	2.074	1.223	1.309	1.121	1.169
1150	2.565	2.103	1.219	1.317	1.128	1.168
1200	2.593	2.132	1.216	1.323	1.134	1.167
1250	2.621	2.159	1.214	1.329	1.140	1.166
1300	2.648	2.186	1.211	1.335	1.146	1.165
1350	2.673	2.212	1.209	1.340	1.151	1.164
1400	2.698	2.236	1.206	1.345	1.156	1.163
1450	2.721	2.260	1.204	1.350	1.161	1.163
1500	2.744	2.283	1.202	1.355	1.166	1.162

Temp. T (°C)	Refrigerant R-22 c_p	c_v	c_p/c_v	Dry Air c_p	c_v	c_p/c_v
−50	0.555	0.459	1.209	1.003	0.713	1.405
0	0.615	0.519	1.185	1.004	0.715	1.404
50	0.674	0.578	1.166	1.006	0.717	1.403
100	0.729	0.633	1.152	1.010	0.721	1.401
150	0.780	0.684	1.141	1.016	0.727	1.398
200	0.826	0.730	1.132	1.025	0.735	1.393
250	0.865	0.768	1.125	1.034	0.745	1.388
300	0.895	0.799	1.120	1.045	0.756	1.383
350	0.923	0.827	1.116	1.056	0.767	1.377
400	0.950	0.854	1.113	1.071	0.782	1.370
450	0.975	0.879	1.109	1.080	0.791	1.365
500	0.997	0.901	1.107	1.092	0.803	1.360
550	1.017	0.921	1.104	1.104	0.815	1.355
600	1.034	0.938	1.102	1.115	0.826	1.337
650	1.050	0.954	1.101	1.126	0.837	1.346
700	1.064	0.968	1.099	1.136	0.847	1.342
750	1.076	0.980	1.098	1.145	0.856	1.338
800	1.087	0.991	1.097	1.154	0.865	1.334
850	1.097	1.001	1.096	1.163	0.873	1.331
900	1.106	1.009	1.095	1.170	0.881	1.328
950	1.114	1.018	1.094	1.178	0.889	1.325
1000	1.122	1.026	1.094	1.185	0.896	1.323
1050	1.129	1.033	1.093	1.191	0.902	1.321
1100	1.137	1.041	1.092	1.197	0.908	1.318
1150	1.142	1.046	1.092	1.203	0.914	1.316
1200	1.149	1.053	1.091	1.208	0.919	1.315
1210	1.157	1.060	1.091	1.213	0.924	1.313
1300	1.164	1.068	1.090	1.218	0.929	1.311
1350	1.173	1.077	1.089	1.223	0.934	1.310
1400	1.182	1.086	1.089	1.227	0.938	1.308
1450	1.193	1.097	1.088	1.231	0.942	1.307
1500	1.205	1.109	1.087	1.241	0.952	1.304

TABLE B.15 (Continued) *Specific Heat Capacities for Ideal Gases (kJ/kg K).*

Temp. T (°C)	Oxygen, O_2 c_p	c_v	c_p/c_v	Nitrogen, N_2 c_p	c_v	c_p/c_v	Hydrogen, H_2 c_p	c_v	c_p/c_v
−50	0.911	0.651	1.399	1.039	0.742	1.400	13.81	9.685	1.426
0	0.915	0.655	1.397	1.039	0.742	1.400	14.19	10.07	1.410
50	0.922	0.717	1.403	1.040	0.743	1.399	14.37	10.25	1.402
100	0.934	0.674	1.386	1.042	0.745	1.398	14.46	10.33	1.399
150	0.948	0.688	1.378	1.046	0.749	1.396	14.49	10.36	1.398
200	0.963	0.703	1.369	1.052	0.755	1.393	14.51	10.38	1.397
250	0.979	0.719	1.361	1.060	0.763	1.380	14.52	10.40	1.397
300	0.995	0.735	1.354	1.069	0.772	1.384	14.54	10.41	1.396
350	1.010	0.750	1.346	1.080	0.783	1.379	14.56	10.43	1.395
400	1.024	0.764	1.340	1.091	0.794	1.374	14.59	10.46	1.394
450	1.037	0.777	1.334	1.103	0.806	1.368	14.62	10.50	1.393
500	1.048	0.789	1.329	1.115	0.819	1.363	14.67	10.54	1.391
550	1.059	0.799	1.325	1.127	0.831	1.357	14.72	10.60	1.389
600	1.069	0.809	1.321	1.139	0.842	1.352	14.78	10.66	1.387
650	1.078	0.818	1.318	1.151	0.854	1.348	14.86	10.73	1.384
700	1.086	0.826	1.315	1.161	0.865	1.343	14.94	10.81	1.381
750	1.093	0.833	1.312	1.172	0.875	1.339	15.02	10.89	1.378
800	1.110	0.840	1.309	1.181	0.885	1.335	15.12	10.99	1.375
850	1.106	0.846	1.307	1.191	0.894	1.332	15.21	11.09	1.372
900	1.112	0.852	1.305	1.199	0.903	1.329	15.32	11.19	1.369
950	1.117	0.858	1.303	1.207	0.911	1.326	14.42	11.30	1.365
1000	1.122	0.863	1.301	1.215	0.918	1.323	15.53	11.41	1.362
1050	1.127	0.867	1.300	1.222	0.925	1.321	15.64	11.52	1.358
1100	1.132	0.872	1.298	1.229	0.932	1.318	15.75	11.62	1.355
1150	1.136	0.876	1.296	1.235	0.938	1.316	15.86	11.73	1.351
1200	1.140	0.881	1.295	1.241	0.944	1.314	15.97	11.84	1.348
1250	1.144	0.885	1.294	1.246	0.949	1.313	16.07	11.95	1.345
1300	1.148	0.889	1.292	1.251	0.955	1.311	16.18	12.05	1.342
1350	1.152	0.893	1.291	1.256	0.959	1.309	16.28	12.15	1.339
1400	1.156	0.896	1.290	1.261	0.964	1.308	16.38	12.26	1.337
1450	1.160	0.900	1.289	1.265	0.968	1.307	16.48	12.36	1.334
1500	1.164	0.904	1.287	1.269	0.972	1.305	16.58	12.46	1.331

TABLE B.16 *Ideal Gas Internal Energies of Various Substances (kJ/kg).*

Temp. T (°C)	Methane, CH_4	Acetylene, C_2H_2	Ethylene, C_2H_4	Ethane, C_2H_6	Propane, C_3H_8	Octane, C_8H_{18}
−50	−121.6	−91.2	−81.5	−97.5	−95.9	
0	−40.4	−30.1	−27.0	−31.3	−34.0	
25	0	0	0	0	0	0
50	45.9	37.4	35.4	39.5	41.0	35.8
100	138.3	112.7	106.9	123.6	125.8	116.9
150	238.5	192.7	186.6	217.6	222.2	206.9
200	347.9	278.3	275.8	323.2	330.7	306.6
250	465.8	367.1	372.1	437.7	449.0	416.2
300	593.5	459.9	476.9	564.8	576.6	535.5
350	729.5	555.4	587.7	700.4	713.1	664.7
400	875.2	654.1	705.6	843.7	858.0	803.6
450	1028.6	755.4	828.9	994.5	1011.2	952.3
500	1190.9	875.4	958.3	1152.8	1171.5	1111.9
550	1360.3	965.7	1092.3	1318.3	1337.3	1281.4
600	1537.9	1074.4	1231.6	1491.2	1508.1	1455.1
650	1722.0	1185.3	1374.9	1668.0	1685.8	1636.5
700	1913.5	1298.1	1523.0	1851.6	1869.6	1826.7
750	2110.7	1413.4	1674.6	2041.4	2058.6	
800	2314.5	1530.3	1830.4	2236.2	2252.7	
850	2523.4	1649.0	1989.3	2436.8	2451.2	
900	2738.1	1769.8	2151.7	2641.5	2652.6	
950	2957.3	1892.0	2316.9	2849.8	2859.7	
1000	3181.8	2016.0	2485.2	3062.3	3068.5	
1050	3410.2	2141.1	2656.0	3278.9	3280.3	
1100	3636.2	2265.1	2829.5	3499.3	3494.1	
1150	3879.5	2396.6	3004.9	3721.8	3712.9	
1200	4119.9	2526.5	3183.0	3948.6	3930.3	
1250	4363.3	2641.4	3362.8	4177.9	4148.5	
1300	4610.4	2789.6	3544.8	4409.6	4378.6	
1350	4860.2	2922.8	3728.4	4644.9	4590.5	
1400	5113.1	3057.1	3913.9	4881.6	4815.7	
1450	5368.3	3192.5	4100.8	5119.5	5042.4	
1500	5626.1	3328.9	4289.3	5358.9	5268.8	

NOTE: Internal energy is taken as zero at 25°C.

Temp. T (°C)	Dry Air	Oxygen, O_2	Nitrogen, N_2	Water Vapor, H_2O	Carbon Dioxide, CO_2	Carbon Monoxide, CO	Hydrogen, H_2
−50	−53.7	−49.8	−55.6	−86.4	−46.4	−55.8	−748.1
0	−17.9	−16.4	−18.6	−34.8	−16.3	−18.7	−253.1
25	0	0	0	0	0	0	0
50	18.0	16.5	18.6	35.1	16.4	18.7	255.5
100	54.1	44.1	55.9	106.1	51.6	56.0	770.2
150	90.4	84.0	93.2	178.1	89.0	93.5	1287.7
200	127.1	118.8	130.9	251.4	128.4	131.4	1806.3
250	164.2	154.3	168.9	326.0	169.6	169.8	2325.7
300	201.7	190.7	206.3	402.0	212.3	208.7	2846.0
350	240.1	227.8	246.1	479.7	256.6	248.2	3366.9
400	277.9	265.7	285.7	558.8	302.2	288.4	3889.2
450	318.3	304.3	325.8	639.6	349.0	329.1	4413.2
500	358.5	343.4	366.3	722.2	397.0	370.6	4997.9
550	399.1	383.1	407.6	806.4	445.8	412.7	5467.9
600	440.3	423.3	449.5	892.5	495.7	455.2	5999.5
650	482.0	464.0	491.8	980.2	546.5	498.5	6534.4
700	524.2	505.2	534.7	1069.6	597.9	542.4	7073.3
750	566.9	546.5	578.2	1169.7	650.2	586.8	7616.0
800	610.1	588.5	622.3	1253.3	703.0	631.7	8163.8
850	653.6	630.7	666.7	1348.8	756.4	677.0	8715.6
900	697.6	673.1	711.7	1445.8	810.7	722.8	9272.4
950	742.1	715.9	757.0	1543.8	865.3	768.8	9834.2
1000	786.7	759.0	802.8	1666.7	920.7	815.4	10,401
1050	831.7	802.2	848.8	1743.4	976.4	862.3	10,975
1100	877.0	845.7	895.3	1843.3	1032.6	909.4	11,553
1150	922.6	889.4	942.1	1952.4	1088.6	956.9	12,136
1200	968.4	933.3	989.1	2058.3	1145.0	1004.7	12,726
1250	1014.4	977.5	1036.4	2165.6	1201.9	1052.7	13,321
1300	1061.3	1021.8	1084.0	2274.3	1259.0	1101.0	13,922
1350	1108.2	1066.3	1131.9	2384.2	1316.5	1149.5	14,526
1400	1170.0	1111.1	1180.0	2495.5	1374.3	1198.1	15,137
1450	1203.4	1156.0	1228.3	2607.4	1432.1	1247.0	15,752
1500	1251.4	1201.1	1276.9	2721.5	1490.3	1296.1	16,372
2000	1735.3	1661.9	1770.4	4860.0	2220.2	1794.7	22,829
2500	2232.2	2140.8	2275.8	5175.2	2685.5	2304.1	29,666
3000	2738.8	2635.9	2788.6	6491.0	3298.2	2820.7	36,810
3500	3251.9	3144.6	3306.6	7841.8	3917.3	3341.3	44,210
4000	3770.6	3664.2	3828.6	9220.1	4541.7	3866.3	51,865
4500	4293.8	4192.4	4354.0	10,618.8	5146.9	4394.5	59,724
5000	4823.4	4726.7	4885.9	12,035.0	5804.4	4925.6	67,776
5500	5354.3	5265.4	5418.6	13,465.9	6442.9	5459.8	76,042
6000	5888.2	5806.7	5954.0	14,907.6	7085.4	5996.6	84,415

TABLE B.17 *Ideal Gas Enthalpies of Various Substances (kJ/kg).*

Temp. T (°C)	Methane, CH_4	Ethane, C_2H_6	Acetylene, C_2H_2	Ethylene, C_2H_4	Propane, C_3H_8	Octane, C_8H_{18}
−50	−160.4	−118.2	−115.1	−103.8	−110.0	
0	−53.3	−38.2	−38.1	−34.4	−38.6	
25	0	0	0	0	0	0
50	58.9	46.4	45.4	42.8	45.8	37.6
100	177.2	144.3	136.6	129.1	140.0	122.4
150	303.3	252.1	232.6	223.6	245.8	216.0
200	438.5	371.5	334.1	327.6	363.7	319.4
250	582.3	499.9	438.9	438.7	491.4	432.6
300	735.9	640.8	547.6	558.3	628.5	555.6
350	897.8	790.2	659.1	683.9	774.4	688.3
400	1069.4	947.3	773.8	816.6	928.7	830.9
450	1248.7	1111.9	891.0	954.7	1091.3	983.3
500	1436.9	1284.0	1011.0	1098.9	1261.0	1146.5
550	1632.2	1463.4	1133.2	1247.7	1436.3	1319.6
600	1835.7	1650.1	1257.9	1401.9	1616.5	1497.0
650	2045.7	1840.7	1384.7	1560.0	1803.6	1682.0
700	2263.1	2038.1	1513.8	1722.9	1996.7	1875.8
750	2486.2	2241.7	1644.7	1889.3	2195.3	
800	2715.9	2450.3	1777.6	2059.9	2398.8	
850	2950.7	2664.8	1912.3	2233.6	2606.7	
900	3191.3	2883.3	2049.0	2410.8	2817.5	
950	3436.4	3105.4	2187.2	2590.8	3034.0	
1000	3686.8	3331.7	2327.1	2773.9	3252.3	
1050	3941.1	3562.1	2468.5	2959.5	3473.5	
1100	4193.0	3796.3	2609.0	3147.8	3696.7	
1150	4462.2	4032.6	2755.6	3338.1	3924.9	
1200	4728.5	4273.2	2901.4	3531.0	4151.7	
1250	4997.8	4516.4	3048.2	3725.6	4379.4	
1300	5270.7	4761.9	3196.4	3922.4	4608.9	
1350	5546.4	5011.0	3345.6	4120.8	4840.2	
1400	5825.2	5261.5	3495.9	4321.1	5074.8	
1450	6106.3	5513.2	3647.2	4522.8	5310.9	
1500	6390.0	5766.4	3799.6	4726.1	5546.8	

NOTE: Enthalpy is taken as zero at 25°C.

Temp. T (°C)	Dry Air	Oxygen, O_2	Nitrogen, N_2	Water Vapor, H_2O	Carbon Dioxide, CO_2	Carbon Monoxide, CO	Hydrogen, H_2
−50	−75.3	−69.3	−77.9	−120.9	−60.6	−78.1	−1057.4
0	−25.2	−22.9	−26.0	−46.3	−21.1	−26.1	−356.2
25	0	0	0	0	0	0	0
50	25.1	23.0	26.0	46.7	21.1	26.1	358.6
100	75.6	63.5	78.1	140.7	65.7	78.2	1079.5
150	126.2	116.4	130.3	235.8	112.6	130.6	1803.2
200	177.3	164.2	182.8	332.2	161.5	183.3	2528.0
250	228.8	212.7	235.6	429.8	212.1	236.5	3253.6
300	280.8	262.1	287.8	528.9	264.3	290.3	3980.1
350	333.3	312.2	342.5	629.6	318.0	344.6	4707.3
400	386.5	363.0	396.9	731.8	373.1	399.6	5435.8
450	440.2	414.6	451.8	835.7	429.3	455.2	6166.0
500	494.6	466.7	507.2	941.3	486.8	511.5	6956.9
550	549.5	519.4	563.3	1048.6	545.0	568.4	7633.1
600	605.0	572.6	620.0	1157.7	604.4	625.9	8370.9
650	661.1	626.3	677.1	1268.5	664.6	683.9	9112.0
700	717.6	680.4	734.9	1381.0	725.5	742.6	9857.1
750	774.7	734.7	793.2	1495.2	787.2	801.8	10,606
800	832.2	789.7	852.1	1610.8	849.5	861.6	11,360
850	890.1	844.9	911.4	1729.3	912.3	921.7	12,118
900	948.4	900.3	971.2	1849.4	975.8	982.3	12,881
950	1007.2	956.1	1031.3	1970.5	1039.9	1043.2	13,649
1000	1066.2	1012.1	1091.9	2093.4	1104.7	1104.6	14,422
1050	1125.5	1068.3	1152.9	2216.2	1169.9	1166.3	15,202
1100	1185.2	1124.8	1214.1	2339.1	1235.5	1228.3	15,986
1150	1245.1	1181.5	1275.7	2471.3	1300.9	1290.6	16,776
1200	1305.2	1238.4	1337.6	2600.3	1366.8	1353.2	17,572
1250	1365.6	1295.5	1399.7	2730.6	1433.1	1416.1	18,373
1300	1426.8	1352.8	1462.1	2862.4	1499.7	1479.2	19,180
1350	1488.0	1410.3	1524.8	2995.4	1566.6	1542.5	19,991
1400	1549.8	1468.1	1587.8	3129.7	1633.8	1606.0	20,808
1450	1611.9	1526.0	1650.9	3265.1	1701.1	1669.7	21,629
1500	1674.3	1584.1	1714.3	3401.8	1768.7	1733.6	22,455
2000	2302.2	2175.1	2356.6	4822.2	2593.3	2381.0	30,974
2500	2942.6	2783.9	3010.4	6317.5	3153.1	3038.8	39,873
3000	3592.7	3408.9	3671.6	7864.1	3860.2	3703.3	49,079
3500	4249.4	4047.5	4338.0	9445.6	4573.8	4372.8	58,542
4000	4911.6	4697.1	5008.4	11,054.7	5292.6	5046.2	68,259
4500	5578.3	5355.2	5682.2	12,684.6	5992.3	5722.9	78,180
5000	6251.4	6019.4	6362.5	14,331.1	6744.3	6402.4	88,294
5500	6925.8	6688.0	7043.6	15,992.7	7477.2	7085.0	98,622
6000	7603.2	7359.2	7727.4	17,665.2	8214.2	7770.2	109,057

TABLE B.17 (Continued) *Ideal Gas Enthalpies of Various Substances (kJ/kg).*

Temp. T (°C)	Hydroxyl, OH	Nitric Oxide, NO	Atomic Hydrogen, H	Atomic Nitrogen, N	Atomic Oxygen, O
0	−44.5	−25.2	−519.6	−37.4	−34.8
25	0	0	0	0	0
500	820.4	489.0	9790.9	704.5	631.4
1000	1718.5	1055.7	20,103	1446.6	1285.2
1500	2679.9	1652.2	30,413	2188.5	1936.5
2000	3691.5	2263.1	40,724	2930.7	2615.0
2500	4738.1	2915.6	51,037	3674.5	3239.3
3000	5860.0	3506.5	61,348	4423.4	3893.9
3500	6899.3	4134.9	71,659	5184.0	4553.3
4000	8003.8	4766.5	81,969	5963.8	5219.5
4500	9120.3	5400.9	92,280	6770.5	5893.2
5000	10,247	6037.6	102,590	7609.7	6574.8
5500	11,384	6676.6	112,901	8485.2	7264.0
6000	12,635	7317.1	123,216	9390.7	7959.1

TABLE B.18 *Heating Values of Some Hydrocarbon Fuels at Constant Pressure.*

Substance	Chemical Formula	Molecular Mass	HHV [kJ/kg(fuel)]	LHV [kJ/kg(fuel)]
Methane	CH_4	16.04	55,518	50,020
Acetylene	C_2H_2	26.04	49,940	48,220
Ethylene	C_2H_4	28.05	50,310	47,170
Ethane	C_2H_6	30.07	51,870	47,480
Propane	C_3H_8	44.10	50,340	46,350
Butane	C_4H_{10}	58.12	49,520	45,740
Octane	C_8H_{18}	114.23	48,255	44,785
Methanol	CH_4O	32.04	23,865	21,110
Ethanol	C_2H_6O	46.07	30,610	27,740

NOTE: Combustion reactants and products are at 25°C. Fuels supplied are in gaseous phase.

TABLE B.19 *Latent Heat of Vaporization of Several Hydrocarbon Fuels.*

Substance	h_{fg} (kJ/kg)
Propane	342.1
Butane	362.5
Octane	363.2
Methanol	1168.1
Ethanol	919.7

NOTE: Vaporization is at atmospheric pressure.

TABLE B.20 *Equilibrium Constant (ln K_p) for Various Reactions.*

Temp. T (°C)	$H_2O \rightleftharpoons H_2 + \frac{1}{2}O_2$	$H_2O \rightleftharpoons \frac{1}{2}H_2 + OH$	$CO_2 \rightleftharpoons CO + \frac{1}{2}O_2$	$CO_2 + H_2 \rightleftharpoons CO + H_2O$
25	−92.214	−106.234	−103.768	−11.554
500	−31.890	−36.137	−33.532	−1.641
1000	−16.754	−18.635	−16.259	0.496
1500	−10.088	−10.957	−8.787	1.301
2000	−6.332	−6.647	−4.639	1.693
2500	−3.924	−3.891	−2.015	1.909
3000	−2.245	−1.976	−0.206	2.039
3500	−1.008	−0.572	1.109	2.117
4000	−0.161	0.506	2.108	2.165
4500	0.698	1.355	2.892	2.193
5000	1.312	2.041	3.518	2.206
5500	1.824	2.612	4.034	2.210
6000	2.253	3.080	4.485	2.210

Temp. T (°C)	$\frac{1}{2}N_2 + \frac{1}{2}O_2 \rightleftharpoons NO$	$H_2 \rightleftharpoons 2H$	$O_2 \rightleftharpoons 2O$	$N_2 \rightleftharpoons 2N$
25	−34.933	−163.999	−186.988	−367.493
500	−12.548	−55.459	−62.976	−132.758
1000	−7.021	−28.316	−32.103	−74.534
1500	−4.610	−16.338	−18.562	−49.041
2000	−3.259	−9.567	−10.942	−34.717
2500	−2.398	−5.206	−6.055	−25.533
3000	−1.802	−2.160	−2.655	−19.138
3500	−1.367	0.088	−0.102	−14.430
4000	−1.038	1.815	1.770	−10.809
4500	−0.780	3.181	3.291	−7.939
5000	−0.574	4.293	4.522	−5.599
5500	−0.404	5.209	5.545	−3.660
6000	−0.272	5.956	6.127	−2.037

NOTE: Better accuracy in the value of the equilibrium constant is obtained by use of linear interpolation of the reciprocal of temperature, similar to that described for the specific volume of superheated substances.

TABLE B.21 *Availability and Entropy of Water*

(*a*) *Saturated Steam*

Abs. Press. (kPa)	Temp. °C	Availability (kJ/kg)			Entropy (kJ/kg K)		
		Sat. Liq.	Evap.	Sat. Vap.	Sat. Liq.	Evap.	Sat. Vap.
p	T	a_f	a_{fg}	a_g	s_f	s_{fg}	s_g
5	32.90	0.2	63.8	64.0	0.4763	7.9197	8.3960
10	45.83	5.4	154.8	160.2	0.7549	7.3962	8.1511
50	81.35	19.8	367.4	387.2	1.0912	6.5035	7.5947
100	99.63	33.7	452.9	486.6	1.3027	6.0571	7.3598
101.3	100.00	34.0	454.5	488.5	1.3069	6.0485	7.3554
150	327.82	53.3	492.0	545.4	1.4336	5.7894	7.2237
200	120.23	53.2	533.8	586.9	1.5300	5.5968	7.1268
250	417.35	60.9	558.6	619.5	1.6070	5.4460	7.0530
300	133.54	67.6	578.3	645.8	1.6720	5.3189	6.9909
500	151.84	90.2	629.6	719.8	1.8600	4.9592	6.8192
1000	179.88	129.9	689.0	819.0	2.1380	4.4448	6.5828
2000	212.37	183.8	729.5	913.3	2.4470	3.8896	6.3366
3000	233.84	224.5	739.5	964.0	2.6450	3.5387	6.1837
4000	250.33	258.2	738.1	996.3	2.7973	3.2712	6.0685
5000	263.91	288.5	730.1	1018.5	2.9210	3.0525	5.9735
6000	275.55	316.1	717.9	1034.0	3.0270	2.8638	5.8908
7000	285.79	341.5	703.2	1044.7	3.1220	2.6942	5.8162
8000	294.97	365.7	686.0	1051.7	3.2070	2.5401	5.7471
9000	303.91	388.7	667.1	1055.8	3.2860	2.3966	5.6820
10000	310.96	410.8	641.6	1052.4	3.3600	2.2598	5.6198
20000	365.70	634.9	315.5	950.3	4.0140	0.9272	4.9412
22120	374.15	787.8	0	787.8	4.4429	0	4.4429

The reference conditions of the surroundings are taken as $T_0 = 298$ K and $p_0 = 101.3$ kPa; hence $h_0 = 104.88$ kJ/kg, $s_0 = 0.3668$ kJ/kg K. Availability is taken as $a = h - h_0 - T_0(s - s_0)$. In this form, availability is also called *exergy*.

(b) Superheated Steam and Compressed Water

Abs. Press. (kPa) p	Sat. Temp. (°C) T_g	Sat. Liq. a_f	Sat. Vapor a_g	Availability (kJ/kg) Temperature (°C) 50	100	150	200	250	300
5	32.90	0.3	64.0	65.7	79.1	103.0	135.3	174.5	219.7
10	45.83	2.8	160.2	160.8	174.2	198.3	230.5	269.7	315.0
50	81.35	19.8	387.1	4.1	391.0	415.5	448.1	487.6	532.9
100	99.63	33.7	486.6	4.1	486.8	511.8	545.0	584.7	630.4
500	151.84	90.2	719.8	4.5	34.5	87.8	755.9	798.4	845.7
1000	179.88	129.9	819.0	5.0	34.9	88.3	837.0	883.5	933.2
5000	263.91	288.5	1018.5	9.0	38.8	92.1	165.3	258.5	1056.9
10000	310.96	410.8	1057.4	14.0	43.8	96.8	169.7	262.0	379.7
20000	365.70	634.9	950.3	23.9	53.5	106.4	178.5	269.4	382.5
30000				34.0	63.3	115.8	187.5	277.1	387.0
40000				43.7	73.2	125.2	196.4	285.2	392.6
50000				53.5	82.7	134.7	205.4	293.0	398.0
60000				63.3	92.3	144.2	214.4	301.3	404.2
70000				72.9	102.0	153.6	223.5	309.5	411.8
80000				82.8	111.7	162.7	232.5	318.0	418.4
90000				92.4	121.1	172.1	241.6	326.5	425.1
100000				102.1	130.9	181.4	250.3	335.0	432.5

p (kPa)	Temperature (°C) 350	400	450	500	550	600	650	700	750
5	270.2	325.2	325.5	385.0	448.5	515.8	586.5	660.5	737.6
10	365.5	420.8	480.3	543.8	611.1	681.8	755.8	833.0	913.0
50	583.6	638.8	698.5	762.1	829.3	900.1	974.2	1051.3	1131.4
100	681.1	736.5	796.2	859.8	928.3	998.0	1072.1	1149.2	1229.4
500	897.5	953.7	1013.9	1078.0	1145.7	1216.9	1291.2	1368.7	1449.0
1000	986.6	1043.8	1104.8	1169.4	1237.6	1309.1	1383.9	1461.7	1542.3
5000	1152.2	1220.8	1289.1	1359.0	1431.2	1506.1	1583.7	1664.1	1747.0
10000	1157.5	1251.3	1333.6	1412.4	1490.8	1570.5	1652.0	1735.8	1821.5
20000	539.9	1168.5	1307.9	1414.1	1510.0	1601.6	1692.2	1783.2	1875.3
30000	528.2	828.3	1206.1	1361.9	1482.3	1589.7	1691.2	1790.4	1889.0
40000	525.2	711.0	1044.6	1279.3	1432.5	1558.8	1672.8	1780.9	1886.7
50000	525.3	687.7	926.1	1184.3	1370.9	1518.2	1645.5	1762.9	1875.6
60000	527.6	678.1	873.0	1103.7	1307.4	1473.0	1613.8	1740.3	1859.8
70000	530.5	674.3	847.5	1050.4	1253.0	1427.9	1580.5	1715.8	1841.7
80000	535.1	673.6	834.0	1016.5	1211.0	1387.6	1547.9	1690.7	1822.6
90000	539.9	674.9	826.5	995.4	1179.0	1354.4	1517.8	1666.2	1803.2
100000	545.0	677.4	822.7	982.1	1156.1	1327.1	1492.6	1642.8	1784.1

TABLE B.22 *Availability and Entropy of R-22 (Chlorodifluoromethane)*

(*a*) *Saturated Refrigerant R-22*

Abs. Press. (kPa)	Temp. (°C)	Availability (kJ/kg) Sat. Liq.	Evap.	Sat. Vap.	Entropy (kJ/kg K) Sat. Liq.	Evap.	Sat. Vap.
p	T	a_f	a_{fg}	a_g	s_f	s_{fg}	s_g
1	−105.1	65.01	−164.11	−99.10	−0.31631	1.57591	1.25960
5	−89.76	58.57	−126.53	−67.96	−0.23908	1.41208	1.17300
10	−80.64	54.62	−106.49	−51.87	−0.19263	1.32311	1.13048
20	−70.41	50.82	−86.56	−35.74	−0.14223	1.23205	1.08982
40	−58.86	47.28	−66.82	−19.54	−0.08703	1.13819	1.05116
60	−51.37	45.37	−55.39	−10.02	−0.05209	1.08158	1.02949
80	−45.68	44.12	−47.38	−3.26	−0.02590	1.04049	1.01459
100	−41.04	43.21	−41.20	2.01	0.04686	0.95638	1.00324
200	−25.12	40.94	−22.54	18.40	0.06649	0.90291	0.96940
300	−14.61	40.11	−12.12	27.99	0.11249	0.83800	0.95049
400	−6.91	39.82	−5.34	34.48	0.14736	0.79060	0.93796
500	0.15	39.81	0.23	40.04	0.17583	0.75147	0.92730
600	5.87	39.97	4.33	44.30	0.20006	0.71907	0.91913
700	10.93	40.23	7.65	47.88	0.22131	0.69092	0.91223
800	15.47	40.56	10.40	50.96	0.24007	0.66612	0.90619
900	19.60	40.89	12.77	53.66	0.25553	0.64531	0.90084
1000	23.41	41.37	14.68	56.05	0.27336	0.62262	0.89598
1500	39.09	43.84	21.12	64.96	0.33813	0.53818	0.87631
2000	51.28	46.58	24.26	70.84	0.38873	0.47169	0.86042
2500	61.38	49.48	25.48	74.96	0.43170	0.41389	0.84559
3000	70.06	52.52	25.31	77.83	0.47004	0.36054	0.83058
3500	77.70	55.72	23.97	79.69	0.50612	0.30791	0.81403
4000	84.52	59.21	21.34	80.55	0.54209	0.25195	0.79404
4500	90.67	63.34	17.11	80.45	0.58216	0.18383	0.76599
4977.39	96.01	72.46	0	72.46	0.67090	0	0.67090

The reference conditions of the surroundings are taken as $T_0 = 273$ K and $p_0 = 101.3$ kPa; hence $h_0 = 257.98$ kJ/kg, $s_0 = 1.10259$ kJ/kg K. Availability is taken as $a = h - h_0 - T_0 (s - s_0)$. In this form, availability is also called *exergy*.

(*b*) *Superheated Refrigerant R-22*

Abs. Press. (kPa) p	Sat. Temp. (°C) T_g	Sat. Vapor Avail. a_g	Availability (kJ/kg) Temperature (°C) −50	0	50	100	150	200
1	−105.1	−99.10	−117.41	−120.41	−117.71	−110.11	−98.32	−82.73
5	−89.76	−67.96	−75.18	−78.17	−75.48	−67.88	−56.09	−40.50
10	−80.64	−51.87	−56.98	−59.99	−57.28	−49.70	−37.90	−22.39
20	−70.41	−35.74	−38.84	−41.87	−39.16	−31.53	−19.72	−4.29
40	−58.86	−19.54	−20.60	−23.64	−20.92	−13.37	−1.54	13.86
60	−51.37	−10.02	−10.14	−13.18	−10.43	−2.84	9.14	24.51
80	−45.68	−3.26		−5.73	−2.98	4.64	16.77	32.11
100	−41.04	2.01		−0.01	2.82	10.35	22.38	37.84
200	−25.12	18.40		17.63	20.42	25.60	37.65	54.26
300	−14.61	27.99		27.62	30.49	38.30	50.41	66.24
400	−6.91	34.48		34.65	37.60	45.48	57.92	73.92
500	0.15	40.04			43.02	50.97	63.14	78.01
600	5.87	44.30			47.28	55.33	67.57	82.60
700	10.93	47.88			50.75	58.87	71.21	86.32
800	15.47	50.96			54.06	62.32	74.72	90.03
900	19.60	53.66			56.43	64.80	77.17	92.58
1000	23.41	56.05			58.51	66.99	79.55	95.01
1500	39.09	64.96			66.38	75.50	88.50	104.88
2000	51.28	70.84				80.76	94.28	111.11
2500	61.38	74.96				83.23	98.29	115.51
3000	70.06	77.83				86.15	101.09	118.74

appendix C

Compressible Flow Tables

LIST OF TABLES

TABLE C.1 *Isentropic Flow Table* ($\gamma = 1.4$)

M	$\frac{p}{p_t}$	$\frac{T}{T_t}$	$\frac{A}{A^*}$	M	$\frac{p}{p_t}$	$\frac{T}{T_t}$	$\frac{A}{A^*}$
0	1.0000	1.0000	∞	0.20	0.9725	0.9921	2.9635
0.01	0.9999	1.0000	57.8738	0.21	0.9697	0.9913	2.8293
0.02	0.9997	0.9999	28.9421	0.22	0.9668	0.9901	2.7076
0.03	0.9994	0.9998	19.3005	0.23	0.9638	0.9895	2.5968
0.04	0.9989	0.9997	14.4815	0.24	0.9607	0.9886	2.4956
0.05	0.9983	0.9995	11.5914	0.25	0.9575	0.9877	2.4027
0.06	0.9975	0.9993	9.6659	0.26	0.9541	0.9867	2.3173
0.07	0.9966	0.9990	8.2915	0.27	0.9506	0.9856	2.2385
0.08	0.9955	0.9987	7.2616	0.28	0.9470	0.9846	2.1656
0.09	0.9944	0.9984	6.4613	0.29	0.9433	0.9835	2.0979
0.10	0.9930	0.9980	5.8218	0.30	0.9395	0.9823	2.0351
0.11	0.9916	0.9976	5.2992	0.31	0.9355	0.9811	1.9765
0.12	0.9900	0.9971	4.8643	0.32	0.9315	0.9799	1.9219
0.13	0.9883	0.9966	4.4969	0.33	0.9274	0.9787	1.8707
0.14	0.9864	0.9961	4.1824	0.34	0.9231	0.9774	1.8229
0.15	0.9844	0.9955	3.9103	0.35	0.9188	0.9761	1.7780
0.16	0.9823	0.9949	3.6727	0.36	0.9143	0.9747	1.7358
0.17	0.9800	0.9943	3.4635	0.37	0.9098	0.9733	1.6961
0.18	0.9776	0.9936	3.2779	0.38	0.9052	0.9719	1.6587
0.19	0.9751	0.9928	3.1123	0.39	0.9004	0.9705	1.6234

TABLE C.1 (*Continued*) *Isentropic Flow Table* ($\gamma = 1.4$)

M	$\frac{p}{p_t}$	$\frac{T}{T_t}$	$\frac{A}{A^*}$	M	$\frac{p}{p_t}$	$\frac{T}{T_t}$	$\frac{A}{A^*}$
0.40	0.8956	0.9690	1.5901	0.70	0.7209	0.9107	1.0944
0.41	0.8907	0.9675	1.5587	0.71	0.7145	0.9084	1.0873
0.42	0.8857	0.9659	1.5289	0.72	0.7080	0.9061	1.0806
0.43	0.8807	0.9643	1.5007	0.73	0.7016	0.9037	1.0742
0.44	0.8755	0.9627	1.4740	0.74	0.6951	0.9013	1.0681
0.45	0.8703	0.9611	1.4487	0.75	0.6886	0.8989	1.0624
0.46	0.8650	0.9594	1.4246	0.76	0.6821	0.8964	1.0570
0.47	0.8596	0.9577	1.4018	0.77	0.6756	0.8940	1.0519
0.48	0.8541	0.9560	1.3801	0.78	0.6691	0.8915	1.0471
0.49	0.8486	0.9542	1.3595	0.79	0.6625	0.8890	1.0425
0.50	0.8430	0.9524	1.3398	0.80	0.6560	0.8865	1.0382
0.51	0.8374	0.9506	1.3212	0.81	0.6495	0.8840	1.0342
0.52	0.8317	0.9487	1.3034	0.82	0.6430	0.8815	1.0305
0.53	0.8259	0.9468	1.2865	0.83	0.6365	0.8789	1.0270
0.54	0.8201	0.9449	1.2703	0.84	0.6300	0.8763	1.0237
0.55	0.8142	0.9430	1.2550	0.85	0.6235	0.8737	1.0207
0.56	0.8082	0.9410	1.2403	0.86	0.6170	0.8711	1.0179
0.57	0.8022	0.9390	1.2263	0.87	0.6106	0.8685	1.0153
0.58	0.7962	0.9370	1.2130	0.88	0.6041	0.8659	1.0129
0.59	0.7901	0.9349	1.2003	0.89	0.5977	0.8632	1.0108
0.60	0.7840	0.9328	1.1882	0.90	0.5913	0.8606	1.0089
0.61	0.7778	0.9307	1.1767	0.91	0.5849	0.8579	1.0071
0.62	0.7716	0.9286	1.1657	0.92	0.5785	0.8552	1.0056
0.63	0.7654	0.9265	1.1552	0.93	0.5721	0.8525	1.0043
0.64	0.7591	0.9243	1.1452	0.94	0.5658	0.8498	1.0031
0.65	0.7528	0.9221	1.1356	0.95	0.5595	0.8471	1.0022
0.66	0.7465	0.9199	1.1265	0.96	0.5532	0.8444	1.0014
0.67	0.7401	0.9176	1.1179	0.97	0.5469	0.8416	1.0008
0.68	0.7338	0.9153	1.1097	0.98	0.5407	0.8389	1.0003
0.69	0.7274	0.9131	1.1018	0.99	0.5345	0.8361	1.0001

M	$\frac{p}{p_t}$	$\frac{T}{T_t}$	$\frac{A}{A^*}$	M	$\frac{p}{p_t}$	$\frac{T}{T_t}$	$\frac{A}{A^*}$
1.00	0.5283	0.8333	1.000	1.05	0.4979	0.8193	1.002
1.01	0.5221	0.8306	1.000	1.06	0.4919	0.8165	1.003
1.02	0.5160	0.8278	1.000	1.07	0.4860	0.8137	1.004
1.03	0.5099	0.8250	1.001	1.08	0.4800	0.8108	1.005
1.04	0.5039	0.8222	1.001	1.09	0.4742	0.8080	1.006

M	$\frac{p}{p_t}$	$\frac{T}{T_t}$	$\frac{A}{A^*}$	M	$\frac{p}{p_t}$	$\frac{T}{T_t}$	$\frac{A}{A^*}$
1.10	0.4684	0.8052	1.008	1.50	0.2724	0.6897	1.176
1.11	0.4626	0.8023	1.010	1.51	0.2685	0.6868	1.183
1.12	0.4568	0.7994	1.011	1.52	0.2646	0.6840	1.190
1.13	0.4511	0.7966	1.013	1.53	0.2608	0.6811	1.197
1.14	0.4455	0.7937	1.015	1.54	0.2570	0.6783	1.204
1.15	0.4398	0.7908	1.017	1.55	0.2533	0.6754	1.212
1.16	0.4343	0.7879	1.020	1.56	0.2496	0.6726	1.219
1.17	0.4287	0.7851	1.022	1.57	0.2459	0.6698	1.227
1.18	0.4232	0.7822	1.025	1.58	0.2423	0.6670	1.234
1.19	0.4178	0.7793	1.026	1.59	0.2388	0.6642	1.242
1.20	0.4124	0.7764	1.030	1.60	0.2353	0.6614	1.250
1.21	0.4070	0.7735	1.033	1.61	0.2318	0.6586	1.258
1.22	0.4017	0.7706	1.037	1.62	0.2284	0.6558	1.267
1.23	0.3964	0.7677	1.040	1.63	0.2250	0.6530	1.275
1.24	0.3912	0.7648	1.043	1.64	0.2217	0.6502	1.284
1.25	0.3861	0.7619	1.047	1.65	0.2184	0.6475	1.292
1.26	0.3809	0.7590	1.050	1.66	0.2151	0.6447	1.301
1.27	0.3759	0.7561	1.054	1.67	0.2119	0.6419	1.310
1.28	0.3708	0.7532	1.058	1.68	0.2088	0.6392	1.319
1.29	0.3658	0.7503	1.062	1.69	0.2057	0.6364	1.328
1.30	0.3609	0.7474	1.066	1.70	0.2026	0.6337	1.338
1.31	0.3560	0.7445	1.071	1.71	0.1996	0.6310	1.347
1.32	0.3512	0.7416	1.075	1.72	0.1966	0.6283	1.357
1.33	0.3464	0.7387	1.080	1.73	0.1936	0.6256	1.367
1.34	0.3417	0.7358	1.084	1.74	0.1907	0.6229	1.376
1.35	0.3370	0.7329	1.089	1.75	0.1878	0.6202	1.386
1.36	0.3323	0.7300	1.094	1.76	0.1850	0.6175	1.397
1.37	0.3277	0.7271	1.099	1.77	0.1822	0.6148	1.407
1.38	0.3232	0.7242	1.104	1.78	0.1794	0.6121	1.418
1.39	0.3187	0.7213	1.109	1.79	0.1767	0.6095	1.428
1.40	0.3142	0.7184	1.115	1.80	0.1740	0.6068	1.439
1.41	0.3098	0.7155	1.120	1.81	0.1714	0.6041	1.450
1.42	0.3055	0.7126	1.126	1.82	0.1688	0.6015	1.461
1.43	0.3012	0.7097	1.132	1.83	0.1662	0.5989	1.472
1.44	0.2969	0.7069	1.138	1.84	0.1637	0.5963	1.484
1.45	0.2927	0.7040	1.144	1.85	0.1612	0.5936	1.495
1.46	0.2886	0.7011	1.150	1.86	0.1587	0.5910	1.507
1.47	0.2845	0.6982	1.156	1.87	0.1563	0.5884	1.519
1.48	0.2804	0.6954	1.163	1.88	0.1539	0.5859	1.531
1.49	0.2764	0.6925	1.169	1.89	0.1516	0.5833	1.543

TABLE C.1 (*Continued*) *Isentropic Flow Table* ($\gamma = 1.4$)

M	$\frac{p}{p_t}$	$\frac{T}{T_t}$	$\frac{A}{A^*}$	M	$\frac{p}{p_t}$	$\frac{T}{T_t}$	$\frac{A}{A^*}$
1.90	0.1492	0.5807	1.555	2.30	0.7997^{-1}	0.4859	2.193
1.91	0.1470	0.5782	1.568	2.31	0.7873^{-1}	0.4837	2.213
1.92	0.1447	0.5756	1.580	2.32	0.7751^{-1}	0.4816	2.233
1.93	0.1425	0.5731	1.593	2.33	0.7631^{-1}	0.4794	2.254
1.94	0.1403	0.5705	1.606	2.34	0.7512^{-1}	0.4773	2.273
1.95	0.1381	0.5680	1.619	2.35	0.7396^{-1}	0.4752	2.295
1.96	0.1360	0.5655	1.633	2.36	0.7281^{-1}	0.4731	2.316
1.97	0.1339	0.5630	1.646	2.37	0.7168^{-1}	0.4709	2.338
1.98	0.1318	0.5605	1.660	2.38	0.7057^{-1}	0.4688	2.359
1.99	0.1298	0.5580	1.674	2.39	0.6948^{-1}	0.4668	2.381
2.00	0.1278	0.5556	1.688	2.40	0.6840^{-1}	0.4647	2.403
2.01	0.1258	0.5531	1.702	2.41	0.6734^{-1}	0.4626	2.425
2.02	0.1239	0.5506	1.716	2.42	0.6630^{-1}	0.4606	2.448
2.03	0.1220	0.5482	1.730	2.43	0.6527^{-1}	0.4585	2.471
2.04	0.1201	0.5458	1.745	2.44	0.6426^{-1}	0.4565	2.494
2.05	0.1182	0.5433	1.760	2.45	0.6327^{-1}	0.4544	2.517
2.06	0.1164	0.5409	1.775	2.46	0.6229^{-1}	0.4524	2.540
2.07	0.1146	0.5385	1.790	2.47	0.6133^{-1}	0.4504	2.564
2.08	0.1128	0.5361	1.806	2.48	0.6038^{-1}	0.4484	2.588
2.09	0.1111	0.5337	1.821	2.49	0.5945^{-1}	0.4464	2.612
2.10	0.1094	0.5313	1.837	2.50	0.5853^{-1}	0.4444	2.637
2.11	0.1077	0.5290	1.853	2.51	0.5762^{-1}	0.4425	2.661
2.12	0.1060	0.5266	1.869	2.52	0.5674^{-1}	0.4405	2.686
2.13	0.1043	0.5243	1.885	2.53	0.5586^{-1}	0.4386	2.712
2.14	0.1027	0.5219	1.902	2.54	0.5500^{-1}	0.4366	2.737
2.15	0.1011	0.5196	1.919	2.55	0.5415^{-1}	0.4347	2.763
2.16	0.9956^{-1}	0.5173	1.935	2.56	0.5332^{-1}	0.4328	2.789
2.17	0.9802^{-1}	0.5150	1.953	2.57	0.5250^{-1}	0.4309	2.815
2.18	0.9649^{-1}	0.5127	1.970	2.58	0.5169^{-1}	0.4289	2.842
2.19	0.9500^{-1}	0.5104	1.987	2.59	0.5090^{-1}	0.4271	2.869
2.20	0.9352^{-1}	0.5081	2.005	2.60	0.5012^{-1}	0.4252	2.896
2.21	0.9207^{-1}	0.5059	2.023	2.61	0.4935^{-1}	0.4233	2.923
2.22	0.9064^{-1}	0.5036	2.041	2.62	0.4859^{-1}	0.4214	2.951
2.23	0.8923^{-1}	0.5014	2.059	2.63	0.4784^{-1}	0.4196	2.979
2.24	0.8785^{-1}	0.4991	2.078	2.64	0.4711^{-1}	0.4177	3.007
2.25	0.8648^{-1}	0.4969	2.096	2.65	0.4639^{-1}	0.4159	3.036
2.26	0.8514^{-1}	0.4947	2.115	2.66	0.4568^{-1}	0.4141	3.065
2.27	0.8382^{-1}	0.4925	2.134	2.67	0.4498^{-1}	0.4122	3.094
2.28	0.8251^{-1}	0.4903	2.154	2.68	0.4429^{-1}	0.4104	3.123
2.29	0.8123^{-1}	0.4881	2.173	2.69	0.4362^{-1}	0.4086	3.153

The superscript ($^{-1}$) denotes $\times 10^{-1}$. For example, 0.9956^{-1} is equal to 0.09956.

M	$\frac{p}{p_t}$	$\frac{T}{T_t}$	$\frac{A}{A^*}$	M	$\frac{p}{p_t}$	$\frac{T}{T_t}$	$\frac{A}{A^*}$
2.70	0.4295^{-1}	0.4068	3.183	3.10	0.2345^{-1}	0.3422	4.657
2.71	0.4229^{-1}	0.4051	3.213	3.11	0.2310^{-1}	0.3408	4.702
2.72	0.4165^{-1}	0.4033	3.244	3.12	0.2276^{-1}	0.3393	4.747
2.73	0.4102^{-1}	0.4015	3.275	3.13	0.2243^{-1}	0.3379	4.792
2.74	0.4039^{-1}	0.3998	3.306	3.14	0.2210^{-1}	0.3365	4.838
2.75	0.3978^{-1}	0.3980	3.338	3.15	0.2177^{-1}	0.3351	4.884
2.76	0.3917^{-1}	0.3963	3.370	3.16	0.2146^{-1}	0.3337	4.930
2.77	0.3858^{-1}	0.3945	3.402	3.17	0.2114^{-1}	0.3323	4.977
2.78	0.3799^{-1}	0.3928	3.434	3.18	0.2083^{-1}	0.3309	5.025
2.79	0.3742^{-1}	0.3911	3.467	3.19	0.2053^{-1}	0.3295	5.073
2.80	0.3685^{-1}	0.3894	3.500	3.20	0.2023^{-1}	0.3281	5.121
2.81	0.3629^{-1}	0.3877	3.534	3.21	0.1993^{-1}	0.3267	5.170
2.82	0.3574^{-1}	0.3860	3.567	3.22	0.1964^{-1}	0.3253	5.219
2.83	0.3520^{-1}	0.3844	3.601	3.23	0.1936^{-1}	0.3240	5.268
2.84	0.3467^{-1}	0.3827	3.636	3.24	0.1908^{-1}	0.3226	5.319
2.85	0.3415^{-1}	0.3810	3.671	3.25	0.1880^{-1}	0.3213	5.369
2.86	0.3363^{-1}	0.3794	3.706	3.26	0.1853^{-1}	0.3199	5.420
2.87	0.3312^{-1}	0.3777	3.741	3.27	0.1826^{-1}	0.3186	5.472
2.88	0.3263^{-1}	0.3761	3.777	3.28	0.1799^{-1}	0.3173	5.523
2.89	0.3213^{-1}	0.3745	3.813	3.29	0.1773^{-1}	0.3160	5.576
2.90	0.3165^{-1}	0.3729	3.850	3.30	0.1748^{-1}	0.3147	5.629
2.91	0.3118^{-1}	0.3712	3.887	3.31	0.1722^{-1}	0.3134	5.682
2.92	0.3071^{-1}	0.3696	3.924	3.32	0.1698^{-1}	0.3121	5.736
2.93	0.3025^{-1}	0.3681	3.961	3.33	0.1673^{-1}	0.3108	5.790
2.94	0.2980^{-1}	0.3665	3.999	3.34	0.1649^{-1}	0.3095	5.845
2.95	0.2935^{-1}	0.3649	4.038	3.35	0.1625^{-1}	0.3082	5.900
2.96	0.2891^{-1}	0.3633	4.076	3.36	0.1602^{-1}	0.3069	5.956
2.97	0.2848^{-1}	0.3618	4.115	3.37	0.1579^{-1}	0.3057	6.012
2.98	0.2805^{-1}	0.3602	4.155	3.38	0.1557^{-1}	0.3044	6.069
2.99	0.2764^{-1}	0.3587	4.194	3.39	0.1534^{-1}	0.3032	6.126
3.00	0.2722^{-1}	0.3571	4.235	3.40	0.1512^{-1}	0.3019	6.184
3.01	0.2682^{-1}	0.3556	4.275	3.41	0.1491^{-1}	0.3007	6.242
3.02	0.2642^{-1}	0.3541	4.316	3.42	0.1470^{-1}	0.2995	6.301
3.03	0.2603^{-1}	0.3526	4.357	3.43	0.1449^{-1}	0.2982	6.360
3.04	0.2564^{-1}	0.3511	4.399	3.44	0.1428^{-1}	0.2970	6.420
3.05	0.2526^{-1}	0.3496	4.441	3.45	0.1408^{-1}	0.2958	6.480
3.06	0.2489^{-1}	0.3481	4.483	3.46	0.1388^{-1}	0.2946	6.541
3.07	0.2452^{-1}	0.3466	4.526	3.47	0.1368^{-1}	0.2934	6.602
3.08	0.2416^{-1}	0.3452	4.570	3.48	0.1349^{-1}	0.2922	6.664
3.09	0.2380^{-1}	0.3437	4.613	3.49	0.1330^{-1}	0.2910	6.727

TABLE C.1 (*Continued*) *Isentropic Flow Table* ($\gamma = 1.4$)

M	$\frac{p}{p_t}$	$\frac{T}{T_1}$	$\frac{A}{A^*}$	M	$\frac{p}{p_t}$	$\frac{T}{T_t}$	$\frac{A}{A^*}$
3.50	0.1311^{-1}	0.2899	6.790	3.90	0.7532^{-2}	0.2474	9.799
3.51	0.1293^{-1}	0.2887	6.853	3.91	0.7431^{-2}	0.2464	9.888
3.52	0.1274^{-1}	0.2875	6.917	3.92	0.7332^{-2}	0.2455	9.977
3.53	0.1256^{-1}	0.2864	6.982	3.93	0.7233^{-2}	0.2446	10.07
3.54	0.1239^{-1}	0.2852	7.047	3.94	0.7137^{-2}	0.2436	10.16
3.55	0.1221^{-1}	0.2841	7.113	3.95	0.7042^{-2}	0.2427	10.25
3.56	0.1204^{-1}	0.2829	7.179	3.96	0.6948^{-2}	0.2418	10.34
3.57	0.1188^{-1}	0.2818	7.246	3.97	0.6855^{-2}	0.2408	10.44
3.58	0.1171^{-1}	0.2806	7.313	3.98	0.6764^{-2}	0.2399	10.53
3.59	0.1155^{-1}	0.2795	7.382	3.99	0.6675^{-2}	0.2390	10.62
3.60	0.1138^{-1}	0.2784	7.450	4.00	0.6586^{-2}	0.2381	10.72
3.61	0.1123^{-1}	0.2773	7.519	4.01	0.6499^{-2}	0.2372	10.81
3.62	0.1107^{-1}	0.2762	7.589	4.02	0.6413^{-2}	0.2363	10.91
3.63	0.1092^{-1}	0.2751	7.659	4.03	0.6328^{-2}	0.2354	11.01
3.64	0.1076^{-1}	0.2740	7.730	4.04	0.6245^{-2}	0.2345	11.11
3.65	0.1062^{-1}	0.2729	7.802	4.05	0.6163^{-2}	0.2336	11.21
3.66	0.1047^{-1}	0.2718	7.874	4.06	0.6082^{-2}	0.2327	11.31
3.67	0.1032^{-1}	0.2707	7.947	4.07	0.6002^{-2}	0.2319	11.41
3.68	0.1018^{-1}	0.2697	8.020	4.08	0.5923^{-2}	0.2310	11.51
3.69	0.1004^{-1}	0.2686	8.094	4.09	0.5845^{-2}	0.2301	11.61
3.70	0.9903^{-2}	0.2675	8.169	4.10	0.5769^{-2}	0.2293	11.71
3.71	0.9767^{-2}	0.2665	8.244	4.11	0.5694^{-2}	0.2284	11.82
3.72	0.9633^{-2}	0.2654	8.320	4.12	0.5619^{-2}	0.2275	11.92
3.73	0.9500^{-2}	0.2644	8.397	4.13	0.5546^{-2}	0.2267	12.03
3.74	0.9370^{-2}	0.2633	8.474	4.14	0.5474^{-2}	0.2258	12.14
3.75	0.9242^{-2}	0.2623	8.552	4.15	0.5403^{-2}	0.2250	12.24
3.76	0.9116^{-2}	0.2613	8.630	4.16	0.5333^{-2}	0.2242	12.35
3.77	0.8991^{-2}	0.2602	8.709	4.17	0.5264^{-2}	0.2233	12.46
3.78	0.8869^{-2}	0.2592	8.789	4.18	0.5195^{-2}	0.2225	12.57
3.79	0.8748^{-2}	0.2582	8.870	4.19	0.5128^{-2}	0.2217	12.68
3.80	0.8629^{-2}	0.2572	8.951	4.20	0.5062^{-2}	0.2208	12.79
3.81	0.8512^{-2}	0.2562	9.032	4.21	0.4997^{-2}	0.2200	12.90
3.82	0.8396^{-2}	0.2552	9.115	4.22	0.4932^{-2}	0.2192	13.02
3.83	0.8283^{-2}	0.2542	9.198	4.23	0.4869^{-2}	0.2184	13.13
3.84	0.8171^{-2}	0.2532	9.282	4.24	0.4806^{-6}	0.2176	13.25
3.85	0.8060^{-2}	0.2522	9.366	4.25	0.4745^{-2}	0.2168	13.36
3.86	0.7951^{-2}	0.2513	9.451	4.26	0.4684^{-2}	0.2160	13.48
3.87	0.7844^{-2}	0.2503	9.537	4.27	0.4624^{-2}	0.2152	13.60
3.88	0.7739^{-2}	0.2493	9.624	4.28	0.4565^{-2}	0.2144	13.72
3.89	0.7635^{-2}	0.2484	9.711	4.29	0.4507^{-2}	0.2136	13.83

M	$\frac{p}{p_t}$	$\frac{T}{T_t}$	$\frac{A}{A^*}$	M	$\frac{p}{p_t}$	$\frac{T}{T_t}$	$\frac{A}{A^*}$
4.30	0.4449^{-2}	0.2129	13.95	4.68	0.2768^{-2}	0.1859	19.26
4.31	0.4393^{-2}	0.2121	14.08	4.69	0.2734^{-2}	0.1852	19.42
4.32	0.4337^{-2}	0.2113	14.20	4.70	0.2701^{-2}	0.1846	19.58
4.33	0.4282^{-2}	0.2105	14.32	4.71	0.2669^{-2}	0.1839	19.75
4.34	0.4228^{-2}	0.2098	14.45	4.72	0.2637^{-2}	0.1833	19.91
4.35	0.4174^{-2}	0.2090	14.57	4.73	0.2605^{-2}	0.1827	20.07
4.36	0.4121^{-2}	0.2083	14.70	4.74	0.2573^{-2}	0.1820	20.24
4.37	0.4069^{-2}	0.2075	14.82	4.75	0.2543^{-2}	0.1814	20.41
4.38	0.4018^{-2}	0.2067	14.95	4.76	0.2512^{-2}	0.1808	20.58
4.39	0.3968^{-2}	0.2060	15.08	4.77	0.2482^{-2}	0.1802	20.75
4.40	0.3918^{-2}	0.2053	15.21	4.78	0.2452^{-2}	0.1795	20.92
4.41	0.3868^{-2}	0.2045	15.34	4.79	0.2423^{-2}	0.1789	21.09
4.42	0.3820^{-2}	0.2038	15.47	4.80	0.2394^{-2}	0.1783	21.26
4.43	0.3772^{-2}	0.2030	15.61	4.81	0.2366^{-2}	0.1777	21.44
4.44	0.3725^{-2}	0.2023	15.74	4.82	0.2338^{-2}	0.1771	21.61
4.45	0.3678^{-2}	0.2016	15.87	4.83	0.2310^{-2}	0.1765	21.79
4.46	0.3633^{-2}	0.2009	16.01	4.84	0.2283^{-2}	0.1759	21.97
4.47	0.3587^{-2}	0.2002	16.15	4.85	0.2255^{-2}	0.1753	22.15
4.48	0.3543^{-2}	0.1994	16.28	4.86	0.2229^{-2}	0.1747	22.33
4.49	0.3499^{-2}	0.1987	16.42	4.87	0.2202^{-2}	0.1741	22.51
4.50	0.3455^{-2}	0.1980	16.56	4.88	0.2177^{-2}	0.1735	22.70
4.51	0.3412^{-2}	0.1973	16.70	4.89	0.2151^{-2}	0.1729	22.88
4.52	0.3370^{-2}	0.1966	16.84	4.90	0.2126^{-2}	0.1724	23.07
4.53	0.3329^{-2}	0.1959	16.99	4.91	0.2101^{-2}	0.1718	23.25
4.54	0.3288^{-2}	0.1952	17.13	4.92	0.2076^{-2}	0.1712	23.44
4.55	0.3247^{-2}	0.1945	17.28	4.93	0.2052^{-2}	0.1706	23.63
4.56	0.3207^{-2}	0.1938	17.42	4.94	0.2028^{-2}	0.1700	23.82
4.57	0.3168^{-2}	0.1932	17.57	4.95	0.2004^{-2}	0.1695	24.02
4.58	0.3129^{-2}	0.1925	17.72	4.96	0.1981^{-2}	0.1689	24.21
4.59	0.3090^{-2}	0.1918	17.87	4.97	0.1957^{-2}	0.1683	24.41
4.60	0.3053^{-2}	0.1911	18.02	4.98	0.1935^{-2}	0.1678	24.60
4.61	0.3015^{-2}	0.1905	18.17	4.99	0.1912^{-2}	0.1672	24.80
4.62	0.2978^{-2}	0.1898	18.32	5.00	0.1890^{-2}	0.1667	25.00
4.63	0.2942^{-2}	0.1891	18.48	6.00	0.6334^{-3}	0.1220	53.18
4.64	0.2906^{-2}	0.1885	18.63	7.00	0.2416^{-3}	0.9259^{-1}	104.1
4.65	0.2871^{-2}	0.1878	18.79	8.00	0.1024^{-3}	0.7246^{-1}	190.1
4.66	0.2836^{-2}	0.1872	18.94	9.00	0.4739^{-4}	0.5814^{-1}	327.2
4.67	0.2802^{-2}	0.1865	19.10	10.00	0.2356^{-4}	0.4762^{-1}	535.9

TABLE C.2 *Normal Shock Table* ($\gamma = 1.4$)

M_1	M_2	$\frac{p_2}{p_1}$	$\frac{\rho_2}{\rho_1}$	$\frac{T_2}{T_1}$	$\frac{p_{t_2}}{p_{t_1}}$	$\frac{p_1}{p_{t_2}}$
1.00	1.000	1.000	1.000	1.000	1.000	0.5283
1.01	0.9901	1.023	1.017	1.007	1.000	0.5221
1.02	0.9805	1.047	1.033	1.013	1.000	0.5160
1.03	0.9712	1.071	1.050	1.020	1.000	0.5100
1.04	0.9620	1.095	1.067	1.026	0.9999	0.5039
1.05	0.9531	1.120	1.084	1.033	0.9999	0.4980
1.06	0.9444	1.144	1.101	1.059	0.9997	0.4920
1.07	0.9360	1.169	1.118	1.016	0.9996	0.4861
1.08	0.9277	1.194	1.135	1.052	0.9994	0.4803
1.09	0.9196	1.219	1.152	1.059	0.9992	0.4746
1.10	0.9118	1.245	1.169	1.065	0.9989	0.4689
1.11	0.9041	1.271	1.186	1.071	0.9986	0.4632
1.12	0.8966	1.297	1.203	1.078	0.9982	0.4576
1.13	0.8892	1.323	1.221	1.084	0.9978	0.4521
1.14	0.8820	1.350	1.238	1.090	0.9973	0.4467
1.15	0.8750	1.376	1.255	1.097	0.9967	0.4413
1.16	0.8682	1.403	1.272	1.103	0.9961	0.4360
1.17	0.8615	0.430	1.290	1.109	0.9953	0.4307
1.18	0.8549	1.458	1.307	1.115	0.9916	0.4255
1.19	0.8485	1.485	1.324	1.122	0.9937	0.4204

M_1	M_2	$\frac{p_2}{p_1}$	$\frac{\rho_2}{\rho_1}$	$\frac{T_2}{T_1}$	$\frac{p_{t_2}}{p_{t_1}}$	$\frac{p_1}{p_{t_2}}$
1.20	0.8422	1.513	1.342	1.128	0.9928	0.4154
1.21	0.8360	1.541	1.359	1.134	0.9918	0.4104
1.22	0.8300	1.570	1.376	1.141	0.9907	0.4055
1.23	0.8241	1.598	1.394	1.147	0.9896	0.4006
1.24	0.8183	1.627	1.411	1.153	0.9884	0.3958
1.25	0.8126	1.656	1.429	1.159	0.9871	0.3911
1.26	0.8071	1.686	1.446	1.166	0.9857	0.3865
1.27	0.8016	1.715	1.463	1.172	0.9842	0.3819
1.28	0.7963	1.745	1.481	1.178	0.9827	0.3774
1.29	0.7911	1.775	1.498	1.185	0.9811	0.3729
1.30	0.7860	1.805	1.516	1.191	0.9794	0.3685
1.31	0.7809	1.835	1.533	1.197	0.9776	0.3642
1.32	0.7760	1.866	1.551	1.204	0.9758	0.3599
1.33	0.7712	1.897	1.568	1.210	0.9738	0.3557
1.34	0.7664	1.928	1.585	1.216	0.9718	0.3516
1.35	0.7618	1.960	1.603	1.223	0.9697	0.3475
1.36	0.7572	1.991	1.620	1.229	0.9676	0.3435
1.37	0.7527	2.023	1.638	1.235	0.9653	0.3395
1.38	0.7483	2.055	1.655	1.242	0.9630	0.3356
1.39	0.7440	2.087	1.672	1.248	0.9607	0.3317
1.40	0.7397	2.120	1.690	1.255	0.9582	0.3280
1.41	0.7355	2.153	1.707	1.261	0.9557	0.3242
1.42	0.7314	2.186	1.724	1.268	0.9531	0.3205
1.43	0.7274	2.219	1.742	1.274	0.9504	0.3169
1.44	0.7235	2.253	1.759	1.281	0.9476	0.3133
1.45	0.7196	2.286	1.776	1.287	0.9448	0.3098
1.46	0.7157	2.320	1.793	1.294	0.9420	0.3063
1.47	0.7120	2.354	1.811	1.300	0.9390	0.3029
1.48	0.7083	2.389	1.828	1.307	0.9360	0.2996
1.49	0.7047	2.423	1.845	1.314	0.9329	0.2962
1.50	0.7011	2.458	1.862	1.320	0.9298	0.2930
1.51	0.6976	2.493	1.879	1.327	0.9266	0.2898
1.52	0.6941	2.529	1.896	1.334	0.9233	0.2866
1.53	0.6907	2.564	1.913	1.340	0.9200	0.2835
1.54	0.6874	2.600	1.930	1.347	0.9166	0.2804
1.55	0.6841	2.636	1.947	1.354	0.9132	0.2773
1.56	0.6809	2.673	1.964	1.361	0.9097	0.2744
1.57	0.6777	2.709	1.981	1.367	0.9061	0.2714
1.58	0.6746	2.746	1.998	1.374	0.9026	0.2685
1.59	0.6715	2.783	2.015	1.381	0.8989	0.2656

TABLE C.2 (*Continued*) *Normal Shock Table* ($\gamma = 1.4$)

M_1	M_2	$\frac{p_2}{p_1}$	$\frac{\rho_2}{\rho_1}$	$\frac{T_2}{T_1}$	$\frac{p_{t_2}}{p_{t_1}}$	$\frac{p_1}{p_{t_2}}$
1.60	0.6684	2.820	2.032	1.388	0.8952	0.2628
1.61	0.6655	2.857	2.049	1.395	0.8915	0.2600
1.62	0.6625	2.895	2.065	1.402	0.8877	0.2573
1.63	0.6596	2.933	2.082	1.409	0.8538	0.2546
1.64	0.6568	2.971	2.099	1.416	0.8799	0.2519
1.65	0.6540	3.010	2.115	1.423	0.8760	0.2493
1.66	0.6512	3.048	2.132	1.430	0.8720	0.2467
1.67	0.6485	3.087	2.148	1.437	0.8680	0.2442
1.68	0.6458	3.126	2.165	1.444	0.8640	0.2417
1.69	0.6431	3.165	2.181	1.451	0.8598	0.2392
1.70	0.6405	3.205	2.198	1.458	0.8557	0.2368
1.71	0.6380	3.245	2.214	1.466	0.8516	0.2344
1.72	0.6355	3.285	2.230	1.473	0.8474	0.2320
1.73	0.6330	3.325	2.247	1.480	0.8431	0.2296
1.74	0.6305	3.366	2.263	1.487	0.8389	0.2273
1.75	0.6281	3.406	2.279	1.495	0.8346	0.2251
1.76	0.6257	3.447	2.295	1.502	0.8302	0.2228
1.77	0.6234	3.488	2.311	1.509	0.8259	0.2206
1.78	0.6210	3.530	2.327	1.517	0.8215	0.2184
1.79	0.6188	3.571	2.343	1.524	0.8171	0.2163
1.80	0.6165	3.613	2.359	1.532	0.8127	0.2142
1.81	0.6143	3.655	2.375	1.539	0.8082	0.2121
1.82	0.6121	3.698	2.391	1.547	0.8038	0.2100
1.83	0.6099	3.740	2.407	1.554	0.7993	0.2080
1.84	0.6078	3.783	2.422	1.562	0.7948	0.2060
1.85	0.6057	3.826	2.438	1.569	0.7092	0.2040
1.86	0.6036	3.870	2.454	1.577	0.7857	0.2020
1.87	0.6016	3.913	2.469	1.585	0.7811	0.2001
1.88	0.5996	3.957	2.485	1.592	0.7765	0.1982
1.89	0.5976	4.001	2.500	1.600	0.7720	0.1963
1.90	0.5956	4.045	2.516	1.608	0.7674	0.1945
1.91	0.5937	4.089	2.531	1.616	0.7627	0.1927
1.92	0.5918	4.134	2.546	1.624	0.7581	0.1909
1.93	0.5899	4.179	2.562	1.631	0.7535	0.1891
1.94	0.5880	4.224	2.577	1.639	0.7488	0.1873
1.95	0.5862	4.270	2.592	1.647	0.7442	0.1856
1.96	0.5844	4.315	2.607	1.655	0.7395	0.1839
1.97	0.5826	4.361	2.622	1.663	0.7349	0.1822
1.98	0.5808	4.407	2.637	1.671	0.7302	0.1806
1.99	0.5791	4.453	2.652	1.679	0.7255	0.1789

M_1	M_2	$\frac{p_2}{p_1}$	$\frac{\rho_2}{\rho_1}$	$\frac{T_2}{T_1}$	$\frac{p_{t_2}}{p_{t_1}}$	$\frac{p_1}{p_{t_2}}$
2.00	0.5774	4.500	2.667	1.688	0.7209	0.1773
2.01	0.5757	4.547	2.681	1.696	0.7162	0.1757
2.02	0.5740	4.594	2.696	1.704	0.7115	0.1741
2.03	0.5723	4.641	2.711	1.712	0.7069	0.1726
2.04	0.5707	4.689	2.725	1.720	0.7022	0.1710
2.05	0.5691	4.736	2.740	1.729	0.6975	0.1695
2.06	0.5675	4.784	2.755	1.737	0.6928	0.1680
2.07	0.5659	4.832	2.769	1.745	0.6882	0.1665
2.08	0.5643	4.881	2.783	1.754	0.6835	0.1651
2.09	0.5628	4.929	2.798	1.762	0.6789	0.1636
2.10	0.5613	4.978	2.812	1.770	0.6742	0.1622
2.11	0.5598	5.027	2.826	1.779	0.6696	0.1608
2.12	0.5583	5.077	2.840	1.787	0.6649	0.1594
2.13	0.5568	5.126	2.854	1.796	0.6603	0.1580
2.14	0.5554	5.176	2.868	1.805	0.6557	0.1567
2.15	0.5540	5.226	2.882	1.813	0.6511	0.1553
2.16	0.5525	5.277	2.896	1.822	0.6464	0.1540
2.17	0.5511	5.327	2.910	1.831	0.6419	0.1527
2.18	0.5498	5.378	2.924	1.839	0.6373	0.1514
2.19	0.5484	5.429	2.938	1.848	0.6327	0.1502
2.20	0.5471	5.480	2.951	1.857	0.6281	0.1489
2.21	0.5457	5.531	2.965	1.866	0.6236	0.1476
2.22	0.5444	5.583	2.978	1.875	0.6191	0.1464
2.23	0.5431	5.636	2.992	1.883	0.6145	0.1452
2.24	0.5418	5.687	3.005	1.892	0.6100	0.1440
2.25	0.5406	5.740	3.019	1.901	0.6055	0.1428
2.26	0.5393	5.792	3.032	1.910	0.6011	0.1417
2.27	0.5381	5.845	3.045	1.919	0.5966	0.1405
2.28	0.5368	5.898	3.058	1.929	0.5921	0.1394
2.29	0.5356	5.951	3.071	1.938	0.5877	0.1382
2.30	0.5344	6.005	3.085	1.947	0.5833	0.1371
2.31	0.5332	6.059	3.098	1.956	0.5789	0.1360
2.32	0.5321	6.113	3.110	1.965	0.5745	0.1349
2.33	0.5309	6.167	3.123	1.974	0.5702	0.1338
2.34	0.5297	6.222	3.136	1.984	0.5658	0.1328
2.35	0.5286	6.276	3.149	1.993	0.5615	0.1317
2.36	0.5275	6.331	3.162	2.002	0.5572	0.1307
2.37	0.5264	6.386	3.174	2.012	0.5529	0.1297
2.38	0.5253	6.442	3.187	2.021	0.5486	0.1286
2.39	0.5242	6.497	3.199	2.031	0.5444	0.1276

TABLE C.2 (*Continued*) *Normal Shock Table* ($\gamma = 1.4$)

M_1	M_2	$\frac{p_2}{p_1}$	$\frac{\rho_2}{\rho_1}$	$\frac{T_2}{T_1}$	$\frac{p_{t_2}}{p_{t_1}}$	$\frac{p_1}{p_{t_2}}$
2.40	0.5231	6.553	3.212	2.040	0.5401	0.1266
2.41	0.5221	6.609	3.224	2.050	0.5359	0.1257
2.42	0.5210	6.666	3.237	2.059	0.5317	0.1247
2.43	0.5200	6.722	3.249	2.069	0.5276	0.1237
2.44	0.5189	6.779	3.261	2.079	0.5234	0.1228
2.45	0.5179	6.836	3.273	2.088	0.5193	0.1218
2.46	0.5169	6.894	3.285	2.098	0.5152	0.1209
2.47	0.5159	6.951	3.298	2.108	0.5111	0.1200
2.48	0.5149	7.009	3.310	2.118	0.5071	0.1191
2.49	0.5140	7.067	3.321	2.128	0.5030	0.1182
2.50	0.5130	7.125	3.333	2.138	0.4990	0.1173
2.51	0.5120	7.183	3.345	2.147	0.4950	0.1164
2.52	0.5111	7.242	3.357	2.157	0.4911	0.1155
2.53	0.5102	7.301	3.369	2.167	0.4871	0.1147
2.54	0.5092	7.360	3.380	2.177	0.4832	0.1138
2.55	0.5083	7.420	3.392	2.187	0.4793	0.1130
2.56	0.5074	7.479	3.403	2.198	0.4754	0.1122
2.57	0.5065	7.539	3.415	2.208	0.4715	0.1113
2.58	0.5056	7.599	3.426	2.218	0.4677	0.1105
2.59	0.5047	7.659	3.438	2.228	0.4639	0.1097
2.60	0.5039	7.720	3.449	2.238	0.4601	0.1089
2.61	0.5030	7.781	3.460	2.249	0.4564	0.1081
2.62	0.5022	7.842	3.471	2.259	0.4526	0.1074
2.63	0.5013	7.903	3.483	2.269	0.4489	0.1066
2.64	0.5005	7.965	3.494	2.280	0.4452	0.1058
2.65	0.4996	8.026	3.505	2.290	0.4416	0.1051
2.66	0.4988	8.088	3.516	2.301	0.4379	0.1043
2.67	0.4980	8.150	3.527	2.311	0.4343	0.1036
2.68	0.4972	8.213	3.537	2.322	0.4307	0.1028
2.69	0.4964	8.275	3.548	2.332	0.4271	0.1021
2.70	0.4956	8.338	3.559	2.343	0.4236	0.1014
2.71	0.4949	8.401	3.570	2.354	0.4201	0.1007
2.72	0.4941	8.465	3.580	2.364	0.4166	0.9998^{-1}
2.73	0.4933	8.528	3.591	2.375	0.4131	0.9929^{-1}
2.74	0.4926	8.592	3.601	2.386	0.4097	0.9860^{-1}
2.75	0.4918	8.656	3.612	2.397	0.4062	0.9792^{-1}
2.76	0.4911	8.721	3.622	2.407	0.4028	0.9724^{-1}
2.77	0.4903	8.785	3.633	2.418	0.3994	0.9658^{-1}
2.78	0.4896	8.850	3.643	2.429	0.3961	0.9591^{-1}
2.79	0.4889	8.915	3.653	2.440	0.3928	0.9526^{-1}

M_1	M_2	$\frac{p_2}{p_1}$	$\frac{\rho_2}{\rho_1}$	$\frac{T_2}{T_1}$	$\frac{p_{t_2}}{p_{t_1}}$	$\frac{p_1}{p_{t_2}}$
2.80	0.4882	8.980	3.664	2.451	0.3895	0.9461^{-1}
2.81	0.4875	9.045	3.674	2.462	0.3862	0.9397^{-1}
2.82	0.4868	9.111	3.684	2.473	0.3829	0.9334^{-1}
2.83	0.4861	9.177	3.694	2.484	0.3797	0.9271^{-1}
2.84	0.4854	9.243	3.704	2.496	0.3765	0.9209^{-1}
2.85	0.4847	9.310	3.714	2.507	0.3733	0.9147^{-1}
2.86	0.4840	9.376	3.724	2.518	0.3701	0.9086^{-1}
2.87	0.4833	9.443	3.734	2.529	0.3670	0.9026^{-1}
2.88	0.4827	9.510	3.743	2.540	0.3639	0.8966^{-1}
2.89	0.4820	9.577	3.753	2.552	0.3608	0.8906^{-1}
2.90	0.4814	9.645	3.763	2.563	0.3577	0.8848^{-1}
2.91	0.4807	9.713	3.773	2.575	0.3547	0.8790^{-1}
2.92	0.4801	9.781	3.782	2.586	0.3517	0.8732^{-1}
2.93	0.4795	9.849	3.792	2.598	0.3487	0.8675^{-1}
2.94	0.4788	9.918	3.801	2.609	0.3457	0.8619^{-1}
2.95	0.4782	9.986	3.811	2.621	0.3428	0.8563^{-1}
2.96	0.4776	10.06	3.820	2.632	0.3398	0.8507^{-1}
2.97	0.4770	10.12	3.829	2.644	0.3369	0.8453^{-1}
2.98	0.4764	10.19	3.839	2.656	0.3340	0.8398^{-1}
2.99	0.4758	10.26	3.848	2.667	0.3312	0.8345^{-1}
3.00	0.4752	10.33	3.857	2.679	0.3283	0.8291^{-1}
3.01	0.4746	10.40	3.866	2.691	0.3255	0.8238^{-1}
3.02	0.4740	10.47	3.875	2.703	0.3227	0.8186^{-1}
3.03	0.4734	10.54	3.884	2.714	0.3200	0.8134^{-1}
3.04	0.4729	10.62	3.893	2.726	0.3172	0.8083^{-1}
3.05	0.4723	10.69	3.902	2.738	0.3145	0.8032^{-1}
3.06	0.4717	10.76	3.911	2.750	0.3118	0.7982^{-1}
3.07	0.4712	10.83	3.920	2.762	0.3091	0.7932^{-1}
3.08	0.4706	10.90	3.929	2.774	0.3065	0.7882^{-1}
3.09	0.4701	10.97	3.938	2.786	0.3038	0.7833^{-1}
3.10	0.4695	11.05	3.947	2.799	0.3012	0.7785^{-1}
3.11	0.4690	11.12	3.955	2.811	0.2986	0.7737^{-1}
3.12	0.4685	11.19	3.964	2.823	0.2960	0.7689^{-1}
3.13	0.4679	11.26	3.973	2.835	0.2935	0.7642^{-1}
3.14	0.4674	11.34	3.981	2.848	0.2910	0.7595^{-1}
3.15	0.4669	11.41	3.990	2.860	0.2885	0.7549^{-1}
3.16	0.4664	11.48	3.998	2.872	0.2860	0.7503^{-1}
3.17	0.4659	11.56	4.006	2.885	0.2835	0.7457^{-1}
3.18	0.4654	11.63	4.015	2.897	0.2811	0.7412^{-1}
3.19	0.4648	11.71	4.023	2.909	0.2786	0.7367^{-1}

TABLE C.2 (*Continued*) *Normal Shock Table* ($\gamma = 1.4$)

M_1	M_2	$\frac{p_2}{p_1}$	$\frac{\rho_2}{\rho_1}$	$\frac{T_2}{T_1}$	$\frac{p_{t_2}}{p_{t_1}}$	$\frac{p_1}{p_{t_2}}$
3.20	0.4643	11.78	4.031	2.922	0.2762	0.7323^{-1}
3.21	0.4639	11.85	4.040	2.935	0.2738	0.7279^{-1}
3.22	0.4634	11.93	4.048	2.947	0.2715	0.7235^{-1}
3.23	0.4629	12.01	4.056	2.960	0.2691	0.7192^{-1}
3.24	0.4624	12.08	4.064	2.972	0.2668	0.7149^{-1}
3.25	0.4619	12.16	4.072	2.985	0.2645	0.7107^{-1}
3.26	0.4614	12.23	4.080	2.998	0.2622	0.7065^{-1}
3.27	0.4610	12.31	4.088	3.011	0.2600	0.7023^{-1}
3.28	0.4605	12.38	4.096	3.023	0.2577	0.6982^{-1}
3.29	0.4600	12.46	4.104	3.036	0.2555	0.6941^{-1}
3.30	0.4596	12.54	4.112	3.049	0.2533	0.6900^{-1}
3.31	0.4591	12.62	4.120	3.062	0.2511	0.6860^{-1}
3.32	0.4587	12.69	4.128	3.075	0.2489	0.6820^{-1}
3.33	0.4582	12.77	4.135	3.088	0.2468	0.6781^{-1}
3.34	0.4578	12.85	4.143	3.101	0.2446	0.6741^{-1}
3.35	0.4573	12.93	4.151	3.114	0.2425	0.6702^{-1}
3.36	0.4569	13.00	4.158	3.127	0.2404	0.6664^{-1}
3.37	0.4565	13.08	4.166	3.141	0.2383	0.6626^{-1}
3.38	0.4560	13.16	4.173	3.154	0.2363	0.6588^{-1}
3.39	0.4556	13.24	4.181	3.167	0.2342	0.6550^{-1}
3.40	0.4552	13.32	4.188	3.180	0.2322	0.6513^{-1}
3.41	0.4548	13.40	4.196	3.191	0.2302	0.6476^{-1}
3.42	0.4544	13.48	4.203	3.207	0.2282	0.6439^{-1}
3.43	0.4540	13.56	4.211	3.220	0.2263	0.6403^{-1}
3.44	0.4535	13.64	4.218	3.234	0.2243	0.6367^{-1}
3.45	0.4531	13.72	4.225	3.247	0.2224	0.6331^{-1}
3.46	0.4527	13.80	4.232	3.261	0.2205	0.6296^{-1}
3.47	0.4523	13.88	4.240	3.274	0.2186	0.6261^{-1}
3.48	0.4519	13.96	4.247	3.288	0.2167	0.6226^{-1}
3.49	0.4515	14.04	4.254	3.301	0.2148	0.6191^{-1}
3.50	0.4512	14.13	4.261	3.315	0.2129	0.6157^{-1}
3.51	0.4508	14.21	4.268	3.329	0.2111	0.6123^{-1}
3.52	0.4504	14.29	4.275	3.343	0.2093	0.6089^{-1}
3.53	0.4500	14.37	4.282	3.356	0.2075	0.6056^{-1}
3.54	0.4496	14.45	4.289	3.370	0.2057	0.6023^{-1}
3.55	0.4492	14.54	4.296	3.384	0.2039	0.5990^{-1}
3.56	0.4489	14.62	4.303	3.398	0.2022	0.5957^{-1}
3.57	0.4485	14.70	4.309	3.412	0.2004	0.5925^{-1}
3.58	0.4481	14.79	4.316	3.426	0.1987	0.5892^{-1}
3.59	0.4478	14.87	4.324	3.440	0.1970	0.5861^{-1}

M_1	M_2	$\frac{p_2}{p_1}$	$\frac{\rho_2}{\rho_1}$	$\frac{T_2}{T_1}$	$\frac{p_{t_2}}{p_{t_1}}$	$\frac{p_1}{p_{t_2}}$
3.60	0.4474	14.95	4.330	3.454	0.1953	0.5829^{-1}
3.61	0.4471	15.04	4.336	3.468	0.1936	0.5798^{-1}
3.62	0.4467	15.12	4.343	3.482	0.1920	0.5767^{-1}
3.63	0.4463	15.21	4.350	3.496	0.1903	0.5736^{-1}
3.64	0.4460	15.29	4.356	3.510	0.1887	0.5705^{-1}
3.65	0.4456	15.38	4.363	3.525	0.1871	0.5675^{-1}
3.66	0.4453	15.46	4.369	3.539	0.1855	0.5645^{-1}
3.67	0.4450	15.55	4.376	3.553	0.1839	0.5615^{-1}
3.68	0.4446	15.63	4.382	3.568	0.1823	0.5585^{-1}
3.69	0.4443	15.72	4.388	3.582	0.1807	0.5556^{-1}
3.70	0.4439	15.81	4.395	3.596	0.1792	0.5526^{-1}
3.71	0.4436	15.89	4.401	3.611	0.1777	0.5497^{-1}
3.72	0.4433	15.98	4.408	3.625	0.1761	0.5469^{-1}
3.73	0.4430	16.07	4.414	3.640	0.1746	0.5440^{-1}
3.74	0.4426	16.15	4.420	3.654	0.1731	0.5412^{-1}
3.75	0.4423	16.24	4.426	3.669	0.1717	0.5384^{-1}
3.76	0.4420	16.33	4.432	3.684	0.1702	0.5356^{-1}
3.77	0.4417	16.42	4.439	3.698	0.1687	0.5328^{-1}
3.78	0.4414	16.50	4.445	3.713	0.1673	0.5301^{-1}
3.79	0.4410	16.59	4.451	3.728	0.1659	0.5274^{-1}
3.80	0.4407	16.68	4.457	3.743	0.1645	0.5247^{-1}
3.81	0.4404	16.77	4.463	3.758	0.1631	0.5220^{-1}
3.82	0.4401	16.86	4.469	3.772	0.1617	0.5193^{-1}
3.83	0.4398	16.95	4.475	3.787	0.1603	0.5167^{-1}
3.84	0.4395	17.04	4.481	3.802	0.1589	0.5140^{-1}
3.85	0.4392	17.13	4.487	3.817	0.1576	0.5114^{-1}
3.86	0.4389	17.22	4.492	3.832	0.1563	0.5089^{-1}
3.87	0.4386	17.31	4.498	3.847	0.1549	0.5063^{-1}
3.88	0.4383	17.40	4.504	3.863	0.1536	0.5038^{-1}
3.89	0.4380	17.49	4.510	3.878	0.1523	0.5012^{-1}
3.90	0.4377	17.58	4.516	3.893	0.1510	0.4987^{-1}
3.91	0.4375	17.67	4.521	3.908	0.1497	0.4962^{-1}
3.92	0.4372	17.76	4.527	3.923	0.1485	0.4938^{-1}
3.93	0.4369	17.85	4.533	3.939	0.1472	0.4913^{-1}
3.94	0.4366	17.94	4.538	3.954	0.1460	0.4889^{-1}
3.95	0.4363	18.04	4.544	3.969	0.1448	0.4865^{-1}
3.96	0.4360	18.13	4.549	3.985	0.1435	0.4841^{-1}
3.97	0.4358	18.22	4.555	4.000	0.1423	0.4817^{-1}
3.98	0.4355	18.31	4.560	4.016	0.1411	0.4793^{-1}
3.99	0.4352	18.41	4.566	4.031	0.1399	0.4770^{-1}

TABLE C.2 (*Continued*) *Normal Shock Table* ($\gamma = 1.4$)

M_1	M_2	$\frac{p_2}{p_1}$	$\frac{\rho_2}{\rho_1}$	$\frac{T_2}{T_1}$	$\frac{p_{t_2}}{p_{t_1}}$	$\frac{p_1}{p_{t_2}}$
4.00	0.4350	18.50	4.571	4.047	0.1388	0.4747^{-1}
4.01	0.4347	18.59	4.577	4.062	0.1376	0.4723^{-1}
4.02	0.4344	18.69	4.582	4.078	0.1364	0.4769^{-1}
4.03	0.4342	18.78	4.588	4.094	0.1353	0.4678^{-1}
4.04	0.4339	18.88	4.593	4.110	0.1342	0.4655^{-1}
4.05	0.4336	18.97	4.598	4.125	0.1330	0.4633^{-1}
4.06	0.4334	19.06	4.604	4.141	0.1319	0.4610^{-1}
4.07	0.4331	19.16	4.609	4.157	0.1308	0.4588^{-1}
4.08	0.4329	19.25	4.614	4.173	0.1297	0.4566^{-1}
4.09	0.4326	19.35	4.619	4.189	0.1286	0.4544^{-1}
4.10	0.4324	19.45	4.624	4.205	0.1276	0.4523^{-1}
4.11	0.4321	19.54	4.630	4.221	0.1265	0.4501^{-1}
4.12	0.4319	19.64	4.635	4.237	0.1254	0.4480^{-1}
4.13	0.4316	19.73	4.640	4.253	0.1244	0.4459^{-1}
4.14	0.4314	19.83	4.645	4.269	0.1234	0.4438^{-1}
4.15	0.4311	19.93	4.650	4.285	0.1223	0.4417^{-1}
4.16	0.4309	20.02	4.655	4.301	0.1213	0.4396^{-1}
4.17	0.4306	20.12	4.660	4.318	0.1203	0.4375^{-1}
4.18	0.4304	20.22	4.665	4.334	0.1193	0.4355^{-1}
4.19	0.4302	20.32	4.670	4.350	0.1183	0.4334^{-1}
4.20	0.4299	20.41	4.675	4.367	0.1173	0.4314^{-1}
4.21	0.4297	20.51	4.680	4.383	0.1164	0.4294^{-1}
4.22	0.4295	20.61	4.685	4.399	0.1154	0.4274^{-1}
4.23	0.4292	20.71	4.690	4.416	0.1144	0.4255^{-1}
4.24	0.4290	20.81	4.694	4.432	0.1135	0.4235^{-1}
4.25	0.4288	20.91	4.699	4.449	0.1126	0.4215^{-1}
4.26	0.4286	21.01	4.704	4.466	0.1116	0.4196^{-1}
4.27	0.4283	21.11	4.709	4.482	0.1107	0.4177^{-1}
4.28	0.4281	21.20	4.713	4.499	0.1098	0.4158^{-1}
4.29	0.4279	21.30	4.718	4.516	0.1089	0.4139^{-1}
4.30	0.4277	21.41	4.723	4.532	0.1080	0.4120^{-1}
4.31	0.4275	21.51	4.728	4.549	0.1071	0.4101^{-1}
4.32	0.4272	21.61	4.732	4.566	0.1062	0.4082^{-1}
4.33	0.4270	21.71	4.737	4.583	0.1054	0.4064^{-1}
4.34	0.4268	21.81	4.741	4.600	0.1045	0.4046^{-1}
4.35	0.4266	21.91	4.746	4.617	0.1036	0.4027^{-1}
4.36	0.4264	22.01	4.751	4.633	0.1028	0.4009^{-1}
4.37	0.4262	22.11	4.755	4.651	0.1020	0.3991^{-1}
4.38	0.4260	22.22	4.760	4.668	0.1011	0.3973^{-1}
4.39	0.4258	22.32	4.764	4.685	0.1003	0.3956^{-1}

M_1	M_2	$\frac{p_2}{p_1}$	$\frac{\rho_2}{\rho_1}$	$\frac{T_2}{T_1}$	$\frac{p_{t_2}}{p_{t_1}}$	$\frac{p_1}{p_{t_2}}$
4.40	0.4255	22.42	4.768	4.702	0.9948^{-1}	0.3938^{-1}
4.41	0.4253	22.52	4.773	4.719	0.9867^{-1}	0.3921^{-1}
4.42	0.4251	22.63	4.777	4.736	0.9787^{-1}	0.3903^{-1}
4.43	0.4249	22.73	4.782	4.753	0.9707^{-1}	0.3886^{-1}
4.44	0.4247	22.83	4.786	4.771	0.9628^{-1}	0.3869^{-1}
4.45	0.4245	22.94	4.790	4.788	0.9550^{-1}	0.3852^{-1}
4.46	0.4243	23.04	4.795	4.805	0.9473^{-1}	0.3835^{-1}
4.47	0.4241	23.14	4.799	4.823	0.9396^{-1}	0.3818^{-1}
4.48	0.4239	23.25	4.803	4.840	0.9320^{-1}	0.3801^{-1}
4.49	0.4237	23.35	4.808	4.858	0.9244^{-1}	0.3785^{-1}
4.50	0.4236	23.46	4.812	4.875	0.9170^{-1}	0.3768^{-1}
4.51	0.4234	23.56	4.816	4.893	0.9096^{-1}	0.3752^{-1}
4.52	0.4232	23.67	4.820	4.910	0.9022^{-1}	0.3735^{-1}
4.53	0.4230	23.77	4.824	4.928	0.8950^{-1}	0.3719^{-1}
4.54	0.4228	23.88	4.829	4.946	0.8878^{-1}	0.3703^{-1}
4.55	0.4226	23.99	4.833	4.963	0.8806^{-1}	0.3687^{-1}
4.56	0.4224	24.09	4.837	4.981	0.8735^{-1}	0.3671^{-1}
4.57	0.4222	24.20	4.841	4.999	0.8665^{-1}	0.3656^{-1}
4.58	0.4220	24.31	4.845	5.017	0.8596^{-1}	0.3640^{-1}
4.59	0.4219	24.41	4.849	4.034	0.8527^{-1}	0.3624^{-1}
4.60	0.4217	24.52	4.853	5.052	0.8459^{-1}	0.3609^{-1}
4.61	0.4215	24.63	4.857	5.070	0.8391^{-1}	0.3593^{-1}
4.62	0.4213	24.74	4.861	5.088	0.8324^{-1}	0.3578^{-1}
4.63	0.4211	24.84	4.865	5.106	0.8257^{-1}	0.3563^{-1}
4.64	0.4210	24.95	4.869	5.124	0.8192^{-1}	0.3548^{-1}
4.65	0.4208	25.06	4.873	5.143	0.8126^{-1}	0.3533^{-1}
4.66	0.4206	25.17	4.877	5.160	0.8062^{-1}	0.3518^{-1}
4.67	0.4204	25.28	4.881	5.179	0.7998^{-1}	0.3503^{-1}
4.68	0.4203	25.39	4.885	5.197	0.7934^{-1}	0.3488^{-1}
4.69	0.4201	25.50	4.889	5.215	0.7871^{-1}	0.3474^{-1}
4.70	0.4199	25.61	4.893	5.233	0.7809^{-1}	0.3459^{-1}
4.71	0.4197	25.71	4.896	5.252	0.7747^{-1}	0.3445^{-1}
4.72	0.4196	25.82	4.900	5.270	0.7685^{-1}	0.3431^{-1}
4.73	0.4194	25.94	4.904	5.289	0.7625^{-1}	0.3416^{-1}
4.74	0.4192	26.05	4.908	5.307	0.7564^{-1}	0.3402^{-1}
4.75	0.4191	26.16	4.912	5.325	0.7505^{-1}	0.3388^{-1}
4.76	0.4189	26.27	4.915	5.344	0.7445^{-1}	0.3374^{-1}
4.77	0.4187	26.38	4.919	5.363	0.7387^{-1}	0.3360^{-1}
4.78	0.4186	26.49	4.923	5.381	0.7329^{-1}	0.3346^{-1}
4.79	0.4184	26.60	4.926	5.400	0.7271^{-1}	0.3333^{-1}

TABLE C.2 (*Continued*) *Normal Shock Table* ($\gamma = 1.4$)

M_1	M_2	$\frac{p_2}{p_1}$	$\frac{\rho_2}{\rho_1}$	$\frac{T_2}{T_1}$	$\frac{p_{t_2}}{p_{t_1}}$	$\frac{p_1}{p_{t_2}}$
4.80	0.4183	26.71	4.930	5.418	0.7214^{-1}	0.3319^{-1}
4.81	0.4181	26.83	4.934	5.437	0.7157^{-1}	0.3365^{-1}
4.82	0.4179	26.94	4.937	5.456	0.7101^{-1}	0.3292^{-1}
4.83	0.4178	27.05	4.941	5.475	0.7046^{-1}	0.3278^{-1}
4.84	0.4176	27.16	4.945	5.494	0.6991^{-1}	0.3265^{-1}
4.85	0.4175	27.28	4.948	5.512	0.6936^{-1}	0.3252^{-1}
4.86	0.4173	27.39	4.952	5.531	0.6882^{-1}	0.3239^{-1}
4.87	0.4172	27.50	4.955	5.550	0.6828^{-1}	0.3226^{-1}
4.88	0.4170	27.62	4.959	5.569	0.6775^{-1}	0.3213^{-1}
4.89	0.4169	27.73	4.962	5.588	0.6722^{-1}	0.3200^{-1}
4.90	0.4167	27.85	4.966	5.607	0.6670^{-1}	0.3187^{-1}
4.91	0.4165	27.96	4.969	5.626	0.6618^{-1}	0.3174^{-1}
4.92	0.4164	28.07	4.973	5.646	0.6567^{-1}	0.3161^{-1}
4.93	0.4163	28.19	4.976	5.665	0.6516^{-1}	0.3149^{-1}
4.94	0.4161	28.30	4.980	5.684	0.6465^{-1}	0.3136^{-1}
4.95	0.4160	28.42	4.983	5.703	0.6415^{-1}	0.3124^{-1}
4.96	0.4158	28.54	4.987	5.723	0.6366^{-1}	0.3111^{-1}
4.97	0.4157	28.65	4.990	5.742	0.6317^{-1}	0.3099^{-1}
4.98	0.4155	28.77	4.993	5.761	0.6268^{-1}	0.3087^{-1}
4.99	0.4154	28.88	4.997	5.781	0.6220^{-1}	0.3075^{-1}
5.00	0.4152	29.00	5.000	5.800	0.6172^{-1}	0.3062^{-1}
6.00	0.4042	41.83	5.268	7.941	0.2965^{-1}	0.2136^{-1}
7.00	0.3974	57.00	5.444	10.47	0.1535^{-1}	0.1574^{-1}
8.00	0.3929	74.50	5.565	13.39	0.8488^{-2}	0.1207^{-1}
9.00	0.3898	94.33	5.651	16.69	0.4964^{-2}	0.9546^{-2}
10.00	0.3876	116.5	5.714	20.39	0.3045^{-2}	0.7739^{-2}

SOURCES OF TABLES

The values of the thermodynamic properties of water were abridged as well as calculated from tabulated data given in "Thermodynamic and Transport Properties of Steam (ASME Steam Tables)," (New York: American Society of Mechanical Engineers, 1967) and in "ASME Steam Tables in SI (Metric) Units," (New York: ASME, 1977). The saturation properties of ice were abridged from "Steam Tables" by Joseph H. Keenan et al. (New York: John Wiley & Sons, 1969) by permission. The values of the thermodynamic properties of carbon dioxide were calculated from tabulated data given in "Tables of Thermal Properties of Gases" by Joseph Hilsenrath et al., National Bureau of Standards Circular 564 (1955) and in "Thermophysical Properties of Carbon Dioxide" by M. P. Vukalovich and V. V. Altunin, (Moscow: Atomizdat, 1965). The properties of Refrigerant R-22 were abridged as well as calculated from tabulated data given in "Thermodynamic Properties of Freon 22 Refrigerant," Technical Bulletin T-22 by E. I. DuPont de Nemours & Company (1964) and are presented with permission of the copyright owner. The thermodynamic properties of dry air were calculated from tabulated data given in NBS Circular 564 (cited above), in "Thermodynamic Properties of Air" by H. D. Baehr and K. Schwier (Berlin: Springer-Verlag, 1961), and in "Thermophysical Properties of Liquid Air and Its Components" by A. A. Vasserman and V. A. Rabinovich (Moscow: Izdatelstro Komiteta Standartov, 1968). Values of specific heat capacities were calculated from tabulated data given in NBS Circular 564 and in "JANAF Thermochemical Tables," NSRDS-NBS Circular 37 (1971). The thermochemical properties of various substances were calculated from tabulated data given in JANAF Thermochemical Tables (cited above). Data in the compressible flow tables are from NACA Report 1135, "Equations, Tables, and Charts for Compressible Flow," Ames Research Staff (1953).

appendix

D

Thermophysical Properties

LIST OF TABLES

TABLE D.1 *Thermal Conductivity (k) of Various Materials at 212°F (100°C)*

Material	k (Btu/ft-h-°R)	k (W/m K)
Metals		
Aluminum	138	239
Copper	228	394
Gold	180	312
Iron	42	73
Lead	20	34
Magnesium	89	154
Nickel	48	83
Silver	243	421
Stainless steel 231	8.3	14
Steel, low carbon	40	70
Titanium	12	21
Zinc	68	117
Semiconductors		
Silicon (pure)	84	146
Silicon (doped)	570	984
Insulation materials		
Air	0.0177	0.0307
Alumina	16	28
Beryllia	114	197
Epoxy	0.114	0.197
Glass	0.455	0.787
Mica	0.410	0.709
Mylar, Teflon	0.114	0.197

TABLE D.2A *Thermal Properties of Saturated Water (English Units)*

T (°F)	p (psia)	k (Btu/ft-h-°R)	ν (ft²/s)	β (1/°R)	c_p (Btu/lbm-°R)	Pr
32		0.322	1.93×10^{-5}	-0.35×10^{-4}	1.007	13.67
50		0.334	1.41×10^{-5}	0.39×10^{-4}	1.003	10.40
100		0.363	0.749×10^{-5}	2.07×10^{-4}	0.998	4.60
150		0.382	0.475×10^{-5}	3.05×10^{-4}	1.000	3.94
200		0.392	0.340×10^{-5}	3.97×10^{-4}	1.006	2.50
212	14.696	0.394	0.319×10^{-5}	4.18×10^{-4}	1.008	1.75
300	67.005	0.395	0.221×10^{-5}	5.56×10^{-4}	1.029	1.18
400	247.259	0.383	0.167×10^{-5}	7.41×10^{-4}	1.081	0.92
500	680.86	0.349	0.141×10^{-5}	10.94×10^{-4}	1.182	0.97
600	1543.2	0.301	0.138×10^{-5}	18.74×10^{-4}	1.510	1.07
700	3094.3	0.191	0.136×10^{-5}	148×10^{-4}	10.33	6.80

TABLE D.2B *Thermal Properties of Saturated Water (SI Units)*

T (°C)	p (kPa)	k (W/m K)	ν (m²/s)	β (1/K)	c_p (kJ/kg K)	Pr
0		0.558	1.80×10^{-6}	-0.63×10^{-4}	4.212	13.67
20		0.597	1.00×10^{-6}	1.82×10^{-4}	4.182	7.02
40		0.633	0.659×10^{-6}	3.87×10^{-4}	4.175	4.31
60		0.658	0.470×10^{-6}	5.11×10^{-4}	4.181	2.98
80		0.673	0.356×10^{-6}	6.32×10^{-4}	4.194	2.21
100	101.325	0.682	0.294×10^{-6}	7.52×10^{-4}	4.211	1.75
150	476.00	0.684	0.203×10^{-6}	10.3×10^{-4}	4.313	1.17
200	1554.9	0.665	0.160×10^{-6}	13.3×10^{-4}	4.501	0.93
250	3977.6	0.618	0.137×10^{-6}	18.1×10^{-4}	4.844	0.86
300	8592.7	0.544	0.128×10^{-6}	29.2×10^{-4}	5.736	0.98
350	16535	0.430	0.126×10^{-6}	66.8×10^{-4}	9.504	1.60

TABLE D.3A *Thermal Properties of Saturated Steam (English Units)*

T (°F)	p (psia)	k (Btu/ft-h-°R)	ν (ft²/s)	β (1/°R)	Pr
212	14.696	0.0139	22.3×10^{-5}	0.00149	1.08
300	67.005	0.0166	6.06×10^{-5}	0.00156	1.15
400	247.259	0.0209	1.89×10^{-5}	0.00173	1.38
500	680.86	0.0278	0.855×10^{-5}	0.00233	1.75
600	1543.2	0.0416	0.404×10^{-5}	0.00411	2.40
700	3094.3	0.102	0.174×10^{-5}	0.0114	11.8

TABLE D.3B *Thermal Properties of Saturated Steam (SI Units)*

T (°C)	p (kPa)	k (W/m K)	ν (m²/s)	β (1/K)	Pr
100	101.325	0.0241	20.7×10^{-6}	0.00268	1.08
150	476.00	0.0288	5.47×10^{-6}	0.00281	1.16
200	1554.9	0.0354	2.03×10^{-6}	0.00305	1.36
250	3977.6	0.0451	0.913×10^{-6}	0.00403	1.68
300	8592.7	0.0627	0.461×10^{-6}	0.00650	2.13
350	16,535	0.107	0.234×10^{-6}	0.0155	4.03

TABLE D.4A *Thermal Properties of Saturated Liquid R-22 (English Units)*

T (°F)	p (psia)	k (Btu/ft-h-°R)	ν (ft²/s)	β (1/°R)	c_p (Btu/lbm-°R)	Pr
−60	8.82	0.0691	2.95×10^{-6}	1.14×10^{-3}	0.256	3.61
−40	15.22	0.0664	2.57×10^{-6}	1.17×10^{-3}	0.262	3.31
−20	24.85	0.0635	2.25×10^{-6}	1.22×10^{-3}	0.266	3.04
0	38.66	0.0607	2.03×10^{-6}	1.29×10^{-3}	0.271	2.81
20	57.73	0.0575	1.85×10^{-6}	1.43×10^{-3}	0.276	2.63
40	83.21	0.0540	1.68×10^{-6}	1.57×10^{-3}	0.283	2.50
60	116.31	0.0512	1.58×10^{-6}	1.74×10^{-3}	0.291	2.41
80	158.33	0.0484	1.43×10^{-6}	1.97×10^{-3}	0.300	2.35
100	210.60	0.0458	1.30×10^{-6}	2.19×10^{-3}	0.313	2.28
120	274.60	0.0431	1.18×10^{-6}	2.49×10^{-3}	0.332	2.25
140	351.94	0.0407	1.12×10^{-6}	2.79×10^{-3}	0.357	2.32

TABLE D.4B *Thermal Properties of Saturated Liquid R-22 (SI Units)*

T (°C)	p (kPa)	k (W/m K)	ν (m²/s)	β (1/K)	c_p (kJ/kg K)	Pr
−50	63.1	0.119	2.75×10^{-7}	2.07×10^{-3}	1.070	3.55
−40	104.9	0.115	2.39×10^{-7}	2.11×10^{-3}	1.097	3.31
−30	163.5	0.110	2.13×10^{-7}	2.20×10^{-3}	1.115	3.06
−20	244.8	0.106	1.93×10^{-7}	2.36×10^{-3}	1.130	2.85
−10	354.3	0.101	1.76×10^{-7}	2.53×10^{-3}	1.151	2.65
0	497.6	0.0957	1.62×10^{-7}	2.71×10^{-3}	1.172	2.53
10	680.7	0.0912	1.51×10^{-7}	2.98×10^{-3}	1.206	2.44
20	909.9	0.0866	1.40×10^{-7}	3.27×10^{-3}	1.235	2.38
30	1191.8	0.0824	1.30×10^{-7}	3.63×10^{-3}	1.277	2.33
40	1533.5	0.0784	1.19×10^{-7}	4.03×10^{-3}	1.344	2.27
50	1942.3	0.0742	1.09×10^{-7}	4.53×10^{-3}	1.382	2.25
60	2426.5	0.0703	1.04×10^{-7}	5.02×10^{-3}	1.495	2.32

TABLE D.5A *Thermal Properties of Saturated Vapor R-22 (English Units)*

T (°F)	p (psia)	k (Btu/ft-h-°R)	ν (ft²/s)	β (1/°R)	c_p (Btu/lbm-°R)	Pr
−60	8.82	0.00337	32.5×10^{-6}	2.72×10^{-3}	0.139	0.892
−40	15.22	0.00384	22.4×10^{-6}	2.74×10^{-3}	0.146	0.887
−20	24.85	0.00431	15.0×10^{-6}	2.77×10^{-3}	0.152	0.879
0	38.66	0.00479	10.4×10^{-6}	2.81×10^{-3}	0.158	0.883
20	57.73	0.00531	7.51×10^{-6}	2.94×10^{-3}	0.165	0.896
40	83.12	0.00586	5.55×10^{-6}	3.11×10^{-3}	0.175	0.916
60	116.31	0.00646	4.19×10^{-6}	3.34×10^{-3}	0.187	0.944
80	158.33	0.00711	3.23×10^{-6}	3.69×10^{-3}	0.204	0.982
100	210.60	0.00784	2.57×10^{-6}	4.20×10^{-3}	0.226	1.04
120	274.60	0.00866	2.03×10^{-6}	4.92×10^{-3}	0.253	1.12
140	351.94	0.00963	1.65×10^{-6}	6.13×10^{-3}	0.288	1.26

TABLE D.5B *Thermal Properties of Saturated Vapor R-22 (SI Units)*

T (°C)	p (kPa)	k (W/m K)	ν (m²/s)	β (1/K)	c_p (kJ/kg K)	Pr
−50	63.1	0.00585	29.1×10^{-7}	4.89×10^{-3}	0.586	0.891
−40	104.9	0.00666	20.8×10^{-7}	4.93×10^{-3}	0.611	0.887
−30	163.5	0.00737	14.4×10^{-7}	4.98×10^{-3}	0.632	0.879
−20	244.8	0.00812	10.3×10^{-7}	5.02×10^{-3}	0.657	0.881
−10	354.3	0.00891	7.59×10^{-7}	5.21×10^{-3}	0.682	0.891
0	497.6	0.00975	5.74×10^{-7}	5.47×10^{-3}	0.716	0.907
10	680.7	0.0107	4.42×10^{-7}	5.77×10^{-3}	0.758	0.928
20	909.9	0.0116	3.47×10^{-7}	6.19×10^{-3}	0.812	0.957
30	1191.8	0.0127	2.77×10^{-7}	6.87×10^{-3}	0.883	0.994
40	1533.5	0.0138	2.28×10^{-7}	7.75×10^{-3}	0.967	1.05
50	1942.3	0.0151	1.84×10^{-7}	9.00×10^{-3}	1.076	1.13
60	2426.5	0.0167	1.53×10^{-7}	11.03×10^{-3}	1.206	1.26

TABLE D.6A *Thermal Properties of Dry Air at Atmospheric Pressure (English Units)*

T (°F)	k (Btu/ft-h-°R)	ν (ft²/s)	β (1/°R)	Pr
32	0.0137	1.26×10^{-4}	0.00203	0.71
50	0.0141	1.52×10^{-4}	0.00196	0.71
100	0.0152	1.80×10^{-4}	0.00179	0.71
150	0.0162	2.09×10^{-4}	0.00164	0.71
200	0.0174	2.40×10^{-4}	0.00152	0.71
400	0.0215	3.77×10^{-4}	0.00116	0.71
600	0.0252	5.02×10^{-4}	0.000943	0.71
800	0.0290	5.71×10^{-4}	0.000794	0.72
1000	0.0321	8.67×10^{-4}	0.000685	0.72
2000	0.0464	20.0×10^{-4}	0.000407	0.74

TABLE D.6B *Thermal Properties of Dry Air at Atmospheric Pressure (SI Units)*

T (°C)	k (W/m K)	ν (m²/s)	β (1/K)	Pr
0	0.0237	1.33×10^{-5}	0.00366	0.71
20	0.0251	1.51×10^{-5}	0.00341	0.71
40	0.0265	1.69×10^{-5}	0.00319	0.71
60	0.0279	1.89×10^{-5}	0.00300	0.71
80	0.0293	2.09×10^{-5}	0.00283	0.71
100	0.0307	2.30×10^{-5}	0.00268	0.71
200	0.0370	3.45×10^{-5}	0.00211	0.71
300	0.0429	4.75×10^{-5}	0.00175	0.71
400	0.0485	6.46×10^{-5}	0.00149	0.72
500	0.0540	8.10×10^{-5}	0.00122	0.72
1000	0.0762	18.1×10^{-5}	0.000786	0.74

TABLE D.7A *Thermal Properties of Steam at Atmospheric Pressure (English Units)*

T (°F)	k (Btu/ft-h-°R)	ν (ft²/s)	β (1/°R)	Pr
212	0.0139	2.23×10^{-4}	0.00149	1.08
300	0.0165	3.20×10^{-4}	0.00132	1.03
400	0.0194	4.21×10^{-4}	0.00116	1.00
500	0.0226	5.36×10^{-4}	0.00104	1.00
600	0.0261	6.65×10^{-4}	0.000943	0.99
700	0.0297	8.07×10^{-4}	0.000862	0.99
800	0.0336	9.62×10^{-4}	0.000794	1.00
900	0.0376	11.3×10^{-4}	0.000735	1.01
1000	0.0418	13.1×10^{-4}	0.000685	1.01
1100	0.0460	15.1×10^{-4}	0.000641	1.02
1200	0.0504	17.2×10^{-4}	0.000602	1.03

TABLE D.7B *Thermal Properties of Steam at Atmospheric Pressure (SI Units)*

T (°C)	k (W/m K)	ν (m²/s)	β (1/K)	Pr
100	0.0241	2.07×10^{-5}	0.00268	1.08
150	0.0278	2.75×10^{-5}	0.00235	1.03
200	0.0317	3.46×10^{-5}	0.00211	1.00
250	0.0357	4.25×10^{-5}	0.00191	0.99
300	0.0399	5.14×10^{-5}	0.00175	0.99
350	0.0446	6.11×10^{-5}	0.00161	0.99
400	0.0483	7.14×10^{-5}	0.00149	1.00
450	0.0525	8.25×10^{-5}	0.00138	1.00
500	0.0569	9.49×10^{-5}	0.00129	1.01
550	0.0612	10.8×10^{-5}	0.00122	1.01
600	0.0658	12.1×10^{-5}	0.00115	1.02

TABLE D.8 *Emissivity (ϵ) of Various Materials*

Aluminum, polished	0.040
rough	0.055
oxidized	0.10
Cast iron, polished	0.21
oxidized	0.64
Copper, polished	0.023
black oxidized	0.78
Flat black lacquer	0.96
Glass, smooth	0.93
Gold, polished	0.018
Nickel, polished	0.045
oxidized	0.37
Silver, polished	0.025
Zinc, polished	0.045
oxidized	0.11

OTTO CYCLE 10-47

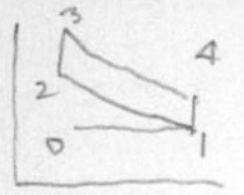

$P_1 = 100$ kPa

$T_1 = 26°C$

Power = 200 kW

N = 3,000 rpm

HV = 44,500 kJ/kg

$\bar{M} = 28$

$\gamma = 1.3$

$\frac{m_a}{m_f} = 17$

SOLN:

$\eta = 1 - CR^{1-\gamma} = 46.2\%$

PT. 3

$\frac{P_2}{P_1} = \left(\frac{V_1}{V_2}\right)^{\gamma}$ — $P_2 = 1468.6$ kPa

$PV = mRT$ (closed system)

$\frac{P_1 V_1}{T_1} = \frac{P_2 V_2}{T_2}$ (because closed so R & m are equal)

$T_2 = T_1 \frac{P_2 V_2}{P_1 V_1} = \frac{(26+273)(1468.6)}{100} \frac{1}{7.9} = T_2 = 555.9$ K $= 282.9$ °C

1st Law: $q_{in} = u_3 - u_2$

$q_{in} = C_v (T_3 - T_2)$

$\frac{44{,}500 \text{ kJ/kg}_f}{17 \frac{m_a}{m_f}} = q_{in} = 2617.65$

$C_v = C_p - R \Rightarrow C_v = C_v\gamma - R$

$C_v = \frac{R}{\gamma - 1}$ $R = \frac{R_u}{\bar{M}} = \frac{8.314}{28} = 0.2969$

$C_v = \frac{.2464}{1.3-1} = 0.990$

$T_3 = 2927.5°C$

$P = \frac{mRT}{V}$ (mR/V const) $\frac{P_2}{T_2} = \frac{P_3}{T_3}$ $P_3 = 8455.2$ kPa

b) $W_{OUT} = mep \cdot D$

$mep = \frac{W_{OUT}}{D} = \frac{\text{Net } W_{OUT}}{V_1 - V_2} = \frac{\text{Net } W_{out}/m}{V_1 - V_2/m} = \frac{\text{Net } w_{OUT}}{v_1 - v_2} = \frac{1209.3}{(.8877 - .1124)} = 1560$ kPa

$pv = RT_1$ $v_1 = \frac{.2969(26+273)}{100} = .8877$ m³/kg

$CR = \frac{V_1}{V_2} = \frac{v_1}{v_2}$ $v_2 = \frac{v_1}{CR} = .1124$ m³/kg

$\eta = \frac{\text{Net } w'_{OUT}}{q_{in}} \Rightarrow \text{Net } w'_{OUT} = \eta q_{in} = .462(2617.65) = 1209.3$ kJ/kg

c) Power = mep N/2 D

$D = \frac{\text{Power}}{mep\, N/2} = \frac{200}{(1560)(3{,}000/2)(1 \text{ min}/60 \text{ sec})} = 0.005128 \text{ m}^3 = 5.128 \text{ l}$

d) Net $\dot{W}_{OUT} = \dot{m}_a \text{ net } w_{out}$

$\dot{m}_{air} = \frac{200 \text{ kW}}{1209.3} = .1654 \text{ kg/s}$

appendix E

Conversion Factors

LENGTH

$$1 \text{ ft} = 12 \text{ in} = 0.3048 \text{ meter (m)}$$
$$= 30.48 \text{ centimeters (cm)} = 304.8 \text{ millimeters (mm)}$$
$$1 \text{ in.} = 2.54 \text{ cm}$$
$$1 \text{ m} = 3.281 \text{ ft}$$
$$1 \text{ cm} = 0.3937 \text{ in}$$

AREA

$$1 \text{ ft}^2 = 144 \text{ in}^2 = 0.092\,90 \text{ m}^2 = 929.0 \text{ cm}^2$$
$$1 \text{ m}^2 = 10\,000 \text{ cm}^2 = 10.764 \text{ ft}^2 = 1550.0 \text{ in}^2$$

VOLUME

$$1 \text{ ft}^3 = 1728 \text{ in}^3 = 0.028\,32 \text{ m}^3 = 28\,317 \text{ cm}^3$$
$$= 7.48 \text{ U.S. gallons} = 29.92 \text{ qt}$$
$$1 \text{ m}^3 = 1\,000\,000 \text{ cm}^3 = 1000 \text{ liters} = 35.315 \text{ ft}^3$$
$$= 61{,}023.7 \text{ in}^3 = 1056.6 \text{ qt}$$
$$1 \text{ liter} = 1.057 \text{ qt}$$

MASS

$$1 \text{ lbm} = 0.453\,59 \text{ kg} = 453.59 \text{ g}$$
$$1 \text{ kg} = 2.2046 \text{ lbm}$$

DENSITY

$$1 \text{ lbm/ft}^3 = 16.02 \text{ kg/m}^3$$
$$1 \text{ kg/m}^3 = 1 \text{ g/liter} = 0.06243 \text{ lbm/ft}^3$$

SPECIFIC VOLUME

$$1 \text{ ft}^3\text{/lbm} = 0.06243 \text{ m}^3\text{/kg}$$
$$1 \text{ m}^3\text{/kg} = 16.02 \text{ ft}^3\text{/lbm}$$

FORCE

$$1 \text{ lbf} = 4.448 \text{ newtons (N)}$$
$$1 \text{ N} = 10^5 \text{ dynes} = 0.2248 \text{ lbf}$$

PRESSURE

$$1 \text{ lbf/ft}^2 \text{ (psf)} = 47.88 \text{ N/m}^2 = 47.88 \text{ Pa} = 0.047\,88 \text{ kPa}$$
$$1 \text{ lbf/in}^2 \text{ (psi)} = 144 \text{ psf} = 6894.8 \text{ Pa} = 6.8948 \text{ kPa}$$
$$1 \text{ N/m}^2 = 1 \text{ pascal (Pa)} = 0.02089 \text{ lbf/ft}^2$$
$$1 \text{ kN/m}^2 = 1 \text{ kPa} = 0.14504 \text{ lbf/in}^2$$
$$1 \text{ atm} = 14.696 \text{ psi} = 2116 \text{ psf} = 1.013 \times 10^5 \text{ Pa}$$
$$= 101.3 \text{ kPa}$$
$$= 29.92 \text{ in. Hg at } 32°\text{F} = 760 \text{ mm Hg at } 0°\text{C}$$

ENERGY

$$1 \text{ lbf-ft} = 0.001285 \text{ Btu} = 0.3240 \text{ cal} = 1.3558 \text{ J}$$

$$1 \text{ Btu} = 778.17 \text{ ft-lbf} = 252 \text{ cal} = 1055 \text{ J}$$

$$1 \text{ joule (J)} = 1 \text{ N-m} = 0.2388 \text{ cal} = 0.7376 \text{ ft-lbf} = 0.9478 \times 10^{-3} \text{ Btu}$$

$$1 \text{ cal} = 4.1868 \text{ J (1 thermochemical cal} = 4.184 \text{ J)}$$

POWER

$$1 \text{ ft-lbf/s} = 0.001818 \text{ hp} = 0.001285 \text{ Btu/s} = 1.3558 \text{ W}$$

$$1 \text{ hp} = 550 \text{ ft-lbf/s} = 0.7068 \text{ Btu/s} = 2544.4 \text{ Btu/h}$$
$$= 1781 \text{ cal/s} = 0.7457 \text{ kW}$$

$$1 \text{ Btu/s} = 778.17 \text{ ft-lbf/s} = 252 \text{ cal/s} = 1.415 \text{ hp} = 1055 \text{ W}$$

$$1 \text{ J/s} = 1 \text{ W} = 0.2388 \text{ cal/s} = 859.85 \text{ cal/h} = 0.7376 \text{ ft-lbf/s}$$
$$= 0.9478 \times 10^{-3} \text{ Btu/s} = 3.412 \text{ Btu/h}$$

SPECIFIC ENERGY

$$1 \text{ Btu/lbm} = 2326 \text{ J/kg} = 2.326 \text{ kJ/kg} = 555.5 \text{ cal/kg}$$

$$1 \text{ kJ/kg} = 0.4299 \text{ Btu/lbm}$$

SPECIFIC ENTROPY, SPECIFIC HEAT

$$1 \text{ Btu/lbm °R} = 4.1868 \text{ kJ/kg K} = 1000 \text{ cal/kg K}$$

$$1 \text{ kJ/kg K} = 0.2388 \text{ Btu/lbm °R} = 238.8 \text{ cal/kg K}$$

SPEED

$$1 \text{ ft/s (fps)} = 0.3048 \text{ m/s}$$

$$1 \text{ mi/h (mph)} = 1.467 \text{ fps} = 0.4470 \text{ m/s} = 1.609 \text{ km/h}$$

$$1 \text{ m/s} = 3.281 \text{ fps} = 3.6 \text{ km/h}$$

TEMPERATURE

$$°\text{F} = 1.8(°\text{C}) + 32 \qquad °\text{R} = °\text{F} + 459.67$$

$$°C = (°F - 32)/1.8 \qquad K = °C + 273.15$$

$$°R = 1.8\ K$$

$$K = °R/1.8$$

SURFACE TENSION

$$1\ \text{lbf/ft} = 14.593\ \text{N/m}$$

$$1\ \text{N/m} = 0.06853\ \text{lbf/ft}$$

VISCOSITY

$$1\ \text{ft}^2/\text{s} = 0.09290\ \text{m}^2/\text{s}$$

$$1\ \text{m}^2/\text{s} = 10.764\ \text{ft}^2/\text{s}$$

HEAT TRANSFER

$$1\ \text{Btu/ft-h} = 0.9613\ \text{W/m}$$

$$1\ \text{W/m} = 1.0403\ \text{Btu/ft-h}$$

$$1\ \text{Btu/ft}^2\text{-h} = 3.154\ \text{W/m}^2$$

$$1\ \text{W/m}^2 = 0.3170\ \text{Btu/ft}^2\text{-h}$$

$$1\ \text{Btu/ft-h-°R} = 1.7307\ \text{W/m K}$$

$$1\ \text{W/m K} = 0.57779\ \text{Btu/ft-h-°R}$$

$$1\ \text{Btu/ft}^2\text{-h-°R} = 5.6786\ \text{W/m}^2\ \text{K}$$

$$1\ \text{W/m}^2\ \text{K} = 0.17610\ \text{Btu/ft}^2\text{-h-°R}$$

Index

O

P

Q

R

S

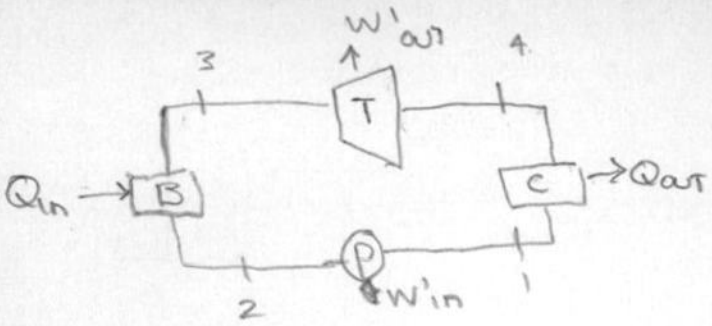

Power = 2 MW

$P_3 = 2.5\,MPa = P_2$

$T_3 = 700\,°C$

$\eta_{Gen} = 0.93$

$\eta_{fuel} = 0.82$

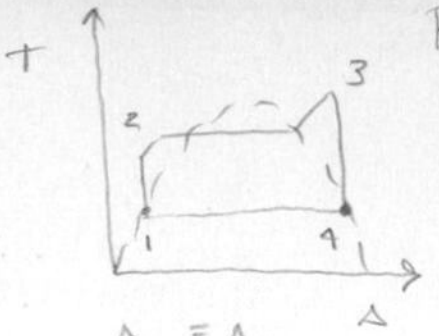

RANKINE CYCLE (IDEAL)

$s_3 = s_4$

$s_2 = s_1$

$T_1 = T_4$

$\dot{m} = ?$

$\dot{W} = \dot{m}\, w'_{OUT} \qquad w'_{OUT} = h_4 - h_3$

$h_3 = 3913.4$ kJ/kg

$s_3 = 7.8431 = s_4$

$7.8431 = 0.8321 + x(7.0773)$

$x = .9906$

$h_4 = 251.09 + 0.9906(2358.4)$

$h_4 = 2587.8$ kJ/kg

$T_4 = 60.09\,°C = T_1$ (sat. liquid for pump)

$w'_{OUT} = 3913.4 - 2587.8 \Rightarrow 1325.6$ kJ/kg

$\frac{2000}{.93} = \dot{m}(1325.6) \qquad \dot{m} = 1.62$ kg/s

$w'_{in,pump} = h_2 - h_1$

$h_1 = 251.45$ kJ/kg

$s_1 = .8321$

$h_2 = 251.45 + (2500 - 20)(.001)$

$h_2 = 253.93$ kJ/kg

Net $w'_{out} = 1325.6 - 2.48 \Rightarrow 1323.12$ kJ/kg

$Q_{in} = \dot{m}\, q_{in} = \dot{m}_f\, HV\, \eta_f \Rightarrow \dot{m}_f = \frac{\dot{m}\, q_{in}}{HV\, \eta_f} \Rightarrow \frac{1.62\,(3659.5)}{40{,}000\,(.82)} = 0.181$ kg/s

$q_{in} = h_3 - h_2 \Rightarrow 3913.4 - 253.93 \Rightarrow 3659.5$ kJ/kg

$P_{OUT,TURB} = (w'_{OUT,TURB})(\dot{m}) = 1.63(1325.6)$

$= 2160.73$ kW

$\eta_{TH} = \frac{\text{net } w'_{OUT}}{q_{in}} \Rightarrow 36.1\%$

RANKINE W/ REHEATER

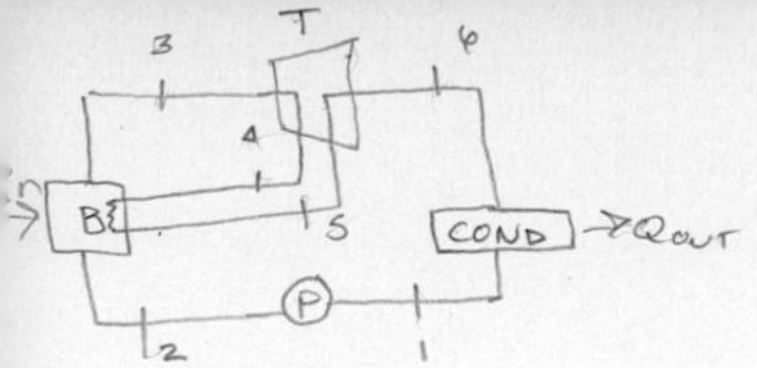

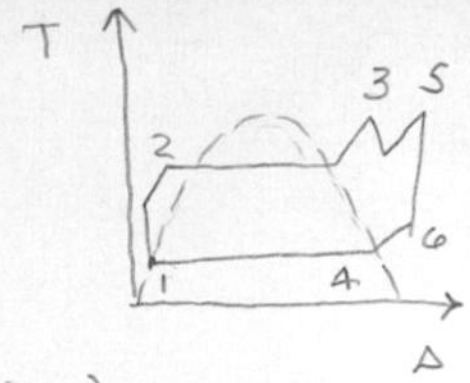

$P_3 = 5.0MPa = P_2$ $P_1 = 20kPa$ (sat. liquid) $= P_6$

$T_3 = 800°C = T_5$ $T_1 = 60.09°C$

$P_5 = 1MPa = P_4$

$\dot{m}_{H_2O} = 2.8$ kg/s

$h_3 = 4135.3$ kJ/kg

$s_3 = 7.7431 = s_4$ $7.7431 = .8321 + x(7.0773)$

$x = .977$

$s_f < s_4 < s_g$

$$\frac{h_4 - 3370.8}{3478.3 - 3370.8} = \frac{7.7431 - 7.6190}{7.7627 - 7.6190}$$ $\underline{h_4 = 3463.64 \text{ kJ/kg}}$

$s_5 = s_4$

$s_5 = 8.4997 = s_6$

$s_6 > s_g$ superheated

$$\frac{8.4997 - 8.5839}{8.5839 - 3.3676} = \frac{h_6 - 2782.3}{2879.2 - 2782.3}$$ $\underline{h_6 = 2841.48}$

$\underline{h_3 = 4135.3 \text{ kJ/kg}}$

$\underline{h_1 = 251.45 \text{ kJ/kg}}$ $\underline{h_5 = 4154.1 \text{ kJ/kg}}$

TURB: $w'_{OUT} = (h_3 - h_4) + (h_5 - h_6)$

$w'_{OUT} = 1934.8$ kJ/kg

Power $= \dot{m}\, w'_{OUT} = 5415.98$ kW

PUMP: $w'_{OUT} = h_1 - h_2 = -v_{f1}(P_2 - P_1) \Rightarrow .00101072(5000 - 20) \Rightarrow -5.07$

$-5.07 = (251.45 - h_2)$

$h_2 = 256.52$ kJ/kg

Power $= 718.26$ kW

c) $\dot{Q} = \dot{m} q_{in}$

$= 2.8(h_3 - h_2) \Rightarrow 10860.34$ kW $q_{in} = 3878.78$

d) $\dot{Q} = \dot{m}(h_5 - h_4)$

$= 1933.29$ kW $q_{in} = 690.46$

e) $$\eta_{TH} = \frac{\text{Net } w'_{OUT}}{q_{in}} = \frac{1934.28 - 5.07}{3878.78 + 690.46} \Rightarrow 42.2\%$$

10-21

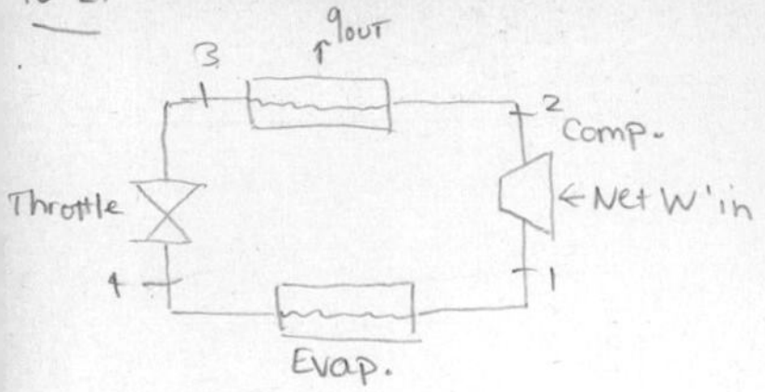

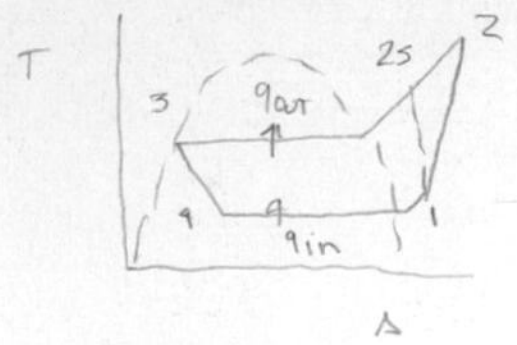

$P_1 = P_4 = 500\text{ kPa}$

$P_2 = P_3 = 1500\text{ kPa}$

$T_1 = T_{g,1} + 5° = 5.15°$ B.11)

$\eta_c = 80\%$

$$COP_{REF} = \frac{q_{in}}{\text{Net } w_{in}} = \frac{q_{in}}{w'_c} = \frac{h_1 - h_4}{w'_c}$$

$$\eta_c = \frac{w'_{c,isen}}{w'_{c,actual}} = \frac{h_{2s} - h_1}{h_2 - h_1} = .8 \qquad w'_c = \frac{h_{2s} - h_1}{\eta_c}$$

$$\dot{w}'_{in} = \dot{m} w'_c$$

$$h_1 = 250 + \frac{5.15 - .15}{50 - .15}(285.96 - 250) = 253.61 \text{ kJ/kg}$$

$$s_1 = .92667 + \frac{\ln\left(\frac{278.30}{273.15}\right)}{\ln\left(\frac{323.15}{273.15}\right)}(1.0479 - .92667) = .94045 \text{ kJ/kg °K}$$

$$s_{2s} = s_1 = .94045$$

$$.94045 = .90752 + \frac{\ln \frac{T_{2s}}{323}}{\ln \frac{373}{323}}(1.03033 - .90752) \qquad T_{2s} = 335.7°K = 62.7°C$$

$P_{2s} = P_2 = 1500$

$$h_{2s} = 271.11 + \frac{12.7}{50}(313.75 - 271.11) = 281.94 \text{ kJ/kg}$$

$$w'_c = \frac{h_{2s} - h_1}{\eta_c} \qquad w'_c = 35.41 \text{ kJ/kg}$$

Throttle: $h_4 = h_3 = h_{f3}$

$P_3 = 1500\text{ kPa}$ B.11) $T_{g3} = 39.10°C$

B.10) $h_3 = h_{f3} = 87.69 + \frac{4.1}{5}(94.26 - 87.69) \Rightarrow 93.08$

$35\text{ kW} = \dot{m}(h_1 - h_4)$

$\dot{W}'_c = \dot{m} w'_c$

$\boxed{\dot{W}_c = 7.72\text{ kW}}$

$$COP_{REF} = \frac{h_1 - h_4}{w'_c} \Rightarrow 4.533$$

If heat pump: $COP_{HP} = \frac{q_{OUT}}{w'_c}$